Lecture Notes in Artificial Intelligence   5866

Edited by R. Goebel, J. Siekmann, and W. Wahlster

Subseries of Lecture Notes in Computer Science

T0180695

Ann Nicholson   Xiaodong Li (Eds.)

AI 2009: Advances in Artificial Intelligence

22nd Australasian Joint Conference
Melbourne, Australia, December 1-4, 2009
Proceedings

 Springer

Series Editors

Randy Goebel, University of Alberta, Edmonton, Canada
Jörg Siekmann, University of Saarland, Saarbrücken, Germany
Wolfgang Wahlster, DFKI and University of Saarland, Saarbrücken, Germany

Volume Editors

Ann Nicholson
Monash University, Clayton School of Information Technology
Clayton, VIC 3800, Australia
E-mail: ann.nicholson@infotech.monash.edu.au

Xiaodong Li
RMIT University, School of Computer Science and Information Technology
Melbourne, VIC 3001, Australia
E-mail: xiaodong.li@rmit.edu.au

Library of Congress Control Number: 2009939152

CR Subject Classification (1998): I.2, H.2.8, I.2.6, K.6, I.5.1, H.5, J.5, H.3

LNCS Sublibrary: SL 7 – Artificial Intelligence

ISSN 0302-9743

ISBN 978-3-642-10438-1 Springer Berlin Heidelberg New York

Typesetting: Camera-ready by author, data conversion by Scientific Publishing Services, Chennai, India
Printed on acid-free paper   SPIN: 12796700   06/3180   5 4 3 2 1 0

Preface

We are pleased to present this LNCS volume, the Proceedings of the 22nd Australasian Joint Conference on Artificial Intelligence (AI 2009), held in Melbourne, Australia, December 1–4, 2009. This long established annual regional conference is a forum both for the presentation of research advances in artificial intelligence and for scientific interchange amongst researchers and practitioners in the field of artificial intelligence. Conference attendees were also able to enjoy AI 2009 being co-located with the Australasian Data Mining Conference (AusDM 2009) and the 4th Australian Conference on Artificial Life (ACAL 2009).

This year AI 2009 received 174 submissions, from authors of 30 different countries. After an extensive peer review process where each submitted paper was rigorously reviewed by at least 2 (and in most cases 3) independent reviewers, the best 68 papers were selected by the senior Program Committee for oral presentation at the conference and included in this volume, resulting in an acceptance rate of 39%. The papers included in this volume cover a wide range of topics in artificial intelligence: from machine learning to natural language systems, from knowledge representation to soft computing, from theoretical issues to real-world applications.

AI 2009 also included 11 tutorials, available through the First Australian Computational Intelligence Summer School (ACISS 2009). These tutorials – some introductory, some advanced – covered a wide range of research topics within artificial intelligence, including data mining, games, evolutionary computation, swarm optimization, intelligent agents, Bayesian and belief networks. There were also four workshops run as part of AI 2009: the 5th Australasian Ontology Workshop (AOW), the Australasian Workshop on Computational Creativity, and the International Workshop on Collaborative Agents – REsearch and Development (CARE), and the First International Workshop on Fictional Prototyping as a Design Tool for Intelligent Environments (FPIE). These tutorials and workshops together provided an excellent start to the week.

The conference featured four distinguished keynote speakers, Ian Witten (University of Waikato, New Zealand), Mark Bedau (Reed College, USA), Eamonn Keogh (University of California - Riverside, USA), and Andries P. Engelbrecht (University of Pretoria, South Africa). Their talks were well received by the attendees.

As with all conferences, the success of AI 2009 depended on its authors, reviewers and organizers. We are very grateful to all the authors for their paper submissions, to all the reviewers for their outstanding work in refereeing the papers within a very tight schedule, and to the senior Program Committee members for their assistance in the paper selection process.

AI 2009 was organized by Clayton School of Information Technology, Monash University, which provided generous financial and organizational support. In particular, we want to thank the conference General Chairs, Damminda Alahakoon and Xinghuo Yu, Advisory Committee Chair, Geoff Webb, Finance Chair, David Albrecht, Workshop Chair, Christian Guttmann, and Publicity Chair, Junbin Gao, for their dedicated efforts that made AI 2009 such a success. We are grateful to the conference coordinator Dianne Nguyen who played such a critical role in managing the conference. Last but not least, AI 2009 relied heavily upon a team of volunteers to keep the conference running smoothly. They were the true heroes working behind the scenes. We are most grateful for their great efforts and contributions.

We would also like to thank our sponsors for their support and financial assistance, including the Centre for Research in Intelligent Systems (CRIS), Monash University, and Platform Technologies Research Institute (PTRI), RMIT University.

September 2009   Ann Nicholson
                                                         Xiaodong Li

Organization

AI 2009 was organized by Clayton School of Information Technology, Monash University, Australia.

AI 2009 Organizing Committee

General Chairs	Damminda Alahakoon (Monash University)
	Xinghuo Yu (RMIT University)
Programme Chairs	Ann Nicholson (Monash University)
	Xiaodong Li (RMIT University)
Advisory Committee Chair	Geoff Webb (Monash University)
Finance Chair	David Albrecht (Monash University)
Workshop Chair	Christian Guttmann (Monash University)
Publicity Chair	Junbin Gao (Charles Sturt University)
Conference Coordinator	Dianne Nguyen (Monash University)

Senior Program Committee

Dan R. Corbett	DARPA, Virginia, USA
Stephen Cranefield	University of Otago, New Zealand
David Dowe	Monash University, Australia
Reinhard Klette	University of Auckland, New Zealand
Kevin Korb	Monash University, Australia
John Lloyd	Australian National University, Australia
Brendan McCane	University of Otago, New Zealand
Mehmet Orgun	Macquarie University, Australia
Chengqi Zhang	University of Technology, Sydney, Australia
Mengjie Zhang	Victoria University of Wellington, New Zealand
Zhi-Hua Zhou	Nanjing University, China

Sponsoring Institutions

The Centre for Research in Intelligent Systems (CRIS), Monash University
Platform Technologies Research Institute (PTRI), RMIT University

AI 2009 Program Committee

Norashikin Ahmad
Sameer Alam
David Albrecht
Lloyd Allison
Peter Andreae
Yun Bai
James Bailey
Timothy Baldwin
Jeewanee Bamunusinghe
Jiri Baum
Ghassan Beydoun
Adrian Bickerstaffe
Tali Boneh
Richard Booth
Sebastian Brand
Thomas Braunl
Lam Thu Bui
Wray Lindsay Buntine
Jinhai Cai
Longbing Cao
Lawrence Cavedon
Chia-Yen Chen
Songcan Chen
Andrew Chiou
Sung-Bae Cho
Vic Ciesielski
Dan R. Corbett
Stephen Cranefield
Michael Cree
Daswin De Silva
Jeremiah D. Deng
Hepu Deng
Grant Dick
Yulin Ding
Trevor Dix
Roland Dodd
Xiangjun Dong
David Dowe
Mark Ellison
Esra Erdem
Daryl Essam
Ling Feng
Alfredo Gabaldon

Marcus Gallagher
Yang Gao
Manolis Gergatsoulis
Chi Keong Goh
Guido Governatori
Charles Gretton
Hans W. Guesgen
Fikret Gurgen
Patrik Haslum
Bernhard Hengst
Jose Hernandez-Orallo
Sarah Hickmott
Philip Hingston
Achim Hoffmann
Geoffrey Holmes
Wei-Chiang Hong
Antony Iorio
Amitay Isaacs
Geoff James
Yu Jian
Warren Jin
Zhi Jin
Ken Kaneiwa
Byeong Ho Kang
George Katsirelos
Michael Kirley
Frank Klawonn
Reinhard Klette
Alistair Knott
Mario Koeppen
Kevin Korb
Rudolf Kruse
Rex Kwok
Gerhard Lakemeyer
Jéróme Lang
Maria Lee
Jimmy Lee
Wei Li
Lily Li
Yuefeng Li
Xiao-Lin Li
Bin Li
Wei Liu

Wanquan Liu
John Lloyd
Abdun Mahmood
Yuval Marom
Eric Martin
Rodrigo Martínez-Béjar
Steven Mascaro
Brendan McCane
Kathryn Merrick
Thomas Meyer
Diego Molla
Saeid Nahavandi
Detlef Nauck
Richi Nayak
David Newth
Michael Niemann
Kouzou Ohara
Mehmet Orgun
Maurice Pagnucco
Andrew Paplinski
Francis Jeffry Pelletier
Nicolai Petkov
Duc-Nghia Pham
Mikhail Prokopenko
Sid Ray
Tapabrata Ray
Jochen Renz
Jeff Riley
Panos Rondogiannis
Malcolm Ryan
Rafal Rzepka
Suzanne Sadedin
Sebastian Sardina
Ruhul Sarker
Torsten Schaub
Daniel Schmidt
Rolf Schwitter
Steven Shapiro
Toby Smith
Andy Song
Maolin Tang
Dacheng Tao
Michael Thielscher

Simon Thompson
John Thornton
Peter Tischer
Andrea Torsello
Charles Twardy
William Uther
Toby Walsh
Kewen Wang
Dianhui Wang
Ian Watson
Peter Whigham
Upali Wickramasinghe
Bill Wilson

Wayne Wobcke
Brendon J. Woodford
Xindong Wu
Roland Yap
Xinghuo Yu
Lean Yu
Mengjie Zhang
Dongmo Zhang
Min-Ling Zhang
Daoqiang Zhang
Chengqi Zhang
Zili Zhang
Shichao Zhang

Xiuzhen Zhang
Haolan Zhang
Jun Zhang
Yanchang Zhao
Fei Zheng
Yi Zhou
Zhi-Hua Zhou
Xingquan Zhu
Li Li
Wei Peng

Additional Reviewers

Bolotov, Alexander
Verden, Andrew
Iorio, Antony
Hengst, Bernhard
Luo, Chao
Li, Chaoming
Lam, Chiou Peng
Moewes, Christian
Zhao, Dengji
Ruÿ, Georg
Singh, Hemant
Qiu, Huining
Varzinczak, Ivan
Deng, Jeremiah

Veness, Joel
Lizier, Joseph
Taylor, Julia
Sung, Ken
Waugh, Kevin
Pipanmaekaporn, Luepol
Newton, M.A. Hakim
Slota, Martin
Knorr, Matthias
Steinbrecher, Matthias
Ptaszynski, Michal
Narodytska, Nina
Obst, Oliver
Ye, Patrick

Dybala, Pawel
Tischer, Peter
Pozos-Parra, Pilar
Li, Ron
Halim, Steven
Patoglu, Volkan
Jin, Warren
Wong, Wilson
Kong, Xiang-Nan
Wang, X. Rosalind
Luo, Xudong
Yu, Yang
Kudo, Yasuo

Table of Contents

Agents

AI Applications

Computer Vision and Image Processing

Data Mining and Statistical Learning

Evolutionary Computing

Game Playing

Knowledge Representation and Reasoning

Natural Language and Speech Processing

Soft Computing

User Modelling

Experimental Market Mechanism Design
for Double Auction*

Masabumi Furuhata[1,2], Laurent Perrussel[2], Jean-Marc Thévenin[2],
and Dongmo Zhang[1]

[1] Intelligent Systems Laboratory, University of Western Sydney, Australia
[2] IRIT, Université de Toulouse, France

Abstract. In this paper, we introduce an experimental approach to the design, analysis and implementation of market mechanisms based on double auction. We define a formal market model that specifies the market policies in a double auction market. Based on this model, we introduce a set of criteria for the evaluation of market mechanisms. We design and implement a set of market policies and test them with different experimental settings. The results of experiments provide us a better understanding of the interrelationship among market policies and also show that an experimental approach can greatly improve the efficiency and effectiveness of market mechanism design.

1 Introduction

Auction has been used for many years as the major trading mechanism for financial markets and electronic markets. The existing researches on auction mostly focus on the theoretical aspects of a market mechanism, such as incentive compatibility, profit optimization, price formation, and so on [1,2,3,4]. From the implementation and market design point of view, "market participants and policy makers would like to know which institution or variant or combination is most efficient, but theoretical and empirical work to date provides little guidance", as Friedman pointed out in [5].

In this paper, we introduce a general approach for the design, analysis, and testing of market mechanisms. The design of a market mechanism involves the development of market policies and evaluation criteria. Different from most existing work in experimental economics, we do not restrict ourselves on specific market policies. Rather, we specify a range of general market policies under the certain trading structure, such as double auction, investigate the properties of market mechanisms with different combinations of market policies. In such a way, we can design a variety of market mechanisms and test them for different purposes.

This paper is organised as follows. Section 2 introduces a market model and specifies the market policies that compose a market mechanism. Section 3 presents a set of evaluation criteria for market mechanism design and testing. Section 4 describes the

* This research was partially supported by the Australian Research Council through Discovery Project DP0988750 and UWS Research Grants Scheme. Also, the work presented in this article is supported by the French Agence Nationale de la Recherche in the framework of the ForTrust (http://www.irit.fr/ForTrust/) project (ANR-06-SETI-006). We thank the anonymous reviewers for their constructive comments.

A. Nicholson and X. Li (Eds.): AI 2009, LNAI 5866, pp. 1–10, 2009.

implementation of market policies. Section 5 presents our experimental results. Finally, we conclude the paper with related work.

2   The Market Model

In this section, we introduce a formal model of market based on double auction market structure. Double auction is a typical market structure in which a set of sellers and buyers trade together for the exchange of certain commodities. Figure 1 illustrates the general structure of double auction markets.

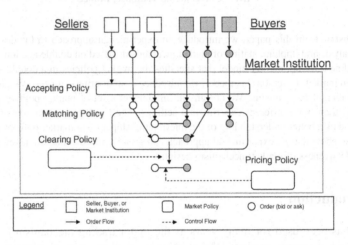

Fig. 1. Structure of Market Institution with Double Auction

In a double auction market, there are three sorts of actors: *sellers*, *buyers* and the *market maker*. The sellers and buyers are called traders. The market maker, who represents the market institution, coordinates the market. During a trading period, the sellers and buyers submit ask orders (sell orders) and bid orders (buy orders) to the market maker of the market institution, respectively. The market maker finds feasible pairs from these incoming orders according to certain market policies, such as accepting policies, matching policies, clearing policies and pricing policies. An *accepting policy* sets criteria for either accepting or rejecting an incoming order. A *matching policy* determines which ask orders match which bid orders. The time for the matched orders to be executed is determined by a *clearing policy*. Meanwhile, the price of the transaction price is determined by a *pricing policy*. According to the structure of double auction markets, the design of market mechanism for a double auction market is to specify each market policy that are to be implemented in the market.

2.1   Market Setting

We consider a double auction market of a single commodity. Let $I = S \cup B$ be a set of traders, where S is the set of sellers; B is the set of buyers. We assume that $S \cap B = \emptyset$[1].

---

[1] In practice, a trader can be both a seller and a buyer for the same commodity. In such a case, we model it as two different roles because the decision making for selling and buying is different.

Each trader $i \in I$ has a fixed valuation for the commodity, which is private information of the trader, denoted by v_i. Let X be the set of incoming orders. An order $x \in X$ consists of two components: the owner of the order, denoted by $\mathfrak{I}(x) \in I$, and the price of the order, denoted by $p(x)$. For any $H \subseteq X$, we write $H^{ask} = \{x \in H : \mathfrak{I}(x) \in S\}$ and $H^{bid} = \{x \in H : \mathfrak{I}(x) \in B\}$. Notice that the meaning of the order prices for sellers and buyers are different. For a seller, sn asking price means that the commodity can be sold with a price no less than this price. For a buyer, a bidding price means that the commodity can be bought with a price no higher than this price.

2.2 Market Policies

Based on the market setting we describe above, we now define the market policies to govern a double auction market.

An *accepting policy* is a function $\mathfrak{A} : X \to \{1, 0\}$ that assigns to each incoming order a value either 1(accepted) or 0 (rejected). Let $A = \{x \in X : \mathfrak{A}(x) = 1\}$ be the set of all the orders that are accepted under the accepting policy $\mathfrak{A}$.

A *matching policy* is a function $\mathfrak{M} : 2^X \to 2^{X \times X}$ such that for any $H \subseteq X$,

1. if $(x, y) \in \mathfrak{M}(H)$, then $x \in H^{ask}$, $y \in H^{bid}$ and $p(x) \leq p(y)$,
2. if (x_1, y_1) and $(x_2, y_2) \in \mathfrak{M}(H)$, then $x_1 = x_2$ if and only if $y_1 = y_2$.

The first condition sets the feasible condition for a match: *the ask price should be less or equal to the bid price*. The second condition specifies that an order can only be matched once. Let M be a set of matched pairs obtained from the accepting policy, that is, $M = \mathfrak{M}(A)$. We use $|M|$ to denote the number of matched pairs in M.

A *pricing policy* on M is a function $\mathfrak{P} : M \to \mathfrak{R}$ that assigns a positive real number, interpreted as the clearing price, to a pair of ask order and bid order such that for any $(x, y) \in M, p(x) \leq \mathfrak{P}(x, y) \leq p(y)$.

Note that any pricing policy is implemented on top of certain accepting policy and matching policy. Without having matched orders, no transactions can be executed.

A clearing policy determines when to clear a matched pair. Formally, let T be the set of time points of a trading period. A *clearing policy* is a function $\mathfrak{C} : T \times M \to \{1, 0\}$ such that for any $t \in T$ and $(x, y) \in M$, if $\mathfrak{C}(t, (x, y)) = 1$, then $\mathfrak{C}(t', (x, y)) = 1$ whenever $t' \in T$ and $t' > t$, which means that once a matched pair is cleared, it can never come back.

With the implementation of all the above policies, a market maker can determine what, when and how the incoming orders being transacted. Briefly speaking, given a set of incoming orders, the accepting policy determines what orders are to be accepted. Among all accepted orders, the matching policy determines whose good can be sold to whom. The pricing policy determines the actual transaction prices and the clearing policy specifies when the transactions should be executed.

3 Evaluation Criteria of Market Mechanisms

In this section, we propose a set of evaluation criteria for the design and evaluation of market mechanisms. We introduce four indicators to measure profiting efficiency, matching efficiency, transaction volume and converging speed, respectively.

Transaction profit (PR) measures the total revenue of the market institution from all transactions that are executed in a given trading period:

$$PR = \sum_{(x,y) \in M} [c_s(\mathfrak{P}(x,y) - p(x)) + c_b(p(y) - \mathfrak{P}(x,y))]$$

where c_s and c_b is the charging rate of the market institution to a seller and a buyer, respectively. The charging rates represent the percentage of profit that a trader has to pay to the market institution for each transaction.

Allocation efficiency (AE) measures the efficiency of matching policies. Matching a given set of orders is usually referred to as allocation. A set $\hat{M} \subseteq X \times X$ is called a potential matching on X if $(x,y) \in \hat{M}$ implies $x \in X^{ask}$, $y \in X^{bid}$ and $v_{\mathfrak{J}}(x) \leq v_{\mathfrak{J}}(y)$. Let $\mathcal{M}$ be the set of all potential matchings on X. Then the indicator AE measures the rate of the total profit that is made by the current matching policy (resulting the matched set M) against the total surplus between buyers' valuation and sellers' valuation given by the optimal matching on all the incoming orders.

$$AE = \frac{\sum\limits_{(x,y) \in M} (p(y) - p(x))}{\max\limits_{\hat{M} \in \mathcal{M}} \sum\limits_{(x,y) \in \hat{M}} (v_{\mathfrak{J}(y)} - v_{\mathfrak{J}(x)})}$$

Note that the value of the denominator is independent to the currently matching policy while the numerator is determined by the current matching policy. Therefore AE measures the quality of a matching policy.

In many situations, the number of transactions indicates the successfulness of a market. We use the *transaction volume* (TV), i.e., the number of transactions $|M|$, to measure the liquidity of a market.

Finally we use *convergence coefficient* (CC), introduced by Smith [6], to measure the efficiency of a clearing policy. Let

$$CC = \frac{100}{\bar{p}} \sqrt{\frac{(\sum\limits_{(x,y) \in M} (\mathfrak{p}(x,y) - \bar{p}))^2}{n}}$$

where $\bar{p}$ is the average market clearing price. Convergence coefficient is the ratio of standard deviation of transaction prices, which measures the spreads of clearing prices.

4   Implementation of Market Policies

In this section, we briefly describe the approaches we have used for the implementation of each market policy we have introduced in Section 2.2.

4.1   Accepting Policy

An accepting policy defines how incoming orders are accepted by the market maker. A widely used accepting policy is *quote-beating accepting policy* (QBA) under which the market maker accepts an incoming order if it surpasses the current best price among the unmatched orders. Let x_{out}^{ask} and x_{out}^{bid} be the best prices among the current unmatched ask orders and bid orders, respectively. For any incoming order x, the QBA satisfies the following accepting rule:

$$\mathfrak{A}(x) = \begin{cases} 1, & \text{if } \left(\mathfrak{I}(x) \in S \text{ and } p(x) < p(x_{out}^{ask})\right) \text{ or } \left(\mathfrak{I}(x) \in B \text{ and } p(x) > p(x_{out}^{bid})\right) \\ 0, & \text{otherwise.} \end{cases}$$

The QBA accepts an order if it is better than the best prices in the current unmatched orders. This rule is known as New York Stock Exchange (NYSE) rule [7].

The QBA above frequently fails to reduce the fluctuation of clearing prices as pointed by Niu et al. [7]. In order to reduce the fluctuation, they propose *equilibrium-beating accepting* policy (EBA). Let $\tilde{p}$ be the price of the expected competitive equilibrium and δ be an adjustment parameter. For any incoming order x, the EBA makes a binary decision based on the following condition:

$$\mathfrak{A}(x) = \begin{cases} 1, & \text{if } \left(\mathfrak{I}(x) \in S \text{ and } p(x) \leq \tilde{p} + \delta\right) \text{ or } \left(\mathfrak{I}(x) \in B \text{ and } p(x) \geq \tilde{p} - \delta\right) \\ 0, & \text{otherwise.} \end{cases}$$

Under the EBA, an incoming order is accepted if it exceeds a threshold which consists of the expected competitive equilibrium $\tilde{p}$ and a slack δ. A key issue in the EBA is how to determine the expected equilibrium price $\tilde{p}$ and the slack δ.

We propose a new accepting policy, namely *learning-based accepting policy* (LBA), which requires less parameter tuning than EBA does. A key concept of LBA is to accept an incoming order at higher chances if it is likely to be successfully transacted according to the historical data. Similarly to *linear reward-inaction algorithm* developed by Hilgard and Bower [8], an LBA policy updates the estimation of successful matches at certain prices according to its history data. Let $L : \Re \rightarrow [0, 1]$ be a learning function that assigns an expected matching success rate (a real number between 0 and 1) to an order price. We use two types of learning functions, $L^{ask}(p)$ and $L^{bid}(p)$, for ask and bid orders, respectively. Let $U = [0, 1]$ be a uniform distribution function. Let $\Pr(U)$ be a probability which is randomly drawn from distribution U. For any incoming order x, an LBA policy determines its acceptance according to the following rule:

$$\mathfrak{A}(x) = \begin{cases} 1, & \text{if } (\mathfrak{I}(x) \in S \text{ and } \Pr(U) \leq L^{ask}(p(x))) \\ & \text{or } (\mathfrak{I}(x) \in B \text{ and } \Pr(U) \leq L^{bid}(p(x))) \\ 0, & \text{otherwise,} \end{cases}$$

If a randomly drawn value $\Pr(U)$ is less than or equal to $L(p)$, the market maker accepts an incoming order. A significant part of LBA is how to update the learning function $L(p)$ from the history data. We set the following three types of update rules. Initially, the learning function $L(p)$ is initialised with a constant value α, where $\alpha \in [0, 1]$. The learning function then updates with generated history data. For each successful matched order $(x, y) \in M$, $L(p)$ is increased by a small value ϵ if $0 \leq p \leq p(x)$ or $p(y) \leq p$. For each unmatched order, $x \in X_{out}^{ask}$ or $y \in X_{out}^{bid}$, $L(p)$ is decreased by ϵ if $p(x) \leq p$ or $0 \leq p \leq p(y)$. In other cases, $L(p)$ stays still.

4.2 Matching Policy

Matching policies determine feasible matched pairs between sell orders and buy orders. Since a matching policy is relevant to all the evaluation criteria we have proposed in Section 3, the design of a matching policy is the most important part of market mechanism design. A well-known algorithm is *4-heap algorithm*, proposed by Wurman et al. in [9], which generates efficient and stable matches. The key idea of the 4-heap algorithm is to make matches between the best prices. In other words, the 4-heap algorithm makes matches from the largest bid-ask spreads. In our implementation of market policies, we use 4-heap algorithm for all the experiments we have done.

4.3 Pricing Policy

A pricing policy rules how to set the clearing price for each feasible matched pair. We have implemented three types of well-known pricing policies.

A *mid-point pricing policy* (MPP) is widely used in clearinghouses. The MPP is a unified pricing under which all the matched pairs are transacted at the same price. Among all the matched orders in M, let $x_{in,l}^{ask}$ be the highest matched ask price and $x_{in,l}^{bid}$ the lowest matched bid price. The MPP sets a unified price $\mathfrak{P}(x,y) = \left(p(x_{in,l}^{ask}) + p(x_{in,l}^{bid}) \right) /2$ for all $(x,y) \in M$. This pricing is the median of all prices in a given matched set.

In contrast to the MPP, k-*pricing policy* (k-PP) discriminates prices between different matched pairs. Under k-PP, the clearing price is a weighted average of the bid price and the ask price of the pair. The k-PP sets a clearing price $\mathfrak{P}(x,y) = k \cdot p(x) + (1 - k) \cdot p(y)$, where k is a weight parameter ranges in $[0,1]$.

Similarly to k-PP, N-*pricing policy* (N-PP) is a pricing policy, proposed by Niu *et al.* in [7]. Let $M' \subseteq M$ be all the transacted matched pairs. The N-PP determines a clearing price according to the average prices of the matched pairs in M': $\mathfrak{P}(x,y) = \sum_{(x',y') \in M'} \frac{(p(x')+p(y'))/2}{|M'|}$ for all $(x,y) \in M$.

In addition, we use a pricing policy which always clears at the competitive equilibrium, denoted by CEPP, for benchmarking purpose.

4.4 Clearing Policy

In general, there are two types of double auction markets classified by clearing timing: continuous clearing and periodic clearing. The former clears all matched pairs immediately and the later clears the matched pairs periodically. We implement two types of clearing policies: *continuous clearing* and *round clearing*, where round is the time unit that the traders submit orders in our experiment as we will explain later.

5   Experimental Analysis

We have implemented all the above mentioned market policies on the JCAT platform [10]. The platform has been used as a game server in the Trading Agent Competition in Market Design (TAC-MD) or known as the CAT competition. We have conducted a set of experiments to test the efficiency of each market policy and their combinations.

Our analysis includes identifying the effects of different types of bidding strategies of trading agents. We evaluate the performance of the market mechanisms by considering three types of bidding strategies implemented in the JCAT platform: *zero-intelligence with constraint* (ZI-C), which sets its price randomly, *zero intelligence plus* (ZIP), which is equipped with a learning-based strategy with reactive to market information, and *Roth and Erev* (RE), which is devised by a reinforcement learning strategy to mimic human-like behaviour.

Our experimental settings are described as follows. Each game consists of 20 trading days with several rounds (either 3 or 5 rounds per day) to give learning time for learning-based bidding strategies. Each trader is able to submit a single order in each round; their demand is up to one order in each day. There are 20 sellers and buyers with fixed valuations assigned as follows: $\{\$50, \$55, \ldots, \$145\}$, respectively. Hence, we have the competitive equilibrium at $\$97.5$ and 10 intra-marginal sellers and buyers. Notice that they are the base values for evaluations of market mechanisms. For all experiments, we use the 4-heap algorithm as a matching policy and 10% for the profit fee. For each type of experiments, we run 10 games with the same settings and we evaluate market performances based on the daily average of the evaluation criteria specified in Section 3.

5.1 Pricing Policy Effect

To test market performance under different pricing policies, we fix accept policy to be AAP, matching policy to be 4-heap and clearing policy to be round-based but vary the pricing policies among MPP, k-PP (with $k = 0.5$) and N-PP (with $N = 20$). The experimental results of these three pricing policies are presented in Table 1.

Table 1. Pricing Effects

Pricing Policy	ZI-C				ZIP				RE			
	PR	AE	CC	TV	PR	AE	CC	TV	PR	AE	CC	TV
MPP	15.75	66.08	3.52	4.92	4.64	64.33	15.09	6.01	8.06	34.13	7.90	2.60
k-PP	16.50	67.18	5.21	5.06	4.61	64.79	15.52	6.09	8.16	33.69	7.56	2.56
N-PP	16.50	67.35	1.84	5.06	4.57	63.70	12.72	6.03	7.25	32.38	4.69	2.45

A key result is that pricing policies have a significant impact on the performance of CC but not for the other indicators for all three types of traders. This result suggests us to focus on CC for the design of pricing policies.

A second observation is that N-PP has smaller CC compared to other two policies for all three types of bidding strategies. This can be explained by the volume of historical data required to determine clearing prices. A higher volume helps to improve CC ratio. We use 20 pairs for N-PP, 1 pair for k-PP, and a few pairs for MPP.

Even though ZI-C traders randomly set their bidding prices, the market performance for ZI-C traders is better than the other two market reactive traders. This is particularly true w.r.t. PR and CC for ZIP traders; PR, AE and CC for RE traders. Therefore, it is interesting whether it is possible to improve these indicators by setting appropriate market policies.

5.2 Accepting Policy Effect

In this experiment, we investigate how accepting policies improve market performance. We compare LBA policy with other three benchmarking accepting policies: AAP, QBA and EBA. We use CEPP for pricing policy, which always clears the market at the middle point of overall traders' valuations. We use continuous clearing for clearing policy. For the LBA policy, we set $\alpha = 1.0$ and dynamically change ϵ in the range between 0.005 and 0.2. We set 3 rounds/day. Table 2 gives the result of our experiments.

Table 2. Accepting Policy Effects

Accepting Policy	ZI-C				ZIP				RE			
	PR	AE	CC	TV	PR	AE	CC	TV	PR	AE	CC	TV
LBA	14.80	91.99	13.23	8.75	11.39	75.00	14.09	6.75	11.06	74.36	14.46	6.74
AAP	13.78	92.14	8.86	8.77	10.80	73.45	10.27	6.72	10.75	72.99	10.34	6.71
QBA	14.76	92.07	8.79	8.78	10.41	72.43	9.77	6.57	10.98	74.59	10.12	6.69
EBA	21.92	84.95	0.44	6.84	15.81	60.15	0.49	4.70	16.10	61.67	0.28	4.80

A key observation is that LBA has better performances than AAP and QBA w.r.t. PR, AE and TV for the market reactive traders (ZIP and RE). This means that LBA signals the market reactive traders in an appropriate way: to accept the orders from the intra-marginal traders and to reject the orders from the extra-marginal traders. This means that LBA can learn the expected matching from history data efficiently. To compare with the results in the previous section, LBA significantly improves AE and TV for all three types of traders. These two indicators are preferable for the traders. However, a disadvantage of LBA is fluctuations of clearing prices as CC indicates.

As a benchmarking accepting policy, we made a set of experiments using EBA policy. We assign the competitive price for the accepting threshold. As a result, the EBA has significant performances on PR and CC. It implies that an EBA policy accepts the intra-marginal traders properly. However, this policy has lower performance on AE and TV, compared to LBA policies. It indicates that there are some unaccepted incoming orders from intra-marginal traders. Hence, the setting of the accepting threshold is too strict for some intra-marginal traders. We detail this point in the next experiment.

In this experiment, we have used a CEPP as pricing policy. In practice, the market institution is not able to obtain the competitive equilibrium. Thus, we consider a case where the market institution uses other pricing policies in the following experiments.

5.3 Robustness of Accepting Policies

Finally, we investigate robustness of accepting policies in a practical setting. We set all the parameters the same as the previous experiment except for the pricing policy. We use N-PP as pricing policy, which has been observed as a well-performing policy in the experiments described in Section 5.1. Notice that N-PP has some fluctuations with clearing prices. Therefore, our aim in this experiment is to investigate the robustness of accepting policies w.r.t. price fluctuation. In other words, we investigate market mechanisms that are trustable w.r.t. different market environmnets. In addition to the accepting policies used in the previous experiment, we use an EBA policy with slack 5, denoted by EBA(5). The reason for using the slack is to relax the rejection range of incoming orders. The results of the experiment is presented in Table 3.

Table 3. Robustness of Accepting Policies

Accepting Policy	ZI-C				ZIP				RE			
	PR	AE	CC	TV	PR	AE	CC	TV	PR	AE	CC	TV
LBA	15.12	92.67	7.89	8.68	2.42	93.82	2.40	9.36	11.27	74.38	9.22	6.64
AAP	14.40	92.31	8.94	8.89	5.92	84.84	6.60	8.49	10.59	73.38	10.08	6.64
QBA	14.09	91.63	9.08	8.78	2.34	92.27	2.95	9.24	10.49	72.79	10.43	6.67
EBA	22.13	85.66	1.33	6.93	1.24	7.84	4.14	0.70	15.80	61.60	1.45	4.77
EBA(5)	20.15	90.85	1.99	7.91	2.33	90.05	2.30	8.99	15.28	67.58	2.24	5.45

According to this result, the LBA policy improves all the market performances for all three bidding strategies compared to AAP and QBA policies. Therefore, the LBA policy is more robust for the volatility of clearing prices relative to other compared accepting policies. This may be because a LBA policy is adaptive to the market situations.

An interesting observation is that EBA is no longer high-performing mechanism for ZIP traders if clearing prices are fluctuated. The failure occurs when the clearing price is unbalanced from the competitive equilibrium. In such a case, the strict rejection and the unbalance of the pricing policy make the trader of one-side stop submitting orders, since there is no way to improve for ZIP traders. We also present a case where the rejection range has $5 slack from the competitive equilibrium. The slack improves all the indicators. However, the issues of EBA are how to determine a proper slack and how to estimate the competitive equilibrium.

6   Conclusions and Related Works

In this paper, we have introduced an experimental approach to the design, analysis, and testing of market mechanisms based on double auction. Firstly, we have defined a formal market model that specifies the market policies in a double auction market. Based on this model, we have introduced a set of criteria for the evaluation of market mechanisms. We have designed and implemented a set of specific market policies, including accepting policy, matching policy, clearing policy and pricing policy. A market mechanism is a combination of these market policies. We have conducted a set of experiments with autonomous trading agents to test the efficiency of different market policies.

There are two key findings from our experiments. First, we have observed that a pricing policy has significant effect on the convergence coefficient but does not have similar effect on the other indicators (Section 5.1). This observation suggests that a mechanism designer should focus on the reduction of fluctuations of transaction prices. Second, an LBA policy can help improving allocation efficiency if pricing policies lead fluctuated clearing prices. This is because an LBA policy rejects orders properly according to the historical data.

The approach that decomposes an auction mechanism into a number of parameterized market policies was introduced by Wurman et al. [11]. In this work, we give a formal definition for each market policies and discussed their properties. Niu et al. [7] presented a set of experimental results on the market mechanisms with double auction. They proposed N-pricing policy that reduces convergence coefficient and the equilibrium beating accepting policy that increased allocation efficiency. However, their experiments were based on only two simple criteria while our experiments have been based on more comprehensive criteria.

Although the system we have used for the experiments has been developed based on the JCAT platform, we have designed market mechanisms in a single market case to focus on the fundamental properties of market policies. This is a major difference from the CAT competition [10] and its analysis [12,13,14], since they have dealt with competitive markets. Nevertheless, the findings from the experiments have provided ideas for us to improve market mechanisms for autonomous trading agents. Especially, some of the ideas have been used in the implementation of our TAC Market Design game entry, **jackaroo**, which is the champion of 2009 TAC Market Design (CAT) Tournament.

In competitive markets like the CAT competition, designing attractive mechanisms for traders is a key issue. In such an environment, it is important for traders to choose the most trustable market. Our aim is to consider these trust aspects to deal with competitive market environments [10].

References

1. Mendelson, H.: Market behavior in clearing house. Econometrica 50, 1505–1524 (1982)
2. McAfee, R.P.: A dominant strategy double auction. JET 56, 434–450 (1992)
3. Krishna, V.: Auction Theory. Academic Press, London (2002)
4. Blum, A., Sandholm, A., Zinkevich, M.: Online algorithms for market clearing. J. of the ACM 53(5), 845–879 (2006)
5. Friedman, D.: How trading institutions affect financial market performance: Some laboratory evidence. Economic Inquiry 31(3), 410–435 (1993)
6. Smith, V.L.: An experimental study of competitive market behavior. J. of Political Economy 70(2), 111–137 (1962)
7. Niu, J., Cai, K., Parsons, S., Sklar, E.: Reducing price fluctuation in continuous double auctions through pricing policy and shout improvement. In: AAMAS, pp. 1143–1150 (2006)
8. Hilgard, E.R., Bower, G.H.: Theories of Learning. Prentice-Hall, NJ (1975)
9. Wurman, P.R., Walsh, W.E., Wellman, M.P.: Flexible double auctions for electronic commerce: Theory and implementation. Decision Support System 24(1), 17–27 (1998)
10. Cai, K., Gerding, E., McBurney, P., Niu, J., Parsons, S., Phelps, S.: Overview of cat: A market design competition (version 2.0). Technical report ulcs-09-005, Department of Computer Science, University of Liverpool, UK (2009)
11. Wellman, M.P., Walsh, W.E., Wurman, P.R.: A parametrization of the auction design space. Games and Economic Behavior 35, 304–338 (2001)
12. Vytelingum, P., Vetsikas, I.A., Shi, B., Jennings, N.R.: Iamwildcat: Thewinning strategy for the tac market design competition. In: ECAI, pp. 428–432 (2008)
13. Niu, J., Cai, K., McBurney, P.: An analysis of entries in the first tac market design competition. In: IEEE/WIC/ACM Int'l Conf. on IAT, vol. 2, pp. 431–437 (2008)
14. Niu, J., Cai, K., Parsons, S., Gerding, E., McBurney, P.: Characterizing effective auction mechanisms: Insights from the 2007 tac market design competition. In: AAMAS, pp. 1079–1086 (2008)

Model Checking Games for a Fair Branching-Time Temporal Epistemic Logic*

Xiaowei Huang and Ron van der Meyden

The University of New South Wales, Australia
{xiaoweih,meyden}@cse.unsw.edu.au

Abstract. Model checking games are instances of Hintikka's game semantics for logic used for purposes of debugging systems verification models. Previous work in the area has developed these games for branching time logic. The paper develops an extension to a logic that adds epistemic operators, and interprets the branching time operators with respect to fairness constraints. The implementation of the extended games in the epistemic model checker MCK is described.

1   Introduction

Model checking is a technique used in computer science for the verification of systems designs. Traditionally, model checkers deal with specifications expressed in a variant of temporal logic — this class of model checkers is now widely applied to the verification of computer hardware and computer network communications protocols. In recent years, a number of model checkers have been developed that are based on modal logics that combine temporal and epistemic modalities [1,2,3,4]. These enable the analysis of systems from the perspective of information theoretic properties, and have been applied to problems such as the verification of security protocols [5] and the verification of knowledge-based programs [1,6]. In the context of Artificial Intelligence, epistemic logic has been the focus of a line of work in the multi-agent systems literature, [7,8], where it is used for reasoning about systems of communicating agents.

One of the reasons for the success of model checking technology is that, at least in the case of linear-time temporal logic specifications, it is possible for a model checker to return to the user a *counter-example*, in the form of an "error-trace" which illustrates a possible execution of the system on which the specification fails. This provides concrete information that helps the user to diagnose the source of the error.

For branching-time temporal logics, the situation is somewhat more complex: while counter-examples can be defined [9], in general, they have a structure that is neither easily presented to the user nor easily comprehended, since, rather than a single execution of the system, one needs to deal with multiple executions, in a complicated branching structure. Once one considers temporal and epistemic logics, it becomes even less clear how to make counter-examples comprehensible, since epistemic operators require even greater flexibility to move between different points of different executions of the system.

---

* Work supported by Australian Research Council Linkage Grant LP0882961 and Defence Research and Development Canada (Valcartier) contract W7701-082453.

A. Nicholson and X. Li (Eds.): AI 2009, LNAI 5866, pp. 11–20, 2009.

In the setting of branching-time temporal logic, the complexity of counter-examples has motivated the application of ideas based on Hintikka's game theoretical semantics for logic [10] as an interactive debugging tool. Game theoretical semantics characterizes the truth of a formula in a model in terms of the existence of a winning strategy in a game constructed from the formula and model. In Hintikka games, there are two players: a verifier, whose objective in the game is to justify that the formula holds in the model, and a refuter, whose objective is to show that the formula is false. The rules of the game are constructed so that steps of the game proceed from a formula to its subformulas, with moves corresponding to cases of the recursive semantics of the logic, with players taking turns depending on the structure of the subformula under consideration. If there exists a winning strategy for verifier, then the assertion holds, otherwise refuter has a winning strategy, and the assertion fails. Playing such a game forces the player to focus on particular subformulas of the specification, and particular states of the model. This provides a useful discipline for helping the user to understand the structure of both, while keeping each step of the process simple enough to be easily comprehended.

Originally developed for simpler logics, Hintikka Games have been adapted to the area of temporal logic model checking [11,12,13,14,15] where they are called model checking games. Our contribution in this paper is to further adapt model checking games to the richer setting of temporal epistemic model checking. We extend previous work on model checking games in two directions. First, we deal with epistemic operators as well as branching-time temporal logic operators. Second, we deal with systems that are subject to *fairness* constraints, which express properties of infinite executions, such as "every action that is always enabled is eventually executed". These are more commonly considered in linear-time temporal logic, but have also been considered in a branching time setting. Fair CTL [16] extends CTL (Computational Tree Logic [17]) models with fairness constraints and thus is strictly more expressive. In this paper, we deal with a language combining Fair CTL and epistemic logic with observational semantics.

The structure of the paper is as follows. Section 2 gives the syntax and semantics of a fair branching time epistemic logic $CTLK_n$. In Section 3, we present a number of variants of the model checking game for this logic. We state the main theoretical results concerning the connection between the semantics and strategies in the game in Section 4. In Section 5 we briefly describe our implementation of the game in the model checker MCK [1]. We make some concluding remarks in Section 6.

2   Syntax and Semantics

We work with a logic $CTLK_n$ that combines CTL and the logic of knowledge and common knowledge for n agents. It will be interpreted with respect to structures representing fairness constraints. Let *Prop* be a set of atomic propositions and $Ags = \{1, \ldots, n\}$ be a set of n agents. The syntax of $CTLK_n$ is given by the following grammar:

$$\phi ::= p \mid \neg\phi \mid \phi_1 \vee \phi_2 \mid EX\phi \mid E[\phi_1 U \phi_2] \mid EG\phi \mid K_i\phi \mid C_G\phi \qquad (1)$$

where $p \in Prop$ and $i \in Ags$ and $G \in \mathcal{P}(Ags) \setminus \{\emptyset\}$.

The formula $K_i\phi$ says that agent i knows ϕ, and $C_G\phi$ says that ϕ is common knowledge to the group of agents G. The operators EX, EU and EG are from the logic CTL, and refer to the branching structure of time. $EX\phi$ says that in some possible future, ϕ will hold at the next moment of time, $E[\phi_1 U\phi_2]$ says that in some possible future, ϕ_1 holds until ϕ_2 does, and $EG\phi$ says that in some possible future, ϕ holds at all future times. The logic CTL contains other operators, but these can be treated as defined. For example, $EF\phi = E[TrueU\phi]$, $AX\phi = \neg EX\neg\phi$, $AF\phi = \neg EG\neg\phi$, etc.

We use a semantics for CTLK$_n$ that is based on a variant of the *interpreted systems* model for the logic of knowledge [18]. Let S be a set, which we call the set of global states. A *run* over S is a function $r : \mathbf{N} \rightarrow S$. An *interpreted system* for n agents is a tuple $\mathcal{I} = (\mathcal{R}, \sim_1, \ldots, \sim_n, \pi)$, where $\mathcal{R}$ is a set of runs over S, each $\sim_i$ is an equivalence relation on S, and $\pi : S \rightarrow \mathcal{P}(Prop)$ is an interpretation function.

A *point* of $\mathcal{I}$ is a pair (r, m) where $r \in \mathcal{R}$ and $m \in \mathbf{N}$. We say that a run r' is *equivalent to a run r up to time $m \in \mathbf{N}$* if $r'(k) = r(k)$ for $0 \le k \le m$. We define the semantics of CTLK$_n$ by means of a relation $\mathcal{I}, (r, m) \models \phi$, where $\mathcal{I}$ is an intepreted system, (r, m) is a point of $\mathcal{I}$ and ϕ is a formula. This relation is defined inductively as follows:

- $\mathcal{I}, (r, m) \models p$ if $p \in \pi(r(m))$,
- $\mathcal{I}, (r, m) \models \neg\phi$ if not $\mathcal{I}, (r, m) \models \phi$
- $\mathcal{I}, (r, m) \models \phi_1 \vee \phi_2$ if $\mathcal{I}, (r, m) \models \phi_1$ or $\mathcal{I}, (r, m) \models \phi_2$
- $\mathcal{I}, (r, m) \models EX\phi$ if there exists a run $r' \in \mathcal{R}$ equivalent to r up to time m such that $\mathcal{I}, (r', m + 1) \models \phi$
- $\mathcal{I}, (r, m) \models E[\phi_1 U\phi_2]$ if there exists a run $r' \in \mathcal{R}$ equivalent to r up to time m and $m' \ge m$ such that $\mathcal{I}, (r', m') \models \phi_2$, and $\mathcal{I}, (r', k) \models \phi_1$ for $m \le k < m'$.
- $\mathcal{I}, (r, m) \models EG\phi$ if there exists a run $r' \in \mathcal{R}$ equivalent to r up to time m such that $\mathcal{I}, (r, k) \models \phi$ for all $k \ge m$
- $\mathcal{I}, (r, m) \models K_i\phi$ if for all points (r', m') of $\mathcal{I}$ such that $r(m) \sim_i r'(m')$ we have $\mathcal{I}, (r', m') \models \phi$
- $\mathcal{I}, (r, m) \models C_G\phi$ if for all sequences of points $(r, m) = (r_0, m_0), (r_1, m_1), \ldots (r_k, m_k)$ of $\mathcal{I}$, such that for each $j = 0 \ldots k - 1$, there exists $i \in G$ such that $r_j(m_j) \sim_i r_{j+1}(m_{j+1})$, we have $\mathcal{I}, (r_k, m_k) \models \phi$.

For the knowledge operators, this semantics is essentially the same as the usual interpreted systems semantics. For the temporal operators, it corresponds to a semantics for branching time known as the *bundle semantics* [19,20].

While they give a clean and coherent semantics to the logic, interpreted systems are not suitable as inputs for a model checking program, since they are infinite structures. We therefore also work with an alternate semantic representation based on transition systems with epistemic indistinguishability relations and fairness constraints. A *(finite) system* is a tuple $M = (S, I, \rightarrow, \sim_1, \ldots, \sim_n, \pi, \alpha)$ where S is a (finite) set of global states, $I \subseteq S$ is the set of initial states, $\rightarrow \subseteq S \times S$ is a serial temporal transition relation, each $\sim_i$ is an equivalence relation representing epistemic accessibility for agent $i \in Ags$, $\pi : S \rightarrow \mathcal{P}(Prop)$ is a propositional interpretation, and α is a set of subsets of S, representing a (generalised Büchi) *fairness condition*. The fairness condition is used to semantically represent constraints such as 'whenever A occurs, B occurs at some later time,' or 'A occurs infinitely often,' that refer to infinite temporal evolutions of the system.

Given a system M over global states S, we may construct an interpreted system $\mathcal{I}(M) = (\mathcal{R}, \sim_1, \ldots, \sim_n, \pi)$ over global states S, as follows. The components $\sim_i$ and π are identical to those in M. The set of runs is defined as follows. We say that a *fullpath* from a state s is an infinite sequence of states $s_0 s_1 \ldots$ such that $s_0 = s$ and $s_i \rightarrow s_{i+1}$ for all $i \geq 0$. We use $Path(s)$ to denote the set of all fullpaths from state s, and $FinPath(s)$ for the set of finite prefixes of fullpaths in $Path(s)$. The fairness condition is used to place an additional constraint on paths. A fullpath $s_0 s_1 \ldots$ is said to be *fair* if for all $Q \in \alpha$, there exists a state $s \in Q$ such that $s = s_i$ for infinitely many i. We write $Path^F(s)$ for the set of all fair fullpaths from s. A *run* of the system is a fair fullpath $s_0 s_1 \ldots$ with $s_0 \in I$. We define $\mathcal{R}$ to be the set of runs of M. A formula is said to *hold* in M, written $M \models \phi$, if $\mathcal{I}(M), (r, 0) \models \phi$ for all $r \in \mathcal{R}$.

We say that a state s is *fair* if it is the initial state of some fair fullpath, otherwise the state is *unfair*. We write $F(M)$ for the set of fair states of M. A state s is *reachable* if there exists a sequence $s_0 \rightarrow s_1 \rightarrow \ldots s_k = s$ where $s_0 \in I$. A state is fair and reachable iff it occurs in some run. We write $FR(M)$ for the set of fair and reachable states of M.

3   Game Semantics for CTLK$_n$

We now reformulate the semantics of CTLK$_n$ on structures M in the form of a Hintikka game. In such a game, there are two *players*, namely system (*Sys*) and user (*Usr*). If p is a player, we write $opp(p)$ for the opponent of p; thus, $opp(Sys) = Usr$ and $opp(Usr) = Sys$. In addition to the two players, we have two *roles*, verifier (**V**) and refuter (**R**). At each game state each player will be in some role, and the opponent will be in the opposite role. Intuitively, a player is in the verifier's (refuter's) role when she believes that the specific formula holds (resp., fails) in current state.

One of the main novelties in our game is that we need to deal with fairness and reachability. In principle, one could avoid this by first restricting systems to the fair reachable states, which would not change the semantics of validity. However, in practice, the existence of unfair reachable states is typically an error that the user will want to be able to diagnose. For this reason, we include unfair and unreachable states in the game, and introduce new propositional constants *Fair*, *Reach* and *Init* to represent that a state is fair (resp., reachable, initial).

Each pair (M, ϕ) consisting of a system M and a formula ϕ determines a game. We assume a fixed system M in what follows, and focus on the role of the formula ϕ (and its subformulas) in determining the states of this game. We call the states of the game *configurations*. There are three types of configuration:

1. Initial configuration: there is a unique initial configuration of the game, denoted $Usr{:}\phi$. Intuitively, this corresponds to the user taking the role of the verifier **V**, and claiming that the formula ϕ is valid in M.
2. Intermediate configurations: these have the form $p : \{(s_1, \phi_1), \ldots, (s_m, \phi_m)\}$ where $p \in \{Sys, Usr\}$ is a player and $\{(s_1, \phi_1), \ldots, (s_m, \phi_m)\}$ is a set of pairs, where each s_k is a state in S and each ϕ_k is either a formula or one of the constants *Fair*, *Reach*, *Init*. Intuitively, such a configuration corresponds to the player p taking the role of the verifier **V**, and claiming that the assertion represented by all of the pairs (s_k, ϕ_k) is true of the system M. If ϕ_k is a formula then pair (s_k, ϕ_k) asserts that $M, s_k \models \phi_k$. If

$\phi_k = Fair$ (*Reach, Init*) then pair (s_k, ϕ_k) means s_k is a fair (resp. reachable, initial) state.

3. Final configuration. Configurations "p wins", where $p \in \{Sys, Usr\}$, are used to denote the completion of play. Intuitively, this means that player p has won the game and $opp(p)$ has lost the game.

Note that each initial and intermediate configuration has the form $p : x$, implying that player p is in the role of verifier and player $opp(p)$ is in the role of the refuter. We write intermediate representations $p : \{(s_1, \phi_1)\}$ with a singleton set of pairs simply as $p : (s_1, \phi_1)$.

At each round of the game, it is the turn of one of the players to make a move, depending on the configuration. Players' roles may exchange during the game. In Table 1, we list the rules of the game. Each rule is in the form

$$\frac{CurrentConfiguration}{NextConfiguration} \quad Role \quad (Condition) \qquad (2)$$

representing that "if the game is in the *CurrentConfiguration* and the *Condition* holds, then it is the turn of the player in role *Role* to move, and one of the choices available to this player is to move the game into configuration *NextConfiguration*." In the rules, *Condition* and *Role* may not be present. If *Condition* is not present, the move can be made unconditionally. If *Role* is not present, there is only one possibility for the next configuration of the game, and the move can be made automatically without any player making a choice. We assume that $M = (S, I, \rightarrow, \sim_1, \ldots, \sim_n, \pi, \alpha)$ and that the fairness condition α has been presented as a set of propositional logic formulas $\{\chi_1, \ldots, \chi_N\}$, where each χ_k represents the set of states $\{s \mid M, s \models \chi_k\}$. The rule for the common knowledge operator uses the set $Kchain(G, s)$, where G is a set of agents and s is a state, defined to be the set of finite sequences of states $s = s_1 \ldots s_m$ such that for all $k = 1 \ldots m - 1$ we have $s_k \sim_j s_{k+1}$ for some $j \in G$.

Note the use of the propositions *Fair* and *Reach* in the cases for the epistemic operators. For example, to refute a claim that $K_i\phi$ holds at a state s, we need not just a state $t \sim_i s$ where $\neg\phi$ holds, but we also need to assure that this state is in fact fair and reachable. We remark that in the rule for $E[\phi_1 U\phi_2]$, of the tuples $(s_k, Fair)$ it would suffice to include only the last $(s_m, Fair)$. However, inclusion of the earlier such tuples allows the refuter at the next move to select the earliest stage in the path at which a transition to an unfair state is made: this is more informative for the user in debugging the system.

In the rules for $EG\phi$ and *Fair*, we make use of a fact concerning generalized Büchi automata, viz., that there exists a fair fullpath from s (satisfying ϕ at every state) iff there exists a cyclic finite path (satisfying ϕ at every state) such that each of the fairness conditions are satisfied on the loop. More precisely, there exists a finite path $s = s_0 \rightarrow s_1 \rightarrow \ldots \rightarrow s_m$ such that $s_m = s_i$ for some $i < m$, and for each fairness constraint χ_k, there exists an index l_k in the loop (i.e.. $i \leq l_k < m$), such that $M, s \models \chi_k$. In the case of $EG\phi$ we also need that $M, s_k \models \phi$ for all $k = 0 \ldots m - 1$. Note that, whereas rules typically include pairs of the form $(s, Fair)$ to represent that only fair states are used in the semantics, we do not need to do this in the case of the rules for EG and *Fair* since these rules already imply fairness of the states introduced.

Table 1. Game Semantics for Fair CTLK$_n$

Initial:	$\dfrac{Usr : \phi}{Usr : (s, \phi)}$ **R**	$(s \in I)$	

AP:

$$\frac{p : (s, q)}{p \ wins} \quad (q \in Prop \wedge q \in \pi(s))$$

$$\frac{p : (s, q)}{opp(p) \ wins} \quad (q \in Prop \wedge q \notin \pi(s))$$

$\phi_1 \vee \phi_2$:

$$\frac{p : (s, \phi_1 \vee \phi_2)}{p : (s, \phi_k)} \quad \textbf{V} \quad (k \in \{1, 2\})$$

$\neg\phi$:

$$\frac{p : (s, \neg\phi)}{opp(p) : (s, \phi)}$$

$EX\phi$:

$$\frac{p : (s, EX\phi)}{p : \{(t, \phi), (t, Fair)\}} \quad \textbf{V} \quad (s \rightarrow t)$$

$E[\phi_1 U \phi_2]$:

$$\frac{p : (s, E[\phi_1 U \phi_2])}{p : \{(s_1, \phi_1), ..., (s_{m-1}, \phi_1), (s_m, \phi_2), (s_1, Fair), ..., (s_m, Fair)\}} \quad \textbf{V}$$

$$(s_1 ... s_m \in FinPath(s))$$

$EG\phi$:

$$\frac{p : (s, EG\phi)}{p : \{(s_1, \phi), ..., (s_{m-1}, \phi), (s_{l_1}, \chi_1), ..., (s_{l_N}, \chi_N)\}} \quad \textbf{V}$$

$$(s_1 ... s_m \in FinPath(s), \ s_m = s_i, \ i \leq l_1, ..., l_N < m)$$

$K_i\phi$:

$$\frac{p : (s, K_i\phi)}{opp(p) : \{(t, \neg\phi), (t, Fair), (t, Reach)\}} \quad \textbf{R} \quad (t \in S, \ s \sim_i t)$$

$C_G\phi$:

$$\frac{p : (s, C_G\phi)}{opp(p) : \{(s_m, \neg\phi), (s_1, Fair), (s_1, Reach), ..., (s_m, Fair), (s_m, Reach)\}} \quad \textbf{R}$$

$$(s_1 ... s_m \in Kchain(G, s))$$

$(s_1, \phi_1), ..., (s_m, \phi_m)$:

$$\frac{p : \{(s_1, \phi_1), ..., (s_m, \phi_m)\}}{p : \{(s_k, \phi_k)\}} \quad \textbf{R} \quad (1 \leq k \leq m)$$

Fair:

$$\frac{p : (s, Fair)}{p : \{(s_{l_1}, \chi_1), ..., (s_{l_N}, \chi_N))\}} \quad \textbf{V}$$

$$(s_1 ... s_m \in FinPath(s), \ s_m = s_k, \ k \leq l_1, ..., l_N < m)$$

Reach:

$$\frac{p : (s, Reach)}{p : (s_1, Init)} \quad \textbf{V} \quad (s_1 ... s_m \in FinPath(s_1), s_m = s)$$

Init:

$$\frac{p : (s, Init)}{p \ wins} \quad (s \in I)$$

$$\frac{p : (s, Init)}{opp(p) \ wins} \quad (s \notin I)$$

4 Main Result

We can now state the main theoretical result of the paper, which is the equivalence of the model checking problem given above with the existence of a winning strategy in the associated game.

A *strategy* of a player is a function mapping the set of configurations in which it is the players' turn, to the set of possible next configurations according to the rules in Table 1.

A *play* of the game for (M, ϕ) according to a pair of strategies $(\sigma_{Usr}, \sigma_{Sys})$ for the user and system, respectively, is a sequence of configurations $C_0 C_1 \ldots$ such that C_0 is the initial configuration $Usr : \phi$, and at each step k, if it is player p's turn on configuration C_k, then there is a successor C_{k+1} in the play, and $C_{k+1} = \sigma_p(C_k)$. Note that it is no player's turn on a final configuration. Thus, a play is either infinite, or ends in a final configuration. In fact, we can show that all plays are finite.

Proposition 1. *If M is a finite state system and ϕ is any $CTLK_n$ formula, then all plays of the game for (M, ϕ) are finite.*

A *winning strategy* for player p is a strategy σ_p, such that for all strategies $\sigma_{opp(p)}$ for the opponent, all plays of the game according to $(\sigma_p, \sigma_{opp(p)})$ are finite and end in the configuration "p wins".

Theorem 1. *For all finite state systems M and formulas ϕ of $CTLK_n$, we have $M \not\models \phi$ iff there exists a winning strategy for Sys in the game for (M, ϕ).*

This theorem forms the basis for our game-based debugging approach. Suppose the user has written a specification ϕ for a system M, and this specification fails to hold in the system. If the user takes the role Usr in the game for (M, ϕ), and plays against a winning strategy for Sys, then the user will lose the game, however they play. In the process of playing the game, and trying different strategies, the user's attention will be drawn to particular states and subformulas. This may help the user to diagnose why their intuitions (concerning either the systems or the specification ϕ) are not in accordance with the facts.

While the game as defined above guarantees termination of any play of the game, it does so at the cost of including rules that involve the construction of rather large game states: viz., the rules for $E[\phi_1 U \phi_2]$, $EG\phi$, $C_G\phi$ and *Reach*, which involve construction of possibly lengthy paths. This creates a cognitive burden for the human player. It is possible to alleviate this in some cases by using more incremental versions of the rules, provided we weaken the correspondence between satisfaction and strategies.

Define the *recursive variant* of the game by replacing the rules for $E[\phi_1 U \phi_2]$, $C_G\phi$ and *Reach* by the rules in Table 2. The recursive variant admits non-terminating plays. E.g., if $s \rightarrow s$ then $Usr : (s, Reach), Usr : (s, Reach), \ldots$ is an infinite play. Thus, Theorem 1 no longer holds for this variant. However, we can recover the result with a slightly different notion of strategy.

Say that a *non-losing* strategy is a strategy σ_p, such that for all strategies $\sigma_{opp(p)}$ for the opponent, all *finite* plays of the game according to $(\sigma_p, \sigma_{opp(p)})$ end in the configuration "p wins".

Table 2. Recursive Game Semantics for Fair CTLK$_n$

$E[\phi_1 U \phi_2]$:

$$\frac{p : (s, E[\phi_1 U \phi_2])}{p : \{(s, \phi_2 \vee (\phi_1 \wedge EX(E[\phi_1 U \phi_2])))\}}$$

$C_G \phi$:

$$\frac{p : (s, C_G \phi)}{p : \{(s, \bigwedge_{i \in G} K_i(\phi \wedge C_G \phi))\}}$$

Reach:

$$\frac{p : (s, Reach)}{p \text{ wins}} \quad (s \in I)$$

$$\frac{p : (s, Reach)}{p : (t, Reach)} \quad \mathbf{V} \ (t \to s)$$

Theorem 2. *For all finite state systems M and formulas ϕ of CTLK$_n$, we have $M \not\models \phi$ iff there exists a non-losing strategy for Sys in the recursive variant of the game for (M, ϕ).*

The recursive version of the game is equally useful for debugging purposes: it enables the user to diagnose the error based on seeing that they cannot win the game, rather than based on seeing that they always lose the game. The key feature that they may explore the most relevant states and subformulas while playing is retained.

Further variants of the game could be constructed: the recursive variant retains the path constructing rules for $EG\phi$, but we could also make this more incremental by recording in the configuration, the state on which we claim the path under construction loops, as well as the fairness constraints already claimed to have been satisfied on previous states in the loop. We could furthermore revise the recursive variant to make it terminating by adding to configurations sufficient information to detect when a play revisits a previously visited configuration (at the cost of admitting very large configurations). We leave the development of such variants to the reader.

5   Implementation

We have implemented the game to provide a debugging facility for the epistemic model checker MCK [1]. MCK provides the ability to model check specifications in both linear and branching time, using a variety of model checking algorithms, depending on the formula in question and a choice of semantics for knowledge: this can be the observational semantics (as in the present paper), a semantics in which local states consist of the current observation plus a clock value, and a synchrononous perfect recall semantics.

Older versions of MCK have been based on algorithms that use symbolic techniques (binary decision diagrams) [21] to do model checking. The implementation of the game adds to MCK a new *explicit-state* model checking facility: this is an algorithm that performs model checking by means of an explicit construction of the reachable states of the system. The approach is essentially an extension of standard explicit-state algorithms for CTL [21] to include epistemic operators. Explicit-state model checking is only feasible for systems with a small state space, but its benefit for our present purposes is that the explicit-state model checking algorithm can be extended to construct during model checking a winning/non-losing strategy for the system. This strategy is then used by the system to play the game against the user in case the debugging game is invoked. An additional benefit of explicit construction of the (reachable) state space is that this allows the state space to be displayed using a graph visualization tool. Our implementation allows the Graphviz tools to be used for this purpose.

The implementation is based on a version of the game that employs the recursive rules of Table 2 only when it is the turn of the user; at system moves the rules of Table 1 are used. In case unfair reachable states exist in the system, the user is offered the choice of playing the game on the system as given (for diagnosing the reason for such states) or a variant of the system in which such states are removed (in case the existence of such states is what the user actually intended).

6   Conclusion and Future Work

We have presented a model checking game for a fair branching time epistemic logic and its implementation in the model checker MCK. Playing the game can help a user to diagnose errors in MCK models.

In future work, we intend to strengthen our tool in two directions: The first is to enable the system to play the game using symbolic model checking algorithms: this will allow the game to be played on models with much larger statespaces. The second is to make the logic supported by the game more expressive: we are presently developing an extension of the game to include μ-calculus operators (these are already supported in the symbolic model checking algorithms in MCK). This will enable notions such as eventual common knowledge [18] to be handled.

Acknowledgements. An initial implementation of the game and explicit state model checking facility for MCK was done by Jeremy Lee; the current implementation is a significant revision by the authors. Cheng Luo has conducted some maintenance on the system.

References

1. Gammie, P., van der Meyden, R.: MCK: Model Checking the Logic of Knowledge. In: Alur, R., Peled, D.A. (eds.) CAV 2004. LNCS, vol. 3114, pp. 479–483. Springer, Heidelberg (2004)
2. Lomuscio, A., Raimondi, F.: MCMAS: A Model Checker for Multi-agent Systems. In: Hermanns, H., Palsberg, J. (eds.) TACAS 2006. LNCS, vol. 3920, pp. 450–454. Springer, Heidelberg (2006)
3. van Ditmarsch, H.P., Ruan, J., Verbrugge, R.: Sum and Product in Dynamic Epistemic Logic. Journal of Logic and Computation 18(4), 563–588 (2008)
4. Nabialek, W., Niewiadomski, A., Penczek, W., Polrola, A., Szreter, M.: VerICS 2004: A model checker for real time and multi-agent systems. In: Proceedings of the International Workshop on Concurrency, Specification and Programming (CS&P 2004), Informatik-Berichte 170, pp. 88–99. Humboldt University (2004)
5. van der Meyden, R., Su, K.: Symbolic Model Checking the Knowledge of the Dinning Cryptographers. In: The Proceedings of 17th IEEE Computer Security Foundations Workshop (CSFW 2004). IEEE Computer Society, Los Alamitos (2004)

6. Baukus, K., van der Meyden, R.: A knowledge based analysis of cache coherence. In: Davies, J., Schulte, W., Barnett, M. (eds.) ICFEM 2004. LNCS, vol. 3308, pp. 99–114. Springer, Heidelberg (2004)
7. Meyer, J.-J.C., van der Hoek, W.: Epistemic Logic for AI and Computer Science. Cambridge University Press, Cambridge (1995)
8. Shoham, Y., Leyton-Brown, K.: Multiagent Systems: Algorithmic, Game-Theoretic, and Logical Foundations. Cambridge University Press, Cambridge (2009)
9. Clark, E., Grumberg, O., Jha, S., Lu, Y., Veith, H.: Counterexample-Guided Abstraction Refinement for Symbolic Model Checking. Journal of the ACM 50(5), 752–794
10. Hintikka, J.: Logic, Language-Games and Information: Kantian Themes in the Philosophy of Logic. Clarendon Press, Oxford (1973)
11. Stirling, C., Walker, D.: Local Model Checking in the Modal mu-Calculus. Theoretical Computer Science 89, 161–177 (1991)
12. Stirling, C.: Local Model Checking Games. In: Lee, I., Smolka, S.A. (eds.) CONCUR 1995. LNCS, vol. 962, pp. 1–11. Springer, Heidelberg (1995)
13. Lange, M., Stirling, C.: Model Checking Games for Branching Time Logics. Journal of Logic Computation 12(4), 623–939 (2002)
14. Fischer, D., Gradel, E., Kaiser, L.: Model Checking Games for the Quantitive μ-calculus. In: Dans Proceedings of the 25th Annual Symposium on the Theoretical Aspects of Computer Science (STACS 2008), pp. 301–312 (2008)
15. Fecher, H., Huth, M., Piterman, N., Wagner, D.: Hintikka Games for PCTL on Labeled Markov Chains (2008)
16. Clark, E.M., Emerson, E.A., Sistla, A.P.: Automatic Verification of Finite-State Concurrent Systems using Temporal Logic Specifications. ACM Transactions on Programming Languages and Systems 8(2), 244–263 (1986)
17. Clark, E.M., Emerson, E.A.: Synthesis of Synchronization Skeletons for Branching Time Temporal Logic. In: Kozen, D. (ed.) Logic of Programs 1981. LNCS, vol. 131. Springer, Heidelberg (1982)
18. Fagin, R., Halpern, J.Y., Moses, Y., Vardi, M.Y.: Reasoning about Knowledge. MIT Press, Cambridge (1995)
19. Burgess, J.: Logic and Time. Journal of Symbolic Logic 44, 556–582 (1979)
20. van der Meyden, R., Wong, K.: Complete Axiomatizations for Reasoning about Knowledge and Branching Time. Studia Logica 75, 93–123 (2003)
21. Clark, E.M., Grumberg, O., Peled, D.A.: Model Checking. MIT Press, Cambridge (1999)

Multistage Fuzzy Decision Making in Bilateral Negotiation with Finite Termination Times

Jan Richter[1], Ryszard Kowalczyk[1], and Matthias Klusch[2]

[1] Centre for Complex Software Systems and Services,
Swinburne University of Technology, Melbourne, Australia
{jrichter,rkowalczyk}@groupwise.swin.edu.au
[2] German Research Center for Artificial Intelligence, Saarbruecken, Germany
klusch@dfki.de

Abstract. In this paper we model the negotiation process as a multistage fuzzy decision problem where the agents preferences are represented by a fuzzy goal and fuzzy constraints. The opponent is represented by a fuzzy Markov decision process in the form of offer-response patterns which enables utilization of limited and uncertain information, e.g. the characteristics of the concession behaviour. We show that we can obtain adaptive negotiation strategies by only using the negotiation threads of two past cases to create and update the fuzzy transition matrix. The experimental evaluation demonstrates that our approach is adaptive towards different negotiation behaviours and that the fuzzy representation of the preferences and the transition matrix allows for application in many scenarios where the available information, preferences and constraints are soft or imprecise.

Keywords: negotiation, multistage, fuzzy, decision, agents, constraints.

1 Introduction

Negotiation is a multistage decision process where each party seeks to find the best course of actions that leads to an agreement which satisfies the requirements of all agents under the presence of conflicting goals and preferences [1]. The decision process is decentralized in the sense that each negotiation partner has its own apparatus for decision making and common knowledge is limited. The interest in automatically resolving conflicts between autonomous software agents has drawn attention to artificial intelligence research, which has the potential to support and automate negotiation in many real world applications including e-commerce, resource allocation and service-oriented computing [2]. The major challenge is the dynamic and distributed environment in which the agents have limited and uncertain information about other agents and the state of the entire system. With the aim to adapt to changes of other agent's behaviours various models for automated negotiation have been proposed ranging from simple If-then rules, heuristic tactics to more advanced learning and reasoning techniques [1]. The agents may explore their environment and the behaviour of other agents

A. Nicholson and X. Li (Eds.): AI 2009, LNAI 5866, pp. 21–30, 2009.

to gain experience, use pre-existing information from past interactions or may have explicit beliefs about other agents parameters, constraints or the underlying decision models. However, many of these assumptions appear to be difficult to fulfil in many scenarios. In this paper, we propose to model the negotiation process as an optimization problem using a multistage fuzzy decision approach where the agent preferences are expressed via a fuzzy goal and fuzzy constraints. The dynamics of the negotiation in the form of offer-response patterns are modelled as a possibilistic Markov decision process which uses fuzzy (possibilistic) state transitions to represent the uncertain knowledge or beliefs about the opponent. The solution is obtained in form of state-action policies by the method of fuzzy dynamic programming by using two reference cases from past interactions. The experimental evaluation demonstrates that we can obtain negotiation strategies that are able to adapt to different negotiation behaviours given the uncertain knowledge about the opponent. The proposed approach is novel in the sense that automated negotiation has not been modelled and solved as a multistage fuzzy decision problem before.

The paper is organized as follows. In Section 2, we present concepts of fuzzy theory and multistage fuzzy decision making. Section 3 presents the negotiation process and the modelling approach. The experimental evaluation in Section 4 demonstrates the improvement compared to heuristic-based negotiation tactics. Section 5 discusses related work, and finally, Section 6 concludes the paper.

2   Multistage Fuzzy Decision Making

This section briefly recalls concepts and notations of multistage fuzzy decision making [3,4]. We assume that the reader is familiar with basic concepts of fuzzy set theory and refer to [4] for a detailed introduction. A fuzzy set A is a set of pairs $A = \{(\mu_A(x), x)\}$ with $\mu_A : X \to [0, 1]$ being the membership function of elements x in the universe of discourse X. A *fuzzy relation* R between two non-fuzzy sets X and Y is defined in the Cartesian product space $X \times Y$: $R = \{(\mu_R(x, y), (x, y))\}$ for each $(x, y) \in X \times Y$ and $\mu_R(x, y) : X \times Y \to [0, 1]$. A binary fuzzy relation is hence a fuzzy set specifying the fuzzy membership of elements in the relation between two non-fuzzy sets. Similarly, any n-ary fuzzy relation is then defined in $X_1 \times \ldots \times X_n$. This allows the formulation of *max-min* and *max-product compositions* of two fuzzy relations R in $X \times Y$ and S in $Y \times Z$ written $R \circ S$ with

$$\mu_{R \circ_{max-min} S}(x, z) = \max_{y \in Y}[\mu_R(x, y) \wedge \mu_S(y, z)] \tag{1}$$

$$\mu_{R \circ_{max-prod} S}(x, z) = \max_{y \in Y}[\mu_R(x, y) \cdot \mu_S(y, z)] \tag{2}$$

for each $x \in X, z \in Z$. Suppose we have a system under control which dynamics are determined by the transition function $f(x_{t+1}|x_t, u_t)$ where $x_t, x_{t+1} \in X = \{\sigma_1, \ldots, \sigma_n\}$ are the states and $u_t \in U = \{\alpha_1, \ldots, \alpha_m\}$ are the controls of the system at stages t, respectively, $t + 1$ with $t = 0, 1, \ldots, N - 1$. Then, according to [3,4] a multistage fuzzy decision problem for the system under control can be defined as $\mu_D(x_0) = \mu_{C^0} \wedge \ldots \wedge \mu_{C^{N-1}} \wedge \mu_{G^N}$ where C^t is a fuzzy constraint at stage t, G^N is the fuzzy goal imposed at the final stage, and D is the fuzzy decision given the initial state x_0. It should be noted that the *min*-operation ($\wedge$) is used as the aggregation operator throughout the paper, but clearly any other t-norm can be employed. The optimal decision $x^* \in X$ then satisfies all constraints and goals such that D is maximized with $\mu_D(x^*) = \max_{x \in X} \mu_D(x)$. In this paper, we consider a system with fuzzy state transitions defined by a conditional fuzzy relation $\mu(x_{t+1}|x_t, u_t)$ with $\mu : X \times U \times X \to [0,1]$, assigning for each $x_t \in X$ and $u_t \in U$ a fuzzy value to the consecutive state $x_{t+1} \in X$. The dynamics of the system can be interpreted as a fuzzy Markov decision process or, from the viewpoint of possibility theory, as assigning a possibility degree to each state transition which determines how plausible the attainment of a succeeding state is [5]. The problem of finding the optimal sequence of controls (or actions) $u_0^*, \ldots, u_{N-1}^*$ for a given intial state x_0 that maximizes the fuzzy decision D can then be written as

$$\mu_D(u_0^*, \ldots, u_{N-1}^* | x_0) = \max_{u_0, \ldots, u_{N-1}} [\mu_{C^0}(u_0) \wedge \ldots \wedge \mu_{C^{N-1}}(u_{N-1}) \wedge E\mu_{G^N}(x_N)], \tag{3}$$

where $E\mu_{G^N}(x_N)$ is the expected goal giving the maximum expected possibility over all next states x_N for controls u_{N-1} and states x_{N-1} with

$$E\mu_{G^N}(x_N) = \max_{x_N \in X} [\mu(x_N | x_{N-1}, u_{N-1}) \wedge \mu_{G^N}(x_N)]. \tag{4}$$

Using the fuzzy transition matrix $\mu(x_N | x_{N-1}, u_{N-1})$ and the expected goal[1] a dynamic programming solution is given by the recurrence equations [4,3]:

$$\mu_{G^{N-i}}(x_{N-i}) = max_{u_{N-i}} [\mu_{C^{N-i}}(u_{N-i}) \wedge E\mu_{G^{N-i+1}}(x_{N-i+1})] \tag{5}$$

$$E\mu_{G^{N-i+1}}(x_{N-i+1}) = \max_{x_{N-i+1} \in X} [\mu(x_{N-i+1} | x_{N-i}, u_{N-i}) \wedge \mu_{G^{N-i+1}}(x_{N-i+1})], \tag{6}$$

for $i = 1, \ldots, N$. Any other s-t norm composition such as the above max-product composition can be used [4]. The optimal solution for Eq. (5,6) for (3,4) is expressed in terms of a policy function with $a_t^* : X \to U$ being the optimal policy at stages $t = 0, 1, \ldots, N-1$ and $u_t^* = a_t^*(x_t)$. The set $A^* = \{a_0^*, \ldots, a_{N-1}^*\}$ then forms the optimal control strategy.

3 Modelling Approach

3.1 Negotiation Model

In this paper we model a bilateral negotiation based on the service-oriented negotiation model by Faratin et al [2] where two agents a and b propose offers

---

[1] Since the expected goal represents a fuzzy relation between states and actions at stage $N - i$ the correct notation for the expected goal is $E\mu_{G^{N-i+1}}(x_{N-i}, u_{N-i})$. In this paper, however, we use the simplified notation introduced by Kacprzyk [4].

$o_a^{k_n}$ and counteroffers $o_b^{k_{n+1}}$ at discrete time points k_n and k_{n+1} on a continuous real-valued issue such as price or delivery time. Each agent has a negotiation interval $[min_a, max_a]$ for the issue j under negotiation defined by the initial value and reservation value of each agent. If the negotiation intervals of both partners overlap an agreement is generally possible. The sequence of all offers exchanged until a time k_n is denoted by the negotiation thread

$$NT_{k_n} = (o_a^{k_1}, o_b^{k_2}, o_a^{k_3}, o_b^{k_4}, \ldots, o_b^{k_n}) \tag{7}$$

with $n \in \mathbb{N}$ and $o_b^{k_n}$ being the last offer of the negotiation thread at time step k_n. Each negotiating agent has a scoring function $V_{aj} : [min_{aj}, max_{aj}] \to [0,1]$ associated to issue j which assigns a score to the current value of the issue within its acceptable interval. The additive scoring function for all issues is $V_a(o) = \sum_{1 \leq j \leq p} w_{aj} * V_{aj}(o)$ where the weight w_a represents the relative importance of issue j for agent a with $\sum_j w_{aj} = 1$. The additive scoring function is either monotonically increasing or decreasing. Utility functions typically correspond to such scoring functions and may include discounts or negotiation costs. Offers are exchanged alternatively during the negotiation until one agent accepts or withdraw from the encounter. An agent's response can hence be defined as

$$response_a(k_{n+1}, o_b^{k_n}) = \begin{cases} withdraw & \text{if } k_{n+1} > k_{max}^a \\ accept(o_b^{k_n}) & \text{if } V_a(o_b^{k_n}) \geq V_a(o_a^{k_{n+1}}) \\ offer(o_a^{k_{n+1}}) & \text{otherwise,} \end{cases}$$

where $o_a^{k_{n+1}}$ is the counterproposal of agent a given agent b's offer $o_b^{k_n}$ at time step k_n. For simplicity, in this paper we consider single-issue negotiation only ($p = 1$). Using this model the agents can have different decision models or negotiation tactics to propose offers and counteroffers [2]. In the next sections we demonstrate how the agents' offers can be used to model negotiation strategies by a multistage fuzzy decision process.

3.2   Negotiation as a Multistage Fuzzy Decision Process

States and Actions. A negotiation according to the model described in above can be modelled using the approach for multistage fuzzy decision making. In the following we refer to actions instead of controls due to the autonomous behaviour of negotiation partners and the limited and uncertain information available. The state space X is created by discretizing the negotiation range, which is generated with the first offer proposals of both agents:

$$X = \left\{ \frac{|o_a^{k_1} - o_b^{k_2}|}{n - 1} * (l - 1) + \min(o_a^{k_1}, o_b^{k_2}) | l = 1, \ldots, n \right\}. \tag{8}$$

An agent's a action is modelled as its response rate to the proposed offer of the opponent agent b and is defined by the relative first order difference,

$$r_a^{k_{n+1}} = \frac{\Delta^1 o_a^{k_{n+1}}}{\Delta^1 o_b^{k_n}}, \tag{9}$$

where $\Delta^1 o_a^{k_{n+1}}$ and $\Delta^1 o_b^{k_n}$ are the previous concessions of agent a and b respectively and $r_a^{k_{n+1}} \in \mathbb{R} : Min(r_a) \leq r_a^{k_{n+1}} \leq Max(r_a)$ with $Max(r_a)$ and $Min(r_a)$ defining the range of acceptable response rates for agent a. The response rate hence specifies the change of the agent's behaviour in relation to the change of the behaviour of the modelled agent [6]. This ensures a monotonic series of offers and at the same time avoids repeated proposals of concessions when the opponent remains in the same state. The action space U is then given by

$$U = \left\{ \frac{Max(r_a) - Min(r_a)}{m-1} * (v-1) + Min(r_a) | v = 1, \ldots, m \right\}, \qquad (10)$$

where a can be a buyer or seller agent. Since the notation of time differs for the negotiation model in 3.1 and the multistage decision model in 2 the alternating sequence of offers in the negotiation thread is mapped to the state-action form where an offer and counteroffer at time k_i and k_{i+1} corresponds to stage t. If, for example, agent a applies this model and agent b proposes the first offer the trajectory TR_t of states x_i and actions u_i until stage t is given by the negotiation thread NT_{k_n} such that TR_t is written as

$$(x_0, u_0, x_1, u_1, \ldots, x_{t-1}, u_{t-1}, x_t) \equiv (o_b^{k_1}, o_a^{k_2}, o_b^{k_3}, r_a^{k_4}, \ldots, o_b^{k_{n-2}}, r_a^{k_{n-1}}, o_b^{k_n}) \quad (11)$$

The optimal response rate $r_a^{k_{n+1}}$ is the action u_t sought by agent a given the state x_t (i.e. offer $o_b^{k_n}$). The offers in the continuous space of the issue under negotiation are mapped to the discrete action and state spaces with $x_t = \arg\min_{\sigma \in X} |o_b^{k_n} - \sigma|$ and $u_t = \arg\min_{\alpha \in U} |r_a^{k_{n+1}} - \alpha|$, where σ and α denote states and actions, respectively. According to Eq. (9) agent a can generate the next counteroffer

$$o_a^{k_{n+1}} = o_a^{k_{n-1}} + r_a^{k_{n+1}}(o_b^{k_n} - o_b^{k_{n-2}}). \qquad (12)$$

The agent needs at least two opponent's offers in order to apply the algorithm. The course of actions proposed by the decision policy can be regarded as a dynamic tit-for-tat strategy where the algorithm adapts the response rate of the agent to the fuzzy preferences given the fuzzy transition matrix.

Transition Matrix Creation. The fuzzy transition matrix represents the agent's beliefs or uncertain knowledge about the opponent's concession behaviour and the agent's responses that may lead to an agreement. The matrix therefore contains the fuzzy state transitions for the range of opponent's offers (states) and possbile response rates (actions) of the agent. In order to obtain the fuzzy transition matrix a small number of reference cases might be used, e.g. in the form of past negotiations. In this paper we focus on the scenario where only a few reference cases are available and their similarity is used to create and update the transition matrix. Let $NT[h]$ be the negotiation thread of reference case h from the set of all cases and $k_{max}[h]$ the negotiation length of h, then the thread can be transformed into the state-action form according to Eq. (14) obtaining the trajectory $TR[h]$ with the number of stages $t_{max}[h]$. The sets of all states and actions of case h are denoted as $X[h]$ and $U[h]$ respectively, where $x_i[h]$ and $u_i[h]$ is a state and action at stage i. The similarity of the current negotiation trajectory TR_{curr} at time t with the reference case h is then defined by

$$sim_t(TR[h], TR_{curr}) = \frac{1}{t+1} \sum_{i=0}^{t} \frac{1 - |x_i[h] - x_i|}{(\max_{h \in H}(x_i[h]) - \min_{h \in H}(x_i[h]))} \qquad (13)$$

for $i \leq t_{max}[h]$. The similarity values provide the necessary fuzzy transitions for states x_i in each case in comparison to the current negotiation. If at the current stage the state exceeds the last offer from a particular case the last similarity value is kept. For each case we can create a state set $X_i'[h]$ for stage i which consists of all states between two consecutive states in the trajectory:

$$X_i'[h] = \begin{cases} \{\sigma_l | x_i[h] \leq \sigma_l \leq \sigma_n\} & \text{for } i = 0 \\ \{\sigma_l | x_{i-1}[h] < \sigma_l \leq x_i[h]\} & \text{for } 0 < i < t_{max}[h] - 1 \\ \{\sigma_l | \sigma_1 < \sigma_l \leq x_i[h]\} & \text{for } i = t_{max}[h] - 1, \end{cases} \qquad (14)$$

where σ_l is a state in the state space X with $l = 1, \ldots, n$. Based on an initial zero transition matrix $\mu(x_{t+1}, x_t, u_t) = 0_{n,m,n}$ for all m actions and n states, the similarity values are created for all cases at each stage to update the fuzzy transition matrix, such that $\mu(x_{i+1}[h], x_i'[h], u_i[h]) = sim_t(TR[h], TR_{curr})$ for all $x_i'[h] \in X_i'[h]$, $x_i[h] \in X[h]$ and $u_i[h] \in U[h]$ with $i = 0, 1, \ldots, t_{max}[h] - 1$. In order to enable fuzzy reasoning over the complete action and state space in the expected goal, we can interpolate over all actions that are zero by using all nonzero state-action pairs. We obtain a fuzzy set over the actions given by

$$E\mu_{G^{t+1}}(x_{t+1} | \sigma_l, \alpha_v) = \frac{E\mu_{G^{t+1}}(\sigma_l, \alpha_{v_2}) - E\mu_{G^{t+1}}(\sigma_l, \alpha_{v_1})}{v_2 - v_1} * (v - v_1) + \alpha_{v_1}, \qquad (15)$$

with $v, v_1, v_2 \in \{1, \ldots, m\}$, $l = 1, \ldots, n$ under the condition that $v_1 < v < v_2$ and $E\mu_{G^{t+1}}(\sigma_l, \alpha_{v_1})$ and $E\mu_{G^{t+1}}(\sigma_l, \alpha_{v_2})$ are greater than zero. In the scenario where the expected goal for actions α_1 and α_m is zero a membership value ε with $\varepsilon > 0$ and $\varepsilon \ll \min_{v | E\mu_{G^{t+1}}(\sigma_l, \alpha_v) > 0}(E\mu_{G^{t+1}}(\sigma_l, \alpha_v))$ is chosen before the interpolation method is applied. The rationale behind is that a limited number of cases is sufficient to estimate an agent's response to the trajectory of offers from the negotiation partner. Since all values in the expected goal are greater than zero for all states and actions after the interpolation, the model can also propose actions not covered by any of the reference cases. Therefore, this approach provides a great flexibility towards the creation of adaptive negotiation strategies.

Fuzzy Goal and Constraints. In the context of negotiation the fuzzy goal and the fuzzy constraints represent the preferences over the opponent's offers (states) and the response rates (actions) of the modelling agent. The degree of membership in the fuzzy goal increases for states closer to the initial value of the agent as they are more preferable to states close to the initial offer of the opponent. Thus, the membership degrees for all states in the fuzzy goal have to be non-zero. Otherwise a state might never be reached as a final or intermediate state. The influence of the (normalized) fuzzy constraint's on the actions depends on the shape and the support of the fuzzy constraints and on the fuzzy set (possibility distribution) over all actions for each state in the expected goal matrix. However, the effect can be increased by normalizing the expected goal distribution for each particular state before the constraint is applied with

$$E\hat{\mu}_{G^i}(x_t|x_{t-1}, u_{t-1}) = \frac{E\mu_{G^i}(x_t|x_{t-1}, u_{t-1})}{\max\limits_{u_{t-1} \in U} E\mu_{G^i}(x_t|x_{t-1}, u_{t-1})} \tag{16}$$

for each $u_{t-1} \in U$ and $x_{t-1} \in X$. As the preference order over all states has to be preserved for all stages $N - i$, the resulting expected goal distribution for each state is scaled back to the height of the subnormal distribution from before the intersection.

4   Evaluation

We evaluate our approach against static mixed strategies in a bilateral, single-issue negotiation environment proposed in [6] and [2] in different deadline scenarios with partial overlap of negotiation intervals. The tactics employed by the mixed strategies are the time- and behaviour-dependent tactics introduced in [2]. The time-dependent tactics use the polynomial decision function with 3 different types of negotiation behaviour: boulware (B), linear (L) and conceder (C) with following settings: $B = \{\beta|\beta \in \{0.1, 0.3, 0.5\}\}$, $L = \{\beta|\beta \in \{1\}\}$, $C = \{\beta|\beta \in \{4, 6, 8\}\}$. Behaviour-dependent tactics are absolute (a) and relative (r) tit-for-tat with $\delta = 1$ and $R(M) = 0$. Weights are classified into 3 groups of small (S), medium (M) and large (L) with $S = \{\gamma|\gamma \in \{0.1, 0.2, 0.3\}$, $M = \{\gamma|\gamma \in \{0.4, 0.5, 0.6\}$ and $L = \{\gamma|\gamma \in \{0.7, 0.8, 0.9\}$. The set of all possible strategies is constructed with $ST = (C \cup L \cup B) \times \{a, r\} \times (S \cup M \cup L)$. We compare our multistage fuzzy decision strategy with a random selected strategy of the set ST from the viewpoint of a buyer agent b while the opponent (seller s) plays a particular subset of static mixed strategies. In that sense the buyer with the random selected strategy plays all strategies ST against one subset of ST (seller) where the average gained utility is compared with the utility of the buyer agent using our adaptive strategy. We use 3 different deadline scenarios where the seller has a shorter, equal or larger deadline than the buyer with $k^b_{max} = 20$ and $k^s_{max} \in \{15, 20, 25\}$. Negotiation intervals of both agents have partial overlap with $min_b = 10$, $max_b = 25$ and $min_s = 15$ and $max_s = 30$. Scoring functions are assumed to be linear and result in a value of 0.1 at the reservation value. Thus, successful negotiations are scored higher as failed negotiations. We use the cost-adjusted utility given by the scoring and cost function $cost(t) = \tanh(t * c)$ with a small communication cost of $c = 0.01$. In order to create a scenario where limited information about the opponent is available we use two reference cases. Figure 1 shows the chosen cases as well as the fuzzy goal and an example of a fuzzy constraint. All fuzzy constraints are specified by an isosceles triangle membership function where the

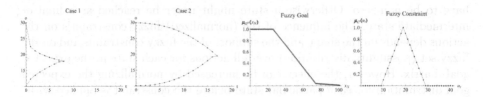

Fig. 1. Reference cases, fuzzy goal and constraints

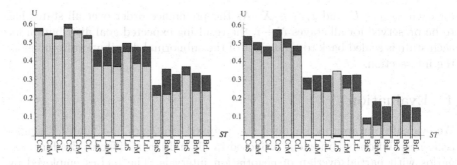

Fig. 2. Utilities without (left) and with communication cost (right)

following set of response rates defines the maximum constraints membership value $\arg\max_{u_t \in U}(u_t) = \{0.4, 0.4, 0.5, 0.5, 0.6, 0.6, 0.8, 0.8, 0.9, 0.9, 1, 1, 1, 1, 1, 1, 1, 1, 1.5, 2\}$. The experimental results without and with communication costs are shown in Figure 2. The light bars represent an average value of utility of a negotiation agent using the random selected strategy from the set of static mixed strategies and the opponent playing a particular strategy group (e.g. CaS). The dark (red) bar demonstrates the gain in utility of the agent using the multistage fuzzy decision approach. As we can see our adaptive strategy performs better than the random strategy selection in almost all scenarios, whereas the improvement is significant when the seller chooses linear or boulware mixed strategies. The reason for that is that the multistage fuzzy decision strategy is able to adapt to the behaviour of the opponent over time whereas the behaviour-dependent part of the static mixed strategy depends on the weight and the pre-defined imitative tactic without considering the opponent's behaviour during the encounter. Hence, boulware and linear tactics often miss the trajectory of opponent's offers in situations where deadlines differ and negotiation intervals overlap only to some degree. It should be noted that the gain in utility depends to a high degree on the choice of the reference cases as they constitute the course of actions that can possibly lead to an agreement. The main advantage is that the agents can adjust or add new negotiation patterns and constraints in future negotiations in order to increase their utility gain or the number of agreements.

5 Related Work

The aim to create negotiation strategies that can adapt to negotiation behaviours of opponents has been a subject of many fields in artificial intelligence. However, many learning and reasoning models require the agents to explore the behaviour of partners, or assume that they have explicit prior knowledge or beliefs about the partner's parameters or underlying decision models [1]. In case-based reasoning, for example, past successful interactions are used to negotiate similar agreements. Matos and Sierra [7] applied case-based reasoning to adjust parameters of combined decision functions. Alongside the negotiation thread of each case the parameter values of the applied strategies are required, which inhibits the use of cases by agents using different decision models. Similiar to our approach Wong et al [8] uses concessions to capture past cases and applies filters to find the best matching case. The major difference to our approach in both methods is that there is no inference on, and interpolation between the cases and the preferences of the agent. The idea of using possibility theory in negotiation has been applied in [9] where potentially good negotiation partners are selected based on the expected qualitative utility. However, the negotiation process is not modelled directly using a fuzzy (or possibilistic) Markov decision process given the limited and uncertain knowledge about the opponent. Relatively few efforts have been reported about using Markov decision processes for modelling negotiation strategies. Narayanan and Jennings [10] model the agent's behaviour by defining the states in terms of resource availability, deadlines and reservation values where counteroffers are proposed based on the opponent's offers and changes in those three realms. The authors show that agreements can be achieved much faster when both agents use this algorithm, but provide no results for cases where only one agent uses this strategy. Similar to our method, Teuteberg [11] models the behaviour of the opponent, but uses a probabilistic approach to generate the transition matrix based on a predefined set of opponent tactics. The major disadvantage is the a large number of negotiations needed to obtain sufficient empirical data for reliable state transitions. Negotiation has also been modelled as a fuzzy constraint satisfaction problem [12] where constraints, preferences and objectives are represented uniformly as fuzzy sets which are distributed among the agents and iteratively relaxed during the exchange of offers [1]. The search process is guided by ordering and pruning the search space but still requires negotiation strategies for proposing offers [13]. Based on the seminal paper of Bellmann and Zadeh [3] decision making in fuzzy environments has been studied and extended by many researchers, such as Kacprzyk [4], Iwamoto [14] and Dubois et al [5], and has been applied in many areas including resource allocation, planning or scheduling [4]. To our knowledge multistage fuzzy decision making has not been used to model agent-based negotiation strategies before.

6 Conclusion

In this paper we have modelled the negotiation process as a multistage fuzzy decision problem where the agents preferences are expressed by a fuzzy goal and fuzzy constraints. A fuzzy Markov decision process represents the uncertain and limited knowledge about the opponent in the form of offer-response patterns whereas two reference cases has been used to create and update the transition matrix. The experimental evaluation has demonstrated that the mechanism is able to adapt to different negotiation behaviours of the opponent and achieves on average higher utilities than the heuristic based negotiation strategies, also in scenarios with different deadlines. In future work we will investigate multistage fuzzy decision models with fuzzy or implicitly specified termination times.

References

1. Braun, P., Brzostowski, J., Kersten, G., Kim, J., Kowalczyk, R., Strecker, S., Vahidov, R.: Intelligent Decision-making Support Systems. In: e-Negotiation Systems and Software Agents: Methods, Models, and Applications, pp. 271–300. Springer, Heidelberg (2006)
2. Faratin, P., Sierra, C., Jennings, N.R.: Negotiation decision functions for autonomous agents. Robotics and Autonomous Systems 24(3-4), 159–182 (1998)
3. Bellman, R., Zadeh, L.: Decision making in a fuzzy environment. Management Science 17(4), 141–164 (1970)
4. Kacprzyk, J.: Multistage Fuzzy Control: A Model-based Approach to Fuzzy Control and Decision Making. John Wiley & Sons, Inc., New York (1997)
5. Dubois, D., Fargier, H., Lang, J., Prade, H., Sabbadin, R.: Qualitative decision theory and multistage decision making - a possibilistic approach. In: Proceedings of the European Workshop on Fuzzy Decision Analysis for Management, Planning and Optimization (1996)
6. Brzostowski, J., Kowalczyk, R.: Predicting partner's behaviour in agent negotiation. In: Proceedings of the fifth international joint conference on Autonomous agents and multiagent systems, pp. 355–361. ACM, New York (2006)
7. Matos, N., Sierra, C.: Evolutionary computing and negotiating agents. In: First International Workshop on Agent Mediated Electronic Trading on Agent Mediated E-Commerce, London, UK, pp. 126–150. Springer, Heidelberg (1999)
8. Wong, W.Y., Zhang, D.M., Kara-Ali, M.: Towards an experience-based negotiation agent. In: Klusch, M., Kerschberg, L. (eds.) CIA 2000. LNCS (LNAI), vol. 1860, pp. 131–142. Springer, Heidelberg (2000)
9. Brzostowski, J., Kowalczyk, R.: On possibilistic case-based reasoning for selecting partners in multi-agent negotiation. In: Webb, G.I., Yu, X. (eds.) AI 2004. LNCS (LNAI), vol. 3339, pp. 694–705. Springer, Heidelberg (2004)
10. Narayanan, V., Jennings, N.R.: An adaptive bilateral negotiation model for e-commerce settings. In: 7th Int. IEEE Conference on E-Commerce Technology, Washington, DC, USA, pp. 34–41. IEEE Computer Society, Los Alamitos (2005)
11. Teuteberg, F., Kurbel, K.: Anticipating agents' negotiation strategies in an e-marketplace using belief models. In: Abramowicz, W. (ed.) Proceeding of the 5th International Conference on Business Information Systems, Poznan, Poland (2002)
12. Luo, X., Jennings, N.R., Shadbolt, N., fung Leung, H., man Lee, J.H.: A fuzzy constraint based model for bilateral, multi-issue negotiations in Semi-Competitive environments. Artificial Intelligence 148(1-2), 53–102 (2003)
13. Kowalczyk, R.: Fuzzy e-negotiation agents. Soft Computing - A Fusion of Foundations, Methodologies and Applications 6(5), 337–347 (2002)
14. Iwamoto, S., Tsurusaki, K., Fujita, T.: Conditional decision-making in fuzzy environment. Journal of the Operations Research 42(2), 198–218 (1999)

Simple Default Reasoning in Theories of Action

Hannes Strass and Michael Thielscher

Department of Computer Science,
Dresden University of Technology
{hannes.strass,mit}@inf.tu-dresden.de

Abstract. We extend a recent approach to integrate action formalisms
and non-monotonic reasoning. The resulting framework allows an agent
employing an action theory as internal world model to make useful default assumptions. While the previous approach only allowed for modeling static defaults, that are independent of state properties, our extension
allows for the expression of dynamic defaults. Problems that arise due
to the interaction of defaults with the solution of the frame problem are
dealt with accordingly: we devise a general method of integrating defaults into the formal representation of action effects and show that the
method prevents counter-intuitive conclusions.

1 Introduction

Recently, [1] proposed a framework for non-monotonic reasoning in theories of
actions and change by embedding them into Raymond Reiter's default logic
[2]. The approach presented there used atomic, normal default rules without
prerequisites to express static world properties. These properties are assumed
once if consistent and then persist over time until supported or refuted by a
definite action effect.

In this paper, we extend that mechanism to atomic, normal default rules
with prerequisites. They allow us to specify dynamic defaults, that is, default
properties that arise and elapse with changing world features. This is, as we shall
argue, most important to capture the fluctuating nature of dynamic worlds that
an intelligent agent might encounter.

As a motivating scenario (and running example of the paper), consider a
very simple domain with an action Fold(x) that turns a sheet of paper x into
a paper airplane. From experience, we might be able to say that in general,
paper airplanes fly. Yet, we don't want to encode this ability to fly as a definite
action effect or general law; we want to retain the possibility of *exceptions*: if
the obtained paper airplane is observed to be unable to fly, we do not want to
get a contradiction. The extension we present here will allow us to use this kind
of defeasible reasoning in theories of actions and change. We show, by means of
an example, that a straightforward generalization of the approach presented in
[1] to normal default rules allows for unintended default conclusions and then
introduce a general, automatic method that is proven to render such conclusions
impossible. Finally, we show how the idea behind this method can also be used
to specify default effects of non-deterministic actions.

A. Nicholson and X. Li (Eds.): AI 2009, LNAI 5866, pp. 31–40, 2009.
© Springer-Verlag Berlin Heidelberg 2009

2 Background

This section presents the formal preliminaries of the paper. In the first subsection we familiarize the reader with a unifying action calculus that we use to logically formalize action domains, and in the second subsection we recall Raymond Reiter's default logic [2].

2.1 The Unifying Action Calculus

The action Fold of our motivating example is characterized by two sets denoting its positive and negative effects, respectively. This is the general method of specifying actions we pursue here: the stated action effects are compiled into an effect axiom that incorporates a solution to the frame problem (similar to that of [3,4]). These effect axioms and action precondition axioms will be formulated in a unifying action calculus (UAC) that was proposed in [5] to provide a universal framework for research in reasoning about actions.

The most notable generalization established by the UAC is its abstraction from the underlying time structure: it can be instantiated with formalisms using the time structure of situations (as the Situation Calculus [6] or the Fluent Calculus [4]), as well as with formalisms using a linear time structure (like the Event Calculus [7]).

The UAC is a sorted logic language which is based on the sorts FLUENT, ACTION, and TIME along with the predicates $<$: TIME $\times$ TIME (denoting an ordering of time points), $Holds$: FLUENT $\times$ TIME (stating whether a fluent evaluates to true at a given time point), and $Poss$: ACTION $\times$ TIME $\times$ TIME (indicating whether an action is applicable for particular starting and ending time points). In this work, we assume a finite number of functions into sorts FLUENT and ACTION and uniqueness-of-names for all of them.

The following definition introduces the most important types of formulas of the unifying action calculus: they allow to express properties of states and applicability conditions and effects of actions.

Definition 1. *Let s be a sequence of variables of sort* TIME.

- *A state formula $\Phi[s]$ in s is a first-order formula with free variables s where*
  - *for each occurrence of $Holds(\varphi, s)$ in $\Phi[s]$ we have $s \in s$ and*
  - *predicate $Poss$ does not occur.*

Let s, t be variables of sort TIME *and A be a function into sort* ACTION.

- *A precondition axiom is of the form*

$$Poss(A(\boldsymbol{x}), s, t) \equiv \pi_A[s] \tag{1}$$

  where $\pi_A[s]$ is a state formula in s with free variables among $s, t, \boldsymbol{x}$.
- *An effect axiom is of the form*

$$Poss(A(\boldsymbol{x}), s, t) \supset (\forall f)(Holds(f, t) \equiv (\gamma_A^+ \vee (Holds(f, s) \wedge \neg \gamma_A^-))) \tag{2}$$

where

$$\gamma_A^+ = \bigvee_{\varphi \in \Gamma_A^+} f = \varphi \quad and \quad \gamma_A^- = \bigvee_{\psi \in \Gamma_A^-} f = \psi$$

and Γ_A^+ and Γ_A^- are sets of terms of sort FLUENT *with free variables among* $\boldsymbol{x}$ *that denote the positive and negative effects of action* $A(\boldsymbol{x})$.

This definition of effect axioms is a restricted version of the original definition of [5]—it only allows for deterministic actions with unconditional effects. Extending the binary *Poss* predicate of the Situation Calculus, our ternary version *Poss*(a, s, t) is to be read as "action a is possible starting at time s and ending at time t".

Definition 2. *A (UAC) domain axiomatization consists of a finite set of foundational axioms Ω (that define the underlying time structure and do not mention the predicates Holds and Poss), a set Π of precondition axioms (1), and a set Υ of effect axioms (2); the latter two for all functions into sort* ACTION.

The domain axiomatizations used here will usually also contain a set Σ_0 of state formulas that characterize the state of the world at the initial time point.

   We illustrate these definitions with the implementation of the action part of our running example.

Example 3. Consider the domain axiomatization $\Sigma = \Omega_{sit} \cup \Pi \cup \Upsilon \cup \Sigma_0$, where Ω_{sit} contains the foundational axioms for situations from [8], Π contains the precondition axiom *Poss*$(\mathsf{Fold}(x), s, t) \equiv t = Do(\mathsf{Fold}(x), s)$, Υ contains effect axiom (2) characterized by $\Gamma_{\mathsf{Fold}(x)}^+ = \{\mathsf{PaperAirplane}(x)\}$ and $\Gamma_{\mathsf{Fold}(x)}^- = \{\mathsf{SheetOfPaper}(x)\}$, and the initial state is $\Sigma_0 = \{Holds(\mathsf{SheetOfPaper}(\mathsf{P}), S_0)\}$. Using the abbreviation $S_1 = Do(\mathsf{Fold}(\mathsf{P}), S_0)$ we can now employ logical entailment to infer that after folding, the object P is no longer a sheet of paper but a paper airplane:

$$\Sigma \models Holds(\mathsf{PaperAirplane}(\mathsf{P}), S_1) \wedge \neg Holds(\mathsf{SheetOfPaper}(\mathsf{P}), S_1)$$

The next definition introduces reachability of a time point as existence of an action sequence leading to the time point. A second order formula expresses this intuition via defining the predicate *Reach* as the least set containing the minimal elements of sort TIME (the initial time points *Init*) and being closed under possible action application (via *Poss*).

Definition 4. *Let Σ be a domain axiomatization and σ be a time point.*

$$Reach(r) \stackrel{\text{def}}{=} (\forall R)(((\forall s)(Init(s) \supset R(s))$$
$$\wedge (\forall a, s, t)(R(s) \wedge Poss(a, s, t) \supset R(t))) \supset R(r))$$

$$Init(t) \stackrel{\text{def}}{=} \neg(\exists s) s < t$$

We say σ is finitely reachable in Σ if $\Sigma \models Reach(\sigma)$.

2.2   Default Logic

Introduced in the seminal work by Reiter [2], default logic has become one of the most important formalisms for non-monotonic reasoning. Its fundamental notion is that of *default rules*, that specify how to extend an incomplete knowledge base with vague, uncertain knowledge.

Definition 5. *A* normal default rule *(or* normal default*) is of the form* $\alpha[s]/\beta[s]$ *where* $\alpha[s]$ *and* $\beta[s]$ *are state formulas in* $s :$ TIME.
   A default rule is called prerequisite-free *or* supernormal *iff* $\alpha = \top$.

Default rules with free (non-TIME) variables are semantically taken to represent their ground instances. By $\mathcal{D}[\sigma]$ we denote the set of defaults in $\mathcal{D}[s]$ where s has been instantiated by the term σ.

Example 3 (continued). The statement "in general, paper airplanes fly" from Section 1 can easily be modeled by the default rule

$$Holds(\mathsf{PaperAirplane}(y), s)/Holds(\mathsf{Flies}(y), s) \tag{3}$$

Definition 6. *A* default theory *is a pair* $(W, \mathcal{D})$ *where* W *is a set of closed formulas and* $\mathcal{D}$ *a set of default rules.*

The set W of a default theory is the set of indefeasible knowledge that we are unwilling to give up under any circumstances.
   The semantics of default logic is defined through extensions: they can be seen as a way of applying to W as many default rules from $\mathcal{D}$ as consistently possible.

Definition 7. *Let* $(W, \mathcal{D})$ *be a default theory. For any set of closed formulas* S, *define* $\Gamma(S)$ *as the smallest set such that:*

- $W \subseteq \Gamma(S)$,
- $Th(\Gamma(S)) = \Gamma(S)^1$, *and*
- *for all* $\alpha/\beta \in \mathcal{D}$, *if* $\alpha \in \Gamma(S)$ *and* $\neg\beta \notin S$, *then* $\beta \in \Gamma(S)$.

A set of closed formulas E *is called an* extension *for* $(W, \mathcal{D})$ *iff* $\Gamma(E) = E$, *that is,* E *is a fixpoint of* Γ.
   The set of generating defaults *of an extension* E *for* $(W, \mathcal{D})$ *is*

$$gd(E) \stackrel{\text{def}}{=} \{\alpha/\beta \in \mathcal{D} \mid \alpha \in E, \neg\beta \notin E\}$$

We denote the set of all extensions for a default theory by $Ex(W, \mathcal{D})$.

By a result from [2], extensions are completely characterized by the consequents of their generating defaults:

---

[1] $Th(F)$ for a set of formulas F denotes the set of its logical consequences, i.e.
   $Th(F) \stackrel{\text{def}}{=} \{\varphi \mid F \models \varphi\}$.

Lemma 8 (Reiter). *Let E be an extension for $(W, \mathcal{D})$.*

$$E = Th(W \cup \{\beta \mid \alpha/\beta \in gd(E)\})$$

Based on extensions, one can define skeptical and credulous conclusions for default theories: skeptical conclusions are formulas that are contained in every extension, credulous conclusions are those that are contained in at least one extension.

Definition 9. *Let $(W, \mathcal{D})$ be a normal default theory and Ψ be a formula.*

$$W \mathrel{\mid\!\approx}_{\mathcal{D}}^{skept} \Psi \stackrel{\text{def}}{\equiv} \Psi \in \bigcap_{E \in Ex(W, \mathcal{D})} E, \quad W \mathrel{\mid\!\approx}_{\mathcal{D}}^{cred} \Psi \stackrel{\text{def}}{\equiv} \Psi \in \bigcup_{E \in Ex(W, \mathcal{D})} E$$

Example 3 (continued). Taking the indefeasible knowledge

$$W = \{Holds(\mathsf{PaperAirplane(P)}, S)\}$$

for a TIME constant S and $\mathcal{D}[s]$ to contain the default rule (3), we can instantiate the default with time point S and skeptically conclude that P flies:

$$W \mathrel{\mid\!\approx}_{\mathcal{D}[S]}^{skept} Holds(\mathsf{Flies(P)}, S)$$

2.3 Domain Axiomatizations with Supernormal Defaults

We recall the notion of a domain axiomatization with supernormal defaults[2] from [1]. It is essentially a supernormal default theory where the set containing the indefeasible knowledge is an action domain axiomatization.

Definition 10. *A domain axiomatization with supernormal defaults is a pair $(\Sigma, \mathcal{D}[s])$, where Σ is a UAC domain axiomatization and $\mathcal{D}[s]$ is a set of default rules of the form*

$$\top / (\neg) Holds(\psi, s)$$

where ψ is a term of sort FLUENT.

3 Domain Axiomatizations with Normal Defaults

As mentioned before, we loosen the restriction to supernormal defaults and allow default rules with prerequisites. The rest of the definition stays the same.

Definition 11. *A domain axiomatization with (normal) defaults is a pair $(\Sigma, \mathcal{D}[s])$, where Σ is a UAC domain axiomatization and $\mathcal{D}[s]$ is a set of default rules of the form*

$$(\neg) Holds(\varphi, s) / (\neg) Holds(\psi, s) \quad or \quad \top / (\neg) Holds(\psi, s)$$

where φ, ψ are terms of sort FLUENT.

---

[2] The endorsement "supernormal" is only used in this work to distinguish the approaches.

For notational convenience, we identify *Holds* statements with the mentioned fluent and indicate negation by overlining: the default $Holds(\varphi, s)/\neg Holds(\psi, s)$, for example, will be written as $\varphi/\overline{\psi}$. Generally, $\overline{\alpha} = \neg\alpha$ and $\overline{\neg\alpha} = \alpha$. We furthermore use $|\cdot|$ to extract the affirmative component of a fluent literal, that is, $|\neg\alpha| = |\alpha| = \alpha$. Both notions generalize to sets of fluents in the obvious way.

We now show the straightforward implementation of our motivating example.

Example 3 (continued). Recall the domain axiomatization Σ from Section 2.1 and let the set of defaults $\mathcal{D}[s]$ contain the single default rule (3). We see that, after applying the action Fold(P), we can indeed infer that P flies:

$$\Sigma \hspace{0.5em} \mathrel{\vbox{\hbox{$\succ$}}}^{skept}_{\mathcal{D}[S_1]} Holds(\mathsf{Flies}(\mathsf{P}), S_1)$$

Note that we need to instantiate the defaults with the resulting situation S_1 (instantiating the defaults with S_0 would not yield the desired result). Now taking a closer look at effect axiom (2) and its incorporated solution to the frame problem, we observe that also

$$\Sigma \hspace{0.5em} \mathrel{\vbox{\hbox{$\succ$}}}^{skept}_{\mathcal{D}[S_1]} Holds(\mathsf{Flies}(\mathsf{P}), S_0)$$

This is because Flies(P) was not a positive effect of the action—according to the effect axiom it must have held beforehand. This second inference is unintended: first of all, the conclusion "the sheet of paper already flew before it was folded" does not correspond to our natural understanding of the example domain. The second, more subtle, reason is that we used defaults about $S_1 = Do(\mathsf{Fold}(\mathsf{P}), S_0)$ to conclude something about S_0 that could not be concluded with defaults about S_0. In practice, it would mean that to make all possible default conclusions about a time point, we had to instantiate the defaults with all future time points (of which there might be infinitely many), which is clearly infeasible.

4 Relaxing the Frame Assumption

We next extend our specification of actions—up to now only via positive and negative effects—with another set of fluents, called *occlusions* (the term first occurred in [9]; our usage of occlusions is inspired by this work). They do not fix a truth value for the respective fluents in the resulting time point of the action and thus allow them to fluctuate freely. In particular, it is then impossible to determine an occluded fluent's truth value at the starting time point employing only information about the ending time point.

Definition 12. *An* effect axiom with unconditional effects and occlusions *is of the form*

$$Poss(A(\boldsymbol{x}), s, t) \supset (\forall f)(\gamma_A^? \vee (Holds(f, t) \equiv (\gamma_A^+ \vee (Holds(f, s) \wedge \neg\gamma_A^-)))) \quad (4)$$

where

$$\gamma_A^+ = \bigvee_{\varphi \in \Gamma_A^+} f = \varphi, \quad \gamma_A^- = \bigvee_{\psi \in \Gamma_A^-} f = \psi, \quad \gamma_A^? = \bigvee_{\chi \in \Gamma_A^?} f = \chi,$$

and Γ_A^+, Γ_A^-, and $\Gamma_A^?$ are sets of terms of sort FLUENT with free variables among $\boldsymbol{x}$ that denote the positive and negative effects and occlusions of action $A(\boldsymbol{x})$.

It is easily seen that effect axiom (2) is a special case of the above effect axiom with $\gamma_A^? = \bot$ (i.e. $\Gamma_A^? = \emptyset$).

4.1 ... to Prevent Default Reasoning Backwards in Time

Example 3 (continued). Set $\Gamma_{\mathsf{Fold}(x)}^? := \{\mathsf{Flies}(x)\}$ and let $\Sigma' = \Omega_{sit} \cup \Pi \cup \Upsilon' \cup \Sigma_0$, where Υ' contains effect axiom (4) for the action $\mathsf{Fold}(x)$. We see that the desired conclusion is preserved, and the undesired one is now disabled:

$$\Sigma \mathrel{\vert\!\approx}^{skept}_{\mathcal{D}[S_1]} Holds(\mathsf{Flies}(P), S_1) \text{ and } \Sigma \mathrel{\not\vert\!\approx}^{skept}_{\mathcal{D}[S_1]} Holds(\mathsf{Flies}(P), S_0)$$

Specifying the occlusions for the action in the example was easy—there was only one default rule, and we had a precise understanding of the desired and undesired inferences. In general, however, defaults might interact and it might become less obvious which of them to exclude from the frame assumption.

Algorithm 1 below implements a general method of identifying the fluents that are to be occluded, taking into account given default rules. It takes as input positive and negative effects Γ_A^+ and Γ_A^- of an action A and a set $\mathcal{D}$ of defaults and computes the set $\Gamma_A^{?\mathcal{D}}$ of default occlusions for A with respect to $\mathcal{D}$. The intuition behind it is simple: it iterates over a set S of fluents potentially influenced by A. This set is initialized with the definite action effects and then extended according to default rules until a fixpoint obtains.

Algorithm 1. Computing the default occlusions

Input: Γ_A^+, Γ_A^-, $\mathcal{D}$
Output: $\Gamma_A^{?\mathcal{D}}$

1: $S := \Gamma_A^+ \cup \{\overline{\gamma} \mid \gamma \in \Gamma_A^-\}$ *// initialization: literals stating the definite effects*
2: **while** there is $\gamma \in S$, $\alpha/\beta \in \mathcal{D}$, a substitution θ with $\alpha\theta = \gamma$; and $\beta\theta \notin S$ **do**
3:      $S := S \cup \{\beta\theta\}$ *// $\beta\theta$ might become default effect of A*
4: **end while**
5: **return** $|S| \setminus (\Gamma_A^+ \cup \Gamma_A^-)$ *// exclude definite effects from occlusions*

Note that prerequisite-free defaults do not contribute to the computation of occlusions: the symbol $\top$ does not unify with any explicitly mentioned action effect. This behavior is semantically perfectly all right: the intended reading of prerequisite-free defaults is that of static world properties that are once assumed (if consistent) and then persist over time until an action effect either refutes or confirms them.

It is easily seen that Algorithm 1 applied to our running example creates the exact set of occlusions that we figured out earlier "by hand".

For the following theoretical results of this paper, let $(\Sigma, \mathcal{D}[s])$ be a domain axiomatization with defaults where all effect axioms are of the form (2), and let Σ' denote the domain axiomatization with effect axioms (4) where the $\Gamma^?$ are constructed by applying Algorithm 1 to each action of Σ. It should be noted that Σ' is consistent whenever Σ is consistent: default occlusions only weaken the restrictions on successor states, thus any model for Σ is a model for Σ'.

The first proposition shows that the default occlusions computed by Algorithm 1 are sound with respect to default conclusions about starting time points of actions: whenever defaults about a resulting time point can be utilized to infer a state property of the starting time point, this state property can also be inferred locally, that is, with defaults about the starting time point itself.

Lemma 13. *Let α be a ground action and σ, τ be terms of sort* TIME *such that $\Sigma' \models Poss(\alpha, \sigma, \tau)$, and let $\Psi[s]$ be a state formula.*

$$\Sigma' \mathrel{\approx\!\!\!|}^{skept}_{\mathcal{D}[\tau]} \Psi[\sigma] \text{ implies } \Sigma' \mathrel{\approx\!\!\!|}^{skept}_{\mathcal{D}[\sigma]} \Psi[\sigma]$$

Proof (Sketch). We prove the contrapositive. Let $\Sigma' \mathrel{\not\approx\!\!\!|}^{skept}_{\mathcal{D}[\sigma]} \Psi[\sigma]$. Then there exists an extension E for $(\Sigma, \mathcal{D}[\sigma])$ with $\Psi[\sigma] \notin E$. We construct an extension F for $(\Sigma, \mathcal{D}[\tau])$ as follows. By Lemma 8, E is characterized by the consequents of its generating defaults (all of which are Holds literals in σ). We determine F's characterizing set of default consequences by removing the ones that are contradicted via action effects and adding consequents of newly applicable normal defaults. All those new default conclusions are, due to the construction of $\Gamma^{?\mathcal{D}}_{\alpha}$ via Algorithm 1, backed by occlusions and do not influence σ. Thus $\Psi[\sigma] \notin F$. $\square$

The absence of unintended inferences about time points connected via a single action then immediately generalizes to time points connected via a sequence of actions and trivially generalizes to disconnected time points. This is the main result of the paper stating the impossibility of undesired default conclusions about the past.

Theorem 14. *Let σ, τ be time points such that σ is reachable and $\sigma \leq \tau$.*

$$\Sigma' \mathrel{\approx\!\!\!|}^{skept}_{\mathcal{D}[\tau]} \Psi[\sigma] \text{ implies } \Sigma' \mathrel{\approx\!\!\!|}^{skept}_{\mathcal{D}[\sigma]} \Psi[\sigma]$$

Another noteworthy property of the presented default reasoning mechanism is the preservation of default conclusions: even if the prerequisite of a default rule is invalidated due to a contradicting action effect, the associated consequent (if not also contradicted) stays intact. This means the algorithm does not occlude unnecessarily many fluents. It would be fairly easy to modify Algorithm 1 such that the resulting effect axioms also "forget" default conclusions whose generating rules have become inapplicable—we would just have to replace all occurrences of literals by their respective affirmative component.

4.2 ... to Model Default Effects of Actions

The usage of occlusions as advocated up to this point is of course not the only way to make use of this concept. When they are specified by the user along with action effects as opposed to computed automatically, occlusions are an excellent means of modeling default effects of non-deterministic actions:

Example 15 (Direct Default Effect). We model the action of tossing a coin via excluding the fluent Heads (whose intention is to denote whether heads is showing upwards after tossing the coin) from the action Toss's frame axiom, that is, $\Gamma^?_{\text{Toss}} := \{\text{Heads}\}$. However, the coin of this example is unbalanced and has a strong tendency towards landing with heads facing upwards. This is modeled by having a default that states the result Heads as "usual outcome":

$$Holds(\text{Heads}, s) \qquad (5)$$

There is another action, Wait, that is always possible and does not change the truth value of any fluent. All $\gamma^{+/-/?}$ not explicitly mentioned are thus to be taken as the empty disjunction, i.e. false. Using the domain axiomatization Σ, that contains the precondition axioms and effect axioms (4) stated above, situations as time structure, and the observation $\Sigma_O = \{\neg Holds(\text{Heads}, Do(\text{Toss}, S_0))\}$ we can draw the conclusion

$$\Sigma \cup \Sigma_O \models \neg Holds(\text{Heads}, Do(\text{Wait}, Do(\text{Toss}, S_0))) \qquad (6)$$

which shows that the observation "the outcome of tossing was tail" persists during Wait, that is, the fluent Heads does not change its truth value during an "irrelevant" action. Tossing the coin again (which results in situation $S_3 = Do(\text{Toss}, Do(\text{Wait}, Do(\text{Toss}, S_0)))$), this time without an observation about the outcome, rule (5) can be applied and yields the default result regardless of previous observations:

$$\Sigma \cup \Sigma_O \models^{skept}_{\mathcal{D}[S_3]} Holds(\text{Heads}, S_3)$$

Hence, Algorithm 1 can also be used to complete a user-specified set of occlusions regarding potential default effects of actions. When trying to achieve the above behavior without specifying the occlusions manually, that is, using a procedure in the spirit of Algorithm 1 that takes as input only definite effects and default rules, one is unlikely to succeed: automatically creating occlusions for all prerequisite-free defaults will cause all these defaults to apply after every action. In the example above, the coin would then magically flip its side (into the default state Heads) after Wait in $Do(\text{Toss}, S_0)$ and we could not infer (6), which contradicts our intuition that Wait has no effects.

5   Conclusions and Future Work

The paper presented a generalization of a recently proposed mechanism for default reasoning in theories of actions and change. Unlike the approach from [1], our work used a logic that allows to express dynamic defaults in addition to static ones. We observed undesired inferences that arose from the interplay of defaults and the solution of the frame problem, and presented an automatic method of adjusting the action effect axioms to preclude the unintended conclusions. Unfortunately, there seems to be a price to pay for being able to express dynamic defaults. The main result of [1] stated the sufficiency of default instantiation in the least time point when restricted to atomic supernormal defaults. This does not apply to our generalization: occlusions may make a previously inapplicable default rule applicable after action execution, therefore defaults need to be locally instantiated to yield a complete picture of the current state of the world.

It is somewhat clear that the syntax-based approach of Algorithm 1, when generalized to formulas rather than single literals, is prone to occlude both too many fluents (for example if the prerequisite is tautological but not $\top$) and too few fluents (for example if the prerequisite is not fulfilled by an action effect alone, but requires some additional state property). In the future, we will therefore be concerned with suitably generalizing the approach for a more expressive class of defaults. The second direction of generalization will be in terms of considered actions: up to now, we allowed only deterministic actions with unconditional effects. Further research will be undertaken to incorporate nondeterminism and conditional effects.

References

1. Strass, H., Thielscher, M.: On Defaults in Action Theories. In: Mertsching, B., Hund, M., Aziz, Z. (eds.) KI 2009. LNCS (LNAI), vol. 5803, pp. 298–305. Springer, Heidelberg (2009)
2. Reiter, R.: A Logic for Default Reasoning. Artificial Intelligence 13, 81–132 (1980)
3. Reiter, R.: The Frame Problem in the Situation Calculus: A Simple Solution (Sometimes) and a Completeness Result for Goal Regression. In: Artificial Intelligence and Mathematical Theory of Computation – Papers in Honor of John McCarthy, pp. 359–380. Academic Press, London (1991)
4. Thielscher, M.: From Situation Calculus to Fluent Calculus: State Update Axioms as a Solution to the Inferential Frame Problem. Artificial Intelligence 111(1-2), 277–299 (1999)
5. Thielscher, M.: A Unifying Action Calculus. Artificial Intelligence (to appear, 2009)
6. McCarthy, J.: Situations and Actions and Causal Laws, Stanford Artificial Intelligence Project: Memo 2 (1963)
7. Kowalski, R.A., Sergot, M.J.: A Logic-based Calculus of Events. New Generation Computing 4(1), 67–95 (1986)
8. Pirri, F., Reiter, R.: Some Contributions to the Metatheory of the Situation Calculus. Journal of the ACM 46(3), 325–361 (1999)
9. Sandewall, E.: Features and Fluents: The Representation of Knowledge about Dynamical Systems. Oxford University Press, Oxford (1994)

From My Agent to Our Agent: Exploring Collective Adaptive Agent via Barnga

Yuya Ushida, Kiyohiko Hattori, and Keiki Takadama

The University of Electro-Communications,
Chofugaoka 1-5-1, Chofu, Tokyo 182-8585, Japan

Abstract. This paper explores the collective adaptive agent that adapts to a *group* in contrast with the individual adaptive agent that adapts to a *single user*. For this purpose, this paper starts by defining the collective adaptive situation through an analysis of the subject experiments in the playing card game, Barnga, and investigates the factors that lead the group to the collective adaptive situation. Intensive simulations using Barnga agents have revealed the following implications: (1) the leader who takes account of other players' opinions contributes to guide players to the collective adaptation situation, and (2) an appropriate role balance among players (*i.e.*, the leader, the claiming and quiet player, which make the most and least number of corrections) is required to derive the collective adaptive situation.

1 Introduction

Owing to the success of AIBO or PLEO, the researches on *HAI (Human Agent Interaction)* [4] [6] [7] have attracted much attention [1]. The HAI, for example, explores the agent that can adapt to a user through the interaction with a user [2]. Most of these researches address the *individual adaptation* that enables an agent to adapt to a *single user* but not the *collective adaptation* that enables an agent to adapt *multiple users*. Considering the fact that our society consists of a lot of people and is based on the complex interaction among them, the agent has to integrate itself into the society. In particular, the group characteristics cannot be described with the summation of the individual characteristics, which suggests that the function of the collective adaptation is indispensable for the agent in addition to that of the individual adaptation.

However, only a few researchers addressed the collective adaptation. Yamada [5], for example, tried to define the collective adaptive situation as the balance between adaptation (*i.e.*, the change of one's mind to others) and diversity (*i.e.*, the recognition of difference from other's mind). However, the result suggested the appropriate balance was just in the middle of them, which can be regarded as ambiguous. Omura, on the other hand, designed the adaptive agent which tried to adapt the society by understanding the preference of all peoples [3], but did not find any specific and numerical conditions that derive the collective adaptation situation.

A. Nicholson and X. Li (Eds.): AI 2009, LNAI 5866, pp. 41–51, 2009.
© Springer-Verlag Berlin Heidelberg 2009

To tackle the issue of the collective adaptive agent, this paper starts by defining the situation of collective adaptation through an analysis of the subject experiments in the playing card game called *Barnga* [5].

This paper is organized as follows: Section 2 introduces the playing card game Barnga. The collective adaptive situation is defined in Section 3 according to the subject experiments on the Barnga. The Barnga agent is designed in Section 4, and the results of two simulations on the Barnga are presented in the Section 5 and 6. Finally, Section 7 gives our conclusion of this research.

2   Barnga: Cross-Cultural Communication Game

Barnga is the playing card game to experience how the players behave when they face the cross-cultural situation. Barnga is played as the following sequence: (1) the players divided into the number of tables have several cards that range from A to 7; (2) they put one card in turn, and decide the winner who puts the strongest card. In the basic rule, the larger a number is, the stronger a card is. Barnga has the trump card (*i.e.*, heart or diamond) and ace-strong/weak rule besides the rule. The trump card becomes the strongest when the card is put, while the ace-strong/weak rule represents that the ace is the strongest/weakest number. In the ace-strong rule, for example, "A" is the strongest and "2" is the weakest; (3) a winner is determined among the players; This is a whole sequence of Barnga and can be counted as one game.

Note that (i) each table has the slightly different rule (except for the basic rule) as shown in the right side of Fig. 1, and (ii) every player does not know such differences among the tables, which represent the cross-cultural situations. Since some players are swapped every several times of the games, the players meet other players who have different rules as shown in the left side of Fig. 1. Plus, the conversation among the players is not allowed during the game, which force the players to interact with non-verbal communication like body languages to determine the winner. Under the circumstances, players are required to play games by adapting the situations.

Fig. 1. The Image of Barnga Game

3   Collective Adaptation in Barnga

We have conducted the subject experiments of Barnga game to define the collective adaptive situation. This experiment was carried out in order to specify the smooth game because the adaptive situation can be regarded as the results of the transition from non-smooth to smooth game.

3.1   Subject Experiments

Subject experiments are explained briefly in this section. The examinees played Barnga game the results were analyzed by finding something in common with the answers by the players written in the questionnaire such as "Do you think the smooth games occurred (smooth or not smooth in general)?" or "Why do you think so?" Most of the examinees answered three important roles of the players in the table (described in Section 3.2). The three group categorizations(described in Section 3.3) can be made from the results, which help abstract the concept of the collective adaptive situation to a theoretical level.

3.2   Roles in a Group

The subject experiments brought three important roles described for the smooth games; (a) a *leader* who decided the winner to reach at consensus decision when two or more different winners were nominated, (b) the *claiming person* who often corrected the different opinions of other players, and (c) the *quiet person* who rarely corrected them. In particular, the leader was selected from the players in each table. Since the smooth games normally appeared in the collective adaptive situation, the appropriate balance among players, (*i.e.*, the leader, claiming, and quiet players) is important to derive such smooth games. As a result of roles' being played, players changed their rules and shared a certain rule such as diamond or heart with others. From the analysis of the subject experiments, it can be concluded that the definition of the collective adaptive situation would be (1) the proper role balance among the players and (2) the smooth game and the rules being shared as a result of it.

3.3   Group Classification

In order to abstract the collective adaptation to a theoretic level, we start by classifying the groups obtained from the experiments into the following categories.

1. **Collective Adaptive Situation**
   As defined previously, most of the players change their rule (*i.e.*, the trump card and ace-strong/weak rules) and share other rules with each other. Such situation derives the smooth games.
2. **Non-collective Adaptive Situation**
   This situation is divided into the following situations.

- **Confused Situation:** Because of the players who are puzzled with the slightly different rules, the players change their rules when they are faced with the rules which are doubtful.
- **Value-Separated Situation:** Players do not change their rules despite of the ones completely different from each other, which results in the non-smooth game.
- **Persisted Situation:** The player persists his rule, which is different from others, in order to force other players to agree with the player for the consensus among the players.

3. **Intermediate Situation**

This situation has the features of both the collective and non-collective adaptive situation, and is divided into the following situations.

- **Compromised Situation:** The players change their rules but do not share one certain rule with each other. This is different from the collective adaptive situation. Such a compromise can be regarded *negatively* because all the players lose their rules, while this can be considered *positively* because the players try to reach the agreement.
- **Absorbed Situation:** One player is absorbed by multiple players. Because of the fact that a rule is shared and a smooth game is played, it seems this situation could be *positively* regarded as collective adaptive situation. However, what makes the collective adaptation different is there are not any corrections in this situation because the minority cannot complain, which brought it being not satisfied with the rule. Thus, it can be classified *negatively* as non-collective adaptive. This analysis found out that the collective adaptive situation requires several number of corrections. Otherwise, it can be regarded as Absorbed situation.

4   Agent Model for Barnga Game

This section describes Barnga agent, who corresponds to the player in our model, considering the results of the subject experiments. The agent is composed of the memory and action modules, each of which has several parameters and actions.

4.1   Memory Module

(1) **Leader Aptitude L:** L(X), which ranges from 0 to 100, shows how the agent X is appropriate as the leader and is used to select the leader. If the agent has the high value it means that it has high possibility to be elected. L(X) is changed by +a or -b, according to the correction made as shown in Eq.(1). When other agents stop to correct the leader's decision, L(X) is reset to the one before the game starts defined as preL(X). L(X) increases because of no correction to the leader's decision means that his action is accepted by the other agents. In this equation, a, and b are coefficients, which are set to 5 and 1 in our simulation, respectively.

$$L(X) = \begin{cases} preL(X) + a & (No\ correction) \\ L(X) - b & (Correction\ to\ the\ leader's\ decision) \end{cases} \quad (1)$$

(2) **Opponent Reliability W:** W(X,Y), which ranges from 0 to 100, shows how much the agent X relies on the agent Y and is updated as shown in Eq. (2). In this equation, R and α are the reward and learning rate respectively, where R is set as shown in Eq.(3), and α is set to 0.2. The parameter n describes the number of corrections to the leader's decision.

$$W(X,Y) = W(X,Y) + \alpha(R - n * 10 - W(X,Y)) \tag{2}$$

$$R = \begin{cases} 100(winner\ that\ agent\ thinks\ =\ winner\ selected\ by\ leader) \\ 0(winner\ that\ agent\ thinks\ \neq\ winner\ selected\ by\ leader) \end{cases} \tag{3}$$

(3) **Rule Strength S(X,r):** S(X,r), which ranges from 0 to 100, indicates the strength of the rule r that the agent X has (*e.g.*, the heart trump or ace-strong rule), in a sense of how the agent X thinks the rule r is correct. S(X,r) decreases whenever the agent X corrects the leader's decision. When the game is over, S(X,r) is reset and is updated as shown in Eqs.(4) and (5), where rNum is denoted for the number of rules. If the game ends according to the heart trump and the ace-strong rule, for example, the heart trump and ace-strong rule(described in *SelectedRule*) are updated according to the Eq. (4), while the diamond trump, the clover, the spade, and the ace-weak rule (described in *OtherRule*) are updated according to Eq. (5).

$$S(X, Selectedrule) = S(X, SelectedRule) + \alpha(R - S(X, SelectedRule)) \tag{4}$$

$$S(X, Otherrule) = S(X, OtherRule) + \triangle S(X, SelectedRule)/rNum \tag{5}$$

4.2   Action Module

The action module of the agent consists of the following functions: (1) rule selection, (2) winner nomination, (3) winner decision (only for the leader) and (4) correction of the leader's decision.

(1) **Rule Selection:** The agent X selects one rule (*e.g.*, the heart trump or ace-strong rule) according to S(X,r) by the roulette-selection.

(2) **Winner Nomination:** Each agent X personally decides the winner according to the selected rule and determines if the decided winner should be nominated according to S(X,r).

(3) **Winner Decision:** The leader decides the winner agent if two or more agents are nominated as winners. To decide one winner, the following two types of the leader are considered in our model: (a) *Self-centered leader*, who determines the winner that the leader thinks as a winner, and (b) *Other-centered leader*, who determines the winner according to W(leader, Y) where Y is denoted for the agent that nominates a certain winner, in a sense that a leader decides which player to rely on.

(4) **Correction of the Leader's Decision:** According to S(X,r), the agent X decides whether he corrects the leader's decision.

4.3 Simulation Sequence

The sequence of Barnga game on the computer simulation shown in Fig. 2 is described as follows.

1: One of the players deals the cards.
2: Each player X puts one card in turn.
3: Each player X selects his own rule (*e.g.*, the heart trump or ace-strong rule) according to S(X,r).
4: Each player X decides the winner, who puts the strongest card, according to the rules selected in 3. Note that the winner agent in this stage has not been nominated yet.
5: The winner agent is nominated according to the value of S(X,r).
6: If the number of nominated agents is one, S(X,r) of the agent X is updated in 6-1, and then go to 10-12.
7: Check whether a leader exists. If not, go to 8-2 to select, and back to 7.
8: Since more than two agents are nominated, the leader X selects one winner, according to the probability of L(X). If X does not select the winner, go to 8-1 to update L(X), to 8-2 to select a new leader, then return to 7.
9: When the leader X decides the winner, the agent Y who think that another agent is the winner corrects X's decision with the probability of S(Y,r). If an agent Y corrects to the leader's decision, go to 9-1 to update S(Y,r) temporarily by Eq. (4), proceeds to 8-1 to update L(X) by Eq. (1), and go to 8-2 to select a leader again. If none of agents correct the leader's decision, go to 10-12.
10-12: Update W, L, and S individually according to Eq. (1) to (5) to go to 13.
13: One game ends, the agents are swapped whenever some numbers of games are played, and then back to 1.

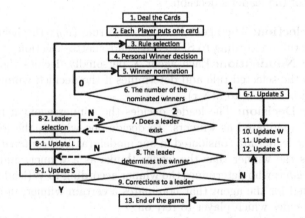

Fig. 2. A sequence of Barnga Simulation

5   Simulation: Characteristic of a Leader

To analyze an effect of the leader, which is one of the potential roles for the collective adaptive situation, the following two types of the agents are tested: (i) self-centered and (ii) other-centered leader agent. The agents are swapped between tables every 21 games and such a cycle is repeated the eight times. (*i.e.*, 8 times of swapping the agents) The results are compared in terms of (1) the number of the collective adaptive situation and (2) the leader aptitude L.

5.1   Simulation Results

We have to mention that the followings are the typical results, although a random seed is changed several times in order to obtain the invariant of the results.

(1) Group Classifications
Fig. 3 shows the result of the group classifications in the two different types of the leaders. The black, gray, and light gray box indicates the collective adaptive situation, non-collective adaptive situation, and intermediate situation, respectively. The horizontal axis indicates the agent types, while the vertical axis indicates the number of the situations. From the simulation result, the following implications have been revealed: (a) the self-centered leader hardly derives the collective adaptive situations, while (b) the other-centered leader agent causes the largest number of collective adaptive situations. According to the definition, the collective adaptive situation requires the smooth games and the shared rule by agents. From the computer point of view, the first component corresponds to the low number of corrections (not 0), and the second one is associated with the change of the value of a particular trump card that results in being shared, which means agents have the same trump card which value is nearly 100. We categorized all the cases into the three types by using the results of correction timing, and the $S(X,r)$ in Fig. 6.

(2) Leader Aptitude L. Fig. 4 and 5 show the change of the leader aptitude L in the self-centered and the other-centered leaders. The horizontal axis indicates the number of games, while the vertical axis indicates the leader aptitude L. The dots in the figure represents the values of L that agents have.

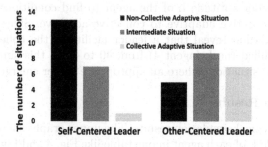

Fig. 3. Group Classification in two types of agent

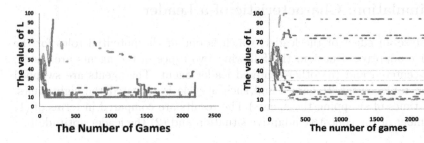

Fig. 4. Distribution of L(self-centered)   **Fig. 5.** Distribution of L(other-centered)

It is not really important because we focus on the overall distribution of the points. Important is the different distribution of the marks between Fig. 4 and 5. The dots of the other-centered leader are distributed in the upper area of the graph, which indicates agents with high values of L, while in the self-centered leader, the points are mapped in the bottom, which suggest that agents have lower values of L.

5.2   Discussion

A few of the collective adaptive situations are observed in the self-centered leader as shown in Fig. 3 and the leader aptitude L decreases as the games proceeds as shown in Fig. 4. These results suggest that the self-centered leader is not perfect to derive the collective adaptive situation. In contrast to the leader, the largest number of the collective adaptation are observed in the other-centered leader agents as shown in Fig. 3 and the high leader aptitude L is kept in comparison with the self-centered leader as shown in Fig. 4. From these points of view, the other-centered leader agents have the great potential of deriving the collective adaptive situation.

6   Simulation: Conditions toward a Collective Adaptive Situation

We modify the leader aptitude L of the agent to find conditions that lead the group from non-collective adaptive to collective adaptive. Since the results of the previous simulation reveal that a leader facilitates the collective adaptive situation, we modified the L (agent 4) from 40 to 80 in this simulation, which corresponds to the situation, where an appropriate leader emerges.

6.1   Simulation Results

Fig. 6 shows the typical simulation results. The upper graphs show the change of the leader aptitude L of each agent in one table like Fig. 4 and Fig. 5. Four types of shapes represent four agents in one table. The middle graph indicates which and when the agent corrects, where the horizontal axis indicates the agent, while the vertical axis describes the number of games. This situation can be regarded as the one, where the game does not proceed smoothly because agent 2, 3, and 4 correct several times, while agent 1 makes few correction, meaning that the situation has unbalanced corrections among the agents. The lower graphs show the change of $S(X,r)$ of the trump card during 21 games, where the each edge of axis corresponds to the maximum value of each trump card. In this case, the left side of the figure illustrates the confused situation because of the different rules among the players. The right side of the figure, on the other hand, shows the rule of each agent is moved toward the heart trump card by modifying their values to share the heart trump card. From this point of view, it can be concluded that these graphs suggest the change of L brings the collective adaptive situation because the correction number decreases (but not 0) and all the agents become to share one rule (the heart trump card), which has consistence with the definition of the collective adaptation described in Section 3.

6.2 Discussion

For the purpose of revealing the specific condition of the collective adaptive situation, all the 7 cases, which are classified as collective adaptive, and the other 14 cases, which are not, are analyzed by (a) finding something in common with them and (b) figuring out what differentiates between collective and non collective adaptation. Although the simulations should be discussed more theoretically, a random seed is changed to obtain variety of results. Through the analysis, following insights that satisfy (a) and (b) are obtained; (i) the leader aptitude L has to be more than 60, and (ii) the ratio of the corrections is less than 3:1 between the claiming and quiet agent. From the computer point of view the claiming agent would be defined as the one who makes the most number of corrections, while the quiet agent could be regarded as the one who corrects the least. In the middle of the left figure agent 1 can be regarded as the quiet agent and agent 4 can be considered as the claiming agent. This case cannot be determined as the smooth game because of the unbalanced corrections, which ratio is 5:1, in a sense that some agents corrects many times and others do not. In the middle of the right figure, however, agent 4 corrects twice over a period of 21 games, while other agents have three times corrections. Thus, the ratio between them is less than 3:1. The following hypothesis can be made in this circumstance; because of no leader, the quiet player has to follow the claiming player, which makes the situation become non-collective adaptive. An appropriate leader, however, tries to bring the opinions together, which results in the decrease of the number of the total corrections and also causes the claiming player to make less corrections, which means that the correction balance among agents is fixed to fit the condition. From these analysis, it can be concluded that the appropriate roles among agents (the leader, quiet and claiming agent) promotes them to share the new rule (*e.g.*, the heart becomes the trump card in Fig. 6).

These results can be applied to a collective adaptive agent. Consider the situation, where humans and an agent play this game together. The agent perceives

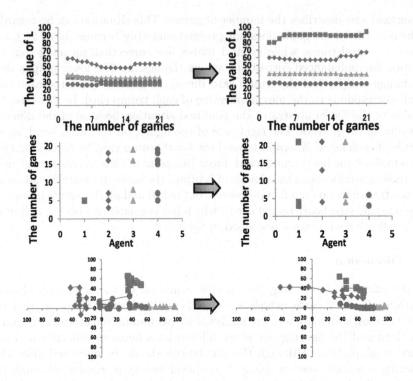

Fig. 6. Transition from non-collective adaptive situation to collective adaptive one

inputs like who the leader is, figures out the best solution by computing, and makes an action such the agent becoming a leader, or a follower. Our project has not revealed specific idea of the agent design. However, the contributions of the project will help to consider modeling collective adaptive agent in the near future.

7 Conclusion

This paper explored the collective adaptive agent that adapts to a *group*. For that purpose, this research defined the collective adaptive situation as the smooth games and the shared rule by players as a result of appropriate role balance through the analysis of the subject experiments in Barnga. The other situations are also classified into the several categories of collective adaptation, which could not be clarified by other previous works. The simulations of Barnga revealed the following insights for the collective adaptive situation: (1) the leader who takes an account of other playersf decisions on the winner contributes to the collective adaptive situation, and (2) the appropriate balance in roles is required, which are the leader, the claiming agent, who makes corrections frequently and the quiet agent who corrects the least times. In detail, the following specific conditions are found out to be indispensable for the collective adaptive situation: (a) the emergence of the leader at least whose value of the Leader Aptitude is more than 60 and (b) the ratio in the correction numbers between the claiming and quiet agents is less than 3:1. Although we should generalize the above conditions and apply it to a real society, an investigation of Barnga, in which both humans and agents play together, a precise analysis of the conditions for leading the collective adaptive situation should be done for the future works as well as (3) the comparisons of the performance with other related works.

References

[1] Miyamoto, E., Lee, M., Okada, M.: Robots as Social Agents: Developing Relationships Between Autistic Children and Robots. The Japanese Journal of Developmental Psychology 18(1), 78–87 (2007) (in Japanese)

[2] Doi, H., Ishizuka, M.: Human-Agent Interaction with a Life-like Character Linked with WWW. The Japanese Society for Artificial Intelligence 17, 693–700 (2002) (in Japanese)

[3] Omura, H., Katagami, D., Nitta, K.: Design of Social Adaptive Agents in Simulation Game of cross-cultural experience. In: Human-Agent Interaction symposium, HAI 2008 (2008) (in Japanese)

[4] Lewis, M.: Designing for Human-Agent Interaction. AI Magazine 19(2), 67–78 (1998)

[5] Thiagarajan, S.: Barnga game: A Simulation Game on Cultural Clashes. Intercultural Press, Inc. (2006)

[6] Yamada, S., Kakusho, K.: Human-Agent Interaction as Adaptation. The Japanese Society for Artificial Intelligence 17, 658–664 (2002) (in Japanese)

[7] Nishida, T.: HAI in Community. The Japanese Society for Artificial Intelligence 17, 665–671 (2002) (in Japanese)

[8] Yamada, Y., Takadama, K.: Exploring Evaluation Criteria for Adaptation of Agents to Groups-Investigation of Correction Behaviors of Agents in BARNGA Game. In: SICE Symposium on Systems and Information (SSI 2004), pp. 19–24 (2004) (in Japanese)

Classification of EEG for Affect Recognition: An Adaptive Approach

Omar AlZoubi[1], Rafael A. Calvo[1], and Ronald H. Stevens[2]

[1] School of Electrical and Information Engineering
The University of Sydney, Australia
http://www.weg.ee.usyd.edu.au
[2] IMMEX Project - University of California Los Angeles

Abstract. Research on affective computing is growing rapidly and new applications are being developed more frequently. They use information about the affective/mental states of users to adapt their interfaces or add new functionalities. Face activity, voice, text physiology and other information about the user are used as input to affect recognition modules, which are built as classification algorithms. Brain EEG signals have rarely been used to build such classifiers due to the lack of a clear theoretical framework. We present here an evaluation of three different classification techniques and their adaptive variations of a 10-class emotion recognition experiment. Our results show that affect recognition from EEG signals might be possible and an adaptive algorithm improves the performance of the classification task.

Keywords: Affective computing, EEG, Classification, Adaptive.

1 Introduction

New brain imaging technologies are opening the windows to new ways of looking at emotions and other affective states (i.e. *affects*). One of the longstanding psychological debates has been between categorical and dimensional models. In the former the assumption is that a discrete number of affects (e.g. 'anger') can be recognized through behavioral (e.g. facial actions or physiological measures) [1]. The latter assumes an underlying set of variables, often two, called valence (going from very positive feelings, to very negative) and arousal (also called activation, going from states like sleepy to excited).

In the studies that use EEG (recently reviewed by Olofsson [2]), most of the focus has been on Event Related Potentials (ERPs). Signal processing [3] and classification algorithms [4] for EEG have been developed in the context of building Brain Computer Interfaces (BCI), and we are seeking ways for developing similar approaches to recognizing affective states from EEG and other physiological signals. Very few of the affect recognition studies based on physiological data use EEG, most use EKG, EMG and skin conductivity [1,5].

These studies used traditional offline classification techniques, compared the performance of different classification algorithms, and evaluated different combinations of feature sets. The ultimate aim is to find an optimal combination of

A. Nicholson and X. Li (Eds.): AI 2009, LNAI 5866, pp. 52–61, 2009.

classifiers and feature sets that could deliver an optimal performance. In addition; offline classification is also useful in evaluating subject's specific features. However, real time affect recognition systems require a real time adaptive classification system that is necessary to cope with non-stationarities of EEG and other physiological data.

Non-stationarities are ubiquitous in EEG signals [6], occurring due to many factors such as 1) user fatigue, 2) electrode drift, 3) changes in the impedance of the electrodes, 4) user cognitive states modulation, such as attention, motivation, and vigilance.

This study provides new data on EEG based affect recognition, and presents a performance comparison of K-Nearest Neighbor (KNN), Support Vector Machines (SVM), and NaiveBayes using an adaptive classification technique. Section 2 discusses some of the literature on classification of EEG signals for affect recognition, and section 3 discusses the need for real time adaptive algorithms for non stationary data. Section 4 discusses the protocol, equipment and subjects used in our data collection. Section 5 presents the performance for both static and adaptive versions of KNN, SVM, and NaiveBayes, and section 6 presents our conclusions.

2   Background

It is hard to compare the results from different studies of affect recognition systems because researchers often use different experimental setups and data preprocessing techniques. Some of these studies [7,8] used a combination of EEG and other physiological signals for this task, while others [9] used EEG solely for affect detection.

In a study to detect the level of arousal from EEG and other physiological signals, Chanel et al [7] formulated this as a classification problem with two classes corresponding to 2 or 3 degree levels of arousal. The performance of two classification methods, NaiveBayes classifier and Fisher Discriminant Analysis (FDA) were evaluated on each EEG and physiological signal separately, and on combination of both. The study used the IAPS protocol for elicitation, 4 subjects, and the EEG was recorded from 64 electrodes with a sampling rate of 1024 Hz.

The EEG was then bandpass filtered between 4-45 Hz, artifacts such as eye blinks were identified and removed from the signals. Using a 6s epoch length, the bandpower at six frequency bands were computed, yielding 6 features from the EEG. According to the authors, most of the EEG features involve the Occipital (O) lobe, since this lobe corresponds to visual cortex and subjects are stimulated with pictures. Using only EEG features and the one leave-out method, a classification accuracy of 72% for NaiveBayes was achieved and 70% for FDA for one subject. Their results suggested that EEG could be used to assess the arousal level of human affects.

In a similar study Khalili et al [8] used EEG recorded from the scalp together with other physiological signals, which was then used to assess subject's arousal and valance levels. Three classes were assessed, Calm (C), Positively Excited (PE), and Negatively Excited (NE). The stimuli to elicit the target affects were IAPS images; each stimulus consists of a block of 5 pictures which assured stability of the emotion over time. Each picture was displayed for 2.5 seconds for a total of 12.5 seconds for each block of pictures. The data was acquired from 5 subjects, with 3 sessions of 30 trials per session from each subject. EEG recorded from 64 electrodes at 1024 sampling rate.

The preprocessing and feature extraction first involved segmenting EEG data into 40s frames. EEG was bandpass filtered between 4-45 HZ, and then applying a local laplacian filter to obtain a better localization of brain activity. The study used a set of features such as, mean, STD, Skewness, Kurtosis, mean of the absolute values of the first difference of raw signals, Mean of the absolute values of the first difference of normalized signal. These six features were computed for each electrode of the 64 electrodes yielding 6*64 = 380 features. This dimension was reduced using genetic algorithms (GA), and classification using KNN and Linear Discriminant analysis (LDA) by applying a leave-one out method. The investigators achieved a classification accuracy of 40% for LDA and 51% for KNN for 3 classes. For a two classes discrimination of PE, and NE they achieved a better results of 50% for LDA and 65% for KNN. However the best classification accuracy according to the authors was achieved using EEG time-frequency features of 70% for KNN.

Horlings et al [9] used EEG alone for classifying affective states into 5 classes on two affective dimensions: valance and arousal. They used the database of the Enterface project [10], and extended it with their own data. 10 subjects were chosen for the task of EEG acquisition using a Truescan32 system; emotion elicitation performed by using the IAPS protocol. The SAM Self-Assessment was also applied where subjects rate their level of emotion on a 2D arousal and valance scale. They performed two recording sessions consisted of 25-35 trials each, with a pause of 5 minutes in between, each trial consists of 5 pictures, and each picture is shown for 2.5 seconds.

The EEG data was then filtered between 2-30 Hz to remove noise and artifacts from the signal. The baseline value was also removed from each EEG signal. Feature extraction involved computing EEG frequency bandpower, Cross-correlation between EEG bandpower, Peak frequency in alpha band and Hjorth parameters, this resulted in 114 features. The best 40 features were selected for each of the valance and arousal dimensions based on the max relevance min redundancy (mRMR) algorithm [11]. Two classifiers were trained on this feature set, one classifier for arousal dimension, and another classifier for valance dimension. According to the authors, each classifier can use different features to obtain optimal performance; using an SVM classifier with 3-fold cross validation performed the best with 32% for the valance and 37% for the arousal dimension.

Most of these studies used offline *non-adaptive* classifiers, and to our knowledge this is the first time adaptive algorithms are evaluated in this context. The next section discusses the need for classifier adaptation, especially if the data source is non-stationary in nature.

3   Online Learning and Adaptation

Most classification methods are based on the hypothesis that data comes from a stationary distribution, this is not particularly true in real life situations, where the underlying concepts of stationarity are violated, by what is known as concept drift in the data mining community [12]. This is particularly the case in EEG signals, where it always changes its nature with time. A stationary signal on the other hand maintains its statistical properties all the time, or over the observation time.

This non-stationary nature of the signals means that a classification model built earlier using a particular set of physiological data is not going to reflect the changes that have already taken place to the signals. Consequently the classification accuracy will degrade with time, unless an update to the classification model is made, or in other words the model is adapted to reflect pattern changes in physiological signals.

The non-stationarity of EEG signals can be seen as a shift in feature space as described by Shenoy et al [6]. The distinguishing patterns of interest of the physiological data are still there, what is really needed is to update or adapt the classification model in real-time to reflect the changes of data distribution. This type of change in the probability distribution of the data is also known as virtual concept drift [13], where the current model error rate is not any more acceptable given the new data distribution.

Online classifier learning and adaptation is particularly important in real time systems based on non stationary data sources in order to maintain the classification accuracy and overall performance of the system. Traditional classification systems learn inefficient models when they assume erroneously that the underlying concept is stationary while in fact it is drifting [14].

One possible solution to the problem is to repeatedly apply a traditional classifier to a fixed sliding window of examples. In this approach a similar number of examples are removed from the end of the window, and the learner is retrained, making sure the classifier is up to date with the most recent examples [15]. Other approaches apply a dynamic training window size strategy, by increasing the window size whenever the concept drift is not detected, and shrinking the window size whenever a concept drift is detected [12]. However, this is a challenging task, especially considering real time systems where memory requirements -especially if the window size is sufficiently large-, and speed/response time are issues [12]. Computationally expensive algorithms are not desired as it might slow the overall performance of the system. Other challenges may exist such as the availability of sufficient real time data as well as the lack of supervised data in actual real life applications. The next section discusses the experimental protocol used here for EEG acquisition.

4   Data and Methods

The system used in the recording was a wireless sensor headset developed by Advanced Brain Monitoring, Inc (Carlsbad, CA). It utilizes an integrated hardware and software solution for acquisition and real-time analysis of the EEG, and it has demonstrated feasibility for acquiring high quality EEG in real-world environments including workplace, classroom and military operational settings. It includes an easily-applied wireless EEG system that includes intelligent software designed to identify and eliminate multiple sources of biological and environmental contamination.

Data was recorded at 256 Hz sampling rate from multiple EEG scalp bi-polar sensor sites: F3-F4, C3-C4, Cz-PO, F3-Cz, Fz-C3, Fz-PO. Bi-polar recordings were selected in order to reduce the potential for movement artifacts that can be problematic for applications that require ambulatory conditions in operational environments. Limiting the sensors (seven) and channels (six) ensures the sensor headset can be applied within 10 minutes, making the tool more feasible in practical scenarios. Further exploratory studies should probably be performed with equipment that allows high density EEG.

Three subjects were asked to self-elicit a sequence of emotions and where recommended to use recollections of real-life incidents. Numerous studies support the notion that this can serve as a sufficient condition for emotion elicitation [16]. Each emotion trial lasted for 3 minutes with a 1 minute rest in between. The power spectral density (PSD) values in each of the 1-Hz bins (from 1 Hz 40 Hz) were calculated from each 1 second epoch. The first and second frequency bins are not considered since they are mostly contaminated by EEG artifacts, which mostly occur at low frequencies.

The end dataset therefore will have (38 frequency bins * 6 EEG Channels) 228 features, and (180 rows * 10 emotions) 1800 instances for each of the three subjects. Based on these datasets, a number of classification techniques were compared, together with an online simulation experiment that incorporated an adaptive classification technique. The next section discusses the classifiers and the results of the two experiments.

5   Results and Discussion

5.1   Offline Analysis

The offline analysis was done using Weka [17], Table 1 lists the classifiers used and their description. The performance of the three classifiers compared in Table 2 are based on 10-fold cross validation. All classifiers are set to their default parameter values as implemented in Weka. The ZeroR classifier represents the baseline accuracy that the 10 affects studied which was 10% (the difference is based on some session being less than the default 3 minutes). The best classification accuracy was achieved using a KNN classifier with k=3 and Euclidian distance measure, and this was nearly uniform across all subjects. An SVM classifier with a linear kernel, which is based on John Platt's sequential minimal optimization algorithm for training a support vector machines classifier [18] was less accurate than KNN; however its performance was comparably better than that of the NaiveBayes classifier. An explanation for the performance of KNN comes from the work done by Cieslak et al [19], where they found that KNN is less sensitive

Table 1. A description of the classifiers used in this study and their parameters

Classifier	Description and Parameters
ZeroR (baseline)	Predicts the majority class in the training data; used as a baseline.
NaiveBayes	A standard probabilistic classifier, the classifier assigns a pattern to the class that has the maximum estimated posterior probability.
KNN	A classical instance-based algorithm; uses normalized Euclidean distance with k=3. KNN assigns the class label by majority voting among nearest neighbors.
SVM	It combines a maximal margin strategy with a kernel method to find an optimal boundary in the feature space, this process is called a kernel machine. The machine is trained according to the structural risk minimization (SRM) criterion [20]. We used Weka's [17] SMO with linear kernel for the offline analysis. The online analysis used a SVM with linear kernel as implemented in PRTools 4.0. [21] Default parameters are used for both methods.

Table 2. Classification accuracy of EEG data using 10-fold cross validation for three subjects A,B,C

Classifier/Subject	A	B	C
ZeroR (baseline)	9.96%	9.93%	9.94%
NaiveBayes	42.83%	28.16%	33.48%
KNN(3)	**66.74%**	39.97%	57.73%
SVM	54.57%	40.80%	33.48%

to non-stationarities than SVM and NaiveBayes. Subject A data showed good separation tendency across all classification methods compared to the other two subjects B,C. The classification performance on subject C data achieved the second best classification accuracy across classifiers except in the case of SVM where subject B data achieved a 40.8% performance compared to 33.48% for subject C. These results suggest that accuracy can change considerably between subjects.

5.2 Online Simulation

This experiment involved comparing the performance of a basic adaptive algorithm [15] in combination with a KNN classifier with k=3 and Euclidian distance measure, SVM with a linear Kernel, and NaiveBayes classifiers as implemented in PRTools 4.0 [21], the classifiers were used with their default parameters. A description of the algorithm is listed in Table 3.

The algorithm was applied with three different training window sizes to compare the effect of window size on classification performance. The static versions

Table 3. Listing of the adaptive algorithm

1. *Choose an initial fixed training window size*
2. *Train a classifier w on the examples of the training window*
3. *On the arrival of new examples, update the training window by:*
   a. *Inserting the new examples into the training window*
   b. *Deleting an equal number of examples from the end of the training window.*
4. *Train the classifier on the new training window*

Table 4. Average error rate and standard deviation for the different classifiers, static and adaptive classifiers over the training sessions, with different window size

Method	Static		Adaptive	
Classifier/window size	AvgErrorRate	STD	AvgErrorRate	STD
Knn/250	0.710	0.140	**0.207**	**0.134**
Knn/450	0.714	0.143	0.247	.145
Knn/900	0.662	0.158	0.288	0.155
NaiveBayes/250	0.694	0.132	0.464	0.153
NaiveBayes/450	0.660	0.124	0.492	0.141
NaiveBayes/900	0.616	0.131	0.507	0.142
SVM/250	0.716	0.129	0.437	0.147
SVM/450	0.704	0.138	0.493	0.159
SVM/900	0.707	0.144	0.542	0.156

of the classifiers were also evaluated with the same window sizes. Static classifiers are those which initially trained on the first training window of examples, but are not updated later on. Training window sizes of 250, 450, and 900 were chosen, which account for 0.15, 0.25, and 0.5 of total dataset size. The training window was updated every 10 examples as it would be inefficient to update the training window and retrain the classifiers on the arrival of every example; 10 is also the number of classes in our dataset. The experiment was done on one subject's data (subject A), and is meant as a demonstration for the need for an adaptive classification technique for real time affect recognition systems, where the physiological data are continuously changing its behavior with time.

Table 4 shows the average error rate and standard deviation over the training sessions of both static and adaptive classifiers, with different window sizes 250, 450, and 900. It can be seen that the adaptive KNN classifier with a window size of 250 samples has the lowest average error rate overall, and the lowest standard deviation among the adaptive classifiers which indicates that the classifier maintained a subtle performance over the training sessions. This can be also inferred from Figure 1 which shows the performance of the KNN classifier with a window size of 250; clearly the adaptive version of the classifier outperforms the static one by nearly 50%. KNN proves to outperform SVM and NaiveBayes with non-stationarity data, and this comes from the way KNN works by voting amongst nearest examples.

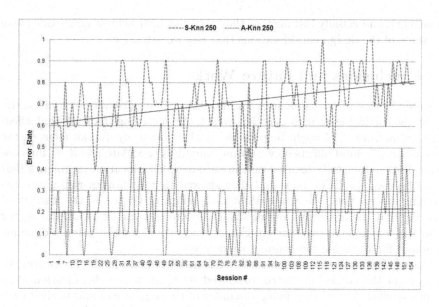

Fig. 1. Adaptive vs. static KNN classifier with a window size of 250 examples, the two solid lines in the middle show linear trend lines

The effect of window size on classifier performance can be inferred from Table 4, adaptive classifiers performance relatively enhanced with a smaller window size. An explanation for this comes particularly from the nature of the non-stationarity data; the smaller the window size, the more is the chance to build a model that can best classify unforeseen examples that are close enough in time, and get more localized information in time from the data, given that the data changes its behavior with time. On the other hand the performance of the adaptive classifiers is degraded with a larger window size, and this is due to the non-stationarity problem mentioned earlier, training the classifiers on a larger window size fails to build an efficient model for the fast changing data.

The average static classification performance was relatively improved with a larger window size, which was not surprising, given the dataset size, and this shouldn't be confused with the earlier discussion as the training and testing was done at different windows in time than the adaptive versions. However, a closer examination of Figure 1 shows the upward trend of the static classifier. That is, as time goes on the error rate goes upwards as well, and the classification performance degrades with time.

It is worth mentioning that the training time for each classifier varied greatly, while NaiveBayes, and KNN training time were relatively small especially if the window size is small, the training time for SVM was considerably higher since the classification task was a multiclass problem. This should be taken in consideration if it is going to affect the response time of the affect recognition system. On the other hand, if the window size is large, the memory requirements for KNN for example becomes larger, since it needs to store its distance matrix in memory, and classify instances as they arrive to their nearest neighbors; these are some of the design considerations that require attention.

6   Conclusions and Future Work

Despite the lack of strong neuroscientific evidence for correlates of brain activity at the cortical level with affective events, our recordings indicate that affect recognition from EEG might be possible. Rather this study did not focus on the neuroscience behind affects so we do not intend to speculate about its implications. Rather the study focused on the automatic classification techniques that could be used for EEG data, and they showed that accuracies well above the baseline are possible. We also evaluated an adaptive version of the algorithms showing that the error rate for the static versions of each algorithm was higher than that of the adaptive version. Future work would look at using a dynamic appraoch for updating the training window size.

Despite the experimental protocol we used is common in the literature, the analysis of the confusion matrix produced by most of the classification algorithms studied showed that fixing the order of the sequence of affects elicitation might be having an effect on their accuracy. Future work should consider using counterbalanced order for the affects elicited, these type of methodological issues can only be solved in larger studies.

References

1. Picard, R.W., Vyzas, E., Healey, J.: Toward machine emotional intelligence: analysis of affective physiological state. IEEE Transactions on Pattern Analysis and Machine Intelligence 23(10), 1175–1191 (2001)
2. Olofsson, J.K., Nordin, S., Sequeira, H., Polich, J.: Affective picture processing: an integrative review of erp findings. Biol. Psychol. 77(3), 247–265 (2008)
3. Bashashati, A., Fatourechi, M., Ward, R.K., Birch, G.E.: A survey of signal processing algorithms in brain-computer interfaces based on electrical brain signals. J. Neural Eng. 4(2), R32–R57 (2007)
4. Lotte, F., Congedo, M., Lécuyer, A., Lamarche, F., Arnaldi, B.: A review of classification algorithms for eeg-based brain-computer interfaces. Journal of Neural Engineering 4(2007) (2007)
5. Calvo, R.A., Brown, I., Scheding, S.: Effect of experimental factors on the recognition of affective mental states through physiological measures. In: Nicholson, A., Li, X. (eds.) AI 2009. LNCS (LNAI), vol. 5866, pp. 61–70. Springer, Heidelberg (2009)
6. Shenoy, P., Krauledat, M., Blankertz, B., Rao, R., Müller, K.: Towards adaptive classification for bci. Journal of Neural Engineering 3(1) (2006)
7. Chanel, G., Kronegg, J., Grandjean, D., Pun, T.: Emotion assessment: Arousal evaluation using eeg's and peripheral physiological signals. In: Gunsel, B., Jain, A.K., Tekalp, A.M., Sankur, B. (eds.) MRCS 2006. LNCS, vol. 4105, pp. 530–537. Springer, Heidelberg (2006)
8. Khalili, Z., Moradi, M.H.: Emotion detection using brain and peripheral signals. In: Cairo International on Biomedical Engineering Conference, CIBEC 2008, pp. 1–4 (2008)

9. Horlings, R., Datcu, D., Rothkrantz, L.: Emotion recognition using brain activity. In: Proceedings of the 9th International Conference on Computer Systems and Technologies and Workshop for PhD Students in Computing. ACM, New York (2008)

10. Savran, A., Ciftci, K., Chanel, G., Mota, J., Viet, L., Sankur, B., Akarun, L., Caplier, A., Rombaut, M.: Emotion detection in the loop from brain signals and facial images (2006)

11. Peng, H., Long, F., Ding, C.: Feature selection based on mutual information: criteria of max-dependency, max-relevance and min-redundancy. IEEE Transactions on Pattern Analysis and Machine Intelligence 27, 1226–1238 (2005)

12. Last, M.: Online classification of nonstationary data streams. Intell. Data Anal. 6(2), 129–147 (2002)

13. Tsymbal, A.: The problem of concept drift: Definitions and related work. Technical report, Department of Computer Science, Trinity College (2004)

14. Hulten, G., Spencer, L., Domingos, P.: Mining time-changing data streams. In: Proceedings of the seventh ACM SIGKDD international conference on Knowledge discovery and data mining, San Francisco, California, pp. 97–106. ACM, New York (2001)

15. Widmer, G., Kubat, M.: Learning in the presence of concept drift and hidden contexts. Mach. Learn. 23(1), 69–101 (1996)

16. Coan, J.A., Allen, J.J.B.: Handbook of emotion elicitation and assessment. Oxford University Press, Oxford (2007)

17. Witten, I.H., Frank, E.: Data Mining: Practical Machine Learning Tools and Techniques, 2nd edn. Morgan Kaufmann Series in Data Management Systems. Morgan Kaufmann, San Francisco (2005)

18. Platt, J.: Machines using sequential minimal optimization. In: Schoelkopf, B., Burges, C., Smola, A. (eds.) Advances in Kernel Methods - Support Vector Learning (1998)

19. Cieslak, D., Chawla, N.: A Framework for Monitoring Classifiers' Performance: When and Why Failure Occurs? Knowledge and Information Systems 18(1), 83–109 (2009)

20. Vapnik, V.: Statistical Learning Theory. John Wiley and Sons, Inc., New York (1998)

21. Heijden, F., Duin, R., de Ridder, D., Tax, D.: Classification, parameter estimation and state estimation. John Wiley & Sons, Chichester (2004)

Effect of Experimental Factors on the Recognition of Affective Mental States through Physiological Measures

Rafael A. Calvo[1], Iain Brown[2], and Steve Scheding[2]

[1] School of Electrical and Information Engineering, The University of Sydney
[2] Australian Centre for Field Robotics, The University of Sydney

Abstract. Reliable classification of an individual's affective state through processing of physiological response requires the use of appropriate machine learning techniques, and the analysis of how experimental factors influence the data recorded. While many studies have been conducted in this field, the effect of many of these factors is yet to be properly investigated and understood. This study investigates the relative effects of number of subjects, number of recording sessions, sampling rate and a variety of different classification approaches. Results of this study demonstrate accurate classification is possible in isolated sessions and that variation between sessions and subjects has a significant effect on classifier success. The effect of sampling rate is also shown to impact on classifier success. The results also indicate that affective space is likely to be continuous and that developing an understanding of the dimensions of this space may offer a reliable way of comparing results between subjects and studies.

Keywords: Emotion recognition, physiological signal processing, data mining, affective computing, human-machine interaction.

1 Introduction

It has been proposed that the next big step in improving the way computers communicate with humans is to adopt an interaction paradigm that imitates aspects of human-human communication [1]; namely, an awareness of a user's affective states (a combination of emotional and other mental states such as boredom or fatigue), so that the system is able to react to these states. Research into affective computing investigates how computers can interpret and simulate emotions to achieve more sophisticated human-computer interaction.

There have been several approaches proposed for determining the affective states of subjects. Some of the more prevalent research techniques are based on the classification and analysis of facial patterns, gestures, speech and posture, as well as studies linking physiological response to emotional state. Each technique has its own challenges. Often, somatic motor expressions of emotion are heavily dependent upon the individual, making any global recognition system impossible. It is hoped that the affective–physiological connection is so rudimentary that strong similarities will be observable independent of the subject.

A. Nicholson and X. Li (Eds.): AI 2009, LNAI 5866, pp. 62–70, 2009.
© Springer-Verlag Berlin Heidelberg 2009

The great challenge of physiological signals is the abundance of available data. Hundreds of features can be extracted by considering all the physiological responses. Heart and muscle activity, brain activity, blood pressure, skin temperature, respiration, and sweat production are all rich sources of information concerning the physiological responses of the human body. Machine learning techniques for processing this data likely hold the key to understanding which responses are correlated to changes in mental and affective state.

This paper contributes a comparison of eight classification techniques and an analysis of the relative effect of a number of experimental factors on the success rate of affect classification. These factors include: number of sessions, number of subjects, sampling rates and classification algorithms used. Affective state is a rich source of information within human communication and learning as it helps clarify both the content and context of the exchange. Indeed, research has shown that along with cognitive processes, affective processes are essential for healthy human functioning [2]. Affect recognition, therefore, is one of the fundamental goals to be achieved in order to develop more *effective* computer systems. While the primary research focus is to investigate affective systems, research in this area has the potential to strongly benefit associated fields such as psychology and teaching by offering quantitative alternatives to affective measures that have traditionally been restricted to qualitative assessment.

Section 2 reviews the literature, focusing on psychophysiological techniques which use the subject's physiological signals as input to a classification algorithm. Section 3 presents an experimental session, and describes the protocol followed for recording physiological signals from three subjects while they elicited a sequence of emotions. The tools used to record and then process the signals for this session are also described. Section 4 provides some details about the eight classification techniques evaluated and some of the many research questions that arise on how different humans elicit emotions. The basic tenet of these open research questions is that the accuracy of the classifiers provides an indication of how complex the emotion identification in a given data set is, and that this complexity is at least partially due to way humans elicit emotions. Section 5 looks at the results obtained for the different classification techniques discussing their accuracy and training time in different situations. Conclusions are presented in Section 6.

2 Background

In recent years several studies started investigating the potential for using biometric data for the classification of affective state [3-7]. Despite a longstanding debate amongst psychologists on the so called 'autonomic specificity', or the possibility of using autonomic nervous system (ANS) recordings to recognize affective state, these recent studies provide some evidence that the discrimination among some affective states is possible.

Emotion recognition is inherently multi-disciplinary, and draws on the fields from psychology, physiology, engineering and computer science. It is not at all surprising, then, that the approaches taken to study in this field also have a tendency to vary greatly. While the research goals of each study overlap there is wide variety in equipment used, signals measured, features extracted, evaluations used and in the format of presented results. These studies have had different levels of success (e.g. Picard, 81%, Kim, 78.4%), and with different limitations. In this volume [8] studies the feasibility of using EEG signals.

Picard [4] built automatic classifiers for the recognition of emotion and showed the relationship physiology and the elicitation of emotions, and that it is consistent within an individual, but it provides no insight as to whether there is any consistency between individuals. The study by Kim [6] uses a large number of subjects (young children, 5-8yrs). Their recognition accuracy was much lower, however this maybe due to the lack of consistency in their sample population. This study also addressed the issue of the inherent subjectivity of the subject-elicited technique (individual understanding of what emotive nouns refer to), by using an immersive, multi-modal environment to trigger the emotion. However it is difficult to create a multi-modal environment where each of the modes is coherently and constructively presented with each other. In cases where this is not achieved, the lack of coherency between triggering stimuli has been shown to heavily reduce the effectiveness and believability of the environment [8], which in turn will influence the clarity and quality of emotions elicited.

This paper investigates some of the effects in classification results by variations in factors such as: number of sessions, number of subjects, sampling rates, and algorithms used for classification. The study also considers the subjective evaluation of each affective elicitation in the three dimensions of arousal, valence and dominance. Until the effects of individual decisions made in the formulation, processing and analysing of the different papers mentioned is properly understood it is hard to see how the results of each study can be effectively viewed together.

3 Subjects and Methods

The signals chosen for this study were the electrocardiograph (ECG), electromyograph (EMG) and galvanic skin response (GSR). The ECG measures the voltage change across the chest due to the electrical activity of the heart. In this case the signal was measured between the wrists and used an electrode connected to the inside of one ankle as a reference node. The EMG measures the electrical impulses across muscle groups that are generated by activation of that muscle group. Electrodes were placed on either end of the masseter muscle group and a reference electrode was placed on the inside of one of the ankles. The masseter muscle group has been used in previous studies [9], [5] and was chosen due to its reliability and ease of measurement. GSR can refer to many different readings; in this study variation in skin conductance was measured. Skin conductance is directly related to sweat production and therefore has been used to measure anxiety levels, however in this study the features extracted from GSR are treated numerically. GSR was measured by subjects placing their index and middle fingers on each of two electrodes in a plastic bar. Subjects were asked to maintain a constant pressure on the electrodes as a variation in pressure affects the results. The equipment used for recording the signals was a Biopac M150 base unit with the appropriate modules for ECG, EMG and GSR. Signals were recorded to a HP Tablet PC using the AcqKnowledge 3.8.2 Software supplied with the equipment.

Following previous studies, here the combination of factors used were: subject-elicited, lab setting, feeling, open-recording and emotion-purpose [4]. A modified version of the Clynes protocol for eliciting emotion [10] was chosen for generating the subject emotion. The Clynes protocol was used in an earlier study by Picard [4] and asks subjects to elicit eight distinct emotions, (*no emotion, anger, hate, grief, platonic love, romantic love, joy, and reverence*). The Clynes protocol typically uses physical expression to give somatosensory feedback, given that the correct equipment was not available, subjects were offered a stress ball to hold in their free hand to use as an object of physical expression. Each emotion was elicited for a three minute period, separated by a period of rest. After subjects were prepared for the study the emotions were elicited in order. In this study subjects were not told exactly what was meant by each emotion (other than its name) allowing individual, subjective, interpretations of each affective label. After each emotion was elicited, subjects were asked to rate the emotion in each terms of Arousal, Valence and Dominance on the Self Assessment Manikin pictorial scale [9]. Three subjects (Male 60, Male 40, Female 30 yrs old) Three sessions were recorded for each subject on different days. The sessions with Subject 1 were recorded at 40Hz, while the sessions of Subjects 2 and 3 were recorded at 1000Hz, after deciding to see the effect of a higher sampling rate on the ability to classify the data. Although the number of subjects is small, the aggregate data is very large. Each of the three sessions for each three subjects contains 24 minute of recordings, for 3 physiological signals at 1000 samples per second.

The raw data was preprocessed using Matlab. The signal data was organised into thirty overlapping 30 second windows for each emotion recording in each session. 120 features were extracted for each 30 second window using the Augsburg Biosignal Toolbox [12]. The features extracted were primarily the mean, median, standard deviation, maxima and minima of several characteristics in each signal. The data was then processed by WEKA, a machine learning toolbox [10].

4 Classification

Eight classification algorithms were evaluated using 10-fold cross validation:

1. ZeroR: predicts the majority class in the training data; used as a baseline.
2. OneR: uses the minimum-error attribute for prediction [11].
3. Function Trees (FT): classification trees that could have logistic regression functions at the inner nodes and/or leaves.
4. Naïve Bayes: A standard probabilistic classifier using estimator classes. Numeric estimator precision values are chosen based on analysis of the training data [10].
5. Bayesian Network: using a hill climbing algorithm restricted by sequential order on the variables, and using Bayes as optimisation criteria.
6. Multilayer Perceptron (MLP): using one hidden layer with 64 hidden units.
7. Linear Logistic Regression (LLR) using boosting.
8. Support Vector Machines: Finds the maximum margin hyperplane between 2 classes. Weka's SMO with polynomial kernel was used [12] with c=1.0, epsilon=1e-12.

An underlying hypothesis of this study is that different emotions manifest themselves in distinct physiological states. Another hypothesis is that the classifiers' performance gives an indication of an internal 'consistency' of the data. If the performance is bad for all algorithms, the data is harder to model. A number of specific problems arise when the classifier performance is used to make other inferences, including:

1. Intra-Subject, Single Session
Subjects might not elicit emotions in the same way on different days. To build a classifier and to test it on data from a single session means excluding the factors of inter-session variation Even for classifiers 'custom' built for a single subject, most applications would require high multisession accuracy.

2. Intra-Subject, All Sessions
A subject specific classifier can be trained and tested by combining data from a number of sessions. By combining the windows from the three sessions for each subject into a single data set, the classifiers' accuracy indicates how variation in affective elicitation deteriorates the accuracy. This is probably caused by differences in the appraisal of emotion, intensity and quality of the elicitation (how close to the emotion the subject was able to elicit).

3. Inter-Subject
The 'universality' of emotions –the assumption that different people elicit emotions in a similar way- has been disputed. Depending on the application, it might be necessary to build a classifier based on data recorded from another subject. For this evaluation, data included both the day-to-day baseline variation in emotion and also the variation in subject interpretation of affective labels. Consequently seeing how the inter-subject data set classification compares to the combined and individual sessions will give insight into how much variation exists between subjects.

5   Results

Table 1 shows the classifiers' accuracy and training time on a PC with an Intel Core 2 Duo Processor (1.83GHz) and 2GB DDR2 RAM. MLP had the highest percentage of correctly classified samples, however data sets take a long time to process (36 minutes to process 9 minutes of recorded signals), making it unsuitable for real time applications. SVM, LLR and Functional Tree (FT) algorithms are faster, and give high accuracy. Of these methods the SVM algorithm gives the most consistent results for the shortest processing time. The FT algorithm also demonstrated unsuitability by failing to be compatible with all data sets. Though often quicker, the remaining algorithms give significantly lower or less consistent results than the SVM algorithm. Hence the SVM algorithm was used as the primary algorithm for comparing the confusion matrices and misclassified sample analysis. Table 1 also shows the processing time of each algorithm for a 3min recording data set.

There is a noticeable decay in classifier accuracy as the data sets become more complex, however even the most complicated data still gives 42% success using the chosen SVM algorithm. This remains three times higher than chance classification.

Table 1. Results of the different classification algorithms used for each data set. S#D# refers to the subject number and session (day) number. 40/1K refers to the sampling rate, (Hz).

	ZeroR	OneR	FT	Naïve Bayes	Bayes Net	MLP	LLR	SVM
S2D1– 40	12.5%	50.4%	89.2%	66.3%	81.3%	92.9%	90%	94.6%
S2D1–1K	12.5%	48.3%	96.7%	61.7%	N/A	97.1%	97.5%	95.8%
S2DA–40	12.5%	55.3%	76.7%	43.6%	64.3%	90.8%	72.6%	74.7%
S2DA-1K	12.5%	59.2%	88.9%	38.6%	N/A	97.8%	86.9%	85.7%
Time to Process	0 s	1 s	1.5min	2 s	N/A	36min	8min	41 s
SADA- 40Hz	12.5%	55.4%	N/A	22.8%	59.3%	70.6%	41.8%	42.2%

5.1 Variation of Results across Different Sample Rates

Table 2 gives a summary of the classifier success for each of the different data sets. Individual sessions displayed strong classifier success across all data sets. For individual sessions the difference in sample rate is fairly small, with classifier success varying by only a few percent in any case. In all but one case the accuracy for 1000Hz is better than the 40Hz equivalent. This is not shown to be true of other algorithms, and is most profoundly noticed where BayesNet failed to process the higher sample rate data set. The results show a progressively increasing difference between the success rates of classification for high and low sample rates as the data sets become more complicated. Although this evidence is far from conclusive it does suggest that the sample rate is a factor to be considered when making physiological recordings.

The different accuracy for the 40Hz and the 1000Hz data sets is not restricted to the SVM classifier. It should also be noted that the effect of sample rate variation was more pronounced in some, but not all techniques. The BayesNet technique for example showed a tendency to fail at higher sample rates, as did the Functional Tree approach. The more consistently correct classifiers, MLP, LLR and SVM, however, all showed classification improvement at higher sample rates. More detailed studies will provide a more complete picture of the effect sample rate has on emotion identification.

Table 2. Percentage of samples correctly classified for data sets using the SVM Algorithm

Subject	1	2	3	Combined Sessions
1 – 40Hz	96.3%	92.1%	95.4%	80.4%
2 – 40Hz	94.2%	97.5%	95.8%	74.7%
2 – 1000Hz	95.8%	97.1%	98.8%	85.7%
3 – 40Hz	90.5%	95%	92.1%	68.1%
3 – 1000Hz	99.2%	99.6%	96.3%	79%
All Subjects (40Hz)	N/A	N/A	N/A	42.2%

5.2 Variation of Results across Different Sessions and Subjects

Comparing the results of the individual sessions, some emotions were consistently poorly classified, others consistently well classified, and others varied from session to session. Platonic love and romantic love stand out as emotions that are often misclassified, while anger and the no emotion baseline were consistently well classified. Table 3 shows percentages of emotions misclassified as other types. For example, Subject 1's romantic love samples are 4% misclassified as 'No emotion' and 3% misclassified as 'Hate'.

The consistency of emotion elicited is better identified from the combined data set of all of a subject's sessions. Subject 3, for example, shows very high classification success in each session individually, but displays the lowest classification success in the combined data set. This suggests that for each individual session, the consistency of emotion elicited for each 3-minute block was very good, but that the character of emotion elicited from session to session was not as consistent. Subject 1, in contrast, shows greater variation in the individual sessions, but better consistency across the three sessions.

In the intra-subject data sets, all three subjects displayed relatively high misclassification in romantic love. The confusion matrices for the three subjects showed one subject with high misclassification towards hate, one with high misclassification towards platonic love and the other with a misclassification split between anger, joy and platonic love. These variations are subject dependent and are likely caused by developed associations as well as variations in mood, concentration and interpretation of the meaning of affective labels.

Table 3. Misclassification results for all data sets

Data Set	Worst Classified			Best Classified Emotion			Most Commonly Misclassified
	1	2	3	1	2	3	
S1–40	J, Re	Re, Ro	P, Ro	A, G, N, P	G	A, G, J, N, Re	Romantic(4%N, 3%H), Reverence(5%J, 3%N), Hate(2%A), Platonic(3%Ro, 2%H)
S2 – 40	J, H, Ro	P	G	A, N, Re	A, H, J, No, Re	H, J	Platonic (3%G, 3%J), Grief (4% P)
S2 – 1K	G, H, J	P, J	-	P	A, H, N, Re, Ro	All	Platonic(4%J), Joy(4%P), Hate(2%A), Grief(2%H, 2%A)
S3 – 40	A, H, Ro	P, G	Re, Ro	Re	A, H, N	G, N, P	Romantic(8%J, 2%Re), Reverence(7%Ro), Platonic(3%G, 2%J), Hate(4%A, 3%G)
S3 – 1K	-	-	Re, Ro	All	All	A, G, H, N, P	Reverence(4%Ro, 1%A), Romantic(4%Re)
S1DA – 40	Ro, H, J			A, No, P			Romantic(10%H, 6%N, 5%J,5%G), Hate(12%A, 5%Re, 4%J), Joy(16%Re, 8%Ro)
S2DA – 40	N, G, P			Re, A			No Emotion(14%J, 10%P, 7%Ro, 5%G), Grief(8%J, 8%N, 4%H, 4%P, 4%Ro), Platonic(13%Ro, 7%J, 7%N)
S2DA – 1k	P, G			A, Re			Platonic(14%Ro, 9%J, 6%N), Grief(10%N, 6%H)
S3DA – 40	Ro, P, Re, G			A, H, N			Romantic(17%J,14%A, 13%P, 7%Re), Platonic(9%Ro, 9%J, 8%G, 7%Re), Reverence(8%G, 6%J, 6%A), Grief(8%P, 6%Re, 6%A)
S3DA – 1k	P, H, Re			N			Plat(19%G, 8%A, 8%J), Hate(12%A, 8%Re), Reverence(4%H,4%N, 4%P, 4%Ro)
SADA	Ro, H, P			A, Re, J			Rom(14%P, 12%A), Hate(18%A, 13%N, 12%Re), Plat(15%Ro, 12%A, 10%N), Grief(11%A, 11%P), No Em(11%J, 11%P), Joy(13%Re, 9%Ro), Rev(12%N, 11%A), Anger(10%H, 8%P)

In the inter-subject data set, romantic love, hate and platonic love showed the worst results for classification, while anger, reverence and joy showed the best classification results. Anger is correctly identified but other emotions tend to be misclassified as anger.

6 Conclusions

The method used in this study utilised a subject-elicited, lab setting, feeling, open-recording and emotion-purpose framework. This particular choice of factors highlighted the individual variation in subject's interpretation of emotive labels. As a consequence, future studies will utilise a detailed description of the emotion to be elicited, or use the induced-emotion approach. Subjects also had preferred techniques for eliciting the emotions, some preferred to visually focus on something, while another preferred to elicit with closed eyes. For this study, the process had the luxury of being flexible and each subject was able to find a way to elicit emotions that they found comfortable.

The strong consistency of classifier success (> 90%) across the nine primary data sets (Table 2) supports the hypothesis of correlation between emotion state and physiological state. Although there is no guarantee that the emotion elicited is an accurate portrayal of the affective label requested, the high success in classification does show that the physiological manifestation caused by each of the eight categories was sufficiently distinct to allow discrimination and classification against the 7 other categories. If further data sets continue to show good discrimination, they will add to the mounting case in support the hypothesis of correlation.

A noteworthy result was the consistency of misclassification within a subject's data sets. Subject 3's romantic love samples were often misclassified as joy, and all subjects showed some misclassification between the negative emotions; anger, hatred and grief. Subjects also showed variation between sessions of which emotions were well classified, and which were relatively poorly classified, this may point to influence from the variation in day-to-day baseline as noted by Picard [2]. It is likely, for example, that on a day where a subject is feeling sad, that many samples might be misclassified as grief, while emotions which are sufficiently distinct, such as joy, might show strong classification success in contrast.

Further studies will continue to use the SAM diagrammatic survey for subject self assessment, but this will be supplemented with a quality assessment rating, ("How well did you feel you elicited the required emotion?"). This rating will help give an understanding of why misclassifications occur within sessions, and whether these misclassifications are predictable.

This study was successful in demonstrating that key factors such as number of sessions, number of subjects, sampling rates, and algorithms used for classification, all play a role in the success of classification. This study also supports the hypothesis that emotions lie in a continuous space. A future challenge will be to identify the axes of this space and determine an appropriate transform from physiological signals into these metrics.

While this study gives an important foundation for recognising the importance of these factors a complete understanding of the ways in which these factors do affect the results can only be properly obtained through more detailed studies.

Acknowledgments

The authors would like to thank the subjects who volunteered their time for this project.

References

[1] Picard, R.: Affective Computing. The MIT Press, Cambridge (1997)

[2] Damasio, A.: Descartes' Error: Emotion, Reason, and the Human Brain. Harper Perennial (1995)

[3] Vyzas, E., Picard, R.: Offline and online recognition of emotion expression from physiological data (1999)

[4] Picard, R.W., Vyzas, E., Healey, J.: Toward machine emotional intelligence: analysis of affective physiological state. IEEE Transactions on Pattern Analysis and Machine Intelligence 23, 1175–1191 (2001)

[5] Wagner, J., Kim, N., Andre, E.: From Physiological Signals to Emotions: Implementing and Comparing Selected Methods for Feature Extraction and Classification. In: IEEE International Conference on Multimedia and Expo., pp. 940–943 (2005)

[6] Kim, K., Bang, S., Kim, S.: Emotion recognition system using short-term monitoring of physiological signals. Medical and Biological Engineering and Computing 42, 419–427 (2004)

[7] Haag, A., Goronzy, S., Schaich, P., Williams, J.: Emotion Recognition Using Bio-sensors: First Steps towards an Automatic System. In: André, E., Dybkjær, L., Minker, W., Heisterkamp, P. (eds.) ADS 2004. LNCS (LNAI), vol. 3068, pp. 36–48. Springer, Heidelberg (2004)

[8] AlZoubi, O., Calvo, R.A.: Classification of EEG for Affect Recognition: An Adaptive Approach. In: Nicholson, A., Li, X. (eds.) AI 2009. LNCS (LNAI), vol. 5866, pp. 51–60. Springer, Heidelberg (2009)

[9] Bradley, M.M., Lang, P.J.: Measuring emotion: the Self-Assessment Manikin and the Semantic Differential. J. Behav. Ther. Exp. Psychiatry 25, 49–59 (1994)

[10] Witten, I., Frank, E.: Data Mining: Practical Machine Learning Tools and Techniques, 2nd edn. Morgan Kaufmann Series in Data Management Systems. Morgan Kaufmann, San Francisco (2005)

[11] Holte, R.C.: Very Simple Classification Rules Perform Well on Most Commonly Used Datasets. Machine Learning 11, 63–90 (1993)

[12] John, C.P.: Fast training of support vector machines using sequential minimal optimization. In: Advances in kernel methods: support vector learning, pp. 185–208. MIT Press, Cambridge (1999)

A Distance Measure for Genome Phylogenetic Analysis

Minh Duc Cao[1], Lloyd Allison[2], and Trevor Dix[1]

[1] Clayton School of Information Technology,
Monash University
[2] National ICT Australia
Victorian Research Laboratory, University of Melbourne
minhduc@infotech.monash.edu.au

Abstract. Phylogenetic analyses of species based on single genes or parts of the genomes are often inconsistent because of factors such as variable rates of evolution and horizontal gene transfer. The availability of more and more sequenced genomes allows phylogeny construction from complete genomes that is less sensitive to such inconsistency. For such long sequences, construction methods like *maximum parsimony* and *maximum likelihood* are often not possible due to their intensive computational requirement. Another class of tree construction methods, namely distance-based methods, require a measure of distances between any two genomes. Some measures such as evolutionary edit distance of gene order and gene content are computational expensive or do not perform well when the gene content of the organisms are similar. This study presents an information theoretic measure of genetic distances between genomes based on the biological compression algorithm *expert model*. We demonstrate that our distance measure can be applied to reconstruct the consensus phylogenetic tree of a number of *Plasmodium* parasites from their genomes, the statistical bias of which would mislead conventional analysis methods. Our approach is also used to successfully construct a plausible evolutionary tree for the γ-Proteobacteria group whose genomes are known to contain many horizontally transferred genes.

1 Introduction

The goal of molecular phylogenetics is to assemble an evolutionary relationship among a set of species from some genetic data such as DNA, RNA or protein sequences. Traditionally, each species is represented by a small molecule sequence such as a gene or a ribosomal RNA, which carries some information of evolutionary history in sequence variations. Closely related organisms generally have a high degree of agreement in these sequences, while the molecules of organisms distantly related usually show patterns of greater dissimilarity. Similar sequences are placed close to each other in the phylogenetic tree to show a probable ancestry relationship of the species. Inferring a phylogenetic tree is a kind of hierarchical clustering of the sequences.

A. Nicholson and X. Li (Eds.): AI 2009, LNAI 5866, pp. 71–80, 2009.
© Springer-Verlag Berlin Heidelberg 2009

The two main classes of phylogenetic tree construction methods are (a) predicting the tree that optimises some criteria such as *maximum parsimony* [1] or *maximum likelihood* [2], and (b) building the tree from a matrix of distances between any pair of sequences. Methods in the former class generally search in the tree space for the best tree and thus can only be applied when the number of sequences is relatively small. Besides, their requirement of multiple alignment of sequences restricts their use to data containing short sequences such as single genes. For data with long sequences or with a large number of species, distance methods such as *neighbour joining*[3] are the methods of choice.

It is well known that phylogenetic analyses of species based on single genes or parts of the genomes are often inconsistent. Many parts of the genomes may have arisen through some forms other than inheritance, for example by viral insertion, DNA transformation, symbiosis or some other methods of horizontal transfer. Furthermore, it can be argued that a single gene hardly possesses enough evolutionary information as some genes may have evolved faster than others [4]. Some genes may even have been deleted from or inserted into the genomes after the separation of species. The availability of more and more sequenced genomes allows phylogeny construction from complete genomes that is less sensitive to such inconsistency because all information is used rather than selective information.

For such long sequences, phylogenetic tree construction methods like *maximum parsimony* and *maximum likelihood* are often not possible due to their intensive computational requirement. The distance methods require a measure of distances between any two genomes. Traditional methods for computing distances based on sequence alignment are also not possible. Some measures such as evolutionary edit distance of gene order [5] and gene content [6] are also computational expensive or do not perform well when the gene content of the organisms are similar.

This paper presents a measure of genetic distances between genomes based on information theory [7,8]. Instead of selecting a gene or an rRNA, the method can use whole genomes. The method does not require sequence alignment, genome annotation and even genome mapping. We apply our distance measure to build the phylogenetic trees for two difficult cases: *Plasmodium* parasite genomes with quite varied AT composition, and a set of bacteria in the γ-Proteobacteria group for which horizontal gene transfer is common.

2   Background

Traditionally, the first step of phylogenetic analysis is to perform a multiple alignment of the sequences involved. The aligned symbol for each sequence is shown in a column for each position. From the alignment, one of tree construction methods is then applied. The maximum parsimony method [1] predicts the phylogenetic tree that minimises the number of steps needed to generate the observed sequences from an ancestral sequence. The method treats all mutations equally whereas in practice, mutation rates are different from positions to positions, and transitions (such as changing A to G) happen more often than transversions (changing A to C or T). As a result, parsimony method may produce misleading information. The problem is overcome by the maximum likelihood method [2], which requires a substitution model to assess the probability of particular mutations, and predicts the most likely tree given the observed data. Both methods require searching the tree space for an optimal tree and apply the criteria on every position of the aligned sequences. They are therefore computationally expensive when applying to data with long sequences or with many sequences.

Distance methods, which are derived from clustering algorithms, are better suited to large data sets. These methods are based on the genetic distances between sequence pairs in the set of sequences. Notable examples of this class are the *neighbour joining* method [3] and the *UPGMA* method [9]. These methods consider sequence pairs that have the smallest distances as "neighbours" and place them under a common ancestor. The goal of distance methods is to find a tree where the sum of the branch lengths between any two nodes closely approximates the distance measurement between the represented sequence pair.

The distances between sequence pairs can be computed from their alignment. Multiple alignment of sequences can only be applied for short, homologous molecules such as genes or ribosomal RNA. However, due to the variation of evolution rates among genes, phylogenetic analysis using different genes may result in different trees [10]. Some genes may even have arisen from means other than inheritance from ancestors and thus are not available in all specices involved. Such inconsistencies raise the question of reliability of the methods.

Since the genomes contain all genetic information of the organisms, it is suggested that phylogenetic analysis using whole genomes would overcome the inconsistency. However, performing alignment of genomes is often impossible due to many factors such as genome rearrangement and DNA transposition. Furthermore, due to their large sizes, genomes cannot be reliably and practically aligned. Early approaches for genome phylogenetics rely on identification of homologies to measure distances. Work by [5] proposes using the number of events needed to rearrange genes in genomes as a measure of genetic dissimilarity. Gene content is considered in [6] to measure genome distances. The similarity of two genomes is defined as the number of genes they share.

Recent years have seen an increasing number of *alignment-free* methods for sequence analysis. These methods are broadly categorised into two main groups, word based and information based [11]. Those in the former group map a sequence to a vector defined by the counts of each *k-mer*, and measure genome distances by some linear algebraic and statistical measures such as the Euclidean distance [12] or covariance distance [13]. These methods are still loosely dependent on local alignment as the comparisons are made for fixed word length. Furthermore, these methods would easily be misled as DNA homologies contain many mutations and indels, and certain genomes show statistical bias toward some skewed composition distributions.

The second group of alignment-free algorithms are founded on information theory [7] and Kolmogorov complexity theory. The advantages of these methods are that they are more elegant and do not require an evolutionary model. These methods are based on the premise that two related sequences would share some information and thus the amount of information in two sequences together would be less than the sum of the amount of information of each sequence. There is no absolute measure of information content in this case. However, information content can be estimated by using a lossless compression algorithm. The better the compression algorithm performs, the closer it can estimate information content of sequences. Nevertheless, compression of biological sequences is very challenging. General text compression algorithms such as Lempel-Ziv [14] and PPM [15] typically fail to compress genomes better than the 2-bits per symbol baseline. A number of biological compression algorithms such as *BioCompress* [16] and *GenCompress* [17] have been developed during the last decade but most of them are too expensive to apply to sequences of size over a million of bases. The GenCompress algorithm, which is used to measure sequence distances in [18], takes about one day to compress the human chromosome 22 of 34 million bases and achieves just 12% compression. In [19], an information measure is developed based on Lempel-Ziv complexity [20], which relates the number of steps in production process of sequence to its complexity. How well the method estimates the complexity is in fact not reported.

To the best of our knowledge, none of the existing methods are sufficiently robust to perform phylogenetics analysis on genome-size sequences. The information theoretic approaches appear to scale well to a large data set, but the existing underlying compression algorithms are either too computationally expensive or do not perform well. To fill in this gap, we introduce here another information theoretic approach to measure genome distances. The method is based on the compression algorithm *expert model* [21], which has been shown to be superior than others in terms of both compression performance and speed. As a rough comparison against GenCompress, the expert model running on a desktop computer can compress the whole human genome of nearly 3 billion bases in about one day and saves about 20%.

3   Genome Distance Measure

Information theory [7] directly relates entropy to the transmission of a sequence under a statistical model of compression. Suppose a sequence X is to be efficiently transmitted over a reliable channel. The sender first compresses X using a compression model and transmits the encoded message to the receiver, who decodes the compressed stream using the same model to recover the original message. The *information content* $\mathcal{I}_X$ of X is the amount of information actually transmitted, i.e. the length of the encoded message.

Suppose a sequence Y *related* to X is available to both parties. The sender needs to transmit only the information of X that is not contained in Y and indicate what was shared. Since the receiver also knows Y, it can recover X correctly. The amount of information actually transmitted in this case is called *conditional information content* of X given Y, denoted as $\mathcal{I}_{X|Y}$. The more related the two sequences are, the more information the two sequences share, and hence the shorter message is transmitted. The shared information of X and Y is called the *mutual information* of X and Y and can be computed as the difference between the information content and the conditional information content: $\mathcal{I}_{X,Y} = \mathcal{I}_X - \mathcal{I}_{X|Y}$.

In this work, we propose using the mutual information of two genomes as an estimation of the genetic similarity between two species. The information content of X is estimated as the message length obtained by compressing X using a lossless compression algorithm. We then concatenate Y and X to have the sequence YX and compress the concatenated sequence to obtain the information content of both sequences together. The conditional information content of X given Y is then computed by taking the difference of the information content of YX and the information content of Y: $\mathcal{I}_{X|Y} = \mathcal{I}_{XY} - \mathcal{I}_Y$.

We use the expert model compression algorithm [21] for estimation of information content because it provides many interesting features. Firstly, the expert model package provides an option to compress one sequence on the background knowledge of the other. Secondly, the expert model has the best performance among the existing biological compression algorithms. Since lossless compression provides an upper bound estimation of the entropy, a better compression gives a better approximation of the information content of sequences. Finally, the expert model runs very quickly on long sequences, and can be used for analysing genome size sequences in practice.

In theory, $\mathcal{I}_{X,Y}$ should be equal to $\mathcal{I}_{Y,X}$ as they both represent the shared information of the two sequences. However, this is not always the case in practice due to arithmetic rounding in compression. We therefore take the average of the two as the similarity measure of the two sequences. Since the two genomes may have different lengths, and they may have nucleotide composition bias, we normalise the similarity measure by a factor of $\mathcal{I}_X + \mathcal{I}_Y$:

$$
\begin{aligned}
S_{X,Y} &= \frac{\mathcal{I}_X - \mathcal{I}_{X|Y} + \mathcal{I}_Y - \mathcal{I}_{Y|X}}{\mathcal{I}_X + \mathcal{I}_Y} \\
&= 1 - \frac{\mathcal{I}_{X|Y} + \mathcal{I}_{Y|X}}{\mathcal{I}_X + \mathcal{I}_Y}
\end{aligned}
\tag{1}
$$

Instead of using the similarity measure, we use the *distance measure*, which is defined as:

$$
D_{X,Y} = 1 - S_{X,Y} = \frac{\mathcal{I}_{X|Y} + \mathcal{I}_{Y|X}}{\mathcal{I}_X + \mathcal{I}_Y}
\tag{2}
$$

4   Experimental Results

We present experimental results for our method on two sets of data. The first data set contains the genomes of 8 malaria parasites and the other contains 13 bacteria genomes. We first applied the expert model to obtain pairwise distances between each pair of genomes in a data set. The phylogenetic trees are then constructed using the *neighbour joining*[3] method from the PHYLIP package [22]. The experiments were carried out on a desktop with Pentium Duo core 2 2.33Ghz CPU and 8GB of memory, running Linux Ubuntu 8.10.

4.1   Plasmodium Phylogeny

Plasmodium species are the parasites that cause malaria in many vertebrates including human. In stages of their life-cycle, *Plasmodium* species interact with a mosquito vector and a vertebrate host. In order to adapt to the environment in the host blood, certain *Plasmodium* genes are under more evolutionary pressure than others, which leads to the variation of evolutionary rates among genes. Most *Plasmodium* species co-evolve with their hosts, and their evolution depends largely on hosts and geographic distributions. Certain species are thought to have emerged as a result of host switches. For example, the human malaria *Plasmodium falciparum* is speculated to have diverted from the chimpanzee malaria *Plasmodium reichenowi* recently and thus is more closely related to *Plasmodium reichenowi* than to other human malaria parasites.

As a result, the study of malaria phylogenetics faces the difficulty of selecting genes or rRNA for analysis. Small subunit rRNA and circumsporozoite protein have been used in many *Plasmodium* phylogenetics analyses [23,24]. However, recent study indicates that these loci are not appropriate for evolutionary studies because *Plasmodium* species possess separate genes, each expressed at a different time in the life cycle[25]. Likewise, the circumsporozoite protein may be problematic as the gene codes for a surface protein is under strong selective pressure from the vertebrate immune system. Indeed, recent phylogeny analyses [26] using these molecules show results that are inconsistent with those of other loci.

We applied our distance measure to construct the phylogenetic tree of eight malaria parasites, namely *P. berghei*, *P. yoelii*, *P. chabaudi* (rodent malaria), *P. falciparum*, *P. vivax*, *P. knowlesi*, *P. reichenowi* (primate malaria) and *P. gallinaceum* (bird malaria). Their genomes were obtained from PlasmoDB release 5.5 (http://www.plasmodb.org/common/downloads/release-5.5/). The genome of *P. reichenowi* has not been completed, only 7.8 megabases out of the estimated 25 megabases are available. The genomes of *P. berghei*, *P. chabaudi*, *P. gallinaceum* and *P. vivax* were completely sequenced, but they have not been fully assembled, each genome consists of several thousand contigs. Only the genomes of three species, *P. falciparum*, *P. knowlesi* and *P. yeolii* have been completely assembled into 14 chromosomes each. Prior to performing analysis, we removed wildcards from the sequences. The characteristics of the genomes are presented in table 1.

Statistical analysis of these *Plasmodium* genomes is very challenging. The composition distributions of these genomes are greatly different. AT content of the *P. falciparum* genome is very high (80% of the genome are A and T) whereas the distribution for the *P. vivax* is more uniform even though both species are human malaria parasites. Conventional analysis tools would be misled by such statistical bias [27]. Because many of the genomes have not been fully assembled, methods taking advantage of gene order or genome rearrangement such as [12,13] cannot be used. Furthermore, due to the size of the dataset, it is not practical to use methods such as [18,19].

Our method took just under 8 hours to process the 150 megabase data set and generate the pairwise distance matrix of the sequences. The neighbour joining method was then applied to produce an unrooted tree. To make the tree rooted,

Table 1. *Plasmodium* genomes characteristics

Species	Host - Geographic Dist.	Genome Size	%AT	Status
P. berghei	Rodent - Africa	18.0 Mb	76.27%	Partly Assembled
P. chabaudi	Rodent - Africa	16.9 Mb	75.66%	Partly Assembled
P. falciparum	Human - Subtropical	23.3 Mb	80.64%	Fully Assembled
P. gallinaceum	Bird - Southeast Asia	16.9 Mb	79.37%	Partly Assembled
P. knowlesi	Macaque - Southeast Asia	22.7 Mb	61.17%	Fully Assembled
P. reichenowi	Chimpanzee - Africa	7.4 Mb	77.81%	Partly Available
P. vivax	Human - Subtropical	27.0 Mb	57.72%	Partly Assembled
P. yoelii	Rodent- Africa	20.2 Mb	77.38%	Fully Assembled

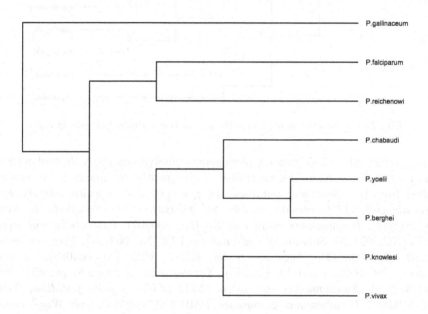

Fig. 1. The generated phylogenetic tree of the *Plasmodium* genus

we selected the *P. gallinaceum* as the outgroup because *P. gallinaceum* is bird malaria, whereas the others are mammal parasites. The tree produced is shown in Fig. 1. The tree is consistent with the majority of the earlier work [26,28]. In particular, it supports the speculation that the species closest to the human malaria parasite *P. falciparum* is in fact the chimpanzee malaria *P. reichenowi*.

4.2 Bacteria Phylogeny

Horizontal gene transfer is found extensively in bacteria genomes. This prevents the establishment of organism relationships based on individual gene phylogenies [10]. In order to perform phylogenetic analysis of such species, typically a number of likely gene orthologs are selected. Resulting phylogenetic hypotheses from these loci are often inconsistent with each other.

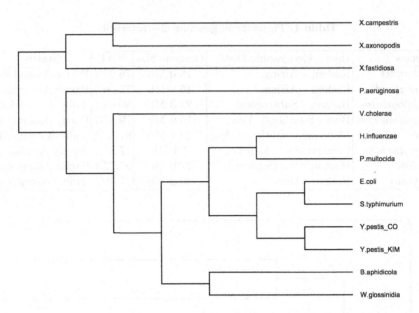

Fig. 2. The generated phylogenetic tree of the γ-Proteobacteria group

We performed a whole-genome phylogenetic analysis on the γ-Proteobacteria group for which horizontal gene transfer is frequently documented. We downloaded from the Genebank database the genomes of 13 species, namely *Escherichia coli* K12 (accession number NC_000913), *Buchnera aphidicola* APS (NC_002528), *Haemophilus influenzae* Rd (NC_000907), *Pasteurella multocida* Pm70 (NC_002663), *Salmonella typhimurium* LT2 (NC_003197), *Yersinia pestis* CO_92 (NC_003143), *Yersinia pestis* KIM5 P12 (NC_004088), *Vibrio cholerae* (NC_002505 and NC_002506), *Xanthomonas axonopodis* pv. citri 306 (NC_003919), *Xanthomonas campestris* (NC_003902), *Xylella fastidiosa* 9a5c (NC_002488), *Pseudomonas aeruginosa* PA01 (NC_002516), and *Wigglesworthia glossinidia brevipalpis* (NC_004344). The sizes of the genomes range from 1.8 megabases to about 7 megabases, and the total size of the data set is 44 megabases.

An earlier phylogenetic analysis of the 13 specices found an inconsistency; the use of different genes resulted in different evolutionary trees. There are 14,158 gene families found on these genomes. The majority of these families contain only one gene. Only 275 families are represented in all 13 species, and 205 families contain exactly one gene per species [10]. The analysis used the alignments of these 205 families and found that the resulting 205 trees are in 13 different topologies. The likelihood tests of 13 topologies reported that four most probable topologies are in agreement with over 180 gene families and that the consensus topology is in agreement with 203 alignments. These four trees differ in regard to the positions of three species, *Wigglesworthia*, *Buchnera* and *Vibrio*.

The tree generated by our method is presented in Fig. 2. Except for the three species, the tree agrees with the four most likely topologies in [10]. Similar to the consensus tree, it also supports the hypothesis that *Wigglesworthia* and *Buchnera* are sister species. It only differs from the consensus tree in positions of the branches involving (*Buchnera, Wigglesworthia*) and (*Haemophilus, Pasteurella*). A close examination of the tree shows that, the distances from these groups to their parent, and the distance between the most recent ancestor of *Vibrio* to its parent are very small. This suggests that, these species split from each other at very similar times. This explains the inconsistency among the four most probable trees generated by [10] and the tree inferred by our approach.

5   Conclusions

We have presented an information theoretic approach to measure genetic distances between genomes for phylogenetic analysis. It is based on the proven *expert model* compression algorithm to estimate mutual information of two genomes which is used as the measure of genetic similarity between the two species. Unlike conventional phylogenetic methods, it does not require alignment and annotation of sequences. In addition, it does not rely on an evolutionary model. Furthermore, the method is able to handle data with considerable bias in genetic composition, which classical statistical analysis approaches often fail to deal with.

We applied our method to generate the phylogenetic trees from the whole genomes of eight *Plasmodium* species, and of 13 species of the γ-Proteobacteria group. The genomes in both data sets are known to contain abundance of horizontal transfer genes. Previous analysis of these species using small molecules showed inconsistencies among the trees inferred from different genes. The trees generated by our approach are largely consistent with the consensus trees from previous work.

As information is the universal measure, the method can be potentially extended for other types of data. Such distance measure can be useful in other data mining applications such as clustering and classification. To the best of our knowledge, our approach is the first to be able to infer reliable phylogenetic trees from whole genomes of eukaryote species, with modest requirements of computation power. Such a tool would be very useful for knowledge discovery from the exponentially increasing databases of genomes resulting from the latest sequencing technology.

References

1. Camin, J., Sokal, R.: A method for deducing branching sequences in phylogeny. Evolution, 311–326 (1965)
2. Felsenstein, J.: Evolutionary trees from DNA sequences: a maximum likelihood approach. J. Mol. Bio., 368–376 (1981)
3. Saitou, N., Nei, M.: The neighbor-joining method: A new method for reconstructing phylogenetic trees. Mol. Biol. Evol., 406–425 (1987)
4. Gogarten, P., Townsend, F.: Horizontal gene transfer, genome innovation and evolution. Nature Reviews Microbiology, 679–687 (2005)
5. Sankoff, D., Leduc, G., Antoine, N., Paquin, B., Lang, B.F., Cedergren, R.: Gene order comparisons for phylogenetic inference: evolution of the mitochondrial genome. PNAS, 6575–6579 (1992)

6. Snel, B., Bork, P., Huynen, M.A.: Genome phylogeny based on gene content. Nat. Genet., 66–67 (1999)
7. Shannon, C.E.: A mathematical theory of communication. The Bell System Technical Journal, 379–423 (1948)
8. Wallace, C.S., Boulton, D.M.: An information measure for classification. Computer Journal, 185–194 (1968)
9. Sokal, R., Michener, C.: A statistical method for evaluating systematic relationships. University of Kansas Science Bulletin, 1409–1438 (1958)
10. Lerat, E., Daubin, V., Moran, N.A.: From gene trees to organismal phylogeny in prokaryotes: the case of the gamma-proteobacteria. PLoS Biology, e19 (2003)
11. Vinga, S., Almeida, J.: Alignment-free sequence comparison - a review. Bioinformatics, 513–523 (2003)
12. Blaisdell, B.E.: A measure of the similarity of sets of sequences not requiring sequence alignment. PNAS, 5155–5159 (1986)
13. Gentleman, J., Mullin, R.: The distribution of the frequency of occurrence of nucleotide subsequences, based on their overlap capability. Biometrics, 35–52 (1989)
14. Ziv, J., Lempel, A.: A universal algorithm for sequential data compression. IEEE Transactions on Information Theory, 337–342 (1977)
15. Cleary, J.G., Witten, I.H.: Data compression using adaptive coding and partial string matching. IEEE Transactions on Communications, 396–402 (1984)
16. Grumbach, S., Tahi, F.: A new challenge for compression algorithms: Genetic sequences. Journal of Information Processing and Management, 875–866 (1994)
17. Chen, X., Kwong, S., Li, M.: A compression algorithm for DNA sequences and its applications in genome comparison. RECOMB, 107 (2000)
18. Li, M., Badger, J.H., Chen, X., Kwong, S., Kearney, P., Zhang, H.: An information-based sequence distance and its application to whole mitochondrial genome phylogeny. Bioinformatics, 149–154 (2001)
19. Otu, H., Sayood, K.: A new sequence distance measure for phylogenetic tree construction. Bioinformatics, 2122–2130 (2003)
20. Lempel, A., Ziv, J.: On the complexity of finite sequences. IEEE Transactions on Information Theory, 75–81 (1976)
21. Cao, M.D., Dix, T.I., Allison, L., Mears, C.: A simple statistical algorithm for biological sequence compression. DCC, 43–52 (2007)
22. Felsenstein, J.: PHYLIP phylogeny inference package. Technical report (1993)
23. Waters, A., Higgins, D., McCutchan, T.: Evolutionary relatedness of some primate models of plasmodium. Mol. Biol. Evol., 914–923 (1993)
24. Escalante, A., Goldman, I.F., Rijk, P.D., Wachter, R.D., Collins, W.E., Qari, S.H., Lal, A.A.: Phylogenetic study of the genus plasmodium based on the secondary structure-based alignment of the small subunit ribosomal RNA. Molecular and Biochemical Parasitology, 317–321 (1997)
25. Corredor, V., Enea, V.: Plasmodial ribosomal RNA as phylogenetic probe: a cautionary note. Mol. Biol. Evol., 924–926 (1993)
26. Leclerc, M.C., Hugot, J.P., Durand, P., Renaud, F.: Evolutionary relationships between 15 plasmodium species from new and old world primates (including humans): an 18s rDNA cladistic analysis. Parasitology, 677–684 (2004)
27. Cao, M.D., Dix, T.I., Allison, L.: Computing substitution matrices for genomic comparative analysis. In: PAKDD, pp. 647–655 (2009)
28. Siddall, M.E., Barta, J.R.: Phylogeny of plasmodium species: Estimation and inference. The Journal of Parasitology, 567–568 (1992)

Pattern Prediction in Stock Market

Saroj Kaushik and Naman Singhal

Department of Computer Science and Engineering
Indian Institute of Technology, Delhi
Hauz Khas, New Delhi – 110019, India
saroj@cse.iitd.ernet.in, singhal.naman@gmail.com

Abstract. In this paper, we have presented a new approach to predict pattern of the financial time series in stock market for next 10 days and compared it with the existing method of exact value prediction [2, 3, and 4]. The proposed pattern prediction technique performs better than value prediction. It has been shown that the average for pattern prediction is 58.7% while that for value prediction is 51.3%. Similarly, maximum for pattern and value prediction are 100% and 88.9% respectively. It is of more practical significance if one can predict an approximate pattern that can be expected in the financial time series in the near future rather than the exact value. This way one can know the periods when the stock will be at a high or at a low and use the information to buy or sell accordingly. We have used Support Vector Machine based prediction system as a basis for predicting pattern. MATLAB has been used for implementation.

Keywords: Support Vector Machine, Pattern, Trend, Stock, Prediction, Finance.

1 Introduction

Mining stock market tendency is regarded as a challenging task due to its high volatility and noisy environment. Prediction of accurate stock prices is a problem of huge practical importance. There are two components to prediction. Firstly, historic data of the firm in consideration needs to be preprocessed using various techniques to create a feature set. This feature set is then used to train and test the performance of the prediction system.

During literature survey we found that in the past, work has been done on the markets around the globe using various preprocessing techniques such as those based on financial indicators, genetic algorithms, principal component analysis and variations of time series models. The better prediction systems are based on Artificial Intelligence techniques such as Artificial Neural Networks [3] and Support Vector Machines (SVM) [2]. More recent work has even tried to come up with hybrid systems [6]. Overall, techniques based on Support Vector Machines and Artificial Neural Networks have performed better than other statistical methods for prediction [1, 2].

A. Nicholson and X. Li (Eds.): AI 2009, LNAI 5866, pp. 81–90, 2009.
© Springer-Verlag Berlin Heidelberg 2009

In all the literature, the emphasis is prediction of the exact value of the stock on the next day using the data of previous N days. We have extended the concept to predict pattern for next M days using the data of previous N days. If we try to predict the actual values of the next M days using the previous N days, the performance is not good. So, we propose a new concept of pattern prediction. The motivation of such an approach is to know the periods of relative highs and lows to be expected rather than knowing the exact values. Here, M and N are the number of days. We analyzed the three cases i.e. when M>N, M=N and M<N. Best results were obtained for M=N.

The data set being considered for the study is based on the real time financial time series of the Reliance Industries of the National Stock Exchange, India. We obtain historic data of the Reliance Industries for the last 8 years from the NSE [9] which contains day-wise closing, high and low prices. Prediction System uses Least Square Support Vector Regression (LS-SVR) based on Support Vector Machines [10]. In the next section we discuss the concepts of LS-SVR followed by implementation, results and observations.

2 Prediction Methods

In attempt to predict the stock markets behavior, study has been done on many prediction methods such as Support Vector Machines and Artificial Neural Networks etc [1,2,7,8]. In our research, we have used SVM based technique and have come up with unique approach to train SVM for prediction. The SVM used in the proposed work is the Least Square Support Vector Regression which is an extension of the Support vector classification proposed by V. Vapnik [1].

2.1 Support Vector Regression

The basic idea of SVM is to use linear model to implement nonlinear class boundaries through some nonlinear mapping of input vector into the high dimensional feature space. A linear model constructed in the new space can represent a nonlinear decision boundary in the original space. In the new space, an optimal separating hyper-plane is constructed. Thus SVM is known as the algorithm that finds a special kind of linear model, the maximum margin hyper-plane. The maximum margin hyper-plane gives the maximum separation between the decision classes. The training examples that are closest to the maximum margin hyper-plane are called support vectors. All other training examples are irrelevant for defining the binary class boundaries. Implementation of SVM is done using Support Vector Regression to predict the output values. Given a set of data points, $\{(x_1, z_1),..,(x_l, z_l)\}$, such that $x_i \in R^n$ is an input and $z_i \in R^1$ is a target output, the standard form of a support vector regression [10] is given below.

$$\min_{w,b,\varepsilon,\varepsilon^*} \frac{1}{2} w^T w + C \sum_{i=1}^{l} \varepsilon_i + C \sum_{i=1}^{l} \varepsilon_i^* \ . \tag{1}$$

subject to,

$$w^T \varphi(x_i) + b - z_i \le +\epsilon + \varepsilon_i .$$

$$z_i - w^T \varphi(x_i) - b \le +\varepsilon_i^* .$$

$$\varepsilon_i, \varepsilon_i^* \ge 0, i = 1,..,l .$$

The dual of (1) is

$$\min_{\alpha,\alpha^*} \frac{1}{2}(\alpha - \alpha^*)^T Q(\alpha - \alpha^*) + \epsilon \sum_{i=1}^{l}(\alpha_i + \alpha_i^*) + \sum_{i=1}^{l} z_i(\alpha_i - \alpha_i^*) . \tag{2}$$

subject to,

$$\sum_{i=1}^{l}(\alpha_i - \alpha_i^*) = 0, 0 \le \alpha_i, \alpha_i^* \le C, i = 1,..,l .$$

where,

$$Q_{ij} = K(x_i, x_j) .$$

The approximate value function is given by the following equation.

$$y = \sum_{i=1}^{l}(-\alpha_i + \alpha_i^*)K(x_i, x) + b . \tag{3}$$

Here w is the weight vector, ϵ, ε, etc are the standard variables used in optimizations, K is the kernel matrix, φ is the kernel function and α_i^*, α_i are the SVM coefficients. From now on let us denote SVM coefficients by $\alpha_i = (-\alpha_i + \alpha_i^*)$ with no restriction on α_i being greater that zero.

3 Feature Set Modeling

In this section, we will discuss the feature set modeling for pattern prediction and value prediction [2] and subsequently, the performance of both the methods will be compared.

3.1 Value Prediction

The basic concept of value prediction is to use previous N days to predict the value of next day [5, 6, 7]. We extend this concept to use values of previous N days to predict the values of next M days. We have analyzed the cases where N > M, N = M and N < M and found that the result comes out to be best if N = M. A total of M SVMs are required to implement the proposed prediction technique.

Let us consider closing price, say, x_i of ith day. Since one SVM is used to predict one day, we have used M SVMs to predict next M days prices $\{ x_{i+1}, x_{i+2}, ..., x_{i+M} \}$ from previous N days prices $\{ x_{i-N+1}, x_{i-N+2}, ..., x_i \}$.

3.2 Pattern Prediction

In the proposed technique for pattern prediction, first we learn all the patterns in the time series then learn to predict a pattern for next M days using closing price of previous N days from training data set and finally predict a pattern on test data.

Learn a Pattern in the Time Series. The pattern is represented as a vector of the coefficients generated by SVM as represented in equation (5). Since we want to predict the pattern for the next M days, first we learn all patterns of size M using same size sliding window in the entire time series. To learn one pattern, we create a training sample consisting of (Day, Price) pair in the current window. Each *Day* in the current window is represented by index from 1 to M and *Price* is represented by x_i, the closing price of ith day (in reference to the complete time series). So to learn a pattern, there are M training pairs required as follows:

$$((1, x_{i+1}), (2, x_{i+2}), ..., (M, x_{i+M})), i \in TrainingSet. \tag{4}$$

We train one SVM corresponding to each pattern in the training set of time series. Once each SVM is trained, we obtain SVM coefficients corresponding to each pattern. Let us represent the coefficients of ith SVM, say, SVM_i by $(\alpha_i, \alpha_{i+1}, ..., \alpha_{i+M})$ and the ith pattern in the time series by coefficients of SVM_i as given below.

$$\bar{\alpha}_i = \{\alpha_{i+1}, \alpha_{i+2}, ..., \alpha_{i+M}\}. \tag{5}$$

Learn to Predict a Pattern. After we have learnt all the patterns in training set, we will learn to predict the pattern of next M days $\{ \alpha_{i+1}, \alpha_{i+2}, ..., \alpha_{i+M} \}$ using the closing price of previous N days $\{ x_{i-N+1}, x_{i-N+2}, ..., x_i \}$ from the training data set. For this, a total of M new SVMs are required. These SVMs have nothing to do with the SVMs that were used to learn the patterns above.

Prediction of a Pattern. For the test set, we compute the coefficients for j^{th} test sample and is represented as follows:

$$\bar{\beta}_j = \{\beta_{j+1}, \beta_{j+2}, ..., \beta_{j+M}\}. \tag{6}$$

To obtain the pattern for j^{th} test sample, we compute the least squared error between $\bar{\beta}_j$ and $\bar{\alpha}_i$'s, $\forall$ $i \in TrainingSet$ and $j \in TestSet$. We consider the predicted pattern of

j^{th} day as the learned pattern of ith day for which least squared error between $\overline{\alpha}_i$ and $\overline{\beta}_j$ is minimum that is computed as follows.

$$error_{\min} = \sum_{k=1}^{M} (\alpha_{i+k} - \beta_{j+k})^2 .$$ (7)

where $i \in TrainingSet$ and $j \in TestSet$.

4   Implementation and Analysis of Results

The financial time series considered is of the Reliance Industries Ltd is its row data is obtained from NSE website [9].

4.1   Implementation

Data consists of closing price, the highest price and the lowest price of the trading day for last 8 years. The latest 1500 values are selected and used for experiment, where 1300 values are used for training and the rest 200 values are used for testing prediction accuracy. We have taken closing price as the feature set for SVM as in general, the prediction is created using the closing price only.

It is assumed the initial training set is large enough sample to represent a complete set of patterns and the process of learning pattern is not repeated when we perform prediction for the same time series.

LS-SVM package [10] is used for implementing Support Vector Machine. For each learning and prediction, first the parameters of the SVM are optimized and then parameters of the features are optimized to obtain the best possible prediction accuracy. During the optimization of the parameters of the feature set, the best results were obtained when N=M and N=10 [8]. Implementation is done using MATLAB.

4.2   Analysis of Results

As already mentioned earlier, the best results for predicting next M days value from previous N days value is obtained when N = M [8], the simulations were done for N = 7,10,14,21. Out of this the best values were obtained for N = 10.

Value Prediction. The graphs are plotted for the actual closing price and the predicted closing price for all the days that form the test set values for the next day and 10th day using value prediction method as discussed in Section 3.1. Values on the Y-axis are the actual closing price of stock in consideration in INR (Indian National Rupee) X-axis goes from 0 to 200 representing each day of the test set. Plot in blue is the predicted price while plot in green is the actual closing price.

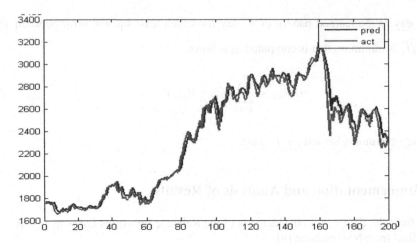

Fig. 1. Actual and predicted closing price for next day

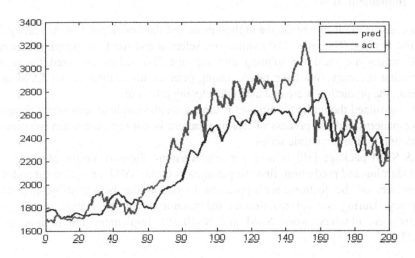

Fig. 2. Actual and predicted closing price for next 10th day

It can be seen that value prediction is good for predicting the next day price. Its performance deteriorates as we predict for more days in the future as shown in Figs 1 and 2.

Pattern Prediction. Now we show the graphs obtained using the proposed technique of pattern prediction. They have been directly compared to the corresponding value prediction plots. Pattern prediction graphs are plotted using the values corresponding to the pattern predicted and the actual values. Learning data and test data is same as above.

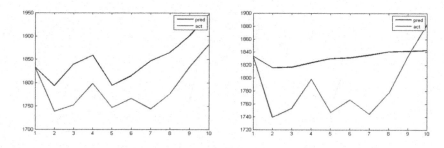

Fig. 3. Pattern Prediction vs Value Prediction at j = 55

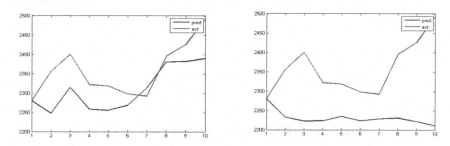

Fig. 4. Pattern Prediction vs Value Prediction at j = 82

We know that $\overline{\alpha}_i$ where $i=1,2,...,1300$ is learned pattern at the close of ith day and $\overline{\beta}_j$ where $j=1,2,...,200$, the predicted coefficients at the close of jth day. Consider such i and j for which error between $\overline{\alpha}_i$ and $\overline{\beta}_j$ is minimum. The pattern to be expected for the next M days at j^{th} day will be similar to the pattern at i^{th} day (refer subsection 3.2). The graph is plotted between $\{x_{i+1}, x_{i+2},...,x_{i+M}\}$ and $\{x_{j+1}, x_{j+2},...,x_{j+M}\}$ where $i \in$ *TrainingSet* and $j \in$ *TestSet*. Graph for the value prediction is the plotted between value predicted by the SVM and the actual value.

The pattern prediction and value prediction graphs shown in the following figures have been compared. Pattern prediction graphs are on the left while value prediction graphs are on the right. Prediction is done for the next M (=10) days following the close of each day of the test set. Out of the total test set of 200 values only a few graphs are shown. Plot in blue is the predicted value while plot in green is the actual closing price.

We can conclude from these graphs that pattern prediction is able to predict the highs and lows that can be expected in the near future more accurately as compared to the approach of prediction based on actual values. In the next section, we compare the results quantitatively.

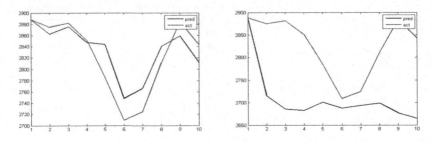

Fig. 5. Pattern Prediction vs Value Prediction at j = 119

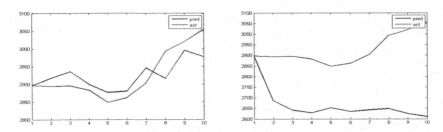

Fig. 6. Pattern Prediction vs Value Prediction at j = 147

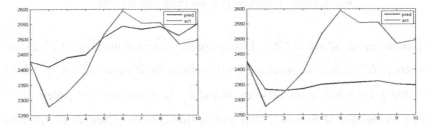

Fig. 7. Pattern Prediction vs Value Prediction at j = 179

4.3 Prediction Accuracy

Prediction accuracy is computed in percentage. For each test sample j, we predict for next M days, where j = 1,..,200. Let us denote the predicted value by p(k), and corresponding actual values by a(k), where k = 1,…,M. The following formula is used to compute the prediction accuracy in %age.

Table 1. Performance of Pattern prediction vs Value prediction

Accuracy ⟶	Mean	Min	Max
Value prediction	51.3%	0%	88.9%
Pattern prediction	58.7%	11.1%	100%

$$acc = \frac{100}{M-1} * \sum_{k=2}^{M} \mathrm{sgn}[(p(k)-p(k-1))*(a(k)-a(k-1))] \,. \tag{8}$$

where,

$$\mathrm{sgn}(x) = \frac{1, x > 0}{0, x <= 0} \,.$$

The table1 shows the performance of both the methods quantitatively using above formula. Here we have shown the minimum, maximum and the average prediction accuracy obtained over the complete test set. We can clearly see that the pattern prediction performs better than value prediction. The average for pattern prediction is 58.7% while that for value prediction is 51.3%. Similarly, maximum for the pattern and value prediction is 100% and 88.9% respectively whereas the minimum is 11.1% and 0% respectively.

5 Conclusion

Value prediction is a good technique for predicting next day price. However, if we want to predict the price for next 10-15 days, we do not get good results by predicting actual value. To tackle such a scenario, we have proposed technique of pattern prediction. Although, it does not attempt to predict the exact value but it predicts the expected trend of the prices for the next 10 days. Pattern-prediction gives better results in prediction of patterns for longer duration.

In the proposed work, we learnt all the patterns present in the time series. Due to this the SVM coefficients corresponding to the patterns obtained while learning are very noisy. As a future study, performance of the pattern prediction can be improved by processing of patterns and using a finite set of patterns rather than all the possible patterns found in the financial time series. To come up with further improved performance of the predicted pattern, we can apply some statistical algorithm between the SVM coefficients and learned pattern coefficients as well.

References

1. Vapnik, V.: The Nature of Statistical Learning Theory. Springer, Heidelberg (1995)
2. Kim, K.: Financial Time Series Forecasting using Support Vector Machines. Neurocomputing 55, 307–319 (2003)
3. Kim, K., Lee, W.B.: Stock Market Prediction using Artificial Neural Networks with Optimal Feature Transformation. Neural Computing and Application 13, 255–260 (2004)
4. Ince, H., Trafalis, T.B.: Kernel Principal Component Analysis and Support Vector Machines for Stock Price Prediction. In: Proceedings of IEEE International Joint Conference on Neural Networks, vol. 3, pp. 2053–2058 (2004)
5. Yu, L., Wang, S., Lai, K.K.: Mining Stock Market Tendency Using GA-based Support Vector Machines. In: Deng, X., Ye, Y. (eds.) WINE 2005. LNCS, vol. 3828, pp. 336–345. Springer, Heidelberg (2005)

6. Li, W., Liu, J., Le, J.: Using GARCH-GRNN Model to Forecast Financial Times Series. In: Yolum, p., Güngör, T., Gürgen, F., Özturan, C. (eds.) ISCIS 2005. LNCS, vol. 3733, pp. 565–574. Springer, Heidelberg (2005)
7. Chen, W.H., Sheh, J.Y., Wu, S.: Comparison of Support Vector Machines and Back Propagation Neural Networks in forecasting the six major Asian Stock Markets. Int. J. Electronic Finance 1(1), 49–67 (2006)
8. Singhal, N.: Stock Price Prediction for Indian Market, Master, Thesis. Department of Computer Science and Engineering. Indian Institute of Technology, Delhi, India (2008)
9. National Stock Exchange of India Ltd., Data Products NCFMOnline ReportCBT, http://www.nseindia.com
10. LS-SVM, package. Home Toolbox Book People Publications Faq Links, http://www.esat.kuleuven.ac.be/sista/lssvmlab

Balancing Workload in Project Assignment

Zhirong Liang[1], Songshan Guo[1], Yanzhi Li[2], and Andrew Lim[2]

[1] Department of Computer Sciences, Sun Yat-sen University, Guangzhou, China
program806@hotmail.com, issgssh@mail.sysu.edu.cn
[2] Department of Management Sciences, City University of Hong Kong, Hong Kong
yanzhili@cityu.edu.hk, lim.andrew@cityu.edu.hk

Abstract. In this paper, we study a project assignment problem. Specifically, a set of projects, each of which needs to be finished over a project development cycle, are to be assigned to a group of identical engineers over a discrete planning horizon. The workload of the projects is different and fluctuates over their development cycles. In any period, an engineer has a maximum allowed workload. The objective of the problem is to assign the projects to engineers with the objective of balancing the total workload among the engineers; the load balance is measured by the difference between the maximum and the minimum total workload. Such a problem is new to the literature. The problem is strongly NP-hard. Therefore, we propose a two-stage heuristic approach to solve it. Extensive numerical experiments show that the proposed approach can achieve optimal or nearly optimal solutions for all test cases; such performance is much better than what can be obtained from an IP model solved with ILOG CPLEX11.

1 Introduction

We study a project assignment problem arising from practices followed by the R&D department of a toys firm. The firm owns its own brand and is headquartered in the U.S. It develops all its products in Hong Kong, and outsources manufacturing to OEM factories in mainland China. Finished products are then shipped to customers around the world; the firm sells over one thousand different types of toys globally every year.

The R&D department is responsible for developing all the firm's products. In the following, we use the term "project" to denote a product that needs to be designed, produced, and delivered by the department. A project typically consists of multiple stages, and lasts from six to thirteen weeks. The workload in each week (stage) is different. The workload also varies with projects, depending on whether a product is totally new, an adapted version from existing products, or an old one. A project is to be assigned to and finished by a single engineer. Considering each year there are over one thousand products to be developed, the R&D department faces a difficult problem in assigning projects to engineers. The current practice is not satisfactory. There are mainly two issues. One is that the workload among engineers is not fairly distributed. Some are loaded much more than others. Engineers who are highly loaded in one round of project

A. Nicholson and X. Li (Eds.): AI 2009, LNAI 5866, pp. 91–100, 2009.

assignments have to be compensated in the next round. The other is that the engineers' workload fluctuates too much over time. An engineer could be very busy during some weeks, and may need to work overtime frequently, while he/she is almost idle during some other weeks. These issues result in the engineers having many complaints and much dissatisfaction. What makes the firm much more concerned, however, is that when an engineer is overloaded, he/she may not be able to finish the project very well, since he/she does not have the time and energy to monitor the project very closely. Sometimes delivery gets delayed, causing a loss of revenue or a penalty cost. Sometimes the finished product may have quality problems, and later on may have to be recalled, which too leads to great loss to the firm.

The problem faced by the firm is by no means unique. Most R&D projects, or generally speaking, most projects, last for different durations of time and the workload fluctuates in different stages. Different projects may have different starting times. In such situations, project assignment has to be done properly, in order to distribute workload fairly among project teams or project managers.

The project assignment problem studied in this paper is new to the literature. It is closely related to, but distinct from, three types of assignment problems. The first type is the generalized assignment problem(GAP) [1]. Our project assignment problem is different from the GAP in two aspects: (1) projects in our setting span over a long time duration, and the required workload fluctuates over time; and (2) the objective of our problem is to balance the workload among the engineers, instead of minimizing cost. The second type of related assignment problems considers the objective of load balancing in the setting of the classical assignment problem, i.e, when agents do not have defined capacities. [4] and [2] studied the problem of balancing load among agents, with the objective to minimize the difference between the maximum workload and the minimum workload of individual agents; they also stipulated that each agent could at the most be assigned one task. The third type of revelant assignment problems studies assignment over a time horizon. [3] considered an assignment problem with time-variant workloads. However, in their setting, a task only lasts one period, whereas in our setting, a task with nonstationary workload lasts multiple periods. An extensive survey of assignment problems can be found in [5].

2   Problem Definition and Formulation

We formally define our problem as follows. A firm has K projects, to be assigned to I engineers, over a planning horizon of T periods. Each project spans over several consecutive periods, and the workload in different periods is different. Each engineer is qualified to work on any project. However, once a project is assigned to an engineer, the engineer needs to work on the project until it is finished, i.e. projects cannot be reassigned while being processed. To ensure quality, and also considering engineers' job satisfaction, it is preferred that the workload of an engineer in any period is restricted to a certain level. Given that all the engineers' period workload has been reasonably confined, the objective of the firm is to assign projects to engineers such that over the planning horizon, the total workload of engineers is balanced. The total workload is considered balanced if the difference between the maximum and minimum total workload of all engineers is minimized. The maximum workload of an engineer in a given period is restricted by certain legal regulations. However, the firm may want to further confine it in order to improve engineers' job satisfaction. Therefore, the firm is interested in finding out the best workload balancing at different levels of maximum period workload, and the final decision about the project assignment is then made with a tradeoff between load balancing and maximum period workload.

We now define some notations:

t: index for period, $t = 1, .., T$
i: index for engineer, $i = 1, .., I$
k: index for project, $k = 1, .., K$
c_{kt}: workload of project k in period t, $k = 1, .., K, t = 1, .., T$. For a project k which spans over period a to b, $c_{kt} = 0, t = 1, 2, .., a - 1, b + 1, .., T$
C: maximum allowed period workload for any engineer

Decision Variable
x_{ik}: 1 if project k is assigned to engineer i, $i = 1, .., I, k = 1, .., K$
U: the maximum total workload of all engineers over the planning horizon
L: the minimum total workload of all engineers over the planning horizon

We can then formulate the problem into the following IP model:

$$\min \quad U - L \tag{1}$$

$$s.t. \quad \sum_{i=1}^{I} x_{ik} = 1, \qquad k = 1, .., K \tag{2}$$

$$\sum_{k=1}^{K} c_{kt} x_{ik} \leq C, \qquad i = 1, .., I; t = 1, .., T \tag{3}$$

$$\sum_{t=1}^{T} \sum_{k=1}^{K} c_{kt} x_{ik} \leq U, i = 1, .., I \tag{4}$$

$$\sum_{t=1}^{T} \sum_{k=1}^{K} c_{kt} x_{ik} \geq L, i = 1, .., I \tag{5}$$

$$U, L \geq 0, x_{ik} \in \{0, 1\}, \quad i = 1, .., I; k = 1, .., K \tag{6}$$

In the above, Constraint (2) makes sure that a project is assigned to one and only one engineer. Constraint (3) ensures period workload of all engineers is upper bounded. Constraints (4) and (5) are used to define the maximum and minimum total workload; due to the objective function, U and L will be exactly equal to the maximum and minimum total workload.

The project assignment problem is strongly NP-hard, which can be proved by showing that the well-known 3-partition problem is its special case. In fact, even to find a feasible solution for the project assignment problem is NP-hard. We omit the proof, which is rather routine, due to page limit.

3   A Two-Stage Solution Approach

To solve the project assignment problem, we propose a two-stage approach. In the first stage, we focus on finding feasible solutions, where feasibility mainly refers to the constraint that period workload is bounded. In the second stage, we focus on improving load balancing while keeping solutions feasible. This two-stage procedure is repeated many times because the initial solutions generated in the first stage have some randomness. As we said earlier, the final project assignment decision is made with a tradeoff between load balancing and maximum period workload. Therefore, we are interested in load balancing at different levels of maximum period workload. For this purpose, we first propose a lower bound for the maximum allowed period workload. We then examine the load balancing as we gradually increase the maximum allowed period workload.

3.1 A Lower Bound for the Maximum Allowed Period Workload

A lower bound can be obtained if we relax the problem by allowing projects to be reassigned during the project development cycle. With this relaxation, for each period t, we can compute a lower bound, $\underline{C_t}$, for the maximum allowed workload. A lower bound for the original problem, $\underline{C}$, would then be the maximum of all the period lower bounds, $\underline{C} = \max\{\underline{C_t}, t = 1, .., T\}$.

For any period t, a lower bound $\underline{C_t}$ can be obtained for period workload, as follows: $\underline{C_t} = \lceil \sum_{k=1,..,K} c_{kt}/I \rceil$ Although this may seem to be an obvious lower bound, and may be quite loose for a specific t, the lower bound for the original problem, $\underline{C}$, after taking the maximum of all the $\underline{C_t}$, tends to be very tight, as shown by experiments.

3.2 Constructing Feasible Solutions

The problem of finding a feasible project assignment, for a given maximum allowed period workload C, can be a challenging task, as it is strongly NP-hard. Therefore, a great deal of effort has been made to construct feasible solutions.

We adopt a two-step procedure. In the first step, we shuffle the projects randomly and then assign them one by one to engineers, following a first-fit or best-fit rule. If we do not end up with a feasible solution in the first step, then, in the second step, we apply branch and bound to subproblems with two engineers, and then, if still infeasible, three engineers. Note that we do not consider performance of solutions in this stage; feasibility is the only concern.

The first step is detailed as follows. First, we randomly permutate projects into a list. Since the two-stage procedure will be repeated many times, such random shuffles are expected to result in different initial solutions. We then assign the projects one by one to engineers, and remove projects from the list as they are assigned. Each time we take the first project from the remaining projects, and try to assign it according to the first-fit rule or the best-fit rule. First-fit rule means the project is to be assigned to the first engineer whose period workload, after taking the current project, will not exceed the maximum allowed period workload C. Best-fit rule means the project is to be assigned to an engineer who, after taking the current project, incurs the minimum penalty; the penalty for engineer i, σ_i, $i = 1, .., I$, is calculated as:

$$\sigma_i = \sum_{t=1}^{T} \delta_{\{W_{it}-C>0\}}(W_{it} - C)$$

Algorithm 1. B&B for Subproblems of Searching for Feasible Solutions

---

Initialization:
  a set of projects to be assigned, $\hat{P}_0$;
  a set of projects assigned to engineer 1, $\hat{P}_1$;
  a set of projects assigned to engineer 2, $\hat{P}_2$; $\hat{P}_1 = \hat{P}_2 := \emptyset$;
  a first-in-last-out queue for B&B nodes, Q, $Q := \{(\hat{P}_0, \hat{P}_1, \hat{P}_2)\}$;
while $Q \neq \emptyset$ **do**
  Take a node from Q, denoted as (P_0, P_1, P_2);
  $Q := Q - \{(P_0, P_1, P_2)\}$;
  Take a project i out of P_0, $P_0 := P_0 - \{i\}$;
  if $P_1 \cup \{i\}$ keeps engineer 1 feasible **then**
    if $P_0 = \emptyset$ **then**
      We stop with a feasible solution $(P_1 \cup \{i\}, P_2)$; ;
    else
      $Q := Q \cup \{(P_0, P_1 \cup \{i\}, P_2)\}$
    end if
  end if
  if $P_2 \cup \{i\}$ keeps engineer 2 feasible **then**
    if $P_0 = \emptyset$ **then**
      We stop with a feasible solution $(P_1, P_2 \cup \{i\})$;
    else
      $Q := Q \cup \{(P_0, P_1, P_2 \cup \{i\})\}$;
    end if
  end if
end while

---

where $\delta_{\{x\}}$ is an indicator function, $\delta_{\{x\}} = 1$, iff $x = true$; W_{it} is the workload of engineer i in period t after taking the current project. The second step is invoked only if the first step does not result in a feasible solution. In the second step, we first sort the engineers in descending order of penalty. Remember that engineers with positive penalty have period workload exceeding C. Our target, then, is to turn these "infeasible" engineers into feasible ones. This is achieved by solving a series of subproblems, each of which takes care of one infeasible engineer. Specifically, we consolidate projects of the engineer with the largest penalty, and the projects of a feasible engineer, and try to reassign the projects such that both engineers get feasible assignments. Given the small scale of the subproblems (only two engineers and a limited number of projects are involved), we apply a branch and bound procedure for reassignment. We formally describe the procedure as follows, in Algorithm 1.

If Algorithm 1, applied to two-engineer subproblems, turns the infeasible engineer feasible, then we continue to apply the algorithm to other infeasible engineers. If Algorithm 1 fails, we try to consolidate the infeasible engineer with a feasible engineer. If this fails, we take up three-engineers subproblems, i.e. pooling the infeasible engineer with two feasible engineers for reassignment. A three-engineers branch and bound procedure is similar to Algorithm 1, but now the number of nodes to be explored will increase dramatically. In the experiment, however, we find that if a feasible solution can be found, it is often found in the early stage of the branch and bound procedure. In view of this, we impose a limit to the number of nodes to be explored in the procedure, so as to save some computational effort, ensuring that we still retain the performance.

Though the branch and bound procedure works quite well in terms of reaching feasible solutions, it often takes a very long time because very often the subproblems do not end up with feasible solutions and therefore many subproblems have to be solved. To address this issue, we use a simple heuristic to eliminate subproblems without a hope of feasibility.

The heuristic follows the same idea as in Section 3.1 for designing the lower bound. For each period, we simply take the sum of workload of all projects, divided by 2 (or 3, for three-engineers subproblems), and this gives a lower bound for the period. We then take the maximum workload over all the periods. The subproblem is infeasible if the maximum thus obtained is greater than the maximum allowed period workload, C. The experiment shows that the heuristic is very effective. Over 80 percent of subproblems are directly shown to be infeasible, eliminating the need of invoking the branch and bound procedure.

3.3 Improving Solution Quality

In the first stage, we pay attention to only feasibility; while in the second stage, we focus on improving the performance of the initial solution. Observing that the performance of a solution, i.e., an assignment, is only determined by the two engineers with the maximum and the minimum total workload, we propose an iterative procedure. During each iteration, focusing on either of the two engineers, we aim to either reduce the maximum workload or to enlarge the minimum total workload.

Specifically, we first sort the engineers in descending order of total workload. We then take the engineer with the maximum total workload, and try to consolidate with another engineer with a lower total workload, starting with the one with the minimum total workload. Denote the maximum total workload as U, and the one used for consolidation, L. The best result would be when reassigning projects between the two engineers, the smaller total workload of the two engineers becomes $\lfloor (U + L)/2 \rfloor$ (the bigger one would be $\lceil (U + L)/2 \rceil$); and the next best is $\lfloor (U + L)/2 \rfloor - 1$, followed by $\lfloor (U + L)/2 \rfloor - 2, .., L$. If the smaller total workload equals L, it means there is no improvement. If there is improvement, we sort the engineers again for the improved solution, and repeat the procedure. Otherwise, we try to consolidate the engineer with the maximum total workload with the engineer with the second lowest total workload. If this does not work, then we continue to choose the engineer with the third lowest, the fourth lowest,.., total workload until we achieve an improvement. With all the above, we attempt to reduce the maximum total workload U. If the maximum total workload cannot be reduced further, we turn to enlarging the minimum total workload. We first consolidate the engineer with the minimum total workload with the one whose total workload is the second largest(since the one with the maximum total workload has been considered before). If there is improvement, we sort the engineers again to continue consolidating engineers with lower

---

Algorithm 2. Dynamic Programming for Second-stage Subproblems

---

Initialization:

  a set of n projects to be assigned, each with total workload $c_k = \sum_{t=1}^{T} c_{kt}, k = 1, .., n$;

  an integer array f of length $\lfloor (U + L)/2 \rfloor$;

  $f[0] := 0; f[j] = -1$, for $j = 1, 2, .., \lfloor (U + L)/2 \rfloor$;{a value of -1 is used to denote infeasibility}

for $k = 1$ to n **do**

  for $j = \lfloor (U + L)/2 \rfloor$ to 1 **do**

    if $j - c_k > 0$ **then**

      if $f[j - c_k] \geq 0$ **then**

        $f[j] := k$;{k means that j is achievable by assigning project k to the less loaded engineer};

      end if

    end if

  end for

end for

---

minimum total workloads with engineers with maximum total workloads, i.e. drawing engineers from the top and bottom of the list; if there is no improvement, we then consolidate with the engineer whose total workload is the third largest, the forth largest, .., etc. The procedure is repeated until we find that no improvement can be made for the minimum total workload. We then start to reduce the maximum total workload again, followed by enlarging the minimum total workload. The second stage stops when no further improvement can be achieved. In the following, we explain the way we consolidate the total workload of two selected engineers. Following notations in the previous paragraph, we denote the larger total workload as U, and the smaller as L. Two approaches are proposed for this subproblem. The first one is a dynamic programming algorithm. As we analyzed earlier, if the two-engineers subproblem can lead to an improvement, the smaller total workload in the new assignment has to be within the range $L+1, .., \lfloor (U+L)/2 \rfloor$. The dynamic programming, therefore, is to check whether this is possible. The dynamic programming is given in Algorithm 2.

To recover a solution with a smaller workload equal to j, we just need to follow $f[j]$: taking the project $f[j]$, and then project $f[j - c_{f[j]}]$, etc. The proposed dynamic programming, however, does not consider feasibility. Therefore, even though the result of Algorithm 2 may say that a value of j, $j = L + 1, .., \lfloor (U + L)/2 \rfloor$, is achievable, the resultant solution may be infeasible. So we need to check backward from $\lfloor (U+L)/2 \rfloor$ to $L+1$ to find out the best possibility. On the other hand, we also need to note that there may exist potential improvements which Algorithm 2 cannot detect because, as shown in Algorithm 2, we only record one path to each state, though there may be multiple paths to the same state. Although there are these limitations to the dynamic programming algorithm, it is still worth applying it because of its simplicity, compared with the next approach.

Algorithm 3. B&B for Second-Stage Subproblems

Initialization:
    a set of projects to be assigned, $\hat{P}_0$;
    a set of projects assigned to engineer 1, $\hat{P}_1$;
    a set of project assigned to engineer 2, $\hat{P}_2$; $\hat{P}_1 = \hat{P}_2 := \emptyset$;
    a first-in-last-out queue for B&B nodes, Q, $Q := \{(\hat{P}_0, \hat{P}_1, \hat{P}_2)\}$
    a current maximum total workload, U_{max};
    the original total workload for the two engineers, U and L;
while $Q \neq \emptyset$ **do**
    Take a node from Q, denoted as (P_0, P_1, P_2);
    $Q := Q - \{(P_0, P_1, P_2)\}$;
    Take a project i out of P_0, $P_0 := P_0 - \{i\}$;
    if $P_1 \cup \{i\}$ keeps engineer 1 feasible && the total workload of engineer $1 \leq \lfloor (U + L)/2 \rfloor$ **then**
        if $P_0 = \emptyset$ **then**
            if The total workload of engineer $1 > L$ **then**
                We stop with an improved assignment $(P_1 \cup \{i\}, P_2)$;
            end if
        else
            $Q := Q \cup \{(P_0, P_1 \cup \{i\}, P_2)\}$
        end if
    end if
    if $P_2 \cup \{i\}$ keeps engineer 2 feasible && the total workload of engineer $2 < U_{max}$ **then**
        if $P_0 = \emptyset$ **then**
            if The total workload of engineer $1 > L$ **then**
                We stop with an improved assignment $(P_1, P_2 \cup \{i\})$;
            end if
        else
            $Q := Q \cup \{(P_0, P_1, P_2 \cup \{i\})\}$;
        end if
    end if
end while

The second approach is a branch and bound algorithm, which is formally described in Algorithm 3. It follows a framework similar to the one used in the first stage. Because the two engineers have the same capability, to reduce the solution space, we have forced a smaller workload on Engineer 1 in Algorithm 3. Note that we apply Algorithm 2 before applying Algorithm 3. This narrows the difference between the maximum and the minimum total workload, compared with the initial solution; U_{max}, which is used in Algorithm 3 as an upper bound, is also smaller than in the initial solution.

4 Experiment

To test the performance of the proposed algorithm, we have generated two sets of test data. One is a "real" data set, generated according to the characteristics of the problem faced by the toys firm which has motivated this study. The

Table 1. Results from the random data set. Time is reported in seconds. K denotes the number of projects; I denotes the number of engineers; and T denotes the length of the planning horizon. "-" denotes that no feasible solution is found.

T	K	I	C	Heuristic U-L	Heuristic time	Cplex U-L	K	I	C	Heuristic U-L	Heuristic time	Cplex U-L
			12	1	5	11			13	1.33	41	-
20	60	10	13	1	0	2	180	30	14	1	0	19
			14	1	0	1			15	1	0	6
			17	1	0	15			17	-	63	-
30	90	10	18	1	0	2	270	30	18	2	81	-
			19	1	0	1			19	1	0	-
			23	1	0	4			21	1	8	-
+ 40	120	10	24	1	0	1	360	30	22	1	0	-
			25	1	0	1			23	1	0	237
			12	1	30	-			12	2.33	93	-
20	120	20	13	1	0	8	240	40	13	1	0	-
			14	1	0	3			14	1	0	10
			17	1	42	-			17	2.67	88	-
30	180	20	18	1	0	-	360	40	18	1	0	-
			19	1	0	6			19	1	0	50
			22	1	1	-			22	1.67	142	-
40	240	20	23	1	0	-	480	40	23	1	0	-
			24	1	0	12			24	1	0	652

other is a "random" data set, aimed to test the robustness of the algorithm. We show only the results from the random data set, which seems more difficult to solve, due to page limit. Remember that our objective is to find the best load balancing at different levels of C, the maximum allowed period workload. Since there is no C given before hand, we have used our lower bound, $\underline{C}$, as defined in Section 3.1. We test three levels of C, i.e., $\underline{C}$, $\underline{C}+1$, and $\underline{C}+2$. Perhaps surprisingly, our algorithm can already find feasible, or even optimal, solutions at these three levels.

The experiments ware conducted on a PC with an 2.66GHz CPU and 2GB memory. Our algorithm is coded in C++. For comparison, we have also tried to solve the test cases with ILOG CPLEX 11.0. Table 1 show the results for the random data set. For each test case, we have run our heuristic three times to test its robustness. Three levels of C are shown in the table, starting from the lower bound $\underline{C}$. CPLEX is allowed to run for at the most three hours. Because we find it is difficult for CPLEX to find a feasible solution, we have set the parameter to let CPLEX focus on feasibility.

As shown by Table 1, our approach has achieved much better performance in a much shorter time. For most cases, our approach can find optimal solutions, even when $C = \underline{C}$. Note that a value of 1 for $U - L$ means that an optimal solution has been found; for such a test case, we can never make $U - L$ equal 0. If an optimum has been found for a smaller C, there is no need to compute for a C larger than that; therefore, in the table, the time required for a larger C

has been set as 0. The results also imply that the lower bound we have proposed tends to be very tight. CPLEX, however, often cannot find feasible solutions within three hours; for test cases for which feasible solutions have been found, the objective function value can be very large. Please see the last test case in Table 1, for example.

References

[1] Cattrysse, D.G., Van Wassenhove, L.N.: A survey of algorithms for the generalized assignment problem. European Journal of Operational Research 60(3), 260–272 (1992)
[2] Duin, C.W., Volgenant, A.: Minimum deviation and balanced optimization: A unified approach. Operations Research Letters 10(1), 43–48 (1991)
[3] Franz, L.S., Miller, J.L.: Scheduling medical residents to rotations: Solving the large-scale multiperiod staff assignment problem. Operations Research 41(2), 269–279 (1993)
[4] Martello, S., Pulleyblank, W.R., Toth, P., de Werra, D.: Balanced optimization problems. Operations Research Letters 3(5), 275–278 (1984)
[5] Pentico, D.W.: Assignment problems: A golden anniversary survey. European Journal of Operational Research 176(2), 774–793 (2007)

Topical Analysis for Identification of Web Communities

Yajie Miao and Chunping Li

Tsinghua National Laboratory for Information Science and Technology (TNList)
Key Laboratory for Information System Security, Ministry of Education
School of Software, Tsinghua University
Beijing, China 100084
miaoyj08@mails.tsinghua.edu.cn, cli@tsinghua.edu.cn

Abstract. Traditional link-based schemes for identification of web communities focus on partitioning the web graph more sophisticatedly, without concerning the topical information inherently held by web pages. In this paper, we give a novel method of measuring the topicality of a hyperlink according to its context. Based on this, we propose a topical maxflow-mincut algorithm which incorporates topical information into the traditional maxflow-mincut algorithm. Experiments show that our algorithm outperforms the traditional algorithm in identifying high-quality web communities.

Keywords: Link Analysis, Topical Analysis, Web Community, Web Structure Mining.

1 Introduction

The vast amount of information pertaining to various topics causes difficulties to web users in finding useful web pages while surfing on the net. To solve such a problem, researchers have been trying to reorganize the web in the form of communities, each of which is related with a single or several related topics. However, it is not an easy task to discover communities on the World Wide Web. Some link analysis based approaches to identification of web communities have been proved to be effective in some cases, one of the most well-known is the *maxflow-mincut* algorithm proposed in [12]. However, this "links-only" method still has its defects because it only utilizes link information and ignores contents of web pages.

In recent years, much attention has been paid to combination of link analysis with topical information. K. Bharat [7] regulated hub and authority scores calculated by HITS [1] using relevance between pages and topics. S. Chakrabarti gave each hyperlink in HITS a weight determined by the similarity of its anchor text with the topic. Topic-Sensitive PageRank [6] and Topical PageRank [3] improved PageRank from a topical perspective and obtained better rank results. However, none of these methods have attempted to combine topical analysis with identification of web communities.

In this paper, we suggest that the weight of a link should embody its topical context. Topical information is incorporated into the traditional method through *Topical Weight* and a *topical maxflow-mincut* algorithm is proposed. Experimental

A. Nicholson and X. Li (Eds.): AI 2009, LNAI 5866, pp. 101–110, 2009.

results show that significant improvements are achieved when using our proposed *topical maxflow-mincut* algorithm.

The remainder of this paper is organized as follows. Related work is introduced in Section 2. The traditional *maxflow-mincut* algorithm is described in Section 3. We present the *topical maxflow-mincut* algorithm in Section 4. In Section 5 the experiment and performance evaluation will be given. We have the concluding remarks and future work in Section 6.

2 Related Work

2.1 Hyperlink Weight

In the early literatures about HITS [1] and PageRank [2], the weight of each link was assumed to be 1 and all the links were treated uniformly. A new metric called *average-click* [5] was then introduced on the basic intuition that users would make a greater effort to find and follow a link among a large number of links than a link among only a couple of links. The weight of a link pointing from p to q was defined to be the probability for a "random surfer" to reach q from p, and therefore was determined by the number of outgoing links of p.

In [4], Chakrabarti et al pointed out that a hyperlink would be more important to the web surfer if the page it pointed to was relevant to this surfer's topic. In order to measure a hyperlink according to a specific topic, they examined the anchor text of a hyperlink and calculated its similarity to the descriptions of the topic. This text similarity was considered as the topical weight for this link.

2.2 Incorporating Topicality in Link Analysis

Bharat et al [7] defined a relevance weight for each web graph node as the similarity of its document to the query topic. Then this weight was used to regulate each node's hub and authority scores computed by HITS. Their experiments showed that adding content analysis could provide appreciable improvements over the basic HITS.

In Havaliwala's Topic-Sensitive PageRank [6], some topics were selected from predefined categories. For each topic, a PageRank vector was computed. A topic-sensitive PageRank score for each page was finally computed by summing up elements of all the PageRank vectors pertaining to various topics. By adopting these topical analysis methods, they captured more accurately the importance of each page with respect to topics.

Nie et al [3] presented a more sophisticated method for incorporating topical information to both HITS and PageRank. For each page, they calculated a score vector to distinguish the contribution from different topics. Using a random walk model, they probabilistically combined page topic distribution with link structure. Experiments showed that their method outperformed other approaches.

2.3 Identification of Web Communities

Identifying communities on the web is a traditional task for web mining, knowledge discovery, graph theory and so on. Gibson et al [9] defined web communities as a core of central authoritative pages interconnected by hub pages and HITS was used to identify the authorities and hubs which formed a tightly knit community.

Kumar [10] represented web communities with community cores, which were identified through a systematic process called *Trawling* during a web crawl. Also with the HITS approach, these cores were used to identify relevant communities iteratively.

Flake et al [12] defined a web community as a set of sites that have more links (in either direction) to members of the community than to non-members. They proposed a *maxflow-mincut* method for identifying web communities. More details about this method will be discussed further in Section 3.

Ino et al [11] introduced a stricter community definition and defined the *equivalence relation* between web pages. After all the equivalence relations have been determined, the web graph can be partitioned into groups using a hierarchical process.

Lee et al [16] proposed to use content similarity between pages to give nodes weights and build new implicit links between nodes in the graph. However, their work focused mainly on viral communities and failed to take topical information and analysis into consideration.

3   Maxflow-Mincut Framework

Flake et al [12] proposed an algorithm for community identification on the World Wide Web. Since our topical approach is mainly based on this method, in this section we elaborate this method and examine its procedures in detail.

The web can be modeled as a graph in which web pages are vertices and hyperlinks are edges. Flake et al defined a web community as a set of websites that have more links to members of the community than to non-members. Also, they devised an iterative algorithm to find web communities. In each iteration, four steps are taken as follows.

First, starting from a set of seed pages, a focus crawler initially proposed in [8] is used to get a number of web pages through a crawl of fixed depth.

Then, a web graph is constructed using these web pages as well as relationships between them. In common cases, these relationships are represented by an adjacency matrix.

In the third step, one of the simplest maximum flow algorithms, i.e., the shortest augmentation path algorithm is run on the web graph and the minimum cut is identified.

The final step involves removing all the edges on the minimum cut found in the third step. All the web vertices which are still reachable from the source form a web community. In the community, all the pages are ranked by the number of links each one has and the highest ranked non-seed web page is added to the seed set.

These four procedures iterate until the desired iteration number is reached. In this paper, we call this algorithm *basic maxflow-mincut*. An example for *basic maxflow-mincut* is shown in Fig. 2(a). The web graph is composed of 10 nodes, each of which is marked by an integer. The seed nodes are 1 and 2. For the limit of space, we do not show the source and sink nodes.

4 Topical Identification of Web Communities

In the following parts, we first define *Topical Weight* (abbreviated as *TW*) to measure the topicality of hyperlinks. Based on this weight metric, we improve the basic algorithm and propose a *topical maxflow-mincut* algorithm.

4.1 Topical Weight for Hyperlinks

Commonly, several seed pages are used for crawling the web for pages used in identification of web communities. We use the TextRank [15] method to automatically extract some keywords from these seed pages. Some noisy contents, like anchor texts and advertisements, are filtered from these seeds manually. All the lexical units, i.e., words in our application, are regarded as vertices in the graph and an edge between two words is assumed to exist in the graph if these two words co-occur in one sentence. The salience score for each word can be calculated through iterative computations similar with PageRank [2]. Using a score threshold, some most important keywords can be selected to form a set, which is denoted as W.

Let's assume that a surfer is browsing page i which has a set of outgoing links $O(i)$. A link $link(i, j)$ is point from page i to page j. The more interesting page j is to the surfer, the more likely he or she is to follow $link(i, j)$ for the next move. So the weight of a hyperlink can be measured by the interestingness of the page it is pointing to.

Since W can be viewed as a representative description of the topic, the interestingness of a page P can be measured by the number of keywords appearing in P. So the topical weight of the hyperlink pointing from page i to page j, denoted as $TW(i, j)$, can be formally formulated as

$$TW(i, j) = \alpha \; Count(W, j), \tag{1}$$

where $Count(W, j)$ is the counting number of keywords appearing in page j and α is a regulating factor. For simplification, we set α as 1.

4.2 Topical Maxflow-Mincut

As discussed above, *basic maxflow-mincut* treats every link between nodes equally. We define *Topical Weight* for links and also give a feasible method for calculating a hyperlink's *Topical Weight*. Therefore, if we have an edge between vertices u and v in the web graph $G=(V,E)$, where both u and v are neither the source s nor the sink t, we can use $TW(u, v)$ as the weight for this edge. The *topical maxflow-mincut* algorithm is shown in Fig. 1.

A subtle problem with this algorithm is that $TW(u, v)$, the keyword number, may have a very broad value spectrum. So we consider normalizing *Topical Weight* to a value in $[1.0, \beta]$. We set the lower bound of this normalization interval to be 1.0 because *Topical Weight* is required to be larger than the weight of links pointing from non-source and non-seed vertices to the sink. Suppose we have TW in an interval $[A, B]$. Using this normalization strategy, TW will be mapped to a value in $[1.0, \beta]$ as

$$normalized \; (\text{TW}) = 1.0 + \frac{TW - A}{B - A} \times (\beta - 1.0) \cdot \tag{2}$$

S: seed pages   d: crawl depth
P: all pages crawled from S with d
G = (V, E): web graph formed from P
Algorithm:
    Create the source s and the sink t and add to V
    for each v ∈ S **do**
        Add (s ,v) to E with c(s ,v) = ∞
    end for
    for all (u ,v) ∈ E **do**
        Set c(u ,v) = *TW(u , v)*
        if (v ,u) ∉ E
        then add (v ,u) to E with c(v ,u) = *TW(v , u)*
    end for
    for each v ∈ V , v ∉ S ∪ {s, t} **do**
        Add (v ,t) to E with c(v ,t) = 1
    end for
    Call **MaxFlow**(G, s, t)
    Remove all the edges on the minimum cut
    Extract a community with v ∈ V connected to s

Fig. 1. Topical maxflow-mincut algorithm

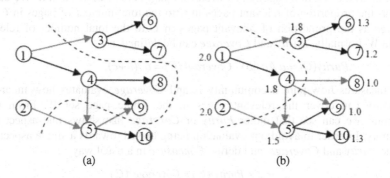

(a)   (b)

Fig. 2. Examples for both basic and topical maxflow-mincut algorithm. (a) An example for basic maxflow-mincut algorithm. (b) An example for topical maxflow-mincut algorithm.

We adopt this normalization operation and use normalized *Topical Weight* as the weight for hyperlinks. The upper bound β is a vital factor for our algorithm and this parameter will be tuned systematically to make the algorithm achieve its best performance.

In Fig. 2(b), the nodes are given different weights showing their varied relevance to the topic. These weights have a range [1.0, 2.0], where 1.0 means "totally irrelevant" and 2.0 means "perfectly relevant". So each edge in the graph also obtains accordingly a weight equal to the weight of the node it is pointing to. We identify a community with *topical maxflow-mincut* algorithm on this web graph.

Similarly, the minimum cut is marked with the dashed curve and the edges in blue are removed. Compared with Fig. 2(a), a community consisting of three members, 1, 2, 4, is identified, excluding two totally irrelevant nodes, 8 and 9. This illustrates the superiority of our algorithm: by utilizing additional topical information, the algorithm can eliminate irrelevant or noise nodes and generate more desirable communities.

5   Experiment and Evaluation

5.1   Metrics for Measuring Web Communities

In order to make our evaluation at a quantitative level, we define some metrics for measuring the quality of web communities. Here are several page sets:

- W : the set of all the pages generated through crawling
- C : the set of pages in the community generated on W
- R : the set of pages which are included in W but excluded from C

So these three page sets have such properties:

$$C \in W \quad R \in W \quad R = W - C. \tag{3}$$

We examine both C and R, and record the number of relevant and irrelevant pages. Suppose after manually counting, we discover a relevant pages and b irrelevant pages in C. Similarly, c relevant pages and d irrelevant pages are discovered in R. We define *Purity* as the proportion of relevant pages in C to the total number of pages in C and *Coverage* as the proportion of relevant pages in C to the total number of relevant pages in W. Formally, *Purity* and *Coverage* can be defined as

$$Purity(C) = a/(a+b) \quad Coverage(C) = a/(a+c). \tag{4}$$

Purity measures how pure a community is and *Coverage* evaluates how much the community can cover the relevant pages in the whole page set W. From their definitions, we can see that either *Purity* or *Coverage* just shows one aspect of a community. In order to avoid our evaluation being biased towards a single aspect, we combine *Purity* and *Coverage*, and define *F-measure* in a usual way

$$F - measure = \frac{2 \times Purity\ (C) \times Coverage\ (C)}{Purity\ (C) + Coverage\ (C)} \tag{5}$$

5.2   Data Set

We select the "data mining conferences" community as our testing case. Our seed set consists of three URLs:

> http://www.kdnuggets.com/meetings/
> http://www.sigkdd.org/kdd2009/
> http://www.kmining.com/info_conferences.html

From the contents of these three pages, we extract 39 keywords, which will be used to calculate *Topical Weight* for edges in the web graph. These keywords include terminologies on data mining, names of academic organizations, well-known data mining conferences, different types of conferences like "workshop", "symposium", "conference", etc., as well as terms frequently used for conference affairs like "registration", "acceptance" and "notification", etc.

After crawling, we totally get 389 distinct pages as the data set. Similar to [3], we rate manually each page as quite relevant, relevant, not relevant, and totally irrelevant, which is assigned the scores of 4, 3, 2 and 1, respectively. We mark pages with scores of 4 or 3 as relevant and pages with scores of 2 or 1 as irrelevant. One point which should be noted is that we are especially interested in a methodology. Therefore, we use a much smaller dataset in our experiment, which makes it feasible for manual rating and close examination of score distributions.

5.3 Parameter Tuning

The parameter β is the upper bound of the normalization interval $[1.0, \beta]$. We tune β with values from 2.0 to 10.0 with a step of 0.5. For each value of β, we run *topical maxflow-mincut* algorithm and generate a web community C_β. For C_β we calculate its *Purity*, *Coverage* and *F-measure* according to their definitions. Fig. 3 shows the values of these metrics with different settings of β. We depict the value level of *basic maxflow-mincut* with the dashed line.

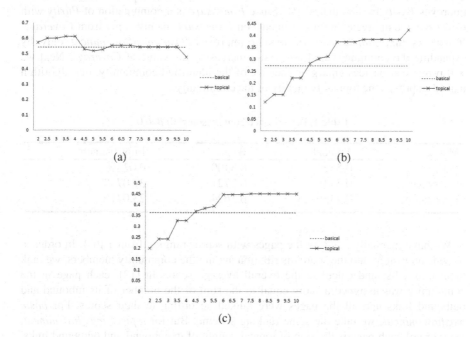

Fig. 3. Values of metrics as β is varied. (a) Community Purity as β changes from 2.0 to 10.0. (b) Community Coverage as β changes from 2.0 to 10.0. (c) Community F-measure as β changes from 2.0 to 10.0.

Fig. 3(a) demonstrates that with β=4.0, *Purity* can get its largest value, which approximately is 0.62 and gains 15% improvements against the basic algorithm. From Fig. 3(c) we can see that when β equals 6.0, the *F-measure* curve reaches its highest point. When β is larger than 6.0, *F-measure* stays unchanged, confirming that increasing β continuously cannot boost the performance anymore. Fig. 3(a) and Fig. 3(b) show that when β is 6.0, both *Purity* and *Coverage* are larger than that of the basic algorithm, though for *Purity* the improvements are not significant. With β=6.0, we get an community containing 66 members, an appropriate size considering the total number of the pages in the data set is 389.

As β varies, *Purity* displays an opposite changing tendency to that of *Coverage*. For a specific application, it is almost impossible to optimize these two competing metrics simultaneously. This demonstrates that both *Purity* and *Coverage* are partial in measuring a community. Therefore it is more reasonable to use *F-measure* as our major evaluation metric.

5.4 Performance Evaluation

Table 1 shows performance comparisons between topical and basic *maxflow-mincut* algorithms. Since *F-measure* should be our major evaluation metric, we can conclude that *topical maxflow-mincut* improves the performance of *basic maxflow-mincut* by 23.041%. However an observation from Table 1 is that *topical maxflow-mincut* improves *Purity* by less than 0.1%. Since *F-measure* is a combination of *Purity* with *Coverage*, the appreciable improvements on *F-measure* mainly come from *Coverage*. It appears that *topical maxflow-mincut* improves *basic maxflow-mincut* only by expanding the community and as a result increasing the value of *Coverage*. Next we will prove that besides enlarging the size of the identified community, our algorithm indeed improves the topicality quality of the community.

Table 1. Performance comparison with β=6.0

Metric	Topical	Basic	Improvement
Purity	0.5523	0.5400	0.023%
Coverage	0.3738	0.2727	37.074%
F-measure	0.4459	0.3624	23.041%

We have manually rated all the pages with scores ranging from 1 to 4. In order to provide an insight into the score distribution among the community members, we rank them into a list and calculate the overall average score. In [12], each page in the community was assigned a score equal to the sum of the number of its inbound and outbound links and all the pages were ranked according to their scores. For *basic maxflow-mincut*, we take the same ranking scheme. But for *topical maxflow-mincut*, the score of each page is the sum of topical weight of its inbound and outbound links. We define $S@n$ as the average score of the first n pages in the ranking result. As the size of the community formed by *basic maxflow-mincut* is 49, we set n to be 5, 10, 15, 20, 25, 30, 35, 40, 45, 49, and for each value we calculate $S@n$ correspondingly for both *topical* and *basic maxflow-mincut*. A comparison is made in Fig. 4. For most

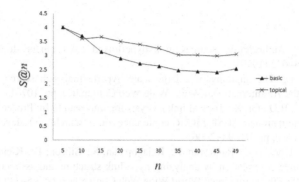

Fig. 4. Comparison of average score as n is varied

values of n, $S@n$ of *topical maxflow-mincut* is higher than that of *basic maxflow-mincut*. This superiority is retained until the last member of the basic community. So we can conclude that besides expanding the community to improve *F-measure*, *topical maxflow-mincut* pulls more high-quality and authoritative pages into the community as well. Also Fig. 4 shows that another advantage of *Topical Weight* is that it can be used as a more effective metric in ranking members of a web community as it can give relevant pages higher positions on the ranking list.

6 Conclusion and Future Work

In this paper, we make a preliminary attempt to incorporate topical analysis into identification of web communities. We use *Topical Weight* to measure the topicality of hyperlinks. By combining *Topical Weight* with the traditional *basic maxflow-mincut* algorithm, we propose *topical maxflow-mincut*, which is an improved algorithm for identification of web communities. Also we define some metrics for measuring the quality of web communities. Experimental results show that our algorithm achieves improvements over *basic maxflow-mincut* and is more capable of finding high-quality web communities.

In our future work, we expect to capture the topicality of hyperlinks more accurately using other methods like TF-IDF or Topic Signature. We would also consider incorporating other types of information, like opinions expressed in the contents of web pages, to the analysis of web communities, which may make it possible for us to identify communities with sentiment characteristics.

Acknowledgments

This work was supported by National Natural Science Funding of China under Grant No. 90718022.

References

1. Kleinberg, J.M.: Authoritative sources in a hyperlinked environment. Journal of the ACM 46(5), 604–632 (1999)
2. Brin, S., Page, L.: The anatomy of a large-scale hypertextual web search engine. In: Proceedings of the 7th International World Wide Web Conference, pp. 107–117 (1998)
3. Nie, L., Davison, B.D., Qi, X.: Topical link analysis for web search. In: Proceedings of the 29th annual international ACM SIGIR conference on research and development in information retrieval, pp. 91–98 (2006)
4. Chakrabarti, S., Dom, B.E., Raghavan, P., Rajagopalan, S., Gibson, D., Kleinberg, J.M.: Automatic resource compilation by analyzing hyperlink structure and associated text. In: Proceedings of the 7th International World Wide Web Conference, pp. 65–74 (1998)
5. Matsuo, Y., Ohsawa, Y., Ishizuka, M.: Average-clicks: a new measure of distance on the World Wide Web. Journal of Intelligent Information Systems, 51–62 (2003)
6. Haveliwala, T.H.: Topic-sensitive PageRank. In: Proceedings of the 11th International World Wide Web Conference, pp. 517–526 (2002)
7. Bharat, K., Henzinger, M.R.: Improved algorithms for topic distillation in hyperlinked environments. In: Proceedings of the 21st International ACM SIGIR conference on research and development in information retrieval, pp. 104–111 (1998)
8. Chakrabarti, S., van der Berg, M., Dom, B.: Focused crawling: a new approach to topic-specific web resource discovery. In: Proceedings of the 8th International World Wide Web Conference (1999)
9. Gibson, D., Klienberg, J., Raghavan, P.: Inferring web communities from link topology. In: Proceedings of the 9th ACM conference on hypertext and hypermedia, pp. 225–234 (1998)
10. Kumar, R., Raghavan, P., Rajagopalan, S., Tomkins, A.: Trawling the web for emerging cyber-communities. In: Proceedings of the 8th International World Wide Web Conference, pp. 65–74 (1999)
11. Ino, H., Kudo, M., Nakamura, A.: Partitioning of web graphs by community topology. In: Proceedings of the 14th International World Wide Web Conference, pp. 661–669 (2005)
12. Flake, G.W., Lawrence, S., Giles, C.L.: Efficient identification of web communities. In: Proceeding of the 14th ACM SIGKDD International Conference on Knowledge Discovery and Data Mining, pp. 150–160 (2000)
13. Flake, G.W., Lawrence, S., Giles, C.L., Coetzee, F.M.: Self-Organization and identification of web communities. IEEE Computer 35(3), 66–71 (2002)
14. Page, L.: PageRank: Bringing order to the web. Stanford Digital Libraries Working Paper (1997)
15. Mihalcea, R., Tarau, P.: TextRank: Bringing Order into Texts. In: Proceedings of the Conference on Empirical Methods in Natural Language Processing, pp. 404–411 (2004)
16. Lee, H., Borodin, A., Goldsmith, L.: Extracting and Ranking Viral Communities Using Seeds and Content Similarity. In: Proceedings of the nineteenth ACM conference on hypertext and hypermedia, pp. 139–148 (2008)

Collaborative-Comparison Learning for Complex Event Detection Using Distributed Hierarchical Graph Neuron (DHGN) Approach in Wireless Sensor Network

Anang Hudaya Muhamad Amin and Asad I. Khan

Clayton School of Information Technology, Monash University,
Clayton, 3168 VIC, Australia
{Anang.Hudaya,Asad.Khan}@infotech.monash.edu.au
http://www.infotech.monash.edu.au

Abstract. Research trends in existing event detection schemes using Wireless Sensor Network (WSN) have mainly focused on routing and localisation of nodes for optimum coordination when retrieving sensory information. Efforts have also been put in place to create schemes that are able to provide learning mechanisms for event detection using classification or clustering approaches. These schemes entail substantial communication and computational overheads owing to the event-oblivious nature of data transmissions. In this paper, we present an event detection scheme that has the ability to distribute detection processes over the resource-constrained wireless sensor nodes and is suitable for events with spatio-temporal characteristics. We adopt a pattern recognition algorithm known as Distributed Hierarchical Graph Neuron (DHGN) with collaborative-comparison learning for detecting critical events in WSN. The scheme demonstrates good accuracy for binary classification and offers low-complexity and high-scalability in terms of its processing requirements.

Keywords: Event detection, distributed pattern recognition, single-cycle learning, Distributed Hierarchical Graph Neuron (DHGN), associative memory.

1 Introduction

Event detection schemes using WSN are being mainly developed with a focus on routing and localisation of sensory data for achieving event sensing and tracking capabilities. Existing event detection infrastructure is mostly reliant on *single-processing* approach, i.e. CPU-centric, where overall analysis of the sensory data is carried out at one point i.e. the *base station*. This architecture introduces two major problems. Firstly, the communication latency between sensor nodes and the base station creates substantial communication overhead due to constant flow of data, re-routing procedures, and relocation of sensor nodes that often occurs in real-time applications. Furthermore, data transmission error is inevitable

A. Nicholson and X. Li (Eds.): AI 2009, LNAI 5866, pp. 111–120, 2009.

in this approach. Secondly, single-processing approach at the base station can add significant delays in detecting critical events owing to the computational bottleneck. It is important for a sensor network that has been equipped with preventive tools such as mobile fire extinguishers to respond in real-time. Furthermore, with the stochastic nature of events, rapid response time is required for critical events. This can only be achieved through on-site detection using some form of distributed processing.

Existing event detection schemes that have been deployed over WSN also suffer from complexity and scalability issues. These schemes mainly apply conventional neural networks or machine learning algorithms. Although these algorithms provide learning mechanisms for event detection, they require extensive retraining as well as large amount of training datasets for generalisation. Hence, this limits the scalability of the scheme for massive sensory data processing.

In this paper, we propose a spatio-temporal event detection scheme using a distributed single-cycle learning pattern recognition algorithm known as Distributed Hierarchical Graph Neuron (DHGN) [1,2] that is deployable in resource-constrained networks such as WSN. Our proposed learning approach works on the principle of *collaborative-comparison learning*, where adjacency comparison technique is applied within a collaborative learning network. We intend to harness the capability of sensor nodes to perform event detection *in situ* and provide instantaneous response only when an event of interest occurs. It therefore eliminates the need for sensor nodes to transmit unnecessary sensory readings to base station and hence reduces the possibility of network congestion.

This paper is structured as follows. Section 2 discusses some of the related works to event detection schemes and distributed associative memory algorithms. Section 3 introduces our proposed spatio-temporal recognition scheme with collaborative-comparison learning for distributed event detection. Section 4 describes the simulation that has been conducted for event detection using DHGN and SVM. Finally, section 5 concludes the paper.

2   Event Detection and Distributed Pattern Recognition

Existing research trends show the tendency to apply classification schemes for event detection by using either artificial neural networks (ANNs) or machine learning approaches. Kulakov and Davcev [3] proposed the implementation of Adaptive Resonance Theory (ART) neural network for event classification and tracking. The scheme reduces the communication overhead by allowing only cluster labels to be sent to the base station. However, the implementation incurs excessive learning cycles to obtain optimum cluster/class matches. Catterall et al. [4] have proposed an implementation of Kohonen Self-Organising Map (SOM) in sensory data clustering for distributed event classification within WSN. SOM implementation provides an avenue for each sensor node to pre-process its readings. Nevertheless, this pre-processing would sometimes incur massive iteration process to obtain an optimum result. Radial-Basis Function (RBF) neural network has been proposed for dynamic energy management within WSN network for particle filter prediction by Wang et al. [5]. RBF has been proven to offer fast learning scheme for neural networks. However, its learning complexity and accuracy is heavily affected by the training method and network generation method being used, e.g. K-means clustering or evolutionary algorithms.

In distributed recognition perspective, Guoqing et al. [6] have proposed a multilayer parallel distributed pattern recognition model using sparse RAM nets. Their scheme performed accurately for non-deterministic patterns. However, the implementation suffered from large memory space allocation. The works of Ikeda et al. [7] and Mu et al. [8] have demonstrated a two-level decoupled Hamming associative memory for recognising binary patterns. Their findings have shown high recall accuracy for noisy pattern recognition. However, the algorithm incurs high complexity due to sequential comparison of input patterns. On machine learning implementation, Cheng et al. [9] have introduced Distributed Support Vector Machines (DSVMs) in their work on chaotic time series prediction. Their work demonstrated SVM implementation as a combination of submodels that could be combined to provide high accuracy outputs. However, this approach has created an expensive computational model to be deployed in resource-constrained networks such as WSN.

3   DHGN Classifier for Event Detection

This section describes the overall structure of DHGN classifier that has been proposed for the spatio-temporal event detection scheme.

3.1   DHGN Network Architecture

DHGN architecture consists of two main components namely DHGN subnets and the Stimulator/Interpreter (SI) Module, as shown in Fig. 1. SI module acts as a network coordinator/manager for DHGN subnets, while each subnet comprises a collection of processing neurons (PNs) within a hierarchical structure [1].

DHGN recognition algorithm works through divide-and-distribute approach. It divides each pattern into smaller subpattern decompositions. These subpatterns are sent to available subnets for recognition process. Given a pattern p with length m and number of possible value v, the number of processing neurons (PNs), P_n required is obtained from (1).

$$P_n = d_n v \left(\frac{\frac{m}{d_n} + 1}{2} \right)^2. \tag{1}$$

d_n represents the number of DHGN subnets allocated within the network. Subsequently, for a given number of PNs within the network, the maximum input size m is derived from the following equation:

$$m = d_n \left(2\sqrt{\frac{P_n}{d_n v}} - 1 \right). \tag{2}$$

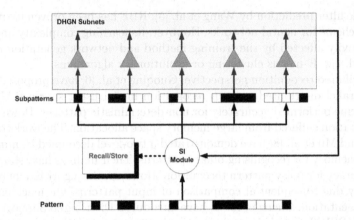

Fig. 1. DHGN network architecture comprises two main components namely DHGN subnets and the SI module. Decomposition of a pattern into subpatterns is done through the SI Module to simulate sensory pattern input operation of a real WSN.

The PNs within DHGN subnets are structured in a hierarchical manner, similar to the Hierarchical Graph Neuron (HGN) [10] approach. Each layer within the DHGN subnet is populated with PNs. The number of layers, L_n, required within each subnet is given by (3):

$$L_n = \frac{\frac{m}{d_n} + 1}{2}. \tag{3}$$

3.2   DHGN Pattern Representation

DHGN adopts pattern representation approach proposed by Nasution and Khan [10]. Each pattern is represented using a two-dimensional *(value, position)* representation. Consider binary patterns given in a memory set $\{p_1, p_2, ..., p_m\}$, where each p_i pattern is an N-bit binary vector, i.e. $p_i \in \{0,1\}^N, i = 1, 2, ..., m$. Each pattern p_i is in the form of $\{(v_1, l_1), (v_2, l_2), ..., (v_N, l_N)\}$ for N-bit pattern, where v and l represent value and position respectively. Each processing neuron handles a specific value and position pair for a given pattern.

Pattern storage scheme for DHGN involves abstract data structure known as *bias array*. This data structure is located on each processing neuron within the DHGN architecture. It stores the information on the unique relationship between the neuron's adjacent neighbours, i.e. a unique composition of all values obtained from its adjacent neurons. These values are retrieved using collaborative-comparison learning approach. For each processing neuron n, its bias array is in the form of vector $\{(i_1, v_1v_2...v_m), (i_2, v_1v_2...v_m), ..., (i_k, v_1v_2...v_m)\}$, where i represents the bias index for each entry $(i_a, v_1v_2...v_m), a = 1, 2, ..., k$. Each bias entry $v_1v_2...v_m$ represents unique v value composition from m adjacent neurons.

3.3 Collaborative-Comparison Single-Cycle Learning

We adopt an adjacency comparison approach in our learning scheme using simple signal/data comparisons. Each PN holds a segment of the overall subpattern. Collectively, these neurons will have an ability to represent the entire subpattern. Consider the following base-level DHGN subnet structure as shown in Fig. 2. The five PNs, where each is responsible to capture its adjacent neurons' values, will be able to store the entire pattern "ABCDE". If we link up these neurons in a one-dimensional structure, we are able to determine collaborative PNs that contain a memory of pattern "ABCDE". We call this approach as *collaborative-comparison learning*.

Our collaborative-comparison learning approach compares external signal from surroundings against the stored entries within each PN's *bias array*, which is a local data structure containing history of adjacent node activation. In this context, each PN learns through comparisons among the signals from its adjacent neighbours and the recorded entries within its memory i.e. the bias array. Consider a bias array $S = (s_1, s_2, ..., s_x)$ which consists of signal entries s with index x. If external signal s_{ext} matches any of the stored entries, i.e. $s_{ext} \in S$, then the respective bias index i of the matched s_i entry will be recalled. Otherwise, the signal will be added into the memory as s_{x+1}. There are two-fold advantages using this approach. Firstly, it minimises data storage requirement. Secondly, the proposed approach accepts all kinds of data. For instance, the signal could be in the form of data vectors or frequency signals, allowing spatial and temporal data to be accommodated. In addition, our proposed learning technique does not require synaptic plasticity rule used by other learning mechanisms, such as the Hebbian learning [11] and incremental learning [12] approaches. Thus new patterns can be learnt without affecting previously stored information.

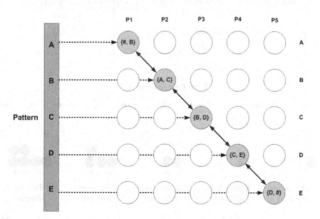

Fig. 2. Collaborative-comparison learning approach for one-dimensional pattern "ABCDE". Each activated processing neuron (PN) stores the signals received by its adjacent neurons.

3.4   DHGN for Event Detection within WSN

In this subsection, we explore the capability of DHGN classifier for event detection within WSN. Our implementation involves matching each DHGN subnet into each sensor node for on-site recognition, while SI Module is allocated at the base station.

DHGN Event Recognition. Our scheme only requires binary input patterns. We consider multiple sensory readings that are used to detect the occurrence of critical events. Given a set of x sensory readings $\{r_1, r_2, ..., r_x\}$ where $r_i \in \mathbb{R}$ and $i = 1, 2, ..., x$, we perform a dimensionality-reduction technique known as threshold-signature that converts each reading value to its respective binary signature. The threshold-signature technique utilises the threshold classes to represent a single data range into a binary format. Given a sensory reading r_i where $i = 1, 2, ..., x$ and H-threshold class, the equivalent binary signature that implies $b_i \rightarrow r_i$ is in the form of $b_i \in \{0, 1\}^H$. Therefore, for x-set sensory readings $\{r_1, r_2, ..., r_x\}$ will be converted into a set of binary signatures $\{b_1, b_2, ..., b_x\}$. The following data in Table 1 shows samples of temperature threshold range with its equivalent binary signature.

If the output index from DHGN subnet matches the stored pattern for the critical event, then a signal is transmitted to the base station in the form of data packet *(node_id, timestamp, class_id)*. The *class_id* parameter is the identification for class of event that has been detected. At a given time t, the base station might receive a number of signals from the network.

Table 1. Temperature threshold ranges with respective binary signatures

Temperature Threshold Range (°C)	Binary Signature
0 - 20	10000
21-40	01000
41-60	00100

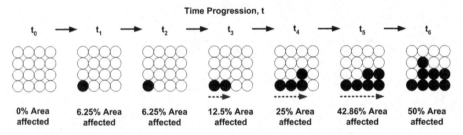

| 0% Area affected | 6.25% Area affected | 6.25% Area affected | 12.5% Area affected | 25% Area affected | 42.86% Area affected | 50% Area affected |

Fig. 3. Spatio-temporal analysis of event data received by base station. Note that the distribution and frequency of critical event detected could be determined through the signals sent by each sensor nodes. The dashed arrow shows the direction in time as the event occurs within the network.

Spatio-temporal Analysis of Event Data. The spatio-temporal analysis involves a process of observing the frequency and distribution of triggered events within WSN network. This process is conducted at base station, since it has a bird's eye view of the overall network. Fig. 3 shows a scenario of spatio-temporal analysis within our proposed scheme.

4   Simulation and Results

In this section, we present the simulation results of an event detection scheme within WSN using our proposed classification approach on the sensory data

Table 2. Threshold classes with respective value range used in the tests

	Threshold Range	Class
Noise Level	≥ 25	Event
Light Exposure Level	≥ 100	
Noise Level	< 25	Non-event
Light Exposure Level	< 100	

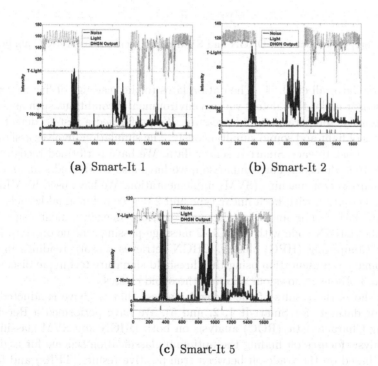

(a) Smart-It 1   (b) Smart-It 2

(c) Smart-It 5

Fig. 4. DHGN event detection results for test using 1690 sensor datasets (x-axis) for Smart-It 1, 2, and 5. Note that Smart-It 3 and Smart-It 4 were not included, since they produced non-events with noise and light exposure readings well below the threshold values (*T-Light* and *T-Noise*).

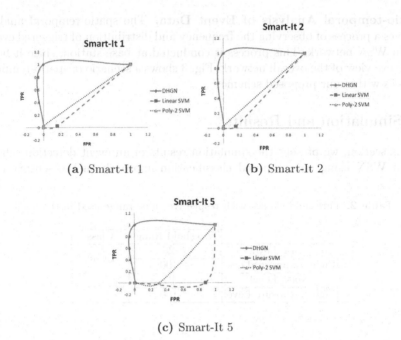

(a) Smart-It 1   (b) Smart-It 2

(c) Smart-It 5

Fig. 5. Comparison between DHGN and SVM recall accuracy using ROC analysis

taken from Catterall et al. [4]. The data relates to the readings of five Smart-It wireless sensor nodes that detect various environmental conditions such as light, sound intensity, temperature, and pressure. We performed a test to detect the occurrences of light and sound simultaneously. The simulation involves assigning a DHGN subnet to each Smart-It sensor data. We have performed recognition tests over 1690 datasets. For comparison, we have also conducted similar tests using support vector machine (SVM) implementation. We have used SVMLight [13] implementation with both linear-type and 2-degree polynomial kernels.

Our simulation runs under parallel and distributed environment using our own parallel DHGN code with MPICH-2 message-passing scheme on High Performance Computing (HPC) cluster. DHGN retrieves sensory readings in the form of binary representation using the threshold-signature technique discussed in Section 3. Table 2 shows our adopted threshold classes.

Fig. 4 shows the results of the recognition test that we have conducted on this sensor dataset (for Smart-It 1, 2, and 5). We have performed a Receiver Operating Characteristic (ROC) analysis on both DHGN and SVM classifiers. ROC analysis focuses on finding the optimum classification scheme for a given problem, based on the trade-off between true positive results (TPRs) and false positive results (FPRs). Fig. 5 shows comparisons between the ROC curves of DHGN and SVM respectively for sensory data from Smart-It 1, 2, and 5 (TPR vs FPR). SVM classifiers use a set of six readings, while DHGN only uses datasets with two entries.

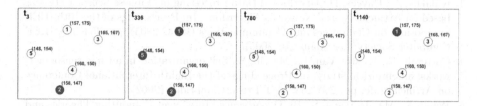

Fig. 6. Changes in the event status detected by SI module with a spatio-temporal perspective. Each node with its spatial coordinates is shown. The highlighted symbols represent the triggered nodes.

From these results, we can see that polynomial SVM produces slightly better results than the linear approach (for Smart-It 1, 2, and 5). This shows that SVM approach depends heavily on the types of kernel being implemented and the nature of data used. This data dependency problem limits the flexibility of SVM for event detection within WSN. DHGN on the other hand, offers efficient dimensionality reduction using our simple threshold-signature technique.

We later extended our simulation to include the proposed spatio-temporal recognition process. Fig. 6 shows some snapshots of the Smart-It network condition for a predefined times. Note the occurrence of events within this network. For instance, at time t_3, only sensor node 2 has sent a signal to the base station, denoting an occurrence of event while at t_{336}, the event has spread out to node 1 and 5. This form of spatio-temporal recognition is still at its preliminary stage.

5   Conclusion

The dual-layer spatio-temporal recognition scheme for distributed event detection has been presented in this paper by utilising DHGN as the single-cycle learning algorithm for event detection within WSN networks. DHGN collaborative-comparison learning approach minimises the need for complex learning and hence it is suitable for lightweight devices such as wireless sensor nodes. We have also shown that DHGN performs better recognition accuracy than SVM for WSN. We presented a preliminary approach for spatio-temporal recognition, which can be further developed for the WSN using the DHGN framework.

References

1. Khan, A.I., Muhamad Amin, A.H.: One Shot Associative Memory Method for Distorted Pattern Recognition. In: Orgun, M.A., Thornton, J. (eds.) AI 2007. LNCS (LNAI), vol. 4830, pp. 705–709. Springer, Heidelberg (2007)
2. Muhamad Amin, A.H., Khan, A.I.: Single-Cycle Image Recognition Using an Adaptive Granularity Associative Memory Network. In: Wobcke, W., Zhang, M. (eds.) AI 2008. LNCS (LNAI), vol. 5360, pp. 386–392. Springer, Heidelberg (2008)

3. Kulakov, A., Davcev, D.: Distributed Data Processing in Wireless Sensor Networks based on Artificial Neural Networks Algorithms. In: Proceedings of the 10th IEEE Symposium on Computers and Communications (ISCC 2005), pp. 353–358. IEEE Computer Society, Los Alamitos (2005)
4. Catterall, E., Kristof Van, L., Martin, S.: Self-organization in ad hoc sensor networks: an empirical study. In: Proceedings of the Eighth International Conference on Artificial Life, pp. 260–263. MIT Press, Cambridge (2002)
5. Wang, X., Ma, J., Wang, S., Bi, D.: Advanced Intelligent Computing Theories and Applications. With Aspects of Theoretical and Methodological Issues. In: Dynamic Energy Management with Improved Particle Filter Prediction in Wireless Sensor Networks, pp. 251–262. Springer, Heidelberg (2007)
6. Guoqing, Y., Songcan, C., Jun, L.: Multilayer parallel distributed pattern recognition system model using sparse RAM nets. Computers and Digital Techniques, IEE Proceedings E 139(2), 144–146 (1992)
7. Ikeda, N., Watta, P., Artiklar, M., Hassoun, M.H.: A two-level Hamming network for high performance associative memory. Neural Networks 14, 1189–1200 (2001)
8. Mu, X., Watta, P., Hassoun, M.H.: A Weighted Voting Model of Associative Memory. IEEE Transaction on Neural Networks 18(3), 756–777 (2007)
9. Cheng, J., Qian, J.-S., Guo, Y.-N.: A Distributed Support Vector Machines Architecture for Chaotic Time Series Prediction. In: King, I., Wang, J., Chan, L.-W., Wang, D. (eds.) ICONIP 2006. LNCS, vol. 4232, pp. 892–899. Springer, Heidelberg (2006)
10. Nasution, B.B., Khan, A.I.: A Hierarchical Graph Neuron Scheme for Real-Time Pattern Recognition. IEEE Transactions on Neural Networks 19(2), 212–229 (2008)
11. Hebb, D.: The Organization of behavior. Wiley, New York (1949)
12. Syed, N.A., Liu, H., Sung, K.K.: Handling concept drifts in incremental learning with support vector machines. In: Proceedings of the Fifth ACM SIGKDD international Conference on Knowledge Discovery and Data Mining, KDD 1999, San Diego, California, United States, August 15-18, pp. 317–321. ACM, New York (1999)
13. Joachims, T.: Making large-Scale SVM Learning Practical. In: Schölkopf, B., Burges, C., Smola, A. (eds.) Advances in Kernel Methods - Support Vector Learning. MIT Press, Cambridge (1999)

Square Root Unscented Particle Filtering for Grid Mapping

Simone Zandara and Ann Nicholson

Monash University, Clayton 3800, VIC

Abstract. In robotics, a key problem is for a robot to explore its environment and use the information gathered by its sensors to jointly produce a map of its environment, together with an estimate of its position: so-called SLAM (Simultaneous Localization and Mapping) [12]. Various filtering methods – Particle Filtering, and derived Kalman Filter methods (Extended, Unscented) – have been applied successfully to SLAM. We present a new algorithm that adapts the Square Root Unscented Transformation [13], previously only applied to feature based maps [5], to grid mapping. We also present a new method for the so-called pose-correction step in the algorithm. Experimental results show improved computational performance on more complex grid maps compared to an existing grid based particle filtering algorithm.

1 Introduction

This paper addresses the classical robotics problem of a robot needing to explore its environment and use the information gathered by its sensors to jointly produce a map of its environment together with an estimate of its position: so-called SLAM (Simultaneous Localization and Mapping) [12]. SLAM is an inherently sequential problem, which suggested the use of Bayesian Filters. Early path tracking methods such as the Kalman Filter (KF) [12] are based on the idea that, given knowledge about the position and heading of a moving object, observed data can be used to track that object; the problem becomes more difficult when the sensors are mounted on the moving object itself. The Extended KF (EKF) [12] is a successful method for modeling the uncertainty of a robot's noisy measurements (e.g. encoders, range finders), however it is unstable and imprecise because of linearization[1]; the Unscented KF (UKF) [8,15] avoids such approximation.

Particle filtering is a popular sequential estimation technique based on the generation of multiple samples from the distribution that is believed to approximate the true distribution. Studies have shown that particle filtering can better approximate a robot's real position than KF techniques, but the method is computationally intense because every particle is updated through a lightweight KF derived technique. Particle filtering has been used successfully to solve SLAM for both grid and feature based maps [12]. Grid maps generate a representation of the surrounding (usually closed) environment through a grid of cells, using raw sensor data for precisation. In contrast, feature based maps describe a (typically

A. Nicholson and X. Li (Eds.): AI 2009, LNAI 5866, pp. 121–130, 2009.
© Springer-Verlag Berlin Heidelberg 2009

open) environment through a set of observed features, usually sensor readings
(e.g. range and bearing).

Unscented Particle Filtering [13] for SLAM [7] has been successfully applied
to feature based mapping. It mixes Particle Filtering and UKF by updating
particles using an unscented transformation, rather than updating the uncertainty through Taylor linearisation of the update functions. Our research draws
from this wide spectrum of KF and particle filtering algorithms; in Section 2 we
provide a brief introduction (see [16] for more details). We present a new algorithm (Section 3) which we call SRUPF-GM (Square Root Unscented Particle
Filtering for Grid Mapping) to adapt Unscented Particle Filtering to *grid* based
maps. We also present a new method for the so-called pose-correction step in
the algorithm. In Section 4 we present experiments comparing its performance
to the well-known GMapping algorithm [2], on three grid environments. Our results show that while SRUPF-GM is slower on simpler maps, it is faster on more
complex maps, and its performance does not degrade as quickly as GMapping
as the number of particles increases.

2   Background

2.1   Particle Filtering for SLAM Problem

The main idea behind Particle Filtering applied to SLAM [11] is to estimate
sequentially the joint posterior $p(x_t, m | x_{t-1}, z_t, u_t)$ for the robot's state x (which
is usually its position X, Y and bearing θ), and the map of the environment m.
This is done using its previous state (x at time $t-1$), odometry information
(u at time t), that is, the robot's own measurements of its movements from its
wheels, and the measurements from sensors (z at time t), e.g. lasers, sonars, etc.

$p(x_t, m | x_{t-1}, z_t, u_t)$ has no closed solution. In Particle Filtering, its estimation
can be decomposed by maintaining a set of n poses, S, that make up a region of
uncertainty (a Monte Carlo method); these poses are called *particles* [10]. Each
particle has its own position, map and uncertainty; the latter is represented by a
Gaussian defined by its position μ (mean) and covariance Σ. The generated distribution is called the *proposal*. The proposal is meant to represent the movement of
the robot and is usually derived from u and z. It is proven that to solve SLAM the
noise that is inherent in the odometry and sensor readings must be modeled; all
SLAM algorithms add a certain amount of noise to do this. The final set of particles becomes the robot's final pose uncertainty ellipse. Another key characteristic
of Particle Filtering is *resampling*, which aims to eliminate those particles which
are believed to poorly represent the true value. This resampling is done using a
weighting mechanism, where each particle has an associated weight. A high-level
description of the particle filtering algorithm is given in Algorithm 1.

2.2   Unscented Transformation

The odometry update in particle filtering can be implemented in different ways;
thus far the most well-known technique is based on an EKF update, which
unfortunately introduces unwanted complexity and error [1]. The Unscented

---

Algorithm 1. Particle filtering algorithm

---

while Robot received data from sensors **do**
    for all x_i in S **do**
        Apply_odometry_to_update_robots_position(x_i,u_t)
        Apply_sensor_measurement_to_correct_pose(x_i,z_t)
        Generate_Map(x_i,z_t)
        x_i updated by sampling new pose
        Update_Weight(x_i)
    end for
    S = resample();
end while

Transformation (UT) [15] aims to avoid Jacobian calculations and has been proved to better approximate the true values. Instead of linearizing odometry and measurement functions, the UT generates a better approximated Gaussian that represents the true distribution through a set of so-called *Sigma points*. It has been used to generate feature based maps using laser or visual sensors [8,5]. Sigma points are generated deterministically using the previous mean and added noise.

Known Problems. The Unscented Transformation is proven to be more accurate than EKF, but its calculation is difficult. During the selection of the Sigma points one needs to calculate the square root of the augmented covariance matrix; this is usually done through a Cholesky Factorization. However, this method needs the matrix to be positive-defined, otherwise the method dramatically fails. Unfortunately, after several updates, the matrix may become non-positive [4]. Different solutions have been proposed; here, we look at one such solution, the Square Root Unscented Transformation method [14]. In this method, the square root of the covariance matrix is propagated during the updating sequence; this requires a number of other changes in the updating algorithm. Complexity is also reduced from $O(N^3)$ to $O(N^2)$ for N Sigma points.

3 The New Algorithm: SRUPF-GM

The Unscented Particle Filter (UPF) as described in [7,5] is an algorithm for feature based maps. Here we combine elements of the UPF with aspects of the GMapping Particle Filtering algorithm [2] to give an improved algorithm, SRUPF-GM (see Algorithm 2), for SLAM with grid based maps.

3.1 Updating Robot's State Using Odometry Information

All SLAM filters add noise when updating the state using the robot's odometry information. While investigating the most recent GMapping implementation [3], we found differences from the method reported in [2]. For example, rather than updating using a KF method, they approximate it, generating the noise around four independent pre-defined zero-mean Gaussians. While fast, this does not

---

Algorithm 2. SRUPF-GM Update Step Algorithm

Input: previous state set $S = < x_{[t-1,0]}, ..., x_{[t-1,n]} >$, sensor z_t and odometry u_t data at time t;

{Cycle through all particles}

for all x_i in S **do**

  {Sigma Point Generation}

  $x_{aug} = [x_i \; 0 \; 0]$

  $cov_{aug} = [cov_i \; Cholesky(Q)]$

  $SP = [x_{aug} \quad x_{aug} + \gamma(cov_{aug})_i \quad x_{aug} - \gamma(cov_{aug})_{i-n}]$ for i=0 to 2n

  {Odometry Update Step, function f}

  for all x_j in SP **do**

    $< v, \delta > = u_t$

    $V = v + x_j[3]$

    $G = \delta + x_j[4]$

    $x_j = \begin{pmatrix} X_{x_j} + V \times cos(G + \theta_{x_j}) \\ Y_{x_j} + V \times sin(G + \theta_{x_j}) \\ \theta_{x_j} + G \end{pmatrix}$

    add(ST,x_j)

  end for

  $x_i = \sum_{i=0}^{2n} w_c x_j$ for all x_j in ST

  $cov_i = QRDecomposition(< x_j - x_i >)$ for j = 1 to 2n

  $cov_i = CholeskyUpdate(cov_i, x_0 - x_i, w_0)$

  {Measurement Update Step, Map Generation and Weight Update}

  if $measurement occurs$ **then**

    $< x_i, cov_i > = $ scanmatch(x_i, z_t)

    Generate_Map(x_i, z_t)

    Update_Weight(x_i)

  end if

  $x_i = $ sample_new_pose(x_i, $cov_i' \times cov_i$) {Final Update Step}

end for

if variance of particle weight is above threshold **then**

  S = resample() {Resampling Step}

end if

---

provide the mean and covariance information we need. Instead, we replace the GMapping odometry update step with a more accurate Square Root Unscented Transformation update step. We augment the mean and the covariance, which are then used to compute the Sigma points (SPs):

$$\mu_{aug} = \begin{bmatrix} \mu_{t-1} \\ 0 \end{bmatrix}, \quad \Sigma_{aug} = \begin{bmatrix} \Sigma_{t-1} & 0 \\ 0 & \sqrt{Q} \end{bmatrix} \tag{1}$$

where Q is the added noise matrix and is constant for every update step. SPs sp_i are then calculated using the augmented covariance matrix:

$$sp_0 = \mu_{aug} \tag{2}$$

$$sp_i = sp_0 + (\gamma \Sigma_{aug})_i \qquad\qquad i = 1, ..., n \tag{3}$$

$$sp_i = sp_0 - (\gamma \Sigma_{aug})_{i-n} \qquad\qquad i = n+1, ..., 2n \tag{4}$$

γ controls how fast uncertainty increases. The SPs are passed through the odometry function f that incorporates the noise and odometry information to generate vectors $\overline{sp}_i$ of size 3 (X, Y, θ).

$$\overline{sp}_i = f(sp_i, u_t) \qquad\qquad i = 0, ..., 2n \qquad (5)$$

The new mean is calculated from the set of SPs. The covariance is calculated with a QR Decomposition of the weighted deltas between the SPs and the new mean.

$$\mu = \sum_{i=0}^{2n} \omega_g \overline{sp}_i \qquad (6)$$

$$\Sigma = QRDecomposition\left[\sqrt{|\omega_c|}(\overline{sp}_i - \mu)\right] \qquad\qquad i = 1, ...2n \qquad (7)$$

$$\Sigma = CholeskyUpdate\left(\Sigma, \overline{sp}_0 - \mu, \omega_0\right) \qquad (8)$$

where ω_g and ω_c are weights on the SPs;[1] our weight system follows [5]. Finally, we add randomness; the particle's new pose is sampled around the Gaussian generated by:

$$x_i \sim \mathcal{N}(\mu, \Sigma^t \Sigma) \qquad (9)$$

3.2   Measurement Update

Gmapping's measurement update step uses laser information both to correct the robot's pose and to generate the map. While we keep the map generation unchanged, we changed the measurement update step. Our method uses the usual UT measurement update step but with a different pose correction method.

Uncertainty Decrement. SPs are passed to the measurement update function h, which acts as a lightweight pose corrector and returns the best SP ν.

$$\nu = h(\overline{sp}_i, z_t) \qquad (10)$$

$$\Sigma^{\mu,\nu} = \sum_{i=0}^{2n} \sqrt{\omega_c}(\overline{sp}_i - \mu)(\overline{sp}_i - \nu)^t \qquad (11)$$

$$\Sigma_\nu = QRDecomposition\left[\sqrt{|\omega_c|}(\overline{sp}_i - \nu)\right] \qquad\qquad i = 1, ...2n \qquad (12)$$

$$\Sigma_\nu = CholeskyUpdate\left(\Sigma_\nu, \overline{sp}_0 - \nu, \omega_0\right) \qquad (13)$$

$\Sigma^{\mu,\nu}$ is the cross covariance between the newly calculated mean and the previously calculated mean (with odometry). The last step is then to calculate the final mean and covariance:

$$K = \Sigma^{\mu,\nu}[\Sigma_\nu \Sigma_\nu^t]^{-1} \qquad (14)$$

$$\mu = \nu \qquad (15)$$

$$\Sigma = CholeskyUpdate\left(\Sigma, K\Sigma_\nu^t, -1\right) \qquad (16)$$

---

[1] Not to be confused with the particle weights.

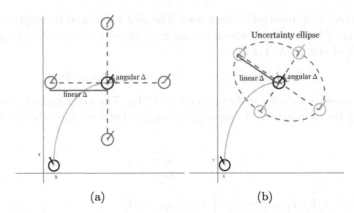

Fig. 1. (a) GMapping pose correction, initial step. Δ is divided by two until a better pose is found. (b) SRUPF-GM pose correction delta uses uncertainty ellipse.

where K is the Kalman Gain. We do not want to perturbate the final mean, as the scan matcher returns the best pose, so μ is just the best Sigma point. The final Σ is decreased with M successive Cholesky updates using all the M columns of the resulting matrix $K\Sigma^t$. No sampling is done after the scan-matching step.

Pose Correction. One reason GMapping works quite well is due to its accurate pose-correction step. The pose corrector generates a number of deterministic samples around a particle's mean and tries to find the best pose within a given δ. Pose correction checks, for every measurement update, if the particle is generating a consistent map. It uses the scan-matching method that incorporates a new scan into particle's existing map to generate a weight. This value is also used to weight the particle. Our square root unscented particle approach works quite well with no accurate pose correction (function h) on simple maps, however more challenging maps do require a pose-correction step. Thus SPs are no longer used to search for the best pose, instead their mean is taken as the starting point.

Our pose correction follows GMapping's with the difference that the δ on which the deterministic samples are generated depends on the covariance cov_t generated during the odometry update. This makes the computation faster than the original GMapping pose-correction, because it searches for the best pose inside the covariance ellipsis, whereas Gmapping version searches over a user pre-defined δ that would eventually include a huge number of improbable poses. Note that (10) in this case is omitted. Figure 1 illustrates the intuition behind our pose-correction approach compared to GMapping pose-correction.

4   Experiments

We tested two versions of the new SRUPF-GM algorithm – with and without pose correction – against the original GMapping implementation. Each algorithm was tested on three different grid-map SLAM problems provided by the Radish

Repository [6] in CARMEN [9] format. Each problem consists of a dataset generated by the sensors of a robot driven around by a human controller, inside a building. The dataset consists of odometry and laser scan readings.

The Radish Repository does not provide a real map neither in an image format nor in an accurate sensor form. Hence it was not possible to compare the algorithms in terms of the quality of the map (for example by generating an error measure). Therefore, we compare the algorithms firstly by whether they achieve the main goal of generating a consistent map. The consistency test assessed whether the overall shape of the map follows the real map, by visually inspecting an image of the results. Secondly, we compared the computation time of the algorithms. Each result reported for the following experiments is the mean of the computational time calculated for 10 runs. To explore the relative computational performance of the algorithms, we also varied two important parameters: (1) the number of particles, and (2) the amount of added noise.

4.1 Experiment 1

The first test was done on a very simple map, generally squared with one single loop and no rooms, as shown in Figure 2(a). For this experiment, we used 30 particles, and the noise parameters were linear 0.1, angular 0.2. (The noise parameters were chosen through preliminary investigations which showed, coincidentally, that these were the best values for all three algorithms.) Table 1 shows the difference in computational time. As one can see, on this simple map, Gmapping is quite fast even using its pose-correction method. SRUPF-GM without pose correction is faster than GMapping, while SRUPF-GM with pose correction is slower, as expected, due to the complexity introduced by the QR and Cholesky computations.

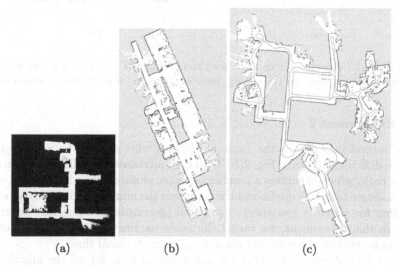

(a)   (b)   (c)

Fig. 2. (a) NSH Building, (b) Cartesium Building, (c) MIT CSAIL

Table 1. Expt 1: Computation time (30 particles, 10 runs)

Algorithm	Average Computation Time (sec)	Std. Deviation
SRUPF-GM without pose correction	17.3140	0.2340
GMapping	21.7994	0.3635
SRUPF-GM with pose correction	27.2739	0.5679

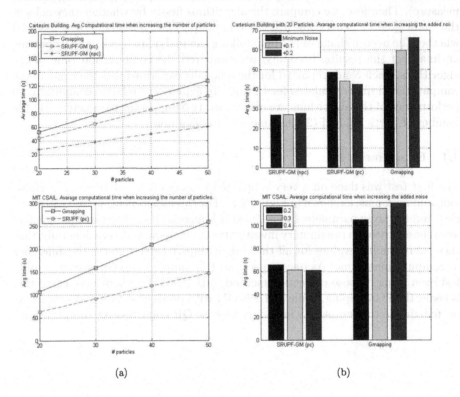

Fig. 3. Expt 2 (above) and Expt 3 (below) comparing SRUPF-GM with and without pose correction (pc/npc) to GMapping, increasing (a) no. of particles (b) added noise

4.2 Experiment 2

In the second experiment the same algorithms were applied to a dataset of medium difficulty, shown in Fig. 2(b). This map increases the difficulty due to the robot's path, which generates a number of loops; closing a loop in the right way is not a simple task. We minimized the error and the number of particles for each algorithm for which it was always successful (generating a consistent accurate map). In this experiment, we varied the noise parameters and the number of particles to explore the resultant changes in computational time.

Figure 3(a)(above) shows that the computation time for all the algorithms increases linearly in the number of particles, however GMapping's results have the largest gradient. Increasing the number of particles is necessary to increase precision. Figure 3 (b)(above) shows the variation of time when increasing the odometry noise. Note that we had to use different minimum added noise for each algorithm; the added noise is algorithm-specific as it depends on how the odometry incorporates this noise into the model function. As the noise is increased, SRUPF-GM with no pose correction shows no significant difference, the computation time for GMapping increases, while the time for SRUPF-GM with pose correction decreases. The explanation for this is found in the pose correction function: by searching inside the uncertainty ellipse SRUPF-GM avoids checking improbable pose. On the other hand, if the ellipse is too small SRUPF-GM may search in vain, hence SRUPF-GM always requires added noise that is not too low. For all the algorithms, increasing the added noise across these values did not decrease accuracy. Of course if the added noise is too high, all the algorithms may no longer find a solution; this 'too high' noise value depends on both the algorithm and the map.

4.3   Experiment 3

In this last experiment, a relatively complicated map was used, still relatively small in area but with an irregular shape and numerous loops in the path. This dataset was taken in the MIT CSAIL building (see Figure 2(c)). On this dataset SRUPF-GM with no pose correction always failed to generate a consistent map (regardless of the number of particles and the amount of noise). Hence it is clear that pose correction *is* actually needed to generate more difficult maps. This fact of course makes SRUPF-GM without pose correction unusable unless the map complexity is known a priori (a rare case). Results are shown in Fig. 3(below). The difference in the computational time is even more pronounced than for the simpler maps, with the computation time again increasing linearly as the number of particles is increased. And again, SRUPF-GM performs better as the noise increases, while GMapping takes longer.

5   Conclusions and Future Works

In this paper, we have presented an improved particle filtering algorithm for solving SLAM on grid based maps. We used as our starting point the GMapping particle algorithm which has been shown (as we confirmed), to generate very accurate maps even for large scale environments. To improve this algorithm, we took aspects from the square root Unscented particle filtering algorithms, previously only applied to feature based maps. We adapted this as required for grid based mapping, increasing the precision during the odometry update as well as decreasing the computation time required for pose correction. One obvious future step is to obtain suitable test datasets that give the real map in a form that allows accurate error measurements to be computed, which will allow us to compare the quality of the resultant maps more accurately. We expect to be able to improve the computation time of SRUPF-GM's pose correction and further optimize the overall algorithm. We envisage these improvements being based on topological hierarchical methods that should decrease the computation time by focusing the accuracy on smaller submaps.

References

1. Bailey, T., Nieto, J., Guivant, J., Stevens, M., Nebot, E.: Consistency of the EKF-SLAM algorithm. In: Proc. of the IEEE/RSJ Int. Conf. on Intelligent Robots and Systems, pp. 3562–3568 (2006)
2. Grisetti, G., Stachniss, C., Burgard, W.: Improving grid-based slam with Rao-Blackwellized particle filters by adaptive proposals and selective resampling. In: Proc. of the IEEE Int. Conf. on Robotics and Automation, pp. 2432–2437 (2005)
3. Grisetti, G., Tipaldi, G., Stachniss, C., Burgard, W.: GMapping Algorithm, http://www.openslam.org/
4. Higham, N.: Analysis of the Cholesky decomposition of a semi-definite matrix. Reliable Numerical Computation (1990)
5. Holmes, S., Klein, G., Murray, D.: A Square Root Unscented Kalman Filter for visual monoSLAM. In: Proc. of IEEE Int. Conf. on Robotics and Automation, pp. 3710–3716 (2008)
6. Howard, A., Roy, N.: The Robotics Data Set Repository (radish), http://radish.sourceforge.net/
7. Kim, C., Sakthivel, R., Chung, W.: Unscented FastSLAM: A robust algorithm for the simultaneous localization and mapping problem. In: Proc. of IEEE Int. Conf. on Robotics and Automation, pp. 2439–2445 (2007)
8. Martinez-Cantin, R., Castellanos, J.: Unscented SLAM for large-scale outdoor environments. In: Proc. of IEEE/RSJ Int. Conf. on Intelligent Robots and Systems, Citeseer, pp. 328–333 (2005)
9. Montemerlo, M., CARMEN-team: CARMEN: The Carnegie Mellon Robot Navigation Toolkit 2002, http://carmen.sourceforge.net
10. Montemerlo, M., Thrun, S., Koller, D., Wegbreit, B.: FastSLAM: A factored solution to the simultaneous localization and mapping problem. In: Proc. of the National Conf. on Artificial Intelligence, pp. 593–598. AAAI Press/MIT Press, Menlo Park, Cambridge (1999/2002)
11. Montemerlo, M., Thrun, S., Koller, D., Wegbreit, B.: FastSLAM 2.0: An improved particle filtering algorithm for simultaneous localization and mapping that provably converges. In: Proc. of Int. Joint Conf. on Artificial Intelligence, vol. 18, pp. 1151–1156 (2003)
12. Thrun, S., Burgard, W., Fox, D.: Probabilistic Robotics. MIT Press, Cambridge (2005)
13. Van der Merwe, R., Doucet, A., De Freitas, N., Wan, E.: The unscented particle filter. In: Adv. in Neural Information Processing Systems, pp. 584–590 (2001)
14. Van Der Merwe, R., Wan, E.: The square-root unscented Kalman filter for state and parameter-estimation. In: Proc. of IEEE Int. Conf. on Acoustics Speech and Signal Processing, vol. 6, pp. 3461–3464 (2001)
15. Wan, E., Van Der Merwe, R.: The unscented Kalman filter for nonlinear estimation. In: Proc. of the IEEE Adaptive Systems for Signal Processing, Communications, and Control Symposium, pp. 153–158 (2000)
16. Zandara, S., Nicholson, A.: Square Root Unscented Particle Filtering for Grid Mapping. Technical report 2009/246, Clayton School of IT, Monash University (2009)

Towards Automatic Image Segmentation Using Optimised Region Growing Technique

Mamoun Alazab[1], Mofakharul Islam[1], and Sitalakshmi Venkatraman[2]

[1] Internet Commerce Security Lab
University of Ballarat, Australia
m.alazab@ballarat.edu.au,
moislam@ballarat.edu.au
[2] Graduate School of ITMS
University of Ballarat, Australia
s.venkatraman@ballarat.edu.au

Abstract. Image analysis is being adopted extensively in many applications such as digital forensics, medical treatment, industrial inspection, etc. primarily for diagnostic purposes. Hence, there is a growing interest among researches in developing new segmentation techniques to aid the diagnosis process. Manual segmentation of images is labour intensive, extremely time consuming and prone to human errors and hence an automated real-time technique is warranted in such applications. There is no universally applicable automated segmentation technique that will work for all images as the image segmentation is quite complex and unique depending upon the domain application. Hence, to fill the gap, this paper presents an efficient segmentation algorithm that can segment a digital image of interest into a more meaningful arrangement of regions and objects. Our algorithm combines region growing approach with optimised elimination of false boundaries to arrive at more meaningful segments automatically. We demonstrate this using X-ray teeth images that were taken for real-life dental diagnosis.

Keywords: Image segmentation, Region growing, False boundary, Automatic diagnosis, Digital forensic.

1 Introduction

A segmentation algorithm partitions a digital image into sets of pixels or segments (objects) and it has wide applications where certain objects in an image are required to be identified and analysed. Hence, several general-purpose image segmentation algorithms have been developed and there is a growing interest in this research area [1]. However, none of these general-purpose techniques could be adopted to any application effectively without modifying it to suit the domain area [2]. In medical diagnosis, factors of image modalities such as noise, bad illumination, low contrast, uneven exposure and complicated 3D structures call for an automated image segmentation that would serve as a significant aid to make real-time decisions quickly and accurately [3]. Even today, X-rays are being used extensively in medical

A. Nicholson and X. Li (Eds.): AI 2009, LNAI 5866, pp. 131–139, 2009.

diagnostics as a safe and affordable imaging technique. Hence, in this study, we concentrate on developing an efficient algorithm for image segmentation in X-ray images for one such application, namely, dental diagnosis.

In dental diagnosis, an image contains several objects and each object contains several regions corresponding to different parts of an object [4]. Hence, it is essential to separate objects of interest from the rest and further, to identify regions of interest (ROI) in each object [5]. With an increasing requirement of diagnosing many dental anomalies that are not seen during visual examination, automatic segmentation of teeth X-ray images would be highly desirable. However, automatic segmentation is quite challenging since there is ambiguities in defining ROI in dental images due to factors such as weak edges and varied non-standard [6]. In this paper, we present a novel segmentation algorithm that combines region-growing method along with optimised elimination of false boundaries using texture features to solve the image segmentation problem in X-ray images of teeth. The proposed algorithm results in meaningful segments that are generated automatically and efficiently to support timely dental diagnosis. The working of the algorithm is demonstrated using X-ray teeth images that were taken for real-life dental diagnosis. The algorithm being generic in nature could be applied for other types of diagnostic applications. For example, in digital forensics, image analysis of teeth structures could be used for identifying a person uniquely. Moreover, since a database of textures is developed automatically for the domain application while performing the iterations of the region-growing process, our proposed generic algorithm could be applied for other domain areas as well.

2   Significance of the Study

In modern dentistry, X-rays are able to show problems that exist below the surface of a tooth and hence X-rays are very essential in doing a complete and thorough dental examination. They are especially important in diagnosing serious conditions early to allow for effective treatment. X-rays show bone anatomy and density, impacted teeth, decay between the teeth, how extensive the decay is or whether an abscess is present. Hence, in digital X-ray images of teeth the most important problem is to find these regions of interest (ROI) that are meaningful arrangements within the objects.

Existing segmentation algorithms for X-ray teeth images predominantly concentrate on applications such as forensics and human identification problems where the ROI are the teeth and their arrangement [5], [7], [8] However, these approaches do not suit the aforementioned problem of dental diagnosis, since they segment only the teeth and not parts of teeth that are decayed or gums with lesions, and even any ROI between teeth or bones that may require dentist's attention. Some recent research have proposed semi-automatic framework for detecting lesions as ROI in X-day teeth images [9], [6]. Such methods require manual interventions as they use a series of filtering operations where the expert inputs the training parameters to introduce some intelligence into the model, and hence could be laborious.

In the segmentation of X-ray teeth images for aiding in dental diagnosis, arriving at a fully automated algorithm is challenging due to the following major inherent problems:

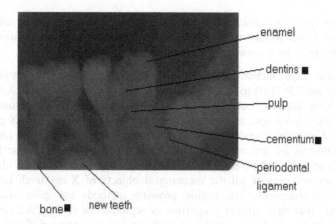

Fig. 1. X-ray image showing different structures of healthy teeth

i) teeth structures (Figure 1) are complicated and hence we cannot have standard set of objects to be matched unlike in other applications,

ii) low contrast and uneven exposure in teeth X-ray images result in the intensities of gums, bones and teeth to be very close and

iii) any noise in data affects the algorithm in identifying the edges between teeth, bones and gums.

To address these problems, we propose a fully automated algorithm that understands how homogeneous regions are formed for identifying meaningful ROI in X-ray teeth images for dental diagnosis. Our algorithm is effective in separating the gum lines, crown and root parts of the teeth and lesions or decay in these regions automatically. We believe that our improved image segmentation algorithm would be a significant contribution in this area of dental diagnosis and other applications that require identification of ROI automatically.

3 Related Works

From literature survey, we observe that traditional image segmentation algorithms fall under three main categories as follows:

- *Histogram thresholding* - assumes that images are composed of regions with different gray ranges, and separates it into a number of peaks, each corresponding to one region. In most cases the threshold is chosen from the brightness histogram of the region or image that we wish to segment and in some cases a fixed threshold of 128 on a scale of 0 to 255 might be sufficiently accurate for high-contrast images [10].

- *Edge-based technique* - uses edge detection filters such as Sobel and Canny [11]. Resulting regions may not be connected, and hence edges need to be joined.

- *Region-based approach* - considers similarity of regional image data. Some of the more widely used approaches in this category are: thresholding, clustering, region growing, splitting and merging [12].

We conducted an experimental study on the ability of these algorithms (implemented in Matlab Version 7.0 [13]) to detect meaningful regions and objects in X-ray teeth images. We limit the scope of this paper by providing here only the summary of observations that have led us towards our proposed algorithm. In X-ray teeth segmentation, histogram thresholding could be mainly used to separate foreground from background as we are unable to find many regions of interest due to single peaks. The Canny method of edge-based detection is quite effective in identifying edges, but does not provide all the meaningful objects of X-ray teeth images for aiding in dental diagnosis. The region growing methods are quite effective in grouping pixels that have similar properties as regions, but finding the appropriate seed and properties of objects in X-ray teeth images are quite difficult. Further, different choices of seed may give different segmentation results and if the seed lies on an edge, ambiguities arise in the growth process.

There is much research work conducted on hybrid algorithms that combine more than one image segmentation technique to bridge the inherent weaknesses of either technique when used separately. In algorithms that combine histogram thresholding and region-based approaches, there is scope for an improved and effective segmentation by introducing textural features as well. Some effective algorithms that adopt boundary relaxation approach on coarsely segmented image using experimental thresholding along with region growing are tailored mainly for colour pixels. Hence, the existing hybrid algorithms that make use of histogram thresholding for region growing are not suitable for X-ray teeth images. Similarly, hybrid algorithms that combine region growing methods with edge-detection techniques are either computationally expensive or suitable more for colour models [14].

To overcome the above mentioned problems, hybrid algorithms are being designed recently, to be semi-automatic with expert intervention for enhancing the filtering process [6], [9]. However, if such algorithms are applied to automatically detect ROI of dental problems, they take long processing time and arrive at results with confusing ROI. Figure 2 shows the X-ray teeth image indicating ROI required for dental diagnosis and Figure 3 shows how the existing semi-automatic algorithms (edge based detection – Figure 3a or region growing based algorithms – Figure 3b) arrive at segmenting the image to different ROI. Hence, they are not effective in detecting dental problems such as tooth decay.

Fig. 2. X-ray image showing ROI

(a) Edge-based technique

(b) Region-based approach

Fig. 3. Results of existing segmentation algorithms

The summary of results from these experiments conducted on X-ray teeth images have highlighted the weaknesses and drawbacks present in existing image segmentation algorithms that are based on histogram thresholding, edge-detection, region-based methods and even hybrid approaches. Hence, to fill this gap, we propose an efficient algorithm that automatically segments X-ray teeth images effectively into meaningful regions that are essential for dental diagnosis.

4 Proposed Image Segmentation Algorithm

The first and foremost problem in dental X-ray images is that the teeth structures are complicated and there are no standard objects to match with. However, our proposed algorithm could build database of objects or textures with intensity ranges for teeth, gums, bones, etc. The second problem of low contrast and uneven exposure in teeth X-ray images that result in very close intensities of gums, bones and teeth could be solved through our algorithm as it adopts morphological enhancement of the contrast in the image to increase texture differentiation among these teeth structures. Finally, any noise in data that could affect in identifying the edges of teeth, bones and gums are overcome in our algorithm as it uses a region growing method based on an adaptive homogeneity measurement rather than a fixed thresholding and further, the region growing is optimised by the elimination of false boundaries. We develop a flag matrix that holds the same value for pixels in one ROI and differs in flag-value from any other ROI. We calculate the horizontal variance and vertical variance of intensities to determine the homogeneity measure within a 4-connected neighbourhood. This homogeneity measure is used in the region growing method where we adopt horizontal and vertical directions of search to find a close region (neighbourhood object). In other words the minimum of the variance is considered for region growing direction with satisfying conditions on the flag property and the homogeneity measure. We also refine the region's boundary using a measure of the texture features and quad tree data structures found along the boundary and its neighbourhood.

In our proposed algorithm, we adopt the following three steps:

Step 1: Contrast Enhancement

Contrast enhancement not only helps in separating the foreground and the background but also solves the problem of low quality dental radiographs where it is difficult to discriminate among the teeth structures such as, bones, gums, crown and root. It also leads to good variability of the intensity degrees among regions for better segmentation. This is because the intensity is enhanced towards the edges and evened out with average values around noisy regions or uneven exposures [15]. We perform a transformation of I (x,y), a given gray-scale image to an enhanced image I'(x,y).of gray-scale range 0 to 255 as follows:

i) I'_{min} (x,y) = 0 and I'_{max} (x,y) = 255 (indicating white and black as extreme contrasts).
ii) I'(i,j) > a such that i $\in$ [i-d, i+d] and j $\in$ [j-d, j+d] (to map grey-scale in the neighbourhood of (i,j) of radius d with same homogeneity measure of value a).
iii) Sum of Histograms of transformed image < Sum of Histograms of original image (guarantees that the contribution of the range of gray scales around (i,j) decreases after enhancement).

Step 2: Region Growing

For region growing, we use the method of arriving at regions by choosing a seed (one or more pixels) as the starting point for a region and, iteratively merging small neighbouring areas of the image that satisfy homogeneity criteria. When the growth of one region stops, the flag matrix and database matrix get updated and another seed that does not yet belong to any region is chosen and the merging iteration starts again. In each iteration, we check if there is any match of any region with the database objects or textures. This whole process is continued until all pixels of the image belong to some region or the other.

We start with a seed as a single pixel and merge regions if the properties (homogeneity measure) of neighbouring pixels are preserved. This is done by exploring the intensity variance horizontally and vertically in the transformed image. Let us denote the gray-scale matrix of the transformed X-ray teeth image by I'(i, j) $\in$ [0, 255] where (i, j) $\in$ [1,m] × [1,n] (with m and n being height and width of the image). Horizontal variance is calculated as |I'(i,j) − I'(i,j+1)| and the vertical variance is computed as |I'(i,j) − I'(i+1,j)|. Then the difference between horizontal and vertical variances is calculated to find the minimum variance. Two matrices called, the flag matrix, F(x,y) with minimum variances in a neighbourhood and the database matrix, D(x,y) with different textures are developed as the algorithm goes through the iterations. The flag matrix, F(x,y) stores same flag value (homogeneity measure) for all the pixels in a region. The flag values are calculated by first taking the homogeneity measure of the seed pixel as the starting value. Subsequently, the flag value gets refined as the iteration progresses by averaging it with that of the neighbouring pixel arrived at. The neighbouring pixel is determined by finding the minimum variance in the horizontal and vertical neighbourhood (as described above). The database matrix,

D(x,y) gets updated and grows from one iteration to the next to contain the segmentation textures of the different teeth structures. Once such a database is developed as a lookup catalog, it becomes very useful in improving the processing time for segmenting X-ray teeth images by matching textures from the database.

Step 3: Optimised Elimination of False Boundaries

Since region growing methods are susceptible to generation of false boundaries that tend to consist of long horizontal and vertical pixels, we adopt a boundary refinement procedure to optimise the iterations in the region growing process of step 2. This is achieved by calculating a measure of textures (α) and a measure of quad tree structures (β) along the boundary (B) as given below:

$$\alpha = (\sum \text{Texture Features along B}) / (\text{Length of B})$$

$$\beta = (\text{\# More than one Pixel across B and in direction of B}) / (\text{Length of B})$$

A merit function is calculated as $\alpha + w\,\beta$, where w is the relative weight associated with these two measures. Boundaries that have small merit function values are discarded as false boundaries. This step optimises the iterations by the elimination of false boundaries. Note that an appropriate weight w is determined through pilot test cases.

5 Experimental Results of Proposed Algorithm

A three-step procedure of our proposed algorithm described above was adopted to conduct an experimental study on 20 X-ray teeth images that had various lesions and dental caries that could be difficult to diagnose such as occlusal caries (decay on upper surface), interproximal caries (decay between teeth), lingual caries (decay of the surface of lower teeth), palatal caries (decay of the inside surface of upper teeth), cementum caries (pulp decay) and recurrent (secondary) caries. As a first step, the contrast enhancement of these images helped in accentuating the texture contrast among dental structures such as teeth, gums and bones and the edges between them. Figure 4a shows the transformed image as a sample X-ray teeth image.

In step 2, the transformed X-ray teeth images went through the region growing iterative algorithm for separating the ROI in the image. The flag and database matrices were used to provide different colour schemes for the various segmented regions in the teeth structures as the final result. Mainly, all normal teeth structures were given shades of blue colour and decayed regions were given red colour. Figure 4b shows the final segmented output of the sample X-ray teeth image indicating the decayed region in red colour. The result demonstrates that our proposed algorithm has automatically and effectively segmented the ROI that are essential for dental diagnosis as these were verified by dentists as well.

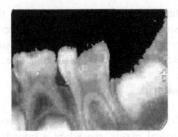

(a) Contrast enhancement

(b) Segmented image showing decay

Fig. 4. Results of our proposed segmentation algorithm

6 Conclusions and Future Work

Image segmentation plays an important role in various diagnostic situations. However, with the increasing number of images to be processed in real-life applications such as digital forensics or medical diagnosis, an automated segmentation of images would aid experts in identifying the problems accurately and quickly to take follow-up actions effectively. This research has taken the first step towards addressing this automatic segmentation problem. In this paper, we have proposed a fully-automatic segmentation using region growing approach with optimised elimination of false boundaries. We conducted an experimental study of our algorithm using dental X-ray images as they provide complex and varied structures with no well-defined contours of teeth problems. The results of the study have demonstrated that our proposed algorithm have automatically segmented X-ray teeth images effectively.

Through the experimental pilot study on X-ray teeth images, we observe that our algorithm was performing better when more images were tested. This is because, the database of textures that was automatically developed during the process had aided in improving the speed of the segmentation process later on. The database of textures formed could serve as a useful lookup table to reduce the processing time significantly.

Our future work would be to analyse the performance of our algorithm on large sample sizes and to ascertain its real-time capability and effectiveness in segmenting all types of teeth defects and even fillings and other types of teeth structures. We also plan to determine its suitability and effectiveness in other areas of medical diagnosis and more importantly in digital forensics, where individual identity could be based on biometric features such as teeth structures.

References

1. Omran, M.G.H., Salman, A., Engelbrecht, A.P.: Dynamic clustering using particle swarm optimization with application in image segmentation. Pattern Analysis and Applications 8, 332–344 (2005)
2. Gonzalez, R.C., Woods, R.E.: Digital Image Processing, 3rd edn. Prentice Hall, NJ (2008)

3. Pham, D.L., Xu, C., Prince, J.L.: Current Methods in Medical Image Segmentation. Annual Review of Biomedical Engineering 2, 315–337 (2000)
4. Jain, A.K., Chen, H.: Matching of Dental X-ray Images for Human Identification. Pattern Recognition 37, 1519–1532 (2004)
5. Said, E.H., Nassar, D.E.M., Fahmy, G., Ammar, H.H.: Teeth Segmentation in Digitized Dental X-Ray Films Using Mathematical Morphology. IEEE Transactions on Information Forensics and Security 1, 178–189 (2006)
6. Li, S., Fevens, T., Krzyzak, A., Jin, C., Li, S.: Semi-automatic Computer Aided Lesion Detection in Dental X-rays Using Variational Level Set. Pattern Recognition 40, 2861–2873 (2007)
7. Nomir, O., Abdel-Mottaleb, M.: A System for Human Identification from X-ray Dental Radiographs. Pattern Recognition 38, 1295–1305 (2005)
8. Zhou, J., Abdel-Mottaleb, M.: A Content-based System for Human Identification Based on Bitewing Dental X-ray Images. Pattern Recognition 38, 2132–2142 (2005)
9. Keshtkar, F., Gueaieb, W.: Segmentation of Dental Radiographs Using a Swarm Intelligence Approach. In: IEEE Canadian Conference on Electrical and Computer Engineering, pp. 328–331 (2006)
10. Hall, E.L.: Computer Image Processing and Recognition. Academic Press, NY (1979)
11. Canny, J.: A Computational Approach to Edge Detection. IEEE Transactions on Pattern Analysis and Machine Intelligence 8(6), 679–698 (1986)
12. Jain, A.K.: Fundamentals of Digital Image Processing. Prentice Hall, Upper Saddle River (1989)
13. Matlab Version 7.0, Mathworks (2005), http://www.matlab.com
14. Shih, F.Y., Cheng, S.: Automatic Seeded Region Growing for Color Image Segmentation. Image and Vision Computing 23, 844–886 (2005)
15. Mukhopadhyay, S., Chanda, B.: A Multiscale Morphological Approach to Local Contrast Enhancement. Signal Processing 80, 685–696 (2000)

Texture Detection Using Neural Networks
Trained on Examples of One Class

Vic Ciesielski and Vinh Phuong Ha

School of Computer Science and Information Technology,
RMIT University, GPO Box 2476V, Melbourne, Vic, 3001, Australia
{vic.ciesielski,vinhphuong.ha}@rmit.edu.au
http://www.cs.rmit.edu.au/~vc

Abstract. We describe an approach to finding regions of a texture of interest in arbitrary images. Our texture detectors are trained only on positive examples and are implemented as autoassociative neural networks trained by backward error propagation. If a detector for texture T can reproduce an $n \times n$ window of an image with a small enough error then the window is classified as T. We have tested our detectors on a range of classification and segmentation problems using 12 textures selected from the Brodatz album. Some of the detectors are very accurate, a small number are poor. The segmentations are competitive with those using classifiers trained with both positive and negative examples. We conclude that the method could be used for finding some textured regions in arbitrary images.

1 Introduction

As digital cameras, mobile phones and computers become more common, more and more people are taking and storing digital images. The problem of finding a particular stored image is becoming more and more difficult. Content Based Image Retrieval (CBIR) systems [1] are attempting to solve this problem using a wide variety of image features. Many images contain textured regions, for example, surf breaking on a beach, a waterfall, a brick building, grass on a sports field and trees in a forest. We are interested in whether using texture can increase the accuracy of retrieval for queries like "Find the picture of the family in front of the waterfall that we took on our holiday a few years ago" or "Find that photo of our son playing hockey when he was in primary school". While some CBIR systems include some texture features in the hundreds of features they use, there is little evidence that the texture features are particularly useful.

Our use of texture will be quite different to the way it is used in CBIR systems. Such systems typically compute a number of Gabor or Wavelet features [2] on the whole image (sometimes image segments) and use feature values in finding a matching database image to the query image. Our approach will be to learn a classifier for a small window of the texture of interest (T), to apply the classifier as a moving window to a database image and report that an image in the data base contains T only if a large enough region of T is detected.

A. Nicholson and X. Li (Eds.): AI 2009, LNAI 5866, pp. 140–149, 2009.

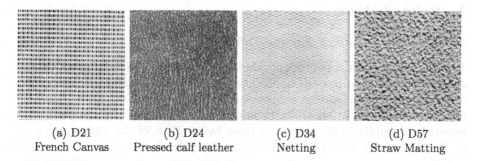

| (a) D21 | (b) D24 | (c) D34 | (d) D57 |
| French Canvas | Pressed calf leather | Netting | Straw Matting |

Fig. 1. Four Brodatz textures. These were used in [4].

There is no generally agreed definition of texture. For our purposes we consider a texture to be an image, or a region in an image, in which a basic visual unit is repeated. In synthetic textures the repeating unit is replicated exactly, for example a checkerboard. In a natural texture the repeating unit is repeated with some variation, for example, blades of grass or pebbles. A considerable amount of work on texture is done with the Brodatz album [3]. The album contains a large variety of natural textures, some highly regular, some very irregular. We have chosen a number of the more regular textures. Some of these are shown in Figure 1.

Most work on learning a classifier requires examples of the class of interest and examples that are not in the class of interest. The learnt classifier is required to distinguish the two classes. A number of texture classification systems of this kind have been built, for example, [2,5]. There is a major problem in using this approach to find a texture T in an arbitrary image since there are many more examples in the non T class. This leads to unbalanced training sets and increased classification error as many of the test/unseen examples are nothing like the ones in the training set. Our goal is to solve these kinds of problems with classifiers by using only examples of the texture of interest in the training data.

1.1 Research Questions

Our overall goal is to determine whether a texture detector for a texture T, trained on only examples of T, is able to find regions of T in arbitrary images, as in Figure 3, for example. In particular, we will address the following research questions:

1. How can a neural network be used to recognize a texture using only examples of the class of interest? What is a suitable neural network architecture?
2. How accurately can the texture of interest be located in a mosaic of other textures? How does this accuracy compare with classifiers that have been trained with examples of all textures?
3. How accurately can the texture of interest be located in arbitrary images?

2 Related Work

2.1 Texture Classification and Detection

Texture classification is the task of taking an image and determining which one of a set of textures it belongs to, for example, distinguishing the 4 classes represented in Figure 1. Texture segmentation is the more difficult task of finding regions of a texture in an arbitrary image, as in Figure 3. There are literally hundreds of publications addressing these tasks. Most of the approaches rely on first computing a set of features such as grey level co-occurrence matrices, Gabor features and wavelet features. Wagner [2] has compared 15 different feature sets on a subset of the Brodatz textures. Recent work that uses features in conjunction with a neural network classifier is described in [6,7,8].

2.2 Learning from Examples of One Class

In recent years there has been increasing interest in learning classifiers from examples of only one class. In the web area, for example, in learning what kinds of pages are of interest to a user, it is very easy to collect positive examples, but there are billions of negative examples. How should they be selected to learn an accurate classifier? Similar problems arise in intrusion detection, document classification and finding objects in images.

In a situation where all the data are numeric, learning a classifier is equivalent to delineating one or more regions in multi-dimensional space. Data points within these regions will be classified as positive and data points outside these regions will be classified as negative. The difficulty with learning from positive only examples is over-generalization, that is, having regions that are too big. This happens because there are no negative examples to clearly mark the boundaries. Neural network classifiers, as they are normally used, need both positive and negative examples.

A variety of approaches have been tried for learning from examples of one class. There is a stream of work using one class support vector machines based on a method originally presented in [9], for example [10]. In [11], this method is applied to a simple texture classification problem. In [12] a co-evolutionary approach is described, one population evolving positive examples, the other negative examples in an attempt to define the boundaries more precisely. A approach in which potential positive and negative examples are evolved a multi-objective algorithm which maximizes separation of clusters and minimizes overlap of clusters is described in [13].

3 Learning a Texture Classifier

Our basic idea is to use a multi-layer neural network, trained by backward error propagation, as an auto associative network (Figure 2b). The input patterns will be $n \times n$ windows of the texture of interest. The desired output is the same

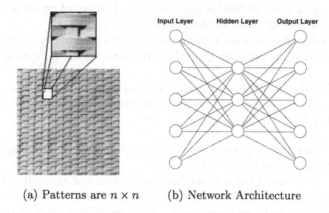

(a) Patterns are $n \times n$ (b) Network Architecture

Fig. 2. Training Patterns and Network Archiecture

as the input. We expect that a network trained for texture T will be able to reproduce unseen examples of T with a small error and that this will not be possible for examples that are not from T. The process of classifying an unseen example E will be to pass E, though the network. If the reproduction error is less than a threshold, $thres_T$ then E is classified as an example of T, otherwise it is classified as 'other'.

3.1 Data Preparation

As noted above, the training examples consist of $n \times n$ windows cut out at random from a large image of the texture of interest, as shown in Figure 2a. A training example is constructed by representing the pixels in the window as a vector, scaling to the range $[0, 1]$ and duplicating the inputs as the outputs. For example, for a window size of 10×10 a training pattern will have 100 inputs and 100 outputs. These training examples need to be generated for all textures of interest.

Clearly the performance of the detector will depend on the value chosen for n. If n is too small, not enough of the variability of the texture will be captured for accurate classification. If n is too big it will not be possible to find a region of this texture in an image unless the region is at least $n \times n$. Furthermore, training and detection times will increase.

3.2 The Training Algorithm

Generating a detector for texture T involves training a neural network in the standard way and then finding a suitable threshold. The methodology is:

1. Generate a training data set of N examples, as described in the previous section.
2. Generate a validation data set of M examples.

3. Choose the number of hidden layers and the number of nodes in each layer. Generate a network with random initial weights.
4. Choose values for learning rate and momentum. Use backward error propagation to train the network. Stop when the error on the validation set begins to rise.
5. Determine the threshold. Compute the error for each example in the validation set, sort the errors, find the value 95% of the way down the list. This is the threshold.

In practice using the validation set was inconvenient and we simply stopped training at a maximum number of epochs.

Choosing a threshold value requires a decision on the tradeoff between true positive and false positive rates. We considered a number of options for choosing the threshold: (1) The largest error on a training example when training stopped. (2) The largest error on a validation example. (3) The average error on the training/validation data. (4) The value which will give true positive rate of 95% and a corresponding false positive rate of 5% on the validation set. We chose option (4) as a reasonable overall compromise. However, further work on optimizing the choice of threshold could improve subsequent detection performance.

4   Segmentation by Texture

A classifier trained by the above algorithm will be a useful detector if it can locate areas of the texture of interest in arbitrary images. We use the following algorithm to generate a segmentation. The segmentation is represented as a binary image in which black pixels represent the texture of interest and white pixels represent other. An example is shown in Figure 3c. Our algorithm to find regions of texture T in image I:

1. For each pixel (except edge pixels) in I
   (a) Centre an $n \times n$ window on this pixel.
   (b) Scale each pixel to $[0, 1]$ and pass to the detector for T.
   (c) Compute error
   (d) If error is less than the threshold for T label this pixel black, else label it white.
2. [Optional] Blob analysis to remove tiny regions.
3. Write the binary image.

More sophisticated strategies are possible to improve the segmentation, for example, removing singletons and very small regions which are most likely false positives, or the kinds voting strategies as used in [4]. We use the above segmentation algorithm because we are primarily interested in detecting the presence of a texture rather than an accurate segmentation.

5   Experiments and Results

In this section we describe a series of experiments to determine a suitable network architecture, to determine whether classifiers trained for one texture can reject examples of other textures, and to determine whether a texture can be detected in an arbitrary image.

For this work we chose twelve textures from the Brodatz album [3]. The full set of electronic images and descriptions can be found in [14]. Selection criteria included a reasonably regular repeating unit and potential to compare with previous work on texture detection/segmentation. All of our experiments used the SNNS system [15].

5.1   Network Architecture

Our goal in this part of the work is to determine a suitable neural network architecture and training strategy (research question 1). We experimented with a wide range hidden layers and different numbers of hidden nodes in the layers. For this set of experiments the window size was 10×10, training set size was 1,600, test set size was 900, the learning rate was 0.4, momentum was not used and training was stopped at 10,000 epochs. Table 1 shows the results for a number of different architectures. Somewhat surprisingly we found that one hidden layer always gave the smallest training and test errors. The number of units in the hidden layer gave varied accuracies; generally choosing approximately the same number of hidden nodes as the number of inputs gave a good trade off between accuracy and training time.

5.2   Discriminating between Different Textures

Our goal in this part of the work is to determine how well our generated classifiers can discriminate different textures (research question 2). For this set of

Table 1. Determining the number of hidden layers and nodes

	D1	D6	D22	D49	D65	D77	D82	D101
100-100-100								
Training MSE	**0.01**	**0.01**	**0.05**	**0.02**	**0.02**	**0.02**	**0.04**	**0.02**
Test MSE	**0.06**	**0.07**	**0.1**	**0.09**	**0.04**	**0.05**	**0.05**	**0.03**
100-50-40-100								
Training MSE	0.03	0.11	0.31	0.02	0.06	0.31	0.33	0.11
Test MSE	0.14	0.16	0.5	0.3	0.15	0.25	0.42	0.17
100-60-60-100								
Training MSE	0.02	0.05	0.19	0.03	0.04	0.15	0.17	0.07
Test MSE	0.12	0.16	0.36	0.24	0.1	0.17	0.27	0.1
100-60-30-60-100								
Training MSE	0.04	0.17	0.31	0.02	0.08	0.31	0.35	0.12
Test MSE	0.21	0.26	0.9	0.42	0.25	0.38	0.6	0.42

Table 2. Detection rates (in percentages) for 12 detectors trained on 12 textures

Thres	D1 2.0	D6 1.23	D21 2.0	D22 7.1	D24 1.595	D34 0.74	D49 2.3	D57 1.9	D65 0.56	D77 1.48	D82 3.9	D101 5.1
D1	**96.13**	0	3.29	94.30	32.95	0	4.91	82.30	0.41	0	28.93	5.52
D6	0	**96.14**	0	0.01	0.03	0	0	0	0	15.46	0	0
D21	0	0	**95.94**	0	0	0	0	10.52	0	0	0	0
D22	0	0	0	**95.56**	0.09	0	0	0.34	0	0	0	0
D24	0	0	0	99.97	**95.92**	0	0	0.07	0	0	99.39	0
D34	1.53	0	3.96	20.91	0.03	**96.25**	17.62	99.9	4.23	0	0	0
D49	0.04	0	3.89	2.3	0	0	**95.08**	60.7	0	0	0	0.56
D57	0	0	0	38.47	0.15	0	0	**94.6**	0	0	0	0
D65	43.32	0	10.61	97.33	71.79	0	53.3	99.9	**96.35**	0	2.86	11.28
D77	0	0	0	2.86	0.55	0	0	0	0	**95.92**	0	0
D82	0	0	0	97.72	11.49	0	0	6.25	0	0	**96.04**	0
D101	0	0	0	0	0	0	0	0.01	0	0	0	**96.04**

experiments the window size was 15×15, training set size was 1,225, validation set size was 2000, test set size was 10,000, the learning rate was 0.4, momentum was not used and training was stopped at 10,000 epochs. Table 2 shows the results of applying the detectors to a new set of test data, generated independently from the training and validation data. Each row shows the performance of a classifier on a test data set. For example, applying the D1 classifier to the D1 test data gives a detection rate of 96.13%. This is expected due the method of setting the threshold described in section 3.2. Applying the D1 classifier to the D6 test data gives a detection rate of 0%. This is a perfect result, none of the D6 examples has been misclassified as D1. Applying the D1 classifier to the D22 test data gives a detection rate of 94.3%. This is a very poor result as most of the D22 examples are being accepted as D1. Interestingly the D22 detector applied to D1 test data is perfect, there are no errors. Overall the picture painted by Table 2 is very promising. All of the diagonal elements are around 95% as expected. Off the diagonal there are many zeros and small percentages indicating accurate rejection of foreign textures by a classifier. However, there are some bad spots. Texture D22, reptile skin, appears to be particularly difficult. While the D22 detector itself is very accurate (horizontal row) many other classifiers falsely accept D22 as their texture. The situation with D57 is somewhat similar. Possibly a larger window size is needed for these textures.

Figure 3 shows a direct comparison between our examples-of-one-class approach (column c) and prior work which used classifiers trained only on examples of the two textures involved (column b) [4] on a texture segmentation problem with irregular boundaries (column a). Our method has captured the texture regions as accurately as [4], however the region boundaries are not as accurate due to our simple approach to segmentation. It is important to note the this is a significant achievement as the method of[4] used a specific classifier trained only to distinguish the two textures of interest. Our method uses a classifier trained to distinguish its own texture from any other image.

The results of this section indicate that our one class method can be competitive with two/multiple class methods.

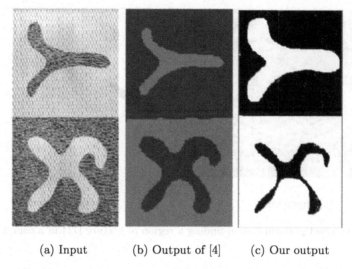

(a) Input   (b) Output of [4]   (c) Our output

Fig. 3. Segmentation of texture in different shapes - comparison with Song's work

5.3   Texture Detection in Arbitrary Images

In this paper we use the Corel Volume 12 image data set. This is a collection of about 2,500 images. They are the kinds of photographs a tourist might take while on holiday in another country. There are images from a number of countries including Nepal, France, Sweden and the USA. The image size is 384×256 pixels.

Our goal in this part of the work is to determine how well our generated classifiers can find regions of their texture in arbitrary images (research question 3). We did this in 2 ways: (1) Classification of image windows cut from randomly selected images, and (2) Segmentation of randomly selected images into which a region of the texture of interest had been pasted.

In the case of (1) we used the Corel Volume 12 image data set [16]. This is a collection of about 2,500 images. They are the kinds of photographs a tourist might take while on holiday in another country. There are images from a number of countries including Nepal, France, Sweden and the USA. The image size is 384 × 256 pixels. We randomly cut 4 windows from each of these images giving a test set of size 10,000. The false detection rates for each of the 12 texture detectors descibed in the previous section are shown in Table 3. This gives us an indication of the kinds of false positive rates that could arise in searching for these textures in arbitrary images. While some of these rates appear high, many

Table 3. False detection rates (in percentages) for detectors on random images

D1	D6	D21	D22	D24	D34	D49	D57	D65	D77	D82	D101
15.25	14.97	3.13	53.68	48.21	0.89	5.5	25.52	7.39	44.46	26.12	5.30

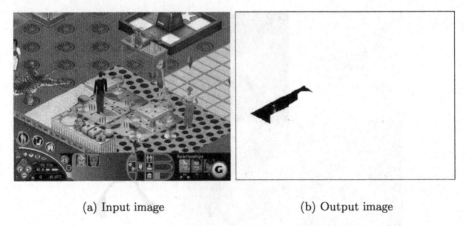

(a) Input image   (b) Output image

Fig. 4. Very good performance in finding a region of texture D24 in a complex image

of these will be eliminated by step 2 of the segmentation algorithm described in Section 4.

In the case of (2) we randomly pasted arbitrary shaped regions of the 12 textures into a number of arbitrarily selected images and then ran the segmentation agorithm. Figure 4 shows one of the outcomes. While we only used a small number of images of this kind, the results suggest that our expectation that many of these false detections would be discarded as singletons or a very small regions by a more sophisticated segmentation algorithm, is a reasonable one.

6   Conclusions

We have shown that it is possible to find regions of a specific texture in arbitrary images, using detectors trained only on positive examples of the texture of interest. The detectors are 3 layer auto associative neural networks, trained by backward error propagation. If a network reproduces an input texture window with a small enough error, the window is classified as a positive example. Texture detectors trained in this way have achieved high accuracy on texture segmentation tasks involving a mosaic of textured regions. The accuracy is competitive with classifiers trained with positive and negative examples. The texture detectors have achieved good performance in detecting the texture class they have been trained for in arbitrary images.

Further work is needed to determine whether better thresholds can be found for the detectors and whether a more sophisticated segmentation algorithm will give better segmentation performance. Our results suggest that learning textures from images selected from data bases such as Google images and Flickr and using the detectors in image retrieval is worth exploring. Also, it is worth exploring whether the method could be used to find objects in images, rather than textures.

References

1. Datta, R., Joshi, D., Li, J., Wang, J.Z.: Image retrieval: Ideas, influences, and trends of the new age. ACM Computing Surveys 40(2), 1–60 (2008)
2. Wagner, P.: Texture analysis. In: Jahne, B., Haussecker, H., Geissler, P. (eds.) Handbook of Computer Vision and Applications, vol. 2, pp. 275–308. Academic Press, San Diego (1999)
3. Brodatz, P.: Textures: A Photographic Album for Artists and Designers. Dover, NY (1966)
4. Song, A., Ciesielski, V.: Fast texture segmentation using genetic programming. In: Sarker, R., et al. (eds.) Proceedings of the 2003 Congress on Evolutionary Computation CEC 2003, Canberra, December 8-12, pp. 2126–2133. IEEE Press, Los Alamitos (2003)
5. Song, A., Ciesielski, V.: Texture segmentation by genetic programming. Evolutionary Computation 16(4), 461–481 (Winter 2008)
6. Feng, D., Yang, Z., Qiao, X.: Texture Image Segmentation Based on Improved Wavelet Neural Network. In: Liu, D., Fei, S., Hou, Z., Zhang, H., Sun, C. (eds.) ISNN 2007. LNCS, vol. 4493, pp. 869–876. Springer, Heidelberg (2007)
7. Teke, A., Atalay, V.: Texture Classification and Retrieval Using the Random Neural Network Model. Computational Management Science 3(3), 193–205 (2006)
8. Avci, E.: An expert system based on Wavelet Neural Network-Adaptive Norm Entropy for scale invariant texture classification. Expert Systems with Applications 32(3), 919–926 (2007)
9. Scholkopf, J., Platt, J., Shawe-Taylor, J., Smola, A., Williamson, R.: Estimating the support of a hige-dimensional distribution. Neural Computation 13, 1443–1471 (2001)
10. Manevitz, L.M., Yousef, M.: One-class SVMs for document classification. J. Mach. Learn. Res. 2, 139–154 (2002)
11. Gondra, I., Heisterkamp, D., Peng, J.: Improving image retrieval performance by inter-query learning with one-class support vector machines. Neural Computing & Applications 13(2), 130–139 (2004)
12. Wu, S.X., Banzhaf, W.: Combatting financial fraud: A coevolutionary anomaly detection approach. In: GECCO 2008: Proceedings of the 10th annual conference on Genetic and evolutionary computation, pp. 1673–1680. ACM, New York (2008)
13. Curry, R., Heywood, M.: One-class learning with multi-objective genetic programming. In: IEEE International Conference on Systems, Man and Cybernetics, ISIC, pp. 1938–1945 (2007)
14. Brodatz Texture Database, http://www.ux.his.no/~tranden/brodatz.html (Visited 3-June-2009)
15. Zell, A.: Stuttgart neural network simulator, http://www.ra.cs.uni-tuebingen.de/SNNS/ (Visted 3-June-2009)
16. Corel Corporation: Corel professional photos CD-ROMs, vol. 12 (1994)

Learning and Recognition of 3D Visual Objects in Real-Time

Shihab Hamid[1,3] and Bernhard Hengst[1,2,3]

[1] School of Computer Science and Engineering, UNSW, Sydney, Australia
[2] ARC Centre of Excellence for Autonomous Systems
[3] NICTA, Sydney, Australia

Abstract. Quickly learning and recognising familiar objects seems almost automatic for humans, yet it remains a challenge for machines. This paper describes an integrated object recognition system including several novel algorithmic contributions using a SIFT feature appearance-based approach to rapidly learn incremental 3D representations of objects as aspect-graphs. A fast recognition scheme applying geometric and temporal constraints localizes and identifies the pose of 3D objects in a video sequence. The system is robust to significant variation in scale, orientation, illumination, partial deformation, occlusion, focal blur and clutter and recognises objects at near real-time video rates.

1 Introduction

The problem of object recognition is a long standing challenge. Changes in scale, illumination, orientation and occlusions can significantly alter the appearance of objects in a scene. Humans easily deal with these subtle variations but machines notice these changes as significant alterations to the matrix of pixels representing the object.

There are two broad approaches to 3D object representation: object-based - 3D geometric modeling, and view-based - representing objects using multiple 2D views. In this paper we report on the development and evaluation of a system that can rapidly learn a robust representation for initially unknown 3D objects and recognise them at interactive frame rates. We have chosen a view-based approach and identified the Scale Invariant Feature Transform (SIFT) [1] as a robust local feature detector, which is used as the computer vision primitive in the object recognition system. In the learning phase a video camera is used to record footage of an isolated object. Objects are learnt as a graph of clustered 'characteristic' views, known as an aspect-graph.

Our contribution is an integrated vision system capable of performing generic 3D object recognition in near real-time. We have developed a fast graphics processing unit (GPU) implementation of SIFT for both building a view-based object representation and later for rapidly recognising learnt objects in images. The system uses an approximate kd-tree search technique, Best-Bin-First (BBF) [2], to significantly speed up feature matching. Additional innovations include: a method to aggregate similar SIFT descriptors based on both a Euclidean distance

A. Nicholson and X. Li (Eds.): AI 2009, LNAI 5866, pp. 150–159, 2009.

Fig. 1. Ten sample objects (left). Part of the 'Mighty-Duck' circular aspect-graph (right).

threshold and a reduced nearest-neighbour ratio; a view clustering algorithm capable of forming rich object representations as aspect-graphs of temporally and geometrically adjacent characteristic views; and an improved view clustering technique using bounding boxes during the learning phase.

The system was evaluated using one to ten objects, each generating about 35 views. Figure 1 shows the ten objects used in our experiments and some of the adjacent views of the generated aspect-graph for the Mighty-Duck object. Recognition speed measurements demonstrate excellent scaling. Our evaluation found the system robust to object occlusion, background clutter, illumination changes, object rotation, scale changes, focal blur and partial deformation.

2   Background and Related Work

Many approaches to 3D object recognition can be found in research literature. A recent survey [3] has concluded that there are as many approaches as there are applications. We only discuss a small subset relevant to our approach.

Local image features usually comprise an interest-point detector and a feature descriptor. One such highly discriminatory feature is the Scale-Invariant Feature Transform (SIFT) [4]. SIFT features are widely adopted because of their ability to robustly detect and invariantly define local patches of an image. Their limitations are that they cannot adequately describe plain untextured objects, their speed of extraction is slow, and their high dimensionality makes the matching process slow. Nevertheless, SIFT is a successful multi-scale detector and invariant descriptor combination. It has been used in object recognition with severe occlusions, detecting multiple deformable objects, panorama stitching and 3D reconstruction.

View-clustering is the process of aggregating similar views into clusters representing the *characteristic views* of the object. Lowe [5] performs view clustering using SIFT features and unlike 2D object recognition, features vote for their object view as well as their neighbour views. There are several limitations to this solution. New features are continually added, potentially creating a scalability problem. The unsupervised clustering cannot recover the orientation of the object. The approach assumes that input object images are a random sequence of images, disregarding any possible temporal link between consecutive frames.

In contrast to view-clustering, view-interpolation explicitly attempts to interpolate the geometric changes caused by changes in the viewing direction. Revaud, et al. [6] apply linear combination theory to the framework of local invariant features. Their model is constructed by forming a homography of features from two nearby object views. On a modest PC, recognition of objects in 800×600 images can be achieved in under half a second. In comparison, the system described here can recognise objects at 12 frames per second using 640×480 images.

The characteristic views of a 3D object can be related using an *aspect-graph*. An aspect-graph is a collection of nodes representing views, connected by edges representing small object rotations. Aspect-graphs can be used to cluster silhouette views of 3D models based on their shape similarity [7]. This method cannot be directly applied to general 3D object recognition as real objects display much more information than just their boundary contours. Possibly closest to our approach is Noor et al.'s [8] aspect-graph for 3D object recognition based on feature matching and temporal adjacency between views.

3   System Description/Implementation

The goal is to rapidly learn the representation of 3D objects with a few training examples, and rapidly recognise the objects and their orientation from unseen test images. The assumptions are that: objects have some texture; images are available for each object without occlusion or background clutter in the training phase; and the camera and objects move slowly relative to each other to produce crisp images. We limit the scope to ten real objects as shown in Figure 1 and view objects around a single plane.

The learning process in the vision system allows for the incremental formation of object representations. The learning algorithm is iterated over all frames in the video stream so that a multi-view object representation can be formed for a single object. This process is repeated for each object to be learnt. Figure 2 depicts the stages of the learning algorithm applied to each training frame.

Capture Frame. A rotating platform captures views of the object at 15 frames per second, producing an image every half-degree of rotation.

Segment Object. Determines a bounding box around the object within the entire image frame by colour separating the distinct fluoro background. This step ensures that the system learns the object within the image frame and not the entire frame itself.

Extract Features. Applies SIFT to identify and describe local key-points in the object image. A SIFT feature is composed of a key-point (x, y, σ, θ) and a 128-dimensional descriptor.

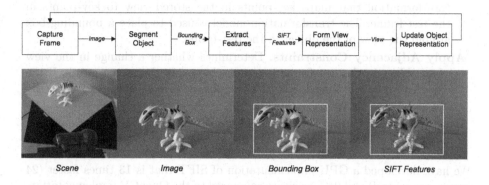

| Scene | Image | Bounding Box | SIFT Features |

Fig. 2. The stages of the object learning algorithm illustrated with the 'dino' object

Form View Representation. Stores a collection of the extracted features. The (x, y) co-ordinates of the features are recorded relative to the segmentation bounding box.

Update Object Representation. Stores the view and associated features in the object database. If the newly formed view significantly matches an existing learnt view, it will be discarded.

The recognition process involves the steps depicted in Figure 3:

Capture Frame. Captures an image from the camera or video.

Extract Features. Finds SIFT features in the entire frame.

Determine View Matches. Matches extracted features to features in the object database. Matches are grouped by views of the object that they are likely to represent. The implementation parameterises affine, perspective and similarity transforms, mapping key-points in a model-view to a test-view. Geometric verification is applied to each of the view-groups to identify the

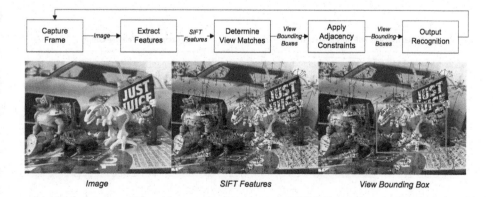

| Image | SIFT Features | View Bounding Box |

Fig. 3. The stages of the object recognition algorithm, illustrated with the 'dino' object

transformation that maps key-points in the stored view to key-points in the test frame. The transformation-model is used to place a bounding-box around the view of the object in the test-frame.

Apply Adjacency Constraints. Determines whether a change in the view of an object in the test frame is permissible using the aspect-graph. If the detected views over time do not correspond to a traversal of the aspect-graph, the views are rejected.

Output Recognition. The final results of the recognition correspond to bounding-boxes around the successfully detected objects.

We have developed a GPU implementation of SIFT that is 13 times faster (24 frames per second) and 94% accurate compared to the OpenCV implementation. It is almost twice as fast as the leading publicly available GPU implementation [9] when run on the same hardware. A full account of the GPU implementation of SIFT is beyond the scope of this paper.

A SIFT feature is said to *correspond to* or *match* another feature if the two descriptors are similar. We use the nearest-neighbour *ratio* strategy [1] that rejects matches if the ratio of the distance between a feature's nearest-neighbour and second nearest-neighbour is high. We also use a Euclidean distance metric to stop distant features from 'matching' during the initial construction of the feature database. To avoid this method from rejecting valid matches to similar SIFT features in different views we store unique view-keypoint pairs in our object model database.

Finding the two nearest neighbours of the test descriptor in the collection of descriptors stored in the database is $O(n)$ and hence prohibitively expensive. We have implemented an approximate kd-tree, Best-Bin-First (BBF) search. The accuracy varies with the number of bins searched, typically 200 [1,4]. We experimentally found a hyperbolic relationship between the accuracy of the search and the maximum number of bins searched, m (Figure 4(a)). By comparing the accuracy results to the speed results (Figure 5 (b)), we trade-off a 200% speed improvement with a 5% loss in accuracy by reducing m from 200 to 50. This

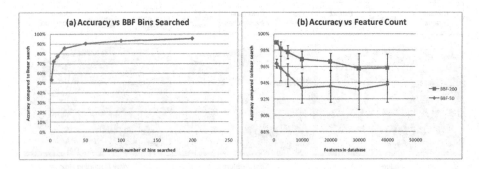

Fig. 4. Left: The accuracy of BBF increases with the number of bins searched. Right: the reduction in accuracy with increased feature numbers.

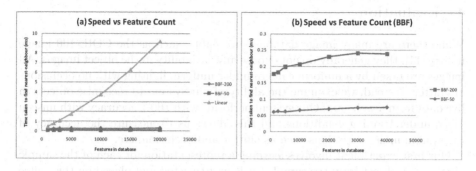

Fig. 5. Left: BBF search speed is almost constant whereas the full nearest-neighbour search is proportional to the number of features. Right: BBF searching 50 bins is significantly faster than searching 200 bins. Error bars are not shown as the standard errors are an order of magnitude smaller than the timing value.

approach is approximately 90% as accurate as linear searching but over 100 times faster for datasets as large as 20,000 features (Figure 5 (a)).

The major difference between our view-clusering approach from [5] is the use of RANSAC [10] to simultaneously develop a geometric model and reject outliers. We cluster views based on how closely they are related by a geometric transform and record view clusters that are *geometrically adjacent* - related by a geometric transform, and *temporally adjacent* - consecutive views in the training video-clip. A view is considered *geometrically identical* to another if it is possible to build a similarity transform to map corresponding key-points using an error tolerance of $\epsilon_{identical}$ pixels. If the error tolerance in pixels is greater than $\epsilon_{identical}$ but less than $\epsilon_{adjacent}$ then the view is part of a geometrically adjacent cluster.

Two views can have many features in common yet not be similar, for example when an object has the same motif on two sides. To avoid clustering these views we use a bounding box matching technique. The error in the view matching bounding box relative to the segmentation bounding box can be measured by considering the distance between each corner of the match-box to the corresponding corner of the segmentation-box. We require each match-box corner to lie within $\beta\%$ of the corresponding segmentation-box corner. This ensures that the tolerance for bounding box matching is proportional to the size of the bounding box. If a view does not sufficiently match the segmentation bounding box it is not clustered and is stored as a new characteristic view.

As recognition is performed on a continuous stream of frames, it is possible to assume that the location and pose of an object through time will remain temporally consistent and there are no rapid changes in the location or pose of the object through time. Restrictions can be imposed to eliminate random view matches which are not temporally consistent. Rather than develop explicit temporal features, temporal consistency is enforced by ensuring consecutive frames from a test video sequence of an object are consistent with the aspect-graph.

4   Evaluation

While there are many image databases available, such as the COIL-100 image library [11], they generally do not expose any variability in the object images. All images are taken by a uniform camera under uniform lighting with no occlusions or clutter. Instead, we examine the ability of our system to recognise objects in live data from a web-cam. Our system performs with 100% accuracy with unseen video under training conditions and this allows us to measure the accuracy of the system under variations such as clutter and occlusion, etc.

Our first experiment involves learning a representation for each of the sample objects. An object from the sample set is first selected and placed on the centre of the rotating platform. The camera is positioned such that the centre of the rotating platform is 20cm from the plane of the camera. A fluorescent light source shines light toward the object from the left of the camera. The 'slow' speed on the rotating platform is used to rotate the object in an automated manner and a video clip of the viewing circle of the object is recorded. The recording commences when the front of the object is facing the camera and ceases after the object has completed a full 360° revolution. Each clip is approximately 60-65 seconds and the entire set of object clips amounts to 5.13GB of loss-less video. We apply the aspect-graph clustering algorithm to each of the training video clips.

Speed of Learning and Recognition. The system is able to construct a representation for an object at 7-9 frames per second. This speed is suitable for learning from live video. Object recognition speed is 11-15 fps in a largely featureless background. In highly cluttered, feature-rich scenes, the recognition speed reduces to 7-10 fps due to the computational cost of feature extraction and matching. This speed is suitable for performing recognition on a live video stream.

Model Size. The view clustering algorithm is able to compress the viewing circle of the sample objects into 20-57 characteristic views (Figure 6). On average, each view consists of 120 features. The more textured and detailed the object, the more features are required to represent it.

Accuracy and Robustness. The recognition system is able to accurately classify, localise and identify the pose of the objects in video streams under considerable variation (Figure 7). The accuracy is 100% for the training clips and diminishes as more variations are introduced. As clutter is added to the background, the accuracy reduces by various amounts for the different objects. Objects that generate more SIFT features, especially features within the body of the object as opposed to the boundary, fare better under heavy clutter.

Most objects can withstand a scale change between −10cm to +15cm. As the resolution of the object decreases with scale change, the accuracy of object recognition rapidly diminishes. Scales changes could be explicitly accommodated in the aspect graph by modelling zooming as well as rotation transitions.

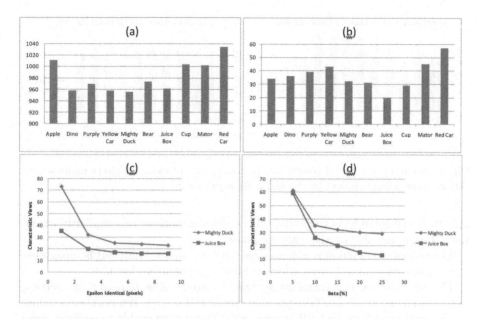

Fig. 6. Learning ten 3D objects. (a) the number of frames in the video sequence, (b) the number of view clusters formed, (c) varying epsilon, (d) varying β.

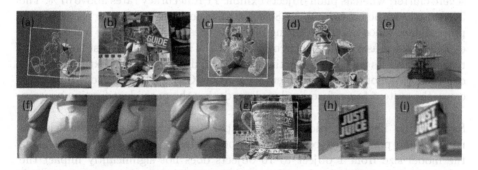

Fig. 7. (a) recognition of significantly occluded objects, (b) object with background clutter, (c) recognition of deformable object, (d-e) examples of scale change, (f) lighting variations, (g) misaligned bounding box with scarcity of cup-handle features, (h-i) blur

Lighting variation can severely affect recognition if the object exhibits few features or a majority of the features are formed as a result of shadows. Objects that are highly textured due to printed or drawn patterns fare better than those that are sculpted.

The recognition system is very robust to partial occlusions. Only 4-5 feature matches are required for accurate recognition. Feature-rich objects can be accurately classified even when the object is more than 80% occluded. The system is

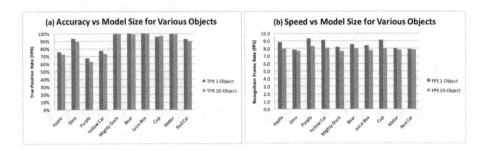

Fig. 8. Accuracy and speed results for the recognition of various objects under severe clutter using a single object model and a 10-object model

able to recognise objects exhibiting minor deformations or can explicitly model the deformations during the learning phase.

The recognition system is able to cope with severe focal blur but is unsuitable to recognise objects in frames which exhibit significant motion blur.

The accuracy of the standard system, without using temporal properties, produces an extremely low false positive rate ($FPR < 1\%$). Temporal constraints can be used to reduce the FPR to near-zero. The system is able to perform more accurate object recognition using objects that are feature-rich. Objects that are highly textured exhibit true positive rate (TPR) accuracy rates above 90% under severe clutter; whereas plain objects exhibit TPR accuracy rates above 67%. The TPR rates can be brought above 80%, for all objects, by exploiting the temporal properties of video and aggregating SIFT features from contiguous frames at the expense of increasing the near-zero FPR to 4%. The recognition system is able to recognise objects with non-reflective surfaces and non-sculpted textures under a considerable amount of environmental variation. Objects that are self-similar can cause the system to confuse similar views during the pose identification process.

Scalability. The recognition system is capable of loading 10 object models into memory and performing recognition at above 7 fps in heavy clutter. Increasing the model size from 1 object to 10 objects does not significantly impact the accuracy of the recognition process (see Figure 8) suggesting that the system should be able to scale to more objects. The system is capable of simultaneous multiple object recognition without showing significant degradation in speed or accuracy, but cannot currently recognize multiple identical objects.

5   Conclusion

The 3D object learning and recognition system described entails several innovations to achieve both speed and high accuracy under varying conditions of occlusion, clutter, lighting, and scale. They include using both geometric and temporal consistency checks from the aspect-graph, comparing estimated bounding boxes between training and test images, and a tuned BBF kd-tree nearest neighbor search. Future work could see the the system benefit from GPU feature matching, active vision, and feature tracking. The system is able to identify object orientation which could be exploited in the robotic manipulation of objects, visual SLAM, and intelligent surveillance applications.

Acknowledgements

NICTA is funded by the Australian Government as represented by the Department of Broadband, Communications and the Digital Economy and the Australian Research Council through the ICT Centre of Excellence program.

References

1. Lowe, D.G.: Distinctive image features from scale-invariant keypoints. International Journal of Computer Vision 60(2), 91–110 (2004)
2. Beis, J.S., Lowe, D.G.: Shape indexing using approximate nearest-neighbour search in high-dimensional spaces. In: CVPR, pp. 1000–1006. IEEE Computer Society, Los Alamitos (1997)
3. Lepetit, V., Fua, P.: Monocular model-based 3d tracking of rigid objects. Found. Trends. Comput. Graph. Vis. 1(1), 1–89 (2005)
4. Lowe, D.G.: Object recognition from local scale-invariant features. In: ICCV, pp. 1150–1157 (1999)
5. Lowe, D.G.: Local feature view clustering for 3D object recognition. In: CVPR, pp. 682–688. IEEE Computer Society, Los Alamitos (2001)
6. Revaud, J., Lavoué, G., Ariki, Y., Baskurt, A.: Fast and cheap object recognition by linear combination of views. In: ACM International Conference on Image and Video Retrieval (CIVR) (July 2007)
7. Cyr, C.M., Kimia, B.B.: 3D object recognition using shape similarity-based aspect graph. In: ICCV, pp. 254–261 (2001)
8. Noor, H., Mirza, S.H., Sheikh, Y., Jain, A., Shah, M.: Model generation for video-based object recognition. In: Proceedings of the 14th ACM International Conference on Multimedia, Santa Barbara, CA, USA, October 23-27, pp. 715–718. ACM, New York (2006)
9. Wu, C.: SiftGPU - a GPU implementation of David Lowe's scale invariant feature transform, SIFT (2007), http://cs.unc.edu/~ccwu/siftgpu/ (accessed: December 18, 2007)
10. Fischler, M.A., Bolles, R.C.: Random sample consensus: A paradigm for model fitting with applications to image analysis and automated cartography. Communications of the ACM 24(6), 381–395 (1981)
11. Nene, S.A., Nayar, S.K., Murase, H.: Columbia object image library (COIL-100). Columbia University (1996) (accessed: May 1, 2008)

Learning Motion Detectors by Genetic Programming

Brian Pinto and Andy Song

School of Computer Science and Information Technology,
RMIT University,
Melbourne, Australia

Abstract. Motion detection in videos is a challenging problem that is
essential in video surveillance, traffic monitoring and robot vision systems. In this paper, we present a learning method based on Genetic
Programming(GP) to evolve motion detection programs. This method
eliminates the need for pre-processing of input data and minimizes the
need for human expertise, which are usually critical in traditional approaches. The applicability of the GP-based method is demonstrated on
different scenarios from real world environments. The evolved programs
can not only locate moving objects but are also able to differentiate
between interesting and uninteresting motion. Furthermore, it is able
to handle variations like moving camera platforms, lighting condition
changes, and cross-domain applications.

Keywords: Genetic Programming, Motion Detection.

1 Introduction

Motion Detection is an important problem in the field of vision systems such as
video surveillance, traffic monitoring and robotic vision systems. Traditionally,
motion detection algorithms involve temporal difference based threshold methods, statistical background modeling and optical flow techniques. For example,
Lipton et al developed a temporal difference based technique, which used a connected component analysis to combine detected pixels into regions of motion[6].
Stauffer et al used the background modeling approach, which represents pixels
in background model by a mixture of Gaussians[10]. By this approach they were
able to handle the "uninteresting" variations of pixels in regions such as swaying
branches of trees and ripples in water. Haritaoglu et al developed a people tracking algorithm, called W4 [3], in which each pixel in background is represented
by its minimum, maximum and the maximum inter-frame difference observed
over a training period.

These traditional motion detection methods, including many methods not
listed here, are often computationally expensive or require intensive development. More critically, they are often highly coupled with domain specific details
and rely on "models" designed by machine vision experts. The performance
of such methods are in turn dependent on fine tuning of parameters in those

A. Nicholson and X. Li (Eds.): AI 2009, LNAI 5866, pp. 160–169, 2009.

models. As a result, these methods are more restrictive. Also variations in environments and changes in task domains will limit the general applicability of these methods. Moreover, real world motion detection problems often contain "uninteresting" motion in given scenes. To address the above issues, a learning method is presented in this study. Motion detection programs are evolved by Genetic Programming (GP)[5] rather than being manually programmed. This method relies less on domain knowledge and human expertise.

GP is a learning paradigm influenced by Darwin's concept of survival of the fittest. It performs a stochastic search to construct a population of computer programs, which are represented as LISP-type of program trees. The leaf nodes on a tree are called *terminals* which usually act like input or parameters of the program, while the internal nodes are called *functions* which usually act like operators. The search is guided by *fitness*: the performance of a program in solving a certain problem. The search starts from a population of randomly generated programs. The programs that have better fitness are more likely to be selected to create new programs for next generation. The fitter programs in the new generation are then used to produce their own offspring. This is an iterative process, by which the best program could improve generation by generation until the solution program is found or a maximum generation is reached.

As a unique machine learning technique, GP has been used on a variety of problems, including image-related areas, such as to detect blood vessels in X-ray coronarograms, and to segment Magnetic Resonance images[8],to detect haemorrhages and micro aneurisms in retina images [12], and to identify regions of field, road or lake in synthetic aperture radar images[1]. GP has also been adapted to detect vehicles on Infrared Line Scan images[4] and to detect interest points in images[11]. All these work show that GP is a powerful learning method. Here we extend GP to learn for motion detection, another important area of machine vision.

The following sections present the GP-based methodology, the real-world scenarios used for testing, the results, discussion and finally, the conclusion.

2   Methodology

The GP-based methodology, initiated in Fang et al[9], can be roughly divided into two phases: the evolution phase and the application phase. The first phase generates a motion detection program based on training samples, while in the application phase this evolved detector is applied on real-time video streams. Both phases use an identical representation, which is discussed below prior to the discussion of the two phases.

2.1   Representation

Video frames in our approach are represented in a multi-frame representation, which is similar to the temporal difference used traditionally in motion detection. This representation is named Multi-frame Accumulate (MF-A) as the motion is accumulated over 'n' number of frames. Each pixel p in MF-A is calculated as in Formula 1.

$$M(p) = \sum_{i=0}^{n-1}(|p_i - p_{i+1}|) \qquad (1)$$

Each pixel p, is the accumulated difference between the intensities at pixel position p for n previous frames. In the formula, p_i refers to pixel p in frame i, $i = 0$ refers to the current frame, $i = 1$ refers to the previous frame and so on. The variation from the n'th previous frame is as relevant as that in the most recent frame since there is no different weight related to n. The value n is set as 3 in this study.

The GP function set, the set of function nodes, in this method is the arithmetic operators $(+, -, \times, /)$ and logic operators $(>, <, ==)$. The division is protected division, in which a division by zero results in zero. An if function is also used, which expects three arguments. If the first argument is true the second argument is returned else the third argument is returned.

Additionally **Average, Minimum** and **Maximum** functions are used to calculate the average, minimum and maximum pixel values of a sub-window in a input image. They expect four arguments, which are the x, y coordinates of the top left and the bottom right corners of the sub-window.

Only two type of terminals are used: a random real number between 0 and 1, and pixel values at position p on the input image (the input image is 20 × 20 in this study). Note a two-dimensional image is represented as a one-dimensional array here. The actual (x, y) coordinate of that pixel p is $(p\%width, p/width)$.

2.2   Evolving Motion Detectors

Evolving a motion detector is effectively learning a classifier, which can differentiate "motion" vs. "no-motion"(or "interesting motion" vs. "anything else"). The performance of a motion detector is measured by how reliably the program can classify the positives from negatives. Therefore, the guide of our search, the fitness function, is the classification accuracy as shown in Formula 2:

$$Fitness = \frac{TP + TN}{TOTAL} \times 100 \qquad (2)$$

Here, TP and TN, refers to the number of true positives and true negatives in classifying "motion" vs. "no-motion", and $TOTAL$ is the total number of classified cases. Note that the fitness function is the classification accuracy, hence, higher the fitness better the program is.

Samples of "motion" and "no-motion" need to be provided as a training set during evolution. Each sample is in the MF-A representation, and is 20×20 pixels in size. These samples are labeled manually as positives, "containing interesting motion" or negatives, as supervising learning is used here. Once the dataset is created, it is divided into a training set and a test set. The fitness, in other words the training accuracy, is the performance of a GP program on the training set. The test set is to evaluate the evolved classifiers. The best performing classifier on the test set is then used in the application phase which is described next.

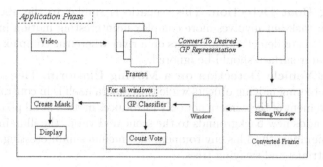

Fig. 1. Application Overview

2.3   Application of Evolved Motion Detectors

The method of applying the best evolved program, the motion detector, is illustrated in Figure 1. The basic steps are 1) retrieving a series of frames from a given video stream. 2) converting them into the MF-A representation. 3) using a sliding window to sample sub-images from the top-left to the bottom-right. 4) Each sub-image, from the frame, is given as the input to the evolved detectors. 5) The detector then classifies the window as either positive or negative. 6) As the sliding windows overlap, pixels are likely to be classified multiple times. Hence, a vote count is maintained to count the number of positive/negative classifications for each pixel. 7) Once all the windows in a frame are classified, each pixel, that has more positive than negative classification votes, is marked as a pixel containing motion.

The size of the sliding window is always at 20 × 20 pixels, which is consistent with the window size in training samples, so the trained program can directly take them as input. Note the overlap between adjacent window positions can be adjusted according to an application. A larger overlap will result faster execution but coarser boundaries. Similarly, a smaller overlap will generate smoother boundaries but requires more time to process each frame. In this study this overlap is always 8 pixels.

3   Test Scenarios

To evaluate the applicability of this GP-based method, described in Section 2, experiments were set up on three different scenarios of increasing complexity. They are described below:

1. **Moving Vehicle Detection with Steady Background.** This scenario involves detecting moving cars on streets in an outdoor environment with steady background. The scenes contain small amounts of uninteresting motion such as pedestrians walking across the street and swaying trees.

2. **Moving Boat Detection.** This scenario is similar to the above vehicle detection task but involves more complex uninteresting motion in the background. It is to detect moving boats on a river where the complex movement of the water surface should be ignored.
3. **Moving Vehicle Detection on a Moving Platform.** This scenario involves detecting moving objects while the camera itself is in constant motion. The movement of the camera not only introduces motion at all pixel positions but also adds new backgrounds to the scene and varies the illumination conditions significantly. This environment also includes uninteresting movement of pedestrians and swaying trees.

4   Experiments and Results

This section, firstly, describes the experiment settings. It then reports on the training and test results achieved in three different scenarios. We also present the results achieved on applying evolved GP motion detection programs on unseen videos of each different scenarios.

4.1   Evolving Motion Detectors

For each scenario, a total of 900 positive and 1800 negative samples were created. Among those samples, 300 positives and 300 negatives were used for training and the remaining were used for test. Note that these positives were marked by us and only contain "interesting motion". For example, in the scenario of detecting moving vehicles, only samples containing moving vehicles are positives. The rest are marked as negatives, which might contain no motion or "uninteresting" motion such as swaying trees. Also these samples were created only from a small portion of the input video. They do not represent the entire video sequence.

The experimental parameters were set as such. The population size, number of programs in each generation, is set to 30. The minimum and the maximum depth of an individual GP program were set at 2 and 12 respectively. The mutation rate was set at 5%, the crossover rate at 85% and the elitism strategy was 10%. In other words, the percentages of new programs generated from mutating programs from parent generation, recombining parent programs and copying parent programs are 5%, 85% and 10% respectively. The low mutation rate, is consistent with the setting in other image related GP studies. The evolution is terminated on either finding a perfect program (100% accuracy) or reaching the maximum number of generations, set at 300.

4.2   Evolution Results

Figure 2 shows both the training accuracies (the fitness) and the test accuracies of the best programs over 300 generations, averaged across 10 runs of training on each of the above mentioned three scenarios. It can be seen that GP can find programs performing well both in training and test for all three scenarios.

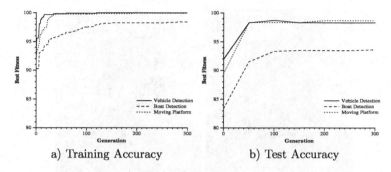

a) Training Accuracy   b) Test Accuracy

Fig. 2. Training and Test Accuracies During the Evolution On Three Scenarios

Program performance is lower in the moving boat detection scenario than in the other two scenarios. This indicates to the relative difficulty of the problem. Although the moving platform scenario is considered to be more complex than the others, high fitness values similar to that in moving vehicle detection were achieved.

Table 1. Training and Test Accuracies of the Best Detector Evolved for Each Scenario

Scenario	Accuracy(%)		True Positives(%)		True Negatives(%)	
	Training	Test	Training	Test	Training	Test
Vehicle Detection	100	99.18	100	99.67	100	98.69
Boat Detection	99.67	96.23	99.67	98.23	96.68	94.24
Moving Platform	100	99.88	100	100	100	99.76

Table 1 presents the training and test accuracies of the best program evolved for each scenario. Each column is divided into two sub-columns, the left showing the training results while the right column showing the test results. The small differences in training and test accuracies indicate the consistency in performance and the absence of over-fitting.

4.3   Application of Evolved GP Programs

The best programs obtained in evolution were used in application phase on videos not seen in training, using the method we described in Section 2. The frames shown in the following figures are the direct output from the programs. They are not modified. Only the white dashed circles/ellipses were added manually to highlight a few points and to aid with the discussion.

Figure 3 shows the output of a program trained for moving vehicle with steady background scenarios. This detector was trained on a completely different video of moving vehicle detection with steady background. The training video was obtained at a higher illumination condition and was situated above a freeway from

Fig. 3. Moving Vehicle Detector on an Unseen Scenario with Steady Background

a) By a Detector Trained   b) By a Detector Trained
for Moving Boat Scenario   for Moving Vehicle Scenario

Fig. 4. Output Frames from the Moving Boat Detection Task

a different city. Despite these unseen factors, the detector performed consistently
and was able to ignore uninteresting motions, such as the pedestrians.

Figure 4 presents two output frames from the moving boat detection task.
The left frame was generated by a detector trained specifically for this boat
detection scenario, whereas the right image was the output of a detector which
was trained for the vehicle detection task and used in Figure 3. Both programs
were able to locate the moving boats and were able to ignore the motion from
water surface, although one is not trained for this task. However the detector
trained for vehicle detection was not able to mark the moving boats thoroughly.

Figure 5 shows the output frames achieved by a detector on a video taken
from a moving platform. The detector was trained for detecting vehicles in such
a scenario. As camera moves steadily the illumination conditions change significantly. This is noticeable in Figure 5, the frame on the right is brighter than
the frame on the left. Despite these challenges consistent detection was achieved
and movement of pedestrians was also ignored successfully.

Fig. 5. Output Frames from the Moving Platform Scenario

On applying the detector used in Figure 3, which was trained for steady background, the results were poor with a large number of false positives. This suggests that although the GP classifiers can be applied on different environments, some similarities between these environments are expected, in this case, classifiers trained on a fixed camera are expected to perform only on fixed camera environments.

5   Discussion

To compare the performance of the GP-based method we test it against a threshold based method. The threshold is applied on the MF-A representation to find the moving pixels. Figure 6 shows the result achieved on applying a manually optimized threshold . Figure 7 shows the output frames achieved on applying the threshold method on the moving boat detection scenario. The frame on the left uses the same threshold as optimized on the moving vehicle detection scenario, the frame on the right uses a manually optimized threshold for the moving boat detection scenario.

Figures 6 and 7 illustrate the problems of threshold as a motion detection method: 1) The threshold must be manually optimized for each different environment. In Figure 7 an un-optimized threshold was not able to attain any significant detection. In contrast, detector evolved by GP can perform reliably on unseen environments as shown in the previous section. 2) The detection is point by point, resulting in partial detection of objects. Further post-processing is necessary to identify whether two separate clusters of detected pixels belong to the same object. Using our method, this process is not often needed. 3) Threshold based detection relies on speed of the movement. As shown in Figure 6, the slow-moving car circled in dashed lines was ignored by threshold but detected by our GP program (see Figure 3).

Fig. 6. Threshold Function on Vehicle Detection with Steady Background

(a) Un-optimized Threshold   (b) Optimized Threshold

Fig. 7. Threshold function on Boat Detection

Similar to thresholds, other traditional approaches usually are also domain specific. For example, a background modeling method designed to be applied on a fixed camera domain would need considerable amount of modifications on the underlying algorithm to be applied on a moving camera domain. Moreover the performance on unseen environments is not readily attained, i.e. parameters need to be optimized again. Additionally, some methods such as background modeling are computationally expensive. On the other hand, the GP-based methodology is readily applicable across different domains. No changes are needed on the methodology itself, but just the new samples for training need to be created. Moreover the evolved motion detectors are small in size and fast in execution. The speed issue will be discussed separately. Additionally this work will be compared with existing traditional approaches separately such as [2] and [7].

6   Conclusion

A general method for learning motion detectors is presented in this paper. This method does not require manual construction of motion detection algorithms and is applicable for producing detection programs on varying scenarios. It can evolve detectors specific for one type of motion while ignore other motions such as ripples, swaying trees, pedestrians and even camera movements. The evolved detectors can perform consistently across different environments without requiring pre-processing of input videos.

References

1. Bhanu, B., Lin, Y.: Object detection in multi-modal images using genetic programming. Applied Soft Computing 4(2), 175–201 (2004)
2. Dalley, G., Migdal, J., Grimson, E.: Background subtraction for temporally irregular dynamic textures. In: IEEE Workshop on Applications of Computer Vision, WACV 2008, pp. 1–7 (2008)
3. Haritaoglu, I., Harwood, D., Davis, L.S.: W4: Real-time surveillance of people and their activities. IEEE Transactions on Pattern Analysis and Machine Intelligence 22(8), 809–830 (2000)
4. Howard, D., Roberts, S.C., Ryan, C.: Pragmatic genetic programming strategy for the problem of vehicle detection in airbourne reconnaissance. Evolutionary Computer Vision and Image Understanding 27(11), 1161–1306 (2006)
5. Koza, J.R.: Genetic Programming: On the Programming of Computers by Means of Natural Selection. Bradford Books (1992)
6. Lipton, A.J., Fujiyoshi, H., Patil, R.S.: Moving target classification and tracking from real-time video. In: Fourth IEEE Workshop on Applications of Computer Vision, WACV 1998, pp. 8–14 (1998)
7. Petkov, N., Subramanian, E.: Motion detection, noise reduction, texture suppression, and contour enhancement by spatiotemporal gabor filters with surround inhibition. Biol. Cybern. 97(5), 423–439 (2007)
8. Poli, R.: Genetic programming for feature detection and image segmentation. Technical report, The University of Birmingham, School of Computer Science, The University of Birmingham, Edgbaston, Birmingham B 15 2TT, UK. R.Poli@cs.bham.ac.uk (2000)
9. Song, A., Fang, D.: Robust method of detecting moving objects in videos evolved by genetic programming. In: GECCO 2008 (July 2008)
10. Stauffer, C., Grimson, W.E.L.: Learning patterns of activity using real-time tracking. IEEE Transactions on Pattern Analysis and Machine Intelligence 22(8), 747–757 (2000)
11. Trujillo, L., Olague, G.: Automated design of image operators that detect interest points. Evolutionary Computation 16(4), 483–507 (2008)
12. Zhang, M., Ciesielski, V.B., Andreae, P.: A domain independent window approach to multiclass object detection using genetic programming. EURASIP Journal on Applied Signal Processing (8), 841–859 (2003)

Information-Theoretic Image Reconstruction and Segmentation from Noisy Projections

Gerhard Visser, David L. Dowe, and Imants D. Svalbe

Monash University, VIC 3800, Melbourne, Australia
gerhardus.visser@infotech.monash.edu.au

Abstract. The minimum message length (MML) principle for inductive inference has been successfully applied to image segmentation where the images are modelled by Markov random fields (MRF). We have extended this work to be capable of simultaneously reconstructing and segmenting images that have been observed only through noisy projections. The noise added to each projection depends on the classes of the pixels (material) that it passes through. The intended application is in low-dose (low-flux) X-ray computed tomography (CT) where irregular projections are used.

1 Introduction

When the members of an observed set of data points each belong to a class from a finite set of classes, the task of inferring those classes is known as classification. When the number of the classes and their properties are not known, this is called clustering (or mixture modelling). If the data points have spatial components, as with images and tomograms, the word segmentation is often used.

This article focuses on segmentation where the pixels are located on a lattice (as with discrete images and tomograms) and the class of each pixel is positively correlated with the classes of its closest neighbours. The pixels are not observed directly but rather noisy projections (summations over subsets of pixels) have been observed.

An example of such a problem is low dose X-ray computed tomography. The pixel intensities are the absorption values at different locations, the classes are the materials (or like regions) and the projections are the sums of the pixel intensities along beams following some path through the object plus noise.

In section 4 we describe what future extensions must be done before a comparison with existing computed tomography (CT) methods can be made.

MML inductive inference has been successfully applied to clustering spatially-correlated data (including image segmentation) in [17], where the images are modelled by Markov random fields (MRF). That work has been extended in [14] to select between different MRF models. Our aim is to extend that work to the image reconstruction problem described above.

One of the advantages of MML inference in image segmentation is that it can infer the number of classes present and the parameters defining those classes. While these two features have not yet been implemented in this work we will

A. Nicholson and X. Li (Eds.): AI 2009, LNAI 5866, pp. 170–179, 2009.

describe how they can be implemented in section 4. For more details on the advantages of MML in classification and clustering see [5, 18, 20, 22].

Our models have been designed for problems where there is a lot of noise present in the observations (as with low-flux X-ray tomography). We will discuss how this work can be extended to infer the nature and degree of the noise from the observed data. For the case where there is little noise, non-probabilistic approaches (such as [13] and [6]) are preferable.

1.1 Minimum Message Length

Minimum message length (MML) [18, 19] is a Bayesian inference method with an information-theoretic interpretation. The minimum message length principle states that the best hypothesis is the one that gives the shortest explanation of the observed data using an optimal two-part encoding scheme.

By minimising the length of a two-part message of Hypothesis (H) followed by Data given Hypothesis ($D|H$), we seek a quantitative trade-off between the desiderata of model simplicity and goodness-of-fit to the data. This can be thought of as a quantitative version of Ockham's razor [10] and is compared to Kolmogorov complexity and algorithmic complexity [2, 9, 12] in [21]. For further discussions of these issues and for contrast with the much later minimum description length (MDL) principle [11], see other articles in that 1999 special issue of the *Computer Journal,* [3, sec. 11.4] and [18, chap. 10].

For our problem the hypothesis H would be some inferred value of x (the class assignments) and y (the pixel intensities) while the detail D is the observed noisy projection values s.

An earlier application of MML to image reconstruction from noisy projections was presented in [4]. While our methods are somewhat different, our aims are similar.

2   Model and Solution

2.1   Image Reconstruction Model

This subsection describes the probabilistic model used. Let $y = (y_1, y_2, ...y_N)$ be a set of pixel intensities, indexed by N locations, with $x_i \in \{1, 2, ..., C\}$ being the class of pixel $y_i \in \{0, 1, ..., B\}$. Here $B + 1$ is the number of intensity values that a pixel can have and C is the number of classes that a pixel can be assigned to. The sequence of projections is $q = (q_1, q_2, ..., q_M)$ where each projection $q_i \subseteq \{1, 2, ..., N\}$ represents a group of locations. For X-ray tomography these groups would be paths followed by the X-ray beams.

The members of x are arranged on a lattice and the a priori distribution over x ($P(x)$) forms a Markov random field (MRF) for a neighbourhood structure defined over the members of x. The neighbourhood structure is defined by a set of neighbours $n_i \subset \{1, 2, .., N\}$ for each location $i \in \{1, 2, .., N\}$. If $i \in n_k$ then $k \in n_i$. For $P(x)$ to be a MRF it is required that $\forall x, P(x) > 0$ and that,

$$P(x_i|x_{\forall j \neq i}) = P(x_i|x_{\forall j \in n_i}) . \tag{1}$$

The second condition (expressed in equation 1) states that the class of a point is conditionally independent of all other points given the classes of its neighbours. We assume that the parameters defining the right hand side of equation (1) are known.

A variety of MRF models that can be used exist in the image segmentation literature, following [17] we will use the auto-logistic model on a toroidal square lattice (see section 2.2).

For now we assume that it is known a priori what the class distributions (defined by $P(y_i|x_i)$) are. For all work presented in this article the classes are defined as Poisson distributions, $P(y_i|x_i) = e^{-\lambda_{x_i}} \lambda_{x_i}^{y_i}/y_i!$. Here x_i is the class that location i is assigned to, so λ_{x_i} is the mean associated with class $x_i \in \{1, 2, ..., C\}$.

The values of the data points y_i are given an upper limit and the probabilities for all values greater than B are added to $P(y_i = B)$. The vector of class parameters λ is also assumed to be known.

Our task is to infer values for x and y given a set $s = \{s_1, s_2, ..., s_M\}$ of noisy summations over the projections q.

$$s_j = \sum_{r \in q_j} (y_r + z_{j,r}) \tag{2}$$

Here s_j is a summation over q_j and $z_{j,r}$ is the noise in s_j contributed by location r. The noise $z_{j,r}$ is assumed to be normally distributed with zero mean and a standard deviation that depends on the class x_r of location r, denoted by σ_{x_r}.

$$P(z_{j,r}|x_r) = \frac{1}{\sigma_{x_r}\sqrt{2\pi}} \exp\left\{-\frac{z_{j,r}^2}{2\sigma_{x_r}^2}\right\} \tag{3}$$

The standard deviation σ_k of each class $k \in \{1, 2, ..., C\}$ is for now assumed known. Let z_j be the sum of noise terms for q_j,

$$z_j = \sum_{r \in q_j} z_{j,r} . \tag{4}$$

2.2   Auto-logistic Model

We have used in our tests the auto-logistic model to express the spatial relations between members of x (the class assignments). The auto-logistic model reduces to the well-known Ising model when the number of classes is equal to two ($C = 2$). When expressed as a Gibbs random field, the prior over x takes the form,

$$P(x) = \frac{1}{U} \exp\left[\sum_{i=1}^{N} \log \alpha_{x_i} - w\beta\right] \tag{5}$$

where w is the number of neighbour pairs (over the entire lattice) that do not have the same class values. Remember that each location has a set of neighbouring locations. For our tests the locations are arranged on a toroidal square-lattice so that each location has four neighbours (left, right, above and below).

The parameters $\alpha = (\alpha_1, \alpha_2, ..., \alpha_C)$ and β are assumed to be known. The vector α is analogous to mixing proportions of the C classes while β determines the degree to which neighbouring classes agree. For example if $x_i = 2$ then $\alpha_{x_i} = \alpha_2$ is the value associated with class 2. Note that the values α_i are equivalent to mixing proportions only when $\beta = 0$. Higher values of β lead to neighbouring class assignments being more likely to have the same value while $\beta = 0$ makes the members of x independent of each other. With w_i defined as the number of neighbours of location i that do not have the same class value as x_i it can be shown that,

$$P(x_i | x_{\forall j \neq i}) = P(x_i | x_{\forall j \in n_i}) \propto \exp\left[\log \alpha_{x_i} - w_i \beta\right] . \tag{6}$$

2.3 The Message Length

This section describes our MML solution to the problem described in section 2.1. All message lengths in this article are measured in nits where 1 bit is equal to $\log_e (2)$ nits.

Given the set of noisy summations s, estimates x (for the class assignments) and y (for the data points) are to be inferred. The optimal estimate is the one that leads to the shortest message length as described in section 1.1. For a given point estimate (a pair $(\hat{x}, \hat{y})$) this code is,

1a. an encoding of $\hat{x}$
1b. an encoding of $\hat{y}$ given $\hat{x}$
2. an encoding of the observed s given $\hat{x}$ and $\hat{y}$

Here the encoding of parts 1a and 1b is known as the assertion and part 2 is known as the detail. The objective function which we try to minimise is the message length, which is equal to the length of assertion plus the length of the detail.

We approximate this message length L as follows,

$$L = T_1 + T_2 + T_3 - T_4 \tag{7}$$

where,

1. T_1 is the length for encoding $\hat{x}$ precisely
2. T_2 is the length for encoding $\hat{y}$ precisely given $\hat{x}$
3. T_3 is the length for encoding s given $\hat{y}$ and $\hat{x}$
4. T_4 is the entropy of the pair $(\hat{y}, \hat{x})$ given s

The first term is equal to the negative log likelihood of $\hat{x}$,

$$T_1 = -\log P(\hat{x}) . \tag{8}$$

We describe in section 2.6 how this can be approximated following [17]. The second term is,

$$T_2 = -\sum_{i=1}^{N} \log P(\hat{y}_i | \hat{x}_i) . \tag{9}$$

To encode s_j given $\hat{y}$ and $\hat{x}$ we need only specify the noise term z_j (see subsection 2.1). These noise terms are normally distributed with mean zero and standard deviation σ_j. The standard deviations depend on $\hat{x}$ and are,

$$\sigma_j^2 = \sum_{i \in q_j} \sigma_{\hat{x}_i}^2 . \tag{10}$$

T_3 is then the sum of the negative log likelihoods of the noise terms z_j, using the above values for σ_j.

Finally, in an optimal code, $\hat{y}$ and $\hat{x}$ do not need to be stated precisely. The number of alternatives that could have been used giving a similar value for $T_1 + T_2 + T_3$ can be approximated by e^{-T_4}. This means that if the pair $(\hat{y}, \hat{x})$ was stated imprecisely it would have cost approximately T_4 nits less. This bits-back approach to calculating the message length was introduced in [15]. The next subsection describes how T_4 can be approximated.

2.4   The Precision of $\hat{y}$ and $\hat{x}$

The entropy of (y, x) given s is defined as,

$$T_4 = - \sum_{\forall (y,x)} P(y,x|s) \log P(y,x|s) \tag{11}$$

Performing the summation over all possible values of y and x is impractical so a numerical approximation for T_4 is used. To explain this approximation we must first express the distribution $P(y, x|s)$ as a Gibbs random field (GRF). First note that applying Bayes's rule,

$$P(y,x|s) \propto P(s|x,y)P(x,y) = P(s|x,y)P(y|x)P(x) . \tag{12}$$

Define the energy V of (y, x) given s as,

$$\begin{aligned}
V(y,x|s) &= - \log \left[P(s|x,y)P(y|x)P(x)U \right] \\
&= - \log P(x)U - \sum_{i=1}^{N} \log P(y_i|x) - \sum_{j=1}^{M} \log P(s_j|y,x) \\
&= - \log P(x)U - \sum_{i=1}^{N} \log P(y_i|x_i) - \sum_{j=1}^{M} \log P(z_j|x) \\
&= - \left[\sum_{i=1}^{N} \log \alpha_{x_i} - w\beta \right] - \sum_{i=1}^{N} \log P(y_i|x_i) - \sum_{j=1}^{M} \log P(z_j|x) .
\end{aligned} \tag{13}$$

Note that given y, x and s this energy $V(y,x|s)$ can be easily calculated. We can now rewrite $P(y, x|s)$ as a GRF,

$$P(y,x|s) = e^{\log P(y,x|s)} = \frac{1}{Z} e^{-V(y,x|s)} \tag{14}$$

where Z is called the partition function and is independent of x and y. Next define,

$$P_T(y,x|s) = \frac{1}{Z_T} e^{-V(y,x|s)/T} \tag{15}$$

as the distribution (over x and y given s) at temperature T. As T increases this distribution reaches its maximum possible entropy. Note that at $T = 1$ this distribution is equivalent to the original distribution $P_1(y, x|s) = P(y, x|s)$. The entropy of this distribution at temperature T is,

$$H_T(y, x|s) = -\sum_{\forall(x,y)} P_T(x, y|s) \log P_T(x, y|s) . \tag{16}$$

The expected energy at temperature T is,

$$Q_T = \sum_{\forall(x,y)} P_T(x, y|s) V(y, x|s) . \tag{17}$$

It can be shown that $\frac{dH_T}{dT} = \frac{dQ_T}{dT}/T$ (hence $dH_T = dQ_T/T$). Gibbs sampling can be used to sample random states of (y, x) given T and s, and hence Q_T can be approximated at any temperature by averaging the energies of those samples.

At $T = \infty$ the entropy H_T attains its maximum value, which is $N \log(CB)$. The entropy of the distribution at temperature $T = 1$ can be calculated as follows. Starting at $T = 1$ and slowly incrementing T up to some value high enough to give a distribution similar to that attained at $T = \infty$, calculate Q_T at each temperature increment. By subtracting the term dQ_T/T at each increment from the maximum entropy, we end with a good estimate of $H_1 = T_4$.

Note that using Gibbs samples from the distribution at each temperature is computationally expensive and to get a good estimate requires that small increments be used [17, Sec. 5.6]. The Gibbs sampling process is discussed in the following subsection.

Q_∞ can be approximated by sampling from the maximum entropy distribution over x and y. It is simple to show then that the error (in calculating H_1) caused by terminating at temperature $T = t$ instead of $T = \infty$ is no greater than $(Q_\infty - Q_t)/t$. This can be used to determine when to terminate the algorithm.

2.5   Estimating $\hat{y}$ and $\hat{x}$ to Optimise the Message Length

For high-dimensional vectors of discrete parameters (such as the one defined by a pair (x, y)) a random selection from the posterior distribution $P(y, x|s)$ can be used as the MML estimate. This type of estimate is discussed in [16] and is also used in [17, sec. 5.1] and [14].

To create such samples we use the Gibbs sampler. This works by repeatedly choosing a random member of (x, y) and changing its value according to its probability distribution given all other values. For example, if x_i is selected then it is re-assigned according to $P(x_i|y, s, x_{\forall k \neq i})$. If this process is repeated for long enough the resulting pair (x, y) can be considered a pseudo-random sample from $P(y, x|s)$.

The same process can be used to sample from $P_T(y, x|s)$ (equation 15) to calculate the approximation for T_4 (equation 11) described in subsection 2.4.

When there is very little noise in the observations s, sampling at temperatures close to $T = 1$ can be difficult due to there being many local minima for the

Gibbs sampler to fall into. This problem can be addressed by using a variation of simulated annealing. In fact, by using simulated annealing the task of finding an estimate $(\hat{x}, \hat{y})$, and the calculation of T_4 (section 2.4), can be done in one step. This is achieved by starting sampling at a high temperature (as described in the last paragraph of section 2.4) and gradually lowering the temperature to $T = 1$. The changes in Q_T are recorded at each decrement and used to calculate T_4 while the final state of (x, y) is used as our MML estimate.

Note that the estimates can be obtained without calculating the message length. There are two uses for calculating the message lengths in our problem. The first is that in some cases multiple runs of the estimation algorithm described above will settle in separate local minima. The message length is a measure of the explanatory value of a hypothesis and can select between these. The second use is for determining the number of classes that can be justified by the data (section 4) [17, 18, 20, 22].

2.6   The Length for Encoding $\hat{x}$

As in section 2.4 equation (15), we define $P_T(x)$ as the distribution over x at temperature T. Since $P(x)$ is a Markov random field (MRF) it can be restated as a Gibbs random field (GRF). This is guaranteed by the Hammersley-Clifford theorem, proven in [1, 7, 8].

This allows us to approximate $H_1(x)$ (the entropy of x at temperature $T = 1$) using the method described in section 2.4. The negative log likelihood of our estimate for $\hat{x}$ is then calculated using,

$$- \log P(\hat{x}) = H_1(x) + V(\hat{x}) - Q(x) \tag{18}$$

where $V(\hat{x})$ is the energy of our estimate $\hat{x}$ and $Q(x)$ is the expected energy for the GRF over x. This type of calculation has also been used to calculate the message lengths for other image models in [17] and [14] and is discussed there in more detail.

3   Test on Artificial Data

For this test there are three classes $C = 3$. The auto-logistic model is used to define the prior $P(x)$ with parameters $\alpha = (1/3, 1/3, 1/3)$ and $\beta = 0.7$.

Similarly the vector of class parameters (class means) is $\lambda = (\lambda_1, \lambda_2, \lambda_3) = (5, 20, 35)$ and the noise parameters (standard deviations) for the three classes are $\sigma = (\sigma_1, \sigma_2, \sigma_3) = (1, 2, 3)$.

From this model, instances of x, y and s were generated with $N = 225$ locations arranged on a 15×15 toroidal square-lattice. The use of a toroidal square-lattice is simply for programming reasons and is not required by our method. The number of projections is $M = 225$ each containing 15 locations. The algorithm was run given s to infer estimates $\hat{y}$ and $\hat{x}$. The true x and inferred $\hat{x}$ class assignments are shown in figure 1. The true y and inferred $\hat{y}$ data point values are also shown in figure 1.

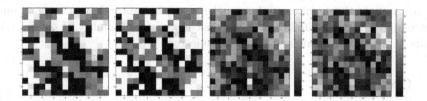

Fig. 1. Far left is the true class assignment vector x with class 1 ($\lambda_2 = 5$) as black, class 2 ($\lambda_2 = 20$) as grey and class 3 ($\lambda_3 = 35$) as white. Centre left is the inferred set of class assignments $\hat{x}$. Centre right is the true value of y and on the far right is the inferred estimate $\hat{y}$. The intensities range from white $y_i = 60$ to black $y_i = 0$ as shown by the bars to the right of the two rightmost images.

The message length calculated as $L = T_1 + T_2 + T_3 - T_4$ was $L = 1740$ with the individual terms being $T_1 = 228$, $T_2 = 664$ $T_3 = 1387$ and $T_4 = 540$. The value of T_4 tells us that there are roughly e^{540} different solutions for the pair (x, y) that are reasonable estimates and gives us some measure of the amount of noise present.

For comparisons to other work to be meaningful our work will have to be developed further. This paper is intended to show how the MML approach to intrinsic classification of spatially correlated data introduced by Wallace [17] can be applied to image reconstruction. The next section discusses what extensions are necessary and how they can be implemented.

4 Further Work

The first problem is with computational expensiveness. Our current implementation is in Java (not a performance language) and little effort was made to make it fast. This Gibbs sampling algorithm is highly parallelisable and specialised hardware is often used in image processing. Before we optimise our implementation we wish to first improve it in other respects.

The earliest applications of the minimum message length principle is in mixture modelling (clustering) [17, 19, 20, 22]. A strong point of MML in this area is the ability to estimate both the class parameters and the number of classes. The next step for our work would be to add those abilities. It should be possible to achieved this using the EM algorithm,

1. initialise all parameters
2. re-estimate x and y using the Gibbs sampler
3. re-estimate the parameters defining the class distributions λ
4. re-estimate the parameters defining $P(x)$
5. re-estimate the projection noise parameters σ
6. if the estimate is stable then stop, else return to step 2

This algorithm should gradually move towards a local minimum in the message length as each individual step reduces it.

To estimate the number of classes, the algorithm above is run several times assuming a different number of classes each time. The number of classes that leads to the shortest message length is preferred.

5 Conclusion

We have shown how Minimum Message Length (MML) can be used to reconstruct and classify (or segment) data sets (images/tomograms) that have been observed only through noisy projections. As a quantitative version of Ockham's razor [10], MML separates noise and pattern information using prior (domain specific) knowledge and it is capable of performing well on noisy data, while being resistant to overfitting. For this reason, applications of MML to low-dose computed tomography are worth exploring.

We have demonstrated how the classification, reconstruction and message length calculations can be done following the approach of [17]. The next step will be to add the ability to infer the class parameters, the noise parameters and the number of classes.

References

1. Besag, J.E.: Spatial interaction and the statistical analysis of lattice systems. Journal of the Royal Statistical Society B36(2), 192–236 (1974)
2. Chaitin, G.J.: On the length of programs for computing finite binary sequences. Journal of the Association of Computing Machinery 13, 547–569 (1966)
3. Comley, J.W., Dowe, D.L.: Minimum message length and generalized Bayesian nets with asymmetric languages. In: Grünwald, P., Pitt, M.A., Myung, I.J. (eds.) Advances in Minimum Description Length: Theory and Applications, pp. 265–294. MIT Press, Cambridge (2005)
4. Dalgleish, A.P., Dowe, D.L., Svalbe, I.D.: Tomographic reconstruction of images from noisy projections - a preliminary study. In: Orgun, M.A., Thornton, J. (eds.) AI 2007. LNCS (LNAI), vol. 4830, pp. 539–548. Springer, Heidelberg (2007)
5. Dowe, D.L.: Foreword re C. S. Wallace. Computer Journal 51(5), 523–560 (2008)
6. Fayad, H., Guedon, J.P., Svalbe, I.D., Bizais, Y., Normand, N.: Applying mojette discrete radon transforms to classical tomographic data. In: Medical Imaging 2008: Physics of Medical Imaging. Proceedings of the SPIE, April 2008, vol. 6913, p. 69132S (2008)
7. Geman, S., Geman, D.: Stochastic relaxations, Gibbs distributions and the Bayesian restoration of images. IEEE Tran. on PAMI PAMI-6, 721–741 (1984)
8. Grimmett, G.R.: A theorem about random fields. Bull. London Math. Soc. 5, 81–84 (1973)
9. Kolmogorov, A.N.: Three approaches to the quantitative definition of information. Problems of Information Transmission 1, 1–17 (1965)
10. Needham, S.L., Dowe, D.L.: Message length as an effective Ockham's razor in decision tree induction. In: Proc. 8th Int. Workshop of Artificial Intelligence and Statistics (AISTATS 2001), Key West, FL, pp. 253–260 (2001)
11. Rissanen, J.: Modeling by the shortest data description. Automatica 14, 465–471 (1978)

12. Solomonoff, R.J.: A formal theory of inductive inference. Information and Control 7, 1–22, 224–254 (1964)
13. Svalbe, I., van der Speck, D.: Reconstruction of tomographic images using analog projections and the digital radon transform. Linear Algebra and its Applications 339, 125–145 (2001)
14. Visser, G., Dowe, D.L.: Minimum message length clustering of spatially-correlated data with varying inter-class penalties. In: Proc. 6th IEEE International Conference on Computer and Information Science (ICIS 2007), Melbourne, Australia, July 2007, pp. 17–22 (2007)
15. Wallace, C.S.: An improved program for classification. In: Proceedings of the Nineteenth Australian Computer Science Conference (ACSC-9), Monash University, Australia, vol. 8, pp. 357–366 (1986)
16. Wallace, C.S.: False Oracles and SMML Estimators. In: Proc. Information, Statistics and Induction in Science conference (ISIS 1996), Was Tech Rept TR 89/128, Monash University, Australia, pp. 304–316. World Scientific, Singapore (1996)
17. Wallace, C.S.: Intrinsic classification of spatially correlated data. Computer Journal 41(8), 602–611 (1998)
18. Wallace, C.S.: Statistical and Inductive Inference by Minimum Message Length. Springer, Heidelberg (2005)
19. Wallace, C.S., Boulton, D.M.: An information measure for classification. Computer Journal 11, 185–194 (1968)
20. Wallace, C.S., Dowe, D.L.: Intrinsic classification by MML - the Snob program. In: Proc. 7th Australian Joint Conf. on Artificial Intelligence, pp. 37–44. World Scientific, Singapore (1994)
21. Wallace, C.S., Dowe, D.L.: Minimum message length and Kolmogorov complexity. Computer Journal 42(4), 270–283 (1999)
22. Wallace, C.S., Dowe, D.L.: MML clustering of multi-state, Poisson, von Mises circular and Gaussian distributions. Statistics and Computing 10, 73–83 (2000)

Belief Propagation Implementation
Using CUDA on an NVIDIA GTX 280

Yanyan Xu[1], Hui Chen[1], Reinhard Klette[2], Jiaju Liu[1], and Tobi Vaudrey[2]

[1] School of Information Science and Engineering, Shandong University, China
[2] The .enpeda.. Project, The University of Auckland, New Zealand

Abstract. Disparity map generation is a significant component of vision-based driver assistance systems. This paper describes an efficient implementation of a belief propagation algorithm on a graphics card (GPU) using CUDA (Compute Uniform Device Architecture) that can be used to speed up stereo image processing by between 30 and 250 times. For evaluation purposes, different kinds of images have been used: reference images from the Middlebury stereo website, and real-world stereo sequences, self-recorded with the research vehicle of the .enpeda.. project at The University of Auckland. This paper provides implementation details, primarily concerned with the inequality constraints, involving the threads and shared memory, required for efficient programming on a GPU.

1 Introduction

The generation of accurate disparity maps for pairs of stereo images is a well-studied subject in computer vision, and is also a major subject in vision-based driver assistance systems (DAS). Within the .enpeda.. project, stereo analysis is used to ensure a proper understanding of distances to potential obstacles (e.g., other cars, people, or road barriers). Recent advances in stereo algorithms involve the use of Markov random field (MRF) models; however, this leads to NP-hard energy minimization problems. Using graph cut (GC) or belief propagation (BP) techniques allows us to generate approximate solutions with reasonable computational costs [5].

Implementations of (potentially) global methods such as GC or BP often generate disparity maps that are closer to (if available) the ground truth than implementations of local methods (e.g., correlation-based algorithms). Obviously, global methods take more time for generating the stereo results [13]. Ideally, one wants to combine the accuracy achieved via global methods with the running time of local methods. One option toward achieving this goal is to speed up, for example, a BP implementation without losing accuracy, by taking advantage of the high performance capabilities of Graphic Processing Units (GPUs); available on most personal computing platforms today.

This report describes a General Purpose GPU (GPGPU) implementation of a BP algorithm using the NVIDIA Compute Uniform Device Architecture (CUDA) language environment [1]. The contemporary graphics processor unit (GPU) has huge computation power and can be very efficient for performing data-parallel tasks [7]. GPUs have recently also been used for many non-graphical applications [8] such

A. Nicholson and X. Li (Eds.): AI 2009, LNAI 5866, pp. 180–189, 2009.

as in Computer Vision. OpenVIDIA [6] is an open source package that implements different computer vision algorithms on GPUs using OpenGL and Cg. Sinha *et al* [14] implemented a feature based tracker on the GPU. Recently, Vineet [15] implemented a fast graph cut algorithm on the GPU. The GPU, however, follows a difficult programming model that applies a traditional graphics pipeline. This makes it difficult to implement general graph algorithms on a GPU.

[9] reports about BP on CUDA. However, that paper does not mention any details about their implementation of BP on CUDA. This paper explains the implementation details clearly. We then go on to detail important pre-processing steps, namely Sobel edge operator (as performed in [10]) and residual images (as performed in [16]), that can be done to improve results on real-world data (with real-world noise).

This paper is structured as follows. Section 2 specifies the used processors and test data. Section 3 describes the CUDA implementation of the BP algorithm. Section 4 presents the experimental results. Some concluding remarks and directions for future work are given in Section 5.

2   Processors and Test Data

For comparison, in our tests we use a normal PC (Intel Core 2 Duo CPU running at 2.13 GHz and 3 GB memory) or a GPU nVidia GTX 280. GPUs are rapidly advancing from being specialized fixed-function modules to highly programmable and parallel computing devices. With the introduction of the Compute Unified Device Architecture (CUDA), GPUs are no longer exclusively programmed using graphics APIs. In CUDA, a GPU can be exposed to the programmer as a set of general-purpose shared-memory Single Instruction Multiple Data (SIMD) multi-core processors, which have been studied since the early 1980*s*; see [11]. The number of threads, that can be executed in parallel on such devices, is currently in the order of hundreds and is expected to multiply soon. Many applications that are not yet able to achieve satisfactory performance on CPUs may have benefit from the massive parallelism provided by such devices.

2.1   Compute Unified Device Architecture

Compute Unified Device Architecture (CUDA) is a general purpose parallel computing architecture, with a new parallel programming model and instruction set architecture. It leverages the parallel compute engine in NVIDIA GPUs to solve many complex computational problems in a more efficient (parallel) way than on a CPU [12]. CUDA comes with a software environment that allows developers to use C as a high-level programming language. A complier named NVCC generates executable code for GPUs.

CUDA Hardware Model.   At the hardware level, the GPU is a collection of multiprocessors (MPs). A multiprocessor consists of eight Scalar Processor (SP) cores, two special function units for transcendentals, a multithreaded instruction unit, and on-chip shared memory, see left of Figure 1 [12]. When a CUDA program on the host CPU invokes a kernel grid, the blocks of the grid are enumerated and distributed to multiprocessors with available execution capacity. The threads of a thread block

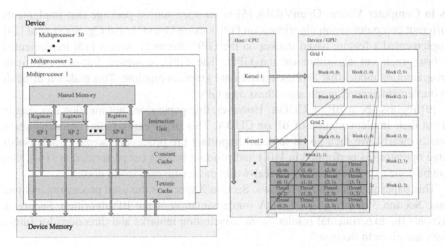

Fig. 1. NVIDIA GeForce GTX 280 CUDA hardware (left) and programming (right) models

Fig. 2. Test data. Left: Tsukuba stereo pair [13]. Right: real image pair captured by HAKA1.

execute concurrently on one multiprocessor. As thread blocks terminate, new blocks are launched on the vacated multiprocessors. SP cores in a multiprocessor execute the same instruction at a given cycle. Each SP can operate on its own data, which identifies each multiprocessor to be of SIMD architecture. Communication between multiprocessors is only through the device memory, which is available to all the SP cores of all the multiprocessors. The SP cores of a multiprocessor can synchronize with one another, but there is no direct synchronization mechanism between the multiprocessors. The GTX 280, used in this paper, has 30 MPs, i.e., 240 SPs. The NVIDIA GeForce GTX 280 graphics card has 1 GB of device memory, and each MP has 16 KB of shared memory (shared amongst all SPs).

CUDA Programming Model. To a programmer, parallel portions of an application are executed on the device as *kernels*. One kernel is executed at a time, with many *thread blocks* in each kernel, which is called a *grid* (see right of Figure 1). A thread block is a batch of threads running in parallel and can cooperate with each other by sharing data through shared memory and synchronizing their execution [12]. For the GTX 280, the maximum number T_{max} of threads equals 512. A *warp* is a collection of threads that are scheduled for execution simultaneously on a multiprocessor. The warp size is fixed for a specific GPU. If the number of threads, that will be executed, is more than the warp size, they are time-shared internally on the multiprocessor. Each thread and block is given a unique ID that can be accessed within the thread during its execution. Using the thread and block IDs, each thread can perform the kernel task on different data. Since the device memory is available to all the threads, it can access any memory location [15].

2.2 Used Test Data

To evaluate our accelerated BP algorithm, different kinds of images have been used: a stereo pair of Tsukuba from the Middlebury Stereo website [13], and real-world stereo sequences, which are captured with HAKA1 (Figure 2), a research vehicle of the .*enpeda.*. project at The University of Auckland [4].

3   Implementation of Belief Propagation

Our GPU implementation is based on the "multiscale" BP algorithm presented by Felzenszwalb and Huttenlocher [5]. If run on the original stereo images, it produces a promising result on high-contrast images such as *Tsukuba*, but the effect is not very satisfying for real-world stereo pairs; [10] shows a way (i.e., Sobel preprocessing) how to improve in the latter case, and [16] provides a general study (i.e., for the use of residual input images) for improving results in correspondence algorithms.

3.1   Belief Propagation Algorithm

Solving the stereo analysis problem is basically achieved by pixel labeling: The input is a set P of pixels (of an image) and a set L of labels. We need to find a labeling $f : P \to L$ (possibly only for a subset of P). Labels are, or correspond to disparities which we want to calculate at pixel positions. It is general assumption that labels should vary only smoothly within an image, except at some region borders. A standard form of an energy function, used for characterizing the labeling function f, is (see [2]) as follows:

$$E(f) = \sum_{p \in P} \left(D_p(f_p) + \sum_{(p,q) \in A} V(f_p - f_q) \right) \tag{1}$$

$D_p(f_p)$ is the cost of assigning label f_p to pixel $p \in \Omega$, where Ω is an M (rows) $\times N$ (columns) pixel grid; the discontinuity term $V(f_p - f_q)$ is the cost of assigning labels f_p and f_q to adjacent pixels p and q. Full details of this equation are found in [5].

As a global stereo algorithm, BP always produces good results (in relation to input data !) when generating disparity images, but also has a higher computational time than local stereo methods. Sometimes we require many iterations to ensure convergence of the message values, and each iteration takes $O(n^2)$ running time to generate each message where n corresponds to the number of possible disparity values (labels). In [5], Felzenszwalb and Huttenlocher present the following methods to speed up the BP calculations.

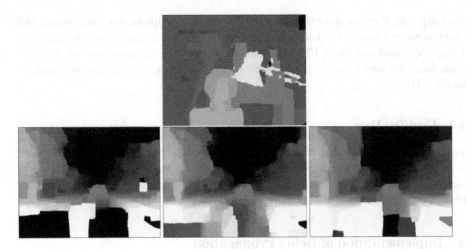

Fig. 3. Example output on test data (see Figure 2 for original images). Top: disparity (colour encoding: dark = far, light = close) for Tsukuba. Bottom: disparity results for a HAKA1 sequence (left to right): on original pair, after applying a Sobel operator, and after mean residual processing.

First, a *red-black method* is provided. Pixels are divided in being either *black* or *red*, at iteration t, messages are sent from black pixels to adjacent red pixels; based on received messages, red pixels sent at iteration $t + 1$ messages to black pixels, and thus the message passing scheme adopts a red-black method, which allows us that only half of all messages are updated at a time.

The *coarse-to-fine algorithm* provides a second method for speeding up BP, and is useful to achieve more reliable results. In the coarse-to-fine method, a Gaussian pyramid is used having L levels (where $L = 0$ is the original image size). Using such a pyramid, long distances between pixels are shortened, this makes message propagation more efficient. This increases the reliability of calculated disparities and reduces computation time without decreasing the disparity search range.

3.2   Belief Propagation on CUDA

BP algorithms have been implemented on the GPU in the past several years: [3] and [17] describe GPU implementations of BP on a set of stereo images. However, each of these implementations uses a graphics API rather than CUDA. Grauer-Gray [9] had implemented BP using CUDA, but did not discuss possible improvements for real-world stereo pairs which always accompany various types of noise, such as different illumination in left and right image, which causes BP to fail.

In our CUDA BP implementation, we define four kernel functions on the GPU, plus the mandatory memory allocation, loading images to GPU global memory, and retrieving the disparity image from the GPU global memory.

1. Allocate GPU global memory
2. Load original images (left and right) to GPU global memory
3. (If real-world image) Pre-process images with Sobel / Residual

4. Calculate data cost
5. Calculate the data (Gaussian) pyramid
6. Message passing using created pyramid
7. Compute disparity map from messages and data-cost
8. Retrieve disparity map to local (host) memory

For the case of image sequences (as for driver assistance applications), Step 1 only needs to be done once. After this, all images can be processed using this memory allocation, thus start at Step 2 (saving computational time). These details are elaborated below. Example output of the algorithm described above can be seen in Figure 3.

The important limitations on the GPU alter the BP GPU kernels. These are the maximum thread limit T_{max} ($= 512$ on the GTX 280), and the shared memory S ($= 16$ KB of memory, which is 4096 single precision 32-bit float data). Keeping functions within the shared memory limit is what makes CUDA programming fast, so we try to keep to these limitations.

The final point to note is that with CUDA programming, you need to divide the threads into j (rows) by k (columns) blocks (omitting the 3rd dimension of 1). You also need to define the q (rows) by r (columns) grid. This defines how the images (M rows by N columns), within each kernel, are processed.

Allocate Global Memory. Memory is allocated and set to zero as follows (all data is single precision, i.e. 32-bit float):

– Left and right image: $2 \times MN$
– Preprocessed left and right images: $2 \times MN$
– Data Pyramid for message passing (central pixel plus: left, right, up and down): $2 \times 5 \times nMN$, where the "$2\times$" part is the upper bound (sum to infinity)
– Disparity image: $1 \times MN$

From this it can be seen that the memory allocation used is $(5 + 10n)MN$ float values. Note: This provides an upper bound on memory allocation dependent on M, N, and n. The limitations of the GTX 280 is 1 GB memory, so we can handle up to around $M = 600$, $N = 800$, and $n = 50$, which is larger than the general resolution in DAS (usually VGA $M = 480$, $N = 640$, and $n = 70$). Since all data is floating point, $S = 4096$ for the remainder of this section (based on the GTX 280).

Load Images to Global Memory. Load the left and right images to allocated memory and smooth with a Gaussian filter of $\sigma = 0{\cdot}7$ before computing the data costs.

Pre-Processing of Images. The images are pre-processed using either Sobel (as done in [10]) or residual images [16]. This is simply done by splitting the image into $j \times k$ overlapping windows. j and k are equal to the height and width (respectively) of the window used in each process. This allows massive parallel windows to be used. The typical window size is $j = 3$ and $k = 3$ (for both Sobel and residual) and only the input, output and two sets of working data are required for both Sobel (gradient in both directions) and one set for residual (smoothed image). This means that we are well within the limits of threads and shared memory, $4jk \leq T_{max} \leq S$.

Calculate Data Cost. This part calculates the data cost for the image; this is a function to calculate the data cost at level $L = 0$ (i.e., D^0). Since we are dealing with only scanline specific information, the image is divided into a grid of $r = \left\lceil \frac{N}{T_{\max}+n} \right\rceil$ (need to store the maximum disparity and the scanline) times $q = M$. Each block will be $j = 1$ times $k = \left\lceil \frac{N}{r} \right\rceil + n$ large, which is under the $T_{\max}$ limit. This means that the total number of blocks is rM. Furthermore, only three data are used (left image, right image, and data-cost image), so $3k \leq 3T_{\max} \leq S$ (in the GTX 280). This means that the shared memory is exploited for fast computation. Comparing this to a conventional CPU, the number of cycles drops from nMN to effectively nrM.

Calculate Gaussian Pyramid. Here we calculate the data pyramid (using the data cost calculated above) in parallel. We employ L levels in the data pyramid to compute the messages and obtain D^1 to D^L. This is done by splitting the image at level $l \in L$ such that the $jk \leq T_{\max}$ limit is satisfied. Only the data from level l and $l + 1$ is needed, so the memory requirement is $2jk$, which is always in the shared memory limit $2jk \leq 2T_{\max} \leq S$. As long as the above limit is kept, any jk can be chosen. A number is ideal that exploits the size of the image (splitting it into even sized blocks) to not waste any threads (i.e., q, r, j, and k are all integers without overlap).

Message Passing. The implementation of the BP algorithm from coarse to fine in L levels is processed here. Here we divide the data cost D^l to a series of tiles whose size is $j \times k$; this is shown in Figure 4. These tiles are overlapping, and the image *apron* (image plus padding from this overlap) is formed. Since both j and k need to have to be odd, we define $j = (2\hat{j} + 1)$ and $k = (2\hat{k} + 1)$, respectively, to represent this. Obviously, one requirement is $jk \leq T_{\max} \Rightarrow (2\hat{j} + 1)(2\hat{k} + 1) \leq T_{\max}$. Since the message passing requires five data pyramids (centre, left, right, up and down pixel) at each disparity, the amount of memory needed is $5jkn$. If we want to exploit the shared

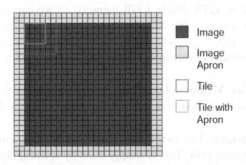

Image

Image Apron

Tile

Tile with Apron

Fig. 4. Image *apron*. This shows how the image is segmented for message passing. Overlapping windows are required, and thus a padding is required to form the *apron*.

memory, the second requirement is: $5jkn \leq S \Rightarrow 5n(2\hat{j} + 1)(2\hat{k} + 1) \leq S$. One last requirement is that $j, k \geq 3 \Rightarrow \hat{j}, \hat{k} \geq 1$. Summarising these requirements:

$$(2\hat{j} + 1)(2\hat{k} + 1) \leq T_{\max} \tag{2}$$

$$5n(2\hat{j} + 1)(2\hat{k} + 1) \leq S \tag{3}$$

$$1 \leq \hat{j}, \hat{k} \in \mathbb{Z} \tag{4}$$

Limit (3) suggests that the maximum disparity calculation is in fact $n \leq \frac{S}{45}$, thus the maximum disparity on the GTX 280 is $n = 91$. If the disparity is above this, then shared memory is not used, thus Limit (3) is no longer enforced.

The grid requires overlap, thus $q = \left\lceil \frac{M^l}{j-2} \right\rceil$ and $r = \left\lceil \frac{N^l}{k-2} \right\rceil$, where M^l and N^l are the rows and columns, respectively, of the data at level $l \in L$.

Compute Disparity Map. Retrieve the estimated disparity map by finding, for each pixel in a block, the disparity that minimizes the sum of the data costs and message values. This process has the same limitations as the message passing for defining the block and grid sizes. However, there is one more piece of memory required (the disparity map), thus Limit (3) becomes $(5n + 1)(2\hat{j} + 1)(2\hat{k} + 1) \leq S$. This reduces the maximum disparity to $n = 90$.

Retrieve Disparity Map. Transport the disparities from the GPU back to the host.

4   Experimental Results

In our experiments we compare high-contrast indoor images (Tsukuba), to real-world images (from HAKA1) which contain various types of noise, such as changes in lighting, out-of-focus lenses, differing exposures, and so forth. So we take two alternative measures to remove this low frequency noise, the first one is using the Sobel edge operator before BP [10], and another method is using the concept of residuals, which is the difference between an image and a smoothed version of itself [16]. Examples of both the Sobel and residual images for real-world images are shown in Figure 5. We use the BP algorithm outlined above with five levels (i.e., $L = 4$) in the Gaussian pyramid, and seven iterations per level.

For the Tsukuba stereo image pair, we have $M = 288$ and $N = 384$; the BP algorithm was run using a data cost of $T_{data} = 15 \cdot 0$, discontinuity cost of $T_{disc} = 1 \cdot 7$, and smoothness parameter of $\lambda = 0 \cdot 07$, and the disparity space runs from 0 to 15 (i.e., $n = 16$). The only other parameters that need to be set explicitly are $\hat{j}, \hat{k} = 3$ (to fit into the shared memory). Figure 3 shows the results of running the implementation on two stereo pairs of images.

The real-world images recorded with HAKA1 are 640×480 ($N \times M$) with an assumed maximum disparity of 32 pixels (i.e., $n = 33$). We use $\hat{j}, \hat{k} = 2$ to fit into shared memory. The BP parameters are $T_{data} = 30 \cdot 0$, $T_{disc} = 11 \cdot 0$, and $\lambda = 0 \cdot 033$. Example results for the HAKA1 sequences, with and without pre-processing, are shown in Figures 3, 5, and 6.

Fig. 5. Top row shows the images (left to right): original, after Sobel processing, mean-residual image. Bottom row shows the disparity results on the images in the top row.

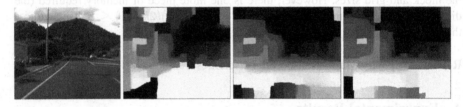

Fig. 6. Left to right: original image and disparity results, first on original images, then Sobel, and finally on residual images

We tested the speed of BP on the NVIDIA Geforce GTX 280 with CUDA compared to the normal PC (outlined in Section 2). The normal CPU implementation runs at 32·42 seconds for the Tsukuba stereo pair, while the running time using CUDA is 0·127 seconds, a speed up by a factor of over 250! The time of BP for real-world images, when running on the normal PC, is 93·98 seconds, compared to 2·75 seconds using the CUDA implementation, and this is (only) a speed improvement by factor 34.

5   Conclusions

We have implemented the belief propagation algorithm (applied to stereo vision) on programmable graphics hardware, which produces fast and accurate results. We have included full details on how to run and implement belief propagation on CUDA. We divided the stereo pairs to a lattice in order to suit the architecture of a GPU. Results for real images are not satisfying as on high-contrast indoor images without pre-processing. Resulting disparity images improve by using either the Sobel edge operator or residual images as input, as suggested in prior work. We have also defined the limitations of this type of BP implementation using limiting inequalities and suggestions for appropriate numbers.

Future work will aim at better speed improvement by exploiting the texture memory. Other ideas for speed improvement are to initialise the BP algorithm with Dynamic Programming Stereo on CUDA, thus speeding up convergence. Finally, the inequalities specified in this paper are ideal for implementation into a linear (quadratic integer) programming optimisation scheme to choose the optimal parameters according to image size.

Acknowledgement

This work is supported by the Natural Science Found of China under No. 60872119.

References

1. CUDA Zone, http://www.nvidia.com/cuda
2. Boykov, Y., Kolmogorov, V.: An experimental comparison of min-cut / max-flow algorithms for energy minimization in vision. IEEE Trans. Pattern Analysis Machine Intelligence 26, 1124–1137 (2004)
3. Brunton, A., Chang, S., Gerhard, R.: Belief Propagation on the GPU for Stereo Vision. In: Proc. Canadian Conf. Computer Robot Vision, p. 76 (2006)
4. .enpeda.. image sequence analysis test site (EISATS), http://www.mi.auckland.ac.nz/EISATS/
5. Felzenszwalb, P.F., Huttenlocher, D.P.: Efficient belief propagation for early vision. Int. J. Computer Vision 70, 41–54 (2006)
6. Fung, J., Mann, S., Aimone, C.: OpenVIDIA: Parallel GPU computer vision. In: Proc. of ACM Multimedia, pp. 849–852 (2005)
7. Fung, J., Mann, S.: Using graphics devices in reverse: GPU-based image processing and computer vision. In: Proc. IEEE Int. Conf. Multimedia Expo., pp. 9–12 (2008)
8. Govindaraju, N.K.: GPUFFTW: High performance GPU-based FFT library. In: Supercomputing (2006)
9. Grauer-Gray, S., Kambhamettu, C., Palaniappan, K.: GPU implementation of belief propagation using CUDA for cloud tracking and reconstruction. In: Proc. PRRS, pp. 1–4 (2008)
10. Guan, S., Klette, R., Woo, Y.W.: Belief propagation for stereo analysis of night-vision sequences. In: Wada, T., Huang, F., Lin, S. (eds.) PSIVT 2009. LNCS, vol. 5414, pp. 932–943. Springer, Heidelberg (2009)
11. Klette, R.: Analysis of data flow for SIMD systems. Acta Cybernetica 6, 389–423 (1984)
12. NVIDIA. NVIDIA CUDA Programming Guide Version 2.1 (2008), http://www.nvidia.com/object/cuda_develop.html
13. Scharstein, D., Szeliski, R.: A taxonomy and evaluation of dense two-frame stereo correspondence algorithms. Int. J. Computer Vision 47, 7–42 (2002)
14. Sinha, S.N., Frahm, J.M., Pollefeys, M., Genc, Y.: Feature tracking and matching in video using graphics hardware. In: Proc. Machine Vision and Applications (2006)
15. Vineet, V., Narayanan, P.J.: CUDA cuts: Fast graph cuts on the GPU. In: CVPR Workshop on Visual Computer Vision on GPUs (2008)
16. Vaudrey, T., Klette, R.: Residual Images Remove Illumination Artifacts for Correspondence Algorithms! In: Proc. DAGM (to appear, 2009)
17. Yang, Q., Wang, L., Yang, R., Wang, S., Liao, M., Nistér, D.: Real-time global stereo matching using hierarchical belief propagation. In: Proc. British Machine Vision Conf., pp. 989–998 (2006)

Face Image Enhancement
via Principal Component Analysis

Deqiang Yang[1], Tianwei Xu[1], Rongfang Yang[1], and Wanquan Liu[2]

[1] College of Computer Science and Information Technology, Yunnan Normal University,
Kunming, 650092, P.R. China
[2] Department of Computing, Curtin University of Technology, Perth, 6002, Australia
qdy75@163.com, w.liu@curtin.edu.au

Abstract. This paper investigates face image enhancement based on the principal component analysis (PCA). We first construct two types of training samples: one consists of some high-resolution face images, and the other includes the low resolution images obtained via smoothed and down-sampling process from the first set. These two corresponding sets form two different image spaces with different resolutions. Second, utilizing the PCA, we obtain two eigenvector sets which form the vector basis for the high resolution space and the low resolution space, and a unique relationship between them is revealed. We propose the algorithm as follows: first project the low resolution inquiry image onto the low resolution image space and produce a coefficient vector, then a super-resolution image is reconstructed via utilizing the basis vector of high-resolution image space with the obtained coefficients. This method improves the visual effect significantly; the corresponding PSNR is much larger than other existing methods.

Keywords: Image enhancement, Principal component analysis (PCA), Hallucinating face.

1 Introduction

Image enhancement has many applications in computer vision, such as blurred image restoration, and the enhancement of low resolution images. Though many techniques for face enhancement have been proposed, many realistic related problems have not been solved satisfactorily. When a digital picture is enlarged many times, blur and the mosaic phenomena often exist; In case of wireless/remote surveillance, the resolution of obtained video is usually quite low due to limited bandwidth requirement in transmission or large data real time transmission, and thus details for the people or object are not clear. In order to solve these problems, as early as 1960s, some researchers proposed some techniques for super-resolution reconstruction. The idea of image super-resolution reconstruction is to reconstruct high-resolution (HR) images from a group of single or multi-frame low-resolution (LR) images. In terms of pixels, the size of HR image is larger than LR image, and the enhance techniques can magnify

A. Nicholson and X. Li (Eds.): AI 2009, LNAI 5866, pp. 190–198, 2009.
© Springer-Verlag Berlin Heidelberg 2009

image and increase image details, making the magnified image close to the origin HR image. This is an important problem in image processing domain, and it has become a popular research topic due to its broad applications in face recognition and surveillance.

The enhancement can be done through interpolation, such as neighbor interpolation, Cubic interpolation etc [1], but its effect is very limited. In order to improve the enhancement effect, many algorithms have been proposed. For example, Schultz *et al.* proposed the Bayesian method [2], Freeman proposed the Markov network method [3], and the method based on neural networks [4]. Inspired by some techniques in machine learning, some techniques based on training samples are proposed recently. In [5] and [6], researchers proposed the famous hallucinating face algorithm. However, this hallucinating face algorithm is only suitable to a single human face image super-resolution reconstruction, in which the training samples include a group of HR face images and their corresponding down-sampling LR images, and in contrast, the candidate face image use the LR samples feature set, the LR sample regions and their corresponding HR sample region's high-frequency information with the smallest feature distance LR sample region to reconstruct the super-resolution face image. Later, researchers proposed a series of algorithms to improve the hallucinating face algorithm based on Baker's theory [7] [8] [9]. A review is conducted in [13]. Most of these algorithms are based on statistical models with local model parameters, and they only improve the quality of high resolution image in a limited extent. The reconstruction effect is not significant when the relationship among the test sample and training samples is not well described. Also the down-sampling function of training images is only by virtue of experience and difficult to identify in practical applications. The proposed algorithms generally require a larger number of training samples, and thus the speed of reconstruction is slow relatively. These human faces constitute a vector space with a certain law after positioning and normalization by taking the similarity of facial structure and appearance into consideration. It is assumed that images with the same resolution in the same vector space, thus, HR and LR samples databases constitute two different vector spaces.

In this paper we design two training image databases by two different resolution face image respectively, one is composed of high resolution face image, known as the HR sample database; the other is called LR sample database, obtained by samples from HR after smoothing and down-sampling. Images in these two libraries have one-to-one relationship based on PCA. We make PCA transform to HR space and LR respectively, to obtain two sets of different orthogonal base vectors, because images in the two spaces have one-one relationship, and it is natural to assume that their basis vectors also have one-one relationship. Projecting the low resolution candidate face image in LR space basis, we can obtain a set of expansion coefficients. By using these coefficients, one can obtain the high resolution face obtained by linear combination of base vectors in HR space.

The rest of this paper is organized as follows. Section 2 presents the theoretical analysis. Section 3 describes the super-resolution reconstruction algorithm based on Principal Component Analysis transform in detail. Experimental results and discussion are drawn in section 4.

2 Theoretical Analysis

Some researchers proved that the relation between intensity distribution and location of human face image satisfies Markov random fields (MRFs) [3], and the pixel value depends on its neighboring pixels. The literature [5] [6] and related image multi-resolution theory suggest that, HR images and their corresponding LR images are closely related, and HR images can be obtained by LR images. Also every LR image has its corresponding HR image. Studies have shown that human faces with the same resolution have a super high-dimensional image subspace [10] after location and normalization, when image samples are up to a certain amount, Extremely most human faces can be expressed by their generating basis of the corresponding vector space. Penev and Sirovich's study found that human face can obtain high quality reconstruction by using PCA method. Different resolution face images are in different image space [11] [12], making PCA transform to one resolution face images is equivalent to look for a set of standard orthogonal basis, which can express all images in this space. Though image features expressed by these basis vectors are irrelevant to each other, the information they contained relates to their corresponding characteristic value. Based on these observations, the PCA projection coefficients can be obtained through projecting image on the basis, and then they can be used for high resolution image reconstruction. In fact, the procedure of down sampling is as follows.

$$L = C_{(\downarrow n)}(H) \tag{1}$$

$$C(H)(x, y) = \frac{1}{9} \sum_{i=-1}^{1} \sum_{j=-1}^{1} H(2 * x + i, 2 * y + j) \tag{2}$$

In this paper we adapt equation (1) and (2) to acquire LR training sample images, where H represents high-resolution images, and L represents low-resolution images obtained by H, operator C is formed by smoothing and down-sampling to H. We use (2) to do down-sampling; a new pixel value will be inferred using mean of its 3×3 neighborhood. Further we can also use the overlapped down-sampling to obtain a smoother LR image.

It is noted that each orthogonal basis vector in human face image space obtained by PCA transform expresses an image feature. The basis vector is also called "feature face", which represents face information in certain frequency band. According to multi-resolution analysis theory, there should be the same low-frequency information between HR image and its corresponding LR image [13] and this motivates us to use the projection coefficient to represent the relationship in HR and LR face space. Experiments also indicate that the PCA projection coefficients for the high-frequency image in HR space are nearly the same to the PCA projection coefficients for the down-sampled low-frequency image.

With the above analysis, the images based on two image space have one-to-one relationship. Assume that their feature vectors also have one-to-one relationship, a feature vector in HR space is corresponding to one in LR space, and both of them contain more low-frequency information. But the feature vectors in HR space contain more high-frequency information, so the details of facial characteristic are clearer. We use the mentioned idea to design a new image enhance technique in next section.

3 The Proposed Face Enhancement Algorithm

In order to describe the algorithm explicitly, we define the following notations. Assume that M represents the number of training samples, H_i ($1 \le i \le M$)   is the column vector for a high resolution human face image i, and the column vector of its corresponding low-resolution face images is represented by L_i. Let μ_h and μ_l represent the mean value of HR image and LR image respectively. Also let I_{in} represent low resolution candidate image, and I_{out} represent reconstructed high resolution image.

In LR image space, we take the column vector of each face image as a random vector L_i, the algorithm for enhancement using PCA method is organized as follows. First, we calculate the mean value of image vector for LR images:

$$\mu_l = \frac{1}{M}\sum_{i=1}^{M} L_i \tag{3}$$

Let L represent the difference set, which is the difference between each image column vector and the mean vector.

$$L = [L_1 - \mu_l, L_2 - \mu_l, \quad L_M - ì_l] \tag{4}$$

The covariance matrix S is constructed as follows:

$$S = \frac{1}{M}\sum_{i=1}^{M}(L_i - ì_l)(L_i - ì_l)T = \frac{1}{M}LL^T \tag{5}$$

The dimension of this matrix is generally very large. If the size of image is N×N, then the dimensions of L and S are N^2×M and N^2×N^2 respectively. In practice, it is difficult to calculate their eigen-values and eigenvectors directly. Instead we use the singular value decomposition approach to calculate these values. Then we construct a matrix R as follows:

$$R = L^T L \tag{6}$$

The dimension of R is M x M, and in general M is much smaller than N. It is much easy to calculate the eigen-values and eigenvectors for R. let λ_i and v_i (i=1,2 ,....,M) represent its eigen-value and eigenvector after orthogonal normalization. In this case, the orthogonal normalized eigenvector of the scatter matrix S can be calculated by equation (7):

$$u_i = \frac{1}{\sqrt{ê_i}}Lv_i, \ i = 1,2, ... M \tag{7}$$

Let each orthogonal normalized eigenvector be a basis vector in LR feature subspace, and, $U_l = [u_1, u_2, \cdots, u_M]$ will be the feature space.

Assume that a set of face samples with same resolution constitute a subspace space and such subspace will represent all face samples if the number of samples is large enough and should include feature details of each face. Finding a set of basis for such subspace and then we expect that all of the faces could be expressed by linear combination of such basis vectors. In this paper we use the PCA to obtain a set of basis vectors for such face space. Such basis vector should also represent a coordinate vector of face in high-dimensional space, because the eigenvector obtained by the PCA transform is orthogonal, these basis vectors are irrelevant to each other in high-dimensional space. Further, each basis vector represents an aspect characteristic of face image; the corresponding coordinate coefficients can be obtained by projecting face on the set of basis vectors.

Next we project the low resolution candidate face I_{in} image on the subspace U_l:

$$W = U_i^T * I_{in} \tag{8}$$

We define $W = [w_1, w_2, \cdots, w_M]$, where w_i (i=1,2, ... ,M) is a set of projection coefficients from (8). The value of w_i represents the weighting of input image I_{in} on basis u_i, i.e., image includes characteristics ratio expressed by u_i. The reconstructed image could be derived from (9) by using these coefficients in LR image subspace.

$$I_{out} = U_i * W + \mathfrak{i}_i \tag{9}$$

where W is coordinate coefficient of input image obtained by projection in LR space, and μ_l is the mean of all faces in LR space. However, $U_l * W$ represent personality characteristic of image. Face image reconstructed using this method is consisting of faces in common in terms μ_l and personality characteristic in terms of $U_l * W$.

The images reconstructed using equation (9) have the same resolution with images in LR space, They are actually without resolution enhancement, and are another expression of images in LR space. Considering the preceding analysis, we noted that the samples in LR space are obtained by smoothing and down-sampling from high resolution images in HR sample space, and there is one-to-one relationship between them. Therefore, their basis vectors also have one-to-one relationship. Actually that the corresponding images in two sample space have similar projection coefficients. In this case, we can use the basis vectors in HR space to substitute the corresponding basis vector in LR space, and use the mean vector in HR space to substitute the mean vector in LR space. Then, we can reconstruct the high-resolution image. Both of basis vectors and the mean vectors in HR space contain more high-frequency information, and more characteristic details of face, so that we can obtain the high-frequency image I_{out} reconstructed through low resolution input images I_{in} in high-resolution space.

$$I_{out} = U_h * W + \mathfrak{i}_h \tag{10}$$

where the feature subspace U_h is formed by basis vectors in HR space, we define $U_h = [u_1, u_2, \cdots, u_M]$, and μ_h represents the mean of HR images. Now the proposed algorithm can be summarized as follows:

Step 1: Smooth and down-sample the images in HR sample space using equation (1) (2), to obtain LR samples.

Step 2: Do PCA transform on LR sample set, and obtain a set of orthogonal basis vectors for LR feature space, which constitute the characteristics subspace of LR space.

Step 3: Project input image I_{in} on the characteristics subspace of LR space U_l, and obtain the projection coefficients of this image in the low-frequency characteristics subspace.

Step 4: Do PCA transform for the HP sample set, and obtain a set of orthogonal basis vectors for HR space, which constitute the characteristics subspace of HR space.

Step 5: Use the projection coefficients obtained in **Step 3** and the feature basis vectors for HR space to reconstruct the high resolution image for the input image via equation (10).

Next we will do some experiments to show the effectiveness of the proposed algorithm.

4   Experimental Results and Analysis

In this section, we conduct experiments using the face database in the high-speed image processing lab in Department of Electronic Engineering at Tsinghua University. In such database, there are more than 1000 face images of different age, illumination conditions and facial expressions. The HR samples consist of high-resolution face images with size 96×128. The corresponding LR sample databases are generated by smoothing and down-sampling using formula (1) and (2), with the reduced size of LR image being 24×32. Two experimental results will be discussed as follows:

In order to compare the performances of different algorithms, we choose the neighbor interpolation algorithm, the cubic interpolation approach, the Baker Hallucinating face approach, and the proposed method in this paper respectively. Some enhanced images are shown in Figure 1. The neighbor interpolation and cubic interpolation algorithms are relatively stable since they do not need any training samples, and do optimal interpolations directly on the input image. They are also faster in implementation, but have very limited effect. Baker proposed the hallucinating algorithm in [6], in which he compares the image features directly and use the typical patch for enhancement. The reconstruction result is closely related to training samples, and the final quality is also related to the down-sampling function on some extent.

In the experiments, 600 images are used as training samples, while the remaining 400 are used for testing. The 600 high-resolution images and the corresponding low-resolution images are composed of HR samples space and LR samples space respectively. We will reconstruct 400 high resolution images for their low resolution images using the proposed algorithm in this paper.

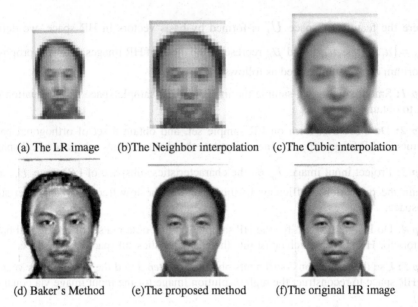

| (a) The LR image | (b)The Neighbor interpolation | (c)The Cubic interpolation |

| (d) Baker's Method | (e)The proposed method | (f)The original HR image |

Fig. 1. Comparisons of different methods (Fig a 24×32, Fig b--f :96×128; Number of training examples:600)

We can see from Figure 1 that the Baker's method, which is patch based, can produce local details of the original image, but have block blur effect. The results of our method are much smoother with more characteristic details, and thus the enhanced images look much closer to the origin HR image. In order to compare the performance numerically instead of visually, we use the common peak signal-noise ratio (PSNR) value for the reconstructed image, and it can be calculated below.

$$PSNR = 10LOG\left(\frac{255^2}{\frac{1}{M*N}\sum_{x=0}^{M-1}\sum_{y=0}^{N-1}\left[I_{out}(x,y)-I_H(x,y)\right]^2}\right).$$

The PSNR values for different algorithms are listed in Table 1 and it can be seen that the proposed algorithm can produce much better results.

Table 1. PSNR Values for different methods (Unit: dB)

methods	Neighbor interpolation	Cubic interpolation	Baker's method	Our method
PSNR	23.1131	22.6393	25.3287	28.2536

In the proposed algorithm, the number of training samples will have impact on the enhancement results. In order to evaluate such impact, we design an experiment, using different number of training samples with HR 96 × 128 pixels with 400 images with 24 × 32, to reconstruct them into HR 96×128 pixels. Figure 2 shows results of a test image under different number of training samples. According to visual effect, the result is quite satisfactory and reconstructed image's quality is very stable when the number of training samples is more than 300.

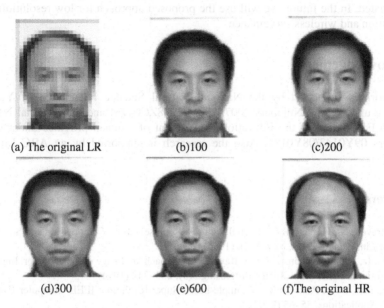

(a) The original LR   (b)100   (c)200

(d)300   (e)600   (f)The original HR

Fig. 2. Reconstructed image with different number of training examples

Table 2. Comparison of time and RMS values for different number of training samples

Number of training samples	50	100	200	300	400	500	600
RMS	18.4	15.0	13.7	12.6	12.3	11.9	11.6
TIME(second)	0.62	1.09	2.13	2.69	3.18	3.74	4.02

Different number of training samples will have an impact on the restoration time and quality in terms of Root Mean Square (RMS), and we do the experiment on a machine (Inter Core 2 CPU and 2.13GHZ) with different training samples and the results are shown in Table 2. We can see that, after number of training samples is above 300, and the values of RMS are in the vicinity of 12. But with the increase of training samples, the time required for the algorithm is enlarged. The formula for RMS is given below.

$$RMS = \frac{1}{M*N} \sum_{x=0}^{M-1} \sum_{y=0}^{N-1} \left\| I_H(x, y) - I_{out}(x, y) \right\| \;.$$

5   Conclusions

In this paper, we proposed a new method for face image enhancement based on PCA transform. The enhanced images obtained by this method are compared with different methods, such as the neighbor interpolation, the Cubic interpolation and the Baker Hallucinating face.

Experimental results show significant improvement both visually and numerically. Also the impact of the number of training samples on the performance is also investigated. In the future, we will use the proposed approach for low resolution face recognition and wireless surveillance.

Acknowledgement

This work is supported by the National Natural Science Foundation of Yunnan province under Grant Numbers 2007F202M/2008ZC047M and the National Natural Science Foundation of the Education Department of Yunnan province under Grant Numbers 09Y0143/08Y0131. Also the research is supported by an ARC Linkage Grant.

References

1. Netravali, N., Haskell, B.G.: Digital Pictures: Representation, Compression and Standards, 2nd edn. Plenum Press, New York (1995)
2. Schultz, R., Stevenson, R.L.: A Bayesian Approach to Image Expansion for Improved Definition. IEEE Trans. Image Processing 3(3), 233–242 (1994)
3. Freeman, W.T., Paztor, E.C.: Example-based Super-resolution. IEEE Computer Graphics and Applications, 55–65 (2002)
4. Frank, M., Candocia, Jose, C.: Principe: Super- Resolution of Images Based on Local Correlation. IEEE Transactions on Neural Networks 10(2), 372–380 (1999)
5. Baker, S., Kanade, T.: Limits on Super-Resolution and How to Break them. IEEE Trans. on PAMI 24(9), 1167–1183 (2002)
6. Baker, S., Kanade, T.: Hallucinating Faces. In: Proc. of IEEE Inter. Conf. on Automatic Face and Gesture Recognition, pp. 83–88 (2003)
7. Liu, W., Lin, D.H., Tang, X.O.: Hallucinating faces: Tensor patch super-resolution and coupled residue compensation. In: Proceedings of the IEEE Computer Society Conference on Computer Vision and Pattern Recognition, vol. 2, pp. 478–484 (2005)
8. Zhuang, Y.T., Zhang, J., Wu, F.: Hallucinating faces: LPH super-resolution and neighbor reconstruction for residue compensation. Pattern Recognition 40, 3178–3194 (2007)
9. Zhao, W., Chellappa, R., Phillips, P.J., Rosenfeld, A.: Face recognition: A literature survey. ACM Computing Surveys 35(4), 399–458 (2003)
10. Zhao, W., Chellapa, W., Philips, P.: Subspace linear discriminate analysis for face recognition. Technical report. CAR-TR-914 (1996)
11. Wang, X.G., Tang, X.O.: Hallucinating face by eigen transformation. IEEE Trans. Syst. Man Cybern. 35(3), 425–434 (2005)
12. Gonzalez, R.C., Woods, R.G.: Digital Image Processing, 2nd edn. Prentice-Hall, Englewood Cliffs (2002)
13. Park, S.C., Park, M.K., Kang, M.G.: Super-resolution image reconstruction: a technical overview. IEEE Signal Processing Magazine (2003)

On Using Adaptive Binary Search Trees to Enhance Self Organizing Maps

César A. Astudillo[1] and B. John Oommen[2]

[1] Universidad de Talca, Curicó, Chile
castudillo@utalca.cl
[2] Carleton University, Ottawa, Canada
oommen@scs.carleton.ca

Abstract. We present a strategy by which a Self-Organizing Map (SOM) with an underlying Binary Search Tree (BST) structure can be adaptively re-structured using conditional rotations. These rotations on the nodes of the tree are *local* and are performed in *constant time*, guaranteeing a decrease in the Weighted Path Length (WPL) of the entire tree. As a result, the algorithm, referred to as the Tree-based Topology-Oriented SOM with Conditional Rotations (TTO-CONROT), converges in such a manner that the neurons are ultimately placed in the input space so as to represent its stochastic distribution, and additionally, the neighborhood properties of the neurons suit the best BST that represents the data.

1 Introduction

Even though numerous researchers have focused on deriving variants of the original Self-Organizing Map (SOM) strategy, few of the reported results possess the ability of modifying the underlying topology, leading to a dynamic modification of the structure of the network by adding and/or deleting nodes and their inter-connections. Moreover, only a small set of strategies use a tree as their underlying data structure [1,2,3,4]. From our perspective, we believe that it is also possible to gain a better understanding of the unknown data distribution by performing *structural* tree-based modifications on the tree, by rotating the nodes within the Binary Search Tree (BST) that holds the whole structure of neurons. Thus, we attempt to use rotations, tree-based neighbors *and* the feature space as an effort to enhance the capabilities of the SOM by representing the underlying data distribution and its structure more accurately. Furthermore, as a long term ambition, this might be useful for the design of faster methods for locating the Best Matching Unit (BMU).

One of the primary goals of Adaptive Data Structures (ADS) is to seek an optimal arrangement of the elements, by automatically reorganizing the structure itself so as to reduce the average access time. The solution to obtain the *optimal* BST is well known when the access probabilities of the nodes are known beforehand [5]. However, our research concentrates on the case when these access probabilities are *not known a priori*. In this setting, the most effective solution

A. Nicholson and X. Li (Eds.): AI 2009, LNAI 5866, pp. 199–209, 2009.
© Springer-Verlag Berlin Heidelberg 2009

is due to Cheetham *et al.* and uses the concept of conditional rotations [6]. The latter paper proposed a philosophy where an accessed element is rotated towards the root if and only if the overall Weighted Path Length (WPL) of the resulting BST decreases.

The strategy that we are presently proposing, namely the Tree-based Topology-Oriented SOM with Conditional Rotations (TTO-CONROT), has a set of neurons, which, like all SOM-based methods, represents the data space in a condensed manner. Secondly, it possesses a connection between the neurons, where the neighbors are based on a learned nearness measure that is tree-based. Similar to the reported families of SOMs, a subset of neurons closest to the BMU are moved towards the sample point using a vector-quantization (VQ) rule. However, unlike most of the reported families of SOMs, the identity of the neurons that are moved is based on the tree-based proximity (and not on the feature-space proximity). Finally, the TTO-CONROT incorporates tree-based mutating operations, namely the above-mentioned conditional rotations.

Our proposed strategy is adaptive, with regard to the migration of the points *and* with regard to the identity of the neurons moved. Additionally, the distribution of the neurons in the feature space mimics the distribution of the sample points. Lastly, by virtue of the conditional rotations, it turns out that the entire tree of neurons is optimized with regard to the overall accesses, which is a unique phenomenon (when compared to the reported family of SOMs) as far as we know.

The contributions of the paper can be summarized as follows. First, we present an integration of the fields of SOMs and ADS. Secondly, the neurons of the SOM are linked together using an underlying tree-based data structure, and they are governed by the laws of the Tree-based Topology-Oriented SOM (TTOSOM) paradigm, and simultaneously by the restructuring adaptation provided by conditional rotations (CONROT). Third, the adaptive nature of TTO-CONROT is unique because adaptation is perceived in two forms: The migration of the codebook vectors in the feature space is a consequence of the SOM update rule, and the rearrangement of the neurons within the tree as a result of the rotations. Finally, we explain how the set of neurons in the proximity of the BMU is affected as a result of applying the rotations on the BST.

The rest of the paper is organized as follows. The next section surveys the relevant literature, which involves both the field of SOMs including their tree-based instantiations, and the respective field of BSTs with conditional rotations. After that, in Section 3, we provide an in-depth explanation of the TTO-CONROT philosophy, which is our primary contribution. The subsequent section shows the capabilities of the approach through a series of experiments, and finally, Section 5 concludes the paper.

2   Literature Review

A number of variants of the original SOM [7] have been presented through the years, attempting to render the topology more flexible, so as to represent complicated data distributions in a better way and/or to make the process faster by, for instance, speeding up the search of the BMU. We focus our attention on a specific family of enhancements in which the neurons are inter-connected using a tree topology [1,2,3,4]. In [1] the authors presented a tree-based SOM called the Growing Hierarchical SOM (GHSOM), in which each node corresponds to an independent SOM, and where dynamic behavior is manifested by adding rows or columns to each SOM depending on a suitable criterion. The authors of [2] have studied a variant of the SOM called the Self-Organizing Tree Map (SOTM), which also utilizes a tree-based arrangement of the neurons, and which uses the distance in the *feature* space to determine the BMU. In [3] the authors proposed a tree-structured neural network called the evolving tree (ET), which takes advantage of a sub-optimal procedure to determine the BMU in $O(\log |V|)$ time, where V is the set of neurons. The ET adds neurons dynamically, and incorporates the concept of a "frozen" neuron, which is a non-leaf node that does not participate in the training process, and which is thus removed from the Bubble of Activity (BoA).

Here, we focus on the TTOSOM [4]. The TTOSOM incorporates the SOM with a tree which has an arbitrary number of children. Furthermore, it is assumed that the user has the ability to describe such a tree, reflecting the *a priori* knowledge about the *structure* of the data distribution[1]. The TTOSOM also possesses a BoA with particular properties, considering the distance in the tree space, where leaves and non-leaf nodes are part of this neighborhood. Another interesting property displayed by the TTOSOM is its ability to reproduce the results obtained by Kohonen [7], when the nodes of the SOM are arranged linearly, i.e., in a list. In this case, the TTOSOM is able to adapt this 1-dimensional grid to a 2-dimensional (or multi-dimensional) object in the same way as the SOM algorithm did [4]. Additionally, if the original topology of the tree followed the overall shape of the data distribution, the results reported in [4] showed that it is also possible to obtain a symmetric topology for the codebook vectors.

We shall now proceed to describe the corresponding relevant work in the field of the tree-based ADS. A BST may be used to store records whose keys are members of an ordered set. In this paper, we are in the domain where the access probability vector is not known *a priori*. We seek a scheme which dynamically rearranges itself and asymptotically generates a tree which minimizes the access cost of the keys.

The primitive tree restructuring operation used in most BST schemes is the well known operation of Rotation [8]. A few memory-less tree reorganizing schemes[2] which use this operation have been presented in the literature. In the Move-to-Root Heuristic [12], each time a record is accessed, rotations are performed on it in an upwards direction until it becomes the root of the tree. On the other hand, the simple Exchange rule [12] rotates the accessed element one level towards the root. Sleator and Tarjan [13] introduced a technique, which also moves the accessed record up to the root of the tree using a restructuring operation. Their

---

[1] The beauty of such an arrangement is that the data can be represented in multiple ways depending on the specific perspective of the user.

[2] This review is necessary brief. A more detailed version is found in [9,10,11].

structure, called the Splay Tree, was shown to have an amortized time complexity of $O(\log N)$ for a complete set of tree operations. The literature also records various schemes which adaptively restructure the tree with the aid of additional memory locations. Prominent among them is the Monotonic Tree [14] and Mehlhorn's D-Tree [15]. In spite of all their advantages, all of the schemes mentioned above have drawbacks, some of which are more serious than others.

This paper focuses on a particular heuristic, namely, the Conditional Rotations (CONROT-BST) [6], which has been shown to reorganize a BST so as to asymptotically arrive at an optimal form. CONROT-BST only requires the maintenance and processing of the values stored at a specific node and its direct neighbors, i.e. its parent and both children, if they exist. Algorithm 1, formally given below, describes the process of the Conditional Rotation for a BST. The algorithm receives two parameters, the first of which corresponds to a pointer to the root of the tree, and the second which corresponds to the key to be searched (assumed to be present in the tree).

Algorithm 1. CONROT-BST(j,k_i)

Input:
i) j, A pointer to the root of a binary search tree T
ii) k_i, A search key, assumed to be in T
Output:
i) The restructured tree T'
ii) A pointer to the record i containing k_i
Method:
1: $\tau_j \leftarrow \tau_j + 1$
2: **if** $k_i = k_j$ **then**
3:     **if** is-left-child(j) = TRUE **then**
4:         $\Psi_j \leftarrow 2\tau_j - \tau_{jR} - \tau_{P(j)}$
5:     **else**
6:         $\Psi_j \leftarrow 2\tau_j - \tau_{jL} - \tau_{P(j)}$
7:     **end if**
8:     **if** $\Psi_j > 0$ **then**
9:         rotate-upwards(j)
10:         recalculate-tau(j)
11:         recalculate-tau$(P(j))$
12:     **end if**
13:     **return** record j
14: **else**
15:     **if** $k_i < k_j$ **then**
16:         CONROT-BST(left-child(j) , k_i)
17:     **else**
18:         CONROT-BST(right-child(j) , k_i)
19:     **end if**
20: **end if**
End Algorithm

We define $\tau_i(n)$ as the total number of accesses to the subtree rooted at node i. CONROT-BST computes the following equation to determine the value of a quantity referred to as Ψ_j, for a particular node j, where:

$$\Psi_j = \begin{cases} 2\tau_j - \tau_{jR} - \tau_{P(j)} & \text{if } j \text{ is a left child of } P(j) \\ 2\tau_j - \tau_{jL} - \tau_{P(j)} & \text{if } j \text{ is a right child of } P(j) \end{cases} \tag{1}$$

When Ψ_j is less than zero, an upward rotation is performed. The authors of [6] have shown that this single rotation yields to a decrease in the WPL of the *entire* tree. Once the rotation takes place, it is necessary to update the corresponding counters, τ. Fortunately this task only involve the updating of τ_i, for the rotated node, and the counter of its parent, $\tau_{P(i)}$. The reader will observe that all the tasks invoked in the algorithm are performed in constant time, and in the worst case, the recursive call is done from the root down to the leaves, leading to a $O(h)$ running complexity, where h is the height of the tree.

3   Merging ADS and TTOSOM

This section concentrates on the details of the integration between the fields of ADS and the SOM. More specifically we shall concentrate on the integration of the CONROT-BST heuristic [6] into a TTOSOM [4], both of which were explained in the previous section. We thus obtain a new species of tree-based SOMs which is self-arranged by performing rotations **conditionally, locally** and in a **constant number of steps**.

As in the case of the TTOSOM [4], the *Neural Distance*, d_N, between two neurons is defined as the minimum number of edges required to go from one to the other. Note however, that in the case of the TTOSOM, since the tree itself was *static*, the inter-node distances can be pre-computed *a priori*, simplifying the computational process. The TTO-CONROT employs a tree which is dynamically modified, where the structure of the tree itself could change, implying that nodes that were neighbors at any time instant may not continue to be neighbors at the next. This renders the resultant SOM to be unique and distinct from the state-of-the-art.

Fig. 1 presents the scenario when the node accessed is B. Observe that the distances are depicted with dotted arrows, with an adjacent numeric index specifying the current distance from node B. Fig. 1a illustrates the situation prior to an access, where nodes H, C and E are all at a distance of 2 from node B, even though they are at different levels in the tree. Secondly, Fig. 1b depicts the configuration of the tree after the rotation is performed. At this time instant, C and E are both at distance of 3 from B, which means that they have increased their distance to B by unity. Moreover, although node H has changed its position, its distance to B remains unmodified. Clearly, the original distances are not necessarily preserved as a consequence of the rotation.

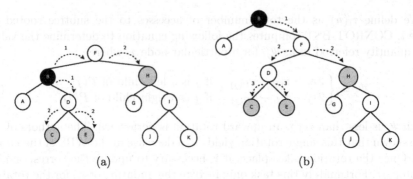

Fig. 1. Example of the Neural Distance before and after a rotation

A concept closely related to the neural distance, is the one referred to as the "Bubble of Activity" (BoA) which is the subset of nodes within a distance of r away from the node currently examined. The BoA can be formally defined as [4]

$$B(v_i; T, r) = \{v | d_N(v_i, v; T) \le r\}, \tag{2}$$

where v_i is the node currently being examined, and v is an arbitrary node in the tree T, whose nodes are V.

Fig. 2 depicts how the BoA differs from the one defined by the TTOSOM as a result of applying a rotation. Fig. 2a shows the BoA around the node B, using the same configuration of the tree as in Fig. 1a, i.e., before the rotation takes place. Here, the BoA when $r = 1$ involves the nodes $\{B, A, D, F\}$, and when $r = 2$ the nodes contained in the bubble are $\{B, A, D, F, C, E, H\}$. Subsequently, considering a radius equal to 3, the resulting BoA contains the nodes $\{B, A, D, F, C, E, H, G, I\}$. Finally, the $r = 4$ case leads to a BoA which includes the whole set of nodes. Now, observe the case presented in Fig. 2b, which corresponds to the BoA around B *after* the rotation upwards has been effected, i.e. the same configuration of the tree used in Fig. 1b. In this case, when the radius is unity, nodes $\{B, A, F\}$ are the *only* nodes within the bubble, which is different from the corresponding bubble before the rotation is invoked. Similarly, when $r = 2$, we obtain a set different from the analogous pre-rotation case, which in

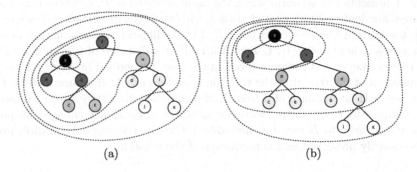

Fig. 2. Example of BoA before and after a rotation is invoked at node B

this case is $\{B, A, F, D, H\}$. Note that coincidentally, for the case of a radius equal to 3, the bubbles are identical before and after the rotation, i.e., they invoke the nodes $\{B, A, D, F, G, I\}$. Trivially, again, when $r = 4$, the BoA invokes the entire tree.

The CONROT-BST heuristic [6] requires that the tree should possess the BST property:

Let x be a node in a BST. If y is a node in the left subtree of x, then $key[y] \leq key[x]$. Further, if y is a node in the right subtree of x, then $key[x] \leq key[y]$.

To satisfy the BST property, first of all we see that, the tree must be binary[3]. The tree trained by the TTOSOM is restricted to contain at most two children per node and a comparison operator between the two children is considered. This comparison can be achieved by defining a unique key that must be maintained for each node in the tree, and which will, in turn, allow a lexicographical arrangement of the nodes.

It happens that the concept of the "just accessed" node in the CONROT-BST is compatible with the corresponding BMU defined for the Competitive Learning (CL) model. During the training phase, when a neuron is a frequent winner of the CL, it gains prominence in the sense that it can represent more points from the original data set. We propose that during the training phase, we can verify if it is worth modifying the configuration of the tree by moving this neuron one level up towards the root as per the CONROT-BST algorithm, and consequently explicitly recording the relevant role of the particular node with respect to its nearby neurons. CONROT-BST achieves this by performing a *local* movement of the node, where only its direct parent and children are aware of the neuron promotion.

Neural Promotion is the process by which a neuron is relocated in a more privileged position[4] in the network with respect to the other neurons in the neural network. Thus, while "all neurons are born equal", their importance in the society of neurons is determined by what they represent. This is achieved, by an explicit advancement of its rank or position.

Initialization, in the case of the BST-based TTOSOM, is accomplished in two main steps which involve defining the initial value of each neuron and the connections among them. The neurons can assume a starting value arbitrarily, for instance, by placing them on randomly selected input samples. On the other hand, a major enhancement with respect to the basic TTOSOM lays in the way the neurons are linked together. The inclusion of the rotations renders this dynamic.

In our proposed approach, the codebooks of the SOM correspond to the nodes of a BST. Apart from the information regarding the codebooks themselves, each neuron requires the maintenance of additional fields to achieve the adaptation.

---

[3] Of course, this is a severe constraint. But we are forced to require this, because the phenomenon of achieving conditional rotations for arbitrary k-ary trees is unsolved. This research, however, is currently being undertaken.

[4] As far as we know, we are not aware of any research which deals with the issue of Neural Promotion. Thus, we believe that this concept, itself, is pioneering.

Also, besides the codebook vectors, each node inherits the properties of a BST Node, and it thus includes a pointer to the left and right children, as well as (to make the implementation easier), a pointer to its parent. Each node also contains a label which is able to uniquely identify the neuron when it is in the "company" of other neurons. This identification index constitutes the lexicographical key used to sort the nodes of the tree and remains static as time proceeds.

The training module of the TTO-CONROT is responsible for determining the BMU, performing restructuring, calculating the BoA and migrating the neurons within the BoA. Basically, what it achieves, is to integrate the CONROT algorithm in the sequence of steps of the Training phase defined by the TTOSOM. Algorithm 2 describes the details of how this integration is fulfilled. Algorithm 2 receives as input a sample point, x, and the pointer to the root of the tree, p. Line No. 1, performs the first task of the algorithm, which involves the determination of the BMU. After that, line No. 2, deals with the call to the CONROT algorithm. The reason why we follow this sequence of steps is that the parameters needed to perform the conditional rotation, as specified in [6], includes the key of the element queried, which, in the present context, corresponds to the key of the BMU. At this stage of the algorithm, the BMU may be rotated or not, and the BoA is determined *after* this restructuring process, which is performed in lines No. 3 and 4 of the algorithm. Finally, lines No. 5 to 7, are responsible for the neural migration, involving the movement of the neurons within the BoA towards the input sample.

Algorithm 2. TTO-CONROT-BST_train(x,p)

Input:
i) x, a sample signal.
ii) p, the pointer to the tree.
Method:
1: $v \leftarrow$ TTOSOM_Find_BMU(p,x,p)
2: cond-rot-bst(p,v.getID())
3: $B \leftarrow \{v\}$
4: TTOSOM_Calculate_Neighborhood($B,v,radius$)
5: **for all** $b \in B$ **do**
6:     update_rule(b.getCodebook(),x)
7: **end for**
End Algorithm

Even though, we have used the advantages of the CONROT algorithm, the architecture that we are proposing allows us to to utilize an alternative restructuring module. Candidates which can be used to perform the adaptation are the ones mentioned in Section 2, and include the splay and the monotonic-tree algorithms, among others [11].

4   Experimental Results

To illustrate the capabilities of our method, the experiments reported in the present work are limited to the two-dimensional feature space. However, it is important to remark that the algorithm is also capable of solving problems in higher dimensions, though a graphical representation of the results is not as illustrative. As per the results obtained in [4], the TTOSOM is capable of inferring the distribution and structure of the data. In our present work, we are interested in knowing the effects of applying the neural rotation as part of the training process. The experiments briefly presented in this section use the same schedule for the learning rate and radius, i.e., no particular refinement of the parameters has been done to each particular data set.

First, we consider the data generated from a triangular-spaced distribution, as shown in Figs. 3a-3d. In this case, the initial tree topology is unidirectional. For the initialization phase a 1-ary tree (i.e., a list) is employed as the special case of the structure, and the respective keys are assigned in an increasing order. At the beginning, the prototype vectors are randomly placed. In the first iteration, the linear topology is lost, which is attributable to the randomness of the data points. As the prototypes are migrated and reallocated (see Figs. 3b and 3c), the 1-ary tree is modified as a consequence of the rotations. Finally, Fig. 3d depicts the case after convergence has taken place. Here, the tree nodes are uniformly distributed over the whole triangular shape. The BST property is still preserved, and further rotations are still possible. This experiment serves as an excellent example to show the differences with respect to the original TTOSOM algorithm

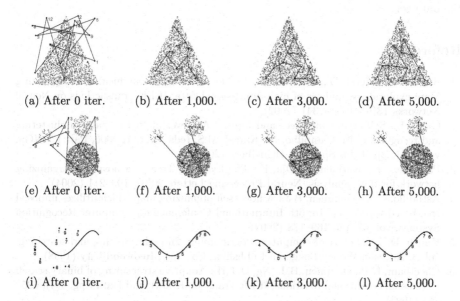

(a) After 0 iter.   (b) After 1,000.   (c) After 3,000.   (d) After 5,000.

(e) After 0 iter.   (f) After 1,000.   (g) After 3,000.   (h) After 5,000.

(i) After 0 iter.   (j) After 1,000.   (k) After 3,000.   (l) After 5,000.

Fig. 3. A 1-ary tree, i.e. a list topology, learns different distributions using the TTO-CONROT algorithm after utilizing the *same* set of parameters

[4], where a similar data set was utilized. In the case of the TTO-CONROT the points effectively represent the whole data set. Here, no particular *a priori* information about the structure of the data distribution is necessary; rather, this is learnt during the training process, as shown in Fig. 3d. In this manner, the specification of the initial tree topology required by the TTOSOM is no longer needed, and an alternative specification, which only requires the number of nodes in the initial 1-ary tree, is sufficient.

Another experiment is the one shown in Figs. 3e-3h, which entails a data set generated from 3 circular-shaped clouds where the circles possess a different size and density. In this experiment, again, in the first iteration, the original structure of 1-ary tree is lost because of the random selection of the codebook vectors. Interestingly, after convergence, and as depicted in Fig. 3h, the algorithm places a proportional number of codebook vectors in each of the three circles according to the density of their data points.

Lastly, Figs. 3i-3l demonstrate the power of the scheme for a linear curve.

5   Conclusions

In this paper, we have proposed a novel integration between the areas of ADS and the SOM. In particular, we have shown how a tree-based SOM can be adaptively transformed by the employment of an underlying BST structure and subsequently, re-structured using rotations that are performed conditionally. Our proposed method is able to infer the topological properties of the stochastic distribution, and at the same time, attempts to build the best BST that represents the data set.

References

1. Rauber, A., Merkl, D., Dittenbach, M.: The Growing Hierarchical Self-Organizing Map: exploratory analysis of high-dimensional data. IEEE Transactions on Neural Networks 13(6), 1331–1341 (2002)
2. Guan, L.: Self-organizing trees and forests: A powerful tool in pattern clustering and recognition. In: Campilho, A., Kamel, M.S. (eds.) ICIAR 2006, Part I. LNCS, vol. 4141, pp. 1–14. Springer, Heidelberg (2006)
3. Pakkanen, J., Iivarinen, J., Oja, E.: The Evolving Tree — a novel self-organizing network for data analysis. Neural Processing Letters 20(3), 199–211 (2004)
4. Astudillo, C.A., Oommen, B.J.: A novel self organizing map which utilizes imposed tree-based topologies. In: 6th International Conference on Computer Recognition Systems, vol. 57, pp. 169–178 (2009)
5. Knuth, D.E.: The art of computer programming, 2nd edn. Sorting and searching, vol. 3. Addison Wesley Longman Publishing Co., Inc., Redwood City (1998)
6. Cheetham, R.P., Oommen, B.J., Ng, D.T.H.: Adaptive structuring of binary search trees using conditional rotations. IEEE Trans. on Knowl. and Data Eng. 5(4), 695–704 (1993)
7. Kohonen, T.: Self-Organizing Maps. Springer-Verlag New York, Inc., Secaucus (2001)

8. Adelson-Velskii, M., Landis, M.E.: An algorithm for the organization of information. Sov. Math. DokL 3, 1259–1262 (1962)
9. Cormen, T.H., Leiserson, C.E., Rivest, R.L., Stein, C.: Introduction to Algorithms, 2nd edn. McGraw-Hill Science/Engineering/Math., New York (2001)
10. Lai, T.W.H.: Efficient maintenance of binary search trees. PhD thesis, University of Waterloo, Waterloo, Ont., Canada (1990)
11. Astudillo, C.A., Oommen, B.J.: Self organizing maps whose topologies can be learnt with adaptive binary search trees using conditional rotations. Journal version of this paper (2009) (submitted for publication)
12. Allen, B., Munro, I.: Self-organizing binary search trees. J. ACM 25(4), 526–535 (1978)
13. Sleator, D.D., Tarjan, R.E.: Self-adjusting binary search trees. J. ACM 32(3), 652–686 (1985)
14. Bitner, J.R.: Heuristics that dynamically organize data structures. SIAM J. Comput. 8, 82–110 (1979)
15. Mehlhorn, K.: Dynamic binary search. SIAM Journal on Computing 8(2), 175–198 (1979)

Topic-Based Computing Model for Web Page Popularity and Website Influence

Song Gao, Yajie Miao, Liu Yang, and Chunping Li

School of Software, Tsinghua University, 100084, Beijing, China
gaosong0329@foxmail.com, bluewillowwind@gmail.com,
miaoyj08@mails.tsinghua.edu.cn, cli@tsinghua.edu.cn

Abstract. We propose a novel algorithm called Popularity&InfluenceCalculator (PIC) to get the most popular web pages and influent websites under certain keywords. We assume that the influence of a website is composed of its own significance and the effects of its pages, while the popularity of a web page is related with the websites and all the other pages. After that, we design a novel algorithm which iteratively computes importance of both websites and web pages. The empirical results show that the PIC algorithm can rank the pages in famous websites and pages with descriptive facts higher. We also find out that those pages contain more popular contents, which is accordant with our previous description of popularity. Our system can help users to find the most important news first, under certain keywords.

Keywords: Popularity of web pages, Influence of websites, Computing model, Topic search.

1 Introduction

The Internet is playing an important role in daily life, with the prevalence of personal computer. There are millions of news and reports propagating over the whole web. Even under a specific topic, users can still find lots of related web pages and corresponding websites. Among those results generated through search engines, it is usually time-consuming for people to find more important news, which is a common but need-to-be-solved problem in the era of information explosion, as well.

According to some researches, most users will only examine the first couple of pages of results returned by any search engine [12]. Under that circumstance, finding important news and related websites which often post such news is becoming more and more essential for users. Actually, it will be beneficial for both personal and enterprise use. For personal use, this technique can save time and make users focus on their original interests. For enterprise use, users can find the negative news or rumors about them, in order to take risk-reduction actions and inspect the websites which have high willingness of posting such important reports or articles.

There have been some currently existing researches on finding out the importance of websites and web pages over the Internet. HITS algorithm [1] is one well-known method to find the hubs and authoritative web pages. It defines two values for each web page: authority and hub property. Hub is a kind of web page pointing to lots of

A. Nicholson and X. Li (Eds.): AI 2009, LNAI 5866, pp. 210–219, 2009.
© Springer-Verlag Berlin Heidelberg 2009

other pages. In HITS, a good hub page is the one connecting with numerous good authoritative pages, and vice versa. In the following years, some people have made some efforts to improve the HITS algorithm. Bharat and Henzinger tried to solve the problems caused by the mutually reinforcing relationships between hosts [6]. Assigning weights is a simple but effective way to improve the original HITS algorithm, as well. Chang considered the user interests of specific document and add more weights to the corresponding web pages [10]. Some people also suggested assigning appropriate weights to in-links of root documents and combining some relevance scoring methods, such as Vector Space Model (VSM) and Okapi, with the original HITS algorithm [9]. According to the web log, the number of users visiting a certain web page can also be used as weights [11]. Lempel and Moran proposed another kind of query-dependent ranking algorithm based on links called SALSA [7, 13], which is inspired by the HITS and PageRank algorithms. SALSA can isolate a particular topological phenomenon called the Tightly Knit Community (TKC) Effect. According to some comparative studies, SALSA is very effective at scoring general queries, but it still exhibits the "sampling anomaly" phenomenon [14]. In addition, some people also proposed to use media focus and user attention information to rank the news topic within a certain news story [15].

However, HITS and these improved algorithms are all based on the link analysis among all the web pages, while major focus of topic ranking research is on web links [4] as well. If the pages are not connected with each other, the results cannot be achieved. Fernandes proposed to use block information in web pages to get a better rank result, which requires some division methods to separate a single page into different blocks [8]. In this paper, we want to eliminate such division process and link information.

The major aim of ranking web pages and websites is to determine which one has more influences on that specific topic. However, this problem is lack of general and efficient solutions. But, there are some similar systems whose major ideas have great value on our problem. Yin and Han proposed an approach to determine the level of trustworthiness of both websites and web pages for a given object [2, 3], which also iteratively calculates the trustworthiness among websites and pages.

In this paper, we define the popularity of a web page as popularity of its contents, as well as design an iteration algorithm called *Popularity&InfluenceCalculator* (PIC), to compute the influence of websites and popularity of web pages. The general thought is that the website which has more popular pages will have more influence correspondingly; while the page which exists on more influent websites will have more popularity, as well. We use two cases to examine the effects of our method and achieve satisfactory ranking results based on values of websites and pages. Our ranking algorithm will not require the existence of web-links among pages and websites. Besides, we are focusing not only on web pages, but also on the websites which have higher possibility to report similar news.

The rest of the paper is organized as follows. The problem description and our proposed model are presented in Section 2. The results of two cases of our algorithm are discussed in Section 3. Finally, we conclude our proposed method in Section 4.

2 Computing Model and PIC Algorithm

2.1 Problem Description

First, we should consider what kind of web pages and websites show higher popularity and influence. In this paper, a certain web page can consist of one or more contents. If a specific content is also mentioned in more web pages, it means that such content does attract more attention from people, which will make it more important. In that case, if a web page contains more important contents, it will become more significant, too. Such definition is meaningful, because in common sense, if a web page is discussed and browsed by more people, it will have greater impacts on the object it discusses, such as a company. Similarly, an important website should contain more web pages which have higher importance. Meanwhile, the web pages published on such a website will have higher possibility to be important in the future as well. The detailed descriptions are as follows.

Importance of contents: The importance of a piece of content represents its appearance in web pages. If a certain piece of content appears on more web pages, it will have higher importance.

Popularity of a web page: The popularity of a web page equals to its contents' importance. If a web page contains more important contents, it will have more popularity. The popularity means the importance of a web page on a certain topic.

Influence of a website: The influence of a website means the average popularities of its pages. If the pages within a website normally have higher popularity, such website will show more influence on a certain topic. Hence, the influence of a website is its importance for a certain topic.

Hence, the aim of our algorithm is to find the web pages having higher popularities and websites showing higher influences under given keywords. The input of the system is the set of all web pages returned through a search engine, while the outputs are the popularities and influences of those web pages and their corresponding websites, which are sorted by the value of their importance, respectively.

2.2 Framework of PIC Algorithm

The major idea of this model is to use an iteration algorithm to compute the importance of websites and pages. In other words, the pages in more influent websites will have more popularity; while the websites which have more pages having high popularity will show higher influence. Before describing the details of the computing model, some variables must be defined first.

P is a set composed of all the pages found in searching step. The size of P is m, which means P has m elements $P_1, ..., P_m$.

W is a set composed of all the websites we found. The size of W is n, which means W has n elements $W_1, ..., W_n$.

Pop is a vector with length of m. Pop_i records the popularity of page P_i.

Inf is a vector with length of n. Inf_i records the influence of website W_i.

Besides, there are also two assumptions about the influence of websites and popularity of web pages in our model.

Assumption 1. The influence of a website can be divided into two parts, its own significance and the popularity of its pages. The own significance of a website can be calculated through several factors, such as the daily visit number, its sensibility to specific domain of news, and so on.

Assumption 2. The popularity of a page can be divided into two parts, influence of the websites it is on and the effects of other pages.

In *PIC* algorithm, the influence of websites will be initialized as their own significance. Then, those values will be used to calculate the popularity of web pages. Meanwhile, the influences among all the pages will also be taken into consideration. Afterwards, the websites' influence will be computed again through their own significance and their pages. The calculation will continue until the convergence of all the importance of both pages and websites. The general computing framework is shown in Fig. 1.

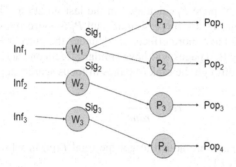

Fig. 1. Computing framework of PIC algorithm

2.3 Computing Process of PIC Algorithm

The computation will start from the websites. In this step, we only consider the inherent significance of a website, such as daily visit number or the popularity in a certain domain. Some people pointed out that there are some common and specific features of websites in different domains [5]. Hence, we believe that there should be some common parameters to determine the inherent importance of the websites. We use Sig_i to represent the significance of W_i.

$$Imp_i = Sig_i .\tag{1}$$

Here, we suppose all the websites have the same significance.

2.3.1 Computing Algorithm for Pages' Popularity

The pages' popularity can be divided into two parts: influence of websites and influence of other pages. First, the influence of websites should be calculated. Suppose the current page P_i exists on r websites, from W_l to W_r. Then, the first part of the importance of P_i will be the average importance of r websites.

$$Pop_i = \frac{\sum_{j=1}^{r} Inf_j}{r} \qquad (2)$$

Second, the effects of other pages should be considered. Except the current page P_i, there still are m-1 pages. In that case, the second part of the popularity of P_i can be achieved through the following equation.

$$\delta Pop_i = rel_i \times \sum_{j=1}^{m-1} rel_j \times sim_{i,j} \times Pop_j \qquad (3)$$

Rel_j means the relevance between page P_j and input keywords; $sim_{i,j}$ represents the similarity between P_i and P_j, which describes how many contents of those two pages are similar. The detailed method of calculating similarity can be found in later section. Pop_j is the popularity of page P_j calculated in the last iteration. This equation means that if two pages P_i and P_j are more similar, and P_j is more relevant with the input keywords, then P_j will have more effects on P_i under the current keywords. The first rel_i is to make sure any two pages will have similar effects on each other.

The rel_j can be got through the rank of pages in focus search step, such as

$$rel_i = \frac{1}{rank_i + 1} \qquad (4)$$

Then, the above two parts are added to get the total Pop, in which b is a regulating factor, ranging from 0 to 1.

$$Pop_i' = Pop_i + b \times \delta Pop_i \qquad (5)$$

Finally, in order to normalize the Pop, a transforming should be done as follows.

$$Pop_i = 1 - e^{-Pop_i'} \qquad (6)$$

2.3.2 Computing Algorithm for Websites' Influence

This step is similar as the above one. The influence of website W_i can also be divided into two parts: the own significance and the effects of its pages. The first part is computed in the last iteration, while the second part is the effects of its pages, which is the average popularity of all the pages on current website.

$$\delta Inf_i = \frac{\sum_{j=1}^{s} Pop_j}{s} \qquad (7)$$

Then, sum the two parts to get the total Inf, in which a is also a regulation factor, ranging from 0 to 1.

$$Inf_i' = Inf_i + a \times \delta Inf_i \qquad (8)$$

Finally, through the normalization, the final *Inf* can be achieved.

$$Inf_i = 1 - e^{-Inf_i'} \qquad (9)$$

Then the calculation will iterate until the vector of *Pop* and *Inf* both converge. Finally, the importance of all pages and websites will be obtained, as well as the rank of them according to their values.

3 Experiments and Evaluation

3.1 Data Collection

Case 1: Suicide case in a company
There was an employee of a company jumped from the fourth floor several years ago. Some people near that building posted pictures and briefly described the event on several BBSs only after about half an hour. During the following days, this suicide event was spread all over the internet.

For this case, we crawl 250 pages given by Google. Also, we find out all the websites these pages "reside" on. This way, we get a dataset composed of 250 web pages and the corresponding 195 websites, together with relationships between them.

Case 2: Event of "South China Tiger"
South China Tiger is a commonly-thought-to-be extinct species, since for decades they haven't been spotted in the wild. On Oct 13, 2007, a local hunter in one province of China claimed that he had managed to take photos of a wild South China Tiger, the evidences of this species' existence. Shortly, a blogger suspected that these photos were likely to be faked and caused a more and more fierce and attention-drawing discussion about the authenticity of these photos to spread over the whole web.

Similarly, for this case, we use Google with a query, "South China Tiger Event", and got 275 pages and 230 websites. In order to accurately measure the similarity between web pages, we process these pages in both cases using html parsing techniques and get the main content of them.

3.2 Results and Analysis

3.2.1 Two Extreme Situations
We run our algorithm on these two datasets and output influence score for each website and popularity score of pages. For each dataset, we rank websites and pages using their values respectively. In order to validate our computational framework, we investigate top-ranked and bottom-ranked websites and pages.

Fig. 2 plots rankings of the 20 most highly ranked websites denoted by the horizontal axis and rankings of web pages these websites correspondingly contain, denoted by the vertical axis. As can be seen, high-ranked websites tend to hold more than one, sometimes four or five pages which generally have relatively competitive positions on the page rank list. The results show that website rankings and page rankings change closely with each other.

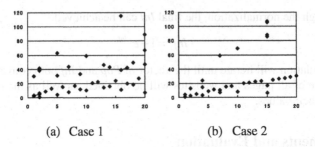

(a) Case 1   (b) Case 2

Fig. 2. Rankings of high-ranked websites and pages

Another observation emerges from the ranking lists of pages for both data sets. That is, if pages contain more descriptive and factual information about the events, they will be given larger scores and higher rankings, while some critical articles with little impact on the spreading of the news will be ranked lowly, such as the ones only discussed by one person or few people. This is accordant with our previous definition of popularity in Section 3, in the sense that one descriptive and factual page is intended to cover objective details of a specific event and can easily find similar pages, whereas critical articles are quite more subjective and individual-dependent, and therefore have few similar ones.

In order to further formulate how websites and pages become high-ranked and low-ranked, we discuss the relationships between websites and pages with two simplified modes. High-ranked websites and pages generally display one-contain-several mode. In this case, because of the process of mutual reinforcement, the website on the left conveys its impact score to the three pages it contains on the right. In return, the three pages convey their popularity to their website, in fact through a summing operation, giving a boost to their website's influence score. Most low-ranked websites and their pages exhibit one-contain-one mode, which makes mutual reinforcement less meaningful. Therefore, the ranking of the single website is mainly and directly determined by its single page and vice versa.

3.2.2 One Moderate Situation
In Section 3.2.1, we have analyzed the results of the most highly and lowly ranked websites and pages. In order to completely evaluate the results and check not-so-extreme situations, we examine supplemental websites and pages at the middle positions on the ranking list.

In Fig. 3, for both case 1 and case 2, we show the websites ranked from 111 to 120 and rankings of all the pages each website contains. Generally, these websites and pages display one-contain-one mode which makes rankings of these websites and their pages highly, and almost linearly, correlated. These results comply with heuristics suggested by our computational algorithm and are consistent with our analysis above.

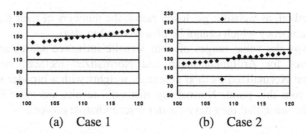

(a) Case 1   (b) Case 2

Fig. 3. Rankings of middle-ranked websites and pages

3.3 Discussion

As elaborated in Section 2, we get each page's popularity score by calculating an arithmetic mean of the influence scores of all the websites this page belongs to. However, an problematic aspect of this procedure is that in practice, as can be seen from the two real data sets in Section 3.1, the mode that one web page resides on several distinct websites rarely exists. One possible way to deal with that is to combine the pages whose similarity is higher than a certain threshold.

3.4 Additional Subtlety

In this section, we start out to handle a subtle problem. As analogously addressed by Bharat and Henzinger in 1998 [6], because of mutual reinforcement, the "cabal" of the websites and pages in Figure 3(a) "conspire" to boost each other and predominate the resulting lists. In order to abbreviate this effect, instead of the straightforward arithmetic-mean method, we divide each importance score evenly among its degree. By summing up all the influence scores endowed by websites containing it, we can get a normalized popularity score of a page. Under such a circumstance, Formula (2) can be formulated as follows:

$$Pop_i = \sum_{j=1}^{r} \frac{Inf_j}{degree_j} \qquad (10)$$

where $degree_j$ is the number of pages on website W_j.

Similarly, the popularity of a page can also be equally divided into several parts, according to the websites it appears on, just as above.

$$\delta Inf_i = \sum_{j=1}^{s} \frac{Pop_j}{degree_j} \qquad (11)$$

In order to examine the effectiveness of this calculation method, we run the modified algorithm on the data set of Case 1 and compare acquired results with previous ones. Fig. 4 shows how the number of pages "nesting" on a website changes with this website's rankings for both the original and modified algorithms. The horizontal axis denotes the top 20 websites and the vertical axis denotes the number of pages each website contains. From this comparison, we can reach the conclusion that when using

Normalized method, as websites get lower ranks, the numbers of their pages decrease accordingly, a tendency which cannot be observed in Fig. 4(a).

Table 1 shows the value ranges of both website influence scores and web page popularity scores for Arithmetic-mean and Normalized method. An observable difference is that Normalized method generates weights with a broader spectrum and wider range. From this point, Normalized method's results are more desirable and reasonable in the sense that they have greater discriminative power.

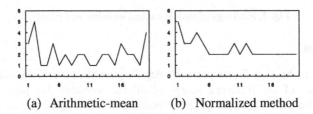

(a) Arithmetic-mean   (b) Normalized method

Fig. 4. Two calculating methods

Table 1. Comparison of value ranges

Method	Rang of website influence	Rang of web page popularity
Arithmetic-mean method	0.75089 ~ 0.7004	0.7533 ~ 0.5040
Normalized method	0.8890 ~ 0.7000	0.5539 ~ 0.1665

4 Conclusion and Future Work

In this paper, we proposed a novel PIC algorithm to get the most popular web pages and the most influent websites under certain keywords. We first define the importance of content, the popularity of web pages and the influence of websites, respectively. Then, we assume that the influence of a website is composed of its own significance and the effects of its pages, while the popularity of a web page is related with the websites and all the other pages' popularity. In that case, we design the iterative algorithm to calculate the value of both websites and web pages, until they converge. Two cases are collected manually as the experiment data. The results show that PIC algorithm can rank the pages in famous websites and pages with descriptive facts higher.

However, there still are some problems. For instance, how to evaluate the inherent significance of a certain website needs thorough consideration. Besides, we also plan to change the current similarity calculation method based on semi-automatic keyword selection into the automatic whole passage comparison method.

References

1. Kleinberg, J.: Authoritative sources in a hyperlinked environment. In: Proceedings of the Ninth Annual ACM-SIAM Symposium on Discrete Algorithms, pp. 668–677. ACM, New York (1998)

2. Yin, X., Han, J., Yu, P.S.: Truth Discovery with Multiple Conflicting Information Providers on the Web. IEEE Transactions on Knowledge and Data Engineering 20, 796–808 (2008)
3. Yin, X., Han, J., Yu, P.S.: Truth Discovery with Multiple Conflicting Information Providers on the Web. In: Proceedings of KDD 2007, pp. 1048–1052. ACM, New York (2007)
4. Wang, J., Chen, Z., Tao, L., Ma, W.-Y., Wenyin, L.: Ranking User's Relevance to a Topic through Link Analysis on Web Logs. In: Proceedings of the 4th international workshop on Web information and data management, pp. 49–54. ACM Press, New York (2002)
5. Zhang, P., von Dran, G.: A Comparison of the Most Important Website Features in Different Domains: An Empirical Study of User Perceptions. In: Proceedings of Americas Conference on Information Systems (AMCIS 2000), pp. 1367–1372 (2000)
6. Bharat, K., Henzinger, M.R.: Improved algorithms for topic distillation in hyperlinked environments. In: Proceedings of the 21st International ACM SIGIR Conference on Research and Development in Information Retrieval, pp. 104–111. ACM, New York (1998)
7. Lempel, R., Moran, S.: The stochastic approach for link-structure analysis (SALSA) and the TKC effect. ACM Transactions on Information Systems, 131–160 (2000)
8. Fernandes, D., de Moura, E.S., Ribeiro-Neto, B.: Computing Block Importance for Searching on Web Sites. In: Proceedings of the sixteenth ACM conference on Conference on information and knowledge management, pp. 165–174. ACM, New York (2007)
9. Li, L., Shang, Y., Zhang, W.: Improvement of HITS-based Algorithms on Web Documents. In: Proceedings of the 11th international conference on World Wide Web, pp. 527–535. ACM, New York (2002)
10. Chang, H., Cohn, D., McCallum, A.: Learning to create customized authority lists. In: Proceedings of the Seventeenth International Conference on Machine Learning, pp. 127–134. ACM, New York (2000)
11. Farahat, A., LoFaro, T., Miller, J.C.: Modication of kleinberg's hits algorithm using matrix exponentiation and web log records. In: Proceedings of the 24th International Conference on Research and Development in Information Retrieval (SIGIR 2001), pp. 444–445. ACM, New York (2001)
12. Jansen, B.J., Spink, A., Bateman, J., Saracevic, T.: Real life information retrieval: a study of user queries on the web. ACM SIGIR Forum 32, 5–17 (1998)
13. Lempel, R., Moran, S.: SALSA: the stochastic approach for link-structure analysis. ACM Transactions on Information Systems (TOIS), 131–160 (2001)
14. Najork, M.: Comparing the Effectiveness of HITS and SALSA. In: Proceedings of the sixteenth ACM conference on Conference on information and knowledge management, pp. 157–164. ACM, New York (2007)
15. Wang, C., Zhang, M., Ru, L., Ma, S.: Automatic Online News Topic Ranking Using Media Focus and User Attention Based on Aging Theory. In: Proceeding of the 17th ACM conference on Information and knowledge management, pp. 1033–1042. ACM, New York (2008)

Classifying Multiple Imbalanced Attributes in Relational Data

Amal S. Ghanem, Svetha Venkatesh, and Geoff West

Curtin University of Technology, GPO Box U1987, Perth 6845, Western Australia

Abstract. Real-world data are often stored as relational database systems with different numbers of significant attributes. Unfortunately, most classification techniques are proposed for learning from balanced non-relational data and mainly for classifying one single attribute. In this paper, we propose an approach for learning from relational data with the specific goal of classifying multiple imbalanced attributes. In our approach, we extend a relational modelling technique (PRMs-IM) designed for imbalanced relational learning to deal with multiple imbalanced attributes classification. We address the problem of classifying multiple imbalanced attributes by enriching the PRMs-IM with the "Bagging" classification ensemble. We evaluate our approach on real-world imbalanced student relational data and demonstrate its effectiveness in predicting student performance.

1 Introduction

Classification is a critical task in many real-world systems, and is a research field in which extensive studies and experiments are conducted to improve the classification results. A wide range of classification techniques, such as Bayesian networks (BNs), decision trees and Support Vector Machines (SVMs), have been successfully employed in many applications to classify various types of objects.

However, most of these classification techniques are usually proposed with specific assumptions, which may not hold in many real-world domains. The classification of a single attribute from flat data files that have balanced data distribution, represent one of these assumptions. However, in many applications, the collected data are stored in relational database systems with highly imbalanced data distribution, where one class of data has a large number of samples as compared with the other classes. Moreover, in many applications, it is often of interest to classify/predict several attributes rather than a single attribute. An example of such a situation is learning from a student relational database to predict the unit results of a second-year undergraduate student given the results of the first-year, in which the unit results are greatly imbalanced.

Studies have shown that learning from imbalanced data usually hinders the performance of the traditional learning techniques [1,2]. This performance degradation is a result of producing more and stronger rules to classify the samples of the majority class in comparison to that of the minority class, and hence incorrectly classify most of the minority samples to be of the majority class.

A. Nicholson and X. Li (Eds.): AI 2009, LNAI 5866, pp. 220–229, 2009.

Several methods have been proposed to handle the general imbalanced class problem [3,4,5,6], and a few attempts have been made to handle the problem particularly in relational data [7,8,9,10]. PRMs-IM [10] has been recently introduced as an extension of a relational learning technique: Probabilistic Relational Models (PRMs) [11,12], to handle the two-class classification of a single imbalanced attribute in relational domains. The idea behind PRMs-IM is to build an ensemble of PRMs on balanced subsets from the original data, in which each subset has an equal number of the minority and majority samples of the imbalanced attribute.

Although the imbalanced class problem is relatively well investigated in both relational and non-relational domains, the classification of several imbalanced attributes in relational domains has not been well addressed. Attempts have been proposed for the special case of classifying two attributes [13,14]. However, these methods did not tackle the imbalanced problem or the relational learning for classifying several attributes.

Therefore, special classification techniques are required to handle the problem of classifying multiple imbalanced attributes in relational domains. In this paper we investigate this problem and review the different proposed approaches. Based on this research, we present a new approach (PRMs-IM2) to handle the problem of classifying multiple imbalanced attributes in relational data. In our approach, we address this problem by combining the balancing concept of PRMs-IM with the "Bagging" classification ensemble [15]. PRMs-IM2 is presented as a Bagging ensemble approach that consists of a set of independent classifiers trained on balanced subsets of the imbalanced data. The subsets are generated using the balancing concept of PRMs-IM for each of the imbalanced attributes. We evaluate our approach on a student relational database with multiple imbalanced attributes, and show the effectiveness of our approach in predicting student results in second semester units.

This paper is organized as follows: section 2 presents a review of the related work. Our methodology is presented in section 3, followed by the experimental results in section 4. Finally, section 5 concludes the paper.

2   Related Work

2.1   Imbalanced Class Problem in Relational Data

Classification techniques such as BNs, decisions trees and SVMs have been shown to perform extremely well in several applications of different domains. However, research has shown that the performance of these techniques is hindered when applied to imbalanced data [1,2], as they get biased to the majority class and hence misclassify most of the minority samples.

Methods proposed to handle the imbalanced class problem can be categorized into three groups [16,17]:

- **Re-sampling:** by either down-sampling the majority class or/and over-sampling the minority class until the two classes have approximately equal numbers of samples. A study of a number of down- and over-sampling methods and their performances is presented by Batista et al. [3].

– **Cost-Sensitive learning:** by assigning a distinct misclassification cost for each class, and particularly increasing that of the minority class [4].
– **Insensitive learning:** by modifying the learning algorithm internally to pay more attention to minority class data, as in building a goal oriented BN [5] and exploring the optimum intervals for the majority and minority classes [6].

However, most of these methods are mainly developed for flat datasets, where all data must be presented in one single file. Therefore, in order to learn from a rich relational database, the data must be first converted into a single file that consists of a fixed set of attributes and the corresponding values. This conversion could result in redundant data and inconsistency. Techniques have been proposed to handle the imbalanced class problem in multi-relational data, including: implementing cost-sensitive learning in structured data [7], combining the classification of multiple flat views of the database [8] and using the G-mean in decision trees [9].

In addition to these methods, PRMs-IM [10] has been recently introduced to handle the imbalanced class problem in relational data. PRMs-IM was introduced as an extension of the relational learning algorithm: Probabilistic Relational Models (PRMs) [11, 12]. PRMs were introduced as an extension of Bayesian Networks (BNs) to satisfy relational learning and inference. PRMs specify a model for probability distribution over the relational domains. The model includes the relational representation of the domain and the probabilistic schema describing the dependencies in the domain. The PRM model learned from the relational data provides a statistical model that can be used to answer many interesting inference queries about any aspect of the domain given the current status and relationships in the database.

Therefore, to handle the imbalanced class problem in relational data, PRMs-IM was presented as an ensemble of independent PRM models built on balanced subsets extracted from the imbalanced training dataset. Each subset is constructed to include all the samples of the minority class and an equal number of randomly selected samples from the majority class. The number of balanced subsets depends on the statistical distribution of the data. Thus, if the number of samples in the majority class is double that of the minority, then two subsets are created. The PRM models of PRMs-IM are then combined using the weighted voting strategy [10], and hence new samples are assigned to the class with the highest weighted score.

2.2 Classifying Multiple Attributes

Most existing pattern classification techniques handle the classification of a single attribute. However, in many real-world applications, it is often the case of being interested in classifying more than one attribute, such as classifying both the activity and location in location-based activity recognition systems. The basic solutions for classifying multiple attributes ($\mathcal{A} = \{A_1, A_2, A_3, \ldots\}$) can be classified as follows [14, 13]:

- The combined method: by considering $\mathcal{A}$ as one complex attribute and hence construct one classifier. In this method, advanced techniques are required to work with the multi-class classification.
- The hierarchal method: in a similar approach to decision trees, by constructing a classifier for a given attributes A_i, and then for each class of A_i construct a specialized classifier for A_j. The performance of this method is hindered by the accuracy of the classifiers at the top of the hierarchy, as any misclassification by the top classifiers can not be corrected later. Moreover, in this method, the top attributes can help to reach conclusions about the lower attributes but not vise versa. In addition, the structure grows rapidly as the number of attributes and classes increases.
- The independent method: for each attribute A_i, construct an independent classifier. This method is based on dealing with each attribute separately, and hence it requires more training and testing phases than the other methods.

In addition to these naïve solutions, other methods were proposed but mostly for the special case of classifying two attributes. One method includes using a bilinear model for solving two-factor tasks [14]. This approach mostly acts as a regression analysis and hence does not provide a graphical model for interpreting the interactions between the attributes in the domain as provided in other classification techniques, such as BNs.

Another method uses the mutual suggestions between a pair of classifiers [13], in which a single classifier is trained for each attribute, and then at the inference phase, the results of each classifier are used as a hint to reach a conclusion in the other classifier. The learning in this approach is similar to that of the independent approach, but differs in obtaining the final classification results in the inference phase, where the hints between the classifiers are used to reach a better result.

3   Methodology

In this paper we aim to develop a classification technique that could handle the problem of classifying multiple imbalanced attributes in relational data by using the concepts of PRMs-IM [10] and the "Bagging" ensemble approach [15]. PRMs-IM are designed specifically to learn from imbalanced relational databases for a single imbalanced attribute. Thus, for classifying N imbalanced attributes, N independent PRMs-IM models must be performed, one model for each attribute. In PRMs-IM2 we aim to extend PRMs-IM to classify the N imbalanced attributes in a single model.

In order to to obtain a single model, we use the idea of the "Bagging" ensemble approach. The Bagging approach uses an ensemble of K classifiers. Each classifier is trained on a different subset randomly sampled, with replacements, from the original data. To classify a new sample, the classification decisions of the K classifiers are combined to reach a final conclusion about the class of the sample. A simple combination technique is to use majority voting, in which the sample is assigned to the class with the largest number of votes.

Our approach relies on the idea of building an ensemble of classifiers, where each classifier is trained on a different relational subset that includes a balanced representation of all the imbalanced attributes. This aim is achieved in PRMs-IM2 by firstly applying the balancing concept of PRMs-IM to build balanced relational subsets for each imbalanced attribute. This results in a separate set of balanced subsets for each imbalanced attribute.

However, to achieve the goal of generating subsets that include all the imbalanced attributes, PRMs-IM2 employs the Bagging concept to further sample the balanced subsets into L datasets. Each of the L datasets is formed by randomly selecting one balanced subset from each imbalanced attribute. At the end of this procedure, L balanced subsets will be generated, each subset including balanced data for each of the imbalanced target attributes. Note that in this paper, we use the same notations to describe the imbalanced situation as those used in [10].

To illustrate our approach, consider a relational dataset S that consists of a set of attributes $(X_1, X_2, ..., X_M, Y_1, Y_2, .., Y_N)$ organized into tables and relationships, where $(Y_1, Y_2, .., Y_N)$ represents the set of the domain imbalanced attributes that we want to classify. Each Y_i has a majority class $Y_{i_{mj}}$ and a minority class $Y_{i_{mr}}$. In addition, $n_{i(mr)}$ represents the number of samples of the minority class of Y_i. The subsets of PRMs-IM2 are constructed as follows:

- For each imbalanced attribute Y_i of the N imbalanced attributes:
  - Compute n_i as the difference between the number of samples of $Y_{i_{mj}}$ and that of $Y_{i_{mr}}$, where n_i is the number of balanced subsets required for Y_i.
  - For each of the n_i iterations, construct a subset $Y_i s_i$, such that it includes:
    * All the $n_{i(mr)}$ samples from $Y_{i_{mr}}$.
    * $n_{i(mr)}$ randomly selected samples with replacements from $Y_{i_{mj}}$.
    * The data of $(X_1, X_2, ..., X_M)$ according to the selected records of Y_i.
- Compute $L = \max_{i=1..N}(n_i)$, where L is the number of datasets required for the bagging approach.
- For L iterations:
  - Construct a database S_i that has the same structure as S.
  - For each Y_j, randomly allocate a subset $Y_j s_k$ from Y_j subsets to S_i.

It is important to note, that when creating the balanced subsets of an imbalanced attribute Y_i, the subsets should include only the data of $(X_1, X_2, ..., X_M)$ and of Y_i. In other words, the data of the other imbalanced attributes rather than Y_i are excluded. This is necessary for creating balanced K databases of all the attributes at the sampling phase. Otherwise, consider the case if Y_i subsets include the related records of Y_j. Then, at the sampling phase, a subset S_k could be generated, such that it includes the random subsets: $Y_i s_l$ and $Y_j s_h$ from Y_i

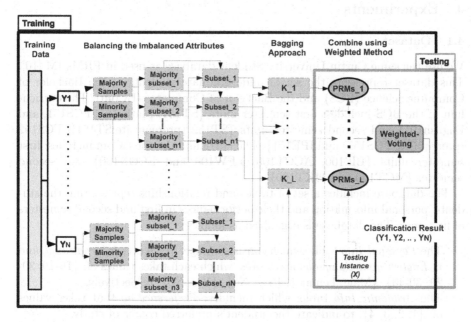

Fig. 1. An illustration of PRMs-IM2 approach for classifying multiple imbalanced attributes

and Y_j, respectively. In this case, the data records of Y_j in S_k will include data from $Y_j s_h$, which are balanced data, and the Y_j records from $Y_i s_l$, which are not balanced.

Having the balanced L relational subsets, an independent PRM model can be learned from each relational subset L_i using the learning techniques described in [12]. Then, these models are combined using the weighted voting strategy as in [10]. In this combination strategy, each PRM model P_i has a different weight $P_{i_w} Y_i$ for each attribute Y_i to be used for the final prediction. The $P_{i_w} Y_i$ is calculated as the average performance accuracy of P_i for classifying Y_i over the training subsets other than the data subset corresponding to P_i. Fig. 1 illustrates the concept of PRMs-IM2.

For a new testing sample x, each P_i outputs the probabilities of each of the imbalanced attributes $(Y_1, Y_2, .., Y_N)$ given the values of $(X_1, X_2, ..., X_M)$. Thus, for each Y_i, each P_i outputs the probabilities $(P_i(x)Y_{i_{mj}}, P_i(x)Y_{i_{mr}})$ for assigning x to $Y_{i_{mj}}$ and $Y_{i_{mr}}$, respectively. Then, for each Y_i, the score of each class equals the summation of the weighted probabilities of the PRM models and hence x is classified to be of the class with the largest weighted score. For example, for Y_i the classification of x is calculated as:

$$F(x) = \text{argmax}_{m \in (Y_{i_{mj}}, Y_{i_{mr}})} (\sum_{\forall P_i} P_i(x)Y_{i_m} * P_{i_w}Y_i) \qquad (1)$$

4   Experiments

4.1   Dataset

We use the same Curtin University Student database as used in PRMs-IM [10]. This dataset represents the set of undergraduate students of the Bachelor of Computer Science (BCS) and the Bachelor of Commerce (BCom). The curriculum of the BCS includes: first semester units {ST151, Maths101, FCS151, and English101} and second semester units {ST152 (prerequisite:ST151), FCS152 (prerequisite:FCS151) and IPE151}. The curriculum of the BCom includes: first semester units {BIS100, ACCT100, LFW100 and ECON100} and second semester {MGT100 and MKT100}.

The database includes a set of tables and relationships representing the students' personal information and their performances in first and second semesters of first year. The database is organized as follows:

- The *Personal_Info* table: which consists of: *age, gender, is_international,* and *is_English_home_language* attributes, which each takes values of: {16-19, 20-29, 30-40}, {Male, Female}, {Yes, No}, {Yes, No}, respectively.
- The *Academic_Info* table: which includes: *Preference_no* that takes values of: {1, 2, 3, 4}, to indicate the student's preferred course of study.
- Semester I units tables: each includes: *grade* of values: {F, 5, 6, 7, 8, 9} representing the marks: {0-49, 50-59, 60-69, 70-79, 80-89, 90-100}.
- Semester II units tables: including the *status* attribute taking values of {Pass, Fail}.

In this dataset, for each of the BCom and BCS degrees, we are interested in predicting a given student's performance in second second semester units based on the personal information and performances in first semester units. However, each of the second semester units represents an imbalanced attribute, in which mot of the data belongs to the majority 'Pass' class compared to few samples belonging to the minority 'Fail' class. Table 1 depicts the data distribution of the training data. For each degree, we perform 5-fold cross validation using the training data for the students enrolled in the period 1999-2005. In addition to the cross validation, we use the data of year 2006 as a separate testing set.

4.2   Experimental Setup

The results of PRMs-IM2 are presented in comparison to the independent and hierarchal approaches discussed earlier in section 2. In this paper, the combined

Table 1. Data distribution of (a) the BCS (b) the BCom training dataset

Unit	$|Fail|$	$|Pass|$		Unit	$|Fail|$	$|Pass|$
ST152	12	58		MGT100	159	1556
FCS152	11	59		MKT100	88	1627
IPE151	7	63				
(a)				(b)		

approach is not evaluated, as it represents a multi-class problem, in which special multi-class algorithms are required. The PRM is used as the learning technique in all the experiments as a result of the relational format of the dataset and the effectiveness of PRMs in relational learning.

For evaluation, we use Receiver Operating Characteristics (ROC) curves and the Area under ROC (AUC) [18], which are often used as a measure for imbalanced classification problems. ROC curves show the trade off between the false positive rate and the true positive rate. The AUC is used to compare several models using the ROC curves, to get a single value of the classifier performance. The closer the AUC value is to the value '1', the better the classifier.

The independent method is represented as the results of PRMs-IM, in which each independent experiment is evaluated for each imbalanced attribute. In the hierarchal method, the imbalanced attributes are first ordered in descending order $(Y_1, Y_2, .., Y_n)$ based on the AUC value of each attribute obtained in PRMs-IM. Thus, the attributes with higher AUCs are listed first. This order is chosen so as to have the most accurate classifiers at the top of the hierarchy to minimize propagating the classification errors to the lower levels. Moreover, to avoid the problem of the imbalanced class problem, each classifier in the hierarchy is build as a PRMs-IM, thus the classifier of Y_i is a PRMs-IM on balanced subsets of Y_i.

4.3   Experimental Results

In this section we present the results obtained from each experiment in terms of: the prediction accuracy in the AUC results and the number of models used for training and inference in each algorithm, as shown in Tables 2 and 3, respectively. For each dataset, the best result is shown in bold. Table 3 presents the normalized number of models of each algorithm for training and inference. The normalized number is the number of models required by each algorithm for a particular dataset divided by the corresponding number of models of PRMs-IM2. Average normalized values greater than one correspond to an algorithm requiring more models than PRMs-IM2.

In terms of the prediction accuracy, the results show that PRMs-IM2 was able to outperform all the other methods except for the *IPE* dataset in the cross validation. In the hierarchal approach, the results are hindered by the misclassification results of the top classifiers in the hierarchy. In the independent method, the models are built independently for each imbalanced attribute and hence the value of one attribute cannot be used to reach any conclusion about the others. However, in real-world applications, usually the information about one attribute can help to reach better understanding about others. Therefore, a model that includes all the attributes can use the different interactions between them to reach better results. This principle could not be achieved using the independent model, as each attribute needs to be modeled separately, and neither can be accomplished in the hierarchical method, as only the top attributes help to reach a conclusion about the lower attributes but not the other way around. Moreover, the combined approach will treat the targeted attributes as one single attribute and thus would not show us the interactions of each attribute by itself.

Table 2. The AUC results (a) Cross validation (b) 2006 testing data

Method	BCom		BCS		
	MGT100	MKT100	ST152	FCS152	IPE151
PRMs-IM	0.914	0.786	0.839	0.901	**0.913**
PRMs-IM2	**0.922**	**0.893**	**0.950**	**0.923**	0.892
Hierarchal	0.914	0.756	0.811	0.897	**0.913**

(a)

Method	BCom		BCS		
	MGT100	MKT100	ST152	FCS152	IPE151
PRMs-IM	**0.921**	0.788	0.875	0.927	0.954
PRMs-IM2	**0.921**	0.840	**0.984**	**0.968**	**0.993**
Hierarchal	**0.921**	0.787	0.785	0.887	0.954

(b)

Table 3. Normalized number of models used for (a) Training (b) Inference

Method	Dataset (DS)		Average	Method	Dataset (DS)		Average
	BCom	BCS	over DS		BCom	BCS	over DS
PRMs-IM	1.53	2.11	1.82	PRMs-IM	1.53	2.11	1.82
PRMs-IM2	1.00	1.00	1.00	PRMs-IM2	1.00	1.00	1.00
Hierarchal	2.32	5.56	3.94	Hierarchal	2.11	3.67	2.89

(a)   (b)

This interaction is achieved in PRMs-IM2, as the final model includes all the attributes and presents all the interactions in the domain. Therefore, PRMs-IM2 offers the opportunity for the imbalanced attributes to be related to each other, and hence the value of one of the imbalanced attributes could strengthen the conclusion of the others. Moreover, PRMs-IM2 could model all the significant imbalanced attributes at once and show the different interactions between the attributes, which can not be achieved by the mutual suggestions approach [13] that learns a separate classifier for each imbalanced attribute, or the bilinear model [14] that uses a linear model.

In terms of the number of models used for training and inference, the results show that PRMs-IM2 requires the least number of models for both training and inference. For example in training, the number of models for PRMs-IM and the hierarchy are about twice and four times, respectively, the number of models of PRMs-IM2, and in inference the number of models are about twice and triple, respectively, those of PRMs-IM.

5   Conclusion

This paper has discussed the problem of classifying multiple imbalanced attributes in relational domains and proposed a technique (PRMs-IM2) to handle this problem. PRMs-IM2 combines the concepts of the relational imbalanced technique (PRMs-IM) and the Bagging ensemble approach. In PRMs-IM2, all the significant imbalanced attributes are modelled in a single model showing the different interactions between the attributes, which can not be achieved by other methods. PRMs-IM2 was evaluated on a student relational database to classify the results of different imbalanced units in semester II. The results show that PRMs-IM2 was able to generally improve over the other naïve methods, while maintaining the least number of models for training and testing.

References

1. Japkowicz, N., Stephen, S.: The class imbalance problem: A systematic study. Intell. Data Anal. 6(5), 429–449 (2002)
2. Kubat, M., Matwin, S.: Addressing the curse of imbalanced training sets: One-sided selection. In: ICML, pp. 179–186 (1997)
3. Batista, G.E., Prati, R.C., Monard, M.C.: A study of the behavior of several methods for balancing machine learning training data. SIGKDD Explor. Newsl. 6(1), 20–29 (2004)
4. Pazzani, M.J., Merz, C.J., Murphy, P.M., Ali, K., Hume, T., Brunk, C.: Reducing misclassification costs. In: ICML, pp. 217–225 (1994)
5. Ezawa, K.J., Singh, M., Norton, S.W.: Learning Goal Oriented Bayesian Networks for Telecommunications Risk Management. In: ICML, pp. 139–147 (1996)
6. Kubat, M., Holte, R.C., Matwin, S.: Machine Learning for the Detection of Oil Spills in Satellite Radar Images. Mach. Learn. 30(2-3), 195–215 (1998)
7. Sen, P., Getoor, L.: Cost-sensitive learning with conditional markov networks. In: ICML, pp. 801–808 (2006)
8. Guo, H., Viktor, H.L.: Mining Imbalanced Classes in Multirelational Classification. In: PKDD/MRDM 2007, Warsaw, Poland (2007)
9. Lee, C.I., Tsai, C.J., Wu, T.Q., Yang, W.P.: An approach to mining the multi-relational imbalanced database. Expert Syst. Appl. 34(4), 3021–3032 (2008)
10. Ghanem, A.S., Venkatesh, S., West, G.: Learning in Imbalanced Relational Data. In: Proc. {ICPR}. IEEE Computer Society, Los Alamitos (December 2008)
11. Koller, D., Pfeffer, A.: Probabilistic frame-based systems. In: AAAI, pp. 580–587 (1998)
12. Friedman, N., Getoor, L., Koller, D., Pfeffer, A.: Learning Probabilistic Relational Models. In: IJCAI, pp. 1300–1309 (1999)
13. Hiraoka, K., Mishima, T.: Classification of double attributes via mutual suggestion between a pair of classifiers. In: ICONIP, November 2002, vol. 4, pp. 1852–1856 (2002)
14. Tenenbaum, J.B., Freeman, W.T.: Separating style and content with bilinear models. Neural Comput. 12(6), 1247–1283 (2000)
15. Breiman, L.: Bagging predictors. Mach. Learn. 24(2), 123–140 (1996)
16. Eavis, T., Japkowicz, N.: A Recognition-Based Alternative to Discrimination-Based Multi-layer Perceptrons. In: Canadian Conference on AI, pp. 280–292 (2000)
17. Barandela, R., Sánchez, J.S., García, V., Rangel, E.: Strategies for learning in class imbalance problems. Pattern Recognition 36(3), 849–851 (2003)
18. Fawcett, T.: An introduction to ROC analysis. PRL 27(8), 861–874 (2006)

Algorithms for the Computation of Reduced Convex Hulls

Ben Goodrich, David Albrecht, and Peter Tischer

Clayton School of IT
Monash University, Australia
ben.goodrich@infotech.monash.edu.au

Abstract. Geometric interpretations of Support Vector Machines (SVMs) have introduced the concept of a reduced convex hull. A reduced convex hull is the set of all convex combinations of a set of points where the weight any single point can be assigned is bounded from above by a constant. This paper decouples reduced convex hulls from their origins in SVMs and allows them to be constructed independently. Two algorithms for the computation of reduced convex hulls are presented – a simple recursive algorithm for points in the plane and an algorithm for points in an arbitrary dimensional space. Upper bounds on the number of vertices and facets in a reduced convex hull are used to analyze the worst-case complexity of the algorithms.

1 Introduction

Reduced convex hulls have been introduced by considering a geometric interpretation of Support Vector Machines (SVMs) [1,2]. Reduced convex hulls provide a method of non-uniformly shrinking a convex hull in order to compensate for noise or outlying points, or to increase the margin between two classes in classification problems. The concept is employed by SVMs to ensure that a pair of hulls does not overlap and can therefore be separated using a hyperplane.

Although reduced convex hulls have been well defined, algorithms which compute them in their entirety are rarely considered. Authors who address the topic of reduced convex hulls generally do so in order to train or understand SVMs [1,2,3,4,5]. By contrast, convex hulls have been applied to problems in many domains such as image processing and pattern classification [6,7] and a range of algorithms for their construction have been proposed. There are also a number of problems such as Delaunay triangulation or Voronoi tessellation which can be either reduced to or interpreted as a convex hull problem [8].

The main contributions of this paper are two algorithms for the construction of reduced convex hulls: a simple divide and conquer algorithm for points in the plane and an algorithm for points in an arbitrary dimensional space. These algorithms are generalizations of the Quickhull [9,10,11,8] and Beneath-Beyond [12,8] algorithms for standard convex hulls. The worst-case complexity of the algorithms is considered by introducing upper bounds on the number of facets and vertices in a reduced convex hull.

A. Nicholson and X. Li (Eds.): AI 2009, LNAI 5866, pp. 230–239, 2009.

2   Reduced Convex Hulls (RCHs)

The μ-reduced convex hull of a set of points $P = \{x_1, x_2, \ldots, x_n\}$ is defined as [2,1]

$$RCH(P, \mu) = \left\{ \sum_i^n \alpha_i x_i \ \middle| \ \sum_i^n \alpha_i = 1, \quad 0 \leq \alpha_i \leq \mu \right\}. \tag{1}$$

The difference between reduced convex hulls and convex hulls is that reduced convex hulls introduce an upper bound $0 < \mu \leq 1$ on α_i values. For $\mu = 1$, the reduced convex hull and the convex hull are identical.

The effect of the upper bound μ is that the impact any single point can have on the hull is limited. Consequently, the convex hull is reduced towards the centroid by an amount controlled by the parameter μ. As μ is decreased, the reduced convex hull shrinks (non-uniformly) towards the centroid and more points will be forced outside the hull. At $\mu = 1/n$, the reduced convex hull contains only the centroid, and for smaller μ the reduced convex hull is empty [2].

The concept of reduced convex hulls arose from geometric interpretations of SVM classifiers. Bennett and Bredensteiner [1] and Crisp and Burges [2] showed that training an SVM classifier on a two-class dataset is equivalent to finding the perpendicular bisector of the line joining the nearest points in the reduced convex hulls of two classes (Figure 1). In this context the amount of reduction in the convex hulls determines the tradeoff between the margin of the classifier and the number of margin errors.

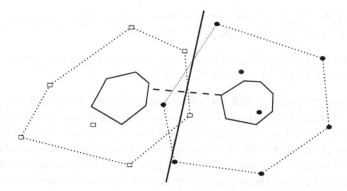

Fig. 1. Convex hulls (dotted lines) are inseparable, whereas the reduced convex hulls (solid lines) are clearly separable. The SVM is the line bisecting the nearest two points of the reduced convex hulls.

3   Theoretical Properties of Reduced Convex Hulls

Like convex hulls, reduced convex hulls form a solid region, the border of which is a convex polytope [3]. Reduced convex hulls can therefore be described in terms of the convex hull of some set of vertices. This suggests that existing convex hull algorithms can be adapted to construct reduced convex hulls, provided there exists a simple means of locating the vertices of a reduced convex hull.

At a glance, locating a vertex of $RCH(P, \mu)$ appears to be a difficult optimization problem, necessitating a search through many possible combinations of weights given to points in P. However, there exists a simple theorem for finding a vertex of $RCH(P, \mu)$ in a particular direction (Theorem 1). This theorem becomes an important part of our reduced convex hull algorithms.

Definition 1. *The* scalar projection *of b onto a is given by $\frac{a \cdot b}{|a|}$.*

Theorem 1. *(Bern and Eppstein [3], Mavroforakis and Theodoridis [5]). For direction v, a vertex $p \in RCH(P, \mu)$ which maximizes $v \cdot p$ satisfies*

$$p = \sum_{i=1}^{m-1} \mu z_i + (1 - (m-1)\mu)z_m, \quad m = \lceil 1/\mu \rceil, \tag{2}$$

where $P = \{z_1, z_2, \ldots, z_n\}$ and points are ordered in terms of scalar projection onto v such that $z_i \cdot v > z_j \cdot v \Rightarrow i < j$ with ties broken arbitrarily.

The process of finding a vertex of the reduced convex hull can be visualized in terms of a plane being 'pushed' into a set of points. If the plane can separate $\lceil 1/\mu \rceil$ points from the set, there is a convex combination of these points (given by Theorem 1) which forms a vertex of the μ-reduced convex hull.

Definition 2. *The $\lceil 1/\mu \rceil$ points which define a vertex of a μ-reduced convex hull are referred to as the* support points *of that vertex.*

Definition 3. *A set of points $P \subset \mathbb{R}^d$ are in* general position *if there exist no $d + 1$ points lying in a $(d-1)$-dimensional plane.*

If the vertices of a convex hull (or reduced convex hull) are in general position, they form a simplicial polytope. It is common practice in convex hull algorithms to assume that vertices are in general position [6]. This means that in $\mathbb{R}^d$ the facets of the polytope are all $(d-1)$-simplices and can each be represented using an identical number of vertices. For example, in two dimensions, facets are 1-simplices (lines) joining the vertices of the convex hull. In three dimensions, facets are 2-simplices (triangles). Our algorithms adopt the convention of constructing convex hulls as simplicial polytopes. The drawback is that when vertices are not in general position, the number of facets used to represent the hull will be sub-optimal. For example, for points in $\mathbb{R}^3$ not in general position, square facets must be represented using several triangles.

For a simplicial polytope in $\mathbb{R}^d$, each facet is itself a $(d-1)$-simplicial polytope with a number of $(d-2)$-simplicial *subfacets* or *ridges*. Each ridge is shared by two adjoining facets. This recursion continues down to 1-simplices and finally 0-simplices, which are generally referred to as *edges* and *vertices* respectively, regardless of the dimensionality of the polytope.

4   An RCH Algorithm for Points in the Plane

Reduced convex hulls are simplest to construct for two-dimensional point sets. For this case our algorithm takes as input a set of points P, a reduction coefficient μ and two initial vertices of the reduced convex hull l and r. Starting vertices can be found using Theorem 1 in conjunction with any direction. The output is an ordered list of vertices V of the reduced convex hull which serves as both a facet list and an adjacency list. Any two consecutive points in V form a facet.

The algorithm itself (Algorithm 1) is a generalization of the Quickhull algorithm for standard convex hulls [9,10,11,8]. The two initial vertices given to the algorithm form a line segment which is treated as two initial facets with opposing normals. Facets are then recursively split into two by using the normal vector of a facet and Theorem 1 to discover new vertices. For the algorithm to function properly, facet normals should be taken to point *away* from the interior of the hull. Once no new facets can be formed, the reduced convex hull is complete.

Algorithm 1. Quickhull for reduced convex hulls

---

function $V = \mathrm{RQH}(P, \mu, l, r)$
$h :=$ a vertex with maximal scalar projection onto the normal vector of the line lr
(computed using Theorem 1)
if $h = l$ or $h = r$ **then**
    return (l, r)
else
    $A :=$ all points in P with a scalar projection onto the normal vector of the line lh
    greater than or equal to that of any of the support points of l and h.
    $B :=$ all points in P with a scalar projection onto the normal vector of the line
    hr greater than or equal to that of any the support points of h and r.
    return $\mathrm{RQH}(A, \mu, l, h) \cup \mathrm{RQH}(B, \mu, h, r)$
end if

The subsets A and B are formed to discard points which can no longer contribute to future iterations of the algorithm (since they are not 'extreme' enough with respect to the current facet being considered). Because the vertices of a reduced convex hull can share multiple support points, A and B are generally not disjoint partitions as they are in the original Quickhull algorithm. The exception is when $\mu = 1$, in which case both algorithms are identical.

5   An RCH Algorithm for Points in an Arbitrary Dimensional Space

For points in an arbitrary dimensional space, reduced convex hulls are more difficult to represent and construct. For d dimensions, hulls consist of a number of d-simplicial facets, each with d neighboring facets. This means that an efficient representation will generally consist of both a facet list and an adjacency list.

The algorithm presented here is a generalization of the Beneath-Beyond algorithm for standard convex hulls [12,8]. The Beneath-Beyond algorithm [12,8] addresses the issue of convex hulls in higher dimensions by starting with a hull containing only a minimal number of points. It then uses Theorem 2 to iteratively update the hull until it contains all points.

Theorem 2 (Barber et al. [6]). *Let H be a convex hull in $\mathbb{R}^d$, and let p be a point in $\mathbb{R}^d - H$. Then F is a facet of the convex hull of $p \cup H$ if and only if*

(a) F is a facet of H and p is below F; or
(b) F is not a facet of H and its vertices are p and the vertices of a sub-facet of H with one incident facet below p and the other incident facet above p.

A point is said to be *above* a facet if it has a scalar projection onto that facet's normal which is greater than that of the facet's vertices. Conversely a point is *below* a facet if it has a scalar projection onto that facet's normal which is less than that of the facet's vertices. Points which are neither below nor above a facet lie on the facet. As in the case of the previous algorithm, facet normals should always be chosen to point *away* from the inside of the hull.

Theorem 2 implies that a convex hull H can be updated to include a new point p as follows. If p lies inside the hull (i.e. it is below all facets of H), no change is required. If p lies outside the hull (i.e. it is above at least one facet of H), any facets of H which are visible from p should be replaced with a cone of new facets (Figure 2) [6,8].

Input to our reduced convex hull algorithm (Algorithm 2) is a set of points P and a reduction coefficient μ. Output consists of:

- A list of vertices V. For d-dimensional input, each vertex is a d-dimensional point.

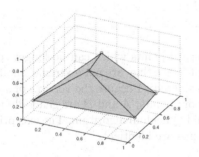 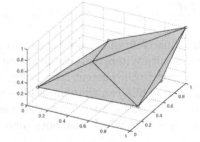

(a) Existing convex hull   (b) Once a new point is added, visible facets are replaced with a cone of new facets

Fig. 2. Updating a convex hull using the Beneath-Beyond Theorem

- A list of facets F. For d-dimensional input, each facet is represented as d indices corresponding to the facet's vertices from V.
- An adjacency list A. Each entry in F has a corresponding entry in A which specifies the indices of neighboring facets.

The starting point for the algorithm (for a d-dimensional point set) is to find any d unique vertices of the reduced convex hull (using Theorem 1 in conjunction with several arbitrary directions). These vertices are used to form two initial facets with opposing normals. For each facet, the algorithm finds a new vertex farthest in the direction the facet's normal vector.[1] The new vertex is added to the hull by 'paving over' any visible facets and removing them as described in the Beneath-Beyond theorem. This process is repeated over all facets until no new vertices can be found.

Algorithm 2. General dimension reduced convex hull algorithm

---

function $[V,F,A] = \mathrm{RH}(P, \mu)$
V = any d vertices of the μ-reduced convex hull (where d is the dimensionality of the points in P)
F = two facets consisting of the vertices in V with opposing normals
for all facets f in F **do**
  v := normal vector of f
  use Theorem 1 to find a vertex p of the RCH with maximal scalar projection onto v
  if p is a new vertex **then**
    X := list of all facets in F which are visible from p (use adjacency list to search facets surrounding f)
    for all edges e along the boundary of X **do**
      g := facet consisting of the vertices in e and p
      add g to F
      update adjacency list
    end for
    remove the facets in X from F
  end if
end for

---

6   Complexity of the Algorithms

There are two main sources of additional complexity in a reduced convex hull as compared to a convex hull. Firstly, a much greater number of vertices and facets can be formed in a reduced convex hull, forcing more iterations of the algorithm. Secondly, support points are shared between multiple vertices, precluding the partitioning methods used in the more efficient convex hull algorithms.

---

[1] In $\mathbb{R}^d$, the normal vector of a d-simplex with vertices $p_1, p_2, \ldots, p_d$ can be calculated by solving the system of linear equations $p_1 \cdot v = p_2 \cdot v = \ldots = p_d \cdot v = \mathrm{const.}$

These additional sources of complexity can be shown by examining the maximum number of facets which can be present in a reduced convex hull. Preparata and Shamos [8] give an upper bound (attributed to Klee [13]) F_n on the number of facets in a standard convex hull of n points in d dimensions as:

$$F_n = \begin{cases} \frac{2n}{d}\binom{n-\frac{d}{2}-1}{\frac{d}{2}-1}, & \text{for } d \text{ even} \\ 2\binom{n-\lfloor\frac{d}{2}\rfloor-1}{\lfloor\frac{d}{2}\rfloor}, & \text{for } d \text{ odd}. \end{cases} \tag{3}$$

This bound corresponds to the case where all of the n points form vertices in the convex hull.

Klee's [13] upper bound can be extended to cover the case of reduced convex hulls by further considering an upper bound on the number of vertices in a reduced convex hull V_n. Letting $m = \lceil 1/\mu \rceil$ and for n points in an arbitrary dimensional space, we can write this bound as

$$V_n = \begin{cases} \binom{n}{m}, & \text{if } \frac{1}{\mu} \text{ is an integer} \\ \binom{n}{m}m, & \text{otherwise.} \end{cases} \tag{4}$$

Equation 4 follows from Theorem 1, which implies that any vertex of a reduced convex hull is a convex combination of exactly $\lceil 1/\mu \rceil$ points. For cases where $1/\mu$ is an integer, the bound V_n is reached if, for all subsets containing $1/\mu$ points, there exists some plane which separates the subset from the rest of the $n - 1/\mu$ points. This can occur, for example, when $n = d+1$. Notice that for cases where $1/\mu$ is not an integer, there are potentially a much larger number of vertices. This is due to the fact that one of the $\lceil 1/\mu \rceil$ support points must receive a smaller weighting than the others, increasing the upper bound by a factor of $\lceil 1/\mu \rceil$.

Combining equations (3) and (4) yields an upper bound $R_n = F_{V_n}$ on the total number of facets in a reduced convex hull of n points in d dimensions. This means that, compared to convex hulls, reduced convex hulls have potentially a much greater number of facets, depending on the value of μ.

For the two dimensional algorithm, the worst case complexity is $O(V_n n)$, or $O(n^{m+1}/(m-1)!)$. This occurs when partitions are highly overlapping and each iteration requires the calculation of dot products for all n points. There is a maximum of V_n iterations since in two dimensions the maximum number of facets is equal to the number of vertices. If $\mu < 1$, this is worse than the Quickhull algorithm, which achieves an $O(n \log n)$ worst case complexity assuming partitions are approximately balanced in each iteration, or $O(n^2)$ otherwise [14].

For the case of points in an arbitrary dimensional space, there are several factors which contribute to the complexity of the algorithm. Again considering the worst case, by far the dominant cost is that of finding the scalar projection of all points onto the normal vectors of all facets, an $O(R_n n)$ operation. There are additional costs such as: finding V_n vertices $(O(V_n n))$; calculating the normal vectors of new facets each time a vertex is added; and iteratively adding V_n

vertices to the hull. However, all such costs are eclipsed by the original cost of $O(R_n n)$, or $O(((\binom{n}{m})m)^{\lfloor d/2 \rfloor} n / \lfloor d/2 \rfloor!)$.

This complexity compares unfavorably to the $O(F_n)$ worst case complexity of efficient convex hull algorithms [6,15]. The introduction of the larger R_n bound instead of F_n is caused by the increased number of facets in a reduced convex hull, whereas the increase by a factor of n is caused by the omission of the 'outside' sets of Barber et al. [6], which cannot be maintained in the reduced case due to vertices sharing support points.

7   Discussion

Reduced convex hulls provide a desirable method of reducing a convex hull because they take into account the density of points. Sparse outlying points are reduced over much more rapidly as μ decreases than dense clusters of points (Figure 3). This is a likely contributor to the high accuracy of SVMs in classification problems, where reduced convex hulls are used to lessen the impact of outlying points and increase the margin between two classes.

The bounds on both the number of vertices and number of facets in a reduced convex hull are at a maximum when $1/\mu \approx n/2$, which highlights an interesting relationship between the cases of $\mu = 1/k$ and $\mu = 1/(n-k)$, for $0 < k < n$. These cases produce reduced convex hulls with an identical number of vertices (Figure 4). This property occurs since a plane separating $(n-k)$ of n points will separate the remaining k points if the orientation of the plane is reversed.

It is also apparent from Figure 4 that reduced convex hulls with μ in a specific range $(1/k, 1/(k-1))$ contain an identical number of vertices and facets. This is because $\lceil 1/\mu \rceil$ has an identical value for all μ values in this range, meaning each vertex has the same number of support points. Notice also the large number of facets which occur in these ranges (much larger than when $1/\mu$ is an integer), as consistent with the bounds given in the previous section.

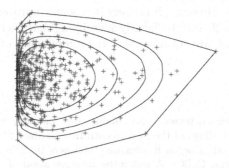

Fig. 3. Reduced convex hulls for $\mu = 1, \frac{1}{10}, \frac{1}{20}, \frac{1}{50}, \frac{1}{100}$. Points are normally distributed across the y-axis and exponentially distributed across the x-axis.

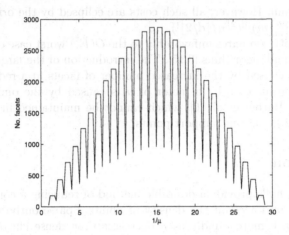

Fig. 4. $1/\mu$ plotted against total number of facets for the reduced convex hull of a set of 30 uniformly distributed random points in $\mathbb{R}^3$

8   Conclusion

To summarize, this paper has decoupled reduced convex hulls from their origins in SVMs. Two reduced convex hull algorithms have been presented: a recursive two-dimensional algorithm for points in the plane and an algorithm for points of arbitrary dimensionality. In future work we hope that applications in which convex hulls act poorly due to noise or outlying points can be identified and improved via the introduction of reduced convex hulls. The high accuracy of SVM classifiers employing the concept of reduced convex hulls suggests that other machine learning tasks could likely also benefit from their use.

Future work could also aim to increase the efficiency of reduced convex hull algorithms. There is a significant gap in efficiency between the reduced convex hull algorithms presented in this paper and state-of-the-art convex hull algorithms. This is partially due to the inherently larger number of vertices and facets in reduced convex hulls. However, it is likely that some improvement could be made by developing methods of intelligently identifying and excluding from calculations any points which are not relevant in the computation of a particular section of a hull.

References

1. Bennett, K.P., Bredensteiner, E.J.: Duality and geometry in SVM classifiers. In: ICML 2000: Proceedings of the Seventeenth International Conference on Machine Learning, pp. 57–64. Morgan Kaufmann Publishers Inc., San Francisco (2000)
2. Crisp, D.J., Burges, C.J.C.: A geometric interpretation of ν-SVM classifiers. In: Solla, S.A., Leen, T.K., Müller, K.R. (eds.) Advances in Neural Information Processing Systems 12, Papers from Neural Information Processing Systems (NIPS), Denver, CO, USA, pp. 244–251. MIT Press, Cambridge (1999)

3. Bern, M., Eppstein, D.: Optimization over zonotopes and training Support Vector Machines. In: Dehne, F., Sack, J.-R., Tamassia, R. (eds.) WADS 2001. LNCS, vol. 2125, pp. 111–121. Springer, Heidelberg (2001)
4. Mavroforakis, M.E., Sdralis, M., Theodoridis, S.: A novel SVM geometric algorithm based on reduced convex hulls. In: 18th International Conference on Pattern Recognition, vol. 2, pp. 564–568 (2006)
5. Mavroforakis, M.E., Theodoridis, S.: A geometric approach to Support Vector Machine (SVM) classification. IEEE Transactions on Neural Networks 17(3), 671–682 (2006)
6. Barber, C.B., Dobkin, D.P., Huhdanpaa, H.: The quickhull algorithm for convex hulls. ACM Transactions on Mathematical Software 22(4), 469–483 (1996)
7. Aurenhammer, F.: Voronoi diagrams – a survey of a fundamental geometric data structure. ACM Computing Surveys 23(3), 345–405 (1991)
8. Preparata, F.P., Shamos, M.I.: Computational Geometry, 2nd edn. Springer, New York (1988)
9. Eddy, W.F.: A new convex hull algorithm for planar sets. ACM Transactions on Mathematical Software 3(4), 398–403 (1977)
10. Bykat, A.: Convex hull of a finite set of points in two dimensions. Information Processing Letters 7, 296–298 (1978)
11. Green, P.J., Silverman, B.W.: Constructing the convex hull of a set of points in the plane. The Computer Journal 22(3), 262–266 (1978)
12. Kallay, M.: Convex hull algorithms in higher dimensions. Department of Mathematics, University of Oklahoma (1981) (unpublished manuscript)
13. Klee, V.: Convex polytopes and linear programming. In: IBM Scientific Computing Symposium: Combinatorial Problems, pp. 123–158. IBM, Armonk (1966)
14. O'Rourke, J.: Computation Geometry in C, 2nd edn. Cambridge University Press, Cambridge (1998)
15. Clarkson, K.L., Mehlhorn, K., Seidel, R.: Four results on randomized incremental constructions. Computational Geometry: Theory and Applications 3(4), 185–212 (1993)

Regularized Kernel Local Linear Embedding on Dimensionality Reduction for Non-vectorial Data

Yi Guo[1], Junbin Gao[2,*], and Paul W. Kwan[3]

[1] yg_au@yahoo.com.au
[2] School of Computing and Mathematics,
Charles Sturt University, Bathurst, NSW 2795, Australia
jbgao@csu.edu.au
[3] School of Science and Technology,
University of New England, Armidale, NSW 2351, Australia
kwan@turing.une.edu.au

Abstract. In this paper, we proposed a new nonlinear dimensionality reduction algorithm called regularized Kernel Local Linear Embedding (rKLLE) for highly structured data. It is built on the original LLE by introducing kernel alignment type of constraint to effectively reduce the solution space and find out the embeddings reflecting the prior knowledge. To enable the non-vectorial data applicability of the algorithm, a kernelized LLE is used to get the reconstruction weights. Our experiments on typical non-vectorial data show that rKLLE greatly improves the results of KLLE.

1 Introduction

In recent years it has been undergoing a large increase in studies on dimensionality reduction (DR). The purpose of DR is mainly to find the corresponding counterparts (or embeddings) of the input data of dimension D in a much lower dimensional space (so-called latent space, usually Euclidean) of dimension d and $d \ll D$ without incurring significant information loss. A number of new algorithms which are specially designed for nonlinear dimensionality reduction (NLDR) have been proposed such as Local Linear Embedding (LLE) [1], Lapacian Eigenmaps (LE) [2], Isometric mapping (Isomap) [3], Local Tangent Space Alignment (LTSA) [4], Gaussian Process Latent Variable Model (GPLVM) [5] etc. to replace the simple linear methods such as Principal Component Analysis (PCA) [6], Linear Discriminant Analysis (LDA) [7] in which the assumption of linearity is essential.

Among these NLDR methods, it is worth mentioning those which can handle highly structured or so-called *non-vectorial* data [8] (for example video sequences, proteins etc which are not readily converted to vectors) directly without vectorization. This category includes the "kernelized" linear methods. Typical methods

---

* Corresponding author.

A. Nicholson and X. Li (Eds.): AI 2009, LNAI 5866, pp. 240–249, 2009.

are Kernel PCA (KPCA) [9], Generalized Discriminant Analysis (GDA or KLDA) [10]. The application of the kernel function not only introduces certain nonlinearity implied by the feature mapping associated with the kernel which enables the algorithms to capture the nonlinear features, but also embraces much broader types of data including the aforementioned non-vectorial data. Meanwhile, kernels can also be regarded as a kind of similarity measurements which can be used in measurement matching algorithms like Multi-Dimensional Scaling (MDS) [11]. A typical example is Kernel Laplacian Eigenmaps (KLE) [12]. Because these methods can directly use the structured data through kernel functions and hence bypass the vectorization procedure which might be a source of bias, they are widely used in complex input patterns like proteins, fingerprints etc.

Because of its simplicity and elegant incarnation of nonlinearity from local linear patches, LLE has attracted a lot of attention. However, it has two obvious drawbacks. Firstly, it can only take vectorial data as input. Secondly, it does not exploit the prior knowledge of input data which is reflected by its somewhat arbitrary constraints on embeddings. As more and more non-vectorial data applications are emerging quickly in machine learning society, it is very desirable to endow LLE the ability to process this type of data. Fortunately, it is not difficult since only inner product is involved in LLE formulation. The "kernel trick " [13] provides an elegant solution to this problem. By introducing kernel functions, LLE can accept non-vectorial data which can be called KLLE. Moreover, another benefit from kernel approaches is its similarity measure interpretation which can be seen as a prior knowledge. We will utilize this understanding to restrict the embeddings and hence provide a solution to the second problem. This is done by incorporating kernel alignment into the current LLE as a regularizer of the embeddings which enforces the similarity contained in kernel function to be duplicated in lower dimensional space. It is equivalent to imposing a preference on the embeddings which favors such configuration that shares the same similarity relation (reflected by kernel function) as that among original input data. We conclude it as a new algorithm called regularized KLLE (rKLLE) as the main contribution of this paper. Our experiments on some typical non-vectorial data show that rKLLE greatly improves the results of KLLE.

This paper is organized as follows. Next section gives a brief introduction to LLE and we deduce the formulation of KLLE in sequel. rKLLE is developed in Section 4, followed by experimental results to show its effectiveness. Finally we conclude this paper in last section with highlight of future research.

2   Local Linear Embedding

We use following notations throughout the paper. $\mathbf{y}_i$ and $\mathbf{x}_i$ are the i-th input datum and its corresponding low-dimensional embedding, and $\mathbf{Y}$ and $\mathbf{X}$ the collection of input data and embeddings respectively. Generally, $\mathbf{X}$ is a matrix with data in rows.

Locally Linear Embedding (LLE) preserves the local linear relations in the input data which is encapsulated in a weight matrix $\mathbf{W}$. The algorithm starts with constructing a neighborhood graph by n nearest neighboring and then $\mathbf{W}$ ($[\mathbf{W}]_{ij} = w_{ij}$) is obtained by

$$\mathbf{W} = \arg\min_{\mathbf{W}} \sum_{i=1}^{N} \|\mathbf{y}_i - \sum_{j=1}^{n} w_{ij}\mathbf{y}_{i_j}\|^2 \qquad (1)$$

subject to $\sum_j w_{ij} = 1$ and $w_{ij} = 0$ if there is no edge between $\mathbf{y}_i$ and $\mathbf{y}_j$ in the neighborhood graph. $\mathbf{y}_{i_j}$ is the j-th neighbor of $\mathbf{y}_i$.

Finally, the lower-dimensional embeddings are estimated by minimizing

$$\sum_i \|\mathbf{x}_i - \sum_j w_{ij}\mathbf{x}_j\|^2. \qquad (2)$$

with respect to $\mathbf{x}_i$'s under the constraints $\sum_i \mathbf{x}_i = \mathbf{0}$ and $\frac{1}{N}\sum_i \mathbf{x}_i\mathbf{x}_i^\top = \mathbf{I}$ to remove arbitrary translations of the embeddings and avoid degenerate solutions. By doing this, the local linearity is reproduced in latent space.

3   Kernelized LLE

Actually, because of the quadratic form (1), w_{ij}'s are solved for each $\mathbf{y}_i$ separately in LLE. So we minimize $\|\mathbf{y}_i - \sum_j w_{ij}\mathbf{y}_{i_j}\|^2$ with respect to w_{ij}'s which is

$$\sum_j \sum_k w_{ij}(\mathbf{y}_{i_k} - \mathbf{y}_i)^\top (\mathbf{y}_{i_j} - \mathbf{y}_i)w_{ik} = \mathbf{w}_i^\top \mathbf{K}_i \mathbf{w}_i \qquad (3)$$

subject to $\mathbf{e}^\top \mathbf{w}_i = 1$ where $\mathbf{e}$ is all 1 column vector. $\mathbf{w}_i$ is the vector of the reconstruction weights of $\mathbf{x}_i$, i.e. $\mathbf{w}_i = [w_{i1}, \ldots, w_{in}]$ and $\mathbf{K}_i$ is the local correlation matrix whose jk-th element is $(\mathbf{y}_{i_k} - \mathbf{y}_i)^\top (\mathbf{y}_{i_j} - \mathbf{y}_i)$.

Apparently, only inner product of input data is involved in (3). By using the "kernel trick" [13], the inner product can be replaced by any other positive definite kernel functions. Hence we substitute every inner product by a kernel $k_y(\cdot, \cdot)$ in the formation of $\mathbf{K}_i$ and have

$$[\mathbf{K}_i]_{jk} = k_y(\mathbf{y}_{i_k}, \mathbf{y}_{i_j}) - k_y(\mathbf{y}_{i_k}, \mathbf{y}_i) - k_y(\mathbf{y}_{i_j}, \mathbf{y}_i) + k_y(\mathbf{y}_i, \mathbf{y}_i).$$

Because the kernel function implies a mapping function ϕ from input data space to feature space and $k_y(\mathbf{y}_i, \mathbf{y}_j) = \phi(\mathbf{y}_i)^\top \phi(\mathbf{y}_j)$, we are actually evaluating the reconstruction weights in feature space. After we get the reconstruction weights, we can go further to solve (2) to obtain the embeddings in latent space.

We proceed to minimize (3) with equality constraint using Lagrange multiplier (we ignore the subscript i for simplicity):

$$J = \mathbf{w}^\top \mathbf{K}\mathbf{w} - \lambda(\mathbf{w}^\top \mathbf{e} - 1)$$

in which stationary point is the solution. $\frac{\partial J}{\partial \mathbf{w}} = 2\mathbf{K}\mathbf{w} - \lambda\mathbf{e} = 0$ leads to $\mathbf{w} = \frac{1}{2}\lambda\mathbf{K}^{-1}\mathbf{e}$. With $\mathbf{w}^\top \mathbf{e} = 1$, we have $\lambda = \frac{2}{\mathbf{e}^\top \mathbf{K}^{-\top}\mathbf{e}}$. What follows is

$$\mathbf{w} = \frac{\mathbf{K}^{-1}\mathbf{e}}{\mathbf{e}^\top \mathbf{K}^{-\top}\mathbf{e}}.$$

The last thing is the construction of the neighborhood graph. It is quite straight-forward in fact. Since kernel function represents similarity, we can just simply choose n most similar input data around $\mathbf{y}_i$. This corresponds to searching n nearest neighbors of $\phi(\mathbf{y}_i)$ in feature space because distance can be converted to inner product easily.

4 Regularized KLLE

Because of introducing kernel functions, not only can we process non-vectorial data in LLE, but also we are provided similarity information among input data which can be further exploited. The constraints used in the KLLE force the embeddings to have standard deviation. Apparently, this preference is imposed artificially on the embeddings which may not reflect the ground truth. This raises a question: can we use something more "natural" instead? Combining these points gives rise to the idea of replacing current constraint in KLLE by similarity matching.

The idea is implemented in following steps. Firstly, we pick up a similarity measure $(k_x(\cdot,\cdot)$, another kernel) in latent space which is matched to its counter-part in input space i.e. $k_y(\cdot,\cdot)$. Secondly, the similarity matching is implemented by kernel alignment. Thirdly, we turn the constrained optimization problem to regularization and therefore, the new regularized KLLE (rKLLE) minimizes the following objective function:

$$L = \sum_i \|\mathbf{x}_i - \sum_j w_{ij}\mathbf{x}_{i_j}\|^2 - \alpha \sum_{ij} k_x(\mathbf{x}_i,\mathbf{x}_j)k_y(\mathbf{y}_i,\mathbf{y}_j) \qquad (4)$$
$$+ \beta \sum_{ij} k_x^2(\mathbf{x}_i,\mathbf{x}_j).$$

The second and third regularization terms are from similarity matching[1]. α and β are positive coefficients which control the strength of the regularization. What is expressed in (4) is that the embedded data will retain the same local linear relationships as input data under the constraint that they should also exhibit the same similarity structure as that in input space.

In rKLLE, the prior knowledge provided by $k_y(\cdot,\cdot)$ is fully used in latent space and hence avoids from introducing other "rigid" assumptions which may be far away from the truth. There is also much room to accommodate additional priors due to its flexible algorithmic structure. For example if we know that the embeddings are from Gaussian distribution, we can add another regularizer on $\mathbf{X}$ (e.g. $\sum_i \mathbf{x}_i^\top \mathbf{x}_i$) at the end to incorporate this. An important issue related to the similarity match is the selection of the $k_x(\cdot,\cdot)$. In practice, we can choose RBF kernel, $k_x(\mathbf{x}_i,\mathbf{x}_j) = \gamma \exp(-\sigma\|\mathbf{x}_i-\mathbf{x}_j\|^2)$, because it has strong connection with Euclidean distance and this connection can be fine tuned by choosing appropriate hyper-parameters. Fortunately, the optimization of hyper-parameters can be done automatically as shown below.

---

[1] The second term is kernel alignment and last term is designed to avoid trivial solution such as infinity.

The computational cost of rKLLE is higher than KLLE since (4) does not have close form solution. The above objective function can be written in simpler matrix form

$$L = \text{tr}[(\mathbf{X} - \mathbf{W}\mathbf{X})^{\top}(\mathbf{X} - \mathbf{W}\mathbf{X})] - \alpha\text{tr}[\mathbf{K}_X^{\top}\mathbf{K}_Y] + \beta\text{tr}[\mathbf{K}_X^{\top}\mathbf{K}_X]$$
$$= \text{tr}[(\mathbf{X}^{\top}\mathbf{M}\mathbf{X})] - \alpha\text{tr}[\mathbf{K}_X^{\top}\mathbf{K}_Y] + \beta\text{tr}[\mathbf{K}_X^{\top}\mathbf{K}_X]$$

where $\mathbf{K}_X$ and $\mathbf{K}_Y$ are the kernel Gram matrices of $k_x(\cdot, \cdot)$ and $k_y(\cdot, \cdot)$ respectively, $\mathbf{M} = (\mathbf{I} - \mathbf{W})^{\top}(\mathbf{I} - \mathbf{W})$ and $\mathbf{I}$ is the identity matrix. We have to employ gradient descent based solver here. For the derivative, we first obtain

$$\frac{\partial L_{2,3}}{\partial \mathbf{K}_X} = -2\alpha\mathbf{K}_Y + 2\beta\mathbf{K}_X,$$

where $L_{2,3}$ is the second and third term of L. Then we get $\frac{\partial L_{2,3}}{\partial X}$ by chain rule (it depends on the form of k_x). The derivative of the first term of L is $2\mathbf{M}\mathbf{X}$. By putting them together, we can obtain $\frac{\partial L}{\partial X}$. The derivative of L with respect to hyper-parameters of k_x, denoted by $\boldsymbol{\Theta}$, can be calculated in the same way. Once we have the current version of $\mathbf{X}$, the gradient can be evaluated. Therefore, optimization process can be initialized by a guess of $\mathbf{X}$ and $\boldsymbol{\Theta}$. The initial $\boldsymbol{\Theta}$ can be arbitrary while starting $\mathbf{X}$ can be provided by other DR methods as long as non-vectorial data are applicable. From the candidates of gradient descent solvers, we choose SCG (scaled conjugate gradient) [14] because of its fast speed.

5   Experimental Results

To demonstrate the effectiveness of the proposed rKLLE algorithm, the experiments of visualizing non-vectorial data (the target latent space is a normal 2-D plane) were conducted on images (MNIST handwritten digits[2] and Frey faces[3]) and proteins (from SCOP database, Structural Classification Of Protein[4]). Proteins are recognized as typical highly structured data. The results of other algorithms are also shown for comparison.

5.1 Parameters Setting

rKLLE has some parameters to be determined beforehand. Through empirical analysis (performing batches of experiments on different data sets varying only the parameters), we found the proposed algorithm is not sensitive to the choice of the parameters, as long as the conjugate gradient optimization can be carried out without immature early stop. So we use the following parameters throughout the experiments which are determined by experiments: $\alpha = 1e-3$ and $\beta = 5e-5$ and $n = 6$ in rKLLE neighborhood graph construction. The minimization will stop after 1000 iterations or when consecutive update of the objective function is less than 10^{-7}. $k_x(\cdot, \cdot)$ is RBF kernel and initialization is done by KPCA[5].

---

[2] MNIST digits are available at http://yann.lecun.com/exdb/mnist/

[3] Available at http://www.cs.toronto.edu/~roweis/data/

[4] SCOP data is available at http://scop.mrc-lmb.cam.ac.uk/scop/

[5] We choose KPCA instead of KLLE because this configuration yields better results in terms of leave-one-out 1NN classification errors.

Table 1. Comparison of leave-one-out 1NN classification errors of different algorithms

Algorithm	rKLLE	KLLE	KLE	KPCA	Isomap	LTSA
error	**110**	171	157	333	222	232

5.2 Handwritten Digits

A subset of handwritten digits images is extracted from the MNIST database. The data set consists of 500 images with 50 images per digit. All images are in grayscale and have a uniform size of 28 × 28 pixels. It is easy to convert them to vectors. So we can also present the results of other DR algorithms for comparison. However, in rKLLE, they were treated as non-vectorial data as bags of pixels [15] and use the shape context based IGV (SCIGV) kernel [16] which is specially designed for shapes in images and robust to the translation of shapes in images.

The experimental results are presented in Figure 1 and legend is in panel (c). Visually, the result of rKLLE is much better than others. The 2D representations of rKLLE reveal clearer clusters of digits than others. To give a quantitative analysis on the quality of the clusters, we use the leave-one-out 1 nearest neighbor (1NN) classification errors as in [5]. The smaller the number of errors, the better the method. The 1NN classification errors of different methods are collected in Table 1 and the result of each method is the best it can achieve by choosing the optimal parameters of the method (as shown in Figure 1) according to this standard. It is interesting to observe that rKLLE is the best regarding this standard. It shows clearly that rKLLE improves the result of KLLE both visually and quantitatively.

5.3 Proteins

Another promising application of DR is in bioinformatics. Experiments were conducted on the SCOP database. This database provides a detailed and comprehensive description of the structural and evolutionary relationships of the proteins of known structure. 292 proteins from different superfamilies and families are extracted for the test. The kernel for proteins is MAMMOTH kernel which is from the family of the so-called alignment kernels whose thorough analysis can be found in [17]. The corresponding kernel Gram matrices are available on the website of the paper and were used directly in our experiments.

Visualizing proteins on the 2D plane is of great importance to facilitate researchers to understand the biological meaning. The representation of proteins on the 2D plane should reflect the relational structure among proteins, that is, proteins having similar structures should be close while those with different structures should be far away.

The results are plotted in Figure 2. The results of other non-vectorial data applicable algorithms are also presented for comparison. Each point (denoted as a shape in the figure) represents a protein. The same shapes with the the same color

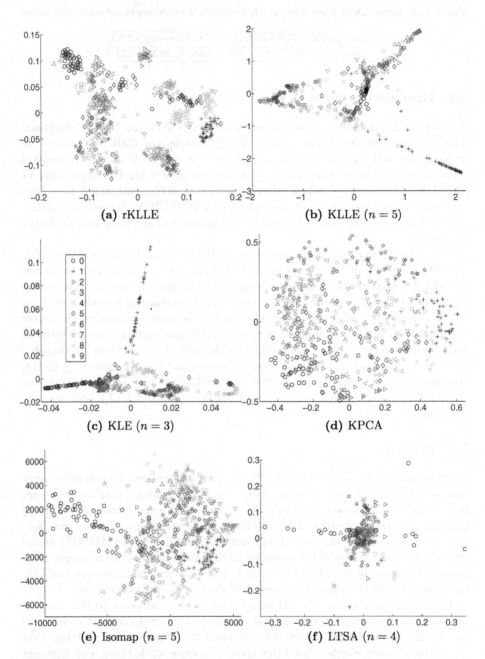

Fig. 1. The result of different algorithms on MNIST handwritten digits database. The parameters of algorithms are chosen to achieve lowest 1NN classification errors.

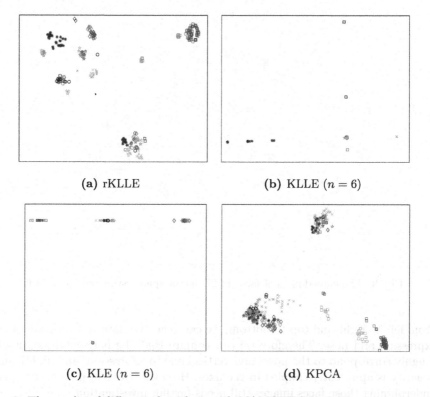

(a) rKLLE   **(b)** KLLE ($n = 6$)

(c) KLE ($n = 6$)   **(d)** KPCA

Fig. 2. The results of different algorithms with MAMOTH kernel. The parameters of algorithms are chosen to achieve lowest 1NN classification errors.

are the proteins from same families while the same shapes with different colors represent the proteins from different families but from the same superfamilies.

rKLLE reveals the fact that proteins from the same families congregate together as tight clusters and hence gains better interpretability. Interestingly, it also reveals the truth that the proteins from the same superfamily but different families are similar in structure, which is reflected by the fact that the corresponding groups (families) are close if they are in the same superfamily (same shape). Others fail to uncover these.

5.4   rKLLE on Image Manifold Learning

Lastly, we present the result of rKLLE on image manifold learning. The objects are 1965 images (each image is 20×28 grayscale) of a single person's face extracted from a digital movie which are also used in [1]. Since the images are well aligned, we simply use the linear kernel as $k_y(\cdot, \cdot)$ in this case.

Two facts can be observed from the result shown in Figure 3. First, it demonstrates group property of the faces. The faces with similar expressions and poses congregated as clusters. Second, The embeddings of the faces in 2D latent space indicate the possible intrinsic dimensionality of the images: expression and pose.

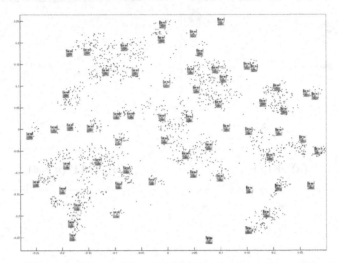

Fig. 3. The embeddings of faces in 2D latent space estimated by rKLLE

From left to right and top to bottom, we can read the natural transition of the expression and pose. Therefore we can conjure that the horizontal axis might roughly correspond to the poses and vertical one to expressions and highly non-linearity is apparently coupled in the axes. However, the real parametric space underpinning those faces images still needs further investigation.

6 Conclusion

In this paper, we proposed the regularized KLLE in which the original constraint of embeddings is replaced by a more natural similarity matching. It exploits the given information of the input data through regularization. It is a showcase of incorporating prior knowledge into dimensionality reduction process. Although the computational cost is higher than KLLE or LLE, the improvement is significant from the results of experiments on typical non-vectorial data.

Due to the flexible structure of rKLLE, it is possible to handle other prior knowledge like class information. So it is likely to extend it to supervised or semisupervised learning setting. This will be our future research.

References

1. Roweis, S.T., Saul, L.K.: Nonlinear dimensionality reduction by locally linear embedding. Science 290(22), 2323–2326 (2000)
2. Belkin, M., Niyogi, P.: Laplacian eigenmaps for dimensionality reduction and data representation. Neural Computation 15(6), 1373–1396 (2003)
3. Tenenbaum, J.B., de Silva, V., Langford, J.C.: A global geometric framework for nonlinear dimensionality reduction. Science 290(22), 2319–2323 (2000)

4. Zhang, Z., Zha, H.: Principal manifolds and nonlinear dimensionality reduction via tangent space. SIAM Journal on Scientific Computing 26(1), 313–338 (2005)
5. Lawrence, N.: Probabilistic non-linear principal component analysis with gaussian process latent variable models. Journal of Machine Learning Research 6, 1783–1816 (2005)
6. Jolliffe, M.: Principal Component Analysis. Springer, New York (1986)
7. Fisher, R.A.: The use of multiple measurements in taxonomic problems. Annals of Eugenics 7, 179–188 (1936)
8. Gärtner, T.: A survey of kernels for structured data. ACM SIGKDD Explorations Newsletter 5(1), 49–58 (2003)
9. Schölkopf, B., Smola, A.J., Müller, K.: Nonlinear component analysis as a kernel eigenvalue problem. Neural Computation 10, 1299–1319 (1998)
10. Baudat, G., Anouar, F.: Generalized discriminant analysis using a kernel approach. Neural Computation 12(10), 2385–2404 (2000)
11. Davison, M.L.: Multidimensional Scaling. Wiley series in probability and mathematical statistics. Applied probability and statistics. Wiley, New York (1983)
12. Guo, Y., Gao, J., Kwan, P.W.: Kernel Laplacian eigenmaps for visualization of non-vectorial data. In: Sattar, A., Kang, B.-h. (eds.) AI 2006. LNCS (LNAI), vol. 4304, pp. 1179–1183. Springer, Heidelberg (2006)
13. Schölkopf, B., Smola, A.: Learning with Kernels: Support Vector Machines, Regularization, Optimization, and Beyond. The MIT Press, Cambridge (2002)
14. Nabney, I.T.: NETLAB: Algorithms for Pattern Recognition. Advances in Pattern Recognition. Springer, London (2004)
15. Jebara, T.: Images as bags of pixels. In: Ninth IEEE International Conference on Computer Vision (ICCV 2003), vol. 1, pp. 265–272 (2003)
16. Guo, Y., Gao, J.: An integration of shape context and semigroup kernel in image classification. In: International Conference on Machine Learning and Cybernetics (2007)
17. Qiu, J., Hue, M., Ben-Hur, A., Vert, J.P., Noble, W.S.: An alignment kernel for protein structures. Bioinformatics 23, 1090–1098 (2007)

Incremental E-Mail Classification and Rule Suggestion Using Simple Term Statistics

Alfred Krzywicki and Wayne Wobcke

School of Computer Science and Engineering
University of New South Wales
Sydney NSW 2052, Australia
{alfredk,wobcke}@cse.unsw.edu.au

Abstract. In this paper, we present and use a method for e-mail categorization based on simple term statistics updated incrementally. We apply simple term statistics to two different tasks. The first task is to predict folders for classification of e-mails when large numbers of messages are required to remain unclassified. The second task is to support users who define rule bases for the same classification task, by suggesting suitable keywords for constructing Ripple Down Rule bases in this scenario. For both tasks, the results are compared with a number of standard machine learning algorithms. The comparison shows that the simple term statistics method achieves a higher level of accuracy than other machine learning methods when taking computation time into account.

1 Introduction

Incremental, accurate and fast automatic e-mail classification methods are important to assist users in everyday e-mail management. In this paper, we propose a solution to an e-mail classification problem in a specific organisational scenario. In this context, users are expected to classify some, but only some, of their messages into one of the folders following a consistent organisational policy, while the rest of the messages remain unclassified and may be deleted or placed in general or private folders. Because of the large volume and flow of messages, any solution must meet stringent computational requirements. Firstly, message categorization must be computed in near real time, typically less than a couple of seconds, so as to avoid delays in the receipt of messages. Secondly, classification must be sufficiently accurate to maintain the consistency of the classification scheme and to reduce the effort required by the user to select the correct category from a number presented, or to decide whether a message should be classified at all. An acceptable level of user interaction can be achieved by flexible selection of accuracy and coverage, in other words, the categorizer must either provide a fairly accurate suggestion or no suggestion at all. The categorizer must be able to distinguish between messages that need to classified and those that do not need to be classified. This makes the classifier's task harder because, apart from predicting the correct class for the classified messages, it also needs to decide which messages are supposed to be classified.

A common way to handle non-classified messages is to treat them as a separate class and to learn a classifier for this class. The problem with this approach is that the

A. Nicholson and X. Li (Eds.): AI 2009, LNAI 5866, pp. 250–259, 2009.

unclassified group of e-mails may cover many different topics and, when treated as a single class, may become highly inconsistent. We found that it is better to separate classified from unclassified messages by using a threshold technique, similar to Yang [6], with the difference that in our work the threshold itself is learned in the process of classification.

In order to test classifiers in the above scenario, we use a set of e-mails from an Australian government organisation, consisting of about 17 000 messages, out of which only 48% are are intended to be classified into 45 categories of interest. Since we did not have access to all messages in users' mailboxes, the messages not to be classified were those messages pre-classified in around 150 other categories and also included a number of messages that were not intended to be classified into any folder. This set, described in more detail in Wobcke, Krzywicki and Chan [5], is referred to as the *gov* data set in this paper.

A second problem we address is the suggestion of keywords from each e-mail in order to construct conditions for rules in a Ripple Down Rule base used to classify e-mails. In previous work (Ho *et al.* [4]), a Ripple Down Rule system (Compton and Jansen [2]) was used to manage e-mail foldering in the E-Mail Management Assistant (EMMA). One aspect to using a Ripple Down Rule system is to provide support to users in defining rules. In the e-mail classification domain, this means finding the most useful keywords in a given e-mail that enable the correct classification of the e-mail into its folder and which also can be applied to future e-mails.

In this paper, we introduce a method based on simple term statistics (STS) that addresses the computational requirements for e-mail classification. We show that this simple method gives a relatively high accuracy for a fraction of the processing cost, when used to predict a folder in an e-mail set, where only a part of the messages are meant to be classified. We also show that the STS method can be applied to keyword generation for e-mail classification rules.

The rest of the paper is structured as follows. In the next section, we briefly describe and discuss other selected research on document categorization. In Section 3 we present definitions of the Simple Term Statistics methods and a description of the STS algorithm. Sections 4 and 5 contain the description of experiments used to evaluate the proposed method on the e-mail classification and term rating tasks respectively. Finally, section 6 summarises and concludes the paper.

2 Related Work

A variety of methods have been researched for document categorization in general and e-mail foldering in particular. Bekkerman *et al.* [1] evaluate a number of machine learning methods on the Enron e-mail data set. In these experiments, methods are evaluated separately for 7 users over all major folders for those users. Messages are processed in batches of 100. For each batch, a model built from all previous messages is tested on the current batch of 100 messages. The overall most accurate SVM method took about half an hour to process about 3000 e-mails. Despite a reasonably high accuracy, the long processing time makes SVM and MaxEnt unsuitable for online applications, especially with larger sets, such as the *gov* data set used in our experiments. Another important difference from our work is that in the Enron e-mail set all e-mails are meant to be classified, whereas in the e-mail set we used only about 48% of messages are supposed to be classification.

Dredze *et al.* [3], use TF-IDF, latent semantic analysis (LSA) and latent Dirichlet allocation (LDA) to pre-select 9 keywords from each message for further use in classification by a perceptron algorithm. Evaluation of the methods was done in batch as well as online modes. Even though the percentage of test messages to training messages was much lower for online training, the results for online evaluation were much lower (about 75% in batch mode compared to about 62% in online mode). As indicated in the paper, this was due to the fact that in online mode only one training iteration was done, presumably because of the long computation time to train the perceptron at each incremental step. In contrast, the STS method does not require multiple iterations as the word statistics calculations are done incrementally.

The above evaluations were done with all e-mails in training and test sets classified into folders. Yang [6] reports a study of text classification algorithms on the Reuters-21450 data set, which includes a number of unlabelled documents, at least some of which should be classified, more similar to our scenario: in this evaluation, k-Nearest Neighbour (k-NN) was shown to perform well. The decision of whether a document needs to be classified or not was made by setting a threshold on an array of values representing learned weights for each document-category pair. A document is classified by the system if the weight is above the threshold, otherwise it is not classified by the system. An optimal threshold is calculated on the training set and used on test documents in a typical batch fashion. We also use the threshold strategy for separating the messages into classified and unclassified, but the threshold is adjusted dynamically in the process of incremental classification, allowing for optimal coverage of classified messages.

3   Simple Term Statistics (STS)

In this section, we present a method of document categorization based on a collection of simple term statistics (STS).

3.1   Method

The STS algorithm maintains an array of weights, one weight for each term-folder pair. Each weight is a product of two numbers: a term ratio and its distribution over folders (for consistency with e-mail classification, we use the term *folder* for the document categories). The prediction of a folder f, given a document d, which is considered to be a set of terms, is made as follows. Each (distinct) term in the document has a weighted "vote" indicating the relevance of the term to a given folder, which in turn, is the product of the term ratio (a "distinctiveness" of term t over the document set, independent of folder-specific measures), and the term distribution (an "importance" of term t for a particular folder). This weighting is analogous to the standard TF-IDF measure, extended to cover multiple folders, and with different types of statistics used. The weighted votes for all terms in a document are simply summed to determine the predicted folder (that with the largest value for the sum).

Term ratios are calculated using the formulas shown in Table 1.

Table 1. Term statistics methods

Term Ratio	$M_a = \frac{1}{N_{dt}}$	$M_b = \frac{1}{N_{ft}}$		$M_c = \frac{C_t}{N_{dt}*C_T}$	$M_d = \frac{1}{(N_{ft})^2}$
Term Distributions	$M_1 = \frac{C_{tf}}{C_T}$	$M_2 = \frac{N_{dtf}}{N_{dt}}$		$M_3 = \frac{N_{dtf}}{N_{df}}$	$M_4 = \frac{N_{dtf}}{N_d}$
	$M_5 = \frac{N_{Tf}}{N_T}$	$M_6 = \frac{N_{Tf}}{N_d}$		$M_7 = \frac{C_{Tf}}{C_T}$	$M_8 = \frac{C_{tf}}{C_{Tf}}$

N_d: number of documents in training set
C_T: count of all terms in training set
N_T: number of distinct terms in training set
N_{dt}: number of documents in training set containing term t
C_t: count of term t in training set
N_{df}: number of documents in folder f
C_{Tf}: count of all terms in folder f
N_{Tf}: number of distinct terms in folder f
N_{ft}: number of folders where term t occurs
C_{tf}: count of term t in folder f
N_{dtf}: number of documents containing term t in folder f

Since the constant function 1, denoted M_0, is both a term ratio and term distribution function, there are 45 STS methods obtained by combining one of the 5 term ratio functions and one of the 9 term distribution functions.

A *term-folder weight* $w_{t,f}$ for a term t and a folder f is calculated using the product of a term ratio and term distribution, for example:

$$M_{c3} = M_c * M_3 = \frac{C_t}{N_{dt} * C_T} * \frac{N_{dtf}}{N_{df}} \ . \tag{1}$$

In addition, if t does not occur in the training set, $w_{t,f}$ is defined to be 0. The predicted category $f_p(d)$ for document d is defined as follows:

$$f_p(d) = argmax_f(w_f), \ where \ w_f = \sum_{t \in d} w_{t,f} \ . \tag{2}$$

4   E-Mail Classification Using Simple Term Statistics

In this section, we present and discuss the results of testing the STS method on the *gov* e-mail data set. The data set consists of 16 998 e-mails pre-classified into 45 folders. Only 48% of all e-mails are classified and the distribution of classified messages over folders is highly uneven.

The training/testing with STS methods was done incrementally, by updating statistics and weights for all terms in the current message d_i, before testing on the next message d_{i+1}. If d_i is unlabelled, no updates were done for its terms. In order to measure the quality of the classification we use accuracy and coverage. Accuracy is the number of messages correctly classified by the learner divided by the number of all messages classified by the learner.

$$Acc = Ncorrect_{learner}/N_{learner} \ . \tag{3}$$

Note that the accuracy calculation does not count unclassified e-mails. The learner's coverage is the number of messages classified by the learner into a category divided by the number of all messages classified by the learner. The classifier's task is to attain high accuracy while keeping the coverage as close as possible to the target coverage.

$$Cov = Nclass_{learner}/N_{learner} \ . \tag{4}$$

In general terms, accuracy corresponds to precision and coverage to recall. In order to separate classified from non-classified messages we extended the STS algorithm in two ways. Firstly, the terms belonging to unclassified e-mails are not counted for statistics. Secondly, a threshold is set on the folder weight in Equation 2. This way, by controlling the threshold, we can flexibly control a tradeoff between the accuracy and coverage. The threshold (denoted by θ) is selected initially on the first 1000 messages to make the learner and the target coverage equal. After that, the threshold is automatically adjusted every 100 messages to keep up with the target coverage. The linear adjustment factor was calculated as follows.

$$\theta_{new} = \theta * Cov_{learner}/Cov_{target} \ . \tag{5}$$

In order to give an approximation on the accuracy of other algorithms, we also obtained results for most accurate general machine learning methods available for experimentation using the Weka toolkit.[1] For all machine learning methods, the top 9 keywords for each of the 45 folders were selected using TF-IDF, thus making 405 attributes. Since these keywords were selected from the whole data set, this somewhat favours the machine learning methods compared to STS. Due to the fact that these methods could not be used in an incremental fashion, we used a setup similar to Bekkerman *et al.* [1], with training on $N * w$ messages and testing on the next w messages for a window w of size 100, repeatedly until the end of the data set. For the k-Nearest Neighbour (k-NN) algorithm, we selected k=1 to achieve the coverage close to the target coverage. Initially, we selected $w = 100$ and observed that, despite good accuracy, the execution times were far too long to be considered for online applications. By selecting $w = 1000$, the execution times for some of these methods were comparable to STS.

Table 2 provides a summary of evaluating the STS and selected machine learning methods on the *gov* message set. The runtimes do not include the time for selecting the top 9 keywords for each folder and preparing the training set, which is needed for the machine learning methods but not for STS. Out of the STS methods, M_{b2} was the most accurate, followed by M_{02} and M_{d2}. All these formulas have a common term distribution component N_{dtf}/N_{dt}, which is similar to TF-IDF applied to folders. Although SVM and Boosted Decision Tree outperform the best STS method when run with a test window of 100 messages, their long execution time is unacceptable for online e-mail

---

[1] http://www.cs.waikato.ac.nz/ml/weka/index_downloading.html

Table 2. Simple Term Statistics accuracy for *gov* data set

Method	Accuracy	Coverage	Threshold	Running Time	Window
$M_{02} = 1 * \frac{N_{dtf}}{N_{dt}}$	0.606	0.460	0.122	2 min 34 sec	1
$M_{b2} = \frac{1}{N_{ft}} * \frac{N_{dtf}}{N_{dt}}$	0.686	0.456	0.071	2 min 34 sec	1
$M_{b8} = \frac{1}{N_{ft}} * \frac{N_{dtf}}{N_{dt}}$	0.509	0.446	0.014	2 min 34 sec	1
$M_{d2} = \frac{1}{(N_{ft})^2} * \frac{N_{dtf}}{N_{dt}}$	0.604	0.456	0.0552	2 min 34 sec	1
$M_{d3} = \frac{1}{(N_{ft})^2} * \frac{N_{dtf}}{N_d}$	0.536	0.456	0.0327	2 min 34 sec	1
$M_{d8} = \frac{1}{(N_{ft})^2} * \frac{C_{tf}}{C_{Tf}}$	0.530	0.468	0.00068	2 min 34 sec	1
Decision Tree	0.663	0.505	N/A	39 min 24 sec	100
Decision Tree	0.568	0.535	N/A	3 min 20 sec	1000
SVM	0.725	0.480	N/A	5 hr 37 min	100
SVM	0.602	0.522	N/A	27 min 32 sec	1000
k-NN (k=1)	0.657	0.481	N/A	39 min	100
k-NN (k=1)	0.540	0.502	N/A	35 min 30 sec	1000
Boost DT	0.735	0.480	N/A	5 hr 57 min	100
Boost DT	0.613	0.510	N/A	37 min 12 sec	1000

categorizers. Even if the SVM or Boosted Decision Tree algorithms are implemented as an off-line component executed overnight, its execution time would be still too long and the window of 100 messages insufficient to handle the flow of messages in a large organisation with a common e-mail foldering system. With the increment of 1000, the Decision Tree runtime is comparable to that of M_{b2}, but the accuracy becomes lower.

k-NN deserves a separate comment, since of all the methods evaluated by Yang [6], this one is the most efficient and scales best to larger data sets. However, one reason this is the case is that the Reuters-21450 data set used by Yang is comparatively small (21450 examples). The k-NN algorithm classifies by comparing each test document to all previously stored cases, so this can be done effectively on the number of cases in the Reuters-21450 data set. However, the classification time increases linearly with the number of stored examples, meaning that the performance will degrade as the size of the data set increases, as does the time for storing the previous cases. For the purpose of suggesting folders in an e-mail management application, it may be important to make fewer suggestions with higher accuracy. In order to do this, the threshold θ in the STS algorithm can be modified to achieve a desired accuracy.

5 Suggesting Rule Keywords

In this section, we evaluate the accuracy of keyword suggestion for rules, based on simulating the process used by an expert to define a rule base for classifying the *gov* data set. The expert is assumed to know the correct folder and their task is to find the best keywords in an e-mail to construct a rule. In the experiments reported here, we retrospectively test a number of learning algorithms to determine which of them could have been used by the expert when defining the rules that were previously defined.

5.1   Methods

For the keyword selection task, five methods were selected: three Simple Term Statistics (STS) methods as defined in Section 3.1, and the Naive Bayes and TF-IDF methods. These methods were selected for the following reasons. STS method M_{d2} provided best overall accuracy when tested on the classification task (Section 4), methods M_{b8} and M_{d8} were selected experimentally for best keyword suggestions, Naive Bayes was originally used in EMMA (Ho *et al.* [4]) for keyword and folder suggestion, and finally TF-IDF was included as a base method commonly used in text mining.

For the purpose of this experiment, the Naive Bayes formula is that used in EMMA defined as follows. For a given folder (which is assumed to be known to an expert), the probability of the folder is calculated for all keywords in the e-mail, using Bayes formula:

$$p(f_c|w) = p(w|f_c) * p(f_c)/p(w) \tag{6}$$

where f_c is the folder into which the e-mail should be classified, and w is a keyword from the e-mail. Approximating probabilities with keyword statistics, we obtain the formula used in EMMA to suggest terms for rules:

$$p(f_c|w) = (C_{tf_c}/C_{Tf_c} * N_{df_c}/N_d)/ \sum_{i \in N_f} (C_{tf_i}/C_{Tf_i} * N_{df_i}/N_d) . \tag{7}$$

where C_{tf} and N_{df} are the counts of term t in a folder f and the number of messages in the folder f respectively.

5.2   Experimental Setup

The experimental setup for keyword suggestion follows the process of defining rules by an expert, described in Wobcke, Krzywicki and Chan [5], in which 368 rules were defined to classify the 16 998 messages with a high degree of accuracy. E-mails were presented to the expert in chronological order in batches of 50. The expert was able to define new rules to correct the rule base, either by refining an existing rule or adding a new rule. In these cases, the viewed message becomes a cornerstone case for the system. In our experiments, e-mails are processed one-by-one in the same order and, whenever a cornerstone case is encountered, a number of keywords are suggested and evaluated against keywords used by the expert in the rule. STS methods provide a keyword rating by summing term-folder weights for each keyword (rather than folder, as in the e-mail classification task) and selects the 9 best keywords for each e-mail as suggested keywords.

The experiments are conducted with the number of suggested keywords chosen as 5, 10 and 15: any greater number of suggestions are likely to be disregarded by the user. Each of these three options are tested in two variants. In the first variant keywords are generated only by the methods described above. These keywords are called *system keywords* and the option is referred to as the *system only* option. In the second variant, suggested keywords are selected from the set of expert keywords used previously in rules, called *expert keywords*, supplemented, if required, with *system keywords*. This option is called the *expert+system* option. For all options, a keyword is included in the suggestion only if it actually occurs in the cornerstone case. The accuracy is calculated as the number of suggested keywords that occur in the rule divided by the total number of keywords occurring in the rule. In order to closely follow the expert selection strategy, suggested keywords come from all parts of the e-mail, including sender, receiver, date/time fields and e-mail addresses.

5.3   Results

Table 3 shows the results for 10 suggested keywords, using both *system only* and textit-system+expert options. The first column of each option shows the accuracy calculated as the number of accurately suggested expert keyword occurrences for all rules divided by the total number of expert keyword occurrences in all rules. The second column for each option is the percentage of rules that have at least one suggested keyword correct.

Comparing the results it is easy to notice that the STS method M_{d2}, which was the most accurate in the e-mail classification task (Section 3), is the least accurate for keyword suggestions. The reason for this is that the classification task is very different from the keyword rating task. In the case of prediction, individual weights of each keyword are summed together to produce a weight for each folder. In this case each individual keyword is not important, as long as it contributes to an overall result indicating a unique folder. This is different for the keyword rating task, where keywords "compete" among themselves for the best description of a given folder.

Looking at the two most accurate STS methods, M_{c8} and M_{d8}, their two common components, are $1/N_{dt}$ and C_{tf}/C_{Tf}, The first component is the inverse document frequency (IDF), which is part of the TF-IDF method. The second component is part of the Bayesian formula. It seems that the best results are obtained by combining these two popular text mining methods.

Increasing the number of suggested keywords from 5 to 10 and 15 (Figure 1) causes a less than proportional increase in accuracy, which suggests that some number of keywords, for example 10, may be optimal.

Figure 2 shows the accuracy and average number of correctly suggested keywords against the number of defined rules for M_{d8} and the *expert+system* option with 10

Table 3. Keyword suggestion for 10 generated keywords

Method	Accuracy	Rules Covered	Accuracy	Rules Covered
M_{b2}	0.179	38.4	0.693	87.3
M_{b8}	0.566	82.1	0.705	89.8
M_{d2}	0.174	37.6	0.694	87.3
M_{d8}	0.462	73.6	0.716	90.3
NB	0.217	43.7	0.692	87.6
TF-IDF	0.329	48	0.655	84

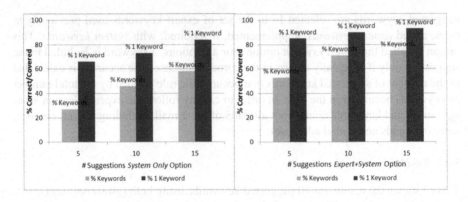

Fig. 1. Keyword suggestion results for method M_{d8}

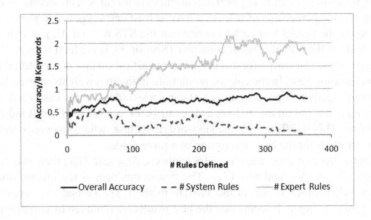

Fig. 2. Accuracy of suggestions for method M_{d8} and Expert+System option

generated keywords. Generally, the accuracy increases with the number of defined rules due to the fact that M_{d8} becomes more accurate as the number of messages and rules increases, and also because the stored expert keywords cover more future rules. It is noticeable that the number of suggested keywords from the "expert set" increases more rapidly, while the number of keywords from the "system set" decreases. This is an expected effect and can be explained by the fact that expert keywords are given priority over system keywords. With the *expert+system* option, the system suggested keywords are most helpful at the beginning, when fewer expert keywords are stored and, at any other time, when no suitable stored keywords are available, for example at rule 200.

6   Conclusion

In this research, we presented a number of methods for document categorization and keyword rating based on Simple Term Statistics (STS). We showed that these methods can be viable alternatives to more complex and resource demanding machine learning methods commonly used in text categorization, such as Decision Tree, SVM and k-NN. STS methods require processing of only the new terms occurring in each step, which makes them truly incremental and sufficiently fast to support online applications. In fact, we discovered that these methods are much faster, while performing well in terms of accuracy, when compared to a range of other methods.

We tested the accuracy of keyword suggestions also using the STS algorithm in the stand alone, *system only* option and in *expert+system* option, where previously used expert keywords were suggested first before system keywords. The suggestions are mostly useful in the initial stage of defining rules, but also in some stages of rule definition where previous expert keywords do not cover the current message sufficiently.

When testing the keyword suggestions, the underlying assumption was that the expert rules are always better than system generated keywords. It would be interesting to test this assumption in a real user environment by presenting a mixture of previous expert and suggested system keywords for the user to make an unbiased selection. The research question to answer in this case would be if greater reliance of the user on system keywords would make the categorization more consistent.

Acknowledgements

This work was supported by Smart Services Cooperative Research Centre.

References

1. Bekkerman, R., McCallum, A., Huang, G.: Automatic Categorization of Email into Folders: Benchmark Experiments on Enron and SRI Corpora. Technical Report IR-418. University of Massachusetts, Amherst (2004)
2. Compton, P.J., Jansen, R.: A Philosophical Basis for Knowledge Acquisition. Knowledge Acquisition 2(3), 241–257 (1990)
3. Dredze, M., Wallach, H.M., Puller, D., Pereira, F.: Generating Summary Keywords for Emails Using Topics. In: Proceedings of the 13th International Conference on Intelligent User Interfaces, pp. 199–206 (2008)
4. Ho, V.H., Wobcke, W.R., Compton, P.J.: EMMA: An E-Mail Management Assistant. In: Proceedings of the 2003 IEEE/WIC International Conference on Intelligent Agent Technology, pp. 67–74 (2003)
5. Wobcke, W., Krzywicki, A., Chan, Y.-W.: A Large-Scale Evaluation of an E-Mail Management Assistant. In: Proceedings of the 2008 IEEE/WIC/ACM International Conference on Web Intelligence and Intelligent Agent Technology, pp. 438–442 (2008)
6. Yang, Y.: An Evaluation of Statistical Approaches to Text Categorization. Information Retrieval 1, 69–90 (1999)

The Positive Effects of Negative Information: Extending One-Class Classification Models in Binary Proteomic Sequence Classification

Stefan Mutter, Bernhard Pfahringer, and Geoffrey Holmes

Department of Computer Science
The University of Waikato
Hamilton, New Zealand
{mutter,bernhard,geoff}@cs.waikato.ac.nz

Abstract. Profile Hidden Markov Models (PHMMs) have been widely used as models for Multiple Sequence Alignments. By their nature, they are generative one-class classifiers trained only on sequences belonging to the target class they represent. Nevertheless, they are often used to discriminate between classes. In this paper, we investigate the beneficial effects of information from non-target classes in discriminative tasks. Firstly, the traditional PHMM is extended to a new binary classifier. Secondly, we propose propositional representations of the original PHMM that capture information from target and non-target sequences and can be used with standard binary classifiers. Since PHMM training is time intensive, we investigate whether our approach allows the training of the PHMM to stop, before it is fully converged, without loss of predictive power.

1 Introduction

Classification of proteins is an important and challenging task in Bioinformatics [1]. It is characterised by imbalanced datasets. Typically, there is only a small number of proteins belonging to the target class compared to a vast number of non-target proteins. The target class is commonly referred to as the positive class and all non-target proteins form the negative class. A solution is to use one-class classification [2], which has been successfully applied in fields with no available negative examples or only a small number of positive training examples [1].

Multiple Sequence Alignments (MSAs) are a standard technique to learn and represent a specific class of proteins. They can be represented using a special kind of Hidden Markov Model (HMM), called Profile Hidden Markov Models (PHMMs) [3,4]. Our fundamental observation is that PHMMs are essentially one-class classifiers. They are trained only on sequences belonging to the positive class and they output a similarity score for unknown test sequences. However, to discriminate between the positive and the negative class, it might be advantageous to use negative information as well.

The advantages have been recognised in the literature, notably by Jaakkola et al. [5] who extract a kernel description of HMMs. They implicitly train the

A. Nicholson and X. Li (Eds.): AI 2009, LNAI 5866, pp. 260–269, 2009.
© Springer-Verlag Berlin Heidelberg 2009

HMM as a one class classifier and introduce negative training information in the subsequent Support Vector Machine classification step. However, this approach is restricted to kernel methods. Additionally, Bánhalmi et al. [1] use one-class classifiers for protein classification. However, they do not use HMMs and work on pre-calculated similarity scores.

In this paper, we explicitly use the PHMM as a one-class classifier. In order to introduce negative information into a subsequent learning step, we propositionalise PHMMs to get a fixed-length feature vector [6]. Propositionalisation is not restricted to kernel approaches nor does it need the definition of an explicit similarity measure as kernel methods do. However, in our research, negative information is not only introduced in a subsequent binary classification step. We also change the original one-class classifier into a binary one and compare the two approaches. Therefore, we propositionalise not only the original PHMM but also the binary classifier.

Previously, results were presented for fully trained HMMs [6]. A further contribution of the paper is to explore whether or not the introduction of negative information can lead to a dramatic decrease in training time. The idea is that propositionalisation of a not fully converged PHMM might compensate for a potential loss of predictive power in the PHMM model.

2   One-Class Classification

An MSA represents a specific group or class of sequences. These can be classes of common evolutionary relationship, function or localisation for example. From a Machine Learning point of view a MSA is a one-class classifier. The MSA is built only using the positively labelled sequences and computes a measure of similarity $s(X)$ for any test sequence X to the positive target class. A threshold Θ on $s(X)$ is used to decide class membership. Often these MSAs are built on a small number of positive instances, whereas there is a huge number of proteins belonging to other classes.

PHMMs represent MSAs. They can be trained from unaligned sequences using the Baum-Welch algorithm which is a special case of the EM (expectation-maximisation) algorithm [4].

A PHMM is, therefore, a generative model of the positive class. We use $P(X|H)$ to denote the probability of an arbitrary sequence X under the PHMM H. In the literature, this probability is often referred to as the forward probability. In the classification step, the log-odds score for $P(X|H)$ measures the similarity of the sequence X to the PHMM H. It is generally a difficult problem to decide on a threshold Θ for these scores. For this research we are not interested in finding the perfect value for Θ, rather the ordering of sequences based on their score. Therefore, Area Under the ROC (AUC) is used to evaluate the performance of our models. This measure is independent of Θ. Later, we will refer to the PHMMs used in the one-class classification setting as one-class PHMMs.

3   Binary Classification

In this paper, we are dealing with binary classification problems of amino acid sequences. Therefore, the question is whether or not a specific amino acid sequence belongs to a class of proteins. As explained in the previous section, these protein classes are represented essentially with a model of a one-class classifier, a PHMM.

Our research focuses on improving these one-class classifiers by adding information from negative training data. There are two starting points to integrate negative information. Firstly, we can extend the original model to be a binary classifier instead of a one-class classifier. Secondly, the negative information can be introduced in a subsequent binary classification step on top of the existing one-class model. In the following two sections, we will introduce our approaches for both scenarios.

3.1   Binary Profile Hidden Markov Models

A binary PHMM consists of two one-class PHMMs: H_{pos} and H_{neg}. The PHMM H_{pos} is trained exclusively on the positive instances, H_{neg} is trained on the negative instances only. Therefore H_{pos} is the one-class classifier we introduced in the previous sections, whereas H_{neg} is a one-class classifier for the negative class. To classify a test sequence X, we calculate the log-odds scores for $P(X|H_{pos})$ and $P(X|H_{neg})$. We predict the class with the higher score. This is a very simple combination of two PHMMs by using the maximum score.

The idea to build a binary PHMM out of two one-class PHMMs ensures that the datasets' imbalance does not negatively influence the binary classification. It allows us to combine two generative one-class classifiers in a discriminative way for binary classification problems. However, training a PHMM is slow and therefore training on a vast amount of (negative) instances is time intensive. On the other hand we expect the negative class to be more diverse. Due to this fact H_{neg} should converge faster.

A binary PHMM calculates for each sequence X two logarithmic scores. These can be easily normalised into probabilities. Thus, no logistic calibration is needed in this setting.

3.2   Propositionalisation

Propositionalisation transforms complex, structured representations of data such as PHMMs into a fixed length representation involving attribute-value pairs [8]. Therefore, it introduces a wide range of possible features to be constructed from the more complex representation. Traditionally Machine Learning has focused on propositional learners. Thus, they are highly optimised. Furthermore, a subsequent propositionalisation step on top of a one-class classifier offers the possibility to make use of negative instances in a discriminative learning task. As Jaakkola et al. [5] point out, discriminative tasks such as classification might benefit from using negative examples. They extract a kernel description of a

HMM and use this for classification by a Support Vector Machine. Feature extraction from the PHMM through propositionalisation and subsequent training of a propositional learner results in a discriminative model. This approach is not restricted to kernel-based classifiers. Whereas Jaakkola et al. [5] only apply their approach to one-class PHMMs, in this paper both one-class PHMMs as well as binary PHMMs are propositionalised.

Mutter et al. [6] introduce the propositionalisation of PHMMs in detail. In this paper, we use two different ways to propositionalise a one-class PHMM which can be applied to general HMMs as well. Two numeric attributes represent scores for one or more paths. The first one is the forward score of the sequence X. We also use the score of the best path through the HMM. This best path is known as the Viterbi path. Additionally, there is a numeric attribute for each state in the HMM. It represents the score of the state given the sequence X. In the remainder, we will refer to this propositionalisation as the logarithmic one. The exponential propositional representation is based on the first one. Given any logarithmic score z, the corresponding exponential score is calculated by e^z. The values of the state scores are then normalised into probabilities. In the case of a binary PHMM there are two scores for the Viterbi and two scores for the forward path. We normalise these values into real probabilities as well. Because in a one-class PHMM there is only one score for the forward and one for the Viterbi path, we do not normalise them.

In case of a binary PHMM, the propositionalisations for each one-class PHMM are combined together. Therefore, propositionalising a binary PHMM leads to twice as many attributes.

4 Experiments

In the experiments we use three amino acid sequence classification datatsets each consisting of a sequence of amino acids represented as a string and a class label. The first two are concerned with protein localisation and the last one classifies enzymes. The protein localisation datasets are from Reinhardt and Hubbart [9]. They have also been used by other researchers [10,11,12]. The first dataset addresses protein localisation in prokaryotes (pro). It consists of 997 instances belonging to three different classes. We transform this multiclass problem into three binary classification tasks. In each binary setting, one class is used as the positive class and all the remaining classes are combined into one negative class. All other datasets are pre-processed in the same way. For the remainder of the paper, we use the datatset name, followed by an index for the positive class, e.g. pro_0 refers to the prokaryote dataset treating class with index 0 as positive and all remaining instances as negative. The eukaryote dataset (euk) consists of 2427 sequences from four different classes, whereas the enzyme dataset has 2765 instances and 16 classes. It was introduced by Chou [13]. We use the protein IDs given in the paper and the Uniprot database to retrieve the primary sequences. We kept all sequences. The major difference between the protein localisation datasets and the enzyme dataset is that the former ones have a high sequence

Table 1. Overview of datasets and their respective size of the positive class. The percentage numbers are rounded.

dataset	positive instances		dataset	positive instances		dataset	positive instances	
	number	percentage		number	percentage		number	percentage
enzyme_1	335	12%	enzyme_9	262	9%	pro_0	688	69%
enzyme_2	225	8%	enzyme_10	92	3%	pro_1	107	11%
enzyme_3	207	7%	enzyme_11	158	6%	pro_2	202	20%
enzyme_4	135	5%	enzyme_12	97	4%	euk_0	684	28%
enzyme_5	120	4%	enzyme_13	259	9%	euk_1	325	13%
enzyme_6	329	12%	enzyme_14	166	6%	euk_2	321	13%
enzyme_7	67	2%	enzyme_15	85	3%	euk_3	1097	45%
enzyme_8	66	2%	enzyme_16	162	6%			

similarity inside classes [14], whereas the latter is characterised by a low sequence similarity inside a class [13]. Table 1 gives an overview of the classes and the size of the positive class in the specific dataset. We learn one-class and binary PHMMs from unaligned sequences using the Baum-Welch algorithm. The transition and emission probabilities in the PHMMs are initialised uniformly. The convergence criterion is a sufficiently small change in the log-odds score relative to a random model. This score is normalized by the number of residues in a sequence[1]. Training a PHMM is time intense, therefore we constrain the model length, by restricting the number of the so-called match states to 35. We do not pre-process the sequences. For more information about the PHMM setup and training we refer to Mutter et al. [6].

The propositional learners are a linear Support Vector Machine (SVM) [15], Random Forests [16] and bagged, unpruned C4.5 decision trees [17]. The complexity parameter for the linear Support Vector Machine and the number of features used in the Random Forests are estimated using an internal 10-fold cross-validation. All Random Forests consist of 100 trees. All experiments are performed using WEKA [7] on 3 GHz Intel© 64-bit machines with 2 GB of memory. Models are evaluated using one 10-fold cross-validation. If not stated explicitly, we report the results for the better way of propositionalising. In this paper, we evaluate the PHMM models after each iteration of the Baum-Welch algorithm. Like the PHMM itself, Bagging is evaluated after each iteration step as well. For the Random Forest and the Support Vector Machine, we perform an evaluation after the first and the final Baum-Welch iteration.

5   Results

Figure 1 shows the results for two datasets: enzyme_1 and enzyme_7. For enzyme_1 there is not a big difference in performance between the one-class PHMM and the binary one. However, the binary PHMM performs better for enzyme_7 during the whole Baum-Welch training process. The figure shows that propositionalisation is able to outperform the pure PHMM based approaches, especially at the start of training. For enzyme_7 the overall best AUC is achieved by propositionalising a one-class PHMM after the first iteration. Towards the end of the

---

[1] The threshold of 0.0001 was proposed by A. Krogh in an e-mail communication.

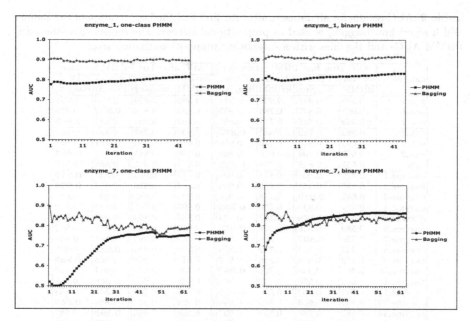

Fig. 1. AUC results for enzyme_1 (on the top) and the enzyme_7 (on the bottom) dataset. The graphs on the left side show the results for a one-class PHMM and subsequent Bagging, whereas the graphs on the right are based on binary PHMM models. We use exponential propositionalisation.

Baum-Welch training the binary PHMM performs better than its propositionalisation.

Table 2 gives an overview of the AUCs for all datasets. We report the AUC for the first and the last iteration. The best AUC for each dataset is printed in bold. The fully trained binary PHMM performs better than the one-class PHMM on all but one dataset. Only enzyme_5 has a slightly higher AUC in the one-class setting after the last iteration. Thus, adding negative information has a positive effect on AUC in our solely PHMM-based experiments. However, training and testing is much more time consuming in the binary case as Table 3 shows. This is the major drawback of methods based on binary PHMMs.

Another important observation is that there is not a big performance difference between propositionalisations based on one-class PHMMs and binary ones, even though the respective PHMMs perform differently. In three cases the one-class based propositionalisation even leads to the best results; all of them after the first iteration. Propositionalising a binary PHMM after the first iteration leads to the highest AUC in seven cases. These findings are important as they reveal the potential of a propositionalisation approach after just one iteration of the Baum-Welch algorithm. For both the one-class and the binary setting, propositionalising after the first iteration leads to a big reduction in training time and can sometimes even improve performance based on AUC.

Table 2. AUC results for all datasets after the first and the last iteration of the Baum-Welch algorithm. Bagging is used as propositional learner. The results show the pure PHMM AUCs and the ones with a subsequent propositionalisation step.

| dataset | AUC after first PHMM iteration | | | | AUC after last PHMM iteration | | | |
| | one-class PHMM | | binary PHMM | | one-class PHMM | | binary PHMM | |
	PHMM	Bagging	PHMM	Bagging	PHMM	Bagging	PHMM	Bagging
pro_0	0.570	0.843	0.920	0.944	0.590	0.859	0.956	**0.961**
pro_1	0.666	0.875	0.784	0.909	0.863	0.873	**0.932**	0.921
pro_2	0.569	0.842	0.711	0.861	0.802	0.861	0.881	**0.903**
euk_0	0.582	0.789	0.823	**0.872**	0.600	0.781	0.831	0.867
euk_1	0.603	0.934	0.750	0.956	0.878	0.931	**0.971**	0.964
euk_2	0.642	0.804	0.675	0.869	0.769	0.845	0.872	**0.885**
euk_3	0.555	0.848	0.774	**0.925**	0.773	0.855	0.888	0.910
enzyme_1	0.777	0.904	0.811	0.908	0.815	0.904	0.832	**0.917**
enzyme_2	0.787	0.884	0.896	0.909	0.863	0.920	0.892	**0.928**
enzyme_3	0.642	**0.910**	0.738	0.906	0.701	0.889	0.805	0.894
enzyme_4	0.631	0.930	0.736	**0.956**	0.805	0.939	0.885	0.942
enzyme_5	0.545	**0.876**	0.707	0.874	0.860	0.874	0.854	0.874
enzyme_6	0.589	0.946	0.809	**0.959**	0.837	0.938	0.877	0.942
enzyme_7	0.520	**0.897**	0.680	0.851	0.753	0.792	0.861	0.840
enzyme_8	0.742	0.854	0.752	0.873	0.892	0.916	**0.932**	0.918
enzyme_9	0.762	0.966	0.846	0.977	0.910	0.971	0.965	**0.983**
enzyme_10	0.467	0.938	0.705	**0.960**	0.728	0.912	0.917	0.917
enzyme_11	0.605	0.936	0.793	0.963	0.936	0.950	**0.968**	0.959
enzyme_12	0.541	0.869	0.686	**0.893**	0.675	0.799	0.846	0.858
enzyme_13	0.807	0.923	0.800	0.946	0.829	0.930	0.894	**0.959**
enzyme_14	0.987	0.985	0.958	0.997	0.988	0.986	**0.999**	0.987
enzyme_15	0.750	0.949	0.884	0.967	0.949	0.928	**0.989**	0.947
enzyme_16	0.736	0.938	0.866	**0.952**	0.842	0.929	0.927	0.950

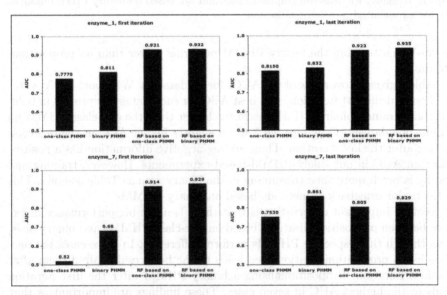

Fig. 2. Comparison of AUCs with and without propositionalisation with a Random Forest (RF). On top are the results for enzyme_1, on the bottom for enzyme_7. The graphs on the left show the AUCs after the first Baum-Welch iteration of the PH-MMs, whereas the right graphs show the ones after the final iteration. The graphs on the left and the binary PHMM for enzyme_7 in the last iteration use logarithmic propositionalisation. Otherwise we use exponential propositionalisation.

Table 3. Execution times for training and evaluation. All learners are evaluated after each Baum-Welch iteration. The execution time for Bagging comprises training and testing time of the propositional step exclusive of the preceding PHMM training time.

dataset	execution time in h				number of iterations	
	PHMM		Bagging		PHMM training	
	one-class	binary	one-class PHMM	binary PHMM	positive class	negative class
enzyme_1	18	40	8	16	44	20
enzyme_7	42	107	7	17	63	20

Table 4. AUC results for all datasets after the first and the last iteration of the Baum-Welch algorithm. Random Forests (RF) and linear Support Vector Machines (SVM) are used as propositional learners. The results show the pure PHMM AUCs and the ones with a subsequent propositionalisation step.

dataset	AUC after first PHMM iteration						AUC after last PHMM iteration					
	one-class PHMM			binary PHMM			one-class PHMM			binary PHMM		
	PHMM	RF	SVM	PHMM	RF	SVM	PHMM	RF	SVM	PHMM	RF	SVM
pro_0	0.570	0.839	0.796	0.920	0.944	0.963	0.590	0.871	0.862	0.956	**0.965**	0.943
pro_1	0.666	0.866	0.862	0.784	0.906	**0.932**	0.863	0.914	0.867	**0.932**	0.931	0.917
pro_2	0.569	0.872	0.801	0.711	0.878	0.898	0.802	0.873	0.832	0.881	**0.904**	0.872
euk_0	0.582	0.815	0.734	0.823	**0.885**	0.875	0.600	0.807	0.691	0.831	0.883	0.849
euk_1	0.603	0.951	0.922	0.750	0.965	0.963	0.878	0.942	0.940	**0.971**	0.969	0.967
euk_2	0.642	0.821	0.812	0.675	0.885	**0.909**	0.769	0.853	0.851	0.872	0.891	0.889
euk_3	0.555	0.862	0.801	0.774	0.922	0.921	0.773	0.868	0.834	0.888	**0.925**	0.913
enzyme_1	0.777	0.931	0.862	0.811	0.932	0.881	0.815	0.923	0.837	0.832	**0.935**	0.867
enzyme_2	0.787	0.929	0.849	0.896	0.932	0.912	0.863	0.921	0.877	0.892	**0.942**	0.892
enzyme_3	0.642	**0.929**	0.721	0.738	0.925	0.846	0.701	0.909	0.781	0.805	0.904	0.828
enzyme_4	0.631	0.945	0.826	0.736	**0.964**	0.908	0.805	0.940	0.900	0.885	0.951	0.914
enzyme_5	0.545	**0.902**	0.856	0.707	0.899	0.880	0.860	0.851	0.837	0.854	0.888	0.844
enzyme_6	0.589	0.962	0.851	0.809	**0.967**	0.915	0.837	0.953	0.886	0.877	0.962	0.905
enzyme_7	0.520	0.914	0.808	0.680	**0.929**	0.858	0.753	0.805	0.794	0.861	0.829	0.820
enzyme_8	0.742	0.910	0.805	0.752	**0.942**	0.908	0.892	0.910	0.917	0.932	0.910	0.923
enzyme_9	0.762	0.979	0.946	0.846	**0.984**	0.969	0.910	0.978	0.964	0.965	**0.984**	0.977
enzyme_10	0.467	0.945	0.706	0.705	**0.965**	0.962	0.728	0.892	0.846	0.917	0.897	0.890
enzyme_11	0.605	0.954	0.879	0.793	0.967	**0.968**	0.936	0.954	0.958	**0.968**	0.966	0.961
enzyme_12	0.541	0.887	0.638	0.686	**0.911**	0.877	0.675	0.853	0.769	0.846	0.886	0.837
enzyme_13	0.807	0.950	0.863	0.800	0.960	0.909	0.829	0.932	0.856	0.894	**0.966**	0.897
enzyme_14	0.987	0.996	0.993	0.958	0.998	0.997	0.988	0.997	0.995	**0.999**	0.997	0.986
enzyme_15	0.750	0.959	0.971	0.884	0.970	**0.995**	0.949	0.934	0.929	0.989	0.952	0.944
enzyme_16	0.736	0.959	0.891	0.866	**0.970**	0.945	0.842	0.936	0.901	0.927	0.959	0.946

Random Forests usually outperform the Bagging results. Figure 2 illustrates their performance for the enzyme_1 and enzyme_7 dataset. Again, there is not a big difference between propositional approaches based on one-class and binary PHMMs. For enzyme_1 the AUC of the Random Forests after the first iteration is only slightly worse than after the last iteration. For enzyme_7 the best results are achieved using a propositional learner after the first Baum-Welch iteration.

Table 4 provides an overview for all datasets. The PHMM results are of course identical to the ones in Table 2 and reported again for a better overall view. For the majority of datasets Random Forests achieve the highest AUC and outperform a purely PHMM-based approach or SVMs. In this setting there are only four datasets for which a pure binary PHMM approach leads to the highest AUC. For two of them the AUC equals the SVM's AUC on a propositional representation after one iteration.

Table 4 shows that Random Forests and SVMs based on attributes from binary PHMMs after the first iteration perform well. Together they achieve the highest AUC in 13 cases. This indicates the strength of propositionalisation. In addition, the table reveals that binary approaches lead to higher AUC. One-class approaches are competitive when used in combination with propositionalisation due to a faster training and test time. In two cases, a propositional one-class approach after the first iteration results in the highest AUC.

6   Conclusions and Future Work

Our research shows that introducing negative information into a discriminative task improves performance. The binary PHMM outperforms the one-class PHMM in almost all cases. However, it is an interesting finding that when propositionalisation is used, the differences between the performance of models built from one-class PHMMs and binary ones is small. There are cases where propositionalisation of the one-class PHMM even outperforms the binary case. In addition the propositional training sets of the one-class PHMM have only half the size. Additionally propositionalisation works extremely well at the start of the training. In more than half of the cases the best AUC resulted from a propositional learner built on top of a one-class or binary PHMM after the first iteration. This fact has the potential to dramatically improve training time for HMM based protein classification tasks but will need further investigation.

In the future we plan to extend our framework in different ways. First of all, we will consider a purely one-class classification approach.

A major strength of propositionalisation is its flexibility to create feature vectors. Thus, secondly, future research will combine propositional representations of different PHMMs in one dataset. These PHMMs can represent different kind of sequences, e.g. primary and secondary structure of a protein.

Bánhalmi et al. [18] extend one-class classification with artificially created negative training examples. This process leads to a binary classification task on positive and artificially created negative instances. In the domain of proteomic sequence classification there is usually an abundance of negative training examples compared to the positive ones. However, it is often not easy to decide which negative training examples are most helpful in building a good decision boundary. Therefore, the third direction of future research will use a PHMM as a generative model, that can create new artificial sequences with a certain score.

Finally, binary approaches perform well in terms of AUC but need more time for testing and especially training. Thus, we will investigate sampling strategies for the negative class.

References

1. Bánhalmi, A., Busa-Fekete, R., Kégl, B.: A one-class classification approach for protein sequences and structures. In: Proceedings of the 5th International Symposium on Bioinformatics Research and Applications (ISBRA), May 2009, pp. 310–322 (2009)

2. Tax, D.: One-class classification: Concept-learning in the absence of counter-examples. PhD thesis, Delft University of Technology (2001)
3. Krogh, A., Brown, M., Mian, I., Sjölander, K., Haussler, D.: Hidden markov models in computational biology: Applications to protein modelling. Journal of Molecular Biology 235(5), 1501–1531 (1994)
4. Durbin, R., Eddy, S., Krogh, A., Mitchison, G.: Biological sequence analysis: probabilistic models of proteins and nucleic acids. Cambridge University Press, Cambridge (1998)
5. Jaakkola, T., Diekhans, M., Haussler, D.: Using the fisher kernel method to detect remote protein homologies. In: Proceedings of the 7th International Conference on Intelligent Systems for Molecular Biology, pp. 149–158 (1999)
6. Mutter, S., Pfahringer, B., Holmes, G.: Propositionalisation of profile hidden markov models for biological sequence analysis. In: Proceedings of the 21st Australasian Joint Conference on Artificial Intelligence, pp. 278–288 (2008)
7. Witten, I.H., Frank, E.: Data Mining: Practical Machine Learning Tools and Techniques, 2nd edn. Morgan Kaufmann, San Francisco (2005)
8. Krogel, M., Rawles, S., Železný, F., Flach, P., Lavrač, N., Wrobel, S.: Comparative evaluation of approaches to propositionalization. In: Horváth, T., Yamamoto, A. (eds.) ILP 2003. LNCS (LNAI), vol. 2835, pp. 197–214. Springer, Heidelberg (2003)
9. Reinhardt, A., Hubbard, T.: Protein subcellular location prediction. Protein Engineering 12, 107–118 (1999)
10. Chou, K., Elrod, D.: Using neural networks for prediction of the subcellular location of proteins. Nucleic Acids Research 26, 2230–2236 (1998)
11. Hua, S., Sun, Z.: Support vector machine approach for protein subcellular localization prediction. Bioinformatics 17(8), 721–728 (2001)
12. Guo, J., Lin, Y., Sun, Z.: A novel method for protein subcellular localization based on boosting and probabilistic neural network. In: Proceedings of the 2nd Conference on Asia-Pacific Bioinformatics (APBC), pp. 21–27 (2004)
13. Chou, K.: Prediction of enzyme family classes. Journal of Proteome Research 2, 183–189 (2003)
14. Nakai, K.: Protein sorting signals and prediction of subcellular localization. Advances in Protein Chemistry 54, 277–343 (2000)
15. Platt, J.: Machines using sequential minimal optimization. In: Schoelkopf, B., Burges, C., Smola, A. (eds.) Advances in Kernel Methods - Support Vector Learning. MIT Press, Cambridge (1998)
16. Breiman, L.: Random forests. Machine Learning 45(1), 5–32 (2001)
17. Breiman, L.: Bagging predictors. Machine Learning 24(2), 123–140 (1996)
18. Bánhalmi, A., Kocsor, A., Busa-Fekete, R.: Counter-example generation-based one-class classification. In: Kok, J.N., Koronacki, J., Lopez de Mantaras, R., Matwin, S., Mladenič, D., Skowron, A. (eds.) ECML 2007. LNCS (LNAI), vol. 4701, pp. 543–550. Springer, Heidelberg (2007)

Using Topic Models to Interpret MEDLINE's Medical Subject Headings

David Newman[1,2], Sarvnaz Karimi[1], and Lawrence Cavedon[1]

[1] NICTA and The University of Melbourne, Victoria, Australia
[2] University of California, Irvine, USA
{david.newman,sarvnaz.karimi,lawrence.cavedon}@nicta.com.au

Abstract. We consider the task of interpreting and understanding a taxonomy of classification terms applied to documents in a collection. In particular, we show how unsupervised topic models are useful for interpreting and understanding MeSH, the Medical Subject Headings applied to articles in MEDLINE. We introduce the resampled author model, which captures some of the advantages of both the topic model and the author-topic model. We demonstrate how topic models complement and add to the information conveyed in a traditional listing and description of a subject heading hierarchy.

1 Introduction

Topic modeling is an unsupervised learning method to automatically discover semantic topics in a collection of documents and allocate a small number of topics to each individual document. But in many collections, documents are already hand-categorised using a human-constructed taxonomy of classification terms or subject headings. We report on a number of experiments that use topic modeling to interpret the meaning of categories, and explain subtle distinctions between related categories, by analysing their use over a document collection. These experiments are performed in the context of the Medical Subject Headings (MeSH) taxonomy.

MeSH are the subject headings used for tagging articles in MEDLINE, the largest biomedical literature database in the world. PubMed – the interface for searching MEDLINE – extensively uses these MeSH headings. Most PubMed queries are mapped to queries that involve MeSH headings, e.g. the query "teen drug use" is mapped to a longer query that searches for the MeSH headings "Adolescent" and "Substance-Related Disorders" (this mapping is explained in [1]). Therefore, it is critical for researchers and health-care professionals using PubMed to understand what is meant by these MeSH headings, since MeSH headings have a direct effect on search results.

One possible approach would be to attempt to understand MeSH headings by analysing how MeSH headings are applied to documents. However, MeSH tagging is a complex procedure performed by a team of expert catalogers at the National Library of Medicine in the US[1]. These catalogers use a range of

---

[1] MeSH tagging is described in detail at http://ii.nlm.nih.gov/mti.shtml

A. Nicholson and X. Li (Eds.): AI 2009, LNAI 5866, pp. 270–279, 2009.
© Springer-Verlag Berlin Heidelberg 2009

Table 1. Most frequent MeSH headings, major MeSH headings, major qualifiers and MeSH-qualifier combinations in articles published since 2000

MeSH heading	Major MeSH heading	Major qualifier	MeSH-qualifier combination
Humans	Brain	metabolism	Signal Transduction (physiology)
Female	Breast Neoplasms	physiology	Antineoplastic Combined Chemotherapy Protocols (therapeutic use)
Male	Neoplasms	genetics	Magnetic Resonance Imaging (methods)
Animals	Apoptosis	methods	Apoptosis (drug effects)
Adult	HIV Infections	chemistry	Neurons (physiology)
Middle Aged	Neurons	pharmacology	DNA-Binding Proteins (metabolism)
Aged	Signal Transduction	therapeutic use	Transcription Factors (metabolism)
Adolescent	Antineoplastic Agents	pathology	Antineoplastic Agents (therapeutic use)
Mice	Magnetic Resonance Imaging	immunology	Anti-Bacterial Agents (pharmacology)
Child	Anti-Bacterial Agents	diagnosis	Brain (metabolism)

techniques, leveraging various biomedical resources and ontologies, and applying machine learning tools that score and suggest MeSH categories for a given document.

We take a statistical approach to this analysis, using topic models of large sets of search results over MEDLINE, to provide a semantic interpretation of MeSH headings. By analyzing large scale patterns of MeSH tagging, and patterns of co-occurring words in titles and abstracts, we independently learn the meaning of MeSH terms in a data-driven way. We argue that this leads to an understanding of the way MeSH headings have been applied to the MEDLINE collection, providing insight into distinctions between headings, and suggesting MeSH terms that can be useful in document search. While this paper focuses on MEDLINE and MeSH, the approach is more broadly useful for any collection of text documents that is tagged with subject headings.

Background on MeSH Headings: MeSH headings are arranged in a large, complex and continually evolving hierarchy. Currently there are over 25,000 MeSH terms arranged in a directed acyclic graph, which includes a root and 11 levels. On average there are 16 MeSH headings attached to a MEDLINE article. All MeSH tags on a given article have an additional attribute Major-TopicYN which can take on the value Y or N, indicating whether the MeSH tag is the primary focus of the article. Furthermore, each application of a MeSH tag on an article may be qualified using zero, one, or more qualifiers, e.g. one could qualify the MeSH tag *Methadone* with the qualifier *therapeutic use*. There are over 80 qualifiers, but only a specific subset of qualifiers may be used with each MeSH heading. Qualifiers applied to articles also always have the attribute MajorTopicYN.

To gain some familiarity with the usage of MeSH headings and qualifiers, we provide lists of most frequent terms in Table 1. Rows in the table do not correspond – the four columns are separate. The first column shows the most frequent MeSH headings, irrespective of MajorTopicYN. We see headings that act as "check tags" (e.g. Human), used to restrict search results to certain classes of interest. The second column shows the most common *major* MeSH headings, where the heading or one of its qualifiers has MajorTopicYN=Y. Here we see a broad range of topics, covering both conditions/diseases (Neoplasms, HIV) and

Table 2. PubMed queries run to produce query resuts sets for experiments. The number of results shown only count search results that contain abstracts.

Label	PubMed query	# results
burns	Burns[MeSH Terms] AND Humans[MeSH Terms]	19,970
dopamine	Dopamine[MeSH Terms]	33,223
drug	Substance-Related Disorders[MeSH Terms] AND Adolescent[MeSH Terms]	22,361
imaging	Imaging, Three-Dimensional[MeSH Terms]	21,858
p53	p53[All Fields]	47,327
smoking	Smoking[MeSH Terms]	63,101

basic research (Neurons, Apoptosis, Signal Transduction). The most frequent qualifiers also span a wide range, and finally we see the top MeSH heading-qualifier combinations partly overlap with the most frequent major MeSH terms, providing more detail about what is prevalent in the published literature. For the rest of this paper, we only consider major MeSH headings (on average, 5 out of 16 MeSH headings applied to an article are major, or have a major qualifier).

2   Interpreting MeSH Headings with Topic Models

2.1   Methods and Data

Topics – learned by topic models – provide a natural basis for representing and understanding MeSH headings. Topic models (also known as Latent Dirichlet Allocation models or Discrete PCA models) are a class of Bayesian graphical models for text document collections represented by bag-of-words (see [2,3,4]). In the standard topic model, each document in the collection of D documents is modeled as a multinomial distribution over T topics, where each topic is a multinomial distributions over W words, and both sets of multinomials are sampled from a Dirichlet.

Rather than learn a single topic model of all of MEDLINE (an impractical task, especially given that we would need to learn thousands of topics), we chose to demonstrate our methodology using six query results sets shown in Table 2. We created PubMed queries that returned a large number of articles (10,000 to 100,000 search results) in a broad area, thus allowing us to obtain a large sample of MeSH tags used in that area. For topic modeling purposes, we only used search results sets that contained abstracts (many pre-1980 MEDLINE citations contain only title and author).

2.2   Topic Model and Author-Topic Model

We start with two topic models appropriate for our task: the *standard topic model*, and the *author-topic model* ([3,5]). In the author-topic model, we use MeSH headings as "authors" of the documents (using the obvious analogy that like an author, a MeSH heading is responsible for generating words in the title and abstract). To learn the model parameters we use Gibbs sampling: the Gibbs sampling equations for the topic model and author-topic model are given by

$$p(z_{id} = t | x_{id} = w, \mathbf{z}^{\neg id}) \propto \frac{N_{wt}^{\neg id} + \beta}{\sum_w N_{wt}^{\neg id} + W\beta} \frac{N_{td}^{\neg id} + \alpha}{\sum_t N_{td}^{\neg id} + T\alpha}, \qquad (1)$$

$$p(z_{id} = t, y_{id} = m, | x_{id} = w, \mathbf{z}^{\neg id}, \mathbf{y}^{\neg id}) \propto \frac{N_{wt}^{\neg id} + \beta}{\sum_w N_{wt}^{\neg id} + W\beta} \frac{N_{tm}^{\neg id} + \gamma}{\sum_t N_{tm}^{\neg id} + T\gamma}, \qquad (2)$$

where $z_{id} = t$ and $y_{id} = m$ are the assignments of the ith word in document d to topic t and author m respectively, and $x_{id} = w$ indicates that the current observed word is word w. $\mathbf{z}^{\neg id}$ and $\mathbf{y}^{\neg id}$ are the vectors of all topic and author assignments not including the current word, N_{wt}, N_{td} and N_{tm} represent integer count arrays (with the subscripts denoting what is counted), and α, β and γ are Dirichlet priors. From the count arrays, we estimate the conditional distributions using

$$p(w|t) = \frac{N_{wt} + \beta}{\sum_w N_{wt} + W\beta}, \, p(t|d) = \frac{N_{td} + \alpha}{\sum_t N_{td} + T\alpha}, \, p(t|m) = \frac{N_{tm} + \gamma}{\sum_t N_{tm} + T\gamma}. \qquad (3)$$

We use a MeSH heading's distribution over topics, $p(t|m)$, as the canonical way to represent a MeSH heading using learned topics. The author-topic model directly estimates this distribution over topics for each MeSH heading. In the topic model, we estimate $p(t|m)$ by summing over the documents using $p(t|m) = \sum_d p(t|d)p(d|m)$, where $p(d|m)$ is the empirical distribution of observed application of MeSH headings to documents.

For each of the six query results sets, we learned topic and author-topic models and computed $p(t|m)$ for all the major MeSH headings that occurred in at least 10 articles in the query results set. The following examples show distributions over topics for three MeSH headings (with query set indicated with 'query='):

Alcohol-Related-Disorders [query="drug"]
(0.27) [t26] drinking alcohol alcohol-use problem alcohol-related drinker women alcohol-consumption heavy ...
(0.11) [t36] dependence use-disorder criteria dsm-iv symptom diagnostic interview treatment adolescent ...

Artificial-Intelligence [query="imaging"]
(0.39) [t62] segmentation shape feature classification detection structure automatic analysis representation ...
(0.32) [t71] image algorithm model object proposed framework approach problem propose estimation ...
(0.17) [t110] approach surface application efficient problem demonstrate texture component computation ...

Tobacco-Smoke-Pollution [query="smoking"]
(0.31) [t98] passive air tobacco-smoke ets pollution environmental exposure active home indoor ...
(0.10) [t129] smoking smoker tobacco smoke consumption tobacco-use daily smoked cigarette current ...
(0.09) [t70] exposure exposed effect level environmental chemical relationship evidence observed dose ...
(0.08) [t120] policies policy ban smoke-free workplace law public restaurant restriction smoking ...

Under each MeSH heading we list the most probable topics according to $p(t|m)$. We denote a topic by a topic ID (e.g. [t26]) which has no external meaning, then the list of most likely words in that topic, followed by an ellipsis to indicate that the cutoff for printing words is arbitrary. The number preceding the topic ID is $p(t|m)$ for that topic. Topics accounting for less than 0.05 probability mass are not shown. This example shows that we learn sensible distributions over topics for these three MeSH headings. Note that the topics learned, and the association of topics with MeSH headings, is not completely independent of the results set returned by the query. For example, the topics associated with the MeSH heading Artificial-Intelligence are clearly oriented towards imaging.

When topic modeling, we want to learn "essential" topics, i.e. topics that are robust (albeit latent) features present in the data, which are reliably and repeatably found by a variety of techniques. However, even with the two closely-related models, the topic model and the author-topic model, we learn topics that are close, but clearly different. For example, for the burns query results set, we learn the following topics relating to children:

(1.1%) [t11] children pediatric year child age **burned** parent month **young** childhood **adult** infant mother **burn** ...

(0.7%) [t25] children child **abuse** parent pediatric **scald** mother year age **physical home** month infant childhood ...

where [t11] is learned by the topic model and [t25] by the author-topic model. While not shown, these topics are highly repeatable over different random initializations. The gist of these topics is clearly different (the different words are bold), with the author-topic model learning an abuse variation of the topic. There is also a difference in the prevalence of the two topics, with the first topic accounting for 1.1% of all words, and the second topic accounting for 0.7% of all words. One may be left wondering which is the better or more correct topic.

In practice, different topic models produce different topics and different statistics, which may not be obvious from the model formulations, but may be revealed by experiments. Figure 1 shows that the distribution of topic sizes for the topic model is flatter than that from the author-topic model for our data sets.

2.3 Resampled Author Model

There may be several reasons for preferring topics learned by the standard topic model. One could argue that the simpler model learns topics that are in some way more fundamental to the collection. Furthermore, even with our MEDLINE abstracts, we have ambiguity over authors: Are they the MeSH headings, or actual authors of the articles? We also may prefer the flatter distribution of topic sizes for better division and faceted searching of the collection.

Here we introduce the *resampled author model*. The resampled author model is the author-topic model run with a fixed word-topic distribution previously learned by the topic model:

$$p(y_{id} = m | x_{id} = w, z_{id} = t, \mathbf{z}^{-id}, \mathbf{y}^{-id}) \propto p(w|t) \ \frac{N_{tm}^{-id} + \gamma}{\sum_t N_{tm}^{-id} + T\gamma} \qquad (4)$$

with $p(w|t)$ given by (3). The idea behind the resampled author model is to keep and use topics learned by the topic model, but learn a better association between topics and authors (in our case MeSH headings) than the naive computation of summing over documents. Indeed, our experimental results shown in Figure 2 show that the resampled author model does produce results that combines the learned topics from the topic model, and the relatively low entropy of topic distributions computed by the author-topic model.

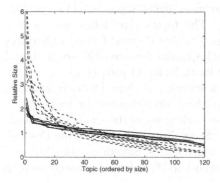

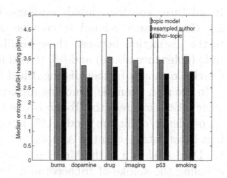

Fig. 1. Spectrum of topic size relative to $\frac{1}{T}$ for topic model (solid) and author-topic model (dashed), showing that the topic model produces a flatter distribution of topics for the six query results sets

Fig. 2. Median entropy of MeSH heading topic distributions, showing that the author-topic model learns clearer associations with learned topics than the topic model, with the resampled author model results falling in between

3   Analysis of MeSH Headings

Our topic representation of MeSH headings is useful for a broad array of tasks. First they are a direct way of explaining or interpreting what is meant by a MeSH heading. Second, topics provide a basis upon which we can compare MeSH headings, and compute quantities related to the MeSH hierarchy. A simple task is to explain differences in closely related MeSH headings. This is useful for educating PubMed users as to the distinctions between MeSH headings, and also suggesting other MeSH headings to use in searches. Below we list topics related to three MeSH headings related to cocaine, and two MeSH headings related to smoking:

Cocaine-Related-Disorders [*query="drug"*]
(0.32) [t114] cocaine user crack drug dependence abuse urine reported day cocaine-dependent ...
(0.05) [t66] drug drug-use substance-use substance substance-abuse drug-abuse illicit-drug alcohol ...
(0.04) [t6] treatment outcome program client outpatient residential abuse-treatment follow-up ...
Cocaine[*query="drug"*]
(0.39) [t114] cocaine user crack drug dependence abuse urine reported day cocaine-dependent ...
(0.04) [t9] urine concentration positive hair sample testing morphine specimen detection test ...
(0.04) [t77] group subject n= found male individual finding examined evaluated test clinical ...
Crack-Cocaine [*query="drug"*]
(0.38) [t114] cocaine user crack drug dependence abuse urine reported day cocaine-dependent ...
(0.07) [t12] sample likely less characteristic multiple recent demographic risk report similar ...
(0.05) [t97] sexual sex partner condom sexually std female transmitted women intercourse risk ...

- - -

Smoking [*query="smoking"*]
(0.23) [t54] smoker smoking cigarette cigarette-smoking effect nonsmoker non-smoker smoked ...
(0.16) [t129] smoking smoker tobacco smoke consumption tobacco-use daily smoked cigarette ...
(0.05) [t108] year age change period young aged pattern related relationship rate ...
Tobacco-Use-Disorder [*query="smoking"*]
(0.17) [t139] dependence measure scale negative addiction questionnaire score positive ...
(0.12) [t129] smoking smoker tobacco smoke consumption tobacco-use daily smoked cigarette ...
(0.09) [t147] nicotine cigarette effect gum smoker patch nrt mg level replacement ...

In all three cocaine-related headings, the most likely topic is [t114], capturing clear content related to cocaine use. The topics that follow give a clue to the distinction between these headings: Cocaine-Related-Disorders features [t66] (substance abuse) and [t6] (treatment), Cocaine features [t9] (testing), and Crack-Cocaine is further distinguished by its inclusion of [t97] (sex).

In the next example, Smoking includes a generic and shared tobacco smoking topic [t129], whereas the Tobacco-Use-Disorder is distinguished by topics [t139] (dependence/addiction) and (t147) nicotine. This usage of these two MeSH headings is consistent with the Annotation and Scope Notes provided in the Descriptor Data as shown in the MeSH browser[2] which states: Smoking = Inhaling and exhaling the smoke of tobacco; and Tobacco-Use-Disorder = Tobacco used to the detriment of a person's health.

3.1   Which MeSH Headings Are Similar?

Knowing topic distributions for MeSH headings allows us to compute distance between two headings using symmetric KL divergence, $KL^*(p(t|m_1)\|p(t|m_2))$. This distance computation provides additional insight into the relationship between related MeSH headings. For example, the MeSH browser page for Substance-Related-Disorders mentions Street-Drugs (under See Also), but does not mention Urban Population or Psychotropic Drugs, which we computed as also being closely related to Substance-Related-Disorders. We display these connections in Figure 3, which shows connections that exist in the MeSH hierarchy (solid and dashed lines), as well as connections that are learned via topics (dotted lines). This type of visualization can immediately convey to a user the *actual* relationships between MeSH headings – possibly even surprising connections – as inferred from their pattern of usage in MEDLINE articles.

3.2   Predicting Major MeSH Headings

We have described several ways in which our topic representation is useful for explaining, interpreting and understanding MeSH headings. But how well do they perform on predictive tasks? We setup the following task: Given all the MeSH tags (major and minor) applied to a test article, predict which tags are major. For each unseen test article, we list by name all the MeSH tags, and indicate the number of major MeSH tags. For example, the article entitled *Effects of caffeine in overnight-withdrawn consumers and non-consumers* has major MeSH tags {Affect, Caffeine, Cognition} and minor MeSH tags {Adolescent, Adult, Attention, Female, Humans, Male, Placebos, Reaction Time, Saliva, Substance Withdrawal Syndrome}. Beyond some of the check-tags like 'Human', it is not immediately obvious (from just looking at the title) which tags would be major. We used the three models to rank MeSH headings in order of $p(m|d) = \sum_t p(m|t)p(t|d)$. The results, shown in Figure 4, show that all models have clear predictive ability that is better than random, with the author-topic model having the best accuracy.

---

[2] http://www.nlm.nih.gov/MeSH/MBrowser.html

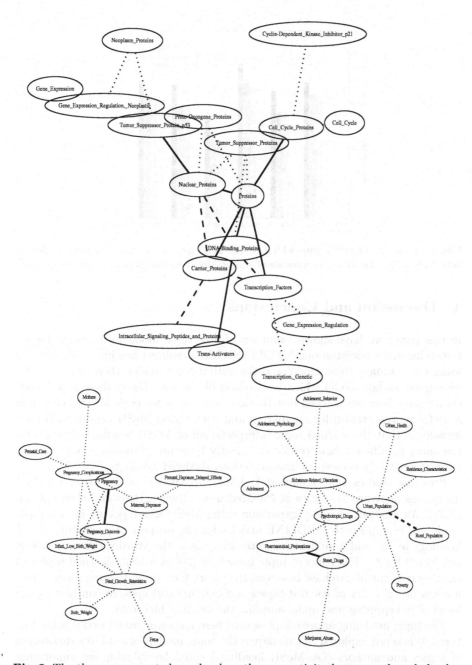

Fig. 3. The three unconnected graphs show the connectivity between selected closely-related MeSH headings for p53, drugs and smoking query data sets (clockwise from top). Solid (blue) lines show parent-child links in the MeSH hierarchy, and dashed (red) lines show sibling links in the MeSH hierarchy. Dotted (black) lines show connections that don't exist in the MeSH hierarchy, but are indicated based on closeness of topic distributions.

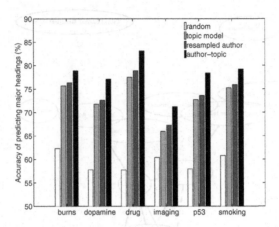

Fig. 4. Accuracy of predicting which MeSH headings are major. Random performs better than 50% because the number of major MeSH headings per document is used.

4 Discussion and Conclusions

In this paper we have shown examples of how topic modeling is useful for interpreting and understanding MEDLINE's MeSH subject headings. We start by using the standard topic model and the author-topic model, then introduce the resampled author model which is a hybrid of the two. Using these topic models we show how the learned distribution over topics for each MeSH heading is a useful representation for comparing and contrasting MeSH headings. We acknowledge that the learned topic interpretation of MeSH headings depends on the query results set; however for sufficiently large query results sets we expect to learn relatively consistent interpretations of MeSH headings.

Previous studies that analyzed PubMed/MEDLINE usage using PubMed query logs analyzed statistics of PubMed users, their actions and their queries ([6,7]). An analysis of query expansion using MeSH was reported in [1]. Topic models were applied to MEDLINE articles for the purpose of predicting MeSH headings in [8], and a similar semantic analysis of the WordNet hierarchy was conducted by [9]. The concept-topic models of [10] also relate learned topics to existing concept hierarchies; however, that work focuses on tagging unseen documents using a mix of learned topics and existing concepts, in contrast to our focus of interpreting and understanding the existing hierarchy.

The topic modeling approach presented here has some useful and flexible features. While not explored in this paper, the topic models' word-level annotation of topics and authors (i.e. MeSH headings) could be valuable for annotating which sections of longer documents are most relevant to each MeSH tag applied. More broadly, this framework is generally useful for any collection that is tagged with subject headings. Extensions to this work could include devising topic models to validate subject heading hierarchies and creating a tool to support ontology maintenance.

Acknowledgement. NICTA is funded by the Australian Government as represented by the Department of Broadband, Communications and the Digital Economy and the Australian Research Council through the ICT Centre of Excellence program.

References

1. Lu, Z., Kim, W., Wilbur, W.J.: Evaluation of query expansion using mesh in pubmed. Inf. Retr. 12(1), 69–80 (2009)
2. Blei, D.M., Ng, A.Y., Jordan, M.I.: Latent Dirichlet allocation. Journal of Machine Learning Research 3, 993–1022 (2003)
3. Griffiths, T., Steyvers, M.: Finding scientific topics. Proceedings of the National Academy of Sciences 101, 5228–5235 (2004)
4. Buntine, W.L., Jakulin, A.: Applying discrete pca in data analysis. In: UAI, pp. 59–66 (2004)
5. Rosen-Zvi, M., Griffiths, T.L., Steyvers, M., Smyth, P.: The author-topic model for authors and documents. In: UAI, pp. 487–494 (2004)
6. Herskovic, J.R., Tanaka, L.Y., Hersh, W., Bernstam, E.V.: A day in the life of pubmed: analysis of a typical day's query log. J. Am. Med. Inform. Assoc. 14(2), 212–220 (2007)
7. Lin, J., Wilbur, W.J.: Modeling actions of pubmed users with n-gram language models. Inf. Retr. 12(4), 487–503 (2009)
8. Mörchen, F., Dejori, M., Fradkin, D., Etienne, J., Wachmann, B., Bundschus, M.: Anticipating annotations and emerging trends in biomedical literature. In: KDD, pp. 954–962 (2008)
9. Snow, R., Jurafsky, D., Ng, A.Y.: Semantic taxonomy induction from heterogenous evidence. In: ACL, pp. 801–808 (2006)
10. Chemudugunta, C., Smyth, P., Steyvers, M.: Combining concept hierarchies and statistical topic models. In: CIKM, pp. 1469–1470 (2008)

A Novel Path-Based Clustering Algorithm Using Multi-dimensional Scaling

Uyen T.V. Nguyen, Laurence A.F. Park, Liang Wang, and Kotagiri Ramamohanarao

Department of Computer Science and Software Engineering
The University of Melbourne, Victoria, Australia 3010
{thivun, lapark, lwwang, rao}@csse.unimelb.edu.au

Abstract. Data clustering is a difficult and challenging task, especially when the hidden clusters are of different shapes and non-linearly separable in the input space. This paper addresses this problem by proposing a new method that combines a path-based dissimilarity measure and multi-dimensional scaling to effectively identify these complex separable structures. We show that our algorithm is able to identify clearly separable clusters of any shape or structure. Thus showing that our algorithm produces model clusters; that follow the definition of a cluster.

Keywords: Unsupervised learning, path-based clustering.

1 Introduction

Data clustering, or cluster analysis, is the process of finding a natural partition of a set of patterns, points or objects [1]. The clustering task plays a very important role in many areas such as exploratory data analysis, pattern recognition, computer vision, and information retrieval. Although cluster analysis has a long history, there are still many challenges, and the goal of designing a general purpose clustering algorithm remains a challenging task [2]. Intuitively, the clustering task can be stated as follows: given a set of n objects, a clustering algorithm tries to partition these objects into k groups so that objects within the same group are alike while objects in different groups are not alike. However, the definition of similarity is application dependent and sometimes unknown, which makes clustering an ill-posed problem.

Despite many clustering algorithms being proposed, K-means is still widely used and is one of the most popular clustering algorithms [2]. This is because it is an efficient, simple algorithm and provides successful results in many practical applications. However, K-means is only good at clustering compact and Gaussian shaped clusters and fails in capturing elongated clusters, or clusters that are non-linearly separable in the input space [3]. In order to tackle this problem, kernel K-means was introduced [4]. This method maps the data into a higher dimensional feature space defined by a non-linear function (intrinsic in the kernel function) so that the possibility of separating the data linearly becomes feasible. However, the task of choosing a suitable kernel function and its parameters for a given dataset is difficult. Another emerging approach is to use a spectral clustering algorithm, which performs

A. Nicholson and X. Li (Eds.): AI 2009, LNAI 5866, pp. 280–290, 2009.

the clustering on a set of eigenvectors of the affinity matrix derived from the data. It has been shown that results obtained by spectral clustering often outperform traditional clustering algorithms like K-means [5]. Although there are many different points of view to explain why spectral clustering works [6-8], it is still not completely understood yet. Moreover, spectral clustering leaves the users many choices and parameters to be set such as the similarity metric and its parameters, the type of graph Laplacian matrix, and the number of eigenvectors to be used [5]. Unfortunately, the success of spectral clustering depends heavily on these choices which make using spectral clustering a difficult task for the user.

In this paper, we propose a new clustering method that is capable of capturing clusters with different shapes that are non-linearly separable in the input space. This is not a new problem and there are two main approaches that address this problem that can be found in the literature [9-12]. In [9-11], a new path-based dissimilarity measure was proposed to embed the connectedness information between objects and a cost function based on this new dissimilarity measure was introduced. Optimization techniques were then used to find a set of clusters that minimizes this cost function. However, finding an optimal partition that minimizes this new cost function is a computationally intensive task and many different optimization techniques have been considered to address this problem. Another approach to this problem [12] is to improve a spectral clustering algorithm by using a path-based similarity measure. Instead of performing the spectral analysis directly on the similarity matrix, they modify the similarity matrix to include the connectedness among the objects. However, this approach is based on spectral clustering, which has many disadvantages as mentioned above.

We address the same problem but approach it differently. Instead of finding a new cost function like [9-11] or improve an existing algorithm like [12], we transform the original data into a new representation that takes into account the connection between objects so that the structures inherent in the data are well represented. This is achieved by a combination of the path-based dissimilarity measure and multi-dimensional scaling as described in Section 3. Compared with other methods, our method is much simpler yet produces very impressive results. The results prove that our new method is able to identify complex and elongated clusters in addition to the compact ones.

2   Background

In this section, we present the two main theories that are used by our algorithm. The first is the path-based dissimilarity measure, which gives a new way to identify the dissimilarity between two objects by taking into account the connection among objects. The second is the multi-dimensional scaling technique which is used to find a set of data points that exhibit the dissimilarities given by a dissimilarity matrix.

2.1   Path-Based Dissimilarity Measure

The path-based dissimilarity measure was first introduced in [9]. The intuitive idea behind this is that if two objects x_i, x_j are very far from each other (reflected by a large distance value d_{ij} with respect to metric m), but there is a path through them consisting of other objects such that the distances between any two successive objects are small, then d_{ij} should be adjusted to a smaller value to reflect this connection. The adjustment of d_{ij} reflects the idea that no matter how far the distance between two objects may be, they should be considered as coming from one cluster if they are connected by a set of successive objects forming density regions. This is reasonable and reflects the characteristic of elongated clusters.

The path-based dissimilarity measure can be described in a more formal way. Suppose that we are given a dataset of n objects X with each object x_i consisting of m features, $x_i = (x_{i1}, x_{i2},... x_{im})$ and an $n \times n$ distance matrix D holding the pair-wise distances of all pairs of objects in X. The objects and their distance matrix D can be seen as a fully connected graph, where each vertex in this graph corresponds to an object and the edge weight between two vertices i and j is the distance between the corresponding objects x_i and x_j, or $d_{ij} = dis(x_i, x_j)$. The path-based distance between x_i and x_j is then defined as follows.

Suppose that P_{ij} is the set of all possible paths from x_i to x_j in the graph, then for each path $p \in P_{ij}$, the effective dissimilarity between x_i and x_j along p is the maximum of all edge weights belonging to this path. The path-based distance d_{ij}' between x_i and x_j ($pbdis(x_i, x_j)$), is then the minimum of effective dissimilarities of all paths in P_{ij}, or:

$$d_{ij}' = pbdis(x_i, x_j) = \min_{p \in P_{ij}}\{\max_{1 \le h < |p|}(dis(p[h], p[h+1]))\} \tag{1}$$

where $p[h]$ denotes the object at the hth position in the path p and $|p|$ denotes the length of path p.

2.2 Multi-dimensional Scaling

Multi-dimensional scaling (MDS) is a technique that allows us to visually explore the data based on its dissimilarity information. In general, given a pair-wise distance matrix of a set of objects, the MDS algorithm finds a new data representation, or a configuration of points, that preserves the given pair-wise distances for a given metric as well as possible. Many MDS algorithms are available [13] and they can be divided into two main categories: metric MDS and non-metric MDS. For the sake of completeness, the theory behind classical multi-dimensional scaling is presented here to show how an MDS algorithm works. Classical multi-dimensional scaling is an attractive MDS method as it provides an analytical solution using an eigen-decomposition.

To start with, the process of deriving the matrix of squared pair-wise distances from a coordinate matrix (also known as data or pattern matrix) in terms of matrix operations is presented. Let X be an $n \times m$ coordinate matrix with each row i containing the coordinates of point x_i on m dimensions ($x_i = [x_{i1}, x_{i2},..., x_{im}]$) and $D^{(2)}(X)$ the squared distance matrix where each element at (i, j) is the squared distance between x_i and x_j. Suppose that Euclidean distance is used, then:

$$d_{ij}^2 = \sum_{k=1}^{m}(x_{ik} - x_{jk})^2 \tag{2}$$

After some simple transformations, the squared distance matrix $D^{(2)}(X)$ can be computed using a compact expression:

$$D^{(2)}(X) = \tilde{c}\tilde{1}^T + \tilde{1}\tilde{c}^T - 2XX^T \tag{3}$$

where $\tilde{c}$ is a vector with the diagonal elements of XX^T, and $\tilde{1}$ is a vector of ones.

Classical multi-dimensional scaling reverses this composition. It takes a dissimilarity matrix Δ (with each element δ_{ij} the dissimilarity between two unknown objects x_i and x_j) as input and finds a set of points $Z = \{z_1, z_2,..., z_n\}$ so that each pairwise distance d_{ij} (the distance between z_i and z_j) is as close to δ_{ij} as possible. This can be done using an eigen-decomposition.

Suppose that Z is the $n \times m'$ coordinate matrix that best matches Δ, then Z and Δ should be related by (3), or:

$$\Delta^{(2)} = \tilde{c}\tilde{1}^T + \tilde{1}\tilde{c}^T - 2ZZ^T \tag{4}$$

where $\Delta^{(2)}$ is the squared dissimilarity matrix, $\tilde{c}$ is now the vector with diagonal elements of ZZ^T.

Because distances are invariant under translation, we assume that Z has column means equal to 0. Then multiplying the left and right sides of (4) by the centering matrix $J (J = I - (1/n)\tilde{1}\tilde{1}^T)$ and $-1/2$, and after some reductions, we have:

$$B = ZZ^T = (-1/2)J\Delta^{(2)}J \tag{5}$$

So the scalar product matrix B of Z can be derived from the dissimilarity matrix Δ as above. From the scalar product matrix B, the coordinate matrix Z is easily computed by using an eigen-decomposition. Let Q and Λ be the eigenvector and eigenvalue matrices of B respectively. Since B is a real and symmetric matrix (because of the symmetry of Δ), we have:

$$B = Q\Lambda Q^T \tag{6}$$

If Δ is a Euclidean distance matrix, which means that it is constructed from the pairwise distances of a set of points, then Λ contains only positive and zero eigenvalues. Otherwise, there might be some negative eigenvalues, and classical scaling ignores them as error. Let Λ_+ be the matrix of positive eigenvalues and Q_+ the matrix of the corresponding eigenvectors, then the coordinate matrix Z is calculated as:

$$Z = Q_+\Lambda_+^{1/2} \tag{7}$$

One point should be noted is if Λ contains only positive and zero eigenvalues, then Z will provide an exact reconstruction of Δ. Otherwise, the distance matrix Δ will be an approximation. Another point is that the relative magnitudes of those eigenvalues in Λ indicate the relative contribution of the corresponding columns of Z in reproducing the original distance matrix Δ. So, if k' eigenvalues in Λ are much larger than the rest, then the distance matrix based on the k' corresponding columns of Z nearly reproduces the original dissimilarity matrix Δ. In this sense, we can reduce the number of dimensions of Z by choosing only the principle eigenvalues with only a small loss of information.

3   A New Algorithm

In this section, the details of our proposed algorithm are presented. With a good data representation, any simple clustering algorithm like K-means can be applied successfully. In order to achieve this goal, we first use the path-based dissimilarity measure to change the dissimilarities (or distances) between all pairs of objects. This transformation is performed once at the beginning on the whole dataset. As the path-based dissimilarity measure takes into account the connection relationships among the objects, this transformation will embed the cluster structure information into the new dissimilarity matrix. We then find a set of objects, or a new data representation, that reflects these new pair-wise dissimilarities by using a multi-dimensional scaling algorithm. After that, K-means is employed to do the clustering on this new data representation.

Algorithm. Path-based clustering using multi-dimensional scaling

- Input: $n \times m$ data matrix X, number of clusters k
- Algorithm:
  1. Compute the $n \times n$ pair-wise distance matrix D from data matrix X
  2. Transform D into D' using path-based dissimilarity measure
  3. Perform classical MDS on D' to get a $n \times m'$ new data matrix Z
  4. Identify k' - the number of principle dimensions of Z.
     Let Y the $n \times k'$ matrix of k' first columns of Z.
  5. Apply K-means on n rows y_i of Y to get a partition $C_1, C_2, ... C_k$
- Output: Clusters $A_1, ... A_k$ with $A_i = \{x_j | y_j \in C_i\}$

In step 2, the path-based distances between all pairs of objects are computed using an algorithm similar to the algorithm of Floyd [14]. In step 3, the classical multi-dimensional scaling algorithm will return a configuration of points whose pair-wise distances approximate the new distance matrix D'. Because of this, the number of dimensions m' of the MDS configuration Z may be very large. However, only some of them are important and the distance matrix can be reconstructed using only these principle dimensions with very small error. So the number of principle dimensions needs to be identified and only those important ones should be used to represent a new data matrix. This can easily be done by an analysis on the eigenvalues of the scalar product matrix which is also returned by classical multi-dimensional scaling algorithm. In our experiments, we choose the number of dimensions as the number of clusters minus one (as done in spectral clustering) and the results showed that this is a suitable setting.

One of the advantages of our algorithm is that it operates based on the dissimilarity matrix which often arises naturally from a data matrix. A distance metric is used to calculate the dissimilarity between objects. Among many distance metrics available,

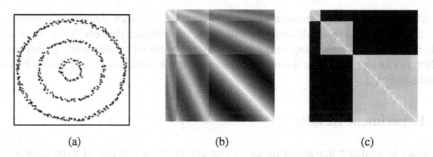

(a)   (b)   (c)

Fig. 1. Distance matrices of three-circle dataset: (a) input data; (b) original distance matrix; (c) transformed distance matrix

Euclidean distance is a simple and popular metric that proves successful in many applications. Another advantage is that our algorithm is more efficient than the original path-based clustering algorithm [9-11] as it calculates the path-based distances of all pairs of objects only one time at the beginning. Moreover, the algorithm requires only one parameter from the user, the number of clusters. Finally, experimental results prove its strong ability to detect complex structures inherent in the data.

To show the effectiveness of our algorithm, we analyze the clustering process on a commonly used three-circle synthetic dataset. Fig. 1 shows the original dataset in two-dimensional space and its distance matrices before and after the path-based transformation. The distance matrices are displayed on gray scale images with white for 0 and darker for higher values. To emphasize the utility of the path-based dissimilarity, the points in this example are ordered so that the points within each circle form a block of successive pixels on the image. It is shown that after the transformation, the distance matrix is nearly block-diagonal with each cluster corresponding to a block. This indicates that applying path-based dissimilarity transformation enhances the cluster structures on the distance matrix.

After performing path-based dissimilarity transformation, the classical MDS is performed on the transformed distance matrix. The new data representation Y with two principle dimensions is obtained and plotted in Fig. 2(a). From this plot, we can see that the data points of the original dataset are transformed into this new space and form three very compact clusters, each of which represents a circle of the original dataset. With this new representation, simple clustering algorithm like K-means can easily detect and correctly identify the three circles as shown in Fig. 2(b).

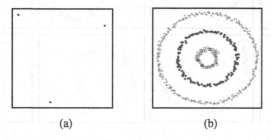

(a)   (b)

Fig. 2. Results obtained on three-circle dataset: (a) new data representation on two-dimensional space; (b) K-means result on three-circle dataset

After the path-based transformation, the distance between data points within a cluster becomes very small compared to those of different clusters. This obeys the rule that the points belonging to the same cluster should be similar while points from different clusters should be dissimilar in the new space, which is clearly shown by the results in Fig. 2.

4 Experiment Results

In order to evaluate the performance of our algorithm, a number of experiments on both synthetic and real datasets were performed. In all these experiments, the results of our algorithm were compared with those of two popular clustering methods, K-means and spectral clustering. With spectral clustering, we used Ncut normalization on the Laplacian. Also, to avoid manually setting the scaling parameter, we employed the local scaling setting proposed in [15] as it has been shown that this setting gives better results than a global scaling parameter.

4.1 Results on Synthetic Datasets

To demonstrate the power of our algorithm on separable data, the first comparison was performed on four synthetic datasets with different data structures: three-circle, face, two-moon, and two-spiral datasets. The results are presented in Fig. 3. We can see that K-means is unable to identify any of the clusters because the clusters are elongated in nature. Spectral clustering is able to identify the clusters in two of the data sets (three-circle and two-moon). Interestingly, our algorithm is able to identify

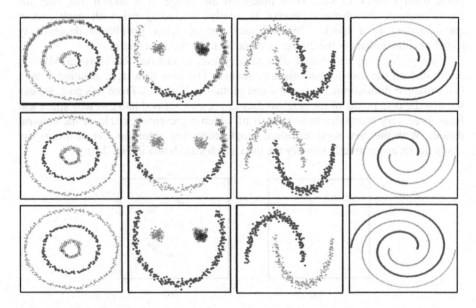

Fig. 3. Results on 4 synthetic datasets: first row: results of K-means; second row: results of spectral clustering; third row: results of our algorithm

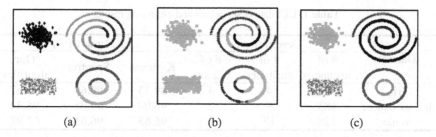

(a)   (b)   (c)

Fig. 4. Results on a complex synthetic dataset: (a) result of K-means; (b) result of spectral clustering; (c) result of our algorithm

each cluster in all of the data sets. To understand these results, we examined the path-based data representation of each dataset and learned that each cluster in the original space forms a very compact cluster in the new space (similar to the case of three-circle dataset explained above). With such good representation, the data space can easily be partitioned using K-means and produce correct results as presented.

The second comparison was performed on a more complex synthetic dataset, which consists of six clusters: two circles, two spirals, a rectangular, and a Gaussian-shaped cluster. The results obtained by K-means, spectral clustering and our algorithm are shown in Fig. 4, which indicate that neither K-means nor spectral clustering can correctly identify all clusters in this dataset. On the contrast, our algorithm detected all the clusters despite of their differences in shape and size.

4.2 Results on Real Datasets

In order to test the performance on real data, we performed a comparison on three commonly used datasets from UCI repository [16]: Iris, Breast-cancer, and Wine. The descriptions of these datasets are presented in Table 1. To measure the performance of each clustering algorithm, the accuracy metric [17] was used. The results are summarized and presented on the same table.

We can see that K-means provides high accuracy on each of the data sets, implying that each of the data sets contain radial clusters. We can also see that our path-based algorithm provides high accuracy on the Breast-cancer data set implying that is has two distinct clusters. Our algorithm gave an accuracy of approximately 2/3 for the other two data sets, which leads us to believe that one cluster is distinct and the other two are overlapping. The overlapping would cause our method to place both clusters in to one, giving an accuracy of 2/3. This overlapping property can be easily seen when each data set is projected into a two-dimensional space.

To deal with the case of overlapping clusters, we will examine a mixed clustering method. We define the new distance as a function of α ($0 \leq \alpha \leq 1$), original distance, and path-based distance as (8). With $\alpha = 0$, the result obtained is equal to the result of K-means while with $\alpha = 1$, the result is of our original algorithm. The remaining values of α give a weighted combination of K-means and our algorithm. The accuracies obtained on three datasets when α changes from 0 to 1 are displayed in Fig. 5.

Table 1. UCI data descriptions and clustering results

Dataset	Descriptions			Accuracy (%)		
	# of instances	# of attributes	# of classes	K-means	Spectral	Our algorithm
Iris	150	4	3	89.33	**90.67**	66.67
Breast-cancer	683	9	2	96.04	69.25	**96.34**
Wine	178	13	3	**96.63**	**96.63**	62.92

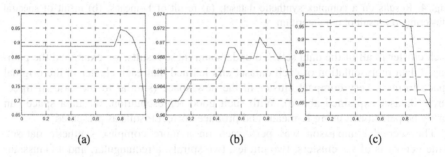

(a)   (b)   (c)

Fig. 5. Accuracies obtained on three datasets when α changes: (a) on Iris dataset; (b) on Breast-cancer dataset; (c) on Wine dataset

$$d_{ij}(\alpha) = (1-\alpha) \times dis(x_i, x_j) + \alpha \times pbdis(x_i, x_j) \qquad (8)$$

We can see that the mixing has increased the accuracy of our algorithm by allowing it to identify overlapped clusters. In two of the three data sets, we can see that the mixing has produced an increase over K-means as well, implying that K-means has also profited from our algorithm identifying the clearly separable clusters.

By applying the mixed algorithm to the synthetic data in Fig. 3 we obtain the greatest accuracy at $\alpha = 1$ (when using our algorithm only), with a descent in accuracy as α is reduced to 0. This result is obvious since our algorithm is suited for separable clusters, and each of the synthetic data sets is clearly separable in Euclidean space.

5 Discussions

As part of our analysis, we examine two cases where our algorithm cannot separate clusters, but K-means is able to provide high accuracy. The first case is when there are overlapping regions between clusters and the second is when separated clusters are connected by small bridges as shown in Fig. 6.

In these cases, our algorithm will consider the data as one cluster since the distances between any two points in different clusters is small due to the path-based transformation. K-means identifies these clusters with only small error. However, these are difficult cases for the clustering task in general. If we removed the class information from the data (remove the color from Fig. 6), there is no reason why we should identify the two cases shown as two clusters. There is also no reason why the

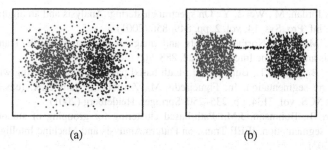

(a)   (b)

Fig. 6. Examples of two cases when the algorithm fails: (a) two Gaussian clusters with an overlapping region; (b) two separated clusters connected by a small bridge

data could not contain one or three clusters. The beauty of our path-based clustering algorithm is that it identifies each clearly separable cluster (independent of shape and structure) and makes no assumptions about inseparable data.

6 Conclusions

In this paper, we have proposed a new clustering method that is capable of capturing complex structures in data. With the combination of the path-based dissimilarity measure and multi-dimensional scaling, we can produce a good data representation for any given dataset, which makes it possible to detect clusters of different shapes that are non-linearly separable in the input space.

We showed that our path-based clustering method clearly identifies separable clusters. We also showed that our algorithm is unable to identify inseparable clusters, but also explained that identifying clusters in such data is in the eye of the beholder. This behavior makes our path-based clustering algorithm produce model clusters; that follow the definition of a cluster.

References

1. Jain, A., Law, M.: Data Clustering: A User's Dilemma. In: Pal, S.K., Bandyopadhyay, S., Biswas, S. (eds.) PReMI 2005. LNCS, vol. 3776, pp. 1–10. Springer, Heidelberg (2005)
2. Jain, A.: Data Clustering: 50 Years Beyond K-means. In: Daelemans, W., Goethals, B., Morik, K. (eds.) ECML PKDD 2008, Part I. LNCS (LNAI), vol. 5211, pp. 3–4. Springer, Heidelberg (2008)
3. Dhillon, I.S., Guan, Y., Kulis, B.: Kernel k-means: spectral clustering and normalized cuts. In: 10th Int. Conf. on Knowledge Discovery and Data Mining, pp. 551–556. ACM, New York (2004)
4. Scholkopf, B., Smola, A., Muller, K.: Nonlinear component analysis as a kernel eigenvalue problem. J. Neu. Com. 10, 1299–1319 (1998)
5. Von Luxburg, U.: A tutorial on spectral clustering. J. Sta. and Com. 17, 395–416 (2007)
6. Meila, M., Shi, J.: A random walks view of spectral segmentation. In: International Conference on AI and Statistics (2001)

7. Ng, A., Jordan, M., Weiss, Y.: On spectral clustering: Analysis and an algorithm. In: Adv. in Neu. Inf. Pro. Sys. 14, vol. 2, pp. 849–856 (2001)
8. Shi, J., Malik, J.: Normalized cuts and image segmentation. IEEE Trans. on Pattern Analysis and Machine Intelligence 22, 888–905 (2000)
9. Fischer, B., Zoller, T., Buhmann, J.: Path based pairwise data clustering with application to texture segmentation. In: Figueiredo, M., Zerubia, J., Jain, A.K. (eds.) EMMCVPR 2001. LNCS, vol. 2134, pp. 235–250. Springer, Heidelberg (2001)
10. Fischer, B., Buhmann, J.M.: Path-based clustering for grouping of smooth curves and texture segmentation. IEEE Trans. on Pattern Analysis and Machine Intelligence 25, 514–519 (2003)
11. Fischer, B., Buhmann, J.M.: Bagging for path-based clustering. IEEE Trans. on Pattern Analysis and Machine Intelligence 25, 1411–1415 (2003)
12. Chang, H., Yeung, D.: Robust path-based spectral clustering. J. Pat. Rec. 41, 191–203 (2008)
13. Borg, I., Groenen, P.: Modern multidimensional scaling: Theory and applications. Springer, Heidelberg (2005)
14. Cormen, T.H., Leiserson, C.E., Rivest, R.L.: Introduction to Algorithms. MIT Press, Cambridge (1989)
15. Zelnik-Manor, L., Perona, P.: Self-tuning spectral clustering. J. Adv. in Neu. Inf. Pro. Sys. 17, 1601–1608 (2004)
16. UCI Machine Learning Repository,
    http://www.ics.uci.edu/~mlearn/MLRepository.html
17. Cai, D., He, X., Han, J.: Document clustering using locality preserving indexing. IEEE Trans. on Knowledge and Data Engineering 17, 1624–1637 (2005)

Ensemble Approach for the Classification of Imbalanced Data

Vladimir Nikulin[1], Geoffrey J. McLachlan[1], and Shu Kay Ng[2]

[1] Department of Mathematics, University of Queensland
v.nikulin@uq.edu.au, gjm@maths.uq.edu.au
[2] School of Medicine, Griffith University
s.ng@griffith.edu.au

Abstract. Ensembles are often capable of greater prediction accuracy than any of their individual members. As a consequence of the diversity between individual base-learners, an ensemble will not suffer from overfitting. On the other hand, in many cases we are dealing with imbalanced data and a classifier which was built using all data has tendency to ignore minority class. As a solution to the problem, we propose to consider a large number of relatively small and balanced subsets where representatives from the larger pattern are to be selected randomly. As an outcome, the system produces the matrix of linear regression coefficients whose rows represent random subsets and columns represent features. Based on the above matrix we make an assessment of how stable the influence of the particular features is. It is proposed to keep in the model only features with stable influence. The final model represents an average of the base-learners, which are not necessarily a linear regression. Test results against datasets of the PAKDD-2007 data-mining competition are presented.

Keywords: ensemble classifier, gradient-based optimisation, boosting, random forest, decision trees.

1 Introduction

Ensemble (including voting and averaged) classifiers are learning algorithms that construct a set of many individual classifiers (called base-learners) and combine them to classify test data points by sample average. It is now well-known that ensembles are often much more accurate than the base-learners that make them up [1], [2]. Tree ensemble called "random forest" was introduced in [3] and represents an example of successful classifier. Another example, bagging support vector machine (SVM) [4] is very important because direct application of the SVM to the whole data set may not be possible. In the case of SVM we are interested to deal with limited sample size which is equal to the dimension of the corresponding kernel matrix. The well known bagging technique [5] is relevant here. According to this technique each base-learner used in the ensemble is trained with data that are randomly selected from the training sample (without replacement).

A. Nicholson and X. Li (Eds.): AI 2009, LNAI 5866, pp. 291–300, 2009.
© Springer-Verlag Berlin Heidelberg 2009

Our approach was motivated by [5], and represents a compromise between two major considerations. On the one hand, we would like to deal with balanced data. On the other hand, we are interested to exploit all available information. We consider a large number n of balanced subsets of available data where any single subset includes two parts 1) all 'positive' instances (minority) and 2) randomly selected 'negative' instances. The method of balanced random sets (RS) is general and may be used in conjunction with different base-learners.

In the experimental section we report test-results against real-world data of the PAKDD-2007 Data Mining Competition[1], which were provided by a consumer finance company with the aim of finding better solutions for a cross-selling problem. The data are strongly imbalanced with significantly smaller proportion of positive cases (1.49%), which have the following practical interpretation: a customer opened a home loan with the company within 12 months after opening the credit card [6].

Regularised linear regression (RLR) represents the most simple example of a decision function. Combined with quadratic loss function it has an essential advantage: using gradient-based search procedure we can optimise the value of the step size. Consequently, we will observe a rapid decline in the target function [7].

By definition, regression coefficients may be regarded as natural measurements of influence of the corresponding features. In our case we have n vectors of regression coefficients, and we can use them to investigate the stability of the particular coefficients.

Proper feature selection may reduce overfitting significantly [8]. We remove features with unstable coefficients, and recompute the classifiers. Note that stability of the coefficients may be measured using different methods. For example, we can apply the t-statistic given by the ratio of the mean to the standard deviation.

The proposed approach is flexible. We do not expect that a single algorithm will work optimally on all conceivable applications and, therefore, an opportunity of tuning and tailoring is a very essential.

Initial results obtained using RLR during PAKDD-2007 Data Mining Competition were reported in [9]. In this paper, using tree-based LogitBoost [10] as a base-learner we improved all results known to us.

This paper is organised as follows: Section 2 describes the method of random sets and mean-variance filtering. Section 3 discusses general principals of the AdaBoost and LogitBoost Algorithms. Section 4 explains the experimental procedure and the most important business insights. Finally, Section 5 concludes the paper.

2 Modelling Technique

Let $\mathbf{X} = (\mathbf{x}_t, y_t)$, $t = 1, \dots, m$, be a training sample of observations where $\mathbf{x}_t \in \mathbb{R}^\ell$ is a ℓ-dimensional vector of features, and y_t is a binary label: $y_t \in \{-1, 1\}$.

---

[1] http://lamda.nju.edu.cn/conf/pakdd07/dmc07/

Boldface letters denote vector-columns, whose components are labelled using a normal typeface.

In a practical situation the label y_t may be hidden, and the task is to estimate it using the vector of features. Let us consider the most simple linear decision function

$$u_t = u(\mathbf{x}_t) = \sum_{j=0}^{\ell} w_j \cdot x_{tj}, \qquad (1)$$

where x_{t0} is a constant term.

We can define a decision rule as a function of decision function and threshold parameter

$$f_t = f(u_t, \Delta) = \begin{cases} 1 & \text{if } u_t \geq \Delta; \\ 0, & \text{otherwise.} \end{cases}$$

We used AUC as an evaluation criterion where AUC is the area under the receiver operating curve (ROC). By definition, ROC is a graphical plot of True Positive Rates (TPR) against False Positive Rates (FPR).

According to the proposed method we consider large number of classifiers where any particular classifier is based on relatively balanced subset with all 'positive' and randomly selected (without replacement) 'negative' data. The final decision function (d.f.) has a form of logistic average of single decision functions.

Definition 1. *We call above subsets as random sets $RS(\alpha, \beta, n)$, where α is a number of positive cases, β is a number of negative cases, and n is the total number of random sets.*

This model includes two very important regulation parameters: 1) n and 2) $q = \frac{\alpha}{\beta} \leq 1$ - the proportion of positive cases where n must be sufficiently large, and q can not be too small.

We consider n subsets of $\mathbf{X}$ with α positive and $\beta = k \cdot \alpha$ negative data-instances, where $k \geq 1, q = \frac{1}{k}$. Using gradient-based optimization [11] we can compute the matrix of linear regression coefficients:

$$W = \{w_{ij}, i = 1, \ldots, n, j = 0, \ldots, \ell\}.$$

The mean-variance filtering (MVF) technique was introduced in [11], and may be efficient in order to reduce overfitting. Using the following ratios, we can measure the consistency of contributions of the particular features by

$$r_j = \frac{|\mu_j|}{s_j}, j = 1, \ldots, \ell, \qquad (2)$$

where μ_j and s_j are the mean and standard deviation corresponding to the j-column of the matrix W.

A low value of r_j indicates that the influence of the j-feature is not stable. We conducted feature selection according to the condition:

$$r_j \geq \gamma > 0.$$

The final decision function,

$$f_t = \frac{1}{n} \sum_{i=1}^{n} \frac{exp\{\tau \cdot u_{ti}\}}{1 + exp\{\tau \cdot u_{ti}\}}, \quad \tau > 0, \tag{3}$$

was calculated as a logistic average of single decision functions,

$$u_{ti} = \sum_{j=0}^{\ell} w_{ij} \cdot x_{tj},$$

where regression coefficients w were re-computed after feature reduction.

Remark 1. It is demonstrated in the Section 4 that performance of the classifier will be improved if we will use in (3) non-linear functions such as decision trees.

3   Boosting Algorithms

Boosting works by sequentially applying a classification algorithm to re-weighted versions of the training data, and then taking a weighted majority vote of the sequence of classifiers thus produced. For many classification algorithms, this simple strategy results in dramatic improvements in performance.

3.1   AdaBoost Algorithm

Let us consider minimizing the criterion [10]

$$\sum_{t=1}^{n} \xi(\mathbf{x}_t, y_t) \cdot e^{-y_t u(\mathbf{x}_t)}, \tag{4}$$

where the weight function is given below

$$\xi(\mathbf{x}_t, y_t) := \exp\{-y_t F(\mathbf{x}_t)\}. \tag{5}$$

We shall assume that the initial values of the ensemble d.f. $F(\mathbf{x}_t)$ are set to zero.

Advantages of the exponential compared with squared loss function were discussed in [9]. Unfortunately, we can not optimize the step-size in the case of exponential target function. We will need to maintain low value of the step-size in order to ensure stability of the gradient-based optimisation algorithm. As a consequence, the whole optimization process may be very slow and time-consuming. The AdaBoost algorithm was introduced in [12] in order to facilitate optimization process.

The following Taylor-approximation is valid under assumption that values of $u(\mathbf{x}_t)$ are small,

$$\exp\{-y_t u(\mathbf{x}_t)\} \approx \frac{1}{2} \left[(y_t - u(\mathbf{x}_t))^2 + 1 \right]. \tag{6}$$

Therefore, we can apply quadratic-minimisation (QM) model in order to minimize (4). Then, we optimize value of the threshold parameter Δ for u_t, and find the corresponding decision rule $f_t \in \{-1, 1\}$.

Next, we will return to (4),

$$\sum_{t=1}^{n} \xi(\mathbf{x}_t, y_t) \cdot e^{-c \cdot y_t \cdot f(\mathbf{x}_t)}, \tag{7}$$

where the optimal value of the parameter c may be easily found

$$c = \frac{1}{2} \log \{\frac{A}{B}\}, \tag{8}$$

and where

$$A = \sum_{y_t = f(\mathbf{x}_t)} \xi(\mathbf{x}_t, y_t), \quad B = \sum_{y_t \neq f(\mathbf{x}_t)} \xi(\mathbf{x}_t, y_t).$$

Finally (for the current boosting iteration), we update the function F:

$$F_{\text{new}}(\mathbf{x}_t) \leftarrow F(\mathbf{x}_t) + c \cdot f(\mathbf{x}_t), \tag{9}$$

and recompute weight coefficients ξ according to (5).

Remark 2. Considering test dataset (labels are not available), we will not be able to optimize value of the threshold parameter Δ. We can use either an average (predicted) value of Δ in order to transform decision function into decision rule, or we can apply direct update:

$$F_{\text{new}}(\mathbf{x}_t) \leftarrow F(\mathbf{x}_t) + c \cdot u(\mathbf{x}_t), \tag{10}$$

where the value of the parameter $c \leq 1$ must be small enough in order to ensure stability of the algorithm.

3.2   LogitBoost Algorithm

Let us parameterize the binomial probabilities by

$$p(\mathbf{x}_t) = \frac{e^{2F(\mathbf{x}_t)}}{1 + e^{2F(\mathbf{x}_t)}}.$$

The binomial log-likelihood is

$$y_t^\star \log \{p(\mathbf{x}_t)\} + (1 - y_t^\star) \log \{1 - p(\mathbf{x}_t)\} = -\log \{1 + \exp \{-2y_t F(\mathbf{x}_t)\}\}, \tag{11}$$

where $y^\star = (y + 1)/2$.

The following relation is valid,

$$\exp \{-2y_t F(\mathbf{x}_t)\} = \xi(\mathbf{x}_t) z_t^2, \tag{12}$$

where
$$z_t = \frac{y_t^* - p(\mathbf{x}_t)}{\xi(\mathbf{x}_t)}, \quad \xi(\mathbf{x}_t) = p(\mathbf{x}_t)(1 - p(\mathbf{x}_t)).$$

We can maximize (11) using a method with Newton's step, which is based on the matrix of second derivatives [11]. As an alternative, we can consider the standard weighted QM-model,

$$\sum_{t=1}^{n} \xi(\mathbf{x}_t)(z_t - u_t)^2. \tag{13}$$

After the solution $u(\mathbf{x}_t)$ was found, we update function $p(\mathbf{x}_t)$ as,

$$p(x_t) = \begin{cases} 1 & \text{if } h_t \geq 1; \\ h_t & \text{if } 0 < h_t < 1; \\ 0, & \text{otherwise,} \end{cases}$$

where $h_t = p(\mathbf{x}_t) + \xi(\mathbf{x}_t)u(\mathbf{x}_t)$. Then, we recompute the weight coefficients ξ and return to the minimization criterion (13).

Let us consider an update of the function F, assuming that $0 < h_t < 1$. By definition,

$$F_{\text{new}}(\mathbf{x}_t) = \frac{1}{2} \log \left\{ \frac{h_t}{1 - h_t} \right\} = \frac{1}{2} \log \left\{ \frac{p(\mathbf{x}_t)}{1 - p(\mathbf{x}_t)} \right\}$$

$$+ \frac{1}{2} \log \left\{ 1 + \frac{u(\mathbf{x}_t)}{1 - p(\mathbf{x}_t)u(\mathbf{x}_t)} \right\} \approx F(\mathbf{x}_t) + \nu \cdot u(\mathbf{x}_t), \quad \nu = 0.5. \tag{14}$$

Remark 3. Boosting trick (similar to the well-known kernel trick): as an alternative to QM-solution, we can apply in (10) or (14) decision function, which was produced by another method, for example, Naïve Bayes, decision trees or random forest.

4   Experimental Results

4.1   Data Preparation

The given home-loan data includes two sets: 1) a training-set with 700 positive and 40000 negative instances, and 2) a test-set with 8000 instances. Any data-instance represents a vector of 40 continuous or categorical features. Using standard techniques, we reduced the categorical features to numerical (dummy) values. Also, we normalized the continuous values to lie in the range $[0, 1]$. As a result of the above transformation we created totally numerical dataset with $\ell = 101$ features. As a result of MVF, the number of features was reduced from 101 to 44 (see Figure 1).

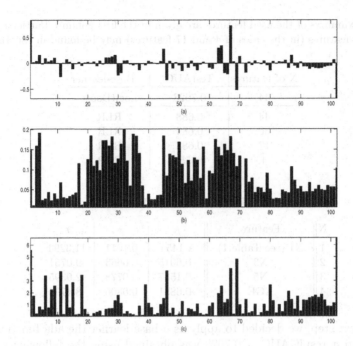

Fig. 1. Mean-variance filtering: (a) means (μ); (b) - standard deviations (s), (c) ratios $r = |\mu|/s$ (see Section 4 for more details)

Table 1. List of 6 the most significant features

N	Feature	μ	r
1	Bureau Enquiries for Morgages last 6 month	0.729	4
2	Age	-0.683	6.6
3	Bureau Enquiries for Loans last 12 month	-0.516	4.8
4	Bureau Enquiries last 3 month	0.342	2.54
5	Number of dependants	-0.322	3.82
6	Bureau Enquiries last month	0.299	1.92

4.2   Test Results

47 participants from various sources including academia, software vendors, and consultancies submitted entries with range of results from 0.4778 to 0.701 in terms of AUC. Our score was 0.688, which resulted in 9th place for us.

Note that the training AUC, which corresponds to the final submission was 0.7253. The difference between training and test results in the case of 44 features appears to be a quite significant. Initial thought (after results were published) was that there are problems with overfitting. We conducted series of experiments with feature selection, but did not make any significant improvement (Table 2).

Table 2. Numbers of the used features are given in the first column. Particular meanings of the features (in the cases of 4 and 17 features) may be found in the Tables 1, 3 and 4.

N of features	TestAUC	Base-learner
44	0.7023	LogitBoost
44	0.688	RLR
30	0.689	RLR
17	0.688	RLR
4	0.6558	RLR

Table 3. Top 4 features

N	Feature	μ	s	r
1	N1 (see Table 1)	1.1454	0.1011	11.3294
2	N3	-0.6015	0.0663	-9.0751
3	N5	-0.1587	0.0778	-2.0395
4	AGE	-0.6831	0.0806	-8.4794

As a next step, we decided to apply as a base-learner the ada-function in R. The best test result AUC = 0.7023 was obtained using the following settings: $loss = e, \nu = 0.3, type = gentle$.

We used in our experiment 100 random balanced sets. In addition, we conducted many experiments with up to 300 random sets, but we did not find any improvement. Also, it is interesting to note that we did not make any changes to the pre-processing technique, which was used before, and conducted our experiments against the same data.

4.3   Discussion and Business Insights

The *RS*-method provides good opportunities to evaluate the significance of the particular features. We can take into account 2 factors: 1) average values μ and 2) *t*-statistic r, which are defined in (2).

Based on the Tables 1 and 4, we can make a conclusion that younger people (AGE: $\mu = -0.654$) with smaller number of dependants (NBR OF DEPENDANTS: $\mu = -0.3298$) who made enquiries for mortgages during last 6 months have higher probability to take up a home loan.

On the other hand, enquiries for loans represent a detrimental factor ($\mu = -0.672$).

Considering a general characteristic such as marital status, we can conclude that "widowed" people are less interested ($\mu = -0.2754$) to apply for home loan.

Also, it is interesting to note that stable job (CURR EMPL MTHS: $\mu = -0.0288$) or long residence (CURR RES MTHS: $\mu = -0.0449$) may be viewed as negative factors. Possibly, these people have already one or more homes and are reluctant to make further investments.

Table 4. Top 17 features, which were selected using MVF

N	Feature	μ	s	r
1	MARITAL STATUS: married	0.0861	0.028	3.0723
2	MARITAL STATUS: single	0.0419	0.0236	1.7786
3	MARITAL STATUS: defacto	0.09	0.0438	2.0572
4	MARITAL STATUS: widowed	-0.2754	0.0766	3.594
5	RENT BUY CODE: mortgage	0.0609	0.0191	3.1838
6	RENT BUY CODE: parents	-0.1285	0.0341	3.7692
7	CURR RES MTHS	-0.0449	0.0101	4.4555
8	CURR EMPL MTHS	-0.0288	0.0111	2.586
9	NBR OF DEPENDANTS	-0.3298	0.0807	4.085
10	Bureau Enquiries last month	0.3245	0.183	1.7736
11	Bureau Enquiries last 3 month	0.1296	0.1338	0.9691
12	Bureau Enquiries for Morgages last 6 month	0.8696	0.1359	6.3982
13	Bureau Enquiries for Loans last 12 month	-0.6672	0.0795	8.3905
14	A DISTRICT APPLICANT=2	-0.1704	0.05	3.4067
15	A DISTRICT APPLICANT=8	-0.1216	0.0397	3.063
16	CUSTOMER SEGMENT=9	-0.0236	0.0317	0.7453
17	AGE	-0.654	0.0962	6.8015

Remark 4. Experiments with 'tree' function (*R*-software, package 'tree') had confirmed that the feature "Bureau enquiries for mortgages during last 6 month" is the most important.

With this model, the company can develop a marketing program such as a direct mail campaign to target customers with highest scores. For example, there are 350 positive cases in the independent test dataset with 8000 instances. We sorted the 8000 customers in a decreasing order according to the decision function with AUC = 0.7023 (see Table 2). As a result, we have found that 50%, 60% and 70% of all positive customers are contained in the field of 1770, 2519 and 3436 top scored customers.

4.4   Computation Time and Used Software

A Dell computer, Duo 2.6GHz, 3GB RAM, was used for computations. It took about 4 hours time in order to complete 100 balanced random sets and produce best reported solution.

5   Concluding Remarks and Further Developments

It is a well known fact that for various reasons it may not be possible to theoretically analyze a particular algorithm or to compute its performance in contrast to another. The results of the proper experimental evaluation are very important as these may provide the evidence that a method outperforms existing approaches. Data mining competitions are very important.

The proposed ensemble method is based on a large number of balanced random sets and includes 2 main steps: 1) feature selection and 2) training. During the PAKDD-2007 Data Mining Competition, we conducted both steps using linear regression. The proposed method is general and may be implemented in conjunction with different base-learners. In this paper we reported results which were obtained using the ADA package in R. These results outperform all known results.

Further improvement may be achieved as a result of more advanced pre-processing technique. Also, it appears to be promising to apply random forest as a single base-learner.

References

[1] Biau, G., Devroye, L., Lugosi, G.: Consistency of random forests and other averaging classifiers. Journal of Machine Learning Research 9, 2015–2033 (2007)
[2] Wang, W.: Some fundamental issues in ensemble methods. In: World Congress on Computational Intelligence, Hong Kong, pp. 2244–2251. IEEE, Los Alamitos (2008)
[3] Breiman, L.: Random forests. Machine Learning 45, 5–32 (2001)
[4] Zhang, B., Pham, T., Zhang, Y.: Bagging support vector machine for classification of SELDI-ToF mass spectra of ovarian cancer serum samples. In: Orgun, M.A., Thornton, J. (eds.) AI 2007. LNCS (LNAI), vol. 4830, pp. 820–826. Springer, Heidelberg (2007)
[5] Breiman, L.: Bagging predictors. Machine Learning 24, 123–140 (1996)
[6] Zhang, J., Li, G.: Overview of PAKDD Competition 2007. International Journal of Data Warehousing and Mining 4, 1–8 (2008)
[7] Hastie, T., Tibshirani, R., Friedman, J.: The Elements of Statistical Learning. Springer, Heidelberg (2001)
[8] Guyon, I., Weston, J., Barnhill, S., Vapnik, V.: Gene selection for cancer classification using support vector machines. Machine Learning 46, 389–422 (2002)
[9] Nikulin, V.: Classification of imbalanced data with random sets and mean-variance filtering. International Journal of Data Warehousing and Mining 4, 63–78 (2008)
[10] Friedman, J., Hastie, T., Tibshirani, R.: Additive logistic regression: a statistical view of boosting. Annals of Statistics 28, 337–374 (2000)
[11] Nikulin, V.: Learning with mean-variance filtering, SVM and gradient-based optimization. In: International Joint Conference on Neural Networks, Vancouver, BC, Canada, July 16-21, pp. 4195–4202. IEEE, Los Alamitos (2006)
[12] Freund, Y., Schapire, R.: A decision-theoretic generalization of online learning and an application to boosting. J. Comput. System Sciences 55, 119–139 (1997)

Adapting Spectral Co-clustering to Documents and Terms Using Latent Semantic Analysis

Laurence A.F. Park, Christopher A. Leckie, Kotagiri Ramamohanarao, and James C. Bezdek

Department of Computer Science and Software Engineering,
The University of Melbourne, 3010, Australia
{lapark,caleckie,kotagiri}@unimelb.edu.au, jcbezdek@gmail.com

Abstract. Spectral co-clustering is a generic method of computing co-clusters of relational data, such as sets of documents and their terms. Latent semantic analysis is a method of document and term smoothing that can assist in the information retrieval process. In this article we examine the process behind spectral clustering for documents and terms, and compare it to Latent Semantic Analysis. We show that both spectral co-clustering and LSA follow the same process, using different normalisation schemes and metrics. By combining the properties of the two co-clustering methods, we obtain an improved co-clustering method for document-term relational data that provides an increase in the cluster quality of 33.0%.

Keywords: co-clustering, spectral graph partitioning, latent semantic analysis, document clustering.

1 Introduction

Spectral co-clustering [1] allows us to partition relational data such that partitions contain more than one type of data. Document-term spectral co-clustering is the application of spectral co-clustering to document-term relational data. It allows us to provide partitions of not only documents, but also the terms that are related to the documents. By using these terms, we can quickly obtain an understanding of how the documents and terms are partitioned. For example, if a set of documents and terms appeared in a partition where the terms were words such as 'Wiggles', 'Teletubbies', and 'Igglepiggle', we would know that the partition is related to children's entertainment. Co-clustering is a useful tool for domains such as information retrieval where many documents and terms can be partitioned into smaller sets and processed by complex algorithms that would have otherwise been infeasible to use.

Much research has gone into the distributions found within document-term relational data [5]. Unfortunately, spectral co-clustering does not make use of this information when applied to document-term relational data.

Latent semantic analysis (LSA) [3] is a method of document analysis for information retrieval that follows a similar process to spectral co-clustering. LSA

A. Nicholson and X. Li (Eds.): AI 2009, LNAI 5866, pp. 301–311, 2009.
© Springer-Verlag Berlin Heidelberg 2009

has been developed for use especially with document-term relational data, but has only been used for dimension reduction and query term expansion.

In this article, we examine the similarities between using spectral co-clustering for co-clustering documents and terms, and using latent semantic analysis for establishing correlations between documents and terms during the information retrieval process. By analysing the components of each method, we are able to produce a spectral co-clustering algorithm that is tailored for use on document-term relational data. The main contributions are:

- an analysis of the similarities between spectral co-clustering and latent semantic analysis
- an outline of a framework for document-term co-clustering
- an improved method of co-clustering document-term relational data by combining ideas from the mentioned co-clustering methods.

The article will proceed as follows: we begin by describing the spectral graph multi-partitioning method in Section 2 and how to compute the spectral co-clusters. We then describe Latent Semantic Analysis and its use for information retrieval in Section 3. This is followed by a comparison of the two methods in Section 4. Experiments are provided in Section 5 showing how each method performs and how we can combine the two methods to improve the computed partitions.

2 Co-clustering Using Spectral Graph Multi-partitioning

Spectral clustering [2] allows us to clearly identify any isolated subgraphs within a graph. In the case where our graph is fully connected, spectral clustering is able to identify subgraphs containing strongly connected components, with weak connections between subgraphs.

In this section, we will examine co-clustering using spectral graph multi-partitioning and examine how to cluster its spectrum.

2.1 Computing Spectral Co-clusters

Given a set of documents vectors A, where each row of A is a document vector containing $a_{d,t}$ (the weight of term t in document d), spectral document clusters are computed by first computing the graph as the document-document relational matrix AA^T. We then compute the graph spectrum, being the eigenvectors of the weighted graph Laplacian matrix of AA^T. The document clusters are obtained by applying K-means to the graph spectrum [2]. If we consider A^T to be a set of term vectors, then to compute the set of term clusters, we perform the same operations on the relational matrix $A^T A$.

To compute the set of document-term co-clusters (rather than only the document clusters or term clusters), we must first compute a common space in which both the documents and terms appear. It has been shown [1] that the spectrum obtained for each of the document and term clusterings are in fact the same

---

Algorithm 1. The spectral co-clustering algorithm. Note that $S(\cdot)$ is defined in equation 2, and $S_d(\cdot)$ and $S_t(\cdot)$ in equation 3.

Data: document-term relational matrix A
Result: documents and terms partitioned into clusters C

```
1  begin
2  │   A_n ← S(A);                           // apply spectral normalisation
3  │   [U, Σ, V] ← svd(A_n) ;                // perform SVD
4  │   U_{2:n+1} ← U[:, 2 : n + 1] ;         // select columns 2 to n + 1
5  │   V_{2:n+1} ← V[:, 2 : n + 1];
6  │   Z_d ← S_d(U_{2:n+1}) ;                // apply spectral weighting
7  │   Z_t ← S_t(V_{2:n+1});
8  │   Z ← [Z_d; Z_t] ;                      // combine to form one matrix
9  │   C ← K-means(Z) ;                      // compute clusters using K-means
10 end
```

---

space as presented above. Therefore, we are able to simply combine the spectra of the documents and terms, and apply K-means to compute document-term co-clusters.

Rather than computing the relational matrices for A and A^T separately and combining the spectra, a more efficient method [1] using singular value decomposition (SVD) is presented.

Spectral graph multi-partitioning algorithm [1] is outlined in Algorithm 1. In the algorithm, we have not defined the functions $S(\cdot)$, $S_d(\cdot)$ and $S_t(\cdot)$, and so we will present them here. Given the matrix:

$$
A = \begin{bmatrix}
a_{d_1,t_1} & a_{d_1,t_2} & \cdots & a_{d_1,t_T} \\
a_{d_2,t_1} & a_{d_2,t_2} & \cdots & a_{d_2,t_T} \\
\vdots & \vdots & \ddots & \vdots \\
a_{d_D,t_1} & a_{d_D,t_2} & \cdots & a_{d_D,t_T}
\end{bmatrix}
\tag{1}
$$

the algorithm consists of first normalising the data:

$$
S(A) = R^{-1/2} A C^{-1/2}
\tag{2}
$$

where the matrices R and C are diagonal matrices containing the elements $R_t = \sum_d a_{d,t}$ and $C_d = \sum_t a_{d,t}$. Then applying the singular value decomposition, giving $A_n = U \Sigma V^T$. Singular vectors $\tilde{u}_i$ and $\tilde{v}_i$ are selected from the columns of U and V and normalised using:

$$
\tilde{e}_{t_i} = S_d(\tilde{u}_i) = R^{-1/2} \tilde{u}_i \qquad \tilde{e}_{d_i} = S_t(\tilde{v}_i) = C^{-1/2} \tilde{v}_i
\tag{3}
$$

giving $\tilde{e}_{t_i}$ and $\tilde{e}_{d_i}$, the eigenvectors of the graph Laplacian for the row and column objects i respectively. The final step of the algorithm is to cluster the combined graph Laplacian eigenvectors.

2.2   Clustering the Spectrum

Given a document-term relational matrix A, where each element $a_{t,d}$ is non-zero if term t appears in document d, it is unlikely that there will be any clear

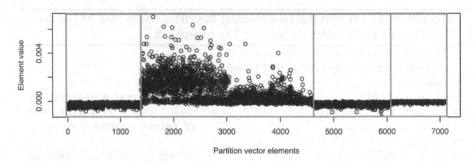

Fig. 1. The elements values of the second partition vector. From the change in element values, we can see four partition of the associated objects (outlined by the gray boxes).

document clusters since each document will have similarity to every other document (due to them containing common terms such as "the"), and it is unlikely that there will be any clear term partitions since every term will be related to a common term.

If there are no clear partitions, the first eigenvector of the Laplacian matrix provides the trivial solution where all elements are equal and the corresponding eigenvalue is zero, implying that the whole data set belongs to one cluster. Therefore, to obtain the document-term co-clusters, we must select a set of *partition vectors* from the remaining set of eigenvectors and apply K-means.

Figure 1 provides us with a plot of the element values within a partition vector from a document collection containing four different topics. The plot shows the vector elements sorted by topic, where the four topic clusters are visible due to the differences in the element values.

Each element in a partition vector represents an object in the data set. If two objects have similar values in many partition vectors, then it is likely that they belong to the same cluster. It is due to this property that Euclidean K-means is used to cluster the objects, based on their similarity in each partition vector.

3   Latent Semantic Analysis

Latent semantic analysis (LSA) is the process of computing the set of hidden topics and their relationship to each of the terms and documents in the document set. Once the relationships are obtained, we are able to represent each of the terms and documents by their latent topics and compute correlations between them.

A document-term index is a $D \times T$ matrix of the form given in equation 1, where D and T are the number of documents and terms in the document collection respectively and $a_{d,t}$ is the count of term t in document d. We can see from this matrix that it is a collection of document vectors, where each document is represented by the set of terms as it features. We can also see that it is a collection of term vectors, where each term is represented by the set of documents as its features. Latent semantic analysis allows us to compute a

---

Algorithm 2. The LSA co-clustering algorithm. Note that element-wise $L(\cdot)$ is defined in equation 4, $L_d(\cdot)$ in equation 6, and $L_t(\cdot)$ in equation 7.

Data: document-term relational matrix A
Result: documents and terms partitioned into clusters C

```
1  begin
2  │   A_n ← L(A);                              // apply LSA normalisation
3  │   [U, Σ, V] ← svd(A_n) ;                             // perform SVD
4  │   U_{1:n} ← U[:, 1 : n] ;                   // select columns 1 to n
5  │   V_{1:n} ← V[:, 1 : n];
6  │   Z_d ← L_d(U_{1:n}) ;                         // apply LSA weighting
7  │   Z_t ← L_t(V_{1:n});
8  │   Z ← [Z_d; Z_t] ;                      // combine to form one matrix
9  │   C ← ∠K-means(Z) ;      // compute clusters using angular K-means
10 end
```

---

mapping for each of the document and term vectors into the topic space, where each document and term are represented by the set of latent topics as their features.

To compute co-clusters in the topic space, we use the algorithm presented in Algorithm 2. In this section we will present the reasoning behind each step and present the functions $L(\cdot)$, $L_d(\cdot)$, and $L_t(\cdot)$ from Algorithm 2.

3.1   Document and Term Normalisation

When performing similarity comparisons of document and term vectors, we must take into account any bias that exists in the data. When examining term frequency counts, bias can exist due to term rarity and document length. For example, when comparing two documents for similarity, they will both contain many common terms such as 'the', 'and', 'is' and 'a'. These terms are not very informative when computing similarity and therefore should not have much impact on the similarity score. On the other hand, if the two documents both contain rare terms, these rare terms should have a greater contribution to the similarity score.

The TF-IDF weighting scheme [4] provides this term rarity normalisation by using the following weights:

$$w_{d,t} = a_{d,t} \log \left(\frac{D}{a_t} + 1 \right) \tag{4}$$

where a_t is the number of documents containing term t, and D is the number of documents. These weights are used in the place of term frequencies to remove the mentioned bias from any similarity computations. By weighting each element of the matrix A, we obtain the normalised matrix $L(A)$.

3.2   Topic Normalisation

To compute the set of latent topics and their relationship to each document and term, we use the singular value decomposition (SVD):

$$A = U\Sigma V^T \tag{5}$$

where A contains the elements $a_{d,t}$.

The SVD is a form of dimension reduction that provides the best n-dimensional fit in the latent semantic space in terms of the l_2 norm when the first n singular values are chosen. The matrices U and V are document to topic space and term to topic space mappings that map the term and document vectors into the latent topic space. Therefore the set of document vectors in the latent topic is given as:

$$L_d(U) = AV = U\Sigma V^T V = U\Sigma \tag{6}$$

and the set of word vectors in the latent topic space is given as:

$$L_t(V) = A^T U = V\Sigma U^T U = V\Sigma \tag{7}$$

By choosing the n greatest singular values, and hence the n associated columns of $U\Sigma$ and $V\Sigma$, we obtain the set of documents and terms in the n-dimensional latent topic space. These document and term vectors allow us to easily compute the correlation between each pair of terms, documents or terms and documents.

3.3   Clustering the Topics

Before we can cluster objects, we must define similarity between them. A simple and effective measure of similarity for document vectors is the cosine measure:

$$\text{sim}(\tilde{d}_i, \tilde{d}_j) = \frac{\tilde{d}_i^T \tilde{d}_j}{\|\tilde{d}_i\|_2 \|\tilde{d}_j\|_2} \tag{8}$$

which is the inner product of the document vectors normalised by their vector length. This measure is independent of the length of the document vectors and therefore computes the similarity based only on the angle between the vectors. If the document vectors point in the same direction, the similarity will be 1, if they point in different directions, the similarity will be less than 1. If the vectors are at right angles to each other (implying that they have no features in common), the similarity will be 0.

Using this knowledge, we employ the use of angular K-means, where the angle between two vectors is used to compute their dissimilarity, rather than Euclidean distance.

4   Comparison of Spectral Co-clustering and LSA

After examining the spectral co-clustering and latent semantic analysis processes, we can see that when applied to a document-term relational matrix, both methods are very similar in that they follow the same generalised process of:

2. Normalise the document-term relational matrix
3. Apply the SVD
4,5. Choose the top n singular vector with largest associated singular values.
6,7. Weight the selected vectors
9. Cluster the weighted vectors

By using the document index, containing term frequencies, as our relational matrix, we apply the following normalisation when using spectral co-clustering:

$$w_{d,t} = \frac{a_{d,t}}{\sqrt{\sum_t a_{d,t} \sum_d a_{d,t}}} \qquad (9)$$

and the normalisation in equation 4 when using LSA, where $w_{d,t}$ is an element of A_n.

Another difference can be seen in the weighting that is used after the SVD is applied. When performing spectral co-clustering, we use the weighting in equation 3, while when using LSA, we use the weighting from equations 6 and 7.

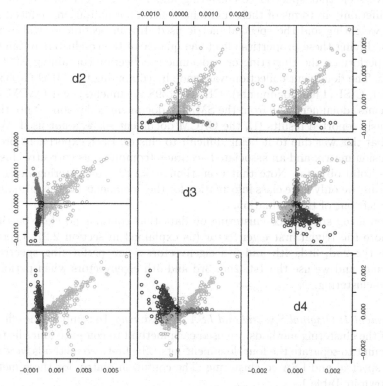

Fig. 2. A plot showing different views of the 2nd, 3rd and 4th spectral dimensions. The different document sets are shown by different shades of gray. The solid horizontal and vertical lines show the x, y axis and the position of (0,0).

The last difference is the metric used with K-means (Euclidean distance vs. vector angle). Given that these two methods are so similar, it seems odd that they should both use different similarity metrics in the reduced space. Spectral co-clustering uses K-means to identify co-clusters, which uses the Euclidean distance as its metric. On the other hand, LSA uses K-means with the cosine measure to compare vectors in the reduced space.

A plot of the 2nd, 3rd and 4th dimensions of the reduced space[1] using the spectral co-clustering normalisation and weighting is shown in Figure 2. The data set consists of the combination of four document sets. Each point in the plot corresponds to one document from the collection. From this set of plots, we can see that each of the four document sets are clustered in a radial fashion about the origin, giving support to the use of the cosine measure as the metric over the Euclidean distance.

5   Experiments

We have seen that spectral co-clustering and LSA are very similar and that they differ only in terms of their relational matrix normalisation, reduced space vector weighting and the spectral metric used. In this section we will examine how modifying these properties affect the quality of the co-clusters obtained.

We performed the clustering on a document collection containing 7,095 documents. The document collection was made by combining the MED (1,033 documents), CISI (1460 documents), CRAN (1398 documents) and CACM (3204 documents) document sets from the SMART document collection[2]. Note that in previously reported results [1], the CACM document set was not used. We believe that this was due to it being difficult to cluster. Each experiment presents a confusion matrix and an associated accuracy (computed as correctly clustered objects/total objects). Note that even though we are co-clustering documents and terms, we only have class information for the documents, so the term objects will be left out of the evaluation.

When using spectral co-clustering on data that contains no clear partitions, we ignore the trivial first eigenvector (as explained in section 2.2). Therefore, we use the 2nd, 3rd, 4th and 5th eigenvectors when performing spectral co-clustering, and we use the 1st, 2nd, 3rd and 4th eigenvectors when performing LSA co-clustering.

Experiment 1: Original Spectral and LSA co-clustering To compare the effectiveness of the clustering methods, we used each method to compute four clusters in an attempt to separate the four document sets. The first two methods were those of the spectral and LSA co-clustering. The confusion matrix for these methods are shown in Table 1.

---

[1] The first dimension is not shown since it contains the trivial eigenvector corresponding to the eigenvalue of value zero.

[2] ftp://ftp.cs.cornell.edu/pub/smart

Table 1. The confusion matrix for spectral co-clustering (left) with 50.6% accuracy and LSA co-clustering (right) with 69.5% accuracy

Cluster	MED	CISI	CRAN	CACM	Cluster	MED	CISI	CRAN	CACM
1	551	0	0	0	1	964	6	1	2
2	240	1460	1398	1622	2	2	0	532	5
3	242	0	0	0	3	8	0	800	51
4	0	0	0	1582	4	59	1454	65	3146

Table 2. The confusion matrix for LSA co-clustering, clustered on the 2nd, 3rd and 4th and 5th dimensions. The table on the left shows the clusters found using Euclidean distance with accuracy of 53.6%, while the table on the right shows those found using the cosine distance with accuracy 68.7%.

Cluster	MED	CISI	CRAN	CACM	Cluster	MED	CISI	CRAN	CACM
1	418	0	0	1	1	988	10	37	6
2	16	964	7	976	2	0	0	566	1
3	0	0	191	0	3	0	0	701	9
4	599	496	1200	2227	4	45	1450	94	3188

We can see from the left confusion matrix that the spectral multi-partitioning method managed to compute two clusters containing only MED documents (cluster 1 and 3) and a cluster containing only CACM documents (cluster 4), and a cluster containing all documents types (cluster 2). We can see that this method performed poorly as it was not able to discriminate the CISI and CRAN documents at all.

From the LSA co-clusters shown in the right confusion matrix of Table 1, we see that cluster 1 clearly contains MED documents and clusters 2 and 3 contain CRAN documents, but cluster 4 contains CISI and CACM documents. So neither the spectral nor LSA co-clustering methods were able to provide clusters centred within each document set.

Experiment 2: LSA without 1st singular vector, using Euclidean and cosine distance Our second set of results show the effect of using LSA co-clustering, where we have ignored the first singular vector, implying that we have used the 2nd, 3rd, 4th and 5th set of singular vectors as our data set (as done in spectral multi-partitioning). The results can be seen in Table 2. From this table, we can see that again the clusters produced are similar to those produced using LSA co-clustering and are also unable to provide clusters that centre on each document set.

Experiment 3: LSA co-clustering using Euclidean distance, Spectral co-clustering using cosine distance Finally, we will examine the effect of switching the

Table 3. The confusion matrix for LSA co-clustering using the Euclidean distance metric (left) with 58.3% accuracy and spectral co-clustering using the cosine measure (right) with 92.4% accuracy

Cluster	MED	CISI	CRAN	CACM	Cluster	MED	CISI	CRAN	CACM
1	399	0	0	1	1	998	0	0	0
2	25	836	67	927	2	34	1457	12	476
3	2	0	659	0	3	1	1	1376	5
4	607	624	672	2276	4	0	2	10	2723

K-means metric. Table 3 contains the confusion matrices for LSA co-clustering using Euclidean distance and spectral co-clustering using the cosine measure.

By modifying the spectral co-clustering method to use LSA's cosine measure we have obtained an improvement over both LSA co-clustering and spectral co-clustering. The right confusion matrix of Table 3 shows each of the clusters containing most of the documents from one of the document sets.

From our analysis, we have produced a spectral co-clustering algorithm that is tailored for use on document-term relational data. The results show that by combining the weighting used with spectral co-clustering with the cosine measure used in LSA co-clustering, we are able to increase the spectral co-clustering accuracy from 69.5% to 92.4% (a 33.0% increase in accuracy) providing a co-clustering method that is more suited to document clustering. It is also interesting to note that the LSA weighting, designed specifically for document-term frequencies, was not as useful as the spectral co-clustering weighting.

6   Conclusion

Co-clustering using spectral multi-partitioning is a generic method of computing co-clusters with relational data. Latent semantic analysis (LSA) is a method of document and term smoothing that follows a similar processes to spectral multi-partitioning.

In this article, we showed that spectral multi-partitioning and LSA both follow the same process but use different normalisation and weighting schemes, and a different metric.

From our analysis, we have produced a spectral co-clustering algorithm that is tailored for use on document-term relational data. By replacing properties of spectral co-clustering with those from LSA, we were able to provide a co-clustering method for documents and terms that provided a 33.0% increase in cluster quality when compared to each method separately. Our results showed that the best clusters were obtained by using the spectral multi-partitioning weighting and normalisation with LSA's similarity metric.

References

1. Dhillon, I.S.: Co-clustering documents and words using bipartite spectral graph partitioning. In: KDD 2001: Proceedings of the Seventh ACM SIGKDD International Conference on Knowledge Discovery and Data Mining, pp. 269–274. ACM, New York (2001)
2. Dhillon, I.S., Guan, Y., Kulis, B.: Kernel k-means: spectral clustering and normalized cuts. In: KDD 2004: Proceedings of the tenth ACM SIGKDD international conference on Knowledge discovery and data mining, pp. 551–556. ACM, New York (2004)
3. Dumais, S.T.: Latent semantic analysis. Annual Review of Information Science and Technology 38(1), 188–230 (2004)
4. Husbands, P., Simon, H., Ding, C.: Term norm distribution and its effects on latent semantic indexing. Inf. Process. Manage. 41(4), 777–787 (2005)
5. Sparck Jones, K., Walker, S., Robertson, S.E.: A probabilistic model of information retrieval: development and comparative experiments, part 1. Information Processing and Management 36(6), 779–808 (2000)

MML Invariant Linear Regression

Daniel F. Schmidt and Enes Makalic

The University of Melbourne
Centre for MEGA Epidemiology
Carlton VIC 3053, Australia
{dschmidt,emakalic}@unimelb.edu.au

Abstract. This paper derives two new information theoretic linear regression criteria based on the minimum message length principle. Both criteria are invariant to full rank affine transformations of the design matrix and yield estimates that are minimax with respect to squared error loss. The new criteria are compared against state of the art information theoretic model selection criteria on both real and synthetic data and show good performance in all cases.

1 Introduction

Consider the linear regression model for explaining data $\mathbf{y}^n \in \mathbb{R}^n$

$$\mathbf{y} = \mathbf{X}_\gamma \boldsymbol{\beta} + \boldsymbol{\varepsilon}$$

where $\boldsymbol{\beta} \in \mathbb{R}^p$ is the vector of linear regression coefficients, $\boldsymbol{\varepsilon} \in \mathbb{R}^n$ are i.i.d. variates distributed as per $\boldsymbol{\varepsilon} \sim N_n(\mathbf{0}, \tau \mathbf{I}_n)$ (where $\mathbf{I}_k$ denotes the $(k \times k)$ identity matrix), $\mathbf{X}_\gamma$ is the design matrix of regressors and $\gamma \subset \{1, \ldots, q\}$ is an index vector determining which regressors comprise the design matrix. Let $\mathbf{X} = (\mathbf{x}_1, \ldots, \mathbf{x}_q)$ be the complete matrix of regressors, where $\mathbf{x}_i \in \mathbb{R}^n$ and q is the maximum number of candidate regressors. Given model structure index γ, the model design matrix is defined as

$$\mathbf{X}_\gamma = (\mathbf{x}_{\gamma_1}, \ldots, \mathbf{x}_{\gamma_p})$$

Denote the full vector of continuous parameters by $\boldsymbol{\theta} = (\boldsymbol{\beta}, \tau) \in \Theta \subset \mathbb{R}^{p+1}$ where Θ is the parameter space. The parameters $\boldsymbol{\theta}$ are considered unknown and must be inferred from the data $\mathbf{y}^n$, along with the optimal regressor subset γ.

This paper considers information theoretic model selection criteria based on Minimum Message Length (MML) [1] for inference of linear regression models. The criteria derived here are: (1) invariant under all full rank affine transformations of the design matrix (2) yield estimates that are minimax with respect to the squared error loss and, (3) require no user specified parameters. Most previous MML criteria for linear regression [1,2,3] are based on ridge regression style priors and none possess all of these attractive properties. In addition, one of the new criteria allows for shrinkage of parameters and selection of relevant

A. Nicholson and X. Li (Eds.): AI 2009, LNAI 5866, pp. 312–321, 2009.

regressors to be performed within a single framework. The resultant criteria are closely related to the linear regression criterion derived using the Normalized Maximum Likelihood (NML) [4] universal model.

2   Minimum Message Length (MML)

Within the MML framework [1,5,6] of inductive inference, the model that best compresses the data resulting in the shortest message length is deemed optimal. The hypothetical message consists of two parts: the assertion, $I_{87}(\boldsymbol{\theta})$, which is a statement of the chosen model, and the detail, $I_{87}(\mathbf{y}^n|\boldsymbol{\theta})$, which denotes the encoding of the data under the assumption that the model named in the assertion is optimal. The Wallace-Freeman, or MML87 approximation [6], states that a model $\boldsymbol{\theta} \in \boldsymbol{\Theta} \subset \mathbb{R}^k$ and data $\mathbf{y}^n = (y_1, \ldots, y_n)$ may be concisely transmitted with a message approximately given by

$$I_{87}(\mathbf{y}^n, \boldsymbol{\theta}) = \underbrace{- \log \pi(\boldsymbol{\theta}) + \frac{1}{2} \log |\mathbf{J}_{\boldsymbol{\theta}}(\boldsymbol{\theta})| + \frac{k}{2} \log \kappa_k + \frac{k}{2}}_{I_{87}(\boldsymbol{\theta})} \underbrace{- \log p(\mathbf{y}^n|\boldsymbol{\theta})}_{I_{87}(\mathbf{y}^n|\boldsymbol{\theta})} \quad (1)$$

where $\pi(\cdot)$ denotes a prior distribution over the parameter space $\boldsymbol{\Theta}$, $\mathbf{J}_{\boldsymbol{\theta}}(\boldsymbol{\theta})$ is the Fisher information matrix, and κ_k is the normalised second moment of an optimal quantising lattice in k-dimensions. In this paper, the need to determine κ_k for arbitrary dimension k is circumvented by using the approximation (pp. 237, [1])

$$\frac{k}{2} (\log \kappa_k + 1) \approx -\frac{k}{2} \log(2\pi) + \frac{1}{2} \log(k\pi) + \psi(1)$$

where $\psi(\cdot)$ is the digamma function. We define log as the natural logarithm, and as such, all message lengths are measured in *nits* (nats), or base-e digits. The MML principle advocates choosing the model $\hat{\boldsymbol{\theta}}_{87}(\mathbf{y}^n)$ that minimises (1) as the most *a posteriori* likely explanation of the data in the light of the chosen priors.

The original Wallace-Freeman approximation requires that the prior be completely specified. Recently, this requirement has been relaxed by the introduction of a Wallace-Freeman like extension to hierarchical Bayes models in which the parameters and hyperparameters are jointly estimated from the data [7]. If $\pi_{\boldsymbol{\theta}}(\cdot|\boldsymbol{\alpha})$ is the prior density over $\boldsymbol{\theta}$ parametrised by hyperparameters $\boldsymbol{\alpha}$, and $\pi_{\boldsymbol{\alpha}}(\cdot)$ is the prior density over $\boldsymbol{\alpha}$, the joint message length of $\mathbf{y}^n$, $\boldsymbol{\theta}$ and $\boldsymbol{\alpha}$ is

$$I_{87}(\mathbf{y}^n, \boldsymbol{\theta}, \boldsymbol{\alpha}) = - \log p(\mathbf{y}^n|\boldsymbol{\theta}) - \log \pi_{\boldsymbol{\theta}}(\boldsymbol{\theta}|\boldsymbol{\alpha})\pi_{\boldsymbol{\alpha}}(\boldsymbol{\alpha}) + \frac{1}{2} \log |\mathbf{J}_{\boldsymbol{\theta}}(\boldsymbol{\theta})||\mathbf{J}_{\boldsymbol{\alpha}}(\boldsymbol{\alpha})| + \text{const}$$

$$(2)$$

where

$$\mathbf{J}_{\boldsymbol{\alpha}}(\boldsymbol{\alpha}) = \mathrm{E} \left[\frac{\partial^2 I_{87}(\mathbf{y}^n, \hat{\boldsymbol{\theta}}_{87}(\mathbf{y}^n|\boldsymbol{\alpha})|\boldsymbol{\alpha})}{\partial \boldsymbol{\alpha} \partial \boldsymbol{\alpha}'} \right]$$

denotes the Fisher information for the hyperparameters $\boldsymbol{\alpha}$, the expectations taken with respect to the marginal distribution $r(\mathbf{y}^n|\boldsymbol{\alpha}) = \int p(\mathbf{y}^n|\boldsymbol{\theta})\pi_{\boldsymbol{\theta}}(\boldsymbol{\theta}|\boldsymbol{\alpha})d\boldsymbol{\theta}$.

3   Linear Regression with a Uniform Prior

Dropping γ for brevity, the negative log-likelihood function for a linear regression model given a set of parameters $\boldsymbol{\theta} = (\boldsymbol{\beta}, \tau) \in \mathbb{R}^{p+1}$ is

$$-\log p(\mathbf{y}^n|\boldsymbol{\theta}) = \frac{n}{2}\log(2\pi) + \frac{n}{2}\log\tau + \frac{1}{2\tau}(\mathbf{y} - \mathbf{X}\boldsymbol{\beta})'(\mathbf{y} - \mathbf{X}\boldsymbol{\beta}) \qquad (3)$$

The Fisher information for the linear regression model is known to be

$$|\mathbf{J}_{\boldsymbol{\theta}}(\boldsymbol{\theta})| = \left(\frac{1}{2\tau^{p+2}}\right)|\mathbf{X}'\mathbf{X}| \qquad (4)$$

To apply the MML87 formula (1) we require a suitable prior distribution $\pi(\boldsymbol{\theta}) = \pi(\boldsymbol{\beta})\pi(\tau)$ for the regression parameters $\boldsymbol{\beta}$ and the noise variance τ. One aim of the paper is to derive model selection criteria that do not require specification of subjective priors which is often difficult in practice in the linear regression setting. Ideally, one wishes to give each set of feasible regression coefficients $\boldsymbol{\beta} \in \mathbb{R}^p$ an equal prior probability. A possible method is to use the uniform prior over each coefficient, which is of course improper and requires a bounding of the parameter space to avoid the Jeffreys-Lindley paradox. The data $\mathbf{y}^n$ are assumed to be generated from the model

$$\mathbf{y} = \mathbf{y}_* + \boldsymbol{\varepsilon}$$

where $\boldsymbol{\varepsilon} \sim N_n(\mathbf{0}, \tau\mathbf{I}_n)$ and $\mathbf{y}_*$ is the 'true' underlying regression curve. Noting that $\mathrm{E}[\boldsymbol{\varepsilon}'\boldsymbol{\varepsilon}] = n\tau$ and $\mathrm{E}[\boldsymbol{\varepsilon}'\mathbf{y}] = 0$, it is clear that

$$\mathrm{E}\left[\mathbf{y}'\mathbf{y}\right] = \mathbf{y}_*'\mathbf{y}_* + n\tau \qquad (5)$$

For a given $\boldsymbol{\beta}$, one can construct an estimate of $\mathbf{y}_*$, say $\mathbf{X}\boldsymbol{\beta}$; since τ is unknown and strictly positive, by (5), we expect this estimate to satisfy

$$\mathbf{y}'\mathbf{y} \geq (\mathbf{X}\boldsymbol{\beta})'(\mathbf{X}\boldsymbol{\beta}) = \boldsymbol{\beta}'(\mathbf{X}'\mathbf{X})\boldsymbol{\beta} \qquad (6)$$

that is, we do not expect the estimate of $\mathbf{y}^*$ to have greater energy than the energy of the observed data $\mathbf{y}$. From (6), the feasible parameter hyper-ellipsoid $\Lambda \subset \mathbb{R}^p$ is then given by

$$\Lambda = \{\boldsymbol{\beta} : \boldsymbol{\beta} \in \mathbb{R}^p, \boldsymbol{\beta}'(\mathbf{X}'\mathbf{X})\boldsymbol{\beta} \leq \mathbf{y}'\mathbf{y}\}$$

A suitable prior for $\boldsymbol{\beta}$ is then a uniform prior over the feasible parameter set Λ

$$\pi(\boldsymbol{\beta}) = \frac{1}{\mathrm{vol}(\Lambda)} = \frac{\Gamma(p/2+1)\sqrt{|\mathbf{X}'\mathbf{X}|}}{(\pi\mathbf{y}'\mathbf{y})^{(p/2)}}, \; \boldsymbol{\beta} \in \Lambda \qquad (7)$$

The prior over τ is chosen to be the standard conjugate prior

$$\pi(\tau) \propto \tau^{-\nu} \qquad (8)$$

where ν is a suitably chosen hyperparameter. The impropriety of (8) is not problematic as τ is a common parameter for all regression models. Using (3), (7), (8) and (4) in (1) and minimising the resulting codelength yields

$$\hat{\beta}_{87}(\mathbf{y}^n) = (\mathbf{X}'\mathbf{X})^{-1}\mathbf{X}'\mathbf{y} = \hat{\beta}_{\mathrm{LS}}(\mathbf{y}^n) \tag{9}$$

$$\hat{\tau}_{87}(\mathbf{y}^n) = \frac{\mathbf{y}'\mathbf{y} - \xi(\mathbf{y}^n)}{n - p + 2\nu - 2} \tag{10}$$

where $\xi(\mathbf{y}^n) = \hat{\beta}_{87}(\mathbf{y}^n)' (\mathbf{X}'\mathbf{X}) \hat{\beta}_{87}(\mathbf{y}^n)$ is the fitted sum of squares, and $\hat{\beta}_{\mathrm{LS}}(\mathbf{y}^n)$ are the usual least squares estimates. The final complete message length, up to a constant, evaluated at the MML estimates (9) and (10) is

$$I_u(\mathbf{y}^n|\gamma) = \left(\frac{n-p}{2}\right) \log 2\pi + \left(\frac{n-p+2\nu-2}{2}\right) (\log \hat{\tau}_{87}(\mathbf{y}^n)+1) + \frac{p}{2} \log(\pi \mathbf{y}'\mathbf{y})$$
$$- \log \Gamma \left(\frac{p}{2}+1\right) + \frac{1}{2} \log(p+1) + \mathrm{const} \tag{11}$$

where $I_u(\mathbf{y}^n|\gamma) \equiv I_u(\mathbf{y}^n, \hat{\beta}_{87}(\mathbf{y}^n), \hat{\tau}_{87}(\mathbf{y}^n)|\gamma)$. The code (11) is henceforth referred to as the MML$_u$ code. We note that as (11) depends on $\mathbf{X}$ only through $\hat{\tau}_{87}(\mathbf{y}^n)$, the message length is invariant under all full rank affine transformations of $\mathbf{X}$. We also note that as the MML87 estmates $\hat{\beta}_{87}(\mathbf{y}^n)$ under the uniform prior coincide with the least-squares estimates they are minimax with respect to squared error loss for all $p > 0$.

The criterion (11) handles the case $p = 0$ (i.e., no signal) gracefully, and is of the same computational complexity as well known asymptotic information criteria such as Akaike's Information Criterion (AIC) [8] or the Bayesian Information Criterion (BIC) [9]. This has the distinct advantage of making it feasible in settings where the complete design matrix may have many thousands of regressors; such problems are becoming increasingly common in bioinformatics, e.g. microarray analysis and genome wide association studies. It remains to select ν; setting $\nu = 1$ renders (10) the minimum variance unbiased estimator of τ, and is therefore one sensible choice.

Remark 1: Coding. To construct the uniform code it has been assumed that the observed signal power $\mathbf{y}'\mathbf{y}$ is known by the receiver. Alternatively, one can design a preamble code to transmit this quantity to the receiver, making the entire message decodable. As all regression models will require this preamble code it may safely be ignored during model selection for moderate sample sizes.

Remark 2: Alternative Prior. An attempt at forming 'uninformative' priors for linear regression models in the MML literature was made in [2]. Here, an additive uniform prior density over the coefficients was chosen to reflect the belief that the higher order coefficients will account for the remaining variance that is unexplained by the already selected lower order coefficients. However, such a prior is not uniform over the feasible set of the regression parameters and depends on an arbritrary ordering of the coefficients.

4   Linear Regression with the g-Prior

Instead of the uniform prior used in the MML_u criterion, we now consider a multivariate Gaussian prior over the regression coefficients. Dropping γ for brevity, this results in the following hierarchy:

$$\mathbf{y} \sim N_n \left(\mathbf{X}\boldsymbol{\beta}, \tau \mathbf{I}_n \right) \tag{12}$$

$$\boldsymbol{\beta} \sim N_p \left(\mathbf{0}_p, m \left(\mathbf{X}'\mathbf{X} \right)^{-1} \right) \tag{13}$$

where both $m > 0$ and τ are hyperparameters that must be estimated from the data. The type of prior considered here is known as Zellner's g-prior [10]. Coding of the assertion now proceeds by first transmitting estimates for $\boldsymbol{\alpha} = (m, \tau)$, and then transmitting the regression parameters $\boldsymbol{\beta}$ given the hyperparameters. This is further detailed in [7].

The negative log-likelihood of the data $\mathbf{y}^n$ given the parameters $(\boldsymbol{\beta}, \tau)$ is given by (3). The Fisher information for $\boldsymbol{\beta}$ now requires a correction as the hyperparameter m is estimated from the data, and may be arbitrarily small leading to problems with the uncorrected MML87 approximation. Following the procedure described in [1] (pp. 236–237), the corrected Fisher information is formed by treating the prior $\pi_\beta(\boldsymbol{\beta}|m)$ as a posterior of some uninformative prior $\pi_0(\boldsymbol{\beta})$ and p prior data samples all set to zero, with design matrix $\mathbf{X}_0$ satisfying $\mathbf{X}_0'\mathbf{X}_0 = (\tau/m)(\mathbf{X}'\mathbf{X})$. The corrected Fisher information is

$$|\mathbf{J}_\beta(\boldsymbol{\beta}|\boldsymbol{\alpha})| = \left(\frac{\tau + m}{\tau m} \right)^p |\mathbf{X}'\mathbf{X}| \tag{14}$$

Substituting (3), (13) and (14) into (1), and minimising the resultant codelength for $\boldsymbol{\beta}$ yields the following MML87 estimates:

$$\hat{\boldsymbol{\beta}}_{87}(\mathbf{y}^n|\boldsymbol{\alpha}) = \left(\frac{m}{m+\tau} \right) (\mathbf{X}'\mathbf{X})^{-1} \mathbf{X}'\mathbf{y} = \left(\frac{m}{m+\tau} \right) \hat{\boldsymbol{\beta}}_{\text{LS}}(\mathbf{y}^n) \tag{15}$$

Using the procedure described in Section 2, the profile message length, say $I_{\hat{\beta}}$, evaluated at $\hat{\boldsymbol{\beta}}_{87}(\mathbf{y}^n|\boldsymbol{\alpha})$ up to constants not depending on $\boldsymbol{\alpha}$ is

$$\frac{n}{2} \log \tau + \frac{p}{2} \log \left(\frac{\tau + m}{\tau} \right) + \left(\frac{1}{2\tau} \right) \mathbf{y}'\mathbf{y} - \left(\frac{m}{2\tau(m+\tau)} \right) \xi(\mathbf{y}^n)$$

where $\xi(\mathbf{y}^n) = \hat{\boldsymbol{\beta}}_{\text{LS}}(\mathbf{y}^n)' (\mathbf{X}'\mathbf{X}) \hat{\boldsymbol{\beta}}_{\text{LS}}(\mathbf{y}^n)$ is the fitted sum of squares. Noting that $\text{E}\left[\mathbf{y}'\mathbf{y}\right] = n\tau + mp$ and $\text{E}\left[\xi(\mathbf{y}^n)\right] = p(\tau+m)$, the entries of the Fisher information matrix for the hyperparameters $\boldsymbol{\alpha}$ are

$$\text{E}\left[\frac{\partial^2 I_{\hat{\beta}}}{\partial m^2} \right] = \text{E}\left[\frac{\partial^2 I_{\hat{\beta}}}{\partial m \partial \tau} \right] = \frac{p}{2(m+\tau)^2} \tag{16}$$

$$\text{E}\left[\frac{\partial^2 I_{\hat{\beta}}}{\partial \tau^2} \right] = \frac{p}{2(m+\tau)^2} + \frac{n-p}{2\tau^2} \tag{17}$$

yielding the Fisher information

$$|\mathbf{J}_{\boldsymbol{\alpha}}(\boldsymbol{\alpha})| = \frac{(n-p)p}{4\tau^2(m+\tau)^2} \tag{18}$$

The hyperparameters m and τ are given the uninformative prior

$$\pi_{\boldsymbol{\alpha}}(\boldsymbol{\alpha}) \propto \tau^{-\nu} \tag{19}$$

where ν is specified *a priori*. Substituting (3), (13), (14), (18) and (19) into (2) and minimising the resultant codelength with respect to $\boldsymbol{\alpha}$ yields

$$\hat{\tau}_{87}(\mathbf{y}^n) = \frac{\mathbf{y}'\mathbf{y} - \xi(\mathbf{y}^n)}{n-p+2\nu-2}$$

$$\hat{m}_{87}(\mathbf{y}^n) = \left(\frac{\xi(\mathbf{y}^n)}{\delta} - \hat{\tau}_{87}(\mathbf{y}^n)\right)_+$$

where $\delta = \max(p-2,1)$ and $(\cdot)_+ = \max(\cdot,0)$ as m may never be negative. When $\hat{m}_{87}(\mathbf{y}^n) > 0$, the complete minimum message length for the data, parameters and the hyperparameters is given by

$$I_g(\mathbf{y}^n|\gamma) = \left(\frac{n-p+2\nu-2}{2}\right)(\log \hat{\tau}_{87}(\mathbf{y}^n)+1) + \frac{p-2}{2}\log\frac{\xi(\mathbf{y}^n)}{\delta}$$

$$+\frac{1}{2}\log(n-p)p^2 + \text{const} \tag{20}$$

where $I_g(\mathbf{y}^n|\gamma) \equiv I_g(\mathbf{y}^n, \hat{\boldsymbol{\beta}}_{87}(\mathbf{y}^n|\hat{\boldsymbol{\alpha}}), \hat{\boldsymbol{\alpha}}_{87}(\mathbf{y}^n)|\gamma)$. Alternatively, when $\hat{m}_{87}(\mathbf{y}^n) = 0$ or $p = 0$, we instead create a 'no effects' design matrix $\mathbf{X}$ comprising a single regressor such that $\mathbf{X}'\mathbf{X} = 0$ which yields the codelength

$$I_g(\mathbf{y}^n|\emptyset) = \left(\frac{n+2\nu-4}{2}\right)(\log \hat{\tau}_{87}(\mathbf{y}^n)+1) + \frac{1}{2}\log(n-1) + \frac{1}{2} + \text{const} \tag{21}$$

for the 'null' model $\gamma = \emptyset$. The codes (20)–(21) are referred to as MML_g; as they depend on the design matrix only through $\hat{\tau}_{87}(\mathbf{y}^n)$ and $\xi(\mathbf{y}^n)$ they are invariant under all full rank affine transformations of $\mathbf{X}$. As in the case of MML_u, the MML_g criterion is of the same computational complexity as AIC and BIC, and is therefore also suitable for application to high dimensional problems.

Remark 3: Minimaxity. The MML87 estimators (15) are minimax with respect to squared error loss for all $p > 2$, assuming the choice of $\nu = 2$ [11]. Furthermore, such estimators dominate the standard least squares estimators for all $p > 2$, and generally outperform them even for $p < 3$.

Remark 4: Given that the least squares estimates are always minimax, it may be preferable to use MML_u when $p < 3$ and MML_g otherwise. This requires a calibration of the two codelengths involving a more efficient coding of the hyperparameter in MML_u and is a topic for future work.

5   Coding the Model Index

The previous criteria ignored the requirement for stating γ, which denotes the regressors included in the model. However, in order to decode the message, a receiver needs to know γ. In nested model selection, such as polynomial regression, a reasonable choice is to use a uniform prior over the maximum number of regressors, $q > 0$, under consideration with codelength

$$I(\gamma) = \log(q+1) \tag{22}$$

If one is instead considering the all subsets regression setting, a prior that treats each combination of subsets of size p as equally likely may be chosen [12]. This prior yields the codelength

$$I(\gamma) = \log \binom{q}{p} + \log(q+1) \tag{23}$$

An alternative prior is to treat *all* combinations of regressors as equally likely; this is equivalent to using a Bernoulli prior with probability of including a regressor set to $(1/2)$. However, it is not difficult to show that for moderate q, such a prior results in codelengths that are longer than those obtained by (23) for almost all combinations of regressors.

Once a suitable code for the model index is chosen, regressor selection is performed by solving

$$\hat{\gamma} = \arg\min_{\gamma} \{ I(\mathbf{y}^n | \gamma) + I(\gamma) \}$$

where $I(\mathbf{y}^n | \gamma)$ is the codelength of the regression model specified by γ; for example, the MML_u (11) or the MML_g (20)–(21) codes.

6   Discussion and Results

The MML_u and MML_g criteria are now compared against two state of the art MDL linear regression criteria, denoted NML [4] and gMDL [13], and the KICc [14] method, on both synthetic and real data. The hyperparameter ν was set to $\nu = 1$ for the MML_u criterion, and to $\nu = 2$ for the MML_g criterion. It has been previously shown that NML, gMDL and KICc regularly outperform the well known AIC and BIC (and allied criteria) and we therefore do not include them in the tests. The NML criterion $I_{\text{NML}}(\mathbf{y}^n)$ for $p > 0$, up to constants, is given by:

$$\left(\frac{n-p}{2} \right) \log \frac{\mathbf{y}'\mathbf{y} - \xi(\mathbf{y}^n)}{n} + \frac{p}{2} \log \frac{\xi(\mathbf{y}^n)}{p} - \log \Gamma \left(\frac{n-p}{2} \right) - \log \Gamma \left(\frac{p}{2} \right) \tag{24}$$

The MML_u and MML_g criteria are clearly similar in form to (24); in fact, using a Jeffrey's prior over $\boldsymbol{\alpha}$ yields a codelength that differs from (24) by $(1/2) \log p + O(1)$. This is interesting given that the NML criterion is derived with the aim of

Table 1. Polynomial order selected by the criteria (expressed as percentages) and squared error in estimated coefficients

	Criterion	Sample Size							
		25	50	75	100	125	150	200	500
	MML_u	15.67	1.64	0.09	0.00	0.00	0.00	0.00	0.00
	MML_g	20.82	2.03	0.12	0.01	0.00	0.00	0.00	0.00
$\hat{p} < p$	NML	15.62	0.87	0.03	0.00	0.00	0.00	0.00	0.00
	gMDL	8.110	0.35	0.01	0.00	0.00	0.00	0.00	0.00
	KICc	32.64	1.50	0.04	0.00	0.00	0.00	0.00	0.00
	MML_u	62.27	86.07	91.24	93.10	94.25	94.90	95.78	97.70
	MML_g	**65.02**	88.83	**93.12**	**94.62**	**95.41**	**95.91**	**96.60**	**98.09**
$\hat{p} = p$	NML	63.38	84.68	89.27	92.53	93.38	94.53	96.97
	gMDL	64.48	82.17	87.19	89.45	91.15	92.15	93.51	96.46
	KICc	62.19	**89.33**	89.94	89.44	89.27	88.94	88.80	88.43
	MML_u	22.07	12.30	8.674	6.896	5.755	5.100	4.219	2.301
	MML_g	14.16	9.141	6.759	5.374	4.589	4.087	3.397	1.909
$\hat{p} > p$	NML	21.00	14.46	10.70	8.741	7.474	6.621	5.468	3.026
	gMDL	27.41	17.48	12.80	10.549	8.847	7.854	6.487	3.536
	KICc	5.170	9.162	10.08	10.56	10.73	11.06	11.20	11.57
	MML_u	113.8	10.38	3.932	2.286	1.625	1.242	0.852	0.294
	MML_g	50.86	7.144	**3.195**	**2.001**	**1.470**	**1.149**	**0.806**	**0.286**
Error	NML	95.38	12.34	4.914	2.871	2.026	1.472	0.957	0.309
	gMDL	136.1	15.69	5.955	3.345	2.302	1.637	1.035	0.319
	KICc	**18.37**	**5.607**	3.614	2.770	2.313	1.890	1.414	0.584

producing minimax regret codes rather than using a formal Bayesian argument. The MML_u criterion may also be rendered even closer to NML by taking the tighter bound $\xi(\mathbf{y}^n)$ when constructing the feasible parameter set Λ. The gMDL approach is derived from the Bayesian mixture code using the g-prior and is thus also closely related to MML_g. The main differences lie in the coding of the hyperparameters, and the fact that the explicit two-part nature of MML_g yields invariant point estimates for β.

6.1 Polynomial Regression

An example application of the newly developed MML criteria is to the problem of polynomial order selection. Following [14] and [4], the simple polynomial basis x^i for $(i = 0, \ldots, q)$ are used. Datasets of various sample sizes $25 \le n \le 500$ were generated from the true model

$$y^* = x^3 - 0.5x^2 - 5x - 1.5$$

with the design points uniformly generated in $[-3, 3]$. The variance τ was chosen to yield a signal-to-noise ratio of one. For every dataset, each criterion was asked to select a nested polynomial model up to maximum degree $q = 10$, and for each sample size the experiment was repeated 10^5 times; the model $p = 0$ was not considered in this test. As the problem is one of nested model selection, the prior (22) was chosen to encode γ. The results are given in Table 1, where the error is the squared ℓ_2 norm of the difference between estimated and true coefficients.

In terms of order selection, MML_g is uniformly superior for all n, although for large n the performance of all the MML/MDL criteria is similar. For small n, the MML_g criterion achieves the best error of all the coding based approaches, followed by MML_u; however, both are slightly inferior to KICc for $n = 25$ and $n = 50$. This is due to the conservative nature of KICc. As the sample size grows, KICc achieves significantly lower correct order selection scores – this is not surprising as KICc is not consistent. Interestingly, even though KICc is asymptotically efficient it still attains a larger error than MML_g at large sample sizes. Of the two MDL criteria, the gMDL criterion appears more prone to overfitting than NML and subsequently attains poorer order selection and error scores for almost all sample sizes.

6.2   Real Datasets

Three real datasets (two from the UCI machine learning repository [15], and one previously analysed in [16] among others) were used to assess the performance of the new MML criteria. The datasets were: (1) the Boston housing data ($q = 14$, $n = 506$), (2) the diabetes data ($q = 10$, $n = 442$) and (3) the concrete compressive strength dataset ($q = 9$, $n = 1030$). Each dataset was randomly divided into a training and testing sample, and the five criteria were asked to choose a suitable subset (including the 'no effects' model) of the candidate regressors from the training sample. The testing sample was subsequently used to assess the predictive performance of the criteria, measured in terms of squared error. Each test was repeated 10^3 times. As this was an all-subsets regression problem, the prior (23) was used for all coding based methods, with q set appropriately depending on the dataset used. The results are presented in Table 2.

Both MML criteria perform well for small sample sizes ($n \leq 25$) and tend to perform marginally worse than the MDL criteria for larger sample sizes. Of

Table 2. Squared prediction errors for three real datasets estimated by cross-validation

	Training Sample	Model Selection Criteria				
		MML_u	MML_g	NML	gMDL	KICc
Housing	25	71.509	**61.922**	69.602	74.842	66.111
	50	36.635	**36.340**	36.918	37.075	36.460
	100	29.383	29.624	29.332	29.135	**29.053**
	200	26.162	26.424	26.031	**25.907**	26.025
	400	24.299	24.304	24.315	24.330	**24.217**
Diabetes	25	4819.2	**4445.0**	4952.5	5136.6	4457.6
	50	3843.8	3851.2	3945.0	3822.6	**3684.0**
	100	3364.2	3385.3	3361.2	3339.3	**3293.8**
	200	3173.3	3199.6	3166.7	3154.2	**3085.7**
	400	3052.7	3052.8	3047.3	3045.2	**3031.8**
Concrete	25	227.41	**221.20**	225.80	225.01	237.02
	50	149.25	**147.46**	148.65	148.52	148.46
	100	123.65	**122.90**	123.82	123.92	123.04
	200	114.50	114.37	114.56	114.62	**114.33**
	400	111.64	**111.59**	111.67	111.67	111.62

the two MML criteria, the MML_g criterion appears slightly superior to MML_u. The KICc criterion is competitive which is not unexpected given that it is an efficient criterion and is expected to perform well for prediction. An interesting point to note is that when MML_g outperforms NML and gMDL, the difference in performance can be relatively large; in contrast, when the MML criteria obtain higher prediction errors than the MDL criteria, the difference in prediction is minimal. The MML_g criterion thus appears to offer better protection against overfitting when there is little signal available (i.e. small sample size or large noise) while trading off little performance as the signal strength increases.

References

1. Wallace, C.S.: Statistical and Inductive Inference by Minimum Message Length, 1st edn. Information Science and Statistics. Springer, Heidelberg (2005)
2. Fitzgibbon, L.J., Dowe, D.L., Allison, L.: Univariate polynomial inference by Monte Carlo message length approximation. In: 19th International Conference on Machine Learning (ICML 2002), Sydney, Australia, pp. 147–154 (2002)
3. Viswanathan, M., Wallace, C.S.: A note on the comparison of polynomial selection methods. In: Uncertainty 1999: The Seventh International Workshop on Artificial Intelligence and Statistics, Fort Lauderdale, Florida, pp. 169–177 (1999)
4. Rissanen, J.: Information and Complexity in Statistical Modeling, 1st edn. Information Science and Statistics. Springer, Heidelberg (2007)
5. Wallace, C.S., Boulton, D.M.: An information measure for classification. Computer Journal 11(2), 185–194 (1968)
6. Wallace, C.S., Freeman, P.R.: Estimation and inference by compact coding. Journal of the Royal Statistical Society (Series B) 49(3), 240–252 (1987)
7. Makalic, E., Schmidt, D.F.: Minimum message length shrinkage estimation. Statistics & Probability Letters 79(9), 1155–1161 (2009)
8. Akaike, H.: A new look at the statistical model identification. IEEE Transactions on Automatic Control 19(6), 716–723 (1974)
9. Schwarz, G.: Estimating the dimension of a model. The Annals of Statistics 6(2), 461–464 (1978)
10. Zellner, A.: Applications of Bayesian analysis in econometrics. The Statistician 32(1–2), 23–34 (1983)
11. Sclove, S.L.: Improved estimators for coefficients in linear regression. Journal of the American Statistical Association 63(322), 596–606 (1968)
12. Roos, T., Myllymäki, P., Rissanen, J.: MDL denoising revisited. IEEE Transactions on Signal Processing 57(9), 3347–3360 (2009)
13. Hansen, M.H., Yu, B.: Model selection and the principle of minimum description length. Journal of the American Statistical Association 96(454), 746–774 (2001)
14. Seghouane, A.K., Bekara, M.: A small sample model selection criterion based on Kullback's symmetric divergence. IEEE Trans. Sig. Proc. 52(12), 3314–3323 (2004)
15. Asuncion, A., Newman, D.: UCI machine learning repository (2007)
16. Efron, B., Hastie, T., Johnstone, I., Tibshirani, R.: Least angle regression. The Annals of Statistics 32(2), 407–451 (2004)

DMCS: Dual-Model Classification System and Its Application in Medicine

Qun Song, Ruihua Weng, and Fengyu Weng

Medical Informatics Technology Development & Research Centre
Jiamusi, Heilongjiang, China
{qsong,rhweng,fyweng}@mitdrc.com

Abstract. The paper introduces a novel dual-model classification method – Dual-Model Classification System (DMCS). The DMCS is a personalized or transductive system which is created for every new input vector and trained on a small number of data. These data are selected from the whole training data set and they are closest to the new vector in the input space. In the proposed DMCS, two transductive fuzzy inference models are taken as the structure functions and trained with different sub-training data sets. In this paper, DMCS is illustrated on a case study: a real medical decision support problem of estimating the survival of hemodialysis patients. This personalized modeling method can also be applied to solve other classification problems.

Keywords: Dual-Model Systems, transductive fuzzy inference system, classification methods, medical decision support problems.

1 Introduction

1.1 Global, Local and Personalized Modeling

Most of learning models and systems in either mathematics or artificial intelligence developed and implemented so far are of global (inductive) models. Such models are trained with all training data and subsequently applied on new data. The derivation of the model in this manner therefore may not optimally account for all of the specific information related to a given new vector in the test data. An error is measured to estimate how well the new data fits into the model. The inductive learning and inference approach is useful when a global model of the problem is needed.

Local models are a type of model ensemble that partitions the problem space into a number of sub-spaces. These sub-spaces can be defined through clustering methods such as k-means, fuzzy c-means and hierarchical clustering. This type of model assumes that each cluster (sub-space) is a unique problem and a unique model is created on this cluster. One or several such model(s) can be used to solve a sub-problem with input vectors belong to the cluster.

Personalized models [13, 14, 15] estimate the value of a potential model (function) only in a single point of the space (the new data vector) utilizing additional information related to this point. A personalized model is created for every new input

A. Nicholson and X. Li (Eds.): AI 2009, LNAI 5866, pp. 322–329, 2009.
© Springer-Verlag Berlin Heidelberg 2009

vector and this individual model is based on the closest data samples to the new samples taken from the training data set.

Global models capture trends in data that are valid for the whole problem space, and local models capture local patterns, valid for clusters of data. Since a personalized model is made for a given new input vector only, the entire modeling process can be specially optimized for it without considering how the model will perform on the problem as a whole.

1.2 Transductive Fuzzy Inference System

Transductive fuzzy inference systems estimate the value of a potential model (function) only in a single point of the space (the new data vector) utilizing additional information related to this point. This approach seems to be more appropriate for clinical and medical applications of learning systems, where the focus is not on the model, but on the individual patient. Each individual data vector (e.g.: a patient in the medical area; a future time moment for predicting a time series; or a target day for predicting a stock index) may need an individual, local model that best fits the new data, rather than - a global model. In the latter case the new data is matched into a model without taking into account any specific information about this data.

The transductive fuzzy inference system is concerned with the estimation of a model in a single point of the space only. For every new input vector that needs to be processed for a prognostic or classificatory task, a number of nearest neighbours, which form a sub-data set, are derived from the whole training data set.

1.3 Dual-Model Systems

Dual-model systems are a type of personalized models that differs from general learning systems. For every new data, it creates two personalized models by use of both training data and the new data and then, the results of such two models on training data are compared to make a classification for the new data. A number of general classifiers can be used in a dual-model systems as structural functions such as linear regression, SVM, RBF and fuzzy inference systems. In our research, we use the transductive fuzzy inference models as structural functions to create a dual-model system for solving a classification problem.

This paper is organized as follows: Section 2 presents the process flow and algorithm of the DMCS method. Section 3 illustrates the DMCS on a case study example. Conclusions are drawn in Section.

2 Process Flow and Algorithm of DMCS

The process flow diagram of a dual-model classification system is shown as Fig.1 and described as follows:

1) Take one new data vector (without class label) and assign different class labels to it, e.g. 0 and 1, to structure two new training samples that have the same input vectors and different labels;

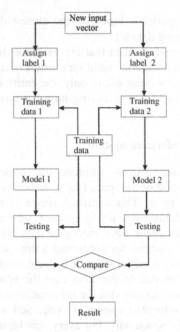

Fig. 1. Process flow diagram of a general dual-model classification system

2) Add such two new training samples into the original training data set respectively to form two training data sets – training data set 1 and training data set 2;
3) Learning with two training data set 1 and set 2, respectively to create two classification models;
4) Apply such two models to the original training data respectively to obtain two classification results;
5) Compare the results and the result with a higher accuracy corresponds with the correct assigned class label.

A DMCS system uses the transductive technology for solving classification problems. The distance between vectors x and y is measured in DMCS in normalized Euclidean distance defined as follows (the values are between 0 and 1):

$$\|x - y\| = \frac{1}{P} \left[\sum_{j=1}^{P} |x_j - y_j|^2 \right]^{\frac{1}{2}} \tag{1}$$

where: $x, y \in R^P$.

Consider the classification problem has two classes and P variables, for each new data vector x_q, the DMCS learning algorithm performs the following steps:

1) Normalize the training data set and the new data (the values are between 0 and 1).
2) Search in the training data set in the whole space to find a cluster D_q that includes N_q training samples closest to x_q. The value of N_q can be pre-defined based on experience, or - optimized through the application of an optimization procedure. Here we assume the former approach.

3) If all training samples in D_q belong to the same class, the new data belongs to this class and the procedure ends. Otherwise,

4) Assign one class label, e.g. 0, to x_q to structure a new training sample and then, add this sample into D_q to form a training data set – T_q;

5) Calculate the distances d_i, $i = 1, 2, ..., N_q +1$, between x_q and each of data samples T_q and calculate the vector weights $w_i = 1 - (d_i - min(d))$, here, $i = 1, 2, ..., N_q +1$, $min(d)$ is the minimum value in the distance vector d, $d = [d_1, d_2, ..., d_{Nq+1}]$.

6) Use a clustering algorithm to cluster and partition the input sub-space that consists of N_{q+1} training samples.

7) Create fuzzy rules and set their initial parameter values according to the clustering procedure results; for each cluster, the cluster centre is taken as the centre of a fuzzy membership function (*Gaussian* function) and the cluster radius is taken as the width.

8) Apply the steepest descent method (back-propagation) to optimize the parameters of the fuzzy rules in the fuzzy inference model following Eq. (2 – 12).

9) Calculate the accuracy *Ar1* of this fuzzy model on the training data T_q.

10) Assign another class label, e.g. 1, to x_q to structure another training sample, add such a sample into D_q to form a training data set – T_q, and repeat step 5 to 9 to obtain the accuracy *Ar2* on the training data.

11) Compare results *Ar1* and *Ar2* and the accuracy with a higher value corresponds with the correct assigned class label.

12) End of the procedure.

The parameter optimization procedure is described as following (for both two fuzzy inference models):

Consider the system having P inputs, one output and M fuzzy rules defined initially through a clustering procedure, the *l*-th rule has the form of:

$$R_l : \qquad \text{If } x_1 \text{ is } F_{l1} \text{ and } x_2 \text{ is } F_{l2} \text{ and } ... x_P \text{ is } F_{lP}, \text{ then } y \text{ is } G_l . \qquad (2)$$

Here, F_{lj} are fuzzy sets defined by the following *Gaussian* type membership function:

$$GaussianMF = \alpha \exp\left[-\frac{(x-m)^2}{2\sigma^2}\right] \qquad (3)$$

and G_l are of a similar type as F_{lj} and are defined as:

$$GaussianMF = \exp\left[-\frac{(y-n)^2}{2\delta^2}\right] \qquad (4)$$

Using the *Modified Centre Average defuzzification* procedure the output value of the system can be calculated for an input vector $x_i = [x_1, x_2, ..., x_P]$ as follows:

$$f(\mathbf{x_i}) = \frac{\sum\limits_{l=1}^{M} \frac{G_l}{\delta_l^2} \prod\limits_{j=1}^{P} \alpha_{lj} \exp\left[-\frac{(x_{ij}-m_{lj})^2}{2\sigma_{lj}^2}\right]}{\sum\limits_{l=1}^{M} \frac{1}{\delta_l^2} \prod\limits_{j=1}^{P} \alpha_{lj} \exp\left[-\frac{(x_{ij}-m_{lj})^2}{2\sigma_{lj}^2}\right]} \qquad (5)$$

Suppose the fuzzy model is given a training input-output data pair $[x_i, t_i]$, the system minimizes the following objective function (a weighted error function):

$$E = \frac{1}{2} w_i [f(x_i) - t_i]^2 \qquad (w_i \text{ are defined in step 5}) \qquad (6)$$

The steepest descent algorithm is used then to obtain the formulas for the optimization of the parameters G_l, δ_l, α_{lj}, m_{lj} and σ_{lj} of the fuzzy model such that the value of E from Eq. (6) is minimized:

$$G_l(k+1) = G_l(k) - \frac{\eta_G}{\delta_l^2(k)} w_i \Phi(\mathbf{x}_i) [f^{(k)}(\mathbf{x}_i) - t_i] \qquad (7)$$

$$\delta_l(k+1) = \delta_l(k) - \frac{\eta_\delta}{\delta_l^3(k)} w_i \Phi(\mathbf{x}_i) [f^{(k)}(\mathbf{x}_i) - t_i][f^{(k)}(\mathbf{x}_i) - G_l(k)] \qquad (8)$$

$$\alpha_{lj}(k+1) = \alpha_{lj}(k) - \frac{\eta_\alpha}{\delta_l^2(k)\alpha_{lj}(k)} w_i \Phi(\mathbf{x}_i) [f^{(k)}(\mathbf{x}_i) - t_i][G_l(k) - f^{(k)}(\mathbf{x}_i)] \qquad (9)$$

$$m_{lj}(k+1) = m_{lj}(k) -$$
$$\frac{\eta_m}{\delta_l^2(k)\sigma_{lj}^2(k)} w_i \Phi(\mathbf{x}_i) [f^{(k)}(\mathbf{x}_i) - t_i][G_l(k) - f^{(k)}(\mathbf{x}_i)][x_{ij} - m_{lj}(k)] \qquad (10)$$

$$\sigma_{lj}(k+1) = \sigma_{lj}(k) -$$
$$\frac{\eta_\sigma}{\delta_l^2(k)\sigma_{lj}^3(k)} w_i \Phi(\mathbf{x}_i) [f^{(k)}(\mathbf{x}_i) - t_i][G_l(k) - f^{(k)}(\mathbf{x}_i)][x_{ij} - m_{lj}(k)]^2 \qquad (11)$$

$$\text{here, } \Phi(x_i) = \frac{\prod_{j=1}^{P} \alpha_{lj} \exp\left\{-\frac{[x_{ij}(k) - m_{lj}(k)]^2}{2\sigma_{lj}^2(k)}\right\}}{\sum_{l=1}^{M} \frac{1}{\delta_l^2} \prod_{j=1}^{P} \alpha_{lj} \exp\left\{-\frac{[x_{ij}(k) - m_{lj}(k)]^2}{2\sigma_{lj}^2(k)}\right\}} \qquad (12)$$

where: η_G, η_δ, η_α, η_m and η_σ are learning rates for updating the parameters G_l, δ_l, α_{lj}, m_{lj} and σ_{lj} respectively.

In the DMSC training algorithm, the following indexes are used:

- Training data samples: $i = 1, 2, \ldots, , N_{q1}$ or N_{q2};
- Input variables: $j = 1, 2, \ldots, P$;
- Fuzzy rules: $l = 1, 2, \ldots, M$;
- Learning epochs: $k = 1, 2, \ldots$.

3   Case Study Example of Applying the DMCS for a Medical Decision Support Problem

A medical dataset is used here for experimental analysis. Data originate from the Dialysis Outcomes and Practice Patterns Study (DOPPS, www.dopps.org) [5]. The DOPPS is based upon the prospective collection of observational longitudinal data from a stratified random sample of hemodialysis patients from the United Sates, 8 European countries (United Kingdom, France, Germany, Italy, Spain, Belgium, Netherlands, and Sweden), Japan, Australia and New Zealand. There have been two phases of data collection since 1996, and a third phase is currently just beginning. To date, 27,880 incident and prevalent patients (approximately 33% and 66% respectively) have been enrolled in the study, which represents approximately 75% of the world's hemodialysis patients. In this study, prevalent patients are defined as those patients who had received maintenance hemodialysis prior to the study period, while incident patients are those who had not previously received maintenance hemodialysis.

The research plan of the DOPPS is to assess the relationship between hemodialysis treatment practices and patient outcomes. Detailed practice pattern data, demographics, cause of end-stage renal disease, medical and psychosocial history, and laboratory data are collected at enrollment and at regular intervals during the study period. Patient outcomes studied include mortality, frequency of hospitalisation, vascular access, and quality of life. The DOPPS aims to measure how a given practice changes patient outcomes, and also determine whether there is any relationship amongst these outcomes, for the eventual purpose of improving treatments and survival of patients on hemodialysis.

The dataset for this case study contains 6100 samples from the DOPPS phase 1 in the United States, collected from 1996-1999. Each record includes 24 patient and treatment related variables (input): demographics (age, sex, race), psychosocial characteristics (mobility, summary physical and mental component scores (sMCS, sPCS) using the Kidney Disease Quality of Life (KD-QOL®) Instrument), co-morbid medical conditions (diabetes, angina, myocardial infarction, congestive heart failure, left ventricular hypertrophy, peripheral vascular disease, cerebrovascular disease, hypertension, body mass index), laboratory results (serum creatinine, calcium, phosphate, albumin, hemoglobin), hemodialysis treatment parameters (Kt/V, hemodialysis angioaccess type, hemodialyser flux), and vintage (years on hemodialysis at the commencement of the DOPPS). The output is survival at 2.5 years from study enrollment (yes or no). All experimental results reported here are based on 10-cross validation experiments [7].

For comparison, several well-known methods of classification are applied to the same problem, such as Support Vector Machine (SVM) and transductive SVM [14], Evolving Classification Function (ECF) [7], Multi-Layer Perceptron (MLP) [12], Radial Basis Function (RBF) [11], and Multiple Linear Regression along with the proposed DMSC, and results are given in Table 1.

The Kappa statistic, K, formally tests for agreement between two methods, raters, or observers, when the observations are measured on a categorical scale. Both methods must rate, or classify, the same cases using the same categorical scale [1]. The degree of agreement is indicated by K, which can be roughly interpreted as follows: $K < 0.20$, agreement quality poor; $0.20 < K < 0.40$, agreement quality fair; $0.40 < K < 0.60$, agreement quality moderate; $0.60 < K < 0.80$, agreement quality good; $K > 0.80$, agreement quality very good. Confidence intervals for K were constructed using the goodness-of-fit approach of Donner & Eliasziw [2]. There is no universally agreed method for comparing K between multiple tests of agreement. In this study, K for different classification methods was compared using the permutation or Monte Carlo resampling routine of McKenzie [9,10].

Table 1. Experimental Results on the DOPPS Data

Model	Kappa(95% Confidence Intervals)*	P-value	Agreement (%)	Specificity (%)	Sensitivity (%)
RBF	0.1675 (0.1268 - 0.2026)	<0.001	60.4	65.3	49.08
ECF	0.1862 (0.1469 - 0.2224)	<0.001	61.5	63.4	51.76
MLP	0.3833 (0.3472 - 0.4182)	<0.001	62.8	65.6	58.72
Multiple Linear Regression	0.4000 (0.3651 - 0.4357)	<0.001	64.9	67.6	63.21
SVM	0.4240 (0.3748 - 0.4449)	<0.001	65.3	68.2	62.3
TSVM	0.4290 (0.3792 - 0.4460)	<0.001	57.2	61.2	52.9
DMCS	0.4612 (0.4498- 0.492)	Reference	76.5	77.0	76.7

* Kappa values and confidence intervals ascertained with Stata Intercooled V 8.2 (StataCorp, College Station, TX), and P-values with KAPCOM [11].

Agreement refers to the quality of the information provided by the classification device and should be distinguished from the usefulness, or actual practical value, of the information. Agreement provides a pure index of accuracy by demonstrating the limits of a test's ability to discriminate between alternative states of health over the complete spectrum of operating conditions. To date, prognostic systems for the prediction of haemodialysis patient survival have published accuracy of 60-70%. The experimental results in Table 1 illustrate that the DMCS in this paper provide incrementally better results, towards a K of > 0.60 and a level of accuracy ~80%, which are generally regarded as thresholds for clinical utility.

4 Conclusions

This paper presents a dual-model classification system – DMCS. The DMCS performs a better local generalization over new data as it develops individual models for each data vector that takes the location of new input vector in the space into account. This approach seems to be more appropriate for clinical and medical applications, where the focus is not on the model, but on the individual patient. At the same time, it is an adaptive model, in the sense that data can be added to the data set continuously and immediately, and made available for DMCS models. This type of modeling can be called "personalized", and it is promising for medical decision support systems. The clinical plausibility of the approach and its results are satisfactory in this study. As the DMCS creates a unique model for each data sample, it usually needs more performing time than inductive models. Further directions for the research include: (1) DMCS system parameter optimization such as optimal number of nearest neighbors; and (2) applying the DMCS method to other decision support systems, such as: cardio-vascular risk prognosis; biological processes modeling and classifications based on gene expression microarray data.

References

1. Altman, D.G.: Practical Statistics for Medical Research. Chapman and Hall, London (1991)
2. Donner, A., Eliasziw, M.: A goodness-of-fit approach to inference procedures for the kappa statistic: confidence interval construction, significance-testing and sample size estimation. Statistics in Medicine 11, 1511–1519 (1992)
3. Gammerman, A., Vovk, V., Vapnik, V.: Learning by transduction. In: Cooper, G.F., Moral, S. (eds.) Proc. of the 14th Conference on Uncertainty in Artificial Intelligence, Madison, Wisconsin, pp. 148–155. Morgan Kaufmann, San Francisco (1998)
4. Golub, C.L., Van Loan, C.: Matrix computations. Jons Hopkins University Press, Baltimore
5. Goodkin, D.A., Mapes, D.L., Held, P.J.: The dialysis outcomes and practice patterns study (DOPPS): how can we improve the care of hemodialysis patients? Seminars in Dialysis 14, 157–159 (2001)
6. Hsia, T.C.: System Identification: Least-Squares Methods. D.C. Heath and Company (1977)
7. Kukar, M.: Transductive reliability estimation for medical diagnosis. Artif. Intell. Med. 29, 81–106 (2003)
8. Marshall, M.R., Song, Q., Ma, T.M., MacDonell, S., Kasabov, N.: Evolving Connectionist System versus Algebraic Formulae for Prediction of Renal Function from Serum Creatinine. Kidney International 67, 1944–1954 (2005)
9. McKenzie, D.P., Mackinnon, A.J., Peladeau, N., Onghena, P., Bruce, P.C., Clarke, D.M., Harrigan, S., McGorry, P.D.: Comparing correlated kappas by resampling: is one level of agreement significantly different from another? Journal of Psychiatric Research 30, 483–492 (1996)
10. McKenzie, D.P., Mackinnon, A.J., Clarke, D.M.: KAPCOM: a program for the comparison of kappa coefficients obtained from the same sample of observations. Perceptual and Motor Skills, 899–902 (1997)
11. Neural Network Toolbox User's Guide. The Math Works Inc., 3 Apple Hill Drive, Natrick, Massachusetts, Ver. 4 (2002)
12. Oja, E.: A simplified neuron model as a principal component analyzer. Journal of Mathematical Biology 16, 267–273 (1982)
13. Song, Q., Kasabov, N.: NFI: A Neuro-Fuzzy Inference Method for Transductive Reasoning. IEEE Trans. on Fuzzy Systems 13(6), 799–808 (2005)
14. Song, Q., Kasabov, N.: TWNFI – Transductive Neuro-Fuzzy Inference System with Weighted Data Normalization for Personalized Modelling. Neural Networks 19, 1591–1596 (2006)
15. Vapnik, V.: The Nature of Statistical Learning Theory. Springer, New York (1995)
16. Xu, L., Oja, E., Suen, C.Y.: Modified Hebbian Learning for Curve and Surface Fitting. Neural Networks 5, 441–457 (1992)

A Graph Distance Based Structural Clustering Approach for Networks

Xin Su and Chunping Li

Tsinghua National Laboratory for Information Science and Technology
School of Software, Tsinghua University, China
sux07@mails.tsinghua.edu.cn,
cli@tsinghua.edu.cn

Abstract. In the era of information explosion, structured data emerge on a large scale. As a description of structured data, network has drawn attention of researchers in many subjects. Network clustering, as an essential part of this study area, focuses on detecting hidden sub-group using structural features of networks. Much previous research covers measuring network structure and discovering clusters. In this paper, a novel structural metric "Graph Distance" and an effective clustering algorithm GRACE are proposed. The graph distance integrates local density of clusters with global structural properties to reflect the actual network structure. The algorithm GRACE generalizes hierarchical and locality clustering methods and outperforms some existing methods. An empirical evaluation demonstrates the performance of our approach on both synthetic data and real world networks.

Keywords: Graph Distance, Structural Clustering, Networks.

1 Introduction

Network is one of the best representations for many complex systems in real world, such as the Internet, social community, and biological systems, etc. It is helpful for people to analyze mass data, model systems and discover underlying features. Many structural properties of real world networks have been revealed, e.g. small world phenomena [17], power-law degree [1] and shrinking diameter [12], etc.

Network clustering is the approach to detect densely connected groups of vertices in a network, with few connections between groups. This issue has drawn a considerable amount of attention in social network [16], physics [14], and bioinformatics [8], etc. Many previous methods [3,7,11,18] reveal that the general process of network clustering has two steps: finding a metric to distinguish the vertices or edges within clusters from those between clusters, and adopting some partitioning or clustering algorithms to discover the clusters.

In this paper we propose a novel structural metric called graph distance. The graph distance combines structural density and edge betweenness. It fixes the length of edges and makes the distance of any pairs of nodes measurable. Then we

A. Nicholson and X. Li (Eds.): AI 2009, LNAI 5866, pp. 330–339, 2009.

propose an algorithm GRACE (GRaph distAnce ClustEring) which guarantees the quality of clustering from a global perspective of network structures. The algorithm can be seen as a generalization of different previous network clustering approaches, such as the partitioning-based and locality-based clustering.

This paper is organized as follows. We review the related works for network clustering methods in Section 2. Then the concept of graph distance is proposed in Section 3. In Section 4, we propose the algorithm GRACE and analyze its computational complexity and generality properties. In Section 5, the experimental results and evaluation of our method are given. Finally, we have the concluding remarks in Section 6.

2   Related Works

The problem of network clustering has been studied for some decades. There is a rich literature about it. Here we focus on undirected and unweighted graph and review some details of recent works.

M. Girvan and M.E.J. Newman [7] introduce the concept of edge betweenness to describe how "between" clusters the edges are. The edge betweenness of an edge, which is derived from Freeman's vertex betweenness [6], is defined as the number of shortest paths between pairs of vertices that run along it. So the inter-cluster edges have higher edge betweenness than the intra-cluster edges. Then the algorithm removes the edges with the highest betweenness to identify clusters. This method defines an important structural metric to discover the edges between clusters.

Recently, W. Hwang et al. [11] propose a unique structural property called bridging centrality. Comparing to edge betweenness, the bridging centrality combines local graph properties called bridging coefficient to edge betweenness. The bridging coefficient is derived from the concept of clustering coefficient [17] which illustrates the local inter-connectivity of a vertex's neighborhood. The experiment shows the bridging centrality is an effective description for network structure.

Both of the methods stated above adopt a cut-based algorithm, which can also be seen as a hierarchical clustering algorithm. They remove the edges with the highest edge betweenness or bridging centrality till no edge remains or isolated modules are identified.

The SCAN [18] method proposed by X. Xu et al. applies the DBSCAN [4] method in text clustering to networks. It finds structure-connected clusters in a two-step approach. First, choose an arbitrary vertex from graph satisfying the core condition as a seed. Second, retrieve all the vertices that are structure reachable from the seed to obtain the cluster grown from the seed. The significant advantage of SCAN is identifying hubs and outliers which play different roles in networks, while the structural similarity based on the number of common neighbors is incompetent to represent the complex structure of networks.

3 Graph Distance

An undirected unweighted graph can be denoted as $G = (V, E)$, V is a set of nodes and E is a set of edges, $E \subseteq V \times V$, an edge $e = (i, j)$ connects two nodes i and j, $i, j \in V$, $e \in E$. The neighbors $N(v)$ of node v are defined to be the set of nodes that directly connect with node v. The degree $deg(v)$ of node v is the number of the neighbors of node v. A path P is defined as a sequence of nodes $(v_1, v_2, \ldots, v_n)$ in which there is an edge from each node to its successor.

Definition 1. *Edge Betweenness*

The edge betweenness of an edge is defined as the number of shortest paths between pairs of vertices that run along it:

$$\Phi(e) = \sum_{s \neq t \in V} \frac{\sigma_{st}(e)}{\sigma_{st}} . \tag{1}$$

where σ_{st} is the number of shortest paths between node s and t, and $\sigma_{st}(e)$ is the number of shortest paths passing through an edge e out of σ_{st}.

Definition 2. *Normalized Edge Betweenness*

The normalized edge betweenness of an edge is the edge betweenness divided by the sum of the edge betweenness from all edges of the graph:

$$\Phi_N(e) = \frac{\sum_{s \neq t \in V} \frac{\sigma_{st}(e)}{\sigma_{st}}}{\sum_{e \in E} \Phi(e)} . \tag{2}$$

If a network contains clusters that are loosely connected by a few intergroup edges, then all the shortest paths between different clusters must go along one of these edges. Thus, the edges connecting clusters have high edge betweenness.

Definition 3. *Structural Density*

The structural density of an edge is the ratio of the number of edges between the nodes in the direct neighborhood to the number of edges that could possibly exist among them. The direct neighborhood of an edge is defined as the union of the neighbors of both ends of the edge. We have:

$$C_{e(v,w)} = \frac{2 | \bigcup_{i,j \in (N(v) \cup N(w))} e(i,j) |}{|N(v) \cup N(w)| (|N(v) \cup N(w)| - 1)} . \tag{3}$$

If there are many edges in the subgraph constructed by the neighbors of the two ends of edge $e(v, w)$, the value of the structural density is high. Further, if the neighbors construct a complete graph, the structural density equals to 1.

Definition 4. *Local Structural Density*

The local structural density of an edge is defined as the product of the structural density and the number of common neighbors of the two ends of the edge.

$$C^w_{e(v,w)} = \frac{2|\bigcup_{i,j \in (N(v) \cup N(w))} e(i,j)|}{|N(v) \cup N(w)|(|N(v) \cup N(w)| - 1)} (|N(v) \cup N(w)| + 1) \quad . \tag{4}$$

To distinguish the edges in a local community with many common neighbors from the bridges that connects many nodes with few common neighbors, we use the number of common neighbors to weight the structural density.

Definition 5. *Graph Distance of an Edge*

The graph distance of an edge is defined as the normalized edge betweenness divided by the local structural density of the edge:

$$Dis(e) = \frac{\Phi_N(e)}{C^w_{e(v,w)}} \quad . \tag{5}$$

Definition 6. *Graph Distance of Arbitrary Pairs of Vertices*

To any pairs of vertices in a network, the graph distance is the sum of the graph distance of every edge that locates along the shortest path between the pairs of vertices. Suppose that (i, j) is an arbitrary pair of vertices, the shortest path from i to j is denoted as $e_1, e_2, \ldots, e_k, \ldots, e_n$, thus:

$$Dis(i,j) = \sum_{k=1}^{n} Dis(e_k) \quad . \tag{6}$$

The graph distance of all pairs of vertices in a network can be computed by Floyd-Warshall algorithm [5] in the time of $O(|V|^3)$.

4 Algorithm GRACE

4.1 Notions

Definition 7. The *influence function* of a vertex $v \in V$ is defined as:

$$f^v_B(w) = f_B(w, v) \quad . \tag{7}$$

where $f_B(v, w)$ can be an arbitrary function of graph distance.

For example:

1. *Square Wave Influence Function*

$$f_{\text{Square}}(w, v) = \begin{cases} 0 & \text{if } Dis(w, v) > \sigma \\ 1 & \text{otherwise} \end{cases} \quad . \tag{8}$$

2. *Gaussian Influence Function*

$$f_{\text{Gaussian}}(w, v) = e^{-\frac{Dis(w,v)^2}{2\sigma^2}} \quad . \tag{9}$$

Definition 8. The *density function* is defined as the sum of the influence functions of all vertices in a network:

$$f_B^V(w) = \sum_{v \in V} f_B^v(w) \ . \tag{10}$$

If the Gaussian influence function is chosen, the density function is:

$$f_{\text{Gaussian}}^V(w) = \sum_{v \in V} e^{-\frac{Dis(w,v)^2}{2\sigma^2}} \ . \tag{11}$$

Definition 9. A vertex $v^* \in V$ is called a *density attractor* for a given influence function, iff v^* is a local maximum of the density function f_B^V. A vertex $v \in V$ is density attracted to a density attractor v^*, iff there is a path $v_0, v_1, \ldots, v_i, \ldots, v_n$, where $v = v_0, v^* = v_n$ and v_i has the largest value of density function in the neighborhood of v_{i-1} where $Dis(v_i, v_{i-1}) < \epsilon$.

Definition 10. A *center-defined cluster* (wrt to σ, ξ) for a density attractor v^* is a subset $C \subseteq V$, with $v \in C$ being density attracted by v^* and $f_B^V(v^*) \geq \xi$. Vertices are called outliers if they are density attracted by a local maximum v_0^* with $f_B^V(v_0^*) < \xi$.

Definition 11. An *arbitrary shape cluster* (wrt to σ, ξ) for the set of density attractors D is a subset $C \subseteq V$, where
1. $\forall v \in C, \exists v^* \in D : f_B^V(v^*) \geq \xi$, v is density attracted to v^* and
2. $\forall v_1^*, v_2^* \in D : \exists$ a path P from v_1^* to v_2^* with $\forall v \in P : f_B^V(v) \geq \xi$.

The basic idea of our algorithm, inspired by [9], is to use a function to describe the impact of each vertex on the rest of the network. The overall structural density of the network can be calculated as the sum of the influence function of all vertices of the network. Clusters can then be determined by identifying density-attractors which are local maxima of the overall density function.

4.2   Algorithm Description

Our algorithm finds the density-attractor of every unclassified vertex in the direction of its neighbor with the greatest differences of density function and put it into the cluster that its density-attractor belongs to.

As Fig. 1 shown, we search for the local maximum of density function for the rear vertex of a queue and push it into the queue if found. Once a local maximum satisfies the condition of density-attractors, the queue stores the "hill-climbing path" in which the vertices should be in the same cluster. Thus, arbitrary shape clusters can be found. The vertices attracted by no density-attractors can be further classified as hubs or outliers.

Given a graph with n vertices and m edges, edge betweenness is computed in $O\left(nm + n^2 \log n\right)$ [2]. The average computational time for local structural density is $O\left(n^2 \log n\right)$ because the average degree of vertices is approximately

```
Algorithm GRACE (G =< V, E >, σ, ξ)

Compute the density function for each vertex v in V;
// all vertices in V are labeled as unclassified
for each unclassified vertex v do
  push v into queue Q;
  while Q is not empty do
    y= the rear of the queue Q;
    if (y has neighbors)
      find the local maximum x in the neighborhood of y;
      if ( x != y)
        if( x is classified)
          label all queue members with the clusterID of x;
        else
          put x into queue Q;
      else
        if (density function of v > Threshold)
          label all queue members with a new clusterID;
          clear the queue;
        else
          label all queue members as non-member;
          clear the queue;
    else
      label y as non-member;
  end for
  for each non-member vertex v do
  if (x and y in N(v): x.clusterID != y.cluserID)
    label v as hub;
  else
    label v as outlier;
  end for
end GRACE
```

Fig. 1. The Algorithm GRACE

equal to $\log n$ in real world networks. Floyd-Warshall algorithm takes $O(n^3)$ [5]. Finding clusters takes $O(deg(v_1) + deg(v_2) + \ldots, deg(v_n))$, which equals to $O(m)$. So the total time cost is bounded by $O(n^3)$.

Our approach can be seen as the generalization of both hierarchical clustering and locality clustering methods. First, we discuss the locality-based clustering algorithm SCAN. Derived from DBSCAN, SCAN has the parameters ϵ and μ replace EPS and MinPts respectively. If we use a square wave influence function with $\sigma = \epsilon$ and an outlier bound $\xi = \mu$, the clustering results are the same as SCAN, because in the case of square wave influence function, the vertex $v : f_B^V(v) > \xi$ satisfies the core vertex condition and each non-core vertex v which is directly density-reachable from a core v^* is attracted by the density attractor of v^*.

In the case of hierarchical clustering algorithm such as edge betweenness cut and bridge cut algorithm, we adopt Gaussian influence function and use different values for σ to generate a hierarchy of clusters. If we start with a very large value for σ, all vertices in the network are in the same cluster. By decreasing the σ, the cluster starts to divide by different density attractors and the dividing line is through the edges with top largest graph distance. It is equivalent to removing edges from network.

5   Experiments and Evaluations

Firstly, we use a synthetic network to illustrate the effectiveness of graph distance. Figure 2 illustrates the comparison between the network with vertices and edges randomly distributed and the network with fixed length of edges which are computed by graph distance. We can easily identify the clusters of the network from the latter one.

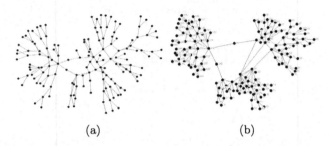

(a)   (b)

Fig. 2. (a) A Synthetic Network with 150 Nodes and 169 Edges, (b) The Network Redrawn with Graph Distance

In Fig. 3, we can see that only a few edges have high graph distance above 0.15 and the distances of most edges are under the level of 0.05. The eight edges between clusters in Fig. 2 (b) are the very edges with top 8 graph distance values of Fig. 3. Moreover, the edges within clusters always have low distance value, which makes the clusters look compact.

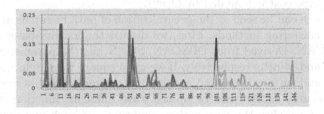

Fig. 3. A Statistical Chart of the Graph Distance of all Edges in the Network

To evaluate the effectiveness of our algorithm, we use two real world networks whose clusters are already known. Adjusted Rand Index (ARI) [10] is adopted as a measure of the quality of clustering to compare our result to the reality. The definition of ARI is:

$$\frac{\sum_{i,j} \binom{n_{ij}}{2} - \left[\sum_i \binom{n_{i.}}{2} \sum_j \binom{n_{.j}}{2} \right] \setminus \binom{n}{2}}{\frac{1}{2} \left[\sum_i \binom{n_{i.}}{2} + \sum_j \binom{n_{.j}}{2} \right] - \left[\sum_i \binom{n_{i.}}{2} \sum_j \binom{n_{.j}}{2} \right] \setminus \binom{n}{2}} . \tag{12}$$

where $n_{i,j}$ is the number of vertices in both cluster x_i and y_j, $n_{i.}$ and $n_{.j}$ is the number of vertices in cluster x_i and y_j respectively.

Political books. This network about books of US politics is compiled by Valdis Kreb [15]. The vertices represent books sold by `Amazon.com`. The edges represent frequent co-purchasing of books by the same buyers. The vertices have been marked as "liberal", "neutral", or "conservative" by Mark Newman [13] based on the descriptions and reviews of the books. The network is illustrated in Fig. 4 (a). The "conservative", "neutral" and "liberal" books are labeled as blue, gray and red respectively.

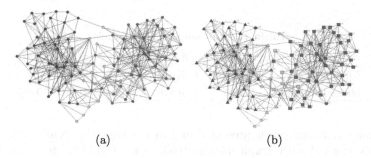

(a)   (b)

Fig. 4. (a) The Network of Books about US Politics, (b) The Result of GRACE on the Network of Political Books

The result of GRACE is shown in Fig. 4 (b). Square, circle and triangle represent "conservative", "neutral" and "liberal" books that our algorithm identifies respectively. We set parameter $\sigma = 10^{-2.5}$, $\xi = 35$ and ARI of the result is 0.678. GRACE finds almost all correct members for "conservative" and "liberal" books, while for "neutral" books, GRACE identifies some of them as hubs because of the inter-connection with two major clusters.

US college football. The network in Fig. 5 (a) is a representation of the schedule of Division I games for the 2000 season. The vertices represent football teams and the edges represent regular-season games between the two teams they connect. This dataset is compiled by M. Girvan and M. Newman [13]. In the network, the teams are divided into conferences containing around 5-13 teams each. There are totally 12 clusters in this network and we label them with different colors.

Applying GRACE on this network, we surprisingly find that 7 clusters out of 12 completely correspond to the real network. Furthermore, it identifies a few independent teams that do not belong to any conference as hubs because they always play games with teams of different conferences. The parameter setting in this case is that $\sigma = 10^{-3}$, $\xi = 8$ and ARI of our result is 0.816. The clustering result is labeled with different shapes in Fig. 5 (b).

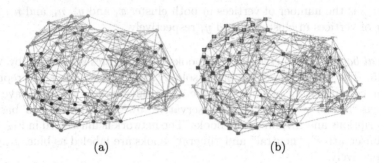

(a)   (b)

Fig. 5. (a) The Network of US College Football, (b) The Result of GRACE on the Network of US College Football

Table 1. ARI Comparative Analysis

	GD+GRACE	GD+Cut	EB+Cut	GD+DB	SCAN
Political Books	**0.678**	0.660	0.677	0.660	0.465
College Football	**0.816**	0.816	0.749	0.816	0.667

A comparative analysis is presented in Table 1. We adopt "cut" method and DBSCAN method with graph distance (GD+Cut, GD+DB) to compare our structural metric with that of edge betweenness cut (EB+Cut) and SCAN. The results show that graph distance generally outperforms edge betweenness and structural similarity proposed in SCAN [18]. Then we compare GRACE with GD+Cut and GD+DB to evaluate our algorithm. The results demonstrate that GRACE performs better than the other methods, espacially in the case of college football, the clustering results are exactly the same as GD+Cut and GD+DB, which validate the generality of our algorithm. All the experiments are conducted in the same environment of our own.

6   Concluding Remarks

Network clustering is a promising topic in mining structured data. This paper proposes a structural metric "Graph Distance" and a clustering algorithm GRACE. The graph distance combines two structural metrics for networks and improves the quality of clustering. The GRACE algorithm builds up the global influence function of the entire network and finds clusters effectively. Experimental results show that GRACE performs better than edge betweenness and SCAN in some scenarios. Depending on the parameter settings, the algorithm provides similar results as locality-based and hierarchical clustering algorithm.

Acknowledgement

This work was supported by National Nature Science Funding of China under Grant No. 90718022.

References

1. Barabási, A.L., Albert, R.: Emergence of Scaling in Random Networks. Science 286, 509–512 (1999)
2. Brandes, U.: A Faster Algorithm for Betweenness Centrality. J. Mathematical Sociology 25, 163–177 (2001)
3. Ding, C., He, X., Zha, H., Gu, M., Simon, H.: A Min-max Cut Algorithm for Graph Partitioning and Data Clustering. In: Proc. of ICDM 2001, pp. 107–114 (2001)
4. Ester, M., Kriegel, H.P., Sander, J., Xu, X.: A Density-based Algorithm for Discovering Clusters in Large Spatial Databases with Noise. In: Proc. of 3rd KDD, pp. 226–231. AAAI Press, Menlo Park (1996)
5. Floyd, R.W.: Algorithm 97 (SHORTEST PATH). Communications of the ACM 5, 345 (1962)
6. Freeman, L.C.: A Set of Measures of Centrality based upon Betweeness. Sociometry 40, 35–41 (1977)
7. Girvan, M., Newman, M.E.J.: Community Structure in Social and Biological Networks. PNAS 99, 7821–7826 (2002)
8. Guimera, R., Amaral, L.A.N.: Functional Cartography of Complex Metabolic Networks. Nature 433, 895–900 (2005)
9. Hinneburg, A., Keim, D.A.: An Efficient Approach to Clustering in Large Multimedia Databases with Noise. In: Proc. KDD 1998, pp. 58–65 (1998)
10. Hubert, L., Arabie, P.: Comparing Partitions. J. Classification. 2, 193–218 (1985)
11. Hwang, W., Kim, T., Ramanathan, M., Zhang, A.: Bridging Centrality: Graph Mining from Element Level to Group Level. In: Proc. of SIGKDD 2008, pp. 336–344. ACM Press, New York (2008)
12. Leskovec, J., Kleinberg, J., Faloutsos, C.: Graphs over Time: Densification Laws, Shrinking Diameters and Possible Explanations. In: Proc. of 11th ACM SIGKDD, pp. 177–187. ACM Press, New York (2005)
13. Network data, http://www-personal.umich.edu/~mejn/netdata
14. Newman, M.E.J., Girvan, M.: Finding and Evaluating Community Structure in Networks. Phys. Rev. E 69, 26113 (2004)
15. Social Network Analysis Software and Services for Organizations, Communities, and their Consultants, http://www.orgnet.com
16. Wasserman, S., Faust, K.: Social Network Analysis. Cambridge Univ. Press, Cambridge (1994)
17. Watts, D.J., Strogatz, S.H.: Collective Dynamics of 'Small World' Networks. Nature 393, 440–442 (1998)
18. Xu, X., Yuruk, N., Feng, Z., Schweiger, T.A.J.: SCAN: a Structural Clustering Algorithm for Networks. In: Proc. of SIGKDD 2007, pp. 824–833. ACM Press, New York (2007)

Constructing Stochastic Mixture Policies for Episodic Multiobjective Reinforcement Learning Tasks

Peter Vamplew, Richard Dazeley, Ewan Barker, and Andrei Kelarev

Graduate School of Information Technology and Mathematical Sciences,
University of Ballarat, University Drive, Mount Helen, Victoria 3353, Australia
{p.vamplew,r.dazeley,e.barker,a.kelarev}@ballarat.edu.au

Abstract. Multiobjective reinforcement learning algorithms extend reinforcement learning techniques to problems with multiple conflicting objectives. This paper discusses the advantages gained from applying stochastic policies to multiobjective tasks and examines a particular form of stochastic policy known as a mixture policy. Two methods are proposed for deriving mixture policies for episodic multiobjective tasks from deterministic base policies found via scalarised reinforcement learning. It is shown that these approaches are an efficient means of identifying solutions which offer a superior match to the user's preferences than can be achieved by methods based strictly on deterministic policies.

Keywords: multiobjective, reinforcement learning, scalarisation, Pareto fronts.

1 Introduction

The vast majority of reinforcement learning (RL) algorithms deal with tasks involving maximising performance on a single objective, as encoded in the scalar reward received from the environment. Whilst many problems can naturally be described by this model, over recent years there has been growing recognition within the optimisation community that many real-world problems require the optimisation of multiple, often conflicting, objectives [1]. This observation is equally true for RL tasks – for example, [2] applied reinforcement learning to simultaneously manage the power consumption and performance of Web application servers. Recently there has been some interest in extending existing single-objective RL methods to handle multiobjective tasks. However one issue which has been largely overlooked in the extension of these approaches is the potential benefits to be gained in the multiobjective domain by moving from deterministic to probabilistic policies.

Section 2 of this paper briefly reviews existing approaches to multiobjective reinforcement learning (MORL). Section 3 discusses the relationship between multiobjective tasks and probabilistic policies, presenting an example to motivate further investigation. Section 4 explores one specific type of probabilistic policy (the mixture policy), and Section 5 investigates means by which mixture policies can be generated from deterministic policies. Finally Section 6 offers suggestions for future directions for research in learning mixture policies for multiobjective tasks.

A. Nicholson and X. Li (Eds.): AI 2009, LNAI 5866, pp. 340–349, 2009.

2   Multiobjective Reinforcement Learning

Before reviewing existing approaches to MORL, it is instructive to examine some fundamental concepts underlying the analysis of multiobjective tasks. Inherently the task in any multiobjective situation is to identify solution(s) which represent a 'good' compromise amongst the objectives. This is often defined via Pareto dominance which allows comparison of a pair of solutions, as shown in Fig 1[1]. A solution dominates another if it is superior on at least one objective, and at least equal on all other objectives. Two solutions are incomparable if each is superior to the other on at least one objective. Any dominated solution is of little value, as clearly the dominating solution is preferable. Therefore the best solutions in a set can be extracted by retaining only those which either dominate or are incomparable with every other member of the set. If this process is applied to the set of all solutions, the resulting set of non-dominated solutions is referred to as the Pareto optimal front (or the Pareto front), and represents the globally optimal set of compromise solutions (see Fig 2). Of course establishing the true front for any problem of significant size is generally impractical, and so the goal of many multiobjective problem-solvers is to produce an approximation to this front. A good approximate front should contain solutions which are accurate (close to the actual front) and well distributed along the front, with a similar extent to that of the actual front.

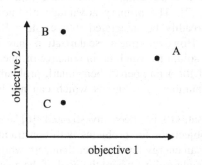

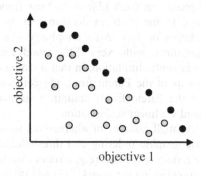

Fig. 1. Solutions A and B dominate solution C; solutions A and B are incomparable to each other

Fig. 2. The black points indicate solutions which form the Pareto front; all grey solutions are dominated by at least one member of the Pareto front

There are several advantages to searching for a set of compromise solutions rather than attempting to find a single 'optimal' solution. Methods which aim for a single solution require *a priori* decisions from the user about the desired nature of that solution (e.g. specifying weights or thresholds for objectives). This requires domain knowledge on the part of the user, and minor variations in these preferences may

---

[1] In multiobjective optimisation, the task is generally to minimise each objective, so a lower value for an objective is superior to a higher value. In contrast in RL the task is to maximise the reward received, and so the notion of superiority is reversed. Given the expected audience for this paper, we chose to frame our discussion in terms of maximisation.

result in significant variations in the solution achieved, which can easily lead to the acceptance of an inferior solution. For example, a slightly higher threshold for one objective may prevent discovery of a solution which provides a significant improvement on all other objectives. Systems which produce sets of solutions allow *a posteriori* decisions about the solution to be accepted, which are easier and better informed as they are based on knowledge of the trade-offs available as encapsulated by the front. Also the presentation of the front to the user may provide better insight into the relationships between the objectives. The primary disadvantage of generating multiple solutions rather than a single solution is the increased computational cost. In addition in the case of on-line learning in a real environment, the losses incurred during the extended learning period may prohibit searching for the complete Pareto set of policies.

The most straightforward and most common means of extending existing RL algorithms to multiobjective problems is to convert the problems into single-objective tasks. The key difference between single-objective and multiobjective RL is that in the former the reward is scalar, whereas in the latter it is a vector, with an element for each objective. Therefore a multiobjective task can be reduced to a single objective via the process of scalarisation, where a function is applied to the reward vector to produce a single, scalar reward. Most commonly this is a linear weighted sum of the individual objective rewards (e.g. [3, 4]). The choice of weights allows the user some control over the nature of the policy found by the system, by placing greater or lesser emphasis on each objective. Less frequently a more complex, non-linear function tuned to the problem domain may be used [2]. The primary advantage of linear scalarisation lies in its simplicity – it can readily be integrated into existing RL algorithms with very little modification. However linear scalarisation has a fundamental limitation, in that it can not find solutions which lie in concave or linear regions of the Pareto front. [5] demonstrated for a number of benchmark problems that the Pareto fronts contain a substantial number of solutions which can not be found via linear scalarisation.

A small number of alternatives to scalarisation have been investigated. [6] used lexicographic ordering and thresholding of objectives for problems with constraints for certain objectives (e.g. a robot maintaining an energy level greater than zero whilst accomplishing some task). [7] and [8] describe algorithms where the goal of the agent is to achieve long-term average rewards which lie in an externally defined 'target' region in objective space. These algorithms produce non-stationary policies in which the actions taken by the agent at any point in time are determined both by the current state and by the position of the current average reward vector relative to the target set. [9] developed a policy-gradient MORL algorithm. The algorithm starts from a policy derived by applying RL independently to each objective. This policy is then improved by following gradients in the policy space which are non-negative with regards to all objectives. An approximate Pareto front is constructed by performing repeated searches with different weightings of the gradient directions.

3 Multiobjective Tasks and Stochastic Policies

With the exception of [9], the majority of MORL work has so far focused on finding deterministic policies (policies where the same action is always selected in any given state, other than when an exploratory action is chosen during learning). This is not surprising as most of these techniques are grounded in single-objective RL algorithms, and in the single-objective case there is little reason to consider stochastic policies as "for any MDP, there exists a stationary, deterministic optimal policy" [9, p288]. However this property of stationary, deterministic policies does not hold true when we consider problems with more than one objective, as illustrated by the simple environment in Fig 3. This environment consists of a single state, with two actions. Both actions lead back to the same state, but their associated rewards correspond to different objectives. Clearly there are only two deterministic policies available: always choosing action a_1 which will maximise the reward on objective 1 but minimise the reward for objective 2, or always choosing action a_2 which will maximise the reward on objective 2 but minimise the reward for objective 1. In any task in which we are considering multiple objectives, inherently we are seeking a solution which is trade-off between these objectives, but in this case neither of these deterministic policies offers any degree of compromise between the two objectives. In contrast, consider a stochastic policy which selects between actions a_1 and a_2 with probabilities p_1 and $(1-p_1)$ respectively. Clearly the average reward received by this policy will be $(p_1, 1-p_1)$. By varying the probability with which each action is selected, a range of policies which offer different compromises between the two objectives can be achieved. For more complex problems usually deterministic policies will exist which do in fact offer a compromise between the different objectives. However these policies will represent discrete, possibly widely-spaced, points in objective-space whereas stochastic policies offer a continuous range of trade-offs between the different objectives. Therefore it likely that a policy which better matches the user's preferences will be found if stochastic policies are considered.

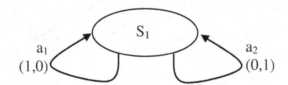

Fig. 3. A single-state environment with 2 actions. Performing action a_1 receives a vector reward of $(1,0)$; performing action a_2 receives a vector reward of $(0,1)$.

4 Mixture Policies for Multiobjective Tasks

Having established the potential benefits of stochastic policies for multiobjective tasks, we need to consider how such policies may be found. As noted earlier, many single-objective RL methods (such as the widely used temporal difference approaches such as Q-learning) do not support stochastic policies. Methods such as policy gradient learning and learning automata can learn the probabilities with which each action should be selected in each state of a stochastic policy, and [9] has pioneered the use of policy gradient methods in finding stochastic multiobjective policies.

Rather than considering more sophisticated approaches, this paper will examine means by which simple methods such as Q-learning and scalarisation can be used to find stochastic policies for episodic multiobjective tasks. [9, p59] describes a special form of stochastic policy known as a mixture policy[2], which is derived from two or more deterministic policies (which we will refer to as base policies). At the start of each episode the mixture policy stochastically selects one of its base policies, which is then followed for the remainder of that episode. Over a large number of episodes, the vector return achieved by this mixture policy will be the mean of that achieved by its base policies, weighted by the probability with which each base policy is selected[3]. In [9], the only base policies considered are those maximising each individual objective, and the mixture policies generated from these base policies are used as the starting point for policy gradient search. Here we consider a more general application of mixture policies in which any set of base policies may be used, and in which the mixture policies themselves are the final outcome of the system.

Consider the generation of mixture policies from a set of deterministic policies. Fig 4 shows the complete set of Pareto optimal deterministic policies for the Deep Sea Treasure task [5]. The line joining policies A and B represents the mixture policies which are derived from that pair of base policies by varying the probability with which the policies are selected. It can be seen that any deterministic policy lying in a concave region of the Pareto front (e.g. policy C) will be dominated by one or more mixture policies derived from policies outside the concave region.

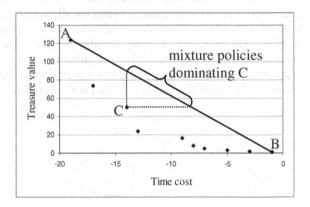

Fig. 4. The Deep Sea Treasure task's objective-space. Points are Pareto-optimal deterministic policies. The line *AB* represents mixture policies derived from those base policies.

The Deep Sea Treasure task is unusual in that only the extremal policies are not in a concave portion of the front. Fig 5 illustrates a more generally representative front – the line segments indicate possible mixture policies which can be generated from this front. It can be seen that as the mixture policies are formed via convex combinations[4]

---

[2] Note that mixture policies are not fully stochastic policies in which actions are chosen stochastically at each state – rather they are a stochastic combination of deterministic policies.

[3] It is important to note that this is only true because the choice between policies A and B is made at the start of each episode. Switching between policies at other time-steps would likely result in erratic and sub-optimal behaviour.

[4] A convex combination is a linear combination of vectors, in which the weights sum to 1.

of the base policies, the non-dominated mixture policies constructed by this process will form the convex hull of the original base-policy points in objective space [11]. All points (deterministic policies) from the original front which are not on this hull will be dominated once the mixture policies are considered. Similarly mixture policies derived from policies which are not neighbouring members of the convex hull will also be dominated by at least one other mixture policy.

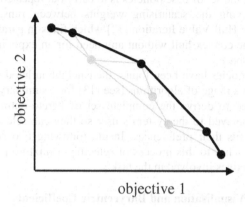

Fig. 5. A hypothetical Pareto front. Points indicate Pareto-optimal deterministic policies (black points indicate policies on the convex hull of the front). Lines indicate mixture policies generated from pairs of deterministic base policies (some combinations of base policies have been omitted for reasons of clarity).

In summary mixture policies provide two potential benefits for episodic tasks when compared to deterministic policies. First they provide a continuous range of trade-offs between the objectives as opposed to the discrete set of trade-offs embodied by the deterministic policies. Hence they are likely to provide a more precise match to the preferences of the decision maker. Second for problems where the Pareto set of deterministic policies contains concave regions (which was shown to be the case for all benchmark problems examined by [5]) mixture policies exist which dominate (in some cases by a significant margin) some otherwise non-dominated deterministic policies.

5   Selecting and Constructing Mixture Policies

In light of the benefits outlined in Section 4, we propose the following general approach to finding suitable policies for episodic multiobjective tasks:

- generate a set of Pareto-optimal deterministic policies
- use these policies as base policies to derive a set of mixture policies
- select a mixture policy which is appropriate to the preferences of the decision-maker who is using this MORL system

As evident in Fig 5, a mixture policy will be non-dominated if and only if it is formed from base policies which are neighbouring points on the convex hull of the Pareto front. Therefore the set of base policies need consist only of Pareto-optimal policies which are not in concave regions of the Pareto front – any computation expended in finding policies in concavities is wasted as they will not be used by any non-dominated mixture policy. Interestingly in this context the inability of scalarised MORL to discover policies other than those on the convex hull (see Section 2) may in fact make it more efficient than other methods which can find such policies. A simple approach to finding the set of base policies is to carry out repeated runs of a scalarised RL algorithm, varying the scalarising weights between runs. A more efficient approach is Convex Hull Value Iteration [12] which finds in parallel all deterministic policies lying on the convex hull without any need for an explicit search through the scalarising weight space.

Once the base policies have been found, their neighbourhood relationships can be established through a range of algorithms (see [13] for a summary). This provides the information required to derive the complete set of Pareto-optimal mixture polices. This set is then displayed to the system's user so they can select the single mixture policy which best fits their preferences. In the following two sub-sections we will describe two approaches to this process of selecting a mixture policy, depending on the number of objectives involved in the task.

5.1 Convex Hull Visualisation and Barycentric Coefficients

For problems with a low dimensionality (two or three objectives) the set of mixture policies can be directly displayed to the user via 2-dimensional, 3-dimensional or stereo graphics. This provides the user with a clear depiction of the relationships between the objectives, and allows them to make an informed choice of the best available policy for their needs. The choice of policy can be indicated by selecting a point anywhere on the surface of the hull. Once the user selects a point in objective-space it is simply a matter of determining the probabilistic weightings of base policies required to construct a mixture policy to achieve that combination of rewards. This can be done by calculating the barycentric coordinates of the target point, and using these as the probability of selection for each base policy. Barycentric coordinates are coordinates defined in terms of the vertices of a simplex (in our case, the points in objective space corresponding to the base policies, which we will label as $V_1, V_2, .., V_n$). If V_0 is the target point (the objective-space position of our desired mixture policy) and $V_{i,j}$ denotes the value of a point V_i for objective j, then the barycentric coordinates are values $b_1..b_n$ such that the following equality holds for all $j=1,...,n$:

$$V_{0,j} = \sum_{i=1}^{n} b_i V_{i,j}$$

If V_0 lies within the simplex defined by $V_1,...,V_n$ (which is always true in our case due to the manner in which V_0 is specified by the user) then the following property holds, and therefore the barycentric coefficient b_i can be directly interpreted as the probability with which base policy i will be selected at the start of each episode:

$$\sum_{i=1}^{n} b_i = 1$$

For a problem with two objectives, each mixture policy lies on the line-segment bounded by the base policy points V_1 and V_2, as shown in Fig 6. In this case the barycentric coefficients can easily be calculated from the ratios of the line-segments V_1V_0 and V_2V_0:

$$b_i = \frac{\|V_i - V_0\|}{\|V_2 - V_1\|}$$

For a problem with three objectives, V_0 will lie within the bounds of a planar triangle defined by V_1, V_2 and V_3. In this case the value of the coefficient for each vertex can be calculated based on the percentage of the area of this triangle which is occupied by the sub-triangle formed by V_0 and the other two vertices (see Fig 7). The area calculations can be efficiently implemented using vector cross-products and length operations as follows:

$$A_1 = \|(V_2 - V_0) \times (V_3 - V_0)\|$$
$$A_2 = \|(V_1 - V_0) \times (V_3 - V_0)\|$$
$$A_3 = \|(V_1 - V_0) \times (V_2 - V_0)\|$$

$$b_i = \frac{A_i}{\sum_{j=1}^{3} A_j}$$

This barycentric approach to constructing mixture policies can be extended to higher dimensions. However as the number of objectives rises beyond three, direct visualisation of the front becomes problematic[5] and the computational cost of establishing hull geometry and calculating the barycentric coefficients increases. Therefore Section 5.2 discusses an interactive approach which does not require visualisation of the hull.

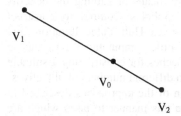

Fig. 6. For the two objective case, each mixture policy V_0 lies on the line-segment formed by the base policies V_1 and V_2

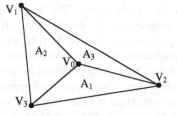

Fig. 7. For the three objective case, each mixture policy V_0 lies in the triangle formed by the base policies V_1, V_2 and V_3, and the barycentric coefficients can be calculated from the relative areas of the sub-triangles A_1, A_2 and A_3

---

[5] Although visualisation of high-dimensional Pareto fronts has been explored in the multiobjective optimisation community – see for example [14].

5.2 Interactive Construction of a Mixture Policy

Where direct selection of mixture policies from a visualisation of the hull is not practical (i.e. where the number of objectives exceeds three), an alternative approach is to allow the user to directly select the base policies and set their probabilities so as to form a suitable mixture policy. This can be achieved via the following process:

1. A complete list of possible base policies is presented to the user textually, augmented by lower-dimensional visualisations.
2. The user selects the base policy which most closely matches their preferences.
3. If the number of selected policies is less than the number of objectives, remove from the list all policies which are not neighbours on the hull of all selected policies and return to Step 2.
4. An initial mixture policy is constructed from an equal weighting of the selected base policies. The user manipulates the base policy probabilities using a linked set of sliders (as one probability is adjusted, the other sliders are adjusted in the opposite direction) whilst the reward vector for the current mixture policy is displayed. The user can explore the trade-offs available based on the currently selected base policies, before settling on a mixture policy, or returning to Step 1 to select a new set of base policies.

6 Conclusions and Future Work

This paper has demonstrated the utility of stochastic policies for multiobjective tasks. Stochastic policies offer a continuous range of solutions, as opposed to the discrete set of solutions offered by deterministic policies which may contain large gaps between neighbouring solutions. Hence it is more likely that a policy closely matching the user's preferences will be discovered if stochastic policies are allowed. In addition it has been shown that for some problems stochastic policies can be superior (in the sense of Pareto dominance) to the best deterministic policies. We have shown that for episodic tasks, mixture policies offer an inexpensive means of gaining the benefits afforded by stochasticity. The base deterministic policies required to construct mixture policies can be found efficiently using Convex Hull Value Iteration, and mixture policies can then be derived with relatively little computational cost and no further interaction with the environment. Two approaches for constructing a suitable mixture policy have been proposed for problems with different numbers of objectives.

It is important to note one fundamental limitation of the approaches described in this paper – mixture policies can only be applied in this manner to tasks which are known to be episodic. The start of a new episode is used as a trigger for stochastically selecting a base policy to follow – switching between base policies at any other time would likely lead to erratic, sub-optimal performance by the agent, such as oscillating between two locations. The benefits of stochastic policies still apply to non-episodic tasks however, and so an important direction for future research is to examine whether suitable switching states can be identified for mixture policies for such tasks. A second possible limitation is that for some tasks consistency of behaviour may itself be a desirable feature, and therefore stochastic policies may be unacceptable. To handle these situations there is still a need for MORL systems which can identify all Pareto-optimal deterministic policies, not just those on the convex hull.

Acknowledgements

We wish to acknowledge the insight into the relationship between Pareto fronts and online learning provided by Dr Peter Andreae from Victoria University of Wellington, and the useful feedback provided by this paper's referees.

References

1. Coello Coello, C.A.: Handling Preferences in Evolutionary Multiobjective Optimization: A Survey. In: 2000 Congress on Evolutionary Computation, vol. 1, pp. 30–37 (2000)
2. Tesauro, G., Das, R., Chan, H., Kephart, J.O., Lefurgy, C., Levine, D.W., Rawson, F.: Managing power consumption and performance of computing systems using reinforcement learning. In: Neural Information Processing Systems (2007)
3. Natarajan, S., Tadepalli, P.: Dynamic preferences in multi-criteria reinforcement learning. In: International Conference on Machine Learning, Bonn, Germany, pp. 601–608 (2005)
4. Castelletti, A., Corani, G., Rizzolli, A., Soncinie-Sessa, R., Weber, E.: Reinforcement learning in the operational management of a water system. In: IFAC Workshop on Modeling and Control in Environmental Issues, Keio University, Yokohama, Japan, pp. 325–330 (2002)
5. Vamplew, P., Yearwood, J., Dazeley, R., Berry, A.: On the Limitations of Scalarisation for Multiobjective Learning of Pareto Fronts. In: Wobcke, W., Zhang, M. (eds.) AI 2008. LNCS (LNAI), vol. 5360, pp. 372–378. Springer, Heidelberg (2008)
6. Gabor, Z., Kalmar, Z., Szepesvari, C.: Multi-criteria reinforcement learning. In: The Fifteenth International Conference on Machine Learning, pp. 197–205 (1998)
7. Mannor, S., Shimkin, N.: The steering approach for multi-criteria reinforcement learning. In: Neural Information Processing Systems, Vancouver, Canada, pp. 1563–1570 (2001)
8. Mannor, S., Shimkin, N.: A geometric approach to multi-criterion reinforcement learning. Journal of Machine Learning Research 5, 325–360 (2004)
9. Shelton, C.R.: Importance sampling for reinforcement learning with multiple objectives, Massachusetts Institute of Technology AI Laboratory Tech Report No. 2001-003 (2001)
10. Mahadevan, S., Ghavamzadeh, M., Theocharous, G., Rohanimanesh, K.: Hierarchical Approaches to Concurrency, Multiagency, and Partial Observability. In: Si, J., Barto, A., Powell, W., Wunsch, D. (eds.) Handbook of Learning and Adaptive Dynamic Programming, pp. 285–310. Wiley-IEEE (2004)
11. Kelley, J.L., Namioka, I.: Linear topological spaces. Graduate Texts in Mathematics, vol. 36. Springer, Heidelberg (1976)
12. Barrett, L., Narayanan, S.: Learning All Optimal Policies with Multiple Criteria. In: Proceedings of the International Conference on Machine Learning (2008)
13. Seidel, R.: Convex Hull Computations. In: Goodman, J.E., O'Rourke, J. (eds.) Handbook of Discrete and Computational Geometry, pp. 361–376. CRC Press, Boca Raton (1997)
14. Agrawal, G., Lewis, K., Chugh, K., Huang, C.-H., Parashar, S., Bloebaum, C.L.: Intuitive Visualization of Pareto Frontier for Multi-Objective Optimization in n-Dimensional Performance Space. In: 10th AIAA/ISSMO Multidisciplinary Analysis and Optimization Conference, Albany, NY (2004)

Enhancing MML Clustering Using Context Data with Climate Applications

Gerhard Visser, David L. Dowe, and Petteri Uotila

Monash University, VIC 3800, Melbourne Australia
gerhardus.visser@infotech.monash.edu.au

Abstract. In Minimum Message Length (MML) clustering (unsupervised classification, mixture modelling) the aim is to infer a set of classes that best explains the observed data items. There are cases where parts of the observed data do not need to be explained by the inferred classes but can be used to improve the inference and resulting predictions. Our main contribution is to provide a simple and flexible way of using such context data in MML clustering. This is done by replacing the traditional mixing proportion vector with a new context matrix. We show how our method can be used to give evidence regarding the presence of apparent long-term trends in climate-related atmospheric pressure records. Akaike Information Criterion (AIC) and Bayesian Information Criterion (BIC) solutions for our model have also been implemented to compare with the MML solution.

1 Introduction

1.1 Minimum Message Length

The Minimum Message Length (MML) [16, 17, 21] principle states that the best explanation for observed data D is the one that minimises the optimal coding length (according to information theory) of a two-part message. The first part encodes the hypothesis H (this is known as the assertion) from Bayesian priors while the second part encodes the data D given the H (this is known as the detail). In practice, we do not actually construct any message but rather strive to infer a hypothesis which minimises some approximation to that code length.

The MML principle can be thought of as a quantitative version of Ockham's razor [5, footnotes 18 and 181-182] and is compared to Kolmogorov complexity and algorithmic complexity [2, 12, 15] in [19]. For a contrast with the much later Minimum Description Length (MDL) principle [14], see [4, sec. 11.4] and [16, chap. 10]. MML inference is statistically invariant (inference is preserved under 1-to-1 transformations of the parameter space) and is in general statistically consistent [5, 6, 16]. Where many methods have been shown to be statistically inconsistent on misspecified models [9], there is as yet no known example of MML having this failing [9, sec. 7.1.5][5, sec. 0.2.5].

MML is capable of selecting between models with varying numbers of parameters without overfitting [21, sec. 6], and outperforms Maximum Likelihood (ML) even when aided by Akaike's Information Criterion (AIC) [6].

A. Nicholson and X. Li (Eds.): AI 2009, LNAI 5866, pp. 350–359, 2009.

MML is a general model selection criterion and as such is intended to replace traditional hypothesis tests and confidence intervals. Instead, message lengths are compared for different hypotheses. Akaike's Information Criterion (AIC) and the Bayesian Information Criterion (BIC) are two comparable and popular model selection criteria. We have implemented both AIC and BIC versions of our method for comparison with the MML solution (section 3).

1.2   Clustering with Minimum Message Length

Given a set of observed items $y = (y_1, y_2, ..., y_N)$ the aim of clustering is to find a set of C classes such that each item can be assigned to a class. The number, C, is often assumed known and the properties of the classes are inferred from the data. These inferred properties of a class describe its typical items.

MML clustering as described in [17, 18, 20] and [16, sec. 6.8] is an unsupervised mixture modelling method which will also select the number of classes present.

The Expectation Maximisation (EM) algorithm is used to infer the class parameters for a fixed number of classes. This is repeated assuming different numbers of classes. For each such EM run a message length (section 1.1) is calculated. The solution with the smallest message length is selected as the best.

Given an inferred hypothesis, the message length is calculated as the length of an optimal code described as follows.

1. The *Assertion* encodes the hypothesis in the following order.
   (a) The number of classes used, C.
   (b) The relative frequencies of all classes.
   (c) The inferred parameters defining each class.
   (d) The partial assignments of items to classes.
2. The *Detail* encodes the data given the hypotheses.
   (a) The observed attributes of each item.

1.3   Our Extension and Some Motivations

Since this work was developed with atmospheric time-series data in mind we will use that as an example throughout this paper but our methods are intended to be general purpose.

MML clustering attempts to capture all regularities that are present in the data. This means that in practice as one adds more attributes to each data item, the number of classes inferred tends to increase. As there is more data to compress the first part of the message can become larger and more complex. Too many classes can be hard to interpret - which in some cases may be undesirable.

In our data set each data item y_i is a set of atmospheric pressure values from several weather stations for a single day. These pressure values $y_{i,j}$ (where i indices a day and j indexes a weather station) are the attributes that we wish to cluster. There may be other attributes that can be associated with each day (data item) which might help with the clustering but which we do not wish to model or explain with the classes inferred. Examples include seasons, extreme weather conditions and global indexes such as those relating to the El Niño cycle.

It helps to notice that with clustering there are often two types of attributes. One can think of them as target attributes and context (known) attributes. We are not interested in discovering regularities in the context attributes and it is not desirable that the number of classes and the complexity of the classes increase to explain those regularities. On the other hand the inferred hypothesis should mention these attributes if they help explain the target attributes.

Our aim is to explain the target attributes while using the context attributes to discriminate and (this aim) is therefore similar to what Jebara [10] describes as combining Discriminative and Generative learning.

Our work provides a simple yet flexible way of dealing with these context attributes differently from target attributes while adhering to the well established MML clustering framework of [17, 20] and [16, sec. 6.8]. By doing that our method inherits the features of MML clustering which has made it successful which includes the ability to select the number of classes without over-fitting.

2   Methods

2.1   A Clustering Model with Context Data

Let y be a set of observed items where $y_{i,j}$ is the value of attribute j for data item i. Let x be the corresponding class assignments where $x_i \in \{1, 2, ..., C\}$ and C is the number of classes. In our atmospheric time-series example $y_{i,j}$ is the measurement on day i at weather station j.

There are other context attributes associated with each day that we can use to improve the clustering but do not wish to model. For our climate example this could include time of year (season) or global indices such as those relating to the El Niño cycle. For this we introduce a context value $z_{i,k}$ where i indexes the item (day) and $k \in \{1, 2, ..., K\}$. Here there are K different *contexts*. Each item i belongs to each context to some degree $z_{i,k}$. The context data z is given as prior knowledge. Each context vector z_i is used much like a fuzzy indicator, however, we interpret them strictly as probability distributions, hence we require that for all i, $\sum_{k=1}^{k=K} z_{i,k} = 1$ and that all $z_{i,k} \geq 0$.

As an example we can divide the days of each year into four seasons, so $K = 4$. A day in the middle of summer (context $k = 1$) can be assigned completely to that season $z_{i,1} = 1$ while a day between summer and autumn can be assigned partially to those two seasons $z_{i,1} = 0.5, z_{i,2} = 0.5$.

In our model we replace the mixing proportion parameter vector with a $K \times C$ mixing proportion matrix S. Now the probability of item i belonging to class c is defined as,

$$\Pr(x_i = c) = \sum_{k=1}^{K} z_{i,k} S_{k,c}. \tag{1}$$

Each of the K rows of matrix S is a relative frequency vector associated with a context. In our example $S_{k,c}$ is the probability of day i belonging to class c if the season is $z_{i,k} = 1$. It follows that $\sum_{c=1}^{C} S_{k,c} = 1$ and all $S_{k,c} \geq 0$. This mixing proportion matrix S will be inferred from the data. In our season example this means that the effective mixing proportions will change gradually according to time of year and we avoid having to use hard boundaries when specifying z.

The matrix S allows the context vector z to be used to provide information about the class assignments x prior to seeing the data y. This means x can be encoded more efficiently given S and z but only if that saving is not outweighed by the cost of stating S, which increases with K and C. Effectively S allows z to inform the classification and inferred model. We are not encoding z at all, one could imagine a separate message fragment encoding z preceding the rest of the message. This imaginary message fragment would be unaffected by y, x, S, C and the class parameters. The idea is that how z is modelled or encoded does not affect the rest of the message.

Each class defines a distribution $\Pr(y_i|x_i)$ for the data items assigned to it. These distributions have parameters associated with them which must be inferred. For our climate example we will consider each weather station to have an independent Gaussian distribution. For details on how these parameters are inferred with MML, for this and other distributions, see [16, 18, 20].

2.2 Coding Approximation and Optimisation Algorithm

In MML inference one usually creates an approximation to the message length of the two-part code described in section 1.1, and then infers a hypothesis which optimises that approximation. We first describe the form of the hypothetical message, then how it is approximated and then the optimisation algorithm. Given a hypothesis the message is made up of the same message fragments as in the list given at the end of section 1.2. For our model part 1b of that list states the matrix S instead of a single mixing proportion vector.

In accordance with MML convention our message lengths are calculated in nits where 1 nit $= \log_2 e$ bits. For item 1a we use the prior distribution 2^{-C} over the number of classes, this message fragment has a length of $C \log_e 2$ nits.

For part 1b the rows of matrix S can be stated using a standard MML multi-state distribution solution (see [20]).

The code length for the class parameters (1c) can be approximated using standard MML solutions for the distributions used (see [20]). Because the order of the classes is arbitrary, $\log_e C!$ nits can be subtracted from this length.

For part 1d the coding length for stating each class assignment x_i precisely is the negative logarithm of the conditional probability $\Pr(x_i|z_i, S)$. Since an optimal code would not state these parameters (x) precisely, a coding trick (see [16, sec. 6.8]) can be used to calculate the message length improvement that can be achieved through imprecisely encoding x. The result of this is that one can subtract form the above described message length the entropy of x given everything else $(z, y, S$ and the class parameters).

Finally the length of the detail (part 2) is simply the negative log likelihood of y given the inferred assignments x and the inferred class parameters.

Because of these imprecise encodings of x one can interpret their assignments as partial (uncertain) and the expectations (over the partial assignments of x) of the code length described above (parts 1b, 1c, 1d and 2) is used.

Now that we have an approximation to the code length for a given hypothesis and data set, a search algorithm which finds an optimal hypothesis is needed. The Expectation Maximisation (EM) algorithm is used.

```
1: initialise partial assignments for x;
2: initialise values for S;
while(not(some termination condition))
{
  3: update class parameters to their optimal values given x and y;
  4: update partial assignments of x given S and y;
  5: update the matrix S given x and z;
}
```

Step 3 is done as with standard MML clustering, see [20] or [16, sec. 6.8]. In step 4 the optimal degree of assignment of item i to class c is equal to its posterior probability $\Pr(x_i = c | S, z, y)$. This type of estimate for discrete parameters like x is discussed in [16, sec. 6]. Step 5 uses the same multi-state distribution solution used in standard MML clustering for the rows of S, however the contribution of item i to the parameter row vector S_k is weighted according to,

$$w_{i,k} = \frac{z_{i,k} \sum_{c=1}^{C} S_{k,c} \Pr(y_i | x_i = c)}{\sum_{t=1}^{K} z_{i,t} \sum_{c=1}^{C} S_{t,c} \Pr(y_i | x_i = c)}. \tag{2}$$

These weights are also used in the message length calculations for S. The individual reassignments (steps 3, 4 and 5) each decrease the overall message length in every iteration and the result is that the solution as a whole moves to a local optimum.

3   Data and Results

3.1   Tests on Artificially Generated Data

Because MML is a Bayesian method, our first claim is that if a *true* hypothesis is generated from the assumed model then our method will on average tend to be good at inferring back that true hypothesis. The hypotheses that were generated for these tests were intended to roughly imitate those one would expect to infer for our atmospheric pressure data.

For the first test the true model has 5 classes with 5 pressure values generated for each day over a 50 year period. The context variable z has been used to divide the 50 years into 4 long term divisions. This simulates how the relative frequencies of our 5 classes change over the long run. Each day is assigned partially to two of these divisions so that the change in relative frequencies occurs slowly and smoothly over time (as with the seasons example in section 2.1).

Data was generated from the assumed hypothesis as described above. Half the attributes from this data were randomly removed as a validation set. From the training set our algorithm was used to infer the number of long term divisions while knowing the true number of classes. This test was repeated for ten such data sets. The results are summarised in figs. 1 and 2. Aside from the MML criterion we have also optimised the Akaike's Information Criterion (AIC) and Bayesian Information Criterion (BIC) for all results. In Fig. 1 the lines titled MML, AIC and BIC show the average resulting criterion values belonging to the left vertical axis. Here we can see all three criteria have their average optimal value at the correct number of divisions ($K = 4$). The predictive performance is measured as the average negative log likelihood of the validation set given the chosen hypotheses, divided by the validation set size. This measure is titled *score* in Fig. 1 and belongs to the right vertical axis. It can be seen that the predictive score reaches its optimal value for $K = 4$ and extra divisions do not improve this.

In Fig. 2 we can see the number of data sets (out of ten) for which MML, AIC and BIC preferred K divisions. The results show that MML and BIC tended to be similarly conservative while AIC sometimes prefers more divisions than the true number $K = 4$.

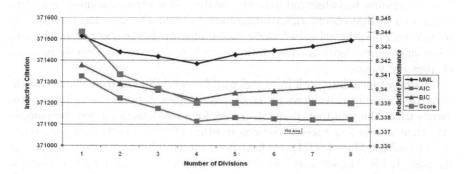

Fig. 1. Average MML, AIC and BIC values inferred for different values of K belong to the left axis. The predictive performance *score* belongs to the right axis. The true value is $K = 4$.

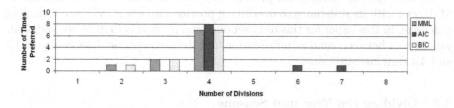

Fig. 2. The number of data sets (out of ten) for which MML, AIC and BIC preferred K divisions, with the true value being $K = 4$

In the next test we have generated data as with the first test but now the algorithm knows the number of long-term divisions ($K = 4$) while the number of classes ($C = 5$) must be inferred. For this test MML and BIC performed similarly and well (preferring either $C = 5$ or $C = 4$) while AIC tended to over-fit (preferring $6 \leq C \leq 8$).

In the final artificial data test we have generated data using only one long term division $K = 1$ (equivalent to no context variable). All three methods were used to infer K as before. Here both MML and BIC chose the correct number $K = 1$ all ten times while AIC chose $K = 1$ seven times but also made estimates as high as $K = 5$.

Our conclusion from these three tests is that both MML and BIC can be expected to either choose the true values for K and C or to choose more conservatively, while AIC will occasionally overestimate these values.

3.2   Atmospheric Time-Series Data and MML Clustering

The meteorological data was derived from historical sub-daily station mean sea level air pressure observations digitised by the Australian Bureau of Meteorology. The air pressure was observed in approximately 50 weather stations across Australia with earliest observations dating back to 1859. The data has been quality controlled. This processing included removal of errors in the observations by mistakes made when digitising observations or when observers incorrectly recorded air pressure values.

Clustering such data both from real world observations or from climate model output is valuable as it allows for large and complex data sets to be interpreted more easily. This can then be used to look for variations in pattern frequencies over time and to link these variations to other climate/weather related events. Self Organising Maps (SOM) [11] have been successfully used for this purpose in the past [1, 13]. The work we present in this paper is an early step in continuing work aimed at providing alternative tools to SOM and k-means clustering specific to atmospheric time-series data.

The existence of multiple atmospheric circulation regimes (classes) in the extratropics is an important, but a controversial, hypothesis in meteorology [3]. Many conflicting results exist and are critically discussed in [3].

We hope that by refining our probabilistic models to fit this problem domain, MML can with its resistance to overfitting provide important evidence regarding this issue. In this paper for this data set our primary goal is to measure and analyse the link between context information and the atmospheric data. Sections 3.3 and 3.4 demonstrate this.

3.3   Dividing the Year into Seasons

It is known that atmospheric pressure states are dependent on time of year (seasons). We demonstrate how our method can be used to determine into how many seasons each year can be divided. We test our method's ability to choose the best predictive model by dividing the data into training and validation sets and comparing predictive performance with the message lengths.

The data from 1865 to 1915 was used. Half the observed measurements were randomly removed as a validation set. We have assumed here that the number of classes is $C = 20$. The algorithm was used to infer the number of seasonal divisions K. Each day is assigned partially to two seasonal divisions as described in section 2.1. In this way we model how the relative frequencies of classes cycle smoothly over time. Fig. 3 compares the resulting MML message lengths and BIC and AIC values for different numbers of seasonal divisions K. For each value of K MML, BIC and AIC inference were repeated 10 times and the solution for which each criterion performed best was selected and is shown on Fig. 3.

For this test, both MML and BIC preferred 8 seasonal divisions while AIC preferred 16. It can be seen that there is no significant improvement on the validation set for more than 8 divisions.

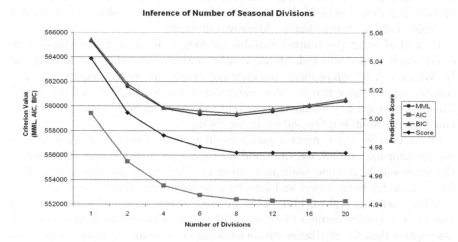

Fig. 3. MML, BIC and AIC values for different numbers of seasonal divisions K belong to the left axis. The *score* measures the performance of the validation set and belongs to the right axis.

3.4 Identifying Long-Term Trends in Atmospheric Time-Series Data

Finally we have used our method with the data from 1865 to 1965 to see how many long-term trends can be justified when assuming $C = 20$ classes. Again half the data was randomly removed as a validation set. The algorithm was used to infer the correct number of long-term divisions K as defined in section 3.1. For each value of K both MML and BIC inference were repeated 10 times and the solution for which each criterion performed best was selected. MML had a clear preference for 4 long term divisions while BIC preferred 7. Both the 4-term and 7-term solutions had the same predictive performance.

4   Conclusion and Further Work

In clustering there is often additional information that can be used to improve the inference but which should not be included as target attributes (attributes to be clustered). The context clustering method that we have presented provides a flexible yet simple extension to standard MML clustering which achieves our goal of using context data. Our results on artificial data show that we can detect the presence of such context divisions and estimate their number if the assumed model is correct. An implementation of our algorithm which uses BIC instead of MML performs similarly while AIC tends to overfit. With the atmospheric pressure time-series data we have demonstrated how our method can be used to give evidence regarding the presence of apparent long term trends in atmospheric pressure patterns and to determine the number of seasonal divisions that can be justified.

It is known for this data set that the class of each day is highly dependant on the class of the previous day and that this can be modelled using a hidden Markov unit model as in [7]. We are currently working on combining our context variable model with that hidden Markov unit model.

Instead of using the context variable for long term divisions or seasons one could use it to try and link global weather indexes, like those measuring the El Niño cycle, to atmospheric pressure patterns. This would require that the context variable have two possible assignments, one for El Niño and one for La Niña, where each day would be (partially) assigned to both with some degree based on the Southern Oscillation Index (SOI).

Other uses for the context variable could include weather extremes such as storms, unusual rainfall, cyclones and hurricanes. Another simple extension of this work will be to allow multiple context variables to be use, this would allow for example for both season and long term trend information to be used.

With clustering real world data the difference between model and reality can lead to an excessive number of classes. One way to address this is to remove the assumption that the attributes within each class can be modelled as independent Gaussian distributions. It should be possible to allow for inter-attribute relations such as latent factors, which have been used in MML clustering in [8].

References

1. Cassano, J.J., Uotila, P., Lynch, A.: Changes in synoptic weather patterns in the polar regions in the twentieth and twenty-first centuries, part 1: Arctic. International Journal of Climatology 26(8), 1027–1049 (2006)
2. Chaitin, G.J.: On the length of programs for computing finite binary sequences. Journal of the Association of Computing Machinery 13, 547–569 (1966)
3. Christainsen, B.: Atmospheric Circulation Regimes: Can Cluster Analysis Provide the Number? Climate Journal 20(10), 2229–2250 (2007)
4. Comley, J.W., Dowe, D.L.: Minimum message length and generalized Bayesian nets with asymmetric languages. In: Grünwald, P., Pitt, M.A., Myung, I.J. (eds.) Advances in Minimum Description Length: Theory and Applications, pp. 265–294. MIT Press, Cambridge (2005)

5. Dowe, D.L.: Foreword re C. S. Wallace. Computer Journal 51(5), 523–560 (2008)
6. Dowe, D.L., Gardner, S., Oppy, G.R.: Bayes not bust! Why simplicity is no problem for Bayesians. British J. Philosophy of Science, 709–754 (December 2007)
7. Edgoose, T., Allison, L.: MML Markov classification of sequential data. Statistics and Computing 9, 269–278 (1999)
8. Edwards, R.T., Dowe, D.L.: Single factor analysis in MML mixture modeling. In: Wu, X., Kotagiri, R., Korb, K.B. (eds.) PAKDD 1998. LNCS (LNAI), vol. 1394, pp. 96–109. Springer, Heidelberg (1998)
9. Grunwald, P., Langford, J.: Suboptimal behavior of Bayes and MDL in classification under misspecification. Machine Learning 66(2-3), 119–149 (2007)
10. Jebara, T.: Discriminative, Generative and Imitative learning. PhD thesis, MIT (2001)
11. Kohonen, T.: Self-Organizing Maps, vol. 30. Springer, Heidelberg (2001)
12. Kolmogorov, A.N.: Three approaches to the quantitative definition of information. Problems of Information Transmission 1, 1–17 (1965)
13. Reusch, D.B., Alley, R.B.: Relative performance of Self-Organizing Maps and Principal Component Analysis in pattern extraction from synthetic climatological data. Polar Geography 29(3), 188–212 (2005)
14. Rissanen, J.: Modeling by the shortest data description. Automatica 14, 465–471 (1978)
15. Solomonoff, R.J.: A formal theory of inductive inference. Information and Control 7, 1–22, 224–254 (1964)
16. Wallace, C.S.: Statistical and Inductive Inference by Minimum Message Length. Springer, Heidelberg (2005)
17. Wallace, C.S., Boulton, D.M.: An information measure for classification. Computer Journal 11, 185–194 (1968)
18. Wallace, C.S., Dowe, D.L.: Intrinsic classification by MML - the Snob program. In: Proc. 7th Australian Joint Conf. on Artificial Intelligence, pp. 37–44. World Scientific, Singapore (1994)
19. Wallace, C.S., Dowe, D.L.: Minimum message length and Kolmogorov complexity. Computer Journal 42(4), 270–283 (1999)
20. Wallace, C.S., Dowe, D.L.: MML clustering of multi-state, Poisson, von Mises circular and Gaussian distributions. Statistics and Computing 10, 73–83 (2000)
21. Wallace, C.S., Freeman, P.R.: Estimation and inference by compact coding. J. Royal Statistical Society B 49, 240–252 (1987)

CoXCS: A Coevolutionary Learning Classifier Based on Feature Space Partitioning

Mani Abedini and Michael Kirley

Department of Computer Science and Software Engineering,
The University of Melbourne, Australia
{mabedini,mkirley}@csse.unimelb.edu.au

Abstract. Learning classifier systems (LCSs) are a machine learning technique, which combine reinforcement learning and evolutionary algorithms to evolve a set of classifiers (or rules) for pattern classification tasks. Despite promising performance across a variety of data sets, the performance of LCS is often degraded when data sets of high dimensionality and relatively few instances are encountered, a common occurrence with gene expression data. In this paper, we propose a number of extensions to XCS, a widely used accuracy-based LCS, to tackle such problems. Our model, CoXCS, is a coevolutionary multi-population XCS. Isolated sub-populations evolve a set of classifiers based on a partitioning of the feature space in the data. Modifications to the base XCS framework are introduced including an algorithm to create the match set and a specialized crossover operator. Experimental results show that the accuracy of the proposed model is significantly better than other well-known classifiers when the ratio of data features to samples is extremely large.

1 Introduction

Learning Classifier Systems (LCSs) are a genetic-based machine learning technique used to solve pattern classification problems [2, 8, 11, 13]. XCS, a well-known Michigan-style model, evolves problem solutions represented by a population of classifiers [19, 20]. Each classifier consists of a *condition-action-prediction* rule, with a fitness value proportional to the accuracy of the prediction of the reward. Evolutionary operators are used to discover better rules that may improve the current population of classifiers. Consequently, XCS is generally able to cover the state space more efficiently than other LCS.

Although there are many papers reporting high accuracy results across a wide spectrum of classification tasks, large state spaces and relatively small sample sizes (a common occurrence with gene expression data [22]) often lead to the evolution of partly-overlapping rules resulting in lower XCS accuracy [15]. Butz and co-workers [3] have shown that by introducing techniques that can efficiently detect building blocks in the condition part of the classifier, it may be possible to improve the performance of XCS. Specific evolutionary operators designed to help avoid the over-generalization phenomena inherent in XCS have also been demonstrated to be useful [14]. However, there is room to further extend these

A. Nicholson and X. Li (Eds.): AI 2009, LNAI 5866, pp. 360–369, 2009.

ideas and introduce modifications/enhancements enabling XCS-based models to solve gene expression classification problems.

A natural way to tackle high dimensional search problems is to adopt a "divide-and-conquer" strategy. To the best of our knowledge, decomposition approaches for XCS has been limited to the models proposed by Gershoff [7] and Richter [17]. Significantly, both of these papers report improved performance when the decomposition approach was used. A cooperative coevolutionary framework [16] may also provide a suitable approach for classification tasks. Zhu and Guan [22] report competitive performance results using a cooperative coevolution LCS. However, if all features are used in the classification process, the excessive computational cost reduces the efficiency/effectiveness of the model. Feature selection provides an alternative approach to help deal with high dimensional data. For gene expression data, techniques that rank genes according to their differential expressions among phenotypes, or techniques based on information gain ranking and principal component analysis can be used [21].

In this paper, we propose a number of extensions to the XCS to solve complex classification tasks. Our model, CoXCS, is fundamentally a coevolutionary model. Here, a number of isolated sub-populations are used to evolve classifiers based on a partitioning of the feature (or attribute) space. A modified version of XCS is used in each of the sub-populations. We introduce a specialization technique for reducing the number of attributes activate during the learning phase and a specificity crossover operator. Detailed computational experiments using a suite of benchmark data sets clearly shows that proposed model is comparable with other classification techniques. Significantly, the performance of the proposed model is better than other models when the ratio of data features to samples is extremely large.

The remainder of this paper is organized as follows: In Section 2 we present background material related to XCS and multi-population implementations. In Section 3 our model is described in detail. This is followed by a list of the experiments and results. We conclude the paper in Section 5 with a discussion of the results and identify future research directions.

2   Background and Related Work

2.1   XCS Overview

XCS is widely accepted as one of the most reliable learning classifier system for data mining. We provide a brief overview of XCS functionality in this subsection. Space constraints preclude us from providing a detailed discussion of XCS. However, further details can be found in Wilson's original paper [19] and related papers (eg.[4, 15, 20]).

XCS maintains a population of classifiers (see Fig 1). Each classifier consists of a *condition-action-prediction* rule, which maps input features to the output signal (or class). A ternary representation of the form 0,1,# (where # is don't care) for the condition and 0,1 for the action can be used. In addition, real encoding can also be used to accurately describe the environment states [20].

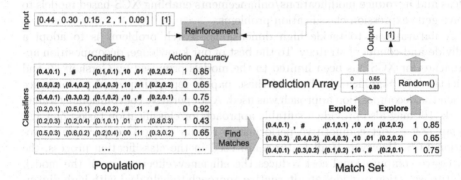

Fig. 1. XCS model overview. The condition segment of the classifier consists of a vector of features, each encoded using real or binary values. The output signal (prediction class) is a binary value in this case. The classifier's fitness value is proportional to the accuracy of the prediction of the reward. See text for further explanation.

Input, in the form of data instances (a vector of features), is passed to the XCS. A match set $[M]$ is created consisting of rules (classifiers) that can be "triggered" by the given data instance. A covering operator is used to create new matching classifiers when $[M]$ is empty. A prediction array is calculated for $[M]$ that contains an estimation of the corresponding rewards for each of the possible actions. Based on the values in the prediction array, an action, a (the output signal), is selected. In response to a, the reinforcement mechanism is invoked and the prediction, p, prediction error, ϵ, accuracy, k, and fitness, F, of the classifier is updated via the following equations:

$$p \leftarrow p + \beta(R - p) \qquad and \qquad \epsilon \leftarrow \epsilon + \beta(|R - p| - \epsilon)$$

where β is the learning rate $(0 < \beta < 1)$. The classifier accuracy is calculated from the following equations:

$$k = \begin{cases} 1 & \text{if } \epsilon < \epsilon_0 \\ \alpha(\frac{\epsilon}{\epsilon_0})^{-\nu} & \text{otherwise} \end{cases} \qquad and \qquad k' = \frac{k}{\sum\limits_{x \in [A]} k_x}$$

Finally, the classifier fitness, F, is updated using the relative accuracy value:

$$F \leftarrow F + \beta(k' - F)$$

It is important to note that the classifier fitness is updated based on the accuracy of the actual reward prediction. This accuracy-based fitness provides a mechanism for XCS to build a complete action map of the inputs space.

A key component of XCS, is the evolutionary computation module. During the evolutionary process, fitness-proportionate selection is used to guide the selection of parents (classifiers in the population), who generate new offspring via crossover and mutation. A bounded population size is typically used. Consequently a form of niching is used to determine if the offspring is added to the population and/or which of the old members of the population are deleted to make room for the new classifier (offspring). The deletion of classifiers is biased towards those with larger action set sizes and lower fitness.

2.2   Multi-population XCS Models

Dam et al., [6] proposed an XCS–based client/server distributed data mining system. Each client had its own XCS, which evolved a set of classifiers using a local data repository. The server then combined the models with its own XCS and attempted to find a set of classifiers to help explain patterns incorrectly classified locally by the clients. The performance of the model was evaluated using benchmark problems, focussing on network load and communication costs. The results suggested that the distributed XCS model was competitive as a distributed data mining system, particularly when the epoch size increased.

In a similar study, a multi-population parallel XCS for classification of electroencephalographic signals was introduced by Skinner et al., [18]. The specific focus of that study was to investigate the effectiveness of migration strategies between sub-populations mapped to ring topologies. They reported that the parameter setting of the multi-population model had a significant effect on the resulting classifier accuracy.

An alternative approach for solving a classification task is to incorporate a decomposition strategy into the model. For example, Gershoff et al., [7] attempted to improve global XCS performance via a hierarchical partitioning scheme. An agent in the model was assigned to each partition, which contained a collection of homogeneous XCS classifiers. The predicted output signal (class) was then estimated using a voting mechanism. This output signal, with a confidence score, was then passed up the hierarchy to a controlling agent. This agent then decided the final output of the system based on the combined output from each of the sub-populations it was responsible for. Gershoff et al., report results with improved performance notes in the limited domain tested.

Richter [17] introduced an extended XCS model, where a series of lower level problems were solved. These results were then combined into a global result for the given problem. Improved performance was noted in a limited range of test problem used in the study. In such an approach, different sub-problem formulations will have a significant impact on the performance of the distributed system.

A recent model employing a cooperative coevolutionary classifier system was introduced by Zhu and Guan [22]. In this fine-grained approach, individuals in isolated sub-populations encoded *if–then* rules for each feature in the data set. As such, the decomposition was taken to the extreme. Individuals were used to classify the partially masked training data corresponding to the feature in focus. However, this particular approach required a two-step process – a concurrent global and local evolutionary process – in order to generate satisfactory accuracy levels. For data sets with a large number of features (attributes) such fine-grain modelling is computationally expensive.

3 Model

CoXCS is fundamentally a coevolutionary parallel learning classifier based on feature space partitioning. Fig. 2 provides a high-level schematic overview of the system.

Within the CoXCS model, the features contained in the data set are partitioned into a set of n sub-populations. The condition segment of each classifier in a given sub-population is initialized using this fixed subset of the features (of size λ) from the data set being processed. Importantly, the sub-populations evolve separately, but with a common objective. As such, each isolated sub-population accumulates and specializes its expertise across a subset of the input space. Bounded sub-population sizes are used as per the standard XCS model. When a new classifier is added to a sub-population, if the size limit is reached, a randomly selected classifier (based on a niching technique) is deleted from the sub-population.

Migration episodes are also used to exchange classifiers between sub populations. After a fixed number of iterations, randomly selected classifiers migrate to a different sub-population based on a random migration topology. It is important to note, that the mutation operator does not destroy the inherent building blocks within the *immigrant* classifiers.

Two important modifications are proposed for the XCS model running in each of the sub-populations:

Firstly, a new covering operator is used to create the match set $[M]$ (see Algorithm 1). This operator builds single feature classifiers (the remaining features are set to #) for each of the features present in the given partition. An important distinction between our model and XCS is the fact that we create λ classifiers, which are added to the bounded population of each partition. In contrast, XCS would create only one classifier. This approach, allows the evolutionary search to slowly build up more specialized classifiers via the genetic operators and reinforcement learning mechanism.

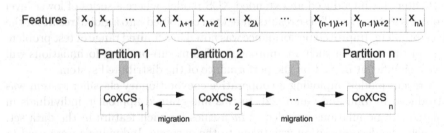

Fig. 2. High level overview of the CoXCS model. Each isolated sub-population evolves solutions based on a partitioning of the feature space (of size λ) using a separate CoXCS. Randomly selected classifiers migrate between sub-populations.

Algorithm 1. CreateMatchSet()

Require: *input*: a vector of features $X \in \{x_0, x_1, \ldots, x_{n\lambda}\}$
            action: a value for the expected output/class (eg. 0 or 1)
            s, e: array index addresses – start (eg. 0) and end (eg. λ)
 1: initialize match set $[M]$
 2: **for** $i = s$ to e **do**
 3:    create new classifier *rule*
 4:    *rule*.setCondition(i,*input*[i])
 5:    *rule*.setAction(*action*)
 6:    $[M]$.add(*rule*)
 7: **end for**
 8: **return** $[M]$

Algorithm 2. SpecCrossover()

Require: $p1$, $p2$: two randomly selected parents (classifiers)
              len: is the length of condition segment in parents $(len = \lambda)$
 1: create new classifier *child*
 2: **for** $i = 0$ to *len* **do**
 3:    **if** $p1$.hasCondition(i) AND ! $p2$.hasCondition(i) **then**
 4:       *child*.setCondition(i, $p1$.getCondition[i])
 5:    **else if** ! $p1$.hasCondition(i) AND $p2$.hasCondition(i) **then**
 6:       *child*.setCondition(i, $p2$.getCondition[i])
 7:    **else if** $p1$.hasCondition(i) AND $p2$.hasCondition(i) **then**
 8:       $\Delta \leftarrow p1$.getCondition[i] $\cap$ $p2$.getCondition[i]
 9:       **if** $\Delta \neq$ null **then**
10:          *child*.setCondition(i, Δ)
11:       **end if**
12:    **end if**
13: **end for**
14: **return** *child* (the new classifier)

Secondly, we introduce a specialized crossover operator, which generates valid offspring (classifiers) across the range of feature encodings used (see Algorithm 2). In the case of nominal and binary features, if the feature appears in either parent, the feature is copied to the child. For real value features, the center-spread and range are examined. The corresponding common range of the feature in both parents is then copied to the child. A standard mutation operator is then applied to the child.

In our implementation, we use a variable length hybrid real-integer encoding for each classifier. A sparse vector representation, indexed to the feature value, is used. The # values are not stored. This approach is used to both speed up computations and minimize memory use. In addition, this approach provides a flexible means to concatenate classifiers generated from different populations (initial partitions of the feature space) after migration episodes and crossover operations. The predicted output class for the classification task is based on majority voting among all CoXCS predictions.

4   Experiments

A series of experiments were conducted to validate our approach. The underlying hypothesis tested was that the classification using the CoXCS model would lead to improved accuracy, particularly for high dimensional data sets.

4.1   Data Sets

A range of data sets displaying different characteristics were used for evaluation. **wbc** (Breast Cancer Wisconsin - Original), **wpbc** (Breast Cancer Wisconsin - Prognostic), **wdbc** (Breast Cancer Wisconsin - Diagnostic) and **hepatitis** were taken from the UC Irvine Machine Learning Repository [1]. Two gene expression datasets were also included: **BRCA** (sporadic breast cancer gene profiles) [10, 12] and **Prostate** (prostate cancer gene profiles) [12].

4.2   Methodology

Model parameters. For all experiments, a hybrid feature encoding scheme was used. The parameter settings for our modified XCS were based on the default XCS settings recommended in [5]. The parameter values that were different include: population sizes of 3000 (UCI data sets) or 5000 (gene expression data sets); the exploration/exploitation rate was set to 0.3, and the reward value set to 1000. The partitioning schemes used was a simply equal linear division of the feature space. In this study, we have employed a simple rule: the number of partitions (and thus sub-populations) $n = \lfloor 0.1 \times \#\text{Features} \rfloor$ for a given data set. The migration ratio was set to 10% of the population size. Five separate migration stages were used, where the number of iterations between migration episodes was fixed at 100.

Validation and performance measures. Ten-fold cross validation was used for the data sets taken from the UCI Repository. The small number of instances (samples) of the gene expression data sets restricted evaluation to two-fold cross validation. In order to compare the performance of our model with other classifier systems, we report results based on the Area Under Curve (AUC) of the Receiver Operating Characteristic (ROC), a widely used technique in machine learning [9]. AUC values vary between 0 and 1, where 0.5 represents a random classification and 1 represents the highest accuracy.

4.3   Results

Table 2 lists the AUC results for each of the data sets considered for a variety of different classifiers. The non-XCS classifier results were generated using the Weka package. The relative performance of the base-line XCS and the other classifiers was very similar. It is interesting to note that accuracy levels were generally very low for the gene expression data sets (**BRCA** and **Prostate**). The accuracy

Table 1. Data set details. The gene expression data sets are characterized by a small number of instances and a very large number of features. All data sets have two output classes.

Data Set	#Instance	#Features	%Majority	%Missing
wbc	699	9	0.65	0.23
wpbc	198	33	0.76	0.06
wdbc	569	30	0.62	–
hepatits	155	19	0.85	5.30
BRCA	22	3226	0.68	–
Prostate	21	12600	0.61	–

Table 2. AUC results. Bold values indicate the the CoXCS model was significantly better when compared to all of the other classifiers.

Classifier	Mode	wbc	wdbc	wpbc	hepatitis	BRCA	Prostate
j48	Train	0.98	0.99	0.93	0.91	0.92	1.00
	Test	0.95	0.93	0.59	0.70	0.35	0.42
NBTree	Train	0.99	0.99	0.79	0.97	1.00	1.00
	Test	0.98	0.95	0.55	0.81	0.45	0.46
Random Forest	Train	1.00	1.00	1.00	1.00	1.00	1.00
	Test	0.98	0.98	0.63	0.84	0.29	0.33
Neural Networks	Train	0.99	0.99	0.98	0.94	0.50	0.50
	Test	0.98	0.99	0.68	0.81	0.50	0.50
Logistic Regression	Train	0.99	1.00	0.94	0.94	1.00	0.50
	Test	0.99	0.97	0.77	0.80	0.56	0.50
Naive Bayes Classifier	Train	0.98	0.98	0.72	0.91	0.99	1.00
	Test	0.98	0.98	0.64	0.83	0.50	0.35
SVM	Train	0.97	0.93	0.50	0.54	1.00	1.00
	Test	0.96	0.93	0.50	0.51	0.53	0.38
XCS	Train	0.99	0.99	0.97	1.00	0.50	0.50
	Test	0.97	0.93	0.70	0.72	0.50	0.50
CoXCS	Train	0.99	1.00	1.00	0.97	1.00	1.00
	Test	**1.00**	0.99	**0.98**	**0.96**	**0.80**	**0.75**

performance of the CoXCS was generally better than other classifiers across data sets. CoXCS performance was significantly better ($p < 0.01$, 15 trials) for problems where the ratio of the number of features to instance was extremely large.

5   Discussion and Conclusion

There are many examples reported in the literature illustrating the effectiveness of LCS, and the accuracy-based XCS in particular, for data mining task. However, there are still many open questions related to improving classification accuracy when confronted with problems of high dimensionality, a small number of data instances, noisy data and multiple classes.

In this paper, we have proposed enhancements for XCS to improve classification accuracy for data sets where the ratio of data features to samples is extremely large in binary classification tasks. In CoXCS, isolated sub-populations were used to evolve classifiers based on a initialization mechanism using a subset of features. Two modifications were made to the base XCS model running in each island: a new algorithm was used to create the match set and a specialized crossover operator was used. This "divide-and-conquer" strategy encourages the evolution of specialized classifiers and allows us to maximize the advantages of the embedded reinforcement learning mechanism in XCS.

Detailed experimental studies show that CoXCS is comparable with, and outpeforms other well-known classifiers in many cases, across the suite of benchmark data sets used for evaluation. The results suggest that the decomposition strategy plays an important role in guiding the trajectory of the evolving populations. Here, we have limited the decomposition to a naive approach. In future work, it would be interesting to examine alternative techniques to detect variable interactions that exist in a problem, and subsequently make use of this "expert knowledge" when partitioning the feature space. There is also scope to examine the effectiveness of distributed deployment and alternative migration policies using a suite of micro-array data classification problems.

References

1. UCI Machine Learning Repository, http://archive.ics.uci.edu/ml/
2. Bull, L., Kovacs, T. (eds.): Foundations of Learning Classifier Systems. Studies in Fuzziness and Soft Computing, vol. 183. Springer, Heidelberg (2005)
3. Butz, M., Pelikan, M., Lloral, X., Goldberg, D.E.: Automated global structure extraction for effective local building block processing in XCS. Evolutionary Computation 14(3), 345–380 (2006)
4. Butz, M.V., Kovacs, T., Lanzi, P.L., Wilson, S.W.: Toward a theory of generalization and learning in XCS. IEEE Transactions on Evolutionary Computation 8(1), 28–46 (2004)
5. Butz, M.V., Wilson, S.W.: An Algorithmic Description of XCS. In: Lanzi, P.L., Stolzmann, W., Wilson, S.W. (eds.) IWLCS 2000. LNCS (LNAI), vol. 1996, pp. 253–274. Springer, Heidelberg (2001)
6. Dam, H.H., Abbass, H.A., Lokan, C.: DXCS: an XCS system for distributed data mining. In: Proceedings of the 2005 conference on Genetic and evolutionary computation (GECCO 2005), pp. 1883–1890. ACM Press, New York (2005)
7. Gershoff, M., Schulenburg, S.: Collective behavior based hierarchical XCS. In: Proceedings of the 2007 Genetic And Evolutionary Computation Conference (GECCO 2007), pp. 2695–2700. ACM Press, New York (2007)
8. Goldberg, D.E.: Genetic Algorithms in Search, Optimization, and Machine Learning. Addison-Wesley, Reading (1989)
9. Hanley, J.A., McNeil, B.J.: The meaning and use of the area under a receiver operating characteristic (ROC) curve. Radiology 143(1), 29–36 (1982)

10. Hedenfalk, I., Duggan, D., Chen, Y., Radmacher, M., Bittner, M., Simon, R., Meltzer, P., Gusterson, B., Esteller, M., Kallioniemi, O.P., Wilfond, B., Borg, A., Trent, J.: Gene-Expression profiles in hereditary breast cancer. N. Engl. J. Med. 344(8), 539–548 (2001)

11. Holland, J.H., Booker, L.B., Colombetti, M., Dorigo, M., Goldberg, D.E., Forrest, S., Riolo, R.L., Smith, R.E., Lanzi, P.L., Stolzmann, W., Wilson, S.W.: What is a Learning Classifier System? In: Lanzi, P.L., Stolzmann, W., Wilson, S.W. (eds.) IWLCS 1999. LNCS (LNAI), vol. 1813, pp. 3–32. Springer, Heidelberg (2000)

12. Hossain, M.M., Hassan, M.R., Bailey, J.: ROC-tree: A Novel Decision Tree Induction Algorithm Based on Receiver Operating Characteristics to Classify Gene Expression Data. In: Proceedings of the SIAM International Conference on Data Mining, Atlanta, Georgia, USA, April 2008, pp. 455–465 (2008)

13. Kovacs, T.: Two views of classifier systems. In: Lanzi, P.L., Stolzmann, W., Wilson, S.W. (eds.) IWLCS 2001. LNCS (LNAI), vol. 2321, pp. 74–87. Springer, Heidelberg (2002)

14. Lanzi, P.L.: A Study of the Generalization Capabilities of XCS. In: Bäck, T. (ed.) Proceedings of the 7th International Conference on Genetic Algorithms, pp. 418–425. Morgan Kaufmann, San Francisco (1997)

15. Lanzi, P.L., Stolzmann, W., Wilson, S.W. (eds.): IWLCS 1999. LNCS (LNAI), vol. 1813. Springer, Heidelberg (2000)

16. Potter, M.A., Jong, K.A.D.: Cooperative Coevolution: An Architecture for Evolving Coadapted Subcomponents. Evolutionary Computation 8(1), 1–29 (2000)

17. Richter, U., Prothmann, H., Schmeck, H.: Improving XCS performance by distribution. In: Li, X., Kirley, M., Zhang, M., Green, D., Ciesielski, V., Abbass, H.A., Michalewicz, Z., Hendtlass, T., Deb, K., Tan, K.C., Branke, J., Shi, Y. (eds.) SEAL 2008. LNCS, vol. 5361, pp. 111–120. Springer, Heidelberg (2008)

18. Skinner, B., Nguyen, H., Liu, D.: Distributed classifier migration in XCS for classification of electroencephalographic signals. In: 2007 IEEE Congress on Evolutionary Computation, pp. 2829–2836. IEEE Press, Los Alamitos (2007)

19. Wilson, S.W.: Classifier Fitness Based on Accuracy. Evolutionary Computation 3(2), 149–175 (1995), http://prediction-dynamics.com/

20. Wilson, S.W.: Get Real! XCS with Continuous-Valued Inputs. In: Lanzi, P.L., Stolzmann, W., Wilson, S.W. (eds.) IWLCS 1999. LNCS (LNAI), vol. 1813, pp. 209–222. Springer, Heidelberg (2000)

21. Zhang, Y., Rajapakse, J.C.: Machine Learning in Bioinformatics, 1st edn. Series in Bioinformatics. Wiley, Chichester (2008)

22. Zhu, F., Guan, S.: Cooperative co-evolution of GA-based classifiers based on input decomposition. Engineering Applications of Artificial Intelligence 21, 1360–1369 (2008)

Multi-Objective Genetic Programming for Classification with Unbalanced Data

Urvesh Bhowan, Mengjie Zhang, and Mark Johnston

School of Engineering and Computer Science,
Victoria University of Wellington, New Zealand

Abstract. Existing learning and search algorithms can suffer a learning bias when dealing with unbalanced data sets. This paper proposes a Multi-Objective Genetic Programming (MOGP) approach to evolve a Pareto front of classifiers along the optimal trade-off surface representing minority and majority class accuracy for binary class imbalance problems. A major advantage of the MOGP approach is that by explicitly incorporating the learning bias into the search algorithm, a good set of well-performing classifiers can be evolved in a single experiment while canonical (single-solution) Genetic Programming (GP) requires some objective preference be *a priori* built into a fitness function. Our results show that a diverse set of solutions was found along the Pareto front which performed as well or better than canonical GP on four class imbalance problems.

1 Introduction

Classification problems arise in a wide range of real world applications; medical diagnosis, fraud detection, and image recognition are just a few examples. Genetic Programming (GP) is a machine learning and search technique based on the principles of Darwinian evolution or natural selection which has been widely successful in solving various classification problems [1]. Some real-world classification problems however, such as working with unbalanced data [2], are difficult to solve. Data sets are unbalanced when they have an uneven distribution of class examples, that is, when at least one class is represented by only a small number of examples (called the *minority class*) while the other class(es) make up the rest (called the *majority class*). Various machine learning approaches have shown that using an uneven distribution of class examples in the learning process can leave the learning algorithm with a performance bias: high accuracy on the majority class(es) but poor performance on the minority class(es) [3].

Addressing this learning bias to more accurately classify examples from both the majority and minority class equally well has shown that these objectives are usually in conflict – increasing the performance of one class results in a trade-off in performance for the other [4]. As class-specific misclassification costs in real-world problems are task-sensitive, objective preference must usually be *a priori* built into the learning system. In many cases, subsequent changes to preference after a solution has been found can require the search to begin afresh with new preference information.

Evolutionary multi-objective optimisation (EMO) is a fast growing area of research which offers a promising solution to the problem of optimising multiple conflicting objectives by simultaneously evolving a Pareto front of solutions along the trade-off

A. Nicholson and X. Li (Eds.): AI 2009, LNAI 5866, pp. 370–380, 2009.

surface in a single experiment [5]. Each objective is treated separately in EMO learning systems allowing the performance trade-off to be incorporated into the learning process. EMO approaches offer the end decision-maker both insights into the performance trade-off for a particular problem and the ability to readily choose preferred solutions along the evolved Pareto front after the search process.

EMO techniques have been successfully applied to a wide range of real world applications [6]. In GP and classification problems specifically, EMO methods have been primarily used for bloat control [7], but research has also involved decomposing the classification accuracy of each class as separate objectives to be optimised [8]. This paper aims to extend this idea by developing a multi-objective GP (MOGP) approach to the class imbalance problem using the classification accuracy of the minority and majority class as separate conflicting objectives to be maximised. Using the MOGP approach, this paper analyses the performance trade-off between these two conflicting objectives along the approximated Pareto fronts for a number of unbalanced data sets with varying levels of class imbalance, and examines the diversity of the Pareto fronts and the types of Pareto-optimal solutions evolved. We also discuss some of the initial problems encountered in developing a multiple-objective approach and present an analysis into the classification performance of the evolved Pareto fronts.

The rest of this paper is organised as follows. Section 2 describes the MOGP approach. Section 3 presents the unbalanced data sets used in our experiments and examines the initial MOGP results. Section 4 introduces a MOGP improvement, presents these new results and analyses the classification performance of the Pareto fronts. Section 5 concludes this paper and gives directions for future work.

2 Multi-Objective GP Approach

2.1 Evolutionary Search Algorithm

In traditional EMO, the evolutionary search is focussed on improving the set of non-dominated solutions until they are Pareto Optimal. Our approach is based on the well known EMO algorithm NSGA-II [9], where the parent and offspring populations are merged together at every generation. This combined parent-child population is then sorted by program fitness where the fittest individuals are copied into a new population, called the Archive population. The Archive population then serves as the parent population in the next generation. The offspring population at every generation is generated using the traditional crossover and mutation genetic operators using binary tournament selection. The Archive population is used to preserve elitism in the population over generations. At the end of the evolutionary cycle, the output of the MOGP system corresponds to the *set* of evolved solutions along the approximated Pareto front.

2.2 Fitness

Every individual in the population has two hierarchical fitness attributes: non-dominance rank and a "crowding" measure. The non-dominance rank is a measure of how well the individual performs on both objectives with respect to every other member in the population. Crowding distance is an estimate of the diversity of a solution with respect to the population. The non-dominance rank serves as the primary fitness attribute – crowding distance is only used to resolve selection when the non-dominance rank is equal between two or more individuals. In other words, between two solutions with differing non-dominance ranks we prefer the solution with the better rank. Otherwise, if both solution have the same rank then we prefer the solution with the better crowding distance.

Non-dominance Rank: In multi-objective optimisation, a single solution *dominates* another solution if it is at least as good as the other solution on all the objectives and better on at least one [6]:

$$S_i \succ S_j \longleftrightarrow \forall m[(S_i)_m \geq (S_j)_m] \wedge \exists k[(S_i)_k > (S_j)_k] \tag{1}$$

In equation (1), if S_i and S_j are two solutions in the population and $(S_i)_m$ denotes the performance of solution S_i on the mth objective, then solution S_i dominates solution S_j if each component of S_i is better than or equal to the corresponding component of S_j and at least one component of S_i is higher. The non-dominance rank for a solution is the number of other solutions in the population that dominate that solution. A solution that is not dominated by any other solution is a non-dominant solution and represents the optimal rank (0). At every generation, all non-dominated individuals form the Pareto-approximated front.

"Crowding" Distance: Crowding techniques use Euclidean distance between solutions in *objective-space* as an estimate of solution diversity. Solutions from sparsely populated regions of objective-space are usually favoured over solutions from densely populated regions (solutions with similar performances across objectives) to promote diversity in the population [6]. Our crowding measure differs from that used in NSGA-II. In NSGA-II, crowding is estimated as the average distance between a given solution's nearest neighbours [9]. As this only takes into account the two immediate solutions surrounding a given solution and not all other solutions in the region, it does not approximate the density of the entire region of solutions in objective-space. For this reason, we use a crowding measure based on the total distance between a solution to *all other solutions* in objective-space, penalising a solution with many close neighbours [10]:

$$d_p = \sum_{j=1}^{n} dist(S_p, S_j) \tag{2}$$

In equation (2), n is the size of the population and $dist(S_p, S_j)$ is the distance in objective space between solutions S_p and S_j, that is, the sum of the differences (absolute-value) between each component of performance vectors S_p and S_j. The smaller the crowding distance d_p (for solution S_p) the less desirable the solution as lower distances indicate densely populated regions of objective-space (many neighbours close together). Equation (2) places equal weighting on all solutions in the population.

2.3 Evaluating Pareto Fronts

To evaluate the evolved Pareto-approximated fronts we calculated the front *hyperarea* as a single measure of the convergence (or classification performance) of the front, and stored all evolved Pareto-front solutions from the series of experiments in a *run-persistent* archive for later analysis. The run-persistent archive allows us to examine trends such as the diversity of the evolved fronts or the kinds of Pareto-front solutions evolved. The run-persistent archive can also be used to track the "global" Pareto front of all evolved Pareto fronts across multiple experiments. The global Pareto front corresponds to the set of all non-dominated solutions with respect to the run-persistent archive, that is, the Pareto-optimal front of all evolved Pareto-approximated fronts.

Front Hyperarea: Hyperarea is typically a measure of front convergence, corresponding to the area under the Pareto front in objective-space [6]. In classification, hyperarea can also represent the area of objective-space correctly classified by the Pareto front, similar to the Area under a ROC curve (AUC) [11]. However, where the AUC represents the classification ability of single classifier at varying classification thresholds, the front hyperarea represents the classification ability of a *set* of classifiers. The hyperarea can be estimated using the sum of the areas of individual trapezoids fitted under each solution in objective-space:

$$hyperarea = \sum_{i=1}^{f-1} \frac{1}{2} \times [(S_{i+1})_{min} - (S_i)_{min}] \times [(S_{i+1})_{maj} + (S_i)_{maj}] \quad (3)$$

In equation (3), f is the number of solutions on the Pareto front, $(S_i)_{min}$ and $(S_i)_{maj}$ represent the performance of Pareto front solution S_i on the two objectives (minority and majority class accuracy), and (S_{i+1}) is the neighbouring solution to S_i. Equation (3) follows from the definition for calculating the area of a trapezoid[1] where the minority and majority class objectives correspond to the width and height of the trapezoid, respectively, in objective-space [11]. Hyperarea values range between 0–1 where 1 is the optimal hyperarea.

2.4 MOGP Configuration

A tree-based structure was used to represent genetic programs [1]. The ramped half-and-half method was used for generating programs in the initial population and for the mutation operator. The population size was 500, crossover and mutation rates were 60% and 40%, respectively, and the maximum program depth was 6 to restrict very large programs in the population. The evolution ran for 50 generations. For the classification strategy we translated the output of a genetic program (floating point number) into two class labels using the division between positive (minority class) and non-positive (majority class) numbers. We used feature terminals (example features) and constant terminals (randomly generated floating point numbers) as the terminal set, and a function set comprising of the four standard arithmetic operators, $+, -, \%$, and $\times$, and the conditional operator if. The $+, -$ and $\times$ operators have their usual meanings (addition, subtraction and multiplication) while % is *protected* division (usual division except that a divide by zero returns zero). Half of each data set was randomly chosen as the training set and the other half as the test set, both preserving the original class imbalance ratio.

---

[1] Trapezoid area is $\frac{1}{2} \times w \times (h+h')$ where w is width, and h and h' are heights of the two sides.

3 Results

3.1 Classification Data Sets

The experiments used three benchmark data sets all chosen based on their uneven distribution of class examples. SPECT and YEAST were from the *UCI Repository of Machine Learning Databases* [12], and FACE, an image data set comprising of a collection of face and non-face images, was from the Center for Biological and Computational Learning at the Massachusetts Institute of Technology [13].

SPECT Heart data: This data set contains 267 data instances (patients) derived from cardiac Single Proton Emmision Computed Tomography (SPECT) images. This is a binary classification task, where patient heart images are classified as *normal* or *abnormal*. The data set has 55 instances of the "abnormal" class (20.6%) and 212 instances of the "normal" class (79.4%), a class imbalance ratio of approximately 1:4. Each SPECT image was processed to extract 44 continuous features, these were further pre-processed to form 22 binary features (F_1–F_{22}) that make up the attributes for each instance [12]. There are no missing attributes.

YEAST data: This data set contains 1482 instances, larger than SPECT data set, generated for the automatic prediction of protein localisation sites in yeast cells. There are eight numeric features calculated from properties of amino acid sequences (F_1–F_8), and nine distinct classes of interest, each with a different degree of class imbalance, making this a multi-class classification problem. For our purposes, we decomposed this data set into binary classification problems with only one "main" (minority) class and *everything else* as the majority class. We used two different "main" classes based on their class imbalance ratios: Yeast$_1$ is reasonably unbalanced with 244 examples (16%) and an imbalance ratio of 1:5, and Yeast$_2$ is highly unbalanced with 44 (3%) examples and an imbalance ratio of approximately 1:35.

FACE image data: This data set contains 30,821 grey-scale PGM-format images of faces and non-faces (background), the largest of the three data sets, each 19×19 pixels in size. There are 2901 face (9.5%) and 28,121 (90.5%) non-face images, an imbalance ratio of approximately 1:10. As image features we used 14 low level pixel statistics F_1–F_{14}, corresponding to the mean and variance of the raw pixel values extracted at eight specific rectilinear regions within each image. These features represent the overall pixel brightness/intensity and the contrast of a given region. For details on the 14 pixel statistical-based features refer to our previous work [4].

Fig. 1. Example face (left two) and non-face images (right two)

3.2 Initial Results

The initial MOGP experiment results are shown in Figure 2. These plots show all evolved Pareto-front solutions generated over 30 experiments for each problem. In

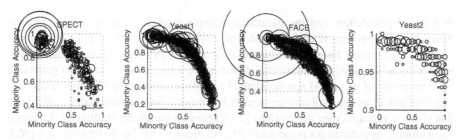

Fig. 2. All evolved Pareto fronts over 30 runs (note that the vertical axis scopes are different)

the plots each circle represents the test performance of a solution in the run-persistent archive (i.e., every evolved Pareto-front solution over all experiments). The size of each circle shows the density of solutions in the run-persistent archive, that is, *how often* a Pareto-front solution with the same performance on both the objectives was evolved across all experiments. The larger the circle, the higher the density (many solutions with the same performance). In Figure 2 the x and y axis correspond to the minority and majority class accuracy, respectively, where 1 indicates optimal accuracy (100%).

Figure 2 shows that a good Pareto front approximation was achieved for each problem as the performance trade-off is clearly visible in each case. Notice that the minority accuracy objective is more sensitive to the performance trade-off than majority accuracy. For example, in all problems solutions with near-optimal majority accuracy (close to 100%) have very poor performance on the minority class (close to 0%) but solutions with near-optimal minority accuracy have some success on the majority class with accuracy varying between 90% ($Yeast_2$), 40% ($Spect$), and 20% ($Face$ and $Yeast_1$).

The density information in Figure 2 reveals that for some problems ($Spect$ and $Face$), there is an noticeably high proportion of solutions in the run-persistent archive with the same sub-optimal performance, that is, 100% majority class accuracy and 0% minority class accuracy, compared to the relatively even distribution of solutions along the rest of Pareto fronts. These are represented by the large circles in the top-left corner of the plots. A closer inspection of these results reveals that the percentage of these sub-optimal or "one-sided" solutions with respect to all evolved solutions is 20% for the $Spect$ problem (675 solutions out of 3338 in the run-persistent archive) and 30% for the $Face$ problem (2457 out of 7934).

This indicates that the diversity of the Pareto fronts can be improved as the proportion of solutions with the same sub-optimal performance should be relatively low for the Pareto fronts to maintain a diverse set of solutions. In addition, as the goal of this MOGP approach is to evolve a diverse set of solutions that generally perform well on both objectives, it is preferable that the evolved Pareto fronts should contain *more* solutions that perform well on both objectives as apposed to large numbers of sub-optimal solutions.

4   Improving Front Diversity

4.1   New Dominance Constraint for Enforcing Diversity

The presence of large sub-optimal solution clusters on the evolved Pareto fronts can be attributed to several factors, such as the method of non-dominance assignment, genetic drift in the population, and the parameter settings used in the experiments. In non-dominance assignment, if multiple solutions with equivalent one-sided performance meet the requirement for non-dominance, they are included on the Pareto front. In genetic drift, as one-sided non-dominated performance is easy to achieve, newly generated child programs of one-sided non-dominated parents are also likely to be one-sided and non-dominated; as the evolutionary search is focussed on non-dominated individuals this effect is repeated over generations [7]. Regarding parameter settings, if the maximum number of generations is too small, the solutions on the Pareto front are not given the chance to spread out along the front using the crowding diversity measure; similarly the crossover and mutation rates also effect how much of the Pareto front is explored [9].

Two potential approaches to limit the presence of one-sided solutions on the Pareto fronts would be to discard such solutions as they are created [8], or keep these solutions but assign them a relatively poor fitness to reduce selection probability. The second approach is advantageous over the first in that it relies on the natural mechanism of selection pressure over manual interference in the evolutionary algorithm, and still allows for *some* one-sided solutions to be included in the learning process – this is generally considered necessary for the diversity of the population [14]. For these reasons, we introduced an additional constraint for non-dominance to reduce selection probability and limit the inclusion of one-sided solutions onto the Pareto front. The new constraint asserts that a given solution is only non-dominant if that solution is not dominated by any other solution *and* achieves at least a minimum performance of P on both objectives:

$$S_i \succ_P S_j \longleftrightarrow S_i \succ S_j \land \forall m[(S_i)_m \geq P) \tag{4}$$

In equation (4), solution S_i P-dominates solution S_j ($S_i \succ_P S_j$) if S_i dominates S_j and achieves at least P on each objective m, where P can be any value between 0 and 1. Under this new definition of non-dominance, one-sided solutions which would otherwise be non-dominant are now assigned a poorer ranking. To avoid significant reductions in the size of Pareto fronts evolved due to too stringent inclusion criteria (high P thresholds), we tested five minimum performance thresholds between 0.05 – 0.2 and present an analysis on the effects of these different thresholds in the next section.

4.2 Results Using New Dominance Constraint

Figure 3 shows the evolved Pareto fronts for three $P\%$ thresholds (the remaining two thresholds are omitted due to space constraints). In these new plots we see that there are fewer and smaller solution clusters around the top-left corner of objective-space as P increases compared to Figure 2. This indicates that the minimum performance constraint was successful in reducing the number of equivalent "one-sided" solutions on the evolved Pareto fronts. Specifically, there are fewer clusters for high thresholds (greater than 10%) compared to low thresholds, but the spread of solutions along the Pareto fronts, that is, the span of solutions along both x and y axis, becomes smaller as P increases. A threshold with a reasonable compromise between the above trade-off lies between 5–10% where the number of solutions clusters is reduced to a reasonable level while still allowing a sufficient span of solutions on the Pareto front.

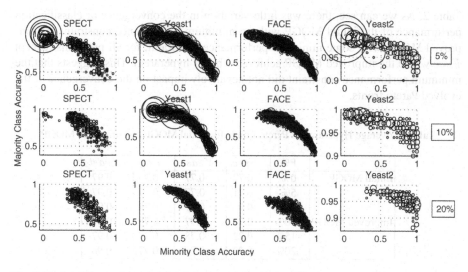

Fig. 3. Evolved Pareto fronts using Minimum-Performance $P\%$ Constraints: 5% (top row), 10% (middle row), and 20% (bottom row)

Diversity Analysis: As a measure of the diversity of the evolved Pareto fronts, we compared the average number of distinctly performing Pareto front solutions from the initial MOGP (no minimum performance constraint) and improved MOGP using the five minimum performance constraints. These results are shown in Table 1 (the average Pareto front sizes are included in parenthesis for comparison). From these results notice that the front diversity is better (higher) in the improved MOGP using *some* minimum performance constraint compared to the initial MOGP for all problems, even though the average Pareto front sizes are smaller in the improved MOGP. This indicates that the minimum performance constraint succeeded in its goal of improving front diversity by reducing the number of "one-sided" solutions allowed to be non-dominant. In terms of the effectiveness of the five different minimum performance thresholds, the best diversity given the size of the Pareto front was achieved using $P\%$ thresholds of 5% and 7.5%. Higher thresholds (greater than 10%) tended to reduce diversity.

Table 1. Average diversity (and size) of Pareto fronts for initial MOGP and improved MOGP

	P%	*Spect*	*Yeast₁*	*Face*	*Yeast₂*
Initial MOGP	0%	18.4 (133.5)	49.7 (239.6)	69.8 (330.6)	12.4 (27.4)
	5%	22.9 (50.4)	46.8 (174.8)	80.9 (111.5)	11.9 (57.6)
	7.5%	19.5 (25.9)	50.2 (171.4)	76.5 (106.3)	13.4 (40.7)
Improved MOGP	10%	18.9 (25.6)	50.2 (185.8)	70.8 (100.3)	12.5 (25.4)
	15%	13.7 (17.1)	45.8 (203.8)	68.0 (112.2)	12.2 (39.6)
	20%	13.3 (16.4)	45.9 (45.9)	57.7 (69.6)	11.1 (17.0)

Front Coverage: A comparison of the average front hyperarea using the initial MOGP and improved MOGP with the five minimum performance thresholds is shown in

Table 2. As we can see, there was little variation in the convergence or classification performance of the initial MOGP with no minimum performance constraint and improved MOGP with some minimum performance constraint, for all problems (except *Face* where the improved MOGP achieved a better hyperarea). This suggests that the minimum performance constraint did not negatively impact on the convergence of the evolved Pareto fronts.

Table 2. Average Hyperarea of Pareto Fronts for initial MOGP and improved MOGP

	P%	*Spect*	*Yeast₁*	*Face*	*Yeast₂*
Initial MOGP	0%	0.72	0.77	0.73	0.97
	5%	0.71	0.77	0.77	0.97
	7.5%	0.71	0.77	0.78	0.97
Improved MOGP	10%	0.71	0.77	0.78	0.97
	15%	0.72	0.78	0.77	0.97
	20%	0.72	0.77	0.77	0.96

Classification performance of MOGP and canonical GP: For a comparison between the classification performance of the evolved Pareto fronts and solutions evolved using canonical GP, we plotted the global evolved Pareto front against the classification results from our previous work which explored the effects of different GP fitness functions on class imbalance problems [4]. These results are shown in Figure 4. The two fitness functions used in this comparison are the overall classification accuracy, and an improved fitness function for class imbalance problems based on the average classification accuracy of the minority and majority classes (for details on these and other new fitness functions see [4]). In Figure 4, the red and black lines in each plot correspond to the *global* Pareto fronts of the initial MOGP and improved MOGP using the optimal minimum performance constraint (5%), respectively, and the two dark shapes correspond to the average performance of the fittest evolved GP solutions using the two fitness functions.

In terms of maximising the performance of both objectives, the MOGP Pareto fronts were shown to perform just as well as canonical GP using both fitness functions on two problems (*Yeast*1 and *Face*), and better than canonical GP on the remaining two

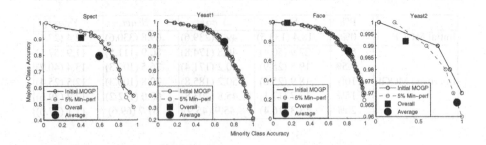

Fig. 4. Comparison of the classification performance between MOGP and canonical GP

problems (*Spect* and *Yeast2*). Another major advantage of the MOGP approach is that by explicitly incorporating the learning bias into the search algorithm, a good set of well-performing solutions was evolved along the optimal trade-off surface in a single experiment while canonical GP required this trade-off be *a priori* built into the fitness function and produced a single solution in a single experiment.

5 Conclusions

The goals of this paper were to develop a multi-objective GP approach to the class imbalance problem using the classification accuracy of the minority and majority class as separate objectives, and examine trends in the evolved Pareto fronts such as the diversity and the kinds of Pareto-optimal solutions evolved. These goals were achieved by examining the evolved Pareto fronts corresponding to the optimal trade-off in objectives for four class imbalance problems. We showed that a large proportion of evolved solutions were optimal only on one objective, the majority class, and that by incorporating a new minimum-performance constraint on each objective into the definition of non-dominance, we were able to increase the diversity of the Pareto fronts.

We also compared the classification performance of the Pareto fronts against canonical GP approaches using different fitness functions for class imbalance problems, and found that the evolved Pareto fronts performed just as well as the solutions evolved using canonical GP on half the problems and better than canonical GP on the rest.

For future work we plan to investigate the effects of different niching schemes specifically designed to penalise solutions in sub-optimal solution clusters and promote diversity. We also plan to develop criteria for extracting single solutions from the evolved Pareto fronts to compare both the classification ability and run-time cost between MOGP solutions and canonical GP solutions.

References

1. Koza, J.R.: Genetic Programming: On the Programming of Computers by Means of Natural Selection. MIT Press, Cambridge (1992)
2. Weiss, G.M., Provost, F.: Learning when training data are costly: The effect of class distribution on tree induction. Journal of Artificial Intelligence Research 19, 315–354 (2003)
3. Chawla, N.V., Japkowicz, N., Kolcz, A.: Editorial: Special issue on learning from imbalanced data sets. ACM SIGKDD Explorations Newsletter 6, 1–6 (2004)
4. Bhowan, U., Johnston, M., Zhang, M.: Differentiating between individual class performance in genetic programming fitness for classification with unbalanced data. In: Proceedings of the 2009 IEEE Congress on Evolutionary Computation (2009)
5. Goldberg, D.E.: Genetic Algorithms in Search, Optimization and Machine Learning. Addison-Wesley, Reading (1989)
6. Coello, C., Lamont, G., Veldhuizen, D.: Evolutionary Algorithms for Solving Multi-Objective Problems, 2nd edn. Springer, US (2007)
7. Jong, E.D., Pollack, J.B.: Multi-objective methods for tree size control. Genetic Programming and Evolvable Machines 4(3), 211–233 (2003)
8. Parrot, D., Li, X., Ciesielski, V.: Multi-objective techniques in genetic programming for evolving classifiers. In: Proceedings of the 2005 Congress on Evolutionary Computation (CEC 2005), September 2005, pp. 1141–1148 (2005)

9. Deb, K., Pratap, A., Agarwal, S., Meyarivan, T.: A fast elitist multi-objective genetic algorithm: NSGA-II. IEEE Transactions on Evolutionary Computation 6, 182–197 (2000)
10. Bot, M.C.J., Boelelaan, D., Langdon, W.B.: Improving induction of linear classification trees with genetic programming. In: Genetic and Evolutionary Computation Conference, pp. 403–410. Morgan Kaufmann, San Francisco (2000)
11. Bradley, A.P.: The use of the area under the ROC curve in the evaluation of machine learning algorithms. Pattern Recognition 30, 1145–1159 (1997)
12. Blake, C., Merz, C.: UCI repository of machine learning databases (1998), http://archive.ics.uci.edu/ml
13. Sung, K.-K.: Learning and Example Selection for Object and Pattern Recognition. PhD thesis, AI Laboratory and Center for Biological and Computational Learning, MIT (1996)
14. Laumanns, M., Thiele, L., Deb, K., Zitzler, E.: Combining convergence and diversity in evolutionary multiobjective optimization. Evolutionary Compututation 10(3), 263–282 (2002)

Scheduling for the National Hockey League Using a Multi-objective Evolutionary Algorithm

Sam Craig, Lyndon While*, and Luigi Barone

Computer Science & Software Engineering,
The University of Western Australia, Australia 6009
{craigs02,lyndon,luigi}@csse.uwa.edu.au

Abstract. We describe a multi-objective evolutionary algorithm that derives schedules for the National Hockey League according to three objectives: minimising the teams' total travel, promoting equity in rest time between games, and minimising long streaks of home or away games. Experiments show that the system is able to derive schedules that beat the 2008–9 NHL schedule in all objectives simultaneously, and that it returns a set of schedules that offer a range of trade-offs across the objectives.

Keywords: Sports scheduling, Multi-objective evolutionary algorithms.

1 Introduction

The National Hockey League (NHL) is the premier ice hockey league in the world. It is one of the four major professional sports leagues in North America, with record game attendance of 21.3 million in 2007–8, and revenues in excess of US$2.6 billion and expected to grow further[1].

The current league has thirty teams that play 1,230 games over six months. Teams play 1–5 games per week, with games happening virtually every day. There is no round structure: at any point in the season, teams will usually have played different numbers of games. Potential schedules are measured according to several criteria, including minimising the travel burden that teams face; avoiding situations where teams have short time periods and long travel between games; and avoiding teams having long streaks of games at home or "on the road". These factors conspire to make fixture scheduling in the NHL a difficult task.

The principal contribution of this paper is the description and analysis of a system that uses a multi-objective evolutionary algorithm to derive good schedules for the NHL. Schedules are assessed on the above three objectives, as well as on a constraint objective that prevents any team having to play multiple games on one day. Experiments show that the system is able to derive schedules that beat the 2008–9 NHL schedule in all objectives simultaneously, and that it returns a set of schedules that offer a range of trade-offs across the objectives.

Section 2 describes some previous applications of evolutionary algorithms to sports scheduling problems, and Section 3 gives some basics of multi-objective optimisation. Section 4 gives the relevant details of the NHL structure. Section 5

---

* Corresponding author.

A. Nicholson and X. Li (Eds.): AI 2009, LNAI 5866, pp. 381–390, 2009.

describes the details of our implementation, and Section 6 describes and analyses some experiments using our system. Section 7 concludes the paper.

2   Evolutionary Approaches to Sports Scheduling

There have been several applications of evolutionary algorithms to sports scheduling problems. Some examples follow.

Yang et al. derived good results for scheduling in Major League Baseball using an evolutionary strategy[2]. Their basic aim was to minimise the travel required (the MLB has 2,430 games!), and also to promote weekend games. They report a significant improvement over the existing schedule.

Schönberger et al. used a memetic algorithm for scheduling a non-professional table tennis tournament[3]. The principal interesting aspect is their use of a repair function to promote feasibility: indeed with no repair function they were unable to produce any feasible solutions.

Barone et al. used an evolutionary strategy to derive good schedules for the Australian Football League[4]. They used the polygon construction method[5] to seed their population, and they optimise against four objectives: minimising interstate travel, maximising the expected revenue for "big games", optimising the geographical distribution of games on each weekend, and balancing the number of home games while minimising streaks of consecutive home games. They were able to derive many solutions that dominate the 2006 AFL schedule.

While and Barone used a broadly similar approach to derive schedules for the Super 14 rugby competition[6], with a fixture template instead of the polygon. They minimised inter-regional travel and maximised the use of prime-time TV slots, again deriving solutions that dominate the pre-existing schedules.

Costa used an evolutionary algorithm combined with tabu search to schedule the NHL[7] (a different league structure was used in 1994, so no direct comparison is possible). This approach proved to be superior to using tabu search alone.

3   Multi-objective Optimisation

In a multi-objective optimisation problem, potential solutions are assessed according to two or more independent quantities. The characteristic of good solutions is that improving in one objective can be achieved only by worsening in at least one other objective. An algorithm for solving such problems returns a set of solutions offering different trade-offs between the various objectives.

Consider a problem where the fitness function maps a solution x into a fitness vector $\overline{f_x}$. A solution x dominates a solution y iff $\overline{f_x}$ is at least as good as $\overline{f_y}$ in every objective, and is better in at least one objective. x is non-dominated wrt a set of solutions X iff there is no solution in X that dominates x. X is a non-dominated set iff every solution in X is non-dominated wrt X. The set of fitness vectors corresponding to a non-dominated set is a non-dominated front.

A solution x is Pareto optimal iff x is non-dominated wrt the set of all possible solutions. Such a solution is characterised by the fact that improvement in one objective can come only at the expense of some other objective(s). The Pareto *optimal set* is the set of all Pareto optimal solutions. The goal in multi-objective optimisation is to find (or approximate) this Pareto optimal set.

Having multiple objectives means there is only a partial order on solutions, which causes problems for selection in an evolutionary algorithm. The usual solution is to define a ranking on solutions: one popular scheme[8] defines the *rank* of a solution x wrt a set X to be the number of solutions in X that dominate x. Selection is then based on ranks: a lower rank implies a better solution.

Precise definitions of all these terms can be found in [9].

4   The National Hockey League

The NHL[10] features twenty-four teams from the USA and six from Canada, divided between two conferences. Each conference is further divided on a regional basis into three divisions of five teams each. The NHL regular season runs from October to April, and each team plays 82 games:

- six games against each of the other four teams in its division;
- four games against each of the other ten teams in its conference;
- one game against each of the fifteen teams in the other conference, and three extra wild-card games against teams (usually traditional rivals) from there.

Half of each set of games are played at home. Thus a season comprises 1,230 games in total. Other season structures have been used in the past[10].

Teams play 1–5 games per week, and games are scheduled virtually every day in the regular season, except for two breaks: Xmas Eve and Xmas Day; and the All Stars week, where the conferences play a game against each other, scheduled approximately half-way through the season.

Thus the NHL schedule has no round structure: at any point teams will usually have played different numbers of games. Coupled with the sheer size of the league, this greatly increases the difficulty of generating good schedules.

5   Our Multi-objective Approach

The principal decisions in constructing a multi-objective evolutionary algorithm are deciding how solutions will be represented; what objectives will be used, and how they will be quantified; how solutions will be selected for survival and reproduction; what crossover and mutation schemes will be used; and how the initial population will be seeded.

5.1   Representation

Our representation of a solution comprises four arrays that describe how the schedule varies from a template, normally one of the existing NHL schedules.

- The *team array* maps logical team IDs to actual teams.
- The *game array* maps logical game IDs to actual games.
- The *game dates array* stores the date for each game in the template.
- The *home team array* identifies the home team for each game.

For example, consider a tournament where three teams play each other home and away, and the following template is used:

T1 vs T2; T2 vs T3; T3 vs T1; T1 vs T3; T3 vs T2; T2 vs T1

Then the solution

{1,3,2}, {2,1,3,4,5,6}, {3,6,8,10,12,16}, {2,1,1,1,1,2}

would represent the schedule

Day 3: T3 vs T2;   Day 6: T3 vs T1;   Day 8: T2 vs T1;
Day 10: T1 vs T2;   Day 12: T2 vs T3;   Day 16: T1 vs T3

The home team array is applied to the template first, followed by the game array and the team array. Finally the game dates array delivers a concrete schedule.

Whilst this representation may not give access to the entire search space of solutions (due to the use of the template), it allows us to efficiently represent differences from a known schedule, and to define a mutation scheme that preserves feasibility.

5.2   Objectives

We assess schedules against three main objectives: minimising travel, minimising inequity in individual games, and minimising long streaks of home or away games. We also use a fourth objective to prevent teams having to play multiple games on one day, but this acts as a so-called "constraint objective": only schedules with a value of zero for this objective are regarded as feasible.

Travel. Travel between game venues is both expensive and detrimental to players' performance. We calculate the travel burden of each team separately by summing the distances that they have to travel between consecutive games. As is the norm, we assume that teams do not return home between games on the road. Thus the travel objective is defined as

$$travel = \sum_{t \in T} \sum_{i=1}^{81} dist(v_{t,i}, v_{t,i+1}) \tag{1}$$

where T is the set of all teams, $v_{t,k}$ is the venue of team t's k^{th} game, and $dist(v_1, v_2)$ returns the distance between venues v_1 and v_2.

Equity. Travel is of course inevitable, but one way to alleviate its effects is to combine long journeys with long time periods between games. The travel burden for a given team in a given game is then some balance between the distance travelled and the number of days since their last game. We defined a penalty function to quantify this burden:

$$P(t,i) = \frac{dist(v_{t,i}, v_{t,i-1})^{0.5}}{(d_{t,i} - d_{t,i-1})^{0.75}} \tag{2}$$

where $d_{t,k}$ is the day on which team t plays its kth game. The sub-linear exponents used in P reflect the fact that the effect of each extra day or distance is smaller than the last.

A problem arises for a game if one team is faced with a significant travel burden while their opponent isn't. The equity objective is thus defined as

$$equity = \sum_{g \in G} |P(h, i) - P(a, j)| \tag{3}$$

where G is the set of all games, and in each game team h is playing its ith game and team a is playing its jth game.

Game Streaks. One way to minimise travel is to play sequences of games at home, or to combine several visits to distant cities into a single "road trip". However, long sequences are regarded as a bad thing: fans like to have the opportunity to watch their team play live on a regular basis. Observing the recent schedules lead us to conclude that game streaks up to length 3 are tolerated by the NHL, but that longer streaks are avoided where possible. We thus defined the streaks objective to increase non-linearly with sequence-lengths over 3:

$$streaks = \sum_{t \in T} \sum_{s \in S} \sum_{i=4}^{s} i - 3 \tag{4}$$

where S is the multi-set of streaks in the schedule for team t. An additional penalty is applied for streaks of identical games.

Feasibility. A schedule is *infeasible* if a team is required to play two games on the same day. We could define our mutation operators to disallow this possibility, but that could limit the search space available and have a negative impact on results. Instead we define a feasibility objective as

$$feasibility = \sum_{t \in T} \sum_{i=1}^{81} f(d_{t,i+1} - d_{t,i}) \tag{5}$$

where $f(0) = 1; f(x) = 0, x > 0$, that counts how many times any team is asked to play two games on one day. The system then optimises this objective in the usual way, and at the end of the run we discard solutions that have a non-zero value for this objective. This approach has been shown to produce good results in previous systems[6, 11, 12].

5.3   Selection

We use a highly elitist approach to selection. In each generation, we allow n parents to produce n children, then promote the best n of these to the next generation. Primary selection is based on a domination ranking[8]. Equally-ranked solutions are separated first by promoting those with lower values in the constraint objective, then using crowding distance[13] to encourage diversity.

5.4 Crossover and Mutation

Given that the representation of a solution has four components, we decided to combine four parents to produce four children using crossover, with each parent's genotype being split between the children in a fixed pattern[15]. We use probabilistic selection to decide which parents were combined.

Mutation is performed on exactly one of the child's components, selected uniformly randomly. The degree of the mutation is determined self-adaptively, allowing for large mutations early in a run and smaller mutations later for fine-tuning of solutions[14]. Mutation operates differently for the different components of the representation. The constraints on the first and the last of these are necessary to preserve feasibility.

- The team array is mutated by exchanging a pair of teams from the same division.
- The game array is mutated by exchanging a pair of games.
- The game dates array is mutated by moving a game either one day forward or one day back.
- The home team array is mutated by flipping the home and away teams in two games H vs A and A vs H.

The adaptive component determines how many such operations are performed.

5.5 Population Seeding

We created our initial population with n mutated copies of the 2008–9 NHL schedule. Given that this is a feasible solution and given the definition of our mutation operators, we know that all solutions derived in the course of a run are feasible, except with regard to the constraint objective from Section 5.2.

Seeding with other existing schedules produced similar results.

6 Experiments and Analysis

We performed a series of experiments to set basic parameters like population size, and to determine which crossover mechanism performed best[15].

The single run of the final system described here had a population of 2,000 and ran for 20,000 generations. It was written using Java v1.6 under Windows Vista: it ran overnight on an Intel 2.66GHz machine with 4GB of RAM.

6.1 Pareto Front Evolution

Figure 1 shows the developing Pareto front, plus the fitness of the 2008–9 NHL schedule. Three objectives can be hard to visualise, so we show three separate projections in each pair of objectives. Note that some solutions which are non-dominated in three objectives may appear to be dominated in the projections.

The graphs show collectively that the system is able to improve all three objectives simultaneously and to quickly derive solutions that dominate the original

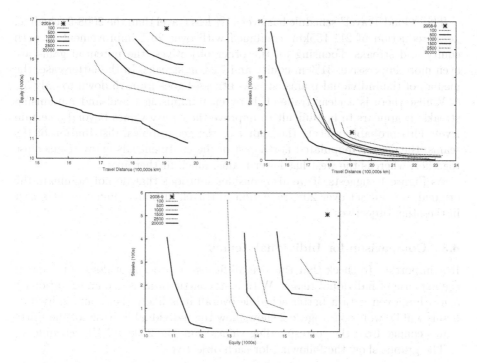

Fig. 1. Front progression at various generation numbers for one run of our system. All objectives are being minimised. Only feasible solutions which are non-dominated in the relevant objectives are shown.

schedule, i.e. that beat it in all three objectives. All of the solutions in the final population were feasible (i.e. the constraint objective was 0), a situation that eventuates fairly early in the run, and it contained 729 solutions that dominate the original. The width of the final front indicates that the system can derive a set of solutions that offer a range of trade-offs across the objectives.

6.2   Comparison with Existing Schedule

Table 1 shows the fitness values of four selected schedules from Figure 1, plus the 2008–9 NHL schedule. We consider only schedules that dominate the original.

Table 1. Fitness values of four schedules from one run of our system, and comparison with the 2008–9 NHL schedule

	Travel (km)	Equity	Streaks
2008–9	1,909,224	16,555	506
Good Travel	1,698,069	13,042	399
Good Equity	1,883,880	11,452	249
Good Streaks	1,851,139	12,148	144
Balanced	1,799,133	12,110	217

The "Good Travel" schedule has over 11% less travel than the 2008-9 schedule, i.e. a reduction of 211,155km, combined with over 20% improvement in both equity and streaks. Focusing on the other objectives, the potential gains are even more impressive: 31% in equity and 72% in streaks. In the latter case, the median of the individual teams' streaks fitnesses goes from 15 down to only 3.

Whilst there is a clear trade-off between minimising travel and minimising streaks, it appears to be difficult to improve the equity values beyond a certain level. This probably reflects the reality of the geographical distribution of the teams: the densely-populated north-east of the continent has many teams close together, whereas the west has fewer teams, much further apart.

As Figure 1 suggests, if we also consider solutions that do not dominate the original, we can get over 20% reduction in travel, although obviously at a cost in the other objectives.

6.3   Comparison for Individual Teams

It is important to check that this overall fitness improvement does not come at the expense of individual teams. With a stochastic process like an evolutionary algorithm, even with a better schedule overall it is likely that some individual teams will be worse off. Figure 2 shows how the individual fitnesses of the thirty teams change from the corresponding values from the 2008–9 NHL schedule.

The graphs show the following for each objective.

Travel: in the "Good Travel" schedule, the biggest increase in travel is around 5,000km, which is small and applies to one of the teams that is doing well originally. The occasional increases in travel in the other schedules should be set against the fact that the teams average 64,000km travel in the 2008–9 NHL schedule.

Equity: in all four schedules, all teams are better off in equity.

Streaks: significantly worse streaks appear only in the "Good Travel" schedule: in the other three schedules, with only one exception losses are small and are confined to teams that were doing well in streaks originally.

Note that for all objectives in all derived schedules, the worst team does better than the worst team in the 2008–9 NHL schedule, and only rarely does a team do worse than its own worst previous schedule under the current NHL structure.

7   Conclusions

We have described a multi-objective evolutionary algorithm that generates fixture schedules for the National Hockey League. Experiments show that the system is able to derive schedules that beat the existing schedule in three objectives simultaneously: minimising the teams' total travel, promoting equity in rest time between games, and minimising long streaks of home or away games. It does this without unduly penalising any individual team. The system returns a set of schedules that offer a range of trade-offs across the different objectives.

Future work could include refining the objectives further, and explicitly assessing the changes for individual teams.

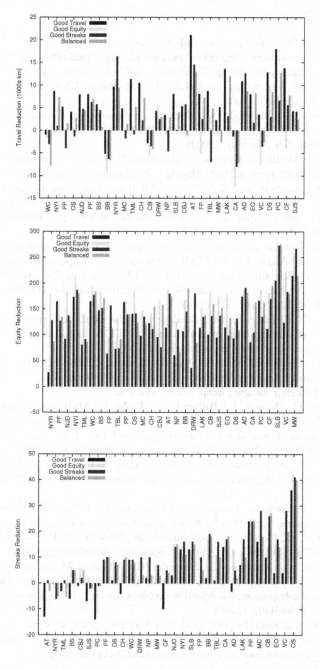

Fig. 2. Histograms of the individual teams' change in each fitness from the 2008–9 NHL schedule for four schedules from one run of our system. In each graph, the thirty teams are ordered best-worst by their fitness in the 2008–9 schedule, and a positive value for a team reflects an improved fitness.

References

1. NHL heads for record revenue and attendance, Reuters, http://www.reuters.com/articles/reutersEdge/idUSTRE5078AI20090108
2. Yang, J., Huang, H., Horng, J.: Devising a Cost-effective Baseball Scheduling by Evolutionary Algorithms. In: IEEE CEC, Honolulu, USA, pp. 1660–1665. IEEE Press, Los Alamitos (2002)
3. Schönberger, J., Mattfield, D., Kopfer, H.: Automated Timetable Generation for Rounds of a Table-tennis League. In: IEEE CEC, La Jolla, USA, pp. 277–284. IEEE Press, Los Alamitos (2000)
4. Barone, L., While, L., Hughes, P., Hingston, P.: Fixture Scheduling for Australian Rules Football using a Multi-objective Evolutionary Algorithm. In: IEEE CEC, Vancouver, Canada, pp. 3377–3384. IEEE Press, Los Alamitos (2006)
5. Dinitz, J., Lamken, E., Wallis, W.: Scheduling a Tournament. In: Dinitz, J., Colbourn, C. (eds.) Handbook of Combinatorial Designs, pp. 578–584. CRC Press, Colbourn (1995)
6. While, L., Barone, L.: Super 14 Rugby Fixture Scheduling using a Multi-objective Evolutionary Algorithm. In: IEEE Symposium on Computational Intelligence and Scheduling, Honolulu, USA, pp. 35–42. IEEE Press, Los Alamitos (2007)
7. Costa, D.: An Evolutionary Tabu Search Algorithm and the NHL Scheduling Problem. INFOR 33(3), 161–178 (1995)
8. Fonseca, C., Fleming, P.: Genetic algorithms for multi-objective optimisation: formulation, discussion, and generalisation. In: ICGA, San Francisco, USA, pp. 416–423. Morgan Kaufmann, San Francisco (1993)
9. Coello Coello, C., Lamont, G., Van Veldhuizen, D.: Evolutionary Algorithms for Solving Multi-objective Problems. Springer, Heidelberg (2007)
10. National Hockey League, http://www.nhl.com
11. Hingston, P., Barone, L., Huband, S., While, L.: Multi-level Ranking for Constrained Multi-objective Evolutionary Optimisation. In: Runarsson, T.P., Beyer, H.-G., Burke, E.K., Merelo-Guervós, J.J., Whitley, L.D., Yao, X. (eds.) PPSN 2006. LNCS, vol. 4193, pp. 563–572. Springer, Heidelberg (2006)
12. Huband, S., Tuppurainen, D., While, L., Barone, L., Hingston, P., Bearman, R.: Maximising Overall Value in Plant Design. Minerals Engineering 19(15), 1470–1478 (2006)
13. Deb, K., Agrawal, S., Pratap, A., Meyarivan, T.: A Fast Elitist Non-dominated Sorting Genetic Algorithm for Multi-objective Optimisation. In: Deb, K., Rudolph, G., Lutton, E., Merelo, J.J., Schoenauer, M., Schwefel, H.-P., Yao, X. (eds.) PPSN 2000. LNCS, vol. 1917, pp. 849–858. Springer, Heidelberg (2000)
14. Laumanns, M., Rudolph, G., Schwefel, H.: Adaptive Mutation Control in Panmictic and Spatially Distributed Multi-objective Evolutionary Algorithms. In: PPSN (workshop on multi-objective problem solving), Paris, France, Paris, France. Springer, Heidelberg (2000)
15. Craig, S.: National Hockey League Scheduling using a Multi-objective Evolutionary Algorithm. Honours thesis, The University of Western Australia (2009), http://www.csse.uwa.edu.au/~lyndon/SamCraigThesis.pdf

Classification-Assisted Memetic Algorithms for Equality-Constrained Optimization Problems

Stephanus Daniel Handoko, Chee Keong Kwoh, and Yew Soon Ong

School of Computer Engineering,
Nanyang Technological University, Singapore 639798
danielhandoko@pmail.ntu.edu.sg

Abstract. *Regressions* has successfully been incorporated into memetic algorithm (MA) to build surrogate models for the objective or constraint landscape of optimization problems. This helps to alleviate the needs for expensive fitness function evaluations by performing local refinements on the approximated landscape. *Classifications* can alternatively be used to assist MA on the choice of individuals that would experience refinements. Support-vector-assisted MA were recently proposed to alleviate needs for function evaluations in the inequality-constrained optimization problems by distinguishing regions of feasible solutions from those of the infeasible ones based on some past solutions such that search efforts can be focussed on some potential regions only. For problems having equality constraints, however, the feasible space would obviously be extremely small. It is thus extremely difficult for the global search component of the MA to produce feasible solutions. Hence, the classification of feasible and infeasible space would become ineffective. In this paper, a novel strategy to overcome such limitation is proposed, particularly for problems having one and only one equality constraint. The raw constraint value of an individual, instead of its feasibility class, is utilized in this work.

1 Introduction

Real-world optimization problems are often constrained. Generally, they can be formulated as finding some vector $\mathbf{x}$ of n real-valued independent variables that minimizes

$$f(\mathbf{x}) \tag{1}$$

subject to

$$\mathbf{g}(\mathbf{x}) \leq \mathbf{0} \tag{2}$$

$$\mathbf{h}(\mathbf{x}) = \mathbf{0} \tag{3}$$

where $\mathbf{x} \in \Re^n$ is often referred to as the *solution*, while $f : \Re^n \to \Re$ the *objective*, whereas $\mathbf{g} : \Re^n \to \Re^{n_g}$ and $\mathbf{h} : \Re^n \to \Re^{n_h}$ the *inequality* and *equality constraints*, respectively. There may also be bound constraints of the form $\mathbf{x}_\ell \leq \mathbf{x} \leq \mathbf{x}_u$ with $\mathbf{x}_\ell$ being the *lower* and $\mathbf{x}_u$ being the *upper bound*.

A. Nicholson and X. Li (Eds.): AI 2009, LNAI 5866, pp. 391–400, 2009.

In addition, real-world problems often involve expensive computation of their objective/constraint functions. Potential energy minimization in computational molecular chemistry or biology, for example, demands minutes to hours in each function evaluation depending on the size of the molecule as well as the fidelity of the model being used. The number of function evaluations required to solve problems in this category is therefore a significant issue.

One method to deal with the situation when only the inequality constraints were present was proposed in [1][2]. The five problems experimented with in [1], all of which have sufficiently reasonable ratio of the feasible to the whole search space, were solved within less amount of function evaluations using the proposed method. Dealing with the equality-constrained optimization problems, however, extremely small ratio of the feasible to the whole search space poses challenges to the global search algorithm to find a feasible solution, deeming the proposed method that separates the regions of feasible from infeasible solutions unsuited. In this paper, a novel strategy designed for the equality-constrained problems is proposed with a primary focus on problems with single equality constraint only.

2   Literature Review

2.1   Deterministic Algorithms

Methods of feasible directions is a class of deterministic algorithms that proceed from one feasible solution to another in order to solve constrained optimization problems [3]. *Zoutendijk* algorithm [4] and *sequential linear programming* (SLP) approaches [5][6][7] employ first-order approximation to both the objective and the constraints and are consequently prone to slow convergence. By employing second-order functional approximation, *sequential quadratic programming* (SQP) technique [8] enjoys quadratic rate of convergence and is the state-of-the-art of nonlinear programming solvers [9].

The following quadratic program is solved for direction $\mathbf{d}$ at the i-th major iteration of the SQP.

$$f(\mathbf{x}^{(i)}) + \nabla f(\mathbf{x}^{(i)})^T \mathbf{d} + \frac{1}{2}\mathbf{d}^T \nabla^2 L(\mathbf{x}^{(i)})\mathbf{d} \tag{4}$$

subject to

$$g_j(\mathbf{x}^{(i)}) + \nabla g_j(\mathbf{x}^{(i)})^T \mathbf{d} \leq 0 \qquad\qquad j = 1,\ldots,n_g \tag{5}$$
$$h_j(\mathbf{x}^{(i)}) + \nabla h_j(\mathbf{x}^{(i)})^T \mathbf{d} = 0 \qquad\qquad j = 1,\ldots,n_h \tag{6}$$

where

$$\nabla^2 L(\mathbf{x}^{(i)}) = \nabla^2 f(\mathbf{x}^{(i)}) + \sum_{i=1}^{n_g} \mu_j^{(i)} \nabla^2 g_j(\mathbf{x}^{(i)}) + \sum_{i=1}^{n_h} \nu_j^{(i)} \nabla^2 h_j(\mathbf{x}^{(i)}) \tag{7}$$

Throughout this work, the gradient vectors are assumed to be readily available while the Hessian matrices are updated using the quasi-Newton approximation.

Although quadratic rate of convergence is achievable, it is well known that deterministic optimization algorithms may not converge to the global optimum. Constrained optimization problems with nonlinear objective or constraints are in general intractable. It is impossible to design a deterministic algorithm that would outperform the exhaustive search in assuring global convergence [10].

2.2   Randomized Algorithms

Genetic algorithm (GA) [11] is a randomized algorithm with ability to overcome the drawback of deterministic optimization algorithms. Belonging to the class of *evolutionary computing*, GA is motivated by the natural inheritance of genes and the natural selection in the course of biological evolution [12] with the crossover, the mutation, and the survival-of-the-fittest being at its very heart. The simplest form of the algorithm assumes only one population evolved from one generation to the next. In dealing with constraints, the ranking scheme in [13] is often used. Summarized in the following three points, it is employed throughout this work.

– The feasible solution is preferred to the infeasible one.
– Between two feasible solutions, the one having better objective is preferred.
– Between two infeasible solutions, the one having less amount of violation to the constraints is preferred.

Research works on constrained evolutionary computing over the last decade include Homomorphous Mapping [14], Stochastic Ranking [15], the ASCHEA [16], Simple Multimembered Evolution Strategy (SMES) [17] that is known to have used the smallest number of fitness function evaluations (FFEs) so far, and some others [18][19][20][21]. Even though specially-designed operators accelerate the search for the global optimum to certain extent, it is a consensus that GA may suffer from excessively slow convergence trying to locate the optimum with sufficient precision because of its failure in exploiting local information [22].

2.3   Hybrid Algorithms

When hybridizing optimization methods, two central yet competing goals meet: *exploration* and *exploitation* [23]. The exploration provides reliable estimates of the global optimum by surveying the search space using global search methods, which are accommodated by the randomized algorithms. The exploitation then enhances each estimate by focussing the search efforts on its local neighborhood in order to produce a sufficiently accurate global optimum. This is accomplished using local search methods, which are facilitated by the deterministic algorithms. Motivated by Dawkins' notion of meme [24] (unit of cultural evolution capable of local refinements), a *memetic algorithm* (MA) exhibits this particular behavior. As the simplest variant, the simple MA simply interleaves global and local search methods one after the other. When compared to its conventional counterparts, the simple MA performs better by converging to a high quality global optimum and searching more efficiently [22].

Local refinements for each individual in the population, unfortunately, need not necessarily be the most efficient strategy. Local refinements of solutions at different locations may end with the same local optimum. Local search methods, such as the SQP, are known to converge quickly only when initialized with an approximate solution close enough to the optimal solution. Thus, the choice of individuals that should undergo local refinements becomes a critical issue in MA.

3   Proposed Approach

3.1   The Global Optimum

The optimization problem constrained by one and only one equality constraint always has global optimum situated at some particular location along the curve defined by $h(\mathbf{x}) = 0$. This curve is the feasible space of the problem. Illustrated in Fig. 1 is the constraint space of benchmark problem **g11**—the objective and constraint functions of which can be found in [25][26]. The solid curve represents the feasible space and the dot the global optimum of the problem. For this type of problems, the feasible space sets the solutions with positive constraint values apart from those with negative ones.

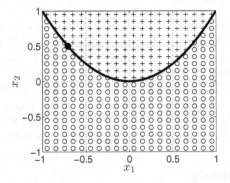

Fig. 1. Constraint Space of Benchmark Problem **g11**
o: constraint space where $h(\mathbf{x}) < 0$; +: constraint space where $h(\mathbf{x}) > 0$

It is understood that local search methods, such as the SQP, converge quickly to local optimum when they are initialized with an approximate solution that is close enough to the optimal solution [3]. Because one of the possibly many local optima must be the global optimum, focussing the search efforts on the regions nearby the feasible space will definitely increase the odds of being more efficient in locating the global optimum of the problem. This is achieved in this work by utilizing one fact that the feasible space of an optimization problem with single equality constraint is the zero-crossing of the constraint value.

3.2   The Neighborhood

Similar to [1], neighborhood $\mathcal{N}$ of the individual $\mathbf{x}$ is defined as the collection of k nearest solutions (to the individual) obtained from database of past solutions. The Euclidean distance in (8) is used throughout as a sparsity measure between any two n-dimensional solutions $\mathbf{p}$ and $\mathbf{q}$.

$$d_{\mathbf{pq}} = \sqrt{\sum_{i=1}^{n}(q_i - p_i)^2} \tag{8}$$

For the k neighbors are infallibly past solutions, no additional FFE is necessary to know quantities associated with these solutions. Freely accessible information include the constraint values based on which two classes can be derived. Should a neighbor and its corresponding class be represented as $\mathbf{x}_i$ and y_i, respectively, $\mathcal{N} = \{(\mathbf{x}_i, y_i) : i = 1, 2, \ldots, k\} = \{(\mathbf{x}_1, y_1), \ldots, (\mathbf{x}_k, y_k)\}$ defines the information contained within this neighborhood with

$$y_i = sign(h(\mathbf{x}_i)) = \begin{cases} +1 & h(\mathbf{x}_i) > 0 \\ -1 & h(\mathbf{x}_i) < 0 \end{cases} \tag{9}$$

Different from [1], the neighbors in this work are restricted to past solutions $\mathbf{x}_i$ for which $h(\mathbf{x}_i) \neq 0$. This means all the neighbors shall be infeasible—which is indeed desirable as the regions surrounding the feasible space must be infeasible. By making use of the signs of the constraint values of an individual's neighbors, it is demanded that the individual relative position can be predicted such that local search will only be executed if the individual is nearby the feasible space of the optimization problem.

Mixed Neighborhood. This type of neighborhood consists of members having positive and negative constraint values. Neighborhood of this type is absolutely of significant interest and importance. A two-class classification subproblem can be formulated out of this scenario. The decision boundary produced as the result of solving the classification subproblem does not only distinguish the regions of positive constraint values from those of the negative ones, but also approximates the feasible space of the optimization problem locally. *Support Vector Machine* (SVM) [27] will be described in the next subsection to serve this purpose.

Positive-only Neighborhood. A positive-only neighborhood, as indicated by its name, consists of members having only positive constraint values. This is of little or no importance as there is no clue about the feasible space of the problem that can be deduced from this type of neighborhood.

Negative-only Neighborhood. A negative-only neighborhood, as revealed by its name, consists of members having only negative constraint values. Similarly, no clue about the feasible space of the problem can be mined from this type of neighborhood, making it of little or no significance.

3.3   The Support Vector Machine (SVM)

SVM is a machine-learning technique initially proposed as a two-class classifier. It is well-known as being characterized by its ability to maximize the geometric margin between the two classes, and simultaneously, minimize the classification error. Upon provision of k training data instances $(\mathbf{x}_i, y_i)$ where $y_i \in \{-1, +1\}$ for all $i = 1, 2, \ldots, k$, the SVM needs to maximize the quadratic program below.

$$\sum_{i=1}^{k} \alpha_i - \frac{1}{2} \sum_{i=1}^{k} \sum_{j=1}^{k} \alpha_i \alpha_j y_i y_j (\mathbf{x}_i \cdot \mathbf{x}_j) \tag{10}$$

subject to

$$\sum_{i=1}^{k} y_i \alpha_i = 0 \tag{11}$$

$$\forall i \; \alpha_i \geq 0 \tag{12}$$

Collection of training data instances having $\alpha > 0$ defines the support vectors. Every one of them is situated at the decision surface $D(\mathbf{x}) = +1$ or $D(\mathbf{x}) = -1$ depending on which class it belongs to. Weight vector $\mathbf{w}$ and bias w_0 are hence computed using (13) and (14) in which SV is the set of support vectors indices.

$$\mathbf{w} = \sum_{i=1}^{k} \alpha_i y_i \mathbf{x}_i = \sum_{i \in SV} \alpha_i y_i \mathbf{x}_i \tag{13}$$

$$w_0 = \frac{1}{|SV|} \sum_{i \in SV} \left(y_i - \sum_{j=1}^{k} \alpha_j y_j (\mathbf{x}_j \cdot \mathbf{x}_i) \right) \tag{14}$$

Upon encountering a mixed neighborhood, the SVM needs training based on the k instances of $\mathcal{N}$. Subsequently, the SVM can be used to predict the relative position of the individual $\mathbf{x}$ with respect to neighborhood $\mathcal{N}$, producing one of the following three possible outcomes.

1. $|D(\mathbf{x})| \leq +1$

   The individual $\mathbf{x}$ has been estimated to be located nearby the feasible space. With no neighbors found within this region, furthermore, local refinement is obviously necessary to exploit this seemingly unexplored search space.

2. $D(\mathbf{x}) > +1$

   Depending on the actual value of $D(\mathbf{x})$, the individual $\mathbf{x}$ may be close enough to the feasible space. With neighbors around, there may not be further need to exploit this previously explored region of the search space.

3. $D(\mathbf{x}) < -1$

   Similar to case 2, the individual $\mathbf{x}$ may be located close to the feasible space depending on the value of $D(\mathbf{x})$. With neighbors around, no exploitation of this previously explored search space would be necessary.

3.4 The Complete Algorithm

Algorithm 1 presents the complete algorithm of the proposed approach in pseudo-code form. An important point worth noting is that the size of the neighborhood is recommended to be some multiple of the dimensionality of the problem being solved such that it would be large enough to capture important information yet small enough to ensure locality and allow the SVM to perform reasonably fast as its complexity depends largely on the number of training data instances. While the cost of running the SVM may not be inexpensive, the efforts required for evaluating the objective and constraint functions may be magnitudes greater for many practical optimization problems. Thus, the additional budget incurred by the SVM will become insignificant when dealing with some computationally-expensive optimization problems.

Algorithm 1. Classification-assisted MA (CaMA)

Initialize a population
Evaluate the population
while no stopping criteria have been fulfilled **do**
    for each individual **x** in the population **do**
        if past solutions are of negative only or positive only constraint values **then**
            Refine **x** using local search
        else
            if neighborhood $\mathcal{N}$ of **x** is a mixed neighborhood **then**
                Train SVM based on $\mathcal{N}$ to obtain decision function $D(\cdot)$
                if $|D(\mathbf{x})| \leq 1$ **then**
                    Refine **x** using local search
                end if
            end if
        end if
    end for
    Evolve the population through crossover, mutation, and elitism
    Evaluate the population
end while

---

4 Results and Discussions

Using GA as the global and SQP as the local search method, an empirical study was carried out with a population size of 100 individuals and a maximum of $2n$ fitness function evaluations (FFEs) for each individual refinement with n being the dimensionality of the problem. Experimented with are benchmark problem **g03** for $n = 2, \ldots, 10$ and **g11**, the objective and constraint functions of which can be found in [25]. These are the only two benchmark problems having single equality constraints among the 24 problems in [26]. When the proposed method was used, a neighborhood size of $2n$ was assumed.

Table 1 tabulates the performance of 30 independent runs of the simple MA (SMA) as well as the classification-assisted MA (CaMA) proposed in this paper.

Table 1. Number of FFEs Required by SMA and CaMA to Locate the Global Optimum

Problem	Statistics	SMA	CaMA	Saving
g03 $(n = 2)$	best	284	124	58.01%
	average	362	152	
	worst	471	187	
g03 $(n = 3)$	best	935	220	72.49%
	average	1,116	307	
	worst	1,244	411	
g03 $(n = 4)$	best	1,543	284	74.18%
	average	1,681	434	
	worst	1,812	552	
g03 $(n = 5)$	best	1,842	258	81.08%
	average	2,083	394	
	worst	2,258	536	
g03 $(n = 6)$	best	2,212	228	85.89%
	average	2,389	337	
	worst	2,639	642	
g03 $(n = 7)$	best	2,499	148	85.00%
	average	2,714	407	
	worst	2,955	2,763	
g03 $(n = 8)$	best	2,763	180	67.11%
	average	3,572	1,175	
	worst	5,747	3,499	
g03 $(n = 9)$	best	3,001	157	45.70%
	average	4,330	2,351	
	worst	6,257	4,441	
g03 $(n = 10)$	best	3,429	401	46.83%
	average	5,539	2,945	
	worst	6,477	5,788	
g11 $(n = 2)$	best	523	116	74.59%
	average	606	154	
	worst	712	225	

As the simplest variant of MA, the simple MA simply interleaves the global with the local search methods one after the other. In other words, each individual in the population would experience local refinements in the context of simple MA. The percentage of saving achievable by the CaMA with respect to the SMA on the average is then calculated as follows.

$$\text{saving} = \frac{\#\text{FFEs}_{(\text{SMA})} - \#\text{FFEs}_{(\text{CaMA})}}{\#\text{FFEs}_{(\text{SMA})}} \times 100\% \tag{15}$$

It is clear from the table that the CaMA consistently outperforms the SMA in the best, worst, and average cases. With average savings ranging from about

45% to 85%, the CaMA shall undoubtedly bring great advantage when solving computationally-expensive optimization problems. For this category of problems, reductions of several hundreds to several thousands of FFEs as exhibited over the benchmark problems could easily translate to savings of days to months of computational time. All these are made possible as search efforts were focussed around the regions that surround the feasible space of the optimization problems, thanks to the classification algorithms, such as the SVM, that enable prediction of local feasible space of the problems. As unnecessary refinements initiated with solutions relatively far away from the feasible space are eliminated, less number of FFEs are required in locating the global optimum of the problems.

5 Conclusion

Raw constraint values—rather than feasibility classes—are utilized in this work to focus search efforts on the regions that surround the minute feasible space of optimization problems with single equality constraint. Savings of up to 85% are achievable in term of the number of fitness function evaluations needed to solve benchmark problem **g03** with dimensionality varied from 2 to 10 as well as **g11**. In the optimization of computationally-expensive problems, such amount could bring significant time reduction. Thus, generalization to problems with multiple equality constraints and possibly some inequality constraints shall be addressed in immediate future.

References

1. Handoko, S.D., et al.: Using Classification for Constrained Memetic Algorithm: A New Paradigm. In: Proceedings of the 2008 IEEE International Conference on Systems, Man and Cybernetics (2008)
2. Handoko, S.D., et al.: Feasibility Structure Modeling: An Effective Chaperone for Constrained Memetic Algorithms. IEEE Transactions on Evolutionary Computation (in press)
3. Bazaraa, M.S., Sherali, H.D., Shetty, C.M.: Nonlinear Programming: Theory and Algorithms. Wiley-Interscience, Hoboken (2006)
4. Zoutendijk, G.: Methods of Feasible Directions. Elsevier, Amsterdam (1960)
5. Baker, T.E., Ventker, R.: Successive Linear Programming in Refinery Logistic Models. Presented at the ORSA/TIMS Joint National Meeting (1980)
6. Baker, T.E., Lasdon, L.S.: Successive Linear Programming at Exxon. Management Science 31(3), 264–274 (1985)
7. Zhang, J.Z., Kim, N.H., Lasdon, L.S.: An Improved Successive Linear Programming Algorithm. Management Science 31(10), 1312–1331 (1985)
8. Wilson, R.B.: A Simplicial Algorithm for Convex Programming. Ph.D. Thesis, Harvard University (1963)
9. Schittkowski, K.: NLPQL: A FORTRAN Subroutine Solving Constrained Nonlinear Programming Problems. Annals of Operations Research 5(2), 485–500 (1986)
10. Wright, S.J.: Nonlinear and Semidefinite Programming. In: Proceedings of Symposia in Applied Mathematics, vol. 61, pp. 115–138 (2004)

11. Holland, J.H.: Adaptation in Natural and Artificial Systems. The University of Michigan Press (1975)
12. Darwin, C.: On the Origin of Species by Means of Natural Selection, or the Preservation of Favoured Races in the Struggle for Life. John Murray (1859)
13. Deb, K.: An Efficient Constraint Handling Method for Genetic Algorithms. Computer Methods in Applied Mechanics and Engineering 186(2-4), 311–338 (2000)
14. Koziel, S., Michalewicz, Z.: Evolutionary Algorithms, Homomorphous Mappings, and Constrained Parameter Optimization. Evolutionary Computation 7(1), 19–44 (1999)
15. Runarsson, T.P., Xin, Y.: Stochastic Ranking for Constrained Evolutionary Optimization. IEEE Transactions on Evolutionary Computation 4(3), 284–294 (2000)
16. Hamida, S.B., Schoenauer, M.: ASCHEA: New Results Using Adaptive Segregational Constraint Handling. In: Proceedings of the 2002 Congress on Evolutionary Computation, pp. 884–889 (2002)
17. Mezura-Montes, E., Coello, C.A.C.: A Simple Multi-Membered Evolution Strategy to Solve Constrained Optimization Problems. IEEE Transactions on Evolutionary Computation 9(1), 1–17 (2005)
18. Barbosa, H.J.C., Lemonge, A.C.C.: A New Adaptive Penalty Scheme for Genetic Algorithms. Information Science 156(3-4), 215–251 (2003)
19. Farmani, R., Wright, J.A.: Self-Adaptive Fitness Formulation for Constrained Optimization. IEEE Transactions on Evolutionary Computation 7(5), 445–455 (2003)
20. Chootinan, P., Chen, A.: Constraint Handling in Genetic Algorithms Using a Gradient-based Repair Method. Computers and Operation Research 33(8), 2263–2281 (2006)
21. Elfeky, E.Z., Sarker, R.A., Essam, D.L.: A Simple Ranking and Selection for Constrained Evolutionary Optimization. In: Wang, T.-D., Li, X., Chen, S.-H., Wang, X., Abbass, H.A., Iba, H., Chen, G.-L., Yao, X. (eds.) SEAL 2006. LNCS, vol. 4247, pp. 537–544. Springer, Heidelberg (2006)
22. Tang, J., Lim, M.H., Ong, Y.S.: Diversity-Adaptive Parallel Memetic Algorithm for Solving Large Scale Combinatorial Optimization Problems. Soft Computing: A Fusion of Foundations, Methodologies, and Applications 11(9), 873–888 (2007)
23. Torn, A., Zilinskas, A.: Global Optimization. Springer, Heidelberg (1989)
24. Dawkins, R.: The Selfish Gene. Oxford University Press, Oxford (1976)
25. Handoko, S.D., et al.: A Study on Constrained MA Using GA and SQP: Analytical vs. Finite-Difference Gradients. In: Proceedings of the 2008 Congress on Evolutionary Computation (2008)
26. Liang, J.J., et al.: Problem Definitions and Evaluation Criteria for the CEC 2006 Special Session on Constrained Real-Parameter Optimization. Technical Report (2006)
27. Vapnik, V.N.: The Nature of Statistical Learning Theory. Springer, Heidelberg (1995)

Unsupervised Text Normalization Approach for Morphological Analysis of Blog Documents

Kazushi Ikeda, Tadashi Yanagihara, Kazunori Matsumoto,
and Yasuhiro Takishima

KDDI R&D Laboratories, Inc., 2-1-15 Ohara Fujimino, Saitama 356-8502, Japan

Abstract. In this paper, we propose an algorithm for reducing the number of unknown words on blog documents by replacing peculiar expressions with formal expressions. Japanese blog documents contain many peculiar expressions regarded as unknown sequences by morphological analyzers. Reducing these unknown sequences improves the accuracy of morphological analysis for blog documents. Manual registration of peculiar expressions to the morphological dictionaries is a conventional solution, which is costly and requires specialized knowledge. In our algorithm, substitution candidates of peculiar expressions are automatically retrieved from formally written documents such as newspapers and stored as substitution rules. For the correct replacement, a substitution rule is selected based on three criteria; its appearance frequency in retrieval process, the edit distance between substituted sequences and the original text, and the estimated accuracy improvements of word segmentation after the substitution. Experimental results show our algorithm reduces the number of unknown words by 30.3%, maintaining the same segmentation accuracy as the conventional methods, which is twice the reduction rate of the conventional methods.

1 Introduction

Internet use becomes more widespread. Blogs are regarded as large linguistic resources where people express their feelings and thoughts. Topics such as news and technologies are also gain on blogs. Studies of blog analysis for information retrieval, topic extraction, blog ranking, and so forth, have attracted much attention in recent years [1, 2]. However, in Japanese blogs, people tend to use peculiar expressions, which greatly reduce the performance of linguistic analysis since morphological analyzers regard them as unknown sequences. Reducing these unknown sequences improves the accuracy of morphological analysis for blog documents. Although manual registration of peculiar expressions to morphological dictionaries is a common solution, registration of unknown words requires much effort and specialized knowledge. For example, information related to the word such as its part of speech and inflected form should be acquired. Compatibility of the dictionary should be maintained. From our experience, only 30,000 unknown words per month are manually registered by an experienced worker. On the other hand, in our pre-examination by using popular Japanese morphological analyzer MeCab [3], 6,000,000

A. Nicholson and X. Li (Eds.): AI 2009, LNAI 5866, pp. 401–411, 2009.

blog sentences contain about 650,000 kinds of unknown words, which shows the difficulty in the manual registration of unknown words.

Most peculiar expressions seen in Japanese blogs are derived from formal expressions and categorized in some typical patterns based on their way of derivation. For example, visually-similar characters tend to be substituted, such as, "わたしは" instead of "わたしは" ("i @m" instead of "I am" in English), where "わ" is substituted by its small character "ゎ" and "し" is substituted by its visually-similar symbol "∪". In the same manner, "かわいい" ("cute") can be substituted by "カわいい" ("cｕтe"). In another example of derivation patterns, words are written to reflect their pronunciation in conversation. "すごい" ("amazing") normally pronounced "sugoi" is emphatically pronounced "sugooi" or "suggoi" in conversation.

In blogs, people exactly describe these pronunciation and create various kinds of expressions, such as, "すごーい", "すごぉい", "すっごい" (pronounced "sugooi", "sugooi" and "suggoi" respectively), and so forth. Japanese is written with several different character types such as kanji, hiragana, katakana, romaji (Roman alphabet), and so forth. People intentionally spell words in different character types. For example, "やばい" ("troublesome" written in hiragana format) sometimes appears in its katakana format "ヤバい" in blogs.

In order to improve the accuracy of morphological analysis of blog documents, we propose an algorithm for reducing the number of unknown words by automatically replacing peculiar expressions with formal expressions. In our algorithm, candidates for the substitution of peculiar expressions are automatically retrieved from formally written documents such as newspapers and stored as substitution rules. In order to replace a peculiar expression with the most suitable expression for the context, substitution rules are scored and selected based on three criteria; the appearance frequency in the retrieval process, the edit distance between substituted sequences and the original text, and the estimated accuracy improvements of word segmentation after the substitution.

We implemented our algorithm and compared its performance to conventional two methods. In our experiments, we modified 100,000 blog sentences and evaluated the number of unknown words and the accuracy of word segmentation. Our evaluation shows our algorithm reduces 30.3% of unknown words in original blog documents, which is twice the reduction rate of the conventional methods, maintaining the same segmentation accuracy. We also evaluated the scalability of the proposed algorithm and the trade-off between the number of unknown words and the ratio of errors in substitution.

2 Related Work

To date, we have found no direct researches which involve automated text normalization for improving morphological analysis of blog documents. However, we have identified a similar work for improving morphological analysis towards casual expressions. For example, the "Dictionary Expansion" algorithm [4] focuses on accuracy improvement of morphological analysis for Japanese colloquial expressions in Web chat applications. The authors find out the rules that generate colloquial expressions from formal expressions registered on morphological dictionaries. For example, they define rules like "adjective words tend to be inserted with a prolonged sound symbol" from the case seen in Web chat applications, where the formal adjective "すごい" tend to be written as "すごーい". Since this method relies heavily on the subjective view and skills of the operator, it is difficult to generate generic rules that can be usefully applied to many expressions. In [5], our experiments with 2,000,000 blogs show 37.2% of all the sentences affected by the algorithm of [4] increase errors in word segmentation. This suggests the algorithm proposed in [4] is not versatile enough to apply to any type of text data, including blogs.

Linguistic analysis of colloquial expressions is reported in [6-8]. The approaches shown in these papers also use manually generated rules, which require specialized knowledge to generate. Considering research to reduce the number of unknown words, the method for fluctuations of words written in katakana format is offered in [9], while the algorithm in [10] automatically obtains new words from Web pages. The estimation algorithm for parts of speech of unknown words is offered in [11].The estimation algorithm for word segmentation in Japanese sentences is offered in [12]. These works are not focused on recognizing peculiar expressions on blogs.

In our previous work [5], we propose the "Initial Rule Expansion" algorithm, which automatically creates highly accurate rules from manually given low-level rules based on the statistics. For example, given rules such as "お⇒"and "ぉ⇒お"will result in specific rules (a) "すごぉい⇒すごい" and (b) "すごぉい⇒すごおい" ("X⇒Y" means substitution of Y for X). Then, (a) "すごぉい⇒すごい"is stored due to its statistically high correctness. Although this method is effective in reducing the number of unknown words of peculiar expression, only a limited number of sentences are modified by the rules. Due to the many kinds of peculiar expressions, it is almost impossible to manually cover all the initial rules.

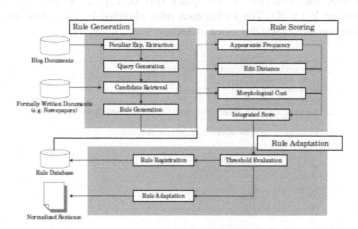

Fig. 1. Overview of the substitution algorithm

3 Algorithm Design

Fig. 1 shows an overview of our algorithm. Inputs of the algorithm are blog documents and formally written documents such as newspapers. Substitution candidates for a peculiar expression are listed up from formally written documents.

Substitution rules are generated from the peculiar expression and its substitution candidates. Substitution rules are scored based on the following three criteria; (1) the appearance frequency in the retrieval process, (2) the edit distance between substituted sequences and original text, and (3) the estimated accuracy improvements of word segmentation after the substitution. A substitution rule with the highest score is selected as the most suitable expression for the context.

3.1 Generation of Substitution Rules

The rule generation algorithm has four steps, (1) extraction of a peculiar expression, (2) generation of a query for substitution candidates of the peculiar expression, (3) retrieval of the substitution candidates, and (4) generation of rules. Fig. 2 shows an example of rule generation. The sentence "できるかどうかゎわかりません" ("I wonder whether it is possible." in English) contains the peculiar expression "かゎ"and this expression should be substituted by the formal expression "かゎ"Most peculiar expressions are detected as unknown words in the morphological analysis because they are not listed in morphological dictionaries ((1) of Fig. 2 shows the peculiar expression "かゎ" is detected as an unknown word). Substitution candidates are retrieved from formally written documents. A query for substitution candidates is created by extracting the unknown word with its adjoining morphemes, and replacing the unknown word with the wild-card symbol. ((2) of Fig. 2 shows the peculiar expression "かゎ"and its adjoining morphemes "どう"and "わかり"are taken out, and "かゎ"is replaced with the wild-card symbol, "*"). As a result of the retrieval, substitution candidates are obtained ((3) of Fig. 2). Substitution rules are generated from the peculiar expression and the parts matched with the wild-card of the query ((4) of Fig. 2, where "X⇒Y" means substitution of Y for X). The substitution rules obtained by the above algorithm are scored based on the criteria described in Section 3.2.

```
(1) Extraction of a peculiar expression
    Blog sentence: 「できるかどう かゎ 分かりません」
    Word segmentation: 「できる/か/どう/かゎ/分かり/ませ/ん」
    Unknown word: かゎ
(2) Generation of a query for substitution candidates
    「どう ＊ 分かり」
(3) Retrieval of the substitution candidates
    「... どう    かは    分かり ...」
    「... どう    か     分かり ...」
    「... どう    したらいいのか    分かり ...」
    「... どう    かは    分かり ...」
    「... どう    か     分かり ...」
    「... どう    かは    分かり ...」
    「... どう    かは    分かり ...」
    「... どう    なっているか    分かり ...」
(4) Generation of substitution rules
    「かゎ ⇒ かは」
    「かゎ ⇒ か」
    「かゎ ⇒ したらいいのか」
    「かゎ ⇒ なっているか」
"X ⇒ Y" means substitution of Y for X
```

Fig. 2. Generation of Substitution Rules

3.2 Scoring Substitution Rules

Substitution rules are scored based on the following criteria.

Scoring Rules Based on the Appearance Frequency. The expressions appearing in the similar contexts to the peculiar expressions are expected to be suitable substitution candidates. The amount of retrievals for each substituted expression represents its appearance frequency. Table 1 is a summary of the appearance frequency of each substitution candidate retrieved from (3) of Fig. 2. The candidate "かは"gains a high score because it often appears in a similar context to the peculiar expression "カ わ". The appearance frequency is divided by the total number of retrievals for normalization so as not to depend on the number of retrievals.

Scoring Rules Based on the Edit Distance Since. Most peculiar expressions are derived from formal expressions, rules which greatly change a peculiar expression are considered not to be the best substitution candidate. The edit distance such as the Levenshtein distance [13] is a criterion for measuring the amount of difference between two character strings. The edit distance between two strings is given by the minimum number of operations needed to transform one string into the other, where an operation is an insertion, deletion, or substitution of a single character. For example, the edit distance of the word " フォーラム " ("forum") and "ファーム"("farm") is 2 since the substitution of "ア"for "オ"and the deletion of "ラ"is the minimum way to change the former into the latter. Table 2 is a summary of the edit distance in each substitution rule generated in the example of Fig. 2. Substitution rules with a large edit distance gain a low score such as, "カわ⇒したらいいのか"and "カわ⇒なっているか".

The weighted edit distance based on the way of derivation is expected to work effectively. For example, as previously stated, people intentionally write a word in katakana format on blogs, which is normally written in hiragana format on formal documents. Giving a small edit distance between katakana and hiragana characters is effective. The visual similarity of two characters and the similarity in the pronunciation of two words should also be reflected on the edit distance, which is our future work.

Scoring Rules Based on the Estimated Accuracy Improvements of Word Segmentation. Mistakes in rule selection lead to the generation of ungrammatical sentences. In our algorithm, morphological analysis cost is used for evaluating the relative unnaturalness of sentences. Morphological analysis cost is calculated from the appearance probability of a word and the joint probability of each word [14]. The

Table 1. Scoring Based on Appearance Frequency

Substitution Rules	Appearance Frequency	Normalized Frequency
カわ ⇒ かは	4	0.5
カわ ⇒ か	2	0.25
カわ ⇒ したらいいのか	1	0.125
カわ ⇒ なっているか	1	0.125

Table 2. Scoring Based on Edit Distance

Substitution Rules	Edit	Edit Distance
カわ ⇒ かわは	2 Substitution	2
カわ ⇒ か	1 Substitution, 1 Deletion	2
カわ ⇒ したらいいのか	2 Substitution, 5 Insertion	7
カわ ⇒ なっているか	2 Substitution, 4 Insertion	6

validity of rule adoption is evaluated by comparing the morphological analysis costs of a sentence with peculiar expressions and substituted sentences.

Fig. 3 shows morphological analysis costs of each substituted sentence obtained in Fig. 2. Each sentence is segmented by a morphological analyzer. Each segment is given a morphological analysis cost (accumulated cost from the beginning of each sentence). The morphological analysis cost at the end of each sentence is considered to show the grammatical correctness of the whole sentence. Morphological analysis cost around peculiar expressions such as " カわ " become higher since appearance probabilities and joint probabilities of peculiar expressions are low compared to formal expressions. The score of a rule is defined as the difference of the morphological analysis cost between a substituted sentence and its original sentence. Although shorter sentences tend to get lower morphological analysis costs, rules deleting many characters get large edit distances. In the following section, we explain the calculation of integrated score of a rule based on the above criteria.

Calculation of Integrated Score. The integrated score of a substitution rule score is generally defined as

$$score = f(freq) + g(dist) + h(cost) \qquad (1)$$

where freq is the appearance frequency of the substituted expression, dist is the edit distance between the substituted expression and original peculiar expression, cost is the difference of the morphological analysis cost between the substituted sentence and the original one, and f; g; h are the functions for weighting each criterion. In this paper, function f; g; h is simply defined as constants α, β and γ as follows.

$$score = \alpha\, freq + \beta\, dist + \gamma\, cost \qquad (2)$$

As an example, Table 3 shows examples of substitution rules that appeared in the experiments in Section 4. The value of constants in Expression (2) is set as $\alpha = 1$, $\beta = -16$, $\gamma = -0.005$. According to the table, the peculiar expression "てたょ"should be substituted by "てた"based on its appearance frequency and morphological analysis cost. The expression "てたといえよう"retrieved as substitution candidates gets a lower score because of its large edit distance from its original expression. In the same manner, "今日ゎ早めに"should be substituted by "今日は早めに"based on its appearance frequency and edit distance. "ぉ金無い" ("no money") should be substituted by "金無い"based on its morphological analysis cost. In this case, however, "税金無い" ("no tax") also gets a relatively high score due to its high appearance frequency and low edit distance. In Section 4, we evaluated the miss ratio of substitution, where the meaning of a sentence has changed.

```
Blog Sentence: できるかどう ｶゎ 分かりません
Segmentation:  できる| か | どう |ｶゎ|分かり|ませ| ん
Morph. cost:   5742|8263|11751|34685|39098|40388|39914
Total cost:    39914

Candidate 1 :  できるかどう かは 分かりません
Segmentation:  できる| か | どう | か | は |分かり|ませ| ん
Morph. cost:   5742|8263|11751|14430|15438|19341|20631|20157
Total cost:    20157
Difference from before substitution: -19757

Candidate 2:   できるかどう か 分かりません
Segmentation:  できる| か |どうか|分かり|ませ| ん
Morph. cost:   5742|8263|16737|20120|21410|20936
Total cost:    20936
Difference from before substitution: -18978

Candidate 3:   できるかどう したらいいのか 分かりません
Segmentation:  できる|か|どう|し|たら|いい|の|か|分かり|ませ|ん
Morph. cost:   5742|8263|11751| ... |26035|27325|26851
Total cost :   26851
Difference from before substitution: -13063

Candidate 4:   できるか どうなっているか 分かりません
Segmentation:  できる|か|どう|なっ|て|いる|か|分かり|ませ|ん
Morph. cost:   5742|8263|11751| ... |22975|24265|23791
Total cost:    23791
Difference from before substitution: -16123
```

Fig. 3. Scoring Based on Morphological Analysis Cost

Table 3. Integrated Scores Based on Each Criterion ($\alpha = 1$, $\beta = -16$, $\gamma = -0.005$)

Substitution Rules	Frequency (%)	Edit Distance	Morph. Cost	Total Score
てたょ⇒てた	20	1	-15757	83.3
てたょ⇒てたよ	0	1	-14037	54.2
てたょ⇒てたい	2	1	-10946	40.7
てたょ⇒てたといえよう	2	5	-9108	-32.5
今日ゎ早めに⇒今日は早めに	33	1	-12721	80.6
今日ゎ早めに⇒今日で、早めに	0	2	-16449	50.2
今日ゎ早めに⇒今日は少し早めに	11	3	-13205	29.0
お金無い⇒金無い	0	1	-13131	49.7
お金無い⇒税金無い	8	1	-9887	41.4
お金無い⇒お金無い	0	1	-10974	38.9
お金無い⇒う金無い	4	1	-6654	21.3

3.3   Adoption and Registration of Substitution Rules

Adoption of a substitution rule is decided depending on whether its score is higher than a given threshold. The number of substitutions and its accuracy are in a trade-off relation. In Section 4, we evaluated the trade-off by monitoring the reduced number of unknown words and word segmentation accuracy on several thresholds.

In our algorithm, rules with higher scores than the given threshold are registered on the database. When few substitution candidates are obtained from formally written documents, additional substitution rules are available from the database. When automatically created queries happen to be poor due to the neighbor characters of peculiar expressions, many unrelated expressions are retrieved. For example, a query generated from the sentence "子供はかわいいよね" ("kids are cute") may be

Table 4. Categorization of Substitution Rules Based on their Effects on the Meanings

Befor Substitution	After Substitution
(a) Small Changes in the Meanings	
今日は猫ちゃん来てたょー	今日は猫ちゃん来てた
("A cat stayed here today")	
とぉつても気持ちいい	とっても気持ちいい
("so good feelings")	
おいしそぉだったんだけどね	おいしそうだったんだけどね
("It seemed delicious.")	
めッちゃ汗かいた	めっちゃ汗かいた
("I was sweating a lot.")	
(b) Large Changes in the Meanings	
じゃあ、② 時に駅前で	じゃあ、七時に駅前で
("See you on the station at 2.")	("See you on the station at 7.")
ばかっぷる〜	ばかっする〜
("couples")	(no meaning)
可愛いすぎぃ	可愛すぎない
("so cute")	("not so cute")
おっはよぉー	おっはよい
("Hello")	(no meaning)
(c) Difficult to Categorize in (a) or (b)	
来ぉへんよっ	来へんよ
("I will not come")	
私も遅くなる時ぁるU	私も遅くなる時もある
("I sometimes late")	
ぜひぉためしあれ	ぜひためしあれ
("Lets try it")	
ハハハよかったね	よかったね
("Thats good")	

"子供は*ね"("kids are *") which retrieves any kind of adjectives as substitution candidates. Calculation and comparison of many substitution candidates require much time, and substitution for them tends to result in errors. In this case, only the rules on the database are used because peculiar expressions are expected to be correctly substituted in other cases and correct rules are stored on the database.

4 Performance Evaluation

We implemented our algorithm and compared its performance with two conventional algorithms, (a) "Dictionary Expansion Algorithm (DEA)"[4] and (b).

"Initial Rule Expansion Algorithm (IREA)"[5]. The problem with using DEA is the high error ratio on the word segmentation due to the over adoption of rules. The problem with using IREA is the lack of scalability, where only limited sentences are normalized by the manually given initial rule sets. Considering these problems, we evaluated the reduction of unknown words and the accuracy of word segmentation. We also evaluated the changes in the meanings of sentences because our substitution algorithm may greatly change the meanings of sentences due to their nature as an unsupervised algorithm. The trade-off relation of our algorithm is shown on several thresholds.

4.1 Experimental Settings

We executed morphological analysis of blog sentences by using DEA, IREA, and our algorithm and compared their performance based on the following four criteria; (1)

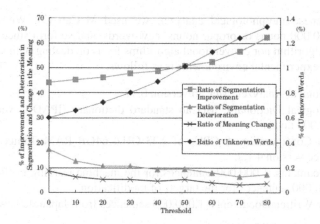

Fig. 4. Ratio of improvement and deterioration in word segmentation accuracy, ratio of change in the meaning of sentences, and ratio of unknown words on the threshold 0 to 80 in our algorithm

Table 5. Performance Evaluation of Each Algorithm

Algorithm	Improvement (%)	Deterioration (%)	Meaning Change (%)	Unknown Words (%)
MeCab	–	–	–	1.619
DEA	48.1	32.1	–	1.458
IREA	52.1	9.7	4.0	1.377
Ours	52.4	8.0	4.0	1.128

The ratio of sentences improving the word segmentation accuracy to the total modified sentences. (2) The ratio of sentences deteriorating the word segmentation accuracy to the total modified sentences. (3) The ratio of sentences changing its meaning to the total modified sentences. (4) The ratio of unknown words to the total morphs.

The accuracy of word segmentation is defined in the same way as the conventional methods [4, 5, 15], where word segmentation is correct when it is divided in the same manner as that performed manually. The segmentation improvement of a sentence is defined as the case where the original sentence has word segmentation errors and the substituted sentence has no errors. The segmentation deterioration of a sentence is defined as the opposite of the improvement case. Changes in the meaning of sentences are defined in three types, (a) almost no change, (b) obviously changed or the meaning of a sentence cannot be understood, and (c) difficult to categorize in category (a) or (b). Table 4 shows examples of substitutions categorized in category (a), (b), and (c). Substitutions in category (a) correctly replace peculiar expressions with formal expressions. They may cause a slight change in the impression of sentences, but no change in the meaning and the facts. Substitutions in category (b) obviously change the facts of sentences or make them incomprehensible.

Substitutions in category (c) do not change the facts of sentences, but their impressions are slightly different. We define that substitutions in category (b) and (c) change the meanings of sentences. In the following experiment, the ratio of improvement and deterioration in word segmentation and the meaning change ratio are manually evaluated by sampling 600 sentences modified in each algorithm.

As a Japanese morphological analyzer, we used MeCab [3]. We additionally registered 180,000 nouns or proper nouns, new words, and so forth since neither our substitution algorithm nor conventional algorithms focus on dealing with those words. The detail of experimental environments is as follows.

- Morphological Analyzer: MeCab version 0.97
- Morphological Dictionary: MeCab standard dictionary (IPADIC version 2.7.0) plus 180,000 nouns.
- Terminal Configuration: 8 CPUs of 2.33 GHz, 64GB RAM, Linux OS version 2.6.24, gcc version 4.1.2.
- Blog Documents: 1,000,000 sentences for the machine learning in IREA and the other 100,000 sentences for the targets of modification.
- Formally Written Documents: 1,000,000 sentences from Japanese newspapers.

4.2 Experimental Results

In our algorithm, the adoption of a substitution rule is decided according to the threshold. First, we evaluated the trade-off relation between the substitution accuracy and the ratio of unknown words. Fig. 4 shows the ratio of improvement and deterioration in word segmentation, the ratio of meaning change in the sentences, and the ratio of unknown words when the threshold changes from 0 to 80. When the threshold is low, many unknown words are replaced and the ratio of unknown words is low. However, manual evaluation of the accuracy in word segmentation and the meaning change ratio shows that most substituted words are recognized in other meanings on morphological analyzer. As the threshold becomes higher, the ratio of unknown words increases, but the accuracy of word segmentation and the meaning change ratio improve.

Table 5 shows the ratio of improvement and deterioration in word segmentation accuracy, the ratio of change in the meaning of sentences and the ratio of unknown words in each algorithm. We tuned the threshold of our algorithm to 60 in order to maintain the same level in word segmentation accuracy and the meaning change ratio as IREA. Our algorithm has higher word segmentation accuracy and smaller ratio of unknown words compared to those in DEA. The ratio of unknown words in our algorithm is 1.128%, where the reduction ratio is 30.3% from 1.619% of MeCab. The reduction ratio of unknown words in our algorithm is twice of that in IREA. This is because, in our algorithm, more substitution rules are obtained by the unsupervised algorithm, and the three criteria for accurate rule selection enable maintaining the word segmentation accuracy and the changes in the meaning of sentences in the same level as IREA. Considering scalability, the execution time of our algorithm was only 1 second to modify 1,000 blog sentences in the case of using 1,000,000 newspaper sentences.

5 Conclusion

In this paper, we proposed an algorithm for reducing the number of unknown words by replacing peculiar expressions seen in blog documents with formal expressions. In our algorithm, candidates for the substitution of peculiar expressions are automatically retrieved from formally written documents and stored as substitution rules. In order to replace a peculiar expression with the most suitable expression for the context, a substitution rule is selected based on three criteria; its appearance frequency in retrieval process, the edit distance between substituted sequences and the original text, and the estimated accuracy improvements of word segmentation after the substitution. The experimental results show our algorithm reduces 30.3% of unknown words in original blog documents at the same segmentation accuracy as conventional ones. This reduction rate is higher than twice of the rate of the conventional algorithm.

References

1. Nakajima, S., Tatemura, J., Hino, Y., Hara, Y., Tanaka, K.: Discovering important bloggers based on analyzing blog threads. In: Proc. of the 2nd Annual Workshop on the Weblogging Ecosystem: Aggregation, Analysis and Dynamics. Workshop at the WWW 2005 (2005)
2. Ni, X., Xue, G.-R., Ling, X., Yu, Y., Yang, Q.: Exploring in the weblog space by detecting informative and affective articles. In: Proc. of the 16th International World Wide Web Conference (WWW 2007), pp. 281–290 (2007)
3. Kudo, T.: Mecab: Yet another part-of-speech and morphological analyzer, http://mecab.sourceforge.net/
4. Kazama, J., Mitsuishi, Y., Makino, T., Kentaro, Torisawa, T., Tsujii, J.: Morphological analysis for Japanese Web chat. In: Proc. of NLP 1999, pp. 509–512 (1999) (in Japanese)
5. Ikeda, K., Yanagihara, T., Matsumoto, K., Takishima, Y.: An automatic rule generation method for modifying informal expression in blog documents. DBSJ Journal, The Database Society of Japan 8(1), 23–28 (2009)
6. Takemoto, Y., Fukushima, S.: Implementation and evaluation of a morphological analysis method for colloquial japanese text. Proc. of IPSJ SIG Notes 94(77), 105–112 (1994)
7. Takeshita, A., Fukunaga, H.: Morphological analysis for spoken language. In: Proc. of the 42nd National Convention of IPSJ, pp. 1–3 (1991)
8. Matsumoto, Y., Den, Y.: Morphological analysis of spoken japanese. Proc. of IPSJ SIG Notes 2001(55), 9–14 (2001)
9. Masuyama, T., Sekine, S., Nakagawa, H.: Automatic construction of japanese katakana variant list from large corpus. In: Proc. of the 20th International Conference on Computational Linguistics (COLING), pp. 1214–1219 (2004)
10. Murawaki, Y., Kurohashi, S.: Online acquisition of japanese unknown morphemes using morphological constraints. In: 2008 Conference on Empirical Methods in Natural Language Processing (EMNLP 2008), pp. 429–437 (2008)
11. Mori, S., Nagao, M.: Word extraction from corpora and its part-of-speech estimation using distributional analysis. In: Proc. of the 11th International Conference on Computational Linguistics (COLING), pp. 1119–1122 (1996)
12. Yanagihara, T., Matsumoto, K., Ikeda, K., Takishima, Y.: Word segmentation estimation using information criteria. In: Proc. of IPSJ NLP 190, pp. 43–47 (2009)
13. Levenshtein, V.I.: Binary codes capable of correcting deletions, insertions and reversals. Journal of Soviet Physics, Doklady, 707–710 (1966)
14. Kudo, T., Yamamoto, K., Matsumoto, Y.: Applying conditional random fields to japanese morphological analysis. In: Proc. of the 2004 Conference on Empirical Methods in Natural Language Processing (EMNLP 2004), pp. 230–237 (2004)
15. Nagata, M.: A stochastic japanese morphological analyzer using a forward-dp backward-a n-best search algorithm. In: Proc. of the 15th International Conference on Computational Linguistics (COLING), pp. 201–207 (1994)

Novel Memetic Algorithm for Protein Structure Prediction

Md. Kamrul Islam and Madhu Chetty

GSIT, Moansh University,
3842, VIC, Australia
{kamrul.islam,madhu.chetty}@infotech.monash.edu.au

Abstract. A novel Memetic Algorithm (MA) is proposed for investigating the complex *ab initio* protein structure prediction problem. The proposed MA has a new fitness function incorporating domain knowledge in the form of two new measures (H-compliance and P-compliance) to indicate hydrophobic and hydrophilic nature of a residue. It also includes two novel techniques for dynamically preserving best fit schema and for providing a guided search. The algorithm performance is investigated with the aid of commonly studied 2D lattice hydrophobic polar (HP) model for the benchmark as well as non-benchmark sequences. Comparative studies with other search algorithms reveal superior performance of the proposed technique.

Keywords: Memetic Algorithm, Pair-wise-interchange, Tabu Search, Modified fitness function, Schema preservation, Guided search space.

1 Introduction

The protein folding problem has remained one of the grand challenges in computational molecular biology. For PSP investigations, hydrophobic-polar (HP) protein model [1] is most commonly applied. The model considers hydrophobic (lacking affinity for water) and hydrophilic (water loving) interactions as the two main dominant forces in protein folding process and amino acids are therefore represented as either hydrophobic (H) or hydrophilic (P). For 2D modeling, these residues are located in square lattice ensuring a self-avoiding walk (SAW) so that the two residues do not occupy same space position. The fitness function measures the energy of the conformation which is obtained by evaluating the topological contacts between two hydrophobic residues (H-H) as -1 (provided they are not neighbors in given sequence) while topological contacts for other possible pairs (H-P, P-H, and P-P) are evaluated as 0. The energy matrix E_{TN} of the HP model [2] is given by *eqn. 1*. Protein conformation can be encoded in various ways such as absolute, relative and so on. In relative encoding, a conformation has three possible moves relative to current position, namely *forward (F), left (L)* and *right (R)*. The first move is always considered as forward (F).

Protein structure prediction (PSP), even in simplified hydrophobic-polar (HP) model, is NP-complete [3]. Hence, not only GA [4,5,6] but a plethora of other

A. Nicholson and X. Li (Eds.): AI 2009, LNAI 5866, pp. 412–421, 2009.

evolutionary algorithms [7] including Ant Colony Optimization (ACO), Tabu Search (TS), Monte Carlo (MC), Memetic Algorithm (MA) are being investigated. Since the PSP problem has a large and complex search space, algorithms which emphasis only on global optimization (e.g. GA) might not be able to perform properly. MA, a powerful combination of GA and local search (LS), due to its ability to combine local search (LS) techniques refines individual population and improves their fitness [8]. Usually, the flexible architecture of MA allows it to include different approaches for local search, i.e. gradient descent, pair wise interchange (PWI), tabu search (TS). In this paper, MA with pair wise interchange is referred as pair-wise MA (PMA) and MA with Tabu search is referred as tabu MA (TMA). Recent studies in various domains [9,10,11] show that MA is both efficient (less computations) and effective (higher accuracy) compared to other EAs. Comparisons between several EAs and MA with pair-wise-interchange (PWI) show better performance for MA [12]. However, limited work has been reported on its application to NP-complete PSP problem. Recently, hybridization of GA with Tabu search on PSP [6] showed a satisfactory performance. With changes in population size based on *complexity* of the protein sequence, its application is, however, limited because it is not always possible to know the complexity upfront. Krasnogor *et al.* [8,13,14,15] and Smith [16,17] applied MA to solve PSP problem using techniques such as fuzzy logic, multi-meme, co-evolution with limited improvement.

$$E_{TN} = \sum_{j=1}^{n-1} \sum_{k=j+1}^{n} N_{jk} \quad \text{where,}$$

$$N_{jk} = \begin{cases} -1 & \text{if } j \text{ and } k \text{ are both H residues and topological neighbour;} \\ 0 & \text{otherwise.} \end{cases} \tag{1}$$

An appropriate fitness function, capturing domain knowledge is very important for enhancing fitness function and improving the accuracy of prediction. For example, Radius of Gyration (RG), measuring radial distance from a given axis, was applied [4] to capture the domain characteristics. In an effort to use the characteristics of the amino acids, hydrophobic property was included in the fitness function [5]. However, there has been no effort to use the equally significant second hydrophilic (P) property of the residues. We propose a novel fitness function which not only maintains the significance of the existing fundamental fitness parameters but also incorporates domain knowledge to bring H type amino acids close to the H-core and pushing P type residues close to the boundary. This is achieved by developing two new measures for H and P characteristics, namely H-compliance and P-compliance. The proposed algorithm also includes a new technique for dynamically preserving the fit schema based on domain knowledge. Further, we also propose a novel approach to add interim individuals in a guided manner (rather than randomly) and also maintain the necessary diversity in population. Experiments are performed using the 2D HP lattice model and using the bench mark as well as non benchmark sequences. Comparisons with other techniques are also carried out which show a superior performance of the proposed algorithm.

2 Proposed Memetic Algorithm

In this section, we present the three novel aspects of the proposed MA which enhances its potential for solving the complex PSP problem.

2.1 Modified Fitness Function

An ideal empirical energy function contains only a few energy terms, is computationally efficient, which can be easily derived from experimental data [18]. Further, for PSP problem it should account for effects such as hydrophobic packing and include penalties for undesirable effects. We will address hydrophobic packing as it can prove to be important in removing the limitations of the existing fitness function *eqn. 1*. This is done by including two new fitness terms for capturing the H and P characteristics of the residues (i) H-compliance factor and (ii) P-compliance factor. The resulting new fitness function obtained by including the two new fitness terms will be referred as 'modified fitness function'.

H-compliance. As we mentioned earlier, the H residues lack affinity for water and tend to be located within the protein fold. We define the H-compliance of a H-type residue as a measure of how compactly (i.e. closely) a residue is located to the H-core centre. It is measured as the radial distance of H residues from H-core centre. The smaller the value of H-compliance, the closer the residue is to the H-core centre. The sum of the distance of all the H-type residues in the sequence gives the H-compliance of the conformation under consideration.

H-compliance of ith H type residue is denoted as h_i. To calculate h_i, we determine the center of a hypothetical rectangle "enclosing" the residues forming the H core as shown in Fig. 1(a). The coordinates (x_{rect}, y_{rect}) of the "center" are obtained as: $x_{rect} = (x_h max - x_h min)/2$, $y_{rect} = (y_h max - y_h min)/2$. Further, if coordinates of any i^{th} hydrophobic residue are given as (x_{hi}, y_{hi}), the overall H-compliance of the jth conformation can be obtained as $H_j = \sum_{i=1}^{n_h} h_i$. That is

$$H_j = \sum_{i=1}^{n_h} h_i = \sum_{i=1}^{n_h} (x_{rect} - x_{hi})^2 + (y_{rect} - y_{hi})^2 \qquad (2)$$

The H-compliance of jth conformation can be added as a fitness term $E_{H-compliance}$ to the function of *eqn. 1*. $E_{H-compliance}$ is the average of the H-compliance of the conformation where n_h is the total number of H type residues in the sequence.

$$E_{H-compliance} = H_j/n_h \qquad (3)$$

P-compliance. The P-compliance of a P type residue is a measure of how close the P residue is to any of the sides ($x_p min$, $x_p max$, $y_p min$ and $y_p max$) of a *P-boundary rectangle* (Fig. 1(b)). P-compliance is defined with the help of P-boundary rectangle rather than H-core because the P residues are located close to the outer periphery of a conformation and it is not possible to measure this from a H-core centre. The smaller the value of P-compliance, the closer it

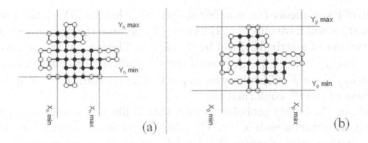

Fig. 1. Boundary rectangles for (a) H residues. (b) P residues.

is to the P-boundary rectangle. The sum of the P-compliance of all the P-type residues gives the P-compliance of the conformation under consideration.

For measuring the P-compliance p_i, we determine the minimum distance of an ith P-type residue from P-boundary rectangle shown in Fig. 1(b).With coordinates of ith P residue given as (x_{pi}, y_{pi}), the P-compliance of the jth conformation is given as follows

$$P_j = \sum_{i=1}^{n_p} p_i = \sum_{i=1}^{n_p} (\min\{|x_pmin - x_{pi}|, |x_pmax - x_{pi}|, \tag{4}$$
$$|y_pmin - y_{pi}|, |y_pmax - y_{pi}|\})$$

Again, to determine the corresponding fitness term $E_{P-compliance}$ to be included in *eqn. 1*, the average P-compliance of the conformation is used as given below. The term n_p is the total number of P residues in the individual.

$$E_{P-compliance} = P_j/n_p \tag{5}$$

Finally, the 'Modified Fitness Function (MFF)' for the jth conformation which is a total fitness is given below

$$E_j^{mff} = aE_{TN} + E_{H-compliance} + E_{P-compliance} \tag{6}$$

Here E_{TN} is fitness for the jth confirmation computed from *eqn. 1*. The original fitness function E_{TN} is multiplied by high integer constant value a so aE_{TN} of *eqn. 6* will remain integer and the later parts of *eqn. 6* will be in decimal and it will ensure that the original fitness term E_{TN} continues to have an influential effect on the MFF.

2.2   Schema Preservation

A conformation in any configuration (2D lattice, FCC etc) can be represented as a two dimensional matrix M_{Ci}. In general, if X is the set of all possible moves and $size|X| = n$, then $X_q \in X$ with $q = 1, 2, \cdots, n$. For relative 2D encoding, $n = 3$ and we have the set of possible moves as $(X_0 = F, X_1 = L, X_2 = R)$. If l is

the length of the sequence (i.e. number of residues), then for 2D relative encoding
a conformation will have only $(l-2)$ moves [19], because the first move is always
F. Thus the size of matrix M_{Ci} will be $(l-2) \times n$. The matrix M_{Ci} is populated
as $M_{Ci} = [a_{rq}]_{r=1,...,l-2,q=1,...,n}$. Now, if the rth position of a conformation C_i is
X_q, then $a_{rq} = \epsilon \times F(C_i)$ otherwise $a_{rq} = 0$. The constant $\epsilon = -1$ and $F(C_i)$
is the fitness of the ith conformation.

To find out the highly probable schema that is likely to occur in subsequent
generation, we obtain a matrix $\pi = \sum_{i=1}^{N} M_{Ci}$ or the entire population, N. Multiplying π with a column vector $[1 \quad 1 \quad 1]^T$, we obtain another column vector
$\Lambda = \pi \times [1 \quad 1 \quad 1]^T = [\rho_1 \quad \cdots \quad \rho_{l-2}]^T$. The rth row of Λ presents the cumulative weight, ρ_r of rth position of all conformations. To obtain the probability
of occurrence of each move at a given position, we multiply each row of matrix
π by $(1/\rho_r)$ to obtain another matrix π'. This matrix π' is important because
it contains the relevant information about the probability of occurrence of a
schema. To establish a move in a given position is highly probable, we define a
cut off value χ (= 0.8). If any element of matrix π' has value greater than χ,
then that position is fixed for finite number (=50) of generations. However, if
the probability of this position changes after 50 iterations, we may get a new
schema. However, based on the Hollands schemata theorem which underpins the
working of MA, we note that the probability of changes in schema will reduce
as the solution converges. This novel technique of schema preservation enables
us to establish the highly probable moves in a conformation. By applying the
technique, if two moves (say, first and third move) are fixed as say F, then for a
sequence of length 10 the conformation would be FxFxxxxx where x is a *dont
care* move. This fixing of schema significantly reduces the search space (hence
computational time) and restrict search to those individuals which contain highly
probable moves.

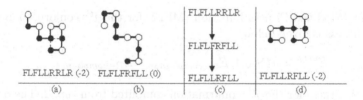

Fig. 2. (a) Conformation from the best individual set with fitness of -2, (b) Newly
generated random conformation with a fitness of 0, (c) Implementing move changes
(F to L) for the confirmation of (b), (d) Modified new individual conformation with a
high fitness of -2

2.3   Guided Search Space

Realizing that best individuals will preserve best schemas, rather than a purely
random generation of new interim individuals, we propose a guided search. For
this, a record is maintained of all those individuals having fitness equal to the

current best fitness value. This set of best individuals is used as templates for generating New Fit Individual (NFI). The strategy is best illustrated by considering an arbitrary toy sequence HPHPPHHPHP. As shown in Fig. 2(a), consider a conformation FLFLLRRLR with fitness -2 from the current best individuals set. Next, we consider a randomly generated conformation to be, say, FLFLFR-FLL with fitness 0 (Fig. 2(b)). If the 5th move (i.e. F) of this conformation is changed with the corresponding move (i.e. 5th move, L) of the best individual, the resulting individual FLFLLRFLL can be seen to improve its fitness equal to the fitness of the best individual (i.e. -2). The whole process is shown in Fig. 2(c) and the conformation is shown in Fig. 2(d).

Simple GA (SGA) is essentially based on fitness function defined by *eqn. 1* is modified incorporating all of the above three features which we refer as enhanced GA (EGA). Its fitness function is given by *eqn. 6* and it incorporates the schema preservation features of 2.2 and the guidance of 2.3.

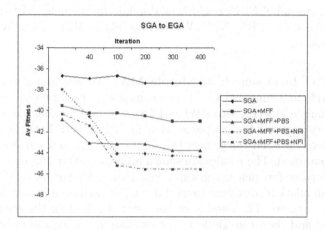

Fig. 3. Effect of enhancement features on SGA applied to benchmark sequence *b7*

3   Results

For investigations, we consider a set of benchmark sequences (*b1, b2, b3, b4, b5, b6, b7*) from [13,2] and also two non-benchmark sequences (*n2, n3*) from [2] given in Appendix. We begin investigations of the performance of the algorithm by selecting one bench sequence (i.e. sequence *b7*) from a set of benchmark protein sequences. Sequence *b7* is chosen as it is a reasonably long sequence with 85 residues possessing necessary level of complexity to discriminate between various techniques.

i)   Enhancements to SGA. For various enhancements to SGA under investigations, we compute the average fitness value of the best 10 conformations in each generation. Fig. 3 shows the variation of average fitness value as a function

Table 1. Comparison of SGA and simple MAs using two local search approaches, i.e.
SGA+PWI and SGA+TS

Test Run	Avg. Iteration					Success Rate
	b1	n2	n3	b2	b3	
SGA	11.4	34.2	29	16	84.2	86.20%
SGA+PWI	13	32.4	18	12.2	20.2	96.15%
SGA+TS	6.4	10.2	8.4	6.4	9	100%

of generation in different cases. We see that SGA (without any enhancement),
has a poor performance. The performance improves by progressively applying
improvements (i) modified fitness function (SGA+ MFF) (ii) preserved best
schema (SGA+MFF+PBS). (iii) Add New Fit Individuals (NFI) using the technique explained in sec 2.3 (SGA+MFF+PBS+NFI). Finally, for the sake of
comparison, instead of NFI we add New Random Individuals (NRI) to the population (SGA+MFF+PBS+NRI). We see that (SGA+MFF+PBS+NFI) has the
best performance.

ii)   Effect of local search on simple MA. We will first study the effect
of local search on a simple GA and then compare its performance with the two
variants of simple MA (LS with PWI and TS). For investigating the performance
of the two variants of the proposed MA, i.e. PMA and TMA, we randomly
consider three benchmark sequences (*b1, b2, b3*) and two new sequences (*n2, n3*)
for experimentation. The results are shown in Table 1. Our aim is to obtain five
values of E* value for each of the sequences. Hence, for each of the 5 sequences,
number of simulation runs were carried out till we achieved 5 successful (which
results in E*) results. The number of iterations required for the successful runs
are averaged and shown in Table 1. For evaluating the algorithm performance
in another manner, we further define a new measure called $SuccessRate =
((25 \div totalnumberofattempts) \times 100)$. The constant 25 appears in the definition
because each of the 5 sequences are successful 5 times. But the total number of
attempts required to achieve this 25 successful runs are different for different
algorithms. In our studies, we found that SGA achieves a success rate of 86.2%
(25 optimum values in 30 attempts) whereas SGA+PWI had a success rate of
96.15% (25 optimum values in 26 attempts) and SGA+TS had 100% of success.
These results are tabulated in Table 1. It shows that although both variants of
simple MA perform better than SGA, simple MA with TS as local search has
best performance.

Enhancement to SGA showed that EGA incorporating: (i) modified fitness
function (ii) preserving schema and (iii) guided search performs better than the
simple GA. Hence we will consider this EGA for further investigations. Effect of
the two local search techniques, PWI and TS on EGA performance is studied
using several benchmark protein sequences given in Appendix. The results of the
studies are presented in Table 2.

Table 2. Results for EGA, PMA and TMA (when an optimal is not reached the number of iteration, each of the algorithms first time reached the suboptimal are given in bracket)

Seq.	EGA Iteration	Fitness	PMA Iteration	Fitness	TMA Iteration	Fitness
b1	11	-9	9	-9	4	-9
n2	8	-4	13	-4	2	-4
n3	7	-8	6	-8	1	-8
b2	6	-9	6	-9	3	-9
b3	19	-8	8	-8	2	-8
b4	36731	-13	3457	-14	10	-14
b5	17251	-21(2687)	14189	-21(1291)	1507	-23
b6	179	-21	441	-21	11	-21

From the Table, it can be observed that while in general, EGA and PMA have a somewhat similar performance, TMA shows a significant improvement over other techniques with regard to both, the optimum value and also the number of iteration required in reaching that optimal value.

iii) Comparison of TMA with other approaches. Since the previous experiments establish that TMA has the best performance, this approach is investigated further by comparing it with five other known approaches, i.e. guided GA (GGA), guided Tabu search (GTB), expected Monte Carlo (EMC), simple GA (SGA), and Monte Carlo (MC), which have been reported in literature. In each of these simulations, for each run, 200 randomly generated individuals were included. The algorithm is set to run up to a maximum of 6 hours if optimum is not reached earlier. The time limit as a termination condition ensures that that all algorithms irrespective of their complexity are compared for similar time duration. Table 3 gives the comparisons. The best results obtained by TMA are compared with the results given in [5,19]. It can be seen that for all smaller

Table 3. TMA compared with other search algorithms (number of iteration required for GGA to reach the fitness are given in bracket and E* denotes optimal fitness value)

Seq.	E*	TMA Iteration	Fitness	GGA	GTB	EMC	GA	MC
b1	-9	4	-9	-9(2)	-9	-9	-9	-9
b2	-9	3	-9	-9(83)	-9	-9	-9	-9
b3	-8	2	-8	-8(124)	-8	-8	-8	-8
b4	-14	10	-14	-14(814)	-14	-14	-12	-13
b5	-23	1507	-23	-23(3876)	-23	-23	-22	-20
b6	-21	11	-21	-21(720)	-21	-21	-21	-21

sequences, TMA outperforms GGA (Guided GA) which is the best result of [5,19] for both, fitness and number of iterations required.

4   Conclusion

In this paper, we show that MA with superior local search proves very useful for PSP prediction. To make MA suitable for the complex PSP problem, the global search algorithm is enhanced by a novel fitness function which includes two new measures: H-compliance and P-compliance for the H and P residues. The enhancements also include novel techniques for schema preservation and guided search. Number of benchmark sequences and new sequences are used for investigations. Comparison with other known search algorithms for PSP problem is also reported. We observe that the enhanced global search (with the three features of novel fitness function, schema preservation and guided search) and incorporating tabu search for local optimization has a superior performance compared to other known algorithms. Experiment with other complex sequences are in progress.

References

1. Dill, K.A., Bromberg, S., Yue, K., Chan, H.S., Ftebig, K.M., Yee, D.P., Thomas, P.D.: Principles of protein folding - a perspective from simple exact models. Protein Science 4(4), 561–602 (1995)
2. Cutello, V., Nicosia, G., Pavone, M., Timmis, J.: An immune algorithm for protein structure prediction on lattice models. IEEE Transaction on Evolutionary Computation 11(1), 101–117 (2007)
3. Berger, B., Leighton, T.: Protein folding in the hydrophobic-hydrophilic (hp) model is np-complete. Journal of Computational Biology 5(1), 27–40 (1998)
4. Lopes, H.S., Scapin, M.P.: An enhanced genetic algorithm for protein structure prediction using the 2d hydrophobic-polar mode. In: Talbi, E.-G., Liardet, P., Collet, P., Lutton, E., Schoenauer, M. (eds.) EA 2005. LNCS, vol. 3871, pp. 238–246. Springer, Heidelberg (2006)
5. Hoque, M., Chetty, M., Dooley, L.: A new guided genetic algorithm for 2d hydrophobic-hydrophilic model to predict protein folding. In: IEEE Congress on Evolutionary Computation, vol. 1, pp. 259–266 (2005)
6. Jiang, T., Cui, Q., Shi, G., Ma, S.: Protein folding simulations of the hydrophobic-hydrophilic model by combining tabu search with genetic algorithms. The Journal of chemical physics 119(8), 4592–4596 (2003)
7. Zhao, X.: Advances on protein folding simulations based on the lattice hp models with natural computing. Applied Soft Computing 8(2), 1029–1040 (2007)
8. Krasnogor, N., Smith, J.: A tutorial for competent memetic algorithms: model, taxonomy, and design issues. IEEE Transactions on Evolutionary Computation 9(5), 474–488 (2005)
9. Tang, M., Yao, X.: A memetic algorithm for vlsi floorplanning. IEEE Transactions on Systems, Man, and Cybernetics, Part B: Cybernetics 37(1), 62–69 (2007)
10. Hasan, S.M.K., Sarker, R., Essam, D., Cornforth, D.: Memetic algorithms for solving job-shop scheduling problems. Memetic Computing 1(1), 69–83 (2008)

11. Fallahi, A.E., Prins, C., Calvo, R.W.: A memetic algorithm and a tabu search for the multi-compartment vehicle routing problem. Computers and Operations Research 35(5), 1725–1741 (2008)
12. Elbeltagi, E., Hegazy, T., Grierson, D.: Comparison among five evolutionary-based optimization algorithms. Advanced Engineering Informatics 19(1), 43–53 (2005)
13. Krasnogor, N., Blackburne, B.P., Burke, E.K., Hirst, J.D.: Multimeme algorithms for protein structure prediction. In: Guervós, J.J.M., Adamidis, P.A., Beyer, H.-G., Fernández-Villacañas, J.-L., Schwefel, H.-P. (eds.) PPSN 2002. LNCS, vol. 2439, pp. 769–778. Springer, Heidelberg (2002)
14. Pelta, D.A., Krasnogor, N.: Recent Advances in Memetic Algorithms. In: Multimeme Algorithms Using Fuzzy Logic Based Memes For Protein Structure Prediction, pp. 49–64. Springer, Berlin (2005)
15. Krasnogor, N., Hart, W., Smith, J., Pelta, D.: Protein structure prediction with evolutionary algorithms. In: Proceedings of the genetic and evolutionary computation (1999)
16. Smith, J.: Protein structure prediction with co-evolving memetic algorithms. In: The 2003 Congress on Evolutionary Computation, vol. 4, pp. 2346–2353 (2003)
17. Smith, J.E.: The Co-Evolution of Memetic Algorithms for Protein Structure Prediction. Studies in Fuzziness and Soft Computing, vol. 166. Springer, Heidelberg (2005)
18. Greenwood, G.W., Shin, J.M.: On the evolutionary search for solutions to the protein folding problem. Morgan Kaufmann, San Francisco (2003)
19. Hoque, M.T.: Genetic algorithm for ab initio protein structure prediction based on low resolution models. PhD thesis, GSIT, Monash University (2007)

Appendix

Benchmark sequences (b1, b2, b3, b4, b5, b6, b7) and non benchmark sequences (n2, n3) (E^* denotes optimal fitness value)

Inst.	Size	Sequence	E^*	Ref
b1	20	2(hp)p2hph2php2hp2(ph)	-9	[2,13]
b2	24	2h2ph2p5(h2p)2h	-9	[2,13]
b3	25	2ph2p3(2h4p)2h	-8	[2,13]
b4	36	3p2h2p2h5p7h2p2h4p2h2ph2p	-14	[2,13]
b5	48	2ph2p2h2p2h5p10h6p2(2h2p)h2p5p	-23	[2,13]
b6	50	2h3(ph)p4hp2(h3p)h4p2(h3p)hp4h3(ph)p2h	-21	[2,13]
b7	85	4h4p12h6p12h3p12h3ph2p2h2p2h2phph	-53	[2]
n2	18	2h5p2h3ph3php	-4	[2,13]
n3	18	hphp3h3p4h2p2h	-8	[2,13]

Interestingness of Association Rules Using Symmetrical Tau and Logistic Regression

Izwan Nizal Mohd Shaharanee, Fedja Hadzic, and Tharam S. Dillon

Digital Ecosystem and Business Intelligence Institute,
Curtin University of Technology, Perth 6102, Australia
izwan.mohdshaharanee@postgrad.curtin.edu.au,
{f.hadzic,tharam.dillon}@cbs.curtin.edu.au

Abstract. While association rule mining is one of the most popular data mining techniques, it usually results in many rules, some of which are not considered as interesting or significant for the application at hand. In this paper, we conduct a systematic approach to ascertain the discovered rules and provide a rigorous statistical approach supporting this framework. The strategy proposed combines data mining and statistical measurement techniques, including redundancy analysis, sampling and multivariate statistical analysis, to discard the non significant rules. A real world dataset is used to demonstrate how the proposed unified framework can discard many of the redundant or non significant rules and still preserve high accuracy of the rule set as a whole.

Keywords: data mining, interesting rules, statistical analysis.

1 Introduction

Various data mining techniques have been successfully employed in acquiring useful rules and patterns from data. The extracted rules offer a proper acknowledgement of potentially useful information that is easily understood by end users. They are considered interesting and useful if they are comprehensible, valid on tests and new data with some degree of certainty, potentially useful, actionable, and novel [1]. Many data mining techniques are available for the acquisition of hidden patterns and rules, and the differences are in terms of objectives, outcomes, and representation techniques. [2] claims that the majority of data mining/machine learning type patterns are rule based in nature with a well defined structure, such as rules derived from decision trees and association rules. The most common patterns that can be evaluated by interestingness measures include association rules, classification rules, and summaries [3]. Association rule mining is one of the most popular data mining techniques widely used to for discovering interesting associations and correlations between data elements in a diverse range of applications [4].

The association rule mining first discovers all the frequent patterns and then constructs the rules from such patterns. Frequent pattern extraction plays an important part in generating good and interesting rules, and is considered the most difficult task. While association rule mining techniques have been successfully used in many applications, in many cases certain aspects of domain knowledge will not be

A. Nicholson and X. Li (Eds.): AI 2009, LNAI 5866, pp. 422–431, 2009.

completely captured by the extracted rules [4]. Another problem is that the rule sets are often too large and complex making it impractical or impossible for a domain expert to analyze them in an efficient manner [5]. Different criteria have been used to limit the nature of rules extracted, such as support and confidence [6], collective strength [7], lift/interest [8], chi-squared test [9], correlation coefficient [10], three alternative interest measure that is; any-confidence, all confidence, and bond [11], log linear analysis [10], leverage [12, 13], and empirical bayes correlation [10].

Several researchers have also anticipated an assessment on pattern discovery by applying a statistical significance test before accepting the patterns. For example, in [9] correlation of rules is used, [14] proposed a pruning and summarizing approach, [15] applies a statistical test with a correction for multiple comparison by using a Bonferroni adjustment, [16] proposed an alternative approach of encountering a rules by change and applying hypothesis testing, [4] contributes to significant statistical quantitative rules and recently [12, 17] summarizes holdout evaluation techniques. While [12] has overviewed the latest development in significance of rules discovery, some areas worth further exploration involve: issues concerning the optimal split between the subset of data used for learning and evaluation, selection of a suitable statistical test and assessment of the rules with more than one itemset in the consequent.

Despite the fact that interesting association rules may be found from a database, by satisfying the various interestingness measures, a problem that remains is that they may only reflect aspects of the database being observed. [12] emphasizes that each assessment of whether a given rule satisfies a certain constraint is accompanied by a risk that the rules will satisfy the constraints with respect to the sample data but not with respect to the whole data distribution. They do not reflect the "real" association rules between the underlying attributes. Even when the rules found pass appropriate statistical tests, it can still be the case that this is caused purely by a statistical coincidence [18]. Since the nature of data mining techniques is data driven, the hypotheses generated by these algorithms must be validated by a statistical methodology for them to be useful in practice [19].

The focus of work presented in this paper, is on developing systematic ways to verify the usefulness of rules obtained from association rules mining using statistical analysis. Some initial ideas and preliminary results using a simple dataset were presented earlier in [20]. This paper presents extensions and refinements for applications to more realistic datasets. A unified framework is proposed, that combines several techniques to access the quality of rules, and remove any redundant and unnecessary rules. In the next section, we explain the problem of ascertaining the discovered rules. In Section 3, we describe our proposed framework. The framework is evaluated using a real world dataset and some experimental findings are given in Section 4. Section 5 concludes the paper and explains our ongoing work in this field of study.

2 Problem Statement

The aim of association rule mining is to discover interesting relationships among items in a given dataset under minimum support and confidence conditions. The problem of finding association rules $x \Rightarrow y$ was first introduced in [6] as a data mining task of finding frequently co-occurring items in a large Boolean transaction database. Commonly used example is in market basket analysis, where an association rule $x \Rightarrow y$ means if consumer buys the set of items X, then he/she probably also buys items Y. These items are typically called as itemsets [5].

2.1 Basic Concepts

Let $I = \{i_1, i_2, ..., i_m\}$ be a set of items. Each transaction T is a set of items, such that $T \subseteq I$. For example, this may correspond to a set of items which a consumer may buy in a basket transaction. An association rule is a condition of the form of $X \Rightarrow Y$ where $X \subseteq I$ and $Y \subseteq I$ are two sets of items. [7] asserts that the idea of the association rule is to develop a systematic method by which user can figure out how to infer the presence of some sets of items, given the presence of other items in a transaction. Mining frequent itemsets or patterns is a fundamental and essential step in the discovery of association rules [6]. The frequent itemset mining problem corresponds to the discovery of all the itemsets that occur in the dataset at least as many times as predetermined minimum support threshold. The association rules that satisfy the minimum support and confidence constraints, are then easily created from these patterns. The support of a rule $X \Rightarrow Y$ is the number of transactions that contain both X and Y, while the confidence of a rule $X \Rightarrow Y$ is the number of transactions containing X, that also contain Y.

2.2 Interestingness Measures

Determining which association rules are useful for the application at hand is a challenging problem. The rules that satisfy the minimum support and confidence threshold are often too numerous to be utilized efficiently and effectively for the application at hand [21]. The numerous patterns are often redundant patterns. [12] define redundant rules as those rules that include items in the antecedent that are entailed by the other elements of the antecedents. Redundant rule constraints discard rule $x \to y$ for which $\exists z \in x : support\ x \to y = support\ (x \to z \to y)$ [12]. In addition to that, [21] define a more dominant minimum improvement constraint in order to discard the redundant rules. The improvement of rule $x \to y$ is defined as

improvement $(x \to y) = confidence\,(x \to y) - \max_{z \subset x}(\,confidence\,(x \to y))$.

This frequent pattern based approach generates a large number of rules and these patterns reflect a strong association between items and carry the underlying information of the data. For a predefined class label dataset, this will contribute to a strong association between occurring items and class labels. [22, 23] suggested and successfully investigated the potential usage of frequent pattern mining for classification problem. This approach directly mined the predefined class label dataset using frequent pattern based classification. This approach offers promising results in terms of the classifier model accuracy and efficiency for classification problem.

Although there are various criteria in determining the usefulness of rules [1, 3, 24], the measures usually just reflect the usefulness of rules with respect to the specific database being observed [12]. It is hard to determine whether the rules produced are useful in practice or are valid in real world problems. Applying a data mining algorithm to practical problems may not be sufficient because we need to ensure that the results have a sound statistical basis. Even data mining algorithms founded on a sound statistical basis are not sufficient, if they cannot solve a practical problem [19]. Therefore, in this paper, we investigate how to combine data mining and statistical measurement techniques to arrive at more reliable and interesting set of rules.

3 Proposed Method

The association rules extracted from an available database will reflect the associations from the observed database and the real world data (implicitly). It is not easily known whether the rules truly represent/reflect the real data, since the mining process occurs at the database level. The data that is used for the rules generation in data mining process, further needs to be verified by the statistical analysis approaches. This will ensure that we identify and discard any coincident and random associations. Generally speaking we interpret interesting rules as those rules that have a sound statistical basis and are not redundant. Such an approach requires sampling process, hypothesis development, model building and finally a measurement using statistical analysis techniques to verify and prove the usefulness and quality of the rules discovered using association rule mining. This statistical approach offers a firm way of identifying significant rules that are statistically valid.

3.1 A Conceptual Framework

The proposed conceptual framework is shown in Figure 1. Initially, the dataset is divided into two partitions. The first partition is used for association rule generation and statistical evaluation, while the second partition acts as a sample data drawn from the database, used to verify the accuracy of discovered rules.

Firstly, we apply preprocessing techniques toward the selected data. This will ensure a clean and consistent data. Secondly, we need to determine the relevance of attributes by classifying their importance to characterize an association. A powerful technique for this purpose is the Symmetrical Tau [25], which is a statistical-heuristic feature selection criterion. It measures the capability of an attribute in predicting the class of another attribute. The measure is based on the probabilities of one attribute value occurring together with the value of the second attribute. To calculate Symmetrical Tau a $c1 \times c2$ contingency table is used, where $c1$ and $c2$ are the values of two attributes. Let there be I rows and J columns in the contingency table for two attributes A and B. The probability that an individual belongs to row category i and column category j is represented as $P(ij)$, and $P(i+)$ and $P(+j)$ are the marginal probabilities in row category i and column category j respectively. The Symmetrical Tau measure for the capability of attribute A in predicting the class of attribute B is defined as [24]:

$$Tau = \frac{\sum_{j=1}^{J} \sum_{i=1}^{I} \frac{P(ij)^2}{P(+j)} + \sum_{i=1}^{I} \sum_{j=1}^{J} \frac{P(ij)^2}{P(i+)} - \sum_{i=1}^{I} P(i+)^2 - \sum_{j=1}^{J} P(+j)^2}{2 - \sum_{i=1}^{I} P(i+)^2 - \sum_{j=1}^{J} P(+j)^2}$$

(1)

For the purpose of feature selection, problem A could be viewed as a feature and B as the target class that needs to be predicted. Higher values of the Tau measure would indicate better discriminating criterions (features) for the class that is to be predicted in the domain. Symmetrical Tau has many more desirable properties in comparison to other feature selection techniques, as was reported in [25]. It is utilized here to provide the relative usefulness of attributes in predicting the value of the class attribute, and discard any of the attributes whose relevance value is fairly low. This would prevent the generation of rules which then would need to be discarded anyway once it was found that they comprise of some irrelevant attributes.

The rules are then generated based on the minimum support and confidence framework. The discovered rules are then ascertained with rigorous statistical techniques. The chi-squared analysis is used to discover the properties of data attributes; principally on the data dependency. The logistic regression analysis is then employed to provide the classification power of the data. The development of logistic regression modeling involves the model building strategies. These statistical analysis results are used to determine and verify the applicability of the association rules to the real world data. We also use some constraint measurement techniques in order to discard the existence of redundant rules. The combination of this information will facilitate the association rule mining framework to determine the right and high quality rules. These rules will have a sound statistical basis and we can be more confident that they reflect the real world situation.

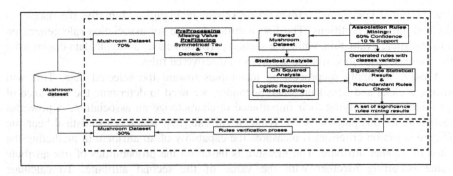

Fig. 1. Framework for analyzing rule interestingness in Association Rules for large database

3.2 Statistical Approaches

A common multivariate statistical analysis is analyzing the association problem [26]. For associations between categorical variables there are several inferential methods involved. Chi-squared analysis is then often used to measure the difference between observed and expected frequencies. The significance used of the chi-squared statistics is for hypothesis testing in tests of independence.

The logistic regression methods have become an integral component of any data analysis concerned with describing the relationship between a response variable and one or more explanatory variables. The logistic regression model has become one of the standard methods of classification problems [26]. Logistics regression is used to estimate the probability that a particular outcome will occur. The dependent variable in logistic regression is the odd ratios and the outcome variable is binary. The coefficients are estimated using a statistical technique called maximum likelihood estimation. The interpretation of regression coefficient in terms of odd ratios is a familiar concept in analysis of categorical data [27].

The selection of logistic regression model involves two competing goals: the model should be complex enough to fit the data well, while at the same time simpler models are preferred since they are easier to interpret [26]. Model building principally involves seeking and determining parsimonious (simple) model that explains the data. The rationale for preferring simple models is that they tend to be numerically more stable, and they are easier to interpret and generalize [27].

4   Experimental Results

Some evaluation of the framework towards unification of data mining and statistical analysis has been performed using the Mushroom dataset, a real world dataset obtained form UCI Machine Learning Repository. Since the Mushroom dataset is a supervised dataset that reflects a classification problem, we have chosen the target variable as the right hand side/consequence of the association rules for the association rule mining analysis.

4.1   Mimicking the Framework

The Mushroom dataset was partitioned randomly. The first partition consists of 70% (5687 records) of the dataset and this partition is to be used for association rule mining and statistical analysis procedures. The remaining 30% (2437 records), is used as a sample data to verify the final rules. We applied the sampling without replacement method [28]. We also make stratification based on the target variable classes. For missing values in the dataset, we applied the distributed based missing value approach. Taking in the whole dataset as input would produce a large number of rules, many of which are caused by the presence of irrelevant attributes. We therefore use the Symmetrical Tau feature selection criterion [25] earlier in the process to remove any irrelevant attributes. This would prevent the generation of rules that comprise of some irrelevant attributes. Hence in this experiment it is not necessary to use Symmetrical Tau to further verify the rules, this is due to the fact that the rules were created from the attribute subset considered as relevant by the measure as was done in [20]. Hence, the Symmetrical Tau [25] was calculated for all attributes in their capability in predicting the value of the class attribute. The attributes were then ranked according to their decreasing Symmetrical Tau value and a relevance cut-off point was picked. The subset of data consists now of 10 attributes: Odor, Spore Print

Color, Gill Size, Ring Type, Bruises, Gill Color, Population, Stalk Color Above Ring, Stalk Color Below Ring and Gill Spacing. Now the necessary pre-processing is done, we proceed with the application of an association rule mining algorithm and verification of the extracted rules through statistical analysis.

We apply association rule mining with minimum support of 10% and confidence of 60%, on the 70% partition of the whole data. This gives us a total of 36474 rules. However, since we have restricted the right-hand set (consequences) of the rules to be either Poisonous or Edibles (i.e. class values), this leaves us with 1094 rules. We next proceed with our statistical analysis to ascertain the discovered rules and reduce them to the most interesting ones. The 70% partition of the dataset is also used for this purpose.

A natural way to express the dependence between antecedent and the consequence of an association rule $x \rightarrow y$ is the correlation based on the chi squared analysis for independence [10]. Based on chi squared analysis for 5687 records, we found that all variables passed the chi squared basic requirement. The requirements are; all cells in the contingency table have expected values greater than 1 and at least 80% of the cells have expected values greater than 5.

The next statistical analysis applied was logistic regression. The relationship between the antecedent and consequent in association rule mining can be presented as a relationship between a target variable and the input variables in logistic regression. As mentioned in Section 3.2, from logistic regression a number of models can be discovered. We have evaluated each model and the most parsimonious model with lowest misclassification rates was selected.

Each model produces a different selection of variables. Such result is possible because different variables may contain different/complementary information that contributes to the prediction of the value of the target variable. Based on the selected logistic regression model, any rules that contain Bruises, Gill Color, Population, Ring Type, Stalk Color Below Ring and Stalk Color Above Ring can be discarded, since they were not significant contributor. Table 1 depicts the examples of pruned rules.

Applying statistical analysis helps determine the usefulness and significance of input variables in predicting the target variables. Hence, this will provide a proper ways to identify and discard rules that are not significant. By applying the combination of the statistical analysis we managed to discard a total of 1050 (96%)

Table 1. Example of prunes rules based on logistic regression analysis

Set Size	Confidence	Support	Count	Rules
7	100	12.55	714	whiteSCBelowRing & whiteSCAboveRing & noneOdor & noBruises & crowdedGspacing & broadGSize => edibleClasses
8	100	10.80	614	whiteSporePrintColor & severalPop & pinkSCAboveRing & noBruises & narrowGSize & evanescentRingType & closeGspacing => poisonousClasses

Table 2. Examples of redundant rules

Set Size	Confidence	Support	Count	Rules
2	100	26.57	1511	foulOdor=>poisonousClasses
*3	100	26.57	1511	foulOdor & closeGspacing =>poisonousClasses
3	100	16.59	942	noneOdor & brownSporePrintColor =>edibleClasses
*4	100	15.93	906	noneOdor & brownSporePrintColor & broadGSize =>edibleClasses

Table 3. Examples of redundant rules (with minimum improvement)

Set Size	Confidence	Support	Count	Rules
2	97.14	41.83	2379	noneOdor => edibleClasses
*3	96.39	28.64	1629	noneOdor & closeGspacing => edibleClasses

* Redundant rules

rules. Of all 44 accepted rules, there are still some redundant rules left, which is due to the incapability of statistical analysis to identify the redundant rules. In Section 2.2 earlier, we have explained the redundant rule constraint [12] and the minimum improvement constraint [21]. These two constraints were used in the final step to discard any redundant rules. Table 2 shows examples of rule discarded according to the redundant rule constraint, while Table 3 shows examples of rules considered as redundant according to the minimum improvement constraint.

Based on the redundant rules analysis performed at 44 rules, we managed to discard another 22 rules. The combination of statistical significance analysis and redundant analysis provided a proper ways in discarding non significant rules. Only 2% (22 rules) from 1094 rules are now considered as significant, which is a significant reduction in the overall complexity of the rule set. However, the question still remains whether this great reduction of rules is at a cost of a significant reduction in accuracy. Hence we evaluate the classification and predictive accuracy of the different sets of rules obtained throughout the process. Table 4 depicts the accuracy results. The 70% of the dataset was used to test the classification accuracy, while the remaining 30% was used for testing the predictive accuracy. The selected input variables for the final rules are Gill Size, Gill Spacing, Odor and Spore Print Color. These are significant contributors in the logistic regression model in predicting the value of the target variable. We investigate this framework on several data partitions to gauge the effect of accuracy test. As can be seen from Table 4 the reduction in the accuracy of the rule set is not significantly large in comparison to the significant reduction in the complexity of the rule set, for all of the data partitions. One can also see that, choosing different data partitions does not cause much difference in the accuracy of the initial and reduced rule set. The reduction of the rules set through statistical analysis was very similar for all of the data partitions. This is due to the similar statistical evaluation result (i.e. significant variables contributing to the target variable in logistic regression) for each partition. These results are caused by the fact that the Mushroom data are uniformly distributed. However, one can expect that when the method is applied on the not uniformly distributed data, the accuracy of the rules at different stages will differ for different data partitions.

Table 4. Accuracy comparison between several data partitions and selected rules

		Rule #	Accuracy	
			Classification	Prediction
70 / 30	[10 Input variables]	1094	93.83%	93.50%
Data	[4 Input] GillSize, GillSpacing ,	44	92.59%	92.16%
Partition	Odor, SporePrintColor	22	91.68%	91.26%
50 / 50	[10 Input variables]	1089	93.71%	93.68%
Data	[4 Input] GillSize, GillSpacing ,	44	92.43%	92.49%
Partition	Odor, SporePrintColor	23	91.38%	91.55%
90 / 10	[10 Input variables]	1099	93.74%	93.42%
Data	[4 Input] GillSize, GillSpacing ,	44	92.48%	92.29%
Partition	Odor, SporePrintColor	21	91.34%	91.09%

5   Conclusions and Future Works

The combination of the approaches used in this method showed a number of ways for ascertaining the extracted data mining rules. In our framework, we focus mainly on the rules discovered from association rule mining algorithm and we integrate the statistical measurement techniques to ascertain the quality of the rules. The experimental results show that, this framework managed to reduce a large number of non significant and redundant rules while at the same time relatively high accuracy was preserved. As part of our ongoing works, the proposed framework will be compared with the current state-of-the-art rule interestingness measures. In addition to that, to fully address the problem of discovering significant/interesting rules, we must consider several factors such as the number of itemsets in the consequent part of the rules, types of variable involved, and selection of appropriated statistical tests and sampling methods.

References

1. Han, J., Kamber, M.: Data mining: concepts and techniques. Morgan Kaufmann Publishers, San Francisco (2001)
2. McGarry, K.: A survey of interestingness measures for knowledge discovery. Knowl. Eng. Rev. 20, 39–61 (2005)
3. Geng, L., Hamilton, H.J.: Interestingness measures for data mining: A survey. ACM Comput. Surv. 38, 9 (2006)
4. Zhang, H., Padmanabhan, B., Tuzhilin, A.: On the discovery of significant statistical quantitative rules. In: Proceedings of the 10th ACM SIGKDD International Conference on Knowledge Discovery and Data Mining. ACM, New York (2004)

5. Philippe, L., Patrick, M., Benoît, V., Stéphane, L.: On selecting interestingness measures for association rules: User oriented description and multiple criteria decision aid. European Journal of Operational Research 184, 610–626 (2008)
6. Agrawal, R., Imieliski, T., Swami, A.: Mining association rules between sets of items in large databases. SIGMOD Rec. 22, 207–216 (1993)
7. Aggarwal, C.C., Yu, P.S.: A new framework for itemset generation. Book A new framework for itemset generation. Series A new framework for itemset generation. ACM, New York (1998)
8. Brin, S., Motwani, R., Ullman, J.D., Tsur, S.: Dynamic itemset counting and implication rules for market basket data. ACM SIGMOD Record 26(2), 255–264 (1997)
9. Silverstein, C., Brin, S., Motwani, R.: Beyond Market Baskets: Generalizing Association Rules to Dependence Rules. Data Min. Knowl. Discov. 2(1), 39–68 (1998)
10. Brijs, T., Vanhoof, K., Wets, G.: Defining interestingness for association rules. International Journal of Information Theories and Applications 10(4), 370–376 (2003)
11. Omiecinski, E.R.: Alternative interest measures for mining associations in databases. IEEE Transactions on Knowledge and Data Engineering 15(1), 57–69 (2003)
12. Webb, G.I.: Discovering Significant Patterns. Machine Learning, 1–33 (2007)
13. Piatetsky-Shapiro, G.: Discovery, analysis; and presentation of strong rules. Knowledge discovery in database 229 (1991)
14. Liu, B., Hsu, W., Ma, Y.: Pruning and summarizing the discovered associations. In: Proceedings of the 5th ACM SIGKDD International Conference on Knowledge Discovery and Data Mining. ACM, New York (1999)
15. Bay, S.D., Pazzani, M.J.: Detecting Group Differences: Mining Contrast Sets. Data Mining and Knowledge Discovery 5, 213–246 (2001)
16. Meggido, N., Srikant, R.: Discovering Predictive Association Rules. In: 4th International Conference on Knowledge Discovery in Databases and Data Mining, pp. 274–278 (1998)
17. Webb, G.I.: Preliminary investigations into statistically valid exploratory rule discovery. In: Australasian Data Mining workshop (AudDM 2003), pp. 1–9 (2003)
18. Aumann, Y., Lindell, Y.: A Statistical Theory for Quantitative Association Rules. J. Intell. Inf. Syst. 20, 255–283 (2003)
19. Goodman, A., Kamath, C., Kumar, V.: Data Analysis in the 21st Century. Stat. Anal. Data Mining 1, 1–3 (2008)
20. Mohd Shaharanee, I.N., Dillon, T.S., Hadzic, F.: Ascertaining Association Rules using Statistical Analysis. In: Proceeding of the 2009 International Symposium on Computing, Communication and Control, Singapore (2009)
21. Roberto, B., Rakesh, A., Dimitrios, G.: Constraint-Based Rule Mining in Large, Dense Databases. Data Mining and Knowledge Discovery 4, 217–240 (2007)
22. Cheng, H., Yan, X., Han, J., Yu, P.S.: Direct discriminative pattern mining for effective classification. IEEE, 169–178 (2008)
23. Cheng, H., Yan, X., Han, J., Hsu, C.-W.: Discriminative frequent pattern analysis for effective classification. IEEE, 10 (2007)
24. Nada, L., Peter, F., Blaz, Z.: Rule Evaluation Measures: A Unifying View. Inductive Logic Programming, 174–185 (1999)
25. Zhou, X.J., Dillon, T.S.: A statistical-heuristic feature selection criterion for decision tree induction. IEEE Transactions on Pattern Analysis and Machine Intelligence 13 (1991)
26. Agresti, A.: An Intro. to Categorical Data Analysis. Wiley-Interscience, New York (2007)
27. Bovas, A., Johannes, L.: Intro. to Regression Modeling. Brooks/Cole, California (2006)
28. Yanrong, L., Raj, G.: Effective Sampling for Mining Association Rules. In: Webb, G.I., Yu, X. (eds.) AI 2004. LNCS (LNAI), vol. 3339, pp. 391–401. Springer, Heidelberg (2004)

Unsupervised Elimination of Redundant Features Using Genetic Programming

Kourosh Neshatian and Mengjie Zhang

School of Engineering and Computer Science,
Victoria University of Wellington, P.O. Box 600, Wellington, New Zealand
{kourosh.neshatian,mengjie.zhang}@ecs.vuw.ac.nz

Abstract. While most feature selection algorithms focus on finding relevant features, few take the redundancy issue into account. We propose a nonlinear redundancy measure which uses genetic programming to find the redundancy quotient of a feature with respect to a subset of features. The proposed measure is unsupervised and works with unlabeled data. We introduce a forward selection algorithm which can be used along with the proposed measure to perform feature selection over the output of a feature ranking algorithm. The effectiveness of the proposed method is assessed by applying it to the output of the Chi-square (χ^2) feature ranker on a classification task. The results show significant improvements in the performance of decision tree and SVM classifiers.

1 Introduction

The goal of *feature selection* (FS), a commonly-used process in machine learning, is to find a minimal subset of features which is sufficient to describe target concepts. *Feature ranking* (FR) is an avenue to FS in which features are ranked based on their relative importance (relevance) with respect to target concepts [1]. Most of the FR methods like information-theoretic and statistical algorithms take a *filter approach* and can only measure the goodness of individual features [2,1]. The output of an FR algorithm is a vector of positive integers called a *ranking vector* whose elements are indexes to the features of a dataset. The elements are sorted in descending order of the importance of their corresponding feature. Usually a small number of features indexed at the beginning of this vector are used to build a learning model.

Since in many FR algorithms features are examined individually, it happens quite often that high-ranked features exhibit redundancy. That is, although a feature at the i-th rank is more relevant than a feature at the $(i+1)$-th rank, the former would not be as useful if it gave the same information as the features at ranks 1 to $i - 1$ did while the latter $((i+1)$-th feature) can provide some more relevant information. Obviously, being limited to use a certain number of features, a feature subset with no redundancy could yield better learning performance. *Redundancy removal* (RR) is the process of correcting a ranking vector by replacing redundant features with non-redundant ones at lower ranks. Cost could be another motivation for RR—for example, in medical domains extracting/measuring extra features might be costly.

A. Nicholson and X. Li (Eds.): AI 2009, LNAI 5866, pp. 432–442, 2009.

If redundancy happened only between two single features following a certain type of functional dependency, then one could use a predefined measure (e.g. linear dependency) for RR. In practice, however, there might be some types of dependency between a group of features which do not necessarily follow any particular functional templates. Genetic programming (GP) has proved to be a powerful search technique for discovering sophisticated relationships between groups of features [3]. In this paper, we propose a GP-based measure for redundancy that extends the univariate linear definition of correlation to a flexible non-linear multivariate form. We then introduce a forward selection algorithm which can be used along with the proposed measure to perform RR over the output of an FR algorithm like χ^2 (Chi-square) [4].

1.1 Syntactic Notation

Throughout this paper we follow a certain mathematical notation. We use capital letters like X for random variables or when we are talking about features in abstract. Lowercase bold letters like $\mathbf{x}$ are used for vectors and features and $\mathbf{x}[i]$ represents the i-th element (or observation) of the vector (or feature). Calligraphic capital letters like $\mathcal{A}$ are used for sets. The unary operator $|.|$ is used to indicate the cardinality of a set. A dataset is a set $\mathcal{F} = \{\mathbf{x}_1, \mathbf{x}_2, \ldots, \mathbf{x}_m\}$ of m features, each of which contains n observations (values). We use $m^\star$ for the desired (or selected) number of features. If $\mathcal{A}$ is an arbitrary subset of features, $\phi(A)$ is a function whose arguments are the features in $\mathcal{A}$.

2 Redundancy Removal Using Genetic Programming

2.1 Fundamental Concepts

Redundancy can be defined based on general consensus:

Definition 1. *A feature X is redundant with respect to a subset $\mathcal{A}$ of features if and only if it can be approximated (reconstructed) by a function of $\mathcal{A}$.*

From an RR perspective, we are interested only in the existence of such a function and not its exact formulation. We also know that in practice there are different degrees of redundancy; that is, X can be partly expressed using $\mathcal{A}$ or up to a certain precision. Thus a measure of the degree of redundancy is required.

A common measure of dependency (an so redundancy) between two random variables is the quotient of their linear relationship. This measure, however, comes with two limitations: I) It is a measure of linear relationships only II) It can be applied to two features at a time. GP has previously been used to measure a wide variety of non-linear correlations between a subset of features and another variable [3]. We use GP to search for a function of features in $\mathcal{A}$ that has the highest linear correlation with X. We prove that as the linear correlation between the two increases, the error of approximating the feature by the GP constructed function decreases.

Proposition 1. *The error of approximating a feature X by a linear function of $\phi(\mathcal{A})$ decreases as $\rho^2 = \text{Cor}^2(X, \phi(\mathcal{A})) \in [0, 1]$, the square of Pearson's product-moment correlation coefficient, approaches one.*

Proof. Let $\hat{X} = \alpha + \beta\phi(\mathcal{A})$, $\alpha, \beta \in \mathbb{R}$, be a linear approximation of X with the error of approximation defined as $\varepsilon = X - \hat{X}$. If α and β are determined by the least squares method, the following properties hold [5]:

(i) $E(\varepsilon) = E(X - \hat{X}) = 0$

(ii) $\text{Cov}(\varepsilon, \hat{X}) = \text{Cov}(X - \hat{X}, \hat{X}) = 0$ (since $\hat{X}$ and ε are orthogonal)

Because of property (ii), the covariance between X and $\hat{X}$ reduces to the variance of $\hat{X}$;

$$\text{Cov}(X, \hat{X}) = \text{Cov}(X - \hat{X}, \hat{X}) + \text{Cov}(\hat{X}, \hat{X}) = \text{Var}(\hat{X}),$$

and hence, the correlation between X and $\hat{X}$ becomes

$$\text{Cor}(X, \hat{X}) = \frac{\text{Cov}(X, \hat{X})}{\sqrt{\text{Var}(X)\text{Var}(\hat{X})}} = \frac{\text{Var}(X)}{\sqrt{\text{Var}(X)\text{Var}(\hat{X})}} = \frac{\sqrt{\text{Var}(\hat{X})}}{\sqrt{\text{Var}(X)}} \quad (1)$$

Alternatively, the correlation between X and $\hat{X}$ can be derived as follows

$$\begin{aligned} \text{Cor}(X, \hat{X}) &= \text{Cor}(X, \alpha + \beta\phi(\mathcal{A})) \\ &= \frac{\beta\,\text{Cov}(X, \phi(\mathcal{A}))}{\sqrt{\text{Var}(X)\beta^2\text{Var}(\phi(\mathcal{A}))}} = \text{Cor}(X, \phi(\mathcal{A})) = \rho \end{aligned} \quad (2)$$

and therefore, from (1) and (2), we get

$$\rho^2 = \frac{\text{Var}(\hat{X})}{\text{Var}(X)} \quad (3)$$

The error of approximation, more specifically, the mean squared error (MSE) of the approximation is a function of this squared correlation coefficient. The MSE of the approximation, using property (i), is

$$MSE(\hat{X}) = E(\varepsilon^2) = E[(X - \hat{X})^2] = \text{Var}(\varepsilon) + E^2(X - \hat{X}) = \text{Var}(\varepsilon)$$

where the variance of ε is

$$\text{Var}(\varepsilon) = \text{Var}(X - \hat{X}) = \text{Var}(X) + \text{Var}(\hat{X}) - 2\text{cov}(X, \hat{X}) = \text{Var}(X) - \text{Var}(\hat{X})$$

Therefore, from (3), it follows that

$$MSE(\hat{X}) = \text{Var}(X) - \rho^2\,\text{Var}(X) = (1 - \rho^2)\,\text{Var}(X)$$

Hence, as ρ^2 increases, $MSE(\hat{X})$ decreases. More precisely

$$\lim_{\rho^2 \to 1} MSE(\hat{X}) = \lim_{\rho^2 \to 1} (1 - \rho^2)\text{Var}(X) = 0 \qquad \square$$

The proposition implies that if a GP search succeeds in maximising $\text{Cor}^2(X, \phi(\mathcal{A}))$, the error of approximating X by some function of $\mathcal{A}$ can be minimised. This can be a ground for defining a redundancy measure.

Definition 2. *The degree of redundancy of a feature X with respect to a set $\mathcal{A}$ of features is*

$$\rho^2_{max} = \max_{\phi \in \Phi}\{\text{Cor}^2(X, \phi(\mathcal{A}))\}$$

where Φ is a finite set of GP-constructable functions of $\mathcal{A}$. The range of ρ^2_{max} is $[0, 1]$ where 0 and 1 correspond to the minimum and maximum possible degree of redundancy. The functions in Φ are constructed (found) by GP using a set of primitive operators and a set of *variable terminals* which correspond to the features in $\mathcal{A}$. GP tries to maximise $\text{Cor}^2(\cdot, \cdot)$ and depending on how successful it is, the degree of redundancy can be determined.

An advantage of this redundancy measure is that GP does not need to find any approximation for X directly, but once it finds any member of the family of functions that optimises $\text{Cor}^2(\cdot, \cdot)$, we would know that X is redundant. In particular, (2) implies that if $\hat{X} = \alpha + \beta\phi(\mathcal{A})$ is the best linear approximation for X, where $\alpha, \beta \in \mathbb{R}$, any function of the form $\alpha' + \beta'\phi(\mathcal{A})$ where $\alpha', \beta' \in \mathbb{R}$ can maximise the fitness function. This is particularly important because GP is not normally equipped with any type of hybrid learning (like gradient descent or least square) to find right values for numeric constants efficiently. This relaxation of the search criteria can significantly improve the probability of success of GP.

2.2   A Synthetic Example

We give an example to illustrate how the proposed measure can detect redundant features that are not normally detectable by ordinary methods and how it can reduce the size of the GP search space. Consider two random variables $X_1 \sim N(\mu = 0, \sigma^2 = 2)$ and $X_2 \sim N(\mu = 0, \sigma^2 = 1)$. We define a third random variable with high functional dependency (redundancy) as $X_3 = -3X_1X_2 + 2 + \xi$ where $\xi \sim U(-1, 1)$ is noise. We create sample vectors $\mathbf{x}_1$, $\mathbf{x}_2$ and $\mathbf{x}_3$ of size 10,000 from the random variables X_1, X_2 and X_3 respectively.

Figure 1(a and b) show these three vectors plotted against each other. There is no mutual linear relationship visible between these vectors as expected. In terms of measurements $\text{Cor}^2(\mathbf{x}_3, \mathbf{x}_1) = 0.00$ and $\text{Cor}^2(\mathbf{x}_3, \mathbf{x}_1) = 0.00$. That is the redundancy of $\mathbf{x}_3$ cannot be detected by measuring its correlation against the other two random variables individually. Using a simple GP search to maximise $\text{Cor}^2(\mathbf{x}_3, \phi(\mathbf{x}_1, \mathbf{x}_2)$, however, results in a variety of solutions like $\phi(\mathbf{x}_1, \mathbf{x}_2) = \mathbf{x}_1\mathbf{x}_2$ as illustrated in 1(c). Using this GP-constructed function, $\text{Cor}^2(\mathbf{x}_3, \mathbf{x}_1\mathbf{x}_2) = 0.99$ which indicates a high redundancy. We see that GP does not need to find the exact formula of $\mathbf{x}_3$; actually any linear combination of $\mathbf{x}_1\mathbf{x}_2$ would yield the same result.

2.3   Measuring Redundancy

Algorithm 1 presents how we use a GP search to measure redundancy. The fitness of a GP individual (program) ϕ is $\text{Cor}^2(\mathbf{x}, \phi(\mathcal{A}))$. The computational

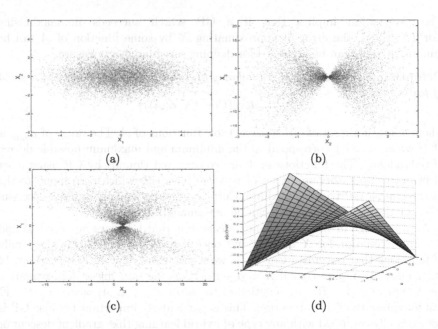

(a)   (b)

(c)   (d)

Fig. 1. An artificial example where x_3 is redundant in the context of x_1 and x_2. The three features are visualised in (a) and (b). A transformation function to detect the redundancy is depicted in (c).

complexity of this fitness function is as low as $O(n)$. The goal of the GP search is to maximise the fitness. The degree of redundancy, ρ^2_{max} in Definition 2, is the maximum fitness obtained during the GP run. The parameter θ determines the maximum acceptable redundancy; the search would stop once the fitness of an individual reaches this threshold.

The algorithm starts with adding all the features in $\mathcal{A}$ to the GP variable terminals set. Lines 2–3 initialise the population and add all the variable terminals as single-node trees to the population. This is to make sure that at all times, the redundancy will be measured against every single feature in the population. Each GP program in the population defines a function $\phi_{program} : \mathbb{R}^{|\mathcal{A}|} \mapsto \mathbb{R}$ which transforms an input vector $(x_1[i], x_2[i], \ldots, x_{|\mathcal{A}|}[i])$ into a scalar value $y[i]$. Lines 7–15 calculate the fitness. Line 16 updates the measured degree of redundancy ρ^2_{max} if the fitness of current program exceeds the current value of ρ^2_{max}.

2.4   Forward Selection

Given a *preliminary* ranking vector, the output of a ranking algorithm, Algorithm 2 introduces a *forward selection* algorithm that removes redundant features from the vector. The algorithm takes the set of all m features in the dataset $\mathcal{F}$, a ranking vector r which is the output of a ranking algorithm, the desired number $m^\star$ of features to be selected and a threshold θ which determines the maximum acceptable redundancy. At the first line the highest-ranked feature

---

Algorithm 1. Measuring redundancy via GP

```
/* The function MeasureRedundancy(x,A,θ) uses GP to measure the
   redundancy of a feature x with respect to a set of features A */
```

Input: $\mathbf{x}$, a feature, where $\mathbf{x} \in \mathcal{F}$
Input: $\mathcal{A}$, a subset of features, where $\mathcal{A} \subseteq \mathcal{F}\backslash\{\mathbf{x}\}$
Input: θ, maximum acceptable redundancy quotient
Output: ρ^2_{max}, the redundancy between $\mathbf{x}$ and $\mathcal{A}$ where $\rho^2_{max} \in [0,1]$

```
1  T ← A ; // variable terminals include all the features in A
2  P ← (a population of randomly-generated GP programs);
3  P ← P ∪ T ; // include all single node (terminal) programs
4  ρ²max ← 0 ; // initialise the measure of redundancy
5  while ¬max-generations ∧ (ρ²max < θ) do
6  |   foreach program ∈ P do
       |   /* calculate the fitness for each program          */
7  |   |   sx, sy, sx², sy², sxy ← 0 ; // initialising the sums
8  |   |   for i ∈ {1, 2, ..., n} do
9  |   |   |   y[i] ← φprogram(x₁[i], x₂[i], ..., x|A|[i]) ; // transformation
10 |   |   |   sx ← sx + x[i]; // updating sum
11 |   |   |   sy ← sy + y[i]; // updating sum
12 |   |   |   sx² ← sx² + (x[i])²; // updating sum of squares
13 |   |   |   sy² ← sy² + (y[i])²; // updating sum of squares
14 |   |   |   sxy ← sxy + x[i]y[i]; // updating sum of products
15 |   |   fitnessprogram ← ( (nsxy − sxsy) / (√(nsx² − sx²)√(nsy² − sy²)) )² ; // calculating Cor²(x,y)
16 |   |   ρ²max ← max(ρ²max, fitnessprogram);
17 |   P ← (new population using genetic operators, keeping the best)
18 return ρ²max;
```

is added to the set $\mathcal{F}^\star$ of selected features. In the loop starting on line 3, the degree of redundancy of the next highest-ranked feature will be measured (on line 4) and the feature will be added to $\mathcal{F}^\star$ only if its redundancy is less than θ. The algorithm stops when $m^\star$ features are selected or all the features in the dataset have been processed.

3   Empirical Results and Analysis

3.1   Dataset

We use the Isolet5 dataset from the UCI machine learning repository [6]. The dataset has been created by recording the voice of 30 people pronouncing the names of the 26 English alphabets twice; there are 52 samples per person and 1559 samples in total (one sample is missing). The task is to classify the alphabets. There are 617 features available in total including spectral coefficients, contour features, sonorant features, pre-sonorant and post-sonorant features. All the features are real-valued, continuous and scaled into the range $[-1, 1]$.

Algorithm 2. Forward Selection

/* Find a (sub)optimal subset of features incrementally by removing
   redundant features found by GP, from a ranking vector.   */

Input: $\mathcal{F} = \{x_1, x_2, \ldots, x_m\}$, the set of all features
Input: $r = (r_1, r_2, \ldots, r_m)$, a ranking vector
Input: m^*, the desired number of selected features
Input: θ, the maximum acceptable redundancy where $\theta \in [0, 1]$
Output: $\mathcal{F}^*$, the set of selected features

1 $\mathcal{F}^* \leftarrow \{x_{r_1}\}$; // adding the feature at the highest rank
2 $i \leftarrow 2$; // the next rank to be processed
3 **while** $(|\mathcal{F}^*| < m^*) \wedge i \leq m)$ **do**
4 $\quad$ $\rho^2 = \texttt{MeasureRedundancy}(x_{r_i}, \mathcal{F}^*, \theta)$; // measure the redundancy
5 $\quad$ **if** $\rho^2 \leq \theta$ **then**
6 $\quad\quad$ $\mathcal{F}^* \leftarrow \mathcal{F}^* \cup \{x_{r_i}\}$;
7 $\quad$ $i \leftarrow i + 1$;
8 **return** $\mathcal{F}^*$;

3.2   GP Settings and Experimental Setup

We adopt the standard tree-based genetic programs. A function node can be one of the four elementary binary arithmetic operators: $+, -, \times, \div$ where $\div$ is *protected*, i.e. returns zero for division by zero. The ramped half-and-half method [7] is used for generating programs in the initial population and for the mutation operator. We use a population size of 1024, however, if the cardinality of $\mathcal{A}$ in Algorithm 1 is very high, using a bigger population is recommended. We take an elitist approach by keeping the best individual at each generation. The initial maximum program tree depth is set to 4, but it can increase to 6 during the evolution process. The probability of the crossover and mutation operators are adapted automatically at runtime. We use Chi-square (χ^2) [4] for FR, and J48 decision tree and SMO-SVM for classification, all from the Weka library [8]. Since no separate test data is available, we use 10-fold cross-validation in our experiments.

3.3   Performance Results and Analysis

Table 1 presents the result of applying the proposed forward selection algorithm on the Isolet5 dataset for different values of θ and m^*. The numbers in each column (except the first column) are indexes to the features in the dataset. By increasing m^* from 1 to 30, one step at a time, new ranking vectors are created for the given values of θ. Each column presents the *corrected ranking* generated by removing redundant features from the preliminary ranking vector for the given value of θ. In the second column $\theta = 1$ which means any level of redundancy is accepted. Therefore the numbers in this column are actually the first 30 elements of the preliminary ranking vector (the output of the χ^2 ranking algorithm, without any changes).

Table 1. Corrected feature ranks based on different redundancy thresholds (θ)

$m^{\star}$ / Rank	$\theta =$ 1.0	$\theta =$ 0.9	$\theta =$ 0.8	$\theta =$ 0.7	$\theta =$ 0.6	$\theta =$ 0.5	$\theta =$ 0.4	$\theta =$ 0.3	$\theta =$ 0.2	$\theta =$ 0.1	$\theta =$ 0.0
1	584	584	584	584	584	584	584	584	584	584	584
2	390	390	390	389	548	548	548	548	548	548	548
3	392	392	548	548	419	419	419	419	413	474	474
4	391	395	419	419	107	358	325	325	325	410	528
5	395	548	73	73	412	411	474	474	474	448	378
6	389	419	413	413	358	171	410	472	448	577	-
7	548	73	517	358	11	474	78	427	214	48	-
8	419	462	10	139	546	387	472	20	480	582	-
9	73	107	358	546	388	425	352	130	48	292	-
10	549	75	458	386	474	12	590	480	437	433	-
11	394	9	325	266	425	522	427	181	435	599	-
12	462	413	139	362	522	545	523	351	164	530	-
13	393	517	515	323	5	472	20	481	528	368	-
14	74	358	546	485	203	352	130	112	463	-	-
15	107	42	386	327	472	69	480	437	347	-	-
16	75	458	134	474	363	589	181	333	428	-	-
17	9	325	362	397	322	214	332	435	334	-	-
18	413	415	173	425	382	427	259	364	331	-	-
19	106	11	360	198	352	446	372	164	370	-	-
20	412	139	474	12	448	451	481	525	600	-	-
21	517	411	397	522	486	20	398	595	473	-	-
22	418	359	387	174	589	130	437	4	-	-	-
23	10	547	110	78	143	321	435	532	-	-	-
24	461	515	233	5	214	480	364	433	-	-	-
25	358	546	425	203	576	577	445	377	-	-	-
26	417	386	198	424	427	493	406	224	-	-	-
27	416	265	76	472	452	16	164	336	-	-	-
28	42	550	426	101	446	541	528	403	-	-	-
29	458	134	298	352	14	24	28	341	-	-	-
30	457	518	12	448	20	332	595	252	-	-	-

For all values of θ, the feature at the first rank is always the same (see the first line of Algorithm 2). However, as the redundancy threshold decreases, features at the lower ranks might be removed due to being redundant with respect to the features at higher ranks. For example by decreasing θ from 1 to 0.9 the feature at the fourth rank (391) is considered redundant with respect to the three feature at higher ranks (584, 390 and 392) and hence, is removed and replaced by the next feature (395) which in this case is not redundant with respect to those three features. As θ decreases, more features are removed due to redundancy. For very low values of θ, like 0.2 and lower, the number of selected features, i.e. the number of remaining features after RR, is even less than 30.

To study the effect of elimination of redundant features we use the selected features in groups of size 5, 10, ..., 30. Table 2 shows the number of redundant (and hence, eliminated) features for different thresholds. For a given θ, the number in each column represents the number of features that have been removed in order to select $m^{\star}$ features. In the first row, where $\theta = 1$, no feature is removed. The hyphens in the table indicate situations where the desired number of selected features cannot be obtained due to the large number of features that have been removed.

It should be noted that although in theory only features with some level of redundancy exhibit a ρ^2_{max} greater than zero, in practice even two independent features may have a ρ^2_{max} greater than zero. The major cause is that the true

Table 2. Number of eliminated redundant features

θ	$m^* = 5$	$m^* = 10$	$m^* = 15$	$m^* = 20$	$m^* = 25$	$m^* = 30$
1.0	0	0	0	0	0	0
0.9	2	6	13	24	34	36
0.8	4	19	45	74	82	86
0.7	4	50	78	96	104	118
0.6	15	84	120	128	141	151
0.5	47	106	138	160	171	199
0.4	89	159	179	241	284	330
0.3	89	184	281	317	494	514
0.2	89	286	468	555	-	-
0.1	143	487	-	-	-	-
0.0	340	-	-	-	-	-

quotient of correlation can only be obtained by an unlimited number of observations. All the measurements obtained from real problems are actually an estimations of the true values. For instance, although X_1 and X_2 in our synthetic example are completely independent, their correlation is not absolute zero which is due to the limited number of observations (in that case 10,000). A minor cause could be the existence of confounding factors or lurking variables [9] which happens when features show some correlation, but their contents are completely different. Therefore, large numbers of eliminated features for low values of θ are not necessarily due to true redundancy.

Table 3 shows the results of applying two classification algorithms, J48 and SVM, on different numbers of selected features. The structure of the table is similar to that of table 2. The numbers in the table are classification test performances which are calculated as the ratio of correctly classified instances to the total number of instances through a 10-fold cross-validation process. The first row, where $\theta = 1$ (with no redundant features being removed), is considered the baseline. Therefore, the first row shows the classification performance using the first 5, 10, ..., 30 top features obtained directly from the χ^2 ranking algorithm. The performance measures on the second and lower rows are obtained using features selected by removing redundant features. In each column, the performance results are obtained based on the same number of features and the highest performance is in bold face.

Table 3. Performance of J48 and SVM using the selected features

θ	$m^* = 5$		$m^* = 10$		$m^* = 15$		$m^* = 20$		$m^* = 25$		$m^* = 30$	
	J48	SVM	J48	SVM	J48	SVM	J48	SVM	J48	SVM	J48	SVM
1.0	0.22	0.23	0.43	0.46	0.54	0.60	0.62	0.69	0.68	0.76	0.68	0.77
0.9	0.31	0.31	0.53	0.58	0.60	0.70	0.66	0.74	0.67	0.76	0.67	0.78
0.8	0.34	0.36	0.53	0.57	0.61	0.66	**0.67**	**0.76**	**0.69**	0.78	**0.70**	0.82
0.7	0.32	0.35	**0.54**	0.57	0.57	0.65	0.63	**0.76**	0.67	**0.82**	0.67	0.84
0.6	**0.38**	**0.43**	**0.54**	**0.59**	**0.64**	**0.72**	0.63	0.73	0.67	0.81	**0.70**	**0.85**
0.5	0.33	0.29	0.52	**0.59**	0.60	0.68	0.64	0.74	0.66	0.81	0.66	0.83
0.4	0.30	0.27	0.48	0.45	0.55	0.64	0.56	0.71	0.58	0.73	0.60	0.75
0.3	0.30	0.27	0.40	0.45	0.52	0.59	0.57	0.69	0.58	0.76	0.58	0.76
0.2	0.34	0.33	0.48	0.52	0.55	0.64	0.56	0.69	-	-	-	-
0.1	0.27	0.29	0.36	0.42	-	-	-	-	-	-	-	-
0.0	0.22	0.25	-	-	-	-	-	-	-	-	-	-

None of the performance results obtained by using the original ranking (the baseline) has achieved the best performance. In fact, compared to the middle rows, the performance of the original ranking is quite low. By decreasing the redundancy threshold on the lower rows, the performance starts rising. The best performances are spread among rows with θ between 0.5 and 0.8. In most cases the difference between the baseline and the highest performance is quite significant. This indicates that replacing redundant features by non-redundant ones has a major effect on the performance. On the other hand, as θ decreases below 0.5, resulting in an aggressive removal of redundant and semi-redundant features, the performance decreases down to the baseline or even less. However, this is not unexpected, since as described earlier, part of the measured redundancy could be due to the limited number of observations or the presence of confounding features. It is also observed that although the two classifiers have different performance results on the same subset of features, their performance trends with respect to the changes in θ and $m^\star$ are similar and they seem to conform with each other on the best selected subset of features.

4   Conclusions and Future Work

FS algorithms that merely rely on FR methods can severely suffer from the redundancy issue. Features at high ranks, although highly related to target concepts, might be redundant with respect to each other. Using GP we were able to devise a method to measure the redundancy of a feature with respect to a group of features. We used this measure with a forward selection algorithm for FS. Our results show that removing redundant features can significantly boost the learning performance. Our proposed algorithm is unsupervised and can, therefore, be quite efficient in applications where labeling data is costly. We tested this method on a classification problem, but it can be generally applied to any problem with numeric features. As future work, one might consider merging the proposed RR method into previous research on using GP for feature subset selection to build up an FS method which is capable of handling complicated relationships between features and target classes while preserving the minimality.

References

1. Jong, K., Mary, J., Cornuéjols, A., Marchiori, E., Sebag, M.: Ensemble feature ranking. In: Boulicaut, J.-F., Esposito, F., Giannotti, F., Pedreschi, D. (eds.) PKDD 2004. LNCS (LNAI), vol. 3202, pp. 267–278. Springer, Heidelberg (2004)
2. Ruiz, R., Riquelme, J.C., Aguilar-Ruiz, J.S.: Fast feature ranking algorithm. In: Knowledge-Based Intelligent Information and Engineering Systems, pp. 325–331 (2003)
3. Neshatian, K., Zhang, M.: Genetic programming for feature subset ranking in binary classification problems. In: Vanneschi, L., et al. (eds.) EuroGP 2009. LNCS, vol. 5481. Springer, Heidelberg (2009)

4. Zheng, Z., Srihari, R., Srihari, S.: A feature selection framework for text filtering. In: Proceedings of the Third IEEE International Conference on Data Mining. IEEE Computer Society, Washington (2003)

5. Johnson, R.A., Wichern, D.W.: Applied Multivariate Statistical Analysis, 5th edn. Prentice Hall, Englewood Cliffs (2002)

6. Asuncion, A., Newman, D.: UCI machine learning repository (2007), http://archive.ics.uci.edu/ml/index.html

7. Koza, J.R.: Genetic Programming: On the Programming of Computers by Means of Natural Selection. MIT Press, Cambridge (1992)

8. Witten, I.H., Frank, E.: Data Mining: Practical Machine Learning Tools and Techniques, 2nd edn. Morgan Kaufmann, San Francisco (2005)

9. Pearl, J.: Causality: Models, Reasoning, and Inference. Cambridge University Press, Cambridge (2000)

A Distance Metric for Evolutionary Many-Objective Optimization Algorithms Using User-Preferences

Upali K. Wickramasinghe and Xiaodong Li

School of Computer Science and Information Technology,
RMIT University, Melbourne, VIC 3001, Australia
{uwickram,xiaodong}@cs.rmit.edu.au
http://www.cs.rmit.edu.au/

Abstract. In this paper we propose to use a distance metric based on user-preferences to efficiently find solutions for many-objective problems. In a user-preference based algorithm a decision maker indicates regions of the objective-space of interest, the algorithm then concentrates only on those regions to find solutions. Existing user-preference based evolutionary many-objective algorithms rely on the use of dominance comparisons to explore the search-space. Unfortunately, this is ineffective and computationally expensive for many-objective problems. The proposed distance metric allows an evolutionary many-objective algorithm's search to be focused on the preferred regions, saving substantial computational cost. We demonstrate how to incorporate the proposed distance metric with a user-preference based genetic algorithm, which implements the reference point and light beam search methods. Experimental results suggest that the distance metric based algorithm is effective and efficient, especially for difficult many-objective problems.

Keywords: Distance metric, User-preference, Many-objective optimization, Multi-objective optimization, Reference point, Light beam search.

1 Introduction

The use of Evolutionary Multi-objective Optimization (EMO) algorithms to find solutions for problems with two or three objectives have been very popular in the recent times [1]. In these studies, the concept of *dominance* plays a major role in the functionality of the algorithms. In many-objective optimization problems (where the number of objectives are greater than three), comparing individuals using dominance becomes less effective [2,3]. Theoretical results in [2] shows that in many-objective search-spaces the number of non-dominated individuals increases to a point where the entire population becomes non-dominated to each other. This severely limits an algorithm's ability to compare and search for solutions in many-objective problems.

A different approach than modifying the dominance concept [3,4] that has been gathering popularity recently in EMO algorithms is *user-preferences* [5,6,7].

A. Nicholson and X. Li (Eds.): AI 2009, LNAI 5866, pp. 443–453, 2009.
© Springer-Verlag Berlin Heidelberg 2009

These preference mechanisms were seen originally in Multi-Criteria Decision Making (MCDM) literature [8]. In user-preference based algorithms, a Decision Maker (DM) is required to first indicate *preferred regions* of the objective-space for an algorithm to find solutions in. This information is extremely valuable and can be used to guide the algorithm to further explore the search-space.

The user-preference based EMO algorithms seen in the literature all use *dominance comparisons* to select their candidate solutions [5,6,7]. Unfortunately, these algorithms suffer from the problem of not being able to distinguish solutions effectively for problems with a large number of objectives, where most solutions are non-dominated to each other. Consequently these algorithms are less effective in search, and inclined to converge prematurely to local Pareto-fronts. To address this issue, we introduce a *distance metric* utilizing the user-preference information which is provided by the DM. This method removes the need to use dominance comparisons. We have presented an EMO algorithm using this distance metric which is capable of handling problems of large number of objectives. This paper focuses on EMO algorithms for many-objective problems, therefore we use the term EMO here onwards to refer to Evolutionary Many-objective Optimization.

This paper is organized as follows. Section 2 briefly describes the user-preference methods used in this study. These include the classical definitions of the *reference point method* and *light beam search*. Section 3 presents the distance metric and an implementation of an EMO algorithm using this metric. The experiments used to evaluate the EMO algorithm are provided in section 4. Finally, in section 5, we present our conclusions and avenues for future research.

2   Background

To better understand the distance metric we first describe user-preference mechanisms used in this study.

2.1   Reference Point Method

The classical reference point method was first described by Wierzbicki [6,9]. It has been included successfully in several EMO algorithms [6,7]. A reference point $\bar{z}$ for a many-objective problem consists of *aspiration values* for each objective. In the classical MCDM literature this reference point is used to construct a single objective function (given by (1)), which is to be minimized over the entire search-space. If $\mathbf{x} = [x_0, \ldots, x_{n-1}]$ is a solution in the search-space of n dimensions,

$$minimize \quad \max_{i=0,\ldots,M-1} \{w_i(f_i(\mathbf{x}) - \bar{z}_i)\} \tag{1}$$

where $\bar{z} = [\bar{z}_0, \ldots, \bar{z}_{M-1}]$ is the reference point and $\mathbf{w} = [w_0, \ldots, w_{M-1}]$ is a set of weights. f_i is the ith objective function, while M denotes the number of objectives. The DM can assign values for weights, which represents any bias toward an objective.

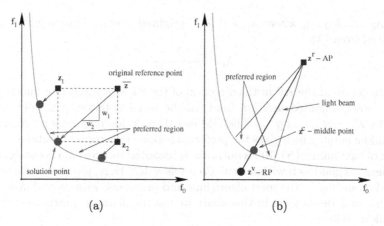

Fig. 1. (a) Classical reference point method (b) Classical light beam search method

Figure 1(a) illustrates the classical reference point method in a two-objective space. The classical MCDM literature [6] shows that several other reference points ($\mathbf{z_1}$ and $\mathbf{z_2}$) can be derived using the original reference point ($\bar{\mathbf{z}}$) and the solution point. In a recursive manner these new reference points can be used to derive more solution points. This traditional approach can be used to define preferred regions in the objective-space.

Using a reference point approach within EMO algorithms is efficient because the entire population can concentrate on finding solutions within this preferred region in a single execution run. We illustrate how preferred regions can be defined with the notion of *outranking* later.

2.2   Light Beam Search Method

The light beam search was first introduced by Jaszkiewicz and Slowinski [10]. The DM first needs to indicate two points in the objective-space, the Aspiration Point (AP), denoted by $\mathbf{z}^r$ and the Reservation Point (RP), denoted by $\mathbf{z}^v$. In situations where the AP and RP are not given, some other points like the *nadir point* and *ideal point* can be used instead. The search direction is given from AP to RP. Metaphorically, this illustrates a light beam originating from AP in the direction of RP. Figure 1(b) illustrates the classical light beam search setup in a two-objective space.

In the classical MCDM literature the light beam search method uses an *achievement scalarizing function* (given by (2)), which is to be minimized. If $\mathbf{x}$ is a solution in the search-space,

$$minimize \quad \max_{i=0,\ldots,M-1} \{\lambda_i \left(f_i(\mathbf{x}) - z_i^r\right)\} + \rho \sum_{i=0}^{M-1} \left(f_i(\mathbf{x}) - z_i^r\right) \qquad (2)$$

where, $\mathbf{z}^r = [z_0^r, \ldots, z_{M-1}^r]$ and $\mathbf{z}^v = [z_0^v, \ldots, z_{M-1}^v]$. ρ is a sufficiently small positive number called the *augmentation coefficient* usually set to 10^{-6}.

$\boldsymbol{\lambda} = [\lambda_0, \ldots, \lambda_{M-1}]$, where $\lambda_i > 0$ is a weighted vector. This weighted vector is derived from (3).

$$\lambda_i = \frac{1}{|z_i^r - z_i^v|} \tag{3}$$

The projection of the AP in the direction of the RP will result in a middle point on the non-dominated solution front. In the usual notation, a middle point is given by $\mathbf{z}^c = [z_0^c, \ldots, z_{M-1}^c]$. The DM can then decide on a region surrounding this middle point, which gives the preferred region. This region is obtained by the notion of outranking (S) [10]. $\mathbf{a}$ outranks $\mathbf{b}$ (denoted by $\mathbf{a}S\mathbf{b}$) if $\mathbf{a}$ is considered to be at least as good as $\mathbf{b}$ within some *threshold* value. Here, the term *better* can be defined according to the used algorithms and problems. *Fitness* and *dominance* are some such definitions. In this study we use the distance metric to define the outranking criteria.

Solutions are obtained in this preferred region *illuminated* by the light beam. We next illustrate how a distance metric is derived from these classical user-preference approaches and how it is used in an EMO algorithm.

3   The Distance Metric

The classical user-preference methods were used to find a single solution on the Pareto front. This single solution would be the closest point to a reference point on the Pareto front or the middle point derived from the light beam search. We introduce a distance metric based on the process of obtaining this solution point.

We define distance of an individual $\mathbf{x}$, to a reference point $\bar{\mathbf{z}}$ using (1) as:

$$dist(\mathbf{x}) = \max_{i=0,\ldots,M-1}\{w_i(f_i(\mathbf{x}) - \bar{z}_i)\} \tag{4}$$

Similarly in the light beam search for any individual $\mathbf{x}$, its distance is defined using (2) as:

$$dist(\mathbf{x}) = \max_{i=0,\ldots,M-1}\{\lambda_i(f_i(\mathbf{x}) - z_i^r)\} + \rho\sum_{i=0}^{M-1}(f_i(\mathbf{x}) - z_i^r) \tag{5}$$

We consider $\mathbf{a}$ to be better than $\mathbf{b}$ if $dist(\mathbf{a}) < dist(\mathbf{b})$. This distance metric will guide the EMO algorithm towards the Pareto front as illustrated in Figure 2(a). We have incorporated both features of the reference point method and light beam search in Figure 2(a) for brevity. Here, it is important to note that RP is defined only if AP is given (for the light beam search). The term *intersection point* ($\mathbf{u}$) is used to identify the closest solution point to the reference point or the middle point of the light beam. It is useful to realize that the many-objective problem is not converted to a single-objective problem with the use of scalarizing functions as seen in traditional MCDM literature. Although (4) and (5) provide a scalar value, the target is not to optimize that value, but to use the value as a metric to guide the population (the closer the individual is to the preferred region, the better it is).

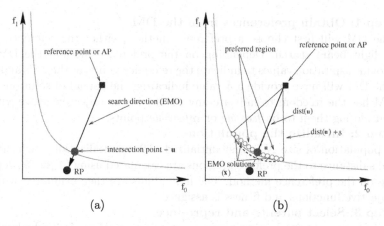

Fig. 2. (a) EMO search yielding a solution point (b) EMO using outranking to define preferred regions

3.1 Controlling the Spread of Solutions

As seen in Figure 2(a) if no control of the spread of solutions is present, the EMO algorithm will explore the search-space along the given direction and converge to the intersection point **u**. To have a control over the spread of solutions we define a threshold value (δ) for the distance metric using the notion of outranking. Here, the aim is to obtain a set of solutions around the intersection point **u**. More specifically, to allow the EMO algorithm to converge to an area of solution points rather than a single solution. If **x** is any solution point we can define **x**S**u** as:

$$\mathbf{x}S\mathbf{u} \Leftrightarrow dist(\mathbf{x}) < dist(\mathbf{u}) + \delta \qquad (6)$$

This relation allows the EMO algorithm to converge not only to the solution point **u**, but to any other solution **x** around **u** as long as **x**S**u**. All the solutions outranking **u** by a given δ defines a preferred region (Figure 2(b)). It is important to note that a preferred region is defined by the search direction, governed by the indicated points, and the spread of solutions. A larger value of δ provides a larger region and a smaller value provides a smaller region. It is also clear that $\delta = 0$ gives **u**. With this δ threshold value the EMO algorithm can have a control of the spread of solutions as required by the DM.

3.2 A Distance Metric Based EMO Algorithm

We now outline how the distance metric can be integrated to an EMO algorithm. Here, a Genetic Algorithm (GA) is used as the search strategy. We have used the original NSGA-II [1] algorithm and replaced the non-dominated sorting procedure by integrating the proposed distance metric approach. Other than GA, the distance metric can be easily integrated into Particle Swarm Optimization (PSO) or Differential Evolution (DE) based EMO algorithms.

- **Step 1: Obtain preferences from the DM**
  The DM will first choose a preference method; either the reference point or light beam search. Depending on the preference method the DM will provide aspiration values to indicate the reference points or the APs and RPs. The DM will next provide a δ value indicating the spread of solutions. The DM has the freedom to indicate any points on the objective-space without considering them to be *feasible* or *infeasible* points.
- **Step 2: Initialize the population**
  A population of size N is first initialized. After initialization, each individual's distance to the preferred regions are calculated using (4) or (5) depending on the preference method. The individuals are then evaluated with the objective functions and fitness is assigned.
- **Step 3: Select parents and reproduce**
  Parents are obtained using tournament selection. Here, individuals closest to the preferred regions are given priority. The parents will crossover and mutate to produce offspring. Here we used the SBX crossover [1] and Polynomial Mutation [1]. The parent population of size N will create N number of offspring.
- **Step 4: Select survivors**
  The parent population of size N is combined with the offspring population of size N to create a population of size $2N$. From this combined population, N number of individuals are selected to move to the next iteration. More specifically, first, the $2N$ population is sorted according to the distance metric. The individuals closest to the preferred regions are selected as the *intersection points*. Next the individuals which outrank these intersection points are selected. If the total number of such selected points are less than N, random individuals are selected from the population to make a final population of size N. If the number of outranked individuals and intersection points are greater than N, the individuals furthest from the preferred regions are removed.

The steps 3 and 4 are repeated until the maximum number of iterations is reached. The distance metric approach guides the population towards the preferred regions such that solutions are found on the Pareto front (Figure 2). At the end of the execution the first non-dominated front is extracted from the population, giving the final solution set.

A dominance comparison based EMO algorithm normally has a computational complexity of $O(MN^2)$ because of the use of the non-dominated sorting procedure [1]. However, the proposed distance metric approach only depends on the sorting procedure using the distance metric. As a result, the computational complexity of using the distance metric (for the entire population) is $O(NlogN)$.

4   Experiments

To evaluate the performance of the EMO algorithm using the distance metric, we used following test problem suits; ZDT [11] for two-objective problems, WFG [12] and DTLZ [13] for two and up to ten objective problems. These test problem suites contain many varieties of multi-objective problems including some with many local optima fronts (multi-modal). The parameter settings were constant throughout the experimentation process because the algorithm was robust, not requiring tweaking depending on the type of problem. The population size was 200 and the maximum number of iterations was 500. The SBX crossover probability was set to 0.9 and the mutation probability was $1/n$, where n was the number of decision variables of each problem. The algorithm executed 50 runs on each problem instance. The proposed algorithm was always able to converge on the global Pareto fronts on the simpler problems and frequently on the more difficult problems. In this section, we only illustrate some of the best results from the more interesting (and difficult) problems from the test problem suites because of the lack of space.

4.1 Two-Objective Problems

Figure 3 shows the solutions fronts obtained for the two-objective multi-modal ZDT4 ($n = 10$) and WFG4 ($n = 6$) test problems. The two preferred regions have spread values of $\delta = 0.01$ and $\delta = 0.05$. Figure 3(a) shows two reference points in the feasible region. ZDT4 is a very challenging problem for EMO algorithms because of its modality (219 local optima fronts). However, the EMO algorithm using reference points is still able to converge onto the global Pareto front. A very interesting result can be seen for the two-objective WFG4 in Figure 3(b). Here, the light beams are located in the infeasible region of the objective-space, because both AP and RPs are infeasible. However, the EMO algorithm with the light beam search still managed to guide the population in the direction of the light beams until solutions are located on the global Pareto front.

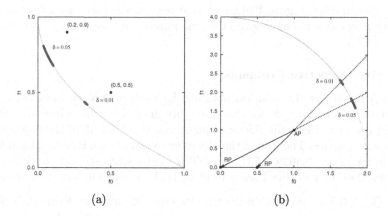

(a)   (b)

Fig. 3. (a) Two-objective ZDT4 with two reference points (b) Two-objective WFG4 with two light beams

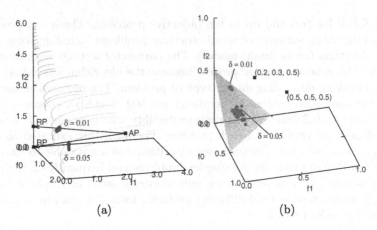

Fig. 4. (a) Three-objective WFG2 with two light beams (b) Three-objective DTLZ1 with two reference points

4.2   Three-Objective Problems

Figure 4 shows the solutions fronts obtained for three-objective WFG2 ($n = 6$) and DTLZ1 ($n = 7$) test problems. Here, WFG2 has disjointed Pareto fronts and DTLZ1 is multi-modal (having $11^5 - 1$ local optima). It is interesting to note that in Figure 4(a) the light beam (with AP $(2.0, 2.0, 2.0)$ and RP $(0.0, 0.0, 0.0)$) goes through the disjoint Pareto front, but the algorithm was still able to locate solutions on the region of the Pareto front which is closest to this light beam. The distance metric guides individuals in the direction given by the vector from AP to RP. This is possible because the algorithm concentrates its search in the direction of this vector. With the population the algorithm has the ability to move in parallel along the direction of this vector until a middle point is found on the Pareto front. Figure 4(b) shows that regardless of the modality, the EMO algorithm was able to converge on the true Pareto with spread values of $\delta = 0.01$ and $\delta = 0.05$.

4.3   Five-Objective Problems

Figure 5 illustrates the solution obtained by each preference mechanism with $\delta = 0.05$. Figure 5(a) shows the result obtained for a five-objective DTLZ1 ($n = 9$) instance. Here, the reference point was at 0.5 for all of the objectives in the objective-space. The sum of the objective values of each individual was found to be in the range $[0.5039, 0.5373]$ for DTLZ1. This suggests that the individuals are very close to the true Pareto front of DTLZ1, since it holds the condition $\sum_{i=0}^{M-1} f_i(\mathbf{x}) = 0.5$ for every $\mathbf{x}$ on the true Pareto optimal front. Figure 5(b) shows the five-objective DTLZ3 ($n = 14$) instances where the AP was set to be the nadir point having the value of 1.0 for all objectives and the RP to be the ideal point having 0.0 for all objectives. DTLZ3 is one of the more difficult multi-modal problems having close to 3^{10} number of local Pareto fronts and one global

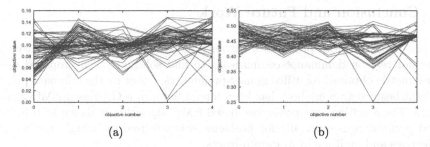

(a)   (b)

Fig. 5. (a) Five-objective DTLZ1 with one reference point (b) Five-objective DTLZ3 with one light beam (each line represents a solution point, where the intersection at the objectives axis represents the value for that objective)

Pareto front. The solutions given for DTLZ3 showed that for each individual **x**, the sum of its squared objective values were in $[1.0475, 1.0671]$. This shows that the solutions are very close to the true Pareto front, because for DTLZ3 a solution **x** is on the true Pareto front if $\sum_{i=0}^{M-1} (f_i(\mathbf{x}))^2 = 1$.

4.4   Ten-Objective Problems

Figure 6 illustrates the solution obtained by each preference mechanism with $\delta = 0.05$. Figure 6(a) shows the result obtained for a ten-objective DTLZ1 ($n = 14$) instance. Here, the reference point was at 0.5 for all of the objectives in the objective-space. The sum of the objective values of each individual was found to be in the range $[0.5084, 0.5662]$ for DTLZ1. Figure 6(b) shows the ten-objective DTLZ3 ($n = 19$) instances where the AP was the nadir point and the RP was the ideal point. The solutions obtained for the DTLZ3 instance indicated that the sum of the squared objective values were in $[1.0837, 1.1322]$, showing that the individuals were very close to the global Pareto front.

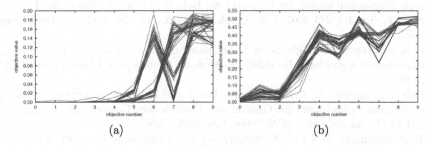

(a)   (b)

Fig. 6. (a) Ten-objective DTLZ1 with one light beam (b) Ten-objective DTLZ3 with one reference point

5   Conclusion and Future Work

In this paper we have proposed a distance metric for EMO algorithms which does not rely on dominance comparisons to find solutions. The proposed distance metric obtained by utilizing user-preferences, either by the reference point or light beam search method, has been integrated into a GA based EMO algorithm. The resulting user-preference based EMO algorithm is shown to provide good performances especially for problems characterized by a high number of objectives and multiple local Pareto-fronts.

Interesting results can be also observed in the behaviour of the proposed EMO algorithm when the preferred regions specified by the DM are in the infeasible regions. In such cases the EMO algorithms are still able to converge to the Pareto front near those specified preferred regions. This property provides an advantage to the DM, since the DM does not have to have the knowledge of where the actual true Pareto optimal front is.

In future we will carry out more comprehensive studies on the distance metric and variations of it. We are also interested in applying EMO algorithms based on this distance metric to solving real world problems.

References

1. Deb, K., Agrawal, S., Pratab, A., Meyarivan, T.: A fast and elitist multiobjective genetic algorithm: NSGA-II. IEEE Transactions on Evolutionary Computation 6, 182–197 (2002)
2. Köppen, M., Vicente-Garcia, R., Nickolay, B.: Fuzzy-pareto-dominance and its application in evolutionary multi-objective optimization. In: Coello Coello, C.A., Hernández Aguirre, A., Zitzler, E. (eds.) EMO 2005. LNCS, vol. 3410, pp. 399–412. Springer, Heidelberg (2005)
3. Kukkonen, S., Lampinen, J.: Ranking-dominance and many-objective optimization. In: IEEE Congress on Evolutionary Computation (CEC), pp. 3983–3990 (2007)
4. Sülflow, A., Drechsler, N., Drechsler, R.: Robust multi-objective optimization in high dimensional spaces. In: Obayashi, S., Deb, K., Poloni, C., Hiroyasu, T., Murata, T. (eds.) EMO 2007. LNCS, vol. 4403, pp. 715–726. Springer, Heidelberg (2007)
5. Deb, K., Kumar, A.: Light beam search based multi-objective optimization using evolutionary algorithms. In: IEEE Congress on Evolutionary Computation, CEC (2007)
6. Deb, K., Sundar, J.: Reference point based multi-objective optimization using evolutionary algorithms. In: Genetic and Evolutionary Computation Conference (GECCO), pp. 635–642. ACM Press, New York (2006)
7. Wickramasinghe, U.K., Li, X.: Integrating user preferences with particle swarms for multi-objective optimization. In: Genetic and Evolutionary Computation Conference (GECCO), pp. 745–752. ACM, New York (2008)
8. Ehrgott, M., Gandibleux, X.: Multiple Criteria Optimization: State of the Art Annotated Bibliographic Survey. International Series in Operations Research and Management Science, vol. 52. Kluwer Academic Publishers, Dordrecht (2002)

9. Miettinen, K.: Some methods for nonlinear multi-objective optimization. In: Zitzler, E., Deb, K., Thiele, L., Coello Coello, C.A., Corne, D.W. (eds.) EMO 2001. LNCS, vol. 1993, pp. 1–20. Springer, Heidelberg (2001)
10. Jaszkiewicz, A., Slowinski, R.: The light beam search approach -an overview of methodology and applications. European Journal of Operational Research 113, 300–314 (1999)
11. Zitzler, E., Deb, K., Thiele, L.: Comparison of multiobjective evolutionary algorithms: Empirical results. Evolutionary Computation 8, 173–195 (2000)
12. Huband, S., Hingston, P., Barone, L., While, R.L.: A review of multiobjective test problems and a scalable test problem toolkit. IEEE Transactions on Evolutionary Computation 10, 477–506 (2006)
13. Deb, K., Thiele, L., Laumanns, M., Zitzler, E.: Scalable test problems for evolutionary multi-objective optimization. In: Evolutionary Multiobjective Optimization (EMO): Theoretical Advances and Applications, pp. 105–145. Springer, Heidelberg (2005)

Balancing Parent and Offspring Selection in Genetic Programming

Huayang Xie and Mengjie Zhang

School of Engineering and Computer Science,
Victoria University of Wellington, New Zealand
{hxie,mengjie}@ecs.vuw.ac.nz

Abstract. In order to drive Genetic Programming (GP) search towards an optimal situation, balancing selection pressure between the parent and offspring selection phases is an important aspect and very challenging. Our previous work showed that stochastic elements cannot be removed from both parent and offspring selections and suggested that maximising diversity in parents and minimising randomness in offspring could provide significantly good performance. This paper conducts additional carefully designed experiments to further investigate how diverse the parent should be if the offspring selection pressure is intensive. This paper shows that any attempt on adding more selection pressure to the parent selection can result in lower GP performance, and the higher the parent selection pressure, the worse the GP performance. The results confirm and strengthen the finding in our previous work.

1 Introduction

Evolutionary Algorithms (EAs) are inspired by biological evolution such as mutation, recombination, natural selection and survival of the fittest, that is, the Darwinian natural selection theory. One form of EAs — Genetic Programming (GP) [1] — started to receive attention from a wide group of researchers from the early 1990s. Since then, it has been rapidly developed into a popular research field of artificial intelligence. To fulfill a certain task, GP starts with a randomly-initialised population of programs. It evaluates each program's performance using a fitness function, which generally compares the program's outputs with the target outputs on a set of training data ("fitness cases"). It assigns each program a fitness value, which in general represents the program's degree of success in achieving the given task. Based on the fitness values, it then chooses some of the programs using a stochastic selection mechanism. After that, it produces a new population of programs for the next generation from these chosen programs using crossover, mutation, and copy operators. The search repeats until it finds an optimal or acceptable solution, or meets certain stopping criteria.

There are many factors that can affect the evolutionary search performance for given problems. These factors include the size of a population, the representation of individuals in a population, the fitness evaluation of individuals,

A. Nicholson and X. Li (Eds.): AI 2009, LNAI 5866, pp. 454–464, 2009.

the selection mechanisms for reproduction and for survival, the genetic operations for modifying individuals, and many more. Amongst these factors, selection mechanisms play an extremely important role.

A selection mechanism consists of a selection scheme and a selection pressure control strategy. According to the configuration of selection pressure, the evolutionary search in GP, as well as in some other instances of EAs, has two extremes [2]. One extreme, when there is no selection pressure, is completely stochastic so that the search acts just like the Monte Carlo method [3], randomly sampling the space of feasible solutions. The other extreme, when the selection pressure is very high, is minimally stochastic so that the search acts like a local hill-climbing search method. It is clear that in general the drawback of the former extreme is its inefficiency and the drawback of the latter extreme is its possible confinement to local optima or *"premature convergence"*. Therefore, an effective and efficient evolutionary search algorithm must balance between these two extremes: removing some stochastic elements in order to distinguish evolutionary search algorithms from a random search algorithm, but on the other hand retaining some stochastic elements in order to prevent the search from being confined in local optima or converging prematurely. In order to obtain the balanced situation, selection pressure, the key element in the selection mechanism, must be properly controlled to maintain the stochastic elements at an optimal level.

As selection in GP and some other forms of EAs consists of two phases — parent selection (selection for reproduction) and offspring selection (selection for survival), selection pressure needs to be controlled in the parent selection and in the offspring selection, as well as be balanced between the parent and offspring selections. The selection of parents has been well explored through the development of EAs. There have been quite a few researches on tuning selection pressure in the parent selection phase [4,5,6,7,8]. However, the selection of offspring (choosing which offspring to put into the next generation) was effectively missing in GP originally: the creation of offspring was a random process and created offspring were directly put into the next generation, meaning that "survival of the fittest" was not applied to offspring. Although later the importance of offspring selection has been noticed and there have been many promising attempts to develop new constructive genetic operators [9,10,11,12], there is little study on tuning selection pressure in the offspring selection phase, and even less on balancing selection pressure between parent and offspring selections. In fact, balancing selection pressure between the parent and offspring selection phases is more challenging than tuning selection pressure in each selection phase alone. How to control selection pressure, consequently the stochastic elements, for driving the evolutionary search towards an optimal situation remains an open and challenging problem in EAs' research.

In our previous work [13], we analysed the selection pressure issue in the context of GP. Although we have found some guidelines to balance the stochastic elements between the parent and offspring selection, to properly achieve that requires further investigations.

1.1 Goals

Our previous work showed that parent diversity is important and recommended to select parents randomly for maximising the parent diversity but select offspring exhaustively for receiving significantly good search performance. The recommendation was derived from a coarse performance comparison where the performance of selecting a pair of parents both randomly was only compared with the performance of selecting a pair of parents both using tournament selection with tournament size 4.

It is likely that choosing one parent randomly while choosing the other one with some selection pressure may also be able to keep parent diversity at an adequate level, and the offspring produced may contain good genetic material with higher chance than that produced by randomly selected parents thus consequently GP search performance may be further improved.

Therefore, this paper further investigates the balance of selection pressure between the parent and offspring selections for driving the GP search towards an optimal situation. In particular, this paper intends to answer if exhaustive offspring selection is chosen, how weak the parent selection pressure is appropriate and whether it is necessary to maximise the parent diversity.

2   The Approach and Experiment Design

This section introduces our previous work for providing sufficient background and describes the experiment design for addressing the research questions.

Constructive genetic operators [9,10,11,12] often replace the standard breeding process, that is, two offspring by crossover and one offspring by mutation, with a many-offspring breeding process. The best offspring is then chosen for survival. The increased offspring selection pressure further reduces the stochastic nature of the GP search. In order to explore the actual effect of offspring selection in combination with parent selection, as well as the balance between parent and offspring selection pressure, we conducted several sets of experiments [13].

The experiments considered six different combinations of selection pressure illustrated in Figure 1. The selection of parents has two options, either without selection pressure by using a random parent selection process, or with selection pressure (using tournament selection with size 4). The selection of offspring has three levels of selection pressure: no selection pressure as in the standard breeding process, or weak selection pressure, or strong selection pressure. The weak offspring selection pressure was implemented through the use of *partial crossover*, which chooses a crossover point randomly in one parent but considers all nodes in the other parent to produce offspring. The strong offspring selection pressure was implemented through the use of *full crossover*, which considers all possible ways of recombining a given pair of parents to produce offspring. Three testing problems were used: the even-6 parity problem, a symbolic regression problem, and a binary classification problem. The ramped half-and-half method was used to create new programs with the maximum depth of four. The population size was 100. The maximum number of generations is 51. The crossover rate and the

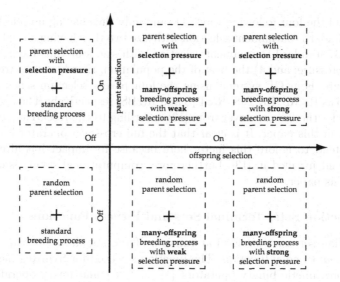

Fig. 1. Six GP systems according to configurations of selection pressure on parent selection and offspring selection (modified from [13])

reproduction rate were 95% and 5%. For ease of analysis, the mutation operator was not used.

Although sometimes increased offspring selection pressure together with parent selection was suggested as effective in the literature [9,10,11,12,14,15], the experimental results showed that stochastic elements cannot be removed from both parent and offspring selections, otherwise premature convergence will appear very often. To obtain a significantly good performance, it was recommended to maximise diversity in parents and minimise randomness in offspring.

2.1   Problem Set

Our experiments considered three problems. Two of them are from [13]: the even-6-parity (EvePar) and the symbolic regression problem (SymReg). Another one is Poly10 as it "is extremely hard" [16]. SymReg and Poly10 are illustrated in Equations 1 and 2.

$$f(x) = exp(1 - x) \times sin(2\pi x) + 50sin(x) \qquad (1)$$

$$f(x_1, x_2, ..., x_{10}) = x_1x_2 + x_3x_4 + x_5x_6 + x_1x_7x_9 + x_3x_6x_{10} \qquad (2)$$

2.2   Parent Selection Pressure Configuration

To answer the above research questions, we investigate four different levels of parent selection pressure configurations when selecting parents for crossover. In these configurations, one parent is selected randomly, the other one is selected using one of the four following ways: 1) randomly thus having no selection pressure; 2) standard tournament selection with size two thus having a weak selection pressure; 3) standard tournament selection with size four thus having a strong selection pressure; and 4) the best of the population, thus having extreme selection pressure. Four GP systems using the four parent selection schemes will be referred to as R&RS, R&T$_2$S, R&T$_4$S and R&BS, respectively. The experiments only consider the full crossover from [13] according to the research questions investigated in this paper. It is clear that the full crossover operator is expensive but we intend to ignore this factor here because its impact has already been analysed and justified in [13]. Furthermore, computational saving is out of the scope of this paper.

2.3 Function Sets, Terminal Sets and Fitness Functions

The function set used for EvePar consists of the standard Boolean operators { and, or, not} and if function. The function set used for SymReg includes the standard arithmetic binary operators { +, -, *, / } and unary operators { abs, exp}. The / function returns zero if it is given invalid arguments. The function set used for Poly10 includes the standard arithmetic binary operators { +, -, * } and a customised {/} operator. The / function returns the denominator if the numerator is zero. The terminal set for EvePar consists of 6 boolean variables. The terminal set for SymReg and Poly10 both consists of a single variable x. In addition, real valued constants in the range [-5.0, 5.0] are also included in the terminal sets for SymReg.

The fitness function for EvePar is the number of wrong outputs (misses) for the 64 combinations of 6-bit length strings. The fitness functions for SymReg and Poly10 are the root-mean-square (RMS) error and the sum of absolute error of the outputs of a program relative to the expected outputs, respectively.

Note that the terminal sets, function sets, and fitness functions are the same as that used in [13] for EvePar and SymReg and that in [16] for Poly10. Other generic parameter configurations are consistent with that in [13].

3   Experiment Results

3.1   Overall Performance

Table 1 shows the effectiveness results over 100 independent runs for the four GP systems on the three problems. The measures for EvePar are the failure rate and the average number of misses. The failure rate shows the fraction of runs that were not able to return the ideal solution. The measures for SymReg and Poly10 are the averages of the RMS error and the sum of absolute errors, respectively. The standard deviation for averaged measures is shown after the ± sign.

The results show the same pattern as that in [13]: overall for all the three problems, the best GP system is the one using random selection for both parents.

We also measured the index of the generation where the best-of-run appeared for the first time in each run (shorten as GenIndex hereafter). In a situation

Table 1. Effectiveness comparison between GP systems using the four parent selection schemes

GP	EvePar		SymReg	Poly10
System	Failure (%)	Miss	RMS Error	Sum Abs Error
R&RS	**20**	**1.3 ± 2.8**	**48.1 ± 6.2**	**8.7 ± 3.2**
R&T$_2$S	23	1.7 ± 3.4	52.2 ± 6.6	9.3 ± 3.2
R&T$_4$S	37	3.3 ± 4.9	53.4 ± 7.4	11.2 ± 3.6
R&BS	97	18.7 ± 7.2	63.6 ± 3.8	17.6 ± 3.0

Table 2. GenIndex comparison between GP systems using the four parent selection schemes

GP System	EvePar	SymReg	Poly10
R&RS	18.0 ± 5.5	37.3 ± 8.1	28.9 ± 13.5
R&T$_2$S	13.2 ± 4.1	24.6 ± 5.8	19.3 ± 8.9
R&T$_4$S	11.0 ± 2.1	17.7 ± 5.6	13.6 ± 7.6
R&BS	6.5 ± 2.8	6.0 ± 4.2	3.4 ± 4.2

where two GP systems have the same effectiveness, the smaller the GenIndex value, the higher the efficiency. On the other hand, if one GP system has a worse average effectiveness than the other but has a smaller average GenIndex, it is likely that the former GP system has encountered a premature convergence problem. Table 2 shows the average GenIndex values for the four GP systems.

From Tables 1 and 2, as the selection pressure increases for the selection of the second parent, the effectiveness of a GP system reduces and the chance of having premature convergence increases. Amongst the four GP systems, the worst performance is given by the GP system in which the best of the population always mates with others, indicating that extremely utilising the known "best" genetic material can easily drive the GP search to local optima. Material in great parents cannot always be considered as good.

3.2 Diversity in Parents

We noticed that the actual performance of the R&RS GP system on the EvePar and SymReg problems was much worse than that reported in [13] (zero percent failure for EvePar and 37.2 RMS error for SymReg). If maximising diversity in parents is the key factor for obtaining the significant performance as reported in [13], then the reason of having worse performance in our R&RS GP system is that our R&RS GP system actually did not maximise the diversity in parents although parents were selected randomly.

In general, random parent selection should give each individual in a population the same chance to be selected as parent. However, since random parent selection can be viewed as a special case of tournament selection with tournament size of 1, the random parent selection in fact has about 36% of the population lost in the parent selection process due to the not-sampled issue [16,17].

Consequently, when the second parent for crossover is selected under some selection pressure, the proportion of a population that can be selected as parents for crossover will be even smaller.

To further verify the necessity of maximising diversity in parents and to investigate the impact of the extra loss of diversity on the performances of the R&T$_2$S, R&T$_4$S and R&BS GP systems, we conducted an additional set of experiments. In this set of experiments, we implemented the round-replacement tournament selection [17], which is designed to ensure every individual in a population can participate into tournaments, for selecting both parents in the R&RS GP system and the first parent in the other three GP systems in order to reduce the loss of parent diversity. Tables 3 and 4 illustrate the effectiveness measures and the GenIndex measure for the four GP systems, respectively.

Table 3. Effectiveness comparison between GP systems using the four parent selection schemes (with round-replacement tournament selection)

GP	EvePar		SymReg	Poly10
System	Failure (%)	Miss	RMS Error	Sum Abs Error
R&RS	**0**	**0**	**38.2 ± 4.3**	**4.3 ± 2.3**
R&T$_2$S	19	1.5 ± 3.5	49.3 ± 6.4	8.9 ± 2.9
R&T$_4$S	35	3.1 ± 3.8	53.0 ± 6.7	11.0 ± 4.1
R&BS	95	17.3 ± 7.1	62.5 ± 4.0	17.6 ± 2.8

Table 4. GenIndex comparison between GP systems using the four parent selection schemes (with round-replacement tournament selection)

GP System	EvePar	SymReg	Poly10
R&RS	14.2 ± 2.7	46.2 ± 5.1	41.7 ± 7.6
R&T$_2$S	12.9 ± 3.2	25.6 ± 6.9	20.9 ± 8.7
R&T$_4$S	10.8 ± 2.3	18.6 ± 6.3	13.0 ± 8.0
R&BS	7.0 ± 2.8	6.5 ± 4.5	3.0 ± 3.6

The performance of the R&RS GP system in the additional experiments is significantly better than that reported in Tables 1 and 2 for all the three problems, and measures for EvePar and SymReg match those in [13]. This observation verified the necessity of maximising diversity in parents when a highly intensive search is conducted in offspring selection.

However, as the selection pressure for the second parent increases, the performance improvement in the additional experiments decreases and becomes even not noticeable in the R&T$_4$S and R&BS GP systems. These additional experimental results show that the negative effect of the increased selection pressure for the second parent overwhelms the positive effect of the extra utilised programs for the first parent.

Overall, our experimental results showed that if exhaustive offspring selection is used, the parent selection pressure should be minimised. Choosing one parent randomly while choosing the other with some selection pressure do not further improve GP performance. It is necessary to maximise the parent diversity. The results confirm and strengthen the work and the recommendation given by [13]: maximising diversity in parents and minimising randomness in offspring can obtain a significantly good performance.

4 Further Discussion

It appears that many researches have been focusing on balancing exploration and exploitation in order to create an optimal search situation for EAs since the 1990s [18,19,20,21,22,23,24,25,26]. Too much exploration results in a pure random search whereas too much exploitation results in a pure local search [18]. If the degree of exploration is too high, search may rapidly degenerate into a random walk where the benefits of evolutionary search are quickly lost. Conversely, if the degree of exploitation is too high, it may result in premature convergence, with significant areas of the search space remaining unexplored [18]. This implies that the key point of balancing exploration and exploitation is to maintain stochastic elements at an appropriate level as required through out the evolutionary process.

Although the terms — *exploration* and *exploitation* — have been used for many years, they have not yet been well defined in EAs' literature. For instance, based on the review in [21], Eshelman *et al.* [19] stated that in GAs selection is commonly seen as the source of exploitation, while the mutation and crossover operators are commonly seen as the source of exploration. However, Eiben and Schippers [21] suggested that mutation and crossover operators can be seen as both exploratory and exploitative. The operators are exploratory because new material is created in mutation and new configurations of material are produced during crossover. They are also exploitative since most of the old genetic material is preserved after applying the operators. Furthermore, Naudts and Schippers [27] interpreted exploitation and exploration in the context of a simple evolutionary algorithm. They defined the process of sampling parents as *neighborhood exploration*, the process of selecting parents as *objective exploitation*, the process of putting selected parents into a mating pool as *generational exploration*, and the process of generating new offspring as *representational exploitation*.

Consequently, due to different interpretations of the two terms, there exist many different views on balancing exploration and exploitation. For instance, in [28], the balance between exploitation and exploration can be adjusted either by the selection pressure of a parent selection operator or by the probability of crossover. In [29], exploitation is encouraged by elitist selection and smaller population sizes, or by using lower mutation rates to promote correlation between parent and offspring. Exploration is encouraged by promoting greater population diversity and selecting parents less discerningly, or by increasing mutation rate. Eiben and Schippers [21] stated (mainly based on research in GAs) that there are three levels at which the phenomena of exploration and exploitation occur: at individual level, at sub-individual level, and at a single gene level. At the first level, individuals are atomic. At the second level, each individual is seen as an instance of the 2^l schemata in the traditional GAs, where l is the length of the individual. At the third level, values of a single gene form the inheritable properties, for instance in a real-valued evolutionary programming that performs a Gaussian perturbation of an allele. It suggests that balancing exploration and exploitation should be considered at the three different levels, but not applicable at all three levels for a given search paradigm. Tackett [9] and Gustafson [25] also suggested that in GP exploration and exploitation are two phases in an evolutionary process, and the exploration phase is followed by the exploitation phase. The two-phase concept in GP treats the maintenance of stochastic elements as a generation-wise process which starts from exploring and ends with exploiting generation by generation.

In order to drive evolutionary search toward an optimal situation, we think it is unnecessarily hard to develop strategies to balance exploration and exploitation when their definitions have not yet been clarified. Instead, it would make more sense to focus on trying to balance parent and offspring selection pressure. This is because not only do they both aim to control stochastic elements through out an evolutionary process, but also the term *selection* is well understood in EAs.

5   Conclusions

This paper further investigated the balance between parent and offspring selection pressure in order to drive GP search towards an optimal situation. It confirmed and strengthened a heuristic in our previous work, that is, when a highly intensive search is conducted in offspring selection, maximising the diversity in parent selection is strongly recommended in order to receive a significantly good performance result. Any attempt on adding more selection pressure to the parent selection can result in lower GP performance, and the higher the parent selection pressure, the worse the GP performance.

Since mutation has not been used in either [13] or this study, further investigation on considering the effect of new material in mutation is necessary in order to move closer to an optimal search situation in GP.

References

1. Koza, J.R.: Genetic Programming — On the Programming of Computers by Means of Natural Selection. MIT Press, Cambridge (1992)
2. Tettamanzi, A., Tomassini, M.: Soft Computing: Integrating Evolutionary, Neural, and Fuzzy Systems. Springer, Heidelberg (2001)
3. Rubinstein, R., Kroese, D.: Simulation and the Monte Carlo Method, 2nd edn. John Wiley and Sons, Chichester (2007)
4. Goldberg, D.E., Deb, K.: A comparative analysis of selection schemes used in genetic algorithms. Foundations of Genetic Algorithms, 69–93 (1991)
5. Julstrom, B.A., Robinson, D.H.: Simulating exponential normalization with weighted k-tournaments. In: Proceedings of the 2000 IEEE Congress on Evolutionary Computation, pp. 227–231. IEEE Press, Los Alamitos (2000)

6. Filipović, V., Kratica, J., Tošić, D., Ljubić, I.: Fine grained tournament selection for the simple plant location problem. In: 5th Online World Conference on Soft Computing Methods in Industrial Applications, pp. 152–158 (2000)
7. Huber, R., Schell, T.: Mixed size tournament selection. Soft Computing - A Fusion of Foundations, Methodologies and Applications 6, 449–455 (2002)
8. Sokolov, A., Whitley, D.: Unbiased tournament selection. In: Proceedings of Genetic and Evolutionary Computation Conference, pp. 1131–1138. ACM Press, New York (2005)
9. Tackett, W.A.: Recombination, selection, and the genetic construction of computer programs. PhD thesis, University of Southern California, Los Angeles, CA, USA (1994)
10. Lang, K.J.: Hill climbing beats genetic search on a boolean circuit synthesis of Koza's. In: Proceedings of the Twelfth International Conference on Machine Learning, Tahoe City, California, USA. Morgan Kaufmann, San Francisco (1995)
11. Harries, K., Smith, P.: Exploring alternative operators and search strategies in genetic programming. In: Proceedings of the Second Annual Conference on Genetic Programming, Stanford University, CA, USA, pp. 147–155. Morgan Kaufmann, San Francisco (1997)
12. Majeed, H., Ryan, C.: A less destructive, context-aware crossover operator for gp. In: Collet, P., Tomassini, M., Ebner, M., Gustafson, S., Ekárt, A. (eds.) EuroGP 2006. LNCS, vol. 3905, pp. 36–48. Springer, Heidelberg (2006)
13. Xie, H., Zhang, M., Andreae, P.: An analysis of constructive crossover and selection pressure in genetic programming. In: Proceedings of Genetic and Evolutionary Computation Conference, pp. 1739–1746 (2007)
14. Mahfoud, S.W.: Crowding and preselection revisited. In: Männer, R., Manderick, B. (eds.) Parallel problem solving from nature, vol. 2, pp. 27–36. North-Holland, Amsterdam (1992)
15. Terrio, M.D., Heywood, M.I.: On naive crossover biases with reproduction for simple solutions to classification problems. In: Deb, K., et al. (eds.) GECCO 2004. LNCS, vol. 3103, pp. 678–689. Springer, Heidelberg (2004)
16. Poli, R., Langdon, W.B.: Backward-chaining evolutionary algorithms. Artificial Intelligence 170, 953–982 (2006)
17. Xie, H., Zhang, M., Andreae, P., Johnston, M.: Is the not-sampled issue in tournament selection critical? In: Proceedings of IEEE Congress on Evolutionary Computation, pp. 3711–3718. IEEE Press, Los Alamitos (2008)
18. Bonham, C.R., Parmee, I.C.: An investigation of exploration and exploitation within cluster oriented genetic algorithms (COGAs). In: Proceedings of the Genetic and Evolutionary Computation Conference, vol. 2, pp. 1491–1497. Morgan Kaufmann, San Francisco (1999)
19. Eshelman, L.J., Caruana, R.A., Schaffer, J.D.: Biases in the crossover landscape. In: Proceedings of the third international conference on Genetic algorithms, pp. 10–19. Morgan Kaufmann Publishers Inc., San Francisco (1989)
20. Eshelman, L., Schaffer, J.: Crossover's niche. In: Proceedings of the Fifth International Conference on Genetic Algorithms, pp. 9–14. Morgan Kaufman, San Francisco (1993)
21. Eiben, A.E., Schippers, C.A.: On evolutionary exploration and exploitation. Fundamenta Informaticae 35, 35–50 (1998)
22. Jong, K.A.D., Spears, W.M.: A formal analysis of the role of multi-point crossover in genetic algorithms. Annals of Mathematics and Artificial Intelligence 5 (1992)

23. Tsutsui, S., Ghosh, A., Corne, D., Fujimoto, Y.: A real coded genetic algorithm with an explorer and an exploiter populations. In: Proceedings of the 7th International Conference on Genetic Algorithms, pp. 238–245. Morgan Kaufmann, San Francisco (1997)

24. Michalewicz, Z., Fogel, D.B.: How to Solve It: Modern Heuristics. Springer, New York (2000)

25. Gustafson, S.M.: An Analysis of Diversity in Genetic Programming. PhD thesis, University of Nottingham (2004)

26. McMahon, A., Scott, D., Browne, W.: An autonomous explore/exploit strategy. In: Proceedings of Genetic and Evolutionary Computation Conference, pp. 103–108. ACM Press, New York (2005)

27. Naudts, B., Schippers, A.: A motivated definition of exploitation and exploration. In: Proceedings of the Genetic and Evolutionary Computation Conference, vol. 1, p. 800. Morgan Kaufmann, San Francisco (1999)

28. Blickle, T., Thiele, L.: A mathematical analysis of tournament selection. In: Proceedings of the Sixth International Conference on Genetic Algorithms, pp. 9–16 (1995)

29. Downing, R.M.: Neutrality and gradualism: encouraging exploration and exploitation simultaneously with binary decision diagrams. In: Proceedings of the 2006 IEEE Congress on Evolutionary Computation, Vancouver, pp. 615–622. IEEE Press, Los Alamitos (2006)

A Memory-Based Approach to Two-Player Texas Hold'em

Jonathan Rubin and Ian Watson

Department of Computer Science,
University of Auckland, New Zealand
jrub001@aucklanduni.ac.nz,
ian@cs.auckland.ac.nz
http://www.cs.auckland.ac.nz/research/gameai

Abstract. A Case-Based Reasoning system, nicknamed SARTRE, that uses a memory-based approach to play two-player, limit Texas Hold'em is introduced. SARTRE records hand histories from strong players and attempts to re-use this information to handle novel situations. SARTRE'S case features and their representations are described, followed by the results obtained when challenging a world-class computerised opponent. Our experimental methodology attempts to address how well SARTRE'S performance can approximate the performance of the expert player, who SARTRE originally derived the experience-base from.

Keywords: Computer Poker, Game-AI, Case-Based Reasoning.

1 Introduction

Poker has been identified as a useful domain for Artificial Intelligence research [1]. As the number of researchers working within the environment of Computer Poker has increased, so too has the development of strong poker robots (or *poker-bots*) which play increasingly more sophisticated strategies [2,4]. A beneficial result of the increased attention paid to computer poker has been the creation of the Annual Computer Poker Competition (CPC) [11], where researchers can evaluate their systems by challenging other computerised opponents to the game of Texas Hold'em poker.

Competitors of past CPC's can typically be characterised into two broad categories. Firstly, those systems that attempt to approximate a *Nash-equilibrium* strategy [2,4]. A *Nash-equilibrium* strategy guarantees that no matter what playing style an opponent adopts, they will never win more than what the *equilibrium* strategy guarantees [2]. This type of strategy can be said to favour not losing, rather than looking for ways to win. At present, *equilibrium* strategies may only be approximated for the game of Texas Hold'em due to the incredibly large size of the game tree. On the other hand, an *exploitative* [3] strategy will attempt to win by maximising profits and exploiting weaker competition. This approach requires a system to model their opponent's play. As this strategy deviates from the *equilibrium*, the system may be prone to exploitation itself.

A. Nicholson and X. Li (Eds.): AI 2009, LNAI 5866, pp. 465–474, 2009.

Our research currently looks into the use of memory in game AI. Rather than relying on game-theoretic principles to construct *near-equilibrium* strategies, the goal of this research is to investigate whether hand histories from strong poker players can be re-used within a Case-Based Reasoning (CBR) framework to achieve a similar performance? CBR is an AI methodology that uses solutions to past problems to solve new problems [7]. A collection of experiences is recorded, which consists of problems and solutions. When a novel situation is encountered a CBR system attempts to retrieve similar experiences from its experience-base and re-use or adapt the solutions to solve the new problem.

SARTRE (Similarity Assessment Reasoning for Texas hold'em via Recall of Experience) is the latest outcome of our research that attempts to address the above question. SARTRE differs from a previous system we developed, CASPER [8], in that it is specifically designed to play 2-player poker, whereas CASPER was more suited to challenge multiple opponents.

SARTRE'S *experience-base* is generated by observing and recording hand histories from the strongest opponents of past CPC's. In 2008 the University of Alberta's Hyperborean-eq took out first place in the limit Hold'em competition [11]. Hyperborean-eq plays a *fixed, near-equilibrium* strategy.

The remainder of this paper proceeds as follows. The game of Texas Hold'em is discussed in Section 2. Section 3 provides an overview of the SARTRE system, followed by Section 4, the experimental results and finally the discussion and conclusion in Section 5.

2   Texas Hold'em

Currently our research focuses around the Texas Hold'em variation of poker. At present, Texas Hold'em is the most popular form of poker as well as being the most strategically complex [5]. In Texas Hold'em play is broken down into four main stages: *preflop, flop, turn* and *river*. For a full description of each stage of the game consult [6].

SARTRE is a heads-up, limit poker-bot. This means SARTRE will only ever challenge one opponent at any one time and betting will be capped at certain limits during each round of play. Factors such as challenging multiple opponents or handling a no-limit betting structure pose extra challenging research problems for poker playing agents. Heads-up, limit poker simplifies these tasks, however, it still preserves the key qualities and structure of other more complicated variants. It also offers its own unique challenges, for example, in heads-up play both players need to play a lot more hands in order to be profitable [5]. Players therefore need to play weaker hands than they would play at a full table (i.e. approx. 9 players), it then becomes more important to determine whether an opponent actually has a valuable hand, or not, more often than would be required at a full table. As only one opponent is available during each game more opportunity exists to model and adapt to your opponents play. It makes sense that two-player, limit poker should be investigated first [2] before focusing effort on more complicated concepts such as no-limit betting and multiple opponents.

3   SARTRE: System Overview

SARTRE makes decisions by retrieving similar cases from its *experience-base*. The authors have hand picked three key factors to represent case features that SARTRE uses to determine a solution for a particular case.

1. The previous betting for the current hand.
2. The current strength of SARTRE'S hand given by combining personal *hole cards* with the publicly available board cards.
3. Information about the state of the current community cards, called the *texture of the board*.

As SARTRE is a computer program the information required needs to be easily recognised and able to be reasoned about algorithmically. Qualitative feature descriptions have been favoured over quantitative descriptions as they are more likely to be used by an expert, human player. Each case feature is described in more detail below, including the representation we have chosen to implement for the SARTRE system.

3.1   The Previous Betting for the Current Hand

The type of betting that can occur at each decision point in a hand consists of a fold *(f)*, check/call *(c)*, or bet/raise *(r)*. A combination of these symbols corresponds to all the decisions made during a particular hand. We have chosen to represent each betting pattern as a path within a betting tree. A betting tree succinctly enumerates all betting combinations up until a certain point in the hand. A path within this tree represents the actual decisions that were made by each player during this hand. Fig. 1. represents a situation where SARTRE'S opponent has made a bet on the *flop* and it is now SARTRE'S turn to act.

Given this representation, we can calculate the similarity between two separate trees (a target tree and a source tree) by comparing the betting path within each tree. If the betting path in the target tree is exactly the same as the betting path within the source tree a similarity value of 1.0 is assigned. Currently, SARTRE will simply assign a value of 0.0 to any betting paths that are not exactly similar, however, we plan to investigate less stringent approaches for future implementations. For example, if one betting path mostly resembles that of another, with a small number of variations, a similarity value close to (but less than) 1.0 could be assigned.

3.2   The Current Strength of SARTRE'S Hand

The second case feature used to determine a betting action is a qualitative category describing SARTRE'S personal hand. During the *pre-flop* SARTRE'S hand simply consists of its personal *hole cards*, whereas for the *post-flop* stages of play SARTRE'S hand is constructed by combining its *hole cards* with the publicly available *community cards*, the best 5 card combination is used.

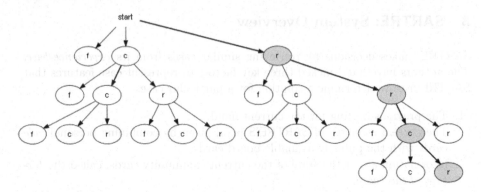

Fig. 1. A tree that describes betting decisions for two players during a hand of Texas Hold'em Poker. The highlighted nodes are the actual decisions that were made by each player.

SARTRE'S best 5 cards are mapped to a category that describes the hand. The classic hand categories in poker include *no-pair, one-pair, two-pair, three-of-a-kind, straight, flush, full-house, four-of-a-kind* and finally a *straight-flush*. Each category has a greater strength than the previous one, where a *straight-flush*, consisting of the cards `Ten, Jack, Queen, King, Ace`, represents the highest rank possible (i.e. a *Royal Flush*).

During the *flop* and the *turn* all the community cards have yet to be dealt and therefore a player's hand has the ability to improve from one category to another, depending on which card is drawn next. It is therefore too simplistic to only consider the current hand category, so further classification is required for hands with the potential to improve. These types of hands are called *drawing hands* (in poker terminology). SARTRE considers two types of drawing hands: *flush draws* & *straight draws*. An example mapping is illustrated in Fig. 2.

The hand categories SARTRE uses to classify cards were decided upon by the authors. Fig. 2. shows a combination of two categories, one which represents the current hand category: `overcards` (i.e. no pair has been made, but both *hole cards* have a higher rank than the community cards). Appended to this category is a separate drawing category: `ace-high-flush-draw-uses-both`, that indicates the strength of the current hand has the ability to improve to a *flush*. The "ace-high" portion of this category further specialises this category by indicating the strength of the possible *flush*.

Currently a simple rule-based system is used to decide which category a combination of cards belongs to. Similarity for this feature is currently either 1.0 when the category of the target case is exactly that of the source case, otherwise it is 0.0 when the categories are distinct.

3.3   The Texture of the Board

The final indexed feature attempts to summarise the state of the community cards without considering the *hole cards* of a player. The *texture of the board*

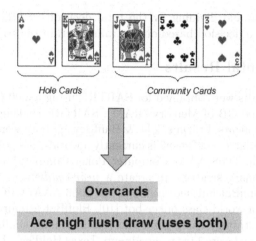

Fig. 2. Mapping a combination of five cards to a category that represents the current hand rank and the drawing strength of this hand

refers to salient information a human poker player would usually notice about the public cards, such as whether a *flush* is possible. Once again a set of qualitative categories were hand-picked by the authors to map various boards into. Some categories used by SARTRE'S current implementation that refer to flush and straight possibilities are *Is-Flush-Possible* (where three cards of the same suit are showing), *Is-Flush-Highly-Possible* (where four cards of the same suit are showing) & *Is-Straight-Possible* (where three consecutive card values are showing), *Is-Straight-Highly-Possible* (where four consecutive card values are showing).

If two boards are mapped into the same category, they are given a similarity value of 1.0, whereas boards that map to separate categories have a similarity of 0.0.

3.4 SARTRE'S Experience-Base

SARTRE'S *experience-base* is generated by analysing information from the logs of previous CPC matches involving Hyperborean-eq. For each hand played in the game log at least one new case is added to SARTRE'S *experience-base*. Each feature described above is assigned into an appropriate category to represent the situation. The decision that Hyperboean-eq made is recorded and acts as the solution for that particular case. The final outcome of that decision is also recorded.

The current version of SARTRE uses just over 1 million cases in total, these are sub-divided into different stages of the game as follows: Preflop cases: 201335, Flop cases: 300577, Turn cases: 281529, River cases: 216597.

When it is time for SARTRE to make a decision, the *experience-base* is consulted and the most similar cases are retrieved, along with their solutions. A *probability triple* is then constructed by summing the number of times each decision was made and dividing by the total decisions. SARTRE then probabilistically selects a decision based on the values within the triple.

4 Experimental Results

Experimental results were obtained for SARTRE using a 3.00 GHz Intel Core 2 Duo CPU with 4.00 GB of Memory (RAM). SARTRE challenged two separate computerized opponents: FellOmen2 [4] & BluffBot [9], both were chosen because they are freely available. FellOmen2 is currently a world-class poker-bot, finishing second equal in the 2008 AAAI Computer Poker Competition [11]. FellOmen2 uses a co-evolutionary strategy, to create a *near-equilibrium* solution [4]. The limit version of BluffBot finished second in the 2006 AAAI CPC and, by today's standards, is not a world-class poker-bot [10]. BluffBot attempts to approach a *Nash-equilibrium* strategy using game-theoretic methods, similar to [2].

All matches played were `limit`, `heads-up`, Texas Hold'em. The betting structure was $2/$4, meaning all bets made during the *preflop* and the *flop* were in increments of $2 and all betting on the *turn* and *river* were in increments of $4. As FellOmen2 and BluffBot were made available in different platforms, two separate poker environments were used to obtain results, described in detail below:

AAAI Computer Poker Competition poker server Version 2.3.1. Using the poker server software, duplicate matches were able to be played. Duplicate matches proceed by playing N hands in a forward direction, then each competitor's memory is reset and the hands are replayed in the reverse direction, i.e. each player now plays the hands that were dealt to their opponent on the forward run. This has the effect of decreasing the inherent variance involved with poker, as one player will not receive a set of better hands than another player. Once the duplicate match is complete the total profit/loss for each direction is summed and the competitor with a positive bankroll is determined the winner. SARTRE challenged FellOmen2 by playing 6 separate duplicate matches, using $N = 3000$, for a total of 36,000 hands.

Poker Academy Pro 2.5. BluffBot was only available to challenge using the commercial application Poker Academy[1]. Poker Academy doesn't allow a duplicate match structure to be played as described above. Instead, all matches played using Poker Academy proceeded in a forward direction and no reduction of variance took place. SARTRE challenged BluffBot by playing a total of 30,000 hands.

4.1 SARTRE vs. FellOmen2

Fig. 3. plots SARTRE'S bankroll for each of the 6 duplicate matches played against FellOmen2. Table 1. provides a summary of the overall outcome. The figures refer to SARTRE'S bankroll.

From the above results we can calculate that, on average, SARTRE loses -2.92 ± 0.5 big bets per 100 hands (BB/100) to FellOmen2. BB/100 is a value

---

[1] http://www.poker-academy.com/poker-software

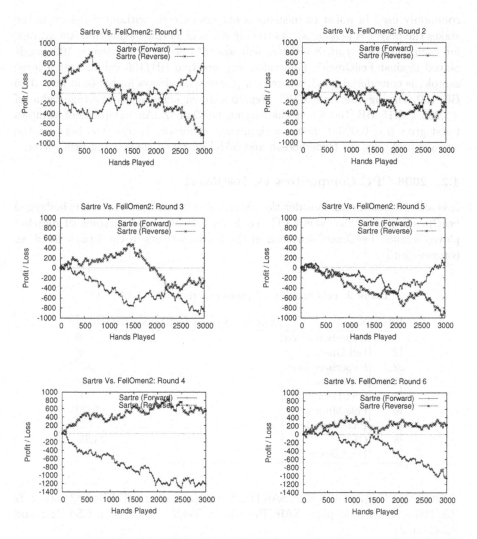

Fig. 3. Sartre vs. FellOmen2

Table 1. Sartre vs. FellOmen2 Summary

	Total Hands	Forward	Reverse	Final Outcome
Round1	6000	532	-827	-295
Round2	6000	-204	-292	-496
Round3	6000	-869	-261	-1130
Round4	6000	549	-1208	-659
Round5	6000	109	-900	-791
Round6	6000	226	-1063	-837
Total	36000	343	-6859	-6516

commonly used in poker to measure a players success, without considering the stakes the player is playing at. As the big bet was $4 this means that, on average for each duplicate run, SARTRE will lose $11.60 ± $2 for every 100 hands played against FellOmen2. Generally, any positive BB/100 value, over a large sample, is considered good, whereas, a player who always folds would lose -37.5 BB/100. During the 2008 CPC, Hyperborean-eq achieved an average value of +1.205 ± 0.15 BB/100 when challenging FellOmen2. An independent samples t-test gives $p < 0.00001$, hence a significant difference is observed between the average profit/loss of Hyperborean and SARTRE when challenging FellOmen2.

4.2   2008 CPC Competitors vs. FellOmen2

It is also interesting to consider the results of other competitors who challenged FellOmen2, during the 2008 CPC. Table 2., lists the final outcome of matches played against FellOmen2 for each of the 9 competitors in the limit Hold'em competition [11,4].

Table 2. FellOmen2 vs. Opponents from 2008 AAAI CPC

Place	Name	Win rate against FellOmen2 (BB/100)
1	Hyperborean-eq	1.2
2	Fell Omen 2	0
2	Hyperborean-on	0.2
2	GGValuta	-0.15
5	GS4-Beta	-1
6	PokeMinn 2	-7.65
7	PokeMinn 1	-7.7
8	GUS	-23.35
9	Dr. Sahbak	-26.6

Our experiments show that SARTRE'S win rate against FellOmen2 was -2.92 BB/100 which would place SARTRE 6th, in Table. 2., between GS4-Beta and PokeMinn2.

4.3   SARTRE vs. BluffBot

The above results only represent SARTRE'S performance against one specific opponent, FellOmen2. To further evaluate the system, SARTRE challenged a separate, computerised opponent. The next opponent SARTRE faced was BluffBot. SARTRE played 30,000 hands against BluffBot and the outcome is illustrated in Fig. 4.

The platform that BluffBot was made available on did not allow for the duplicate match structure that was used when SARTRE challenged FellOmen2, so caution must be used in interpreting the results. However, it is safe to say that Fig. 4. clearly illustrates a profitable trend for SARTRE. SARTRE achieves a win rate of +7.48 BB/100 against BluffBot.

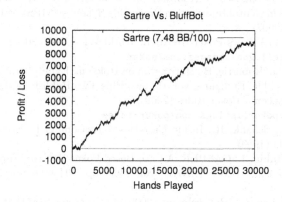

Fig. 4. Sartre vs. BluffBot

5 Discussion and Conclusion

From the results it is clear that SARTRE has not reached the quality of performance that Hyperborean-eq exhibits, as Hyperborean-eq is profitable against FellOmen2, but SARTRE is unprofitable. Some possible reasons for this include:

– The hand strength feature needs to be improved. Presently, a large combination of dissimilar hands are mapped into one category. This results in detailed information being lost which could degrade the level of play.
– There are still many situations where case retrieval is sparse. For one match (chosen at random) against FellOmen2 the results indicated that out of a total of 3769 river decisions made by SARTRE, for 357 (9.47%) of these, SARTRE was unable to retrieve any similar cases. When SARTRE cannot retrieve a similar case a crude strategy of always *checking/calling* is adopted.

However, while SARTRE does not yet achieve the level of play of Hyperborean-eq, the system still appears to play reasonably strong poker. SARTRE was profitable against BluffBot and appears to perform better than four other competitors of the 2008 CPC when challenging FellOmen2.

Acknowledgements. Thanks to Teppo Salonen & Ian Fellows for supplying BluffBot and FellOmen2.

References

1. Billings, D., Davidson, A., Schaeffer, J., Szafron, D.: The challenge of poker. Artificial Intelligence (134), 201–240 (2002)
2. Billings, D., Burch, N., Davidson, A., Holte, R., Schaeffer, J., Schauenberg, T., Szafron, D.: Approximating Game-Theoretic Optimal Strategies for Full-scale Poker. In: IJCAI 2003, pp. 661–668. Morgan Kaufmann, San Francisco (2003)

3. Johanson, M., Zinkevich, M., Bowling, M.: Computing Robust Counter-Strategies. In: Advances in Neural Information Processing Systems (NIPS), vol. 20. MIT Press, Cambridge (2007)
4. Fellows, I.: Ian Fellows Poker Research. World Wide Web (2008), http://thefell.googlepages.com/poker
5. Harrington, D., Robertie, B.: Harrington on Hold'em: Expert Strategy for No-Limit Tournaments. The Endgame, vol. II. Two Plus Two Publishing, Las Vegas (2005)
6. PokerListings.com.: Poker Rules (2009), http://www.pokerlistings.com/poker-rules
7. Riesbeck, C., Schank, R.: Inside Case-Based Reasoning. L. Erlbaum Associates Inc., Hillsdale (1989)
8. Watson, I., Rubin, J.: CASPER: A Case-Based Poker-Bot. In: Wobcke, W., Zhang, M. (eds.) AI 2008. LNCS (LNAI), vol. 5360, pp. 594–600. Springer, Heidelberg (2008)
9. Salonen, T.: The BluffBot webpage (2009), http://www.bluffbot.com/
10. Salonen, T.: Personal communication (2009)
11. University of Alberta.: The Annual Computer Poker Competition (2009), http://www.cs.ualberta.ca/~pokert/

Decomposition of Multi-player Games

Dengji Zhao[1], Stephan Schiffel[2], and Michael Thielscher[2]

[1] Intelligent Systems Laboratory
University of Western Sydney, Australia
[2] Department of Computer Science
Dresden University of Technology, Germany

Abstract. Research in General Game Playing aims at building systems that learn to play unknown games without human intervention. We contribute to this endeavour by generalising the established technique of decomposition from AI Planning to multi-player games. To this end, we present a method for the automatic decomposition of previously unknown games into independent subgames, and we show how a general game player can exploit a successful decomposition for game tree search.

1 Introduction

Research in General Game Playing is concerned with the development of systems that understand the rules of previously unknown games and learn to play well without human intervention. Identified as a new Grand AI Challenge, this endeavour requires to combine methods from a variety of a sub-disciplines including Knowledge Representation, Search, Planning, and Learning [1,2,3,4]. An annual AAAI Contest has been established in 2005 to foster research in this area by evaluating general game playing systems in a competitive setting [5].

With this paper we contribute to the science of General Game Playing by tackling an important and open sub-problem: how can game tree search be improved by automatically decomposing a game into independent parts? The general value of decomposition has been widely recognised in AI Planning, where it is used to help solve large, complex problems arising in practical settings using a divide-and-conquer strategy [6,7,8]. In [9] we have shown how this method can be directly adapted to the special case of single-player games. This previous result provides the starting point for our generalisation to multi-player games. Specifically, we address the following two issues in the present paper: Given its mere rules, how can a previously unknown multi-player game be automatically decomposed into independent subgames? And how can a successful decomposition be exploited for a significant improvement of game tree search during play?

We begin (Section 2) with a brief introduction to the formal basis for our analysis, the Game Description Language [5]. In Section 3, we present a general decomposition method for multi-player games. This result is used in Section 4 to obtain a significant improvement of game tree search for decomposable games. In Section 5, we further improve our method in the special case of so-called impartial games. This is accompanied by both a formal complexity analysis and an overview of experimental results. We conclude in Section 6.

A. Nicholson and X. Li (Eds.): AI 2009, LNAI 5866, pp. 475–484, 2009.

2   Preliminaries

The Game Description Language (GDL) [5,10] is the standard language to communicate the rules of an arbitrary game to each player. It is a variant of first-order logic enhanced by distinguished keywords for the conceptualisation of games. GDL is purely axiomatic, that is, no algebra or arithmetics is included in the language; if a game requires this, the relevant portions of arithmetics have to be axiomatized in the game description.

The class of games that can be expressed in GDL can be classified as *n-player* ($n \geq 1$), *deterministic*, *perfect information* games with *simultaneous moves*. "Deterministic" excludes all games that contain any element of chance, while "perfect information" prohibits that any part of the game state is hidden from some players, as is common in most card games. "Simultaneous moves" allows to describe games like Roshambo, where the players move at the same time, while still permitting to describe games with alternating moves, like chess or checkers, by restricting all players except one to a single "noop" move. Also, GDL games are *finite* in several ways: All reachable states are composed of finitely many fluents; there is a finite, fixed number of players; each player has finitely many possible actions in each game state, and the game has to be formulated such that it leads to a terminal state after a finite number of moves. Each terminal state has an associated goal value for each player, not necessarily zero-sum.

A game state is defined by a set of atomic properties, the *fluents*, that are represented as ground terms. The leading function symbol of a fluent will be called a *fluent symbol*. One game state is designated as the initial state. The transitions are determined by the combined actions of all players. The game progresses until a terminal state is reached.

Example 1. Figure 1 shows the GDL rules[1] of "Double-Tictactoe". This game consists of two instances of the well-known Tic Tac Toe played in parallel.

The `role` keyword (lines 1–2) declares the players in the game. The initial state of the game is described by the keyword `init` (lines 3–7). The two Tic Tac Toe boards are described by fluent functions `cell1` and `cell2`, respectively. Constant `b` indicates a blank cell. The fluent function `control` defines whose turn it is.

The keyword `legal` (lines 8–13) defines what actions (i.e., moves) are possible for each player depending on the properties of the current state, which in turn are encoded using the keyword `true`. The game designer has to ensure that each player always has at least one legal action in every game state. In turn-taking games, players typically have "noop" as their only legal move if it is not their turn. In Double-Tictactoe, the player whose turn it is has to choose one of the two boards and a cell on this board to mark.

The keyword `next` (lines 14–23) defines the effects of the players' actions. For example, lines 14–16 declare that cell (M, N) on the first board is marked with constant `x` if `xplayer` executes action `mark1(M, N)`. The reserved keyword `does` refers to the actions executed by the players. GDL also requires the game

---

[1] We use Prolog notation with variables denoted by uppercase letters.

1 **role**(xplayer).
2 **role**(oplayer).
3 **init**(cell1 (1 ,1 ,b)).
   ...
4 **init**(cell1 (3 ,3 ,b)).
5 **init**(cell2 (1 ,1 ,b)).
   ...
6 **init**(cell2 (3 ,3 ,b)).
7 **init**(control (xplayer)).
8 **legal**(W, mark1 (X,Y)) :−
9   **true**(cell1 (X,Y,b)) ,
10   **true**(control (W)) ,
11   **not** terminal1 .
     ...
12 **legal**(xplayer , noop) :−
13   **true**(control (oplayer)).
14 **next**(cell1 (M,N, x)) :−
15   **does**(xplayer , mark1 (M,N)) ,
16   **true**(cell1 (M,N,b)).
     ...

17 **next**(cell1 (M,N,X)) :−
18   **does**(P, mark2 (X2,Y2)) ,
19   **true**(cell1 (M,N,X)).
20 **next**(control (xplayer)) :−
21   **true**(control (oplayer)).
22 **next**(control (oplayer)) :−
23   **true**(control (xplayer)).
24 open1 :− **true**(cell1 (M,N,b)).
25 **goal**(xplayer ,100) :−
26   line1 (x) , line2 (x).
27 **goal**(oplayer ,75) :−
28   line1 (o) , **not** line2 (x) ,
29   **not** line2 (o).
     ...
30 **terminal** :−
31   terminal1 , terminal2 .
32 terminal1 :− line1 (x).
33 terminal1 :− line1 (o).
34 terminal1 :− **not** open1.
     ...

Fig. 1. Some GDL rules of the game Double-Tictactoe

designer to specify the non-effects of actions by *frame axioms*; e.g., lines 17–19 say that marking a cell on the second board does not affect the first board.

The **goal** predicate (lines 25–29) assigns a number between 0 (loss) and 100 (win) to each role in a terminal state. It is defined with the help of the *auxiliary* predicates line1(W) and line2(W). Auxiliary predicates are not part of the pre-defined language, but are defined in the game description itself. The game is over when a state is reached that implies **terminal** (lines 30–34).[2]

3 Subgame Detection

In our previous work [9], we have developed an algorithm to detect independent subgames and applied this algorithm to single-player games. The basic idea is to build a *dependency graph* for a given GDL description of a game, consisting of the actions and fluents as vertices and edges between them if a fluent is a precondition or an effect of an action. The *connected components* of this graph then correspond to independent subgames of that particular game.

While in principle this idea can be applied to multi-player games, some improvements are necessary in order to extend the range of decomposable games. One problem arises from the fact that in [9] the dependency graph is composed of the mere fluent and action symbols of a game. This does not allow to decompose a game based on different *instances* of these fluents and actions.

Example 2. Consider the following rules of the well-known game Nim with four heaps (a,b,c,d), where the size of the heaps is represented by the fluent **heap**.

---
[2] For a complete definition of syntax and semantics of GDL we refer to [11].

```
 1  init ( heap ( a , 1 )).
 2  init ( heap ( b , 2 )).
 3  init ( heap ( c , 3 )).
 4  init ( heap ( d , 5 )). ...
 5  legal (W, reduce (X,N))  :−
 6     true ( control (W)),
 7     true ( heap (X,M)),
 8     smaller (N,M).
 9  ...
10  next ( heap (X,N))  :−
11     does (W, reduce (X,N)).
12  next ( heap (X,N))  :−
13     true ( heap (X,N)),
14     does (W, reduce (Y,M)),
15     distinct (X,Y).
16  ...
```

Identifying each heap as an independent game is not possible with a dependency graph that does not allow to distinguish different (partial) instances of **heap**.[3]

To overcome this restriction, we base subgame detection for multi-player games on partially instantiated fluent and action terms instead of the mere fluent and action symbols. Considering fully instantiated (i.e., ground) fluents and actions would yield the best results for subgame detection, but this is practically infeasible except for very simple games. Therefore, we instantiate fluents and actions according to the following heuristics:

- The i-th argument of a fluent **f** is instantiated with all possible values iff for every rule that matches $\text{next}(\text{f}(\ldots, X_i, \ldots)) : - B$ the *call graph* (see below) of B contains $\text{true}(\text{f}(\ldots, X_i, \ldots))$ and does not contain $\text{true}(\text{f}(\ldots, X'_i, \ldots))$ with $X'_i \neq X_i$.
- The j-th argument of a move **m** is instantiated with all possible values iff the i-th argument of **f** is instantiated and there is a rule $\text{next}(\text{f}(\ldots, X_i, \ldots)) : - B$ where the call graph of B contains $\text{does}(\text{r}, \text{m}(\ldots, Y_j, \ldots))$ with $Y_j = X_i$.

A *call graph* [9] of a formula is the least set of atoms containing all atoms in the formula as well as all atoms that occur in a rule whose head matches an atom in the call graph. For computing the call graph, we replace every **does**(R, M) in the rules by **legal**(R, M), in order to reflect the fact that every executed move must be a legal one.

The idea behind the heuristics is that an argument of a fluent is instantiated if its value does not change from one state to the next and if instances of the fluent that differ in that argument do not interact. If the different instances do not interact they are likely to belong to separate subgames. Arguments of moves are instantiated if they refer to an instantiated argument of a fluent. In example 2, for instance, the first argument of **heap** is instantiated because the rules for **next** always refers directly (line 13) or indirectly (line 11 along with lines 5, 7) to the first argument of **heap** in the current state and there is no other heap referred to. The first argument of the move **reduce** is instantiated because in the first next-rule the first argument of **reduce** is identical to that of **heap**.

Another problem we face in multi-player games is to determine which individual action of a joint move (by all players) is responsible for a positive or negative effect. To this end, we extended the definition of potential effects from [9] by the following notion of *a role affecting some fluent*.

---

[3] See www.general-game-playing.de for the complete Nim rules.

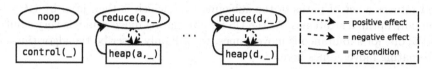

Fig. 2. Dependency graph for the game Nim

Definition 1. *A role r **affects** a fluent f iff there is a game rule unifiable with*
next(f) : $-B$, *where the call graph of B contains* does(r', m) *and r and r' are unifiable.*

*A move m is called a **noop move** if it is the only legal move of a player when not in control and if m does not occur in the call graph of any* next *rule.*

*Fluent f is a **potential positive effect** of move m if there is a game rule unifiable with* next(f) : $-B$ *such that m is not a noop move, B does not imply* true(f), *and B is compatible (see below) with* $\exists r, \vec{y}.$ does(r, m) *where r affects f and $\vec{y}$ are the free variables in m.*

*Fluent f is a **potential negative effect** of move m if m is not a noop move and there is no game rule unifiable with* next(f) : $-B$ *such that for all r that affect f we have* $\forall \vec{y}.$ (true(f) $\wedge$ does(r, m) $\Rightarrow$ B) *where $\vec{y}$ are the free variables in m.*

*Fluent f is a **potential precondition** of move m if f occurs in the call graph of the body of a game rule with head* legal(p, m), *or head* next(f') *where f' is a potential positive or negative effect of m.*

Compatibility means logical consistency under the constraint that each player can do only one action at a time. Thus a fluent is a potential positive effect if there is a non-frame axiom compatible with the action in question, and it is a potential negative effect if there is no frame axiom for this fluent that applies whenever the action is executed. The potential preconditions of a move include all fluents occurring in a legal rule for that move and also the fluents that are preconditions of its (conditional) effects.

The control-fluent that is used to encode turn-taking in multi-player games typically occurs in the legal rules of all actions. We identify and subsequently ignore the control-fluent as precondition during the subgame detection in turn-taking games. Otherwise, it would connect all actions in the dependency graph, effectively rendering subgame detection for turn-taking games impossible.

Applying the above definitions to the Nim game in example 2, we obtain the dependency graph in figure 2 with six subgames: one for each heap consisting of the respective heap-fluent and reduce-action, one consisting of the control-fluent, and one for the noop-action.

4 Solving Decomposable Games

Once a multi-player game has been successfully decomposed, it needs to be solved by what we call *decomposition search* (DS). DS is composed of *subgame search* (SGS) and *global game search* (GGS). SGS searches each subgame independently and returns a set of paths of the subgame tree, which is then used by GGS to compute optimal strategies. As the DS for alternating move games is very similar to the one for simultaneous move games, we will only describe the algorithm for alternating move games here.

Subgame Search (SGS). In each state of an alternating move game, we know whose turn it is next. However, the turn in each state of the subgames is unknown, because in each turn, a player can only choose one subgame to play. Thus SGS needs to consider all players' legal moves during each subgame state expansion and to return a set of paths of the subgame tree, called *turn-move sequences* (TMSeqs).

Definition 2. *A **turn-move sequence** is a tuple* (Ts, Ms, Es) *where*

- Ts *is a list of roles (**turn sequence**), indicated by* $T_1 \circ T_2 \circ ... \circ T_n$,
- Ms *is a list of moves (**move sequence**), indicated by* $M_1 \circ M_2 \circ ... \circ M_n$,
- Es *is a set of evaluations of **local concepts** (see below),*

where $n \geq 0$ *is the length of the sequence. If* $n = 0$*, we call it **empty turn-move sequence**.*

We extend our notion of *local concepts* from single-player games [9] by recording a sign (positive or negative) for each concept: A *local concept* is a ground literal that occurs in the call graph of a `goal` or `terminal` rule, and the local concept's call graph is only related to the fluents of one subgame. For the `goal` rule in lines 27–29 of figure 1, for example, we get three local concepts: `line1(o)`, `not line2(x)`, and `not line2(o)`. The sign for each concept is determined by whether an even or odd number of negations occurs in the path from the root of the `goal` or `terminal` rule's call graph to the concept. In this way we know that a rule is satisfied if all its local concepts (with signs) are true.

With the help of the extended notion of local concepts, we can not only check the equality of two TMSeqs but also find if one is better than another one by using the following definition.

Definition 3. *A turn move sequence* $s_1 = (Ts_1, Ms_1, Es_1)$ *is **evaluation dominated** by* $s_2 = (Ts_2, Ms_2, Es_2)$ *(written* $s_1 \preceq_{Cs} s_2$*) **under** local concepts* Cs *iff* $Ts_1 = Ts_2$ *and* $\forall_{C \in Cs}(Es_1 \models C \Rightarrow Es_2 \models C)$*, where* $Es \models C$ *means* C *is satisfied after playing the moves in the* $TMSeq$*. If* $\exists_{C \in Cs}(Es_1 \not\models C \wedge Es_2 \models C)$*, we call* s_1 ***strongly evaluation dominated*** *(* $s_1 \prec_{Cs} s_2$*) by* s_2 ***under** local concepts* Cs*. This extends to sets of turn move sequences in the following way:* $T_1 \preceq_{Cs} T_2 \equiv \forall_{s_1 \in T_1} \exists_{s_2 \in T_2} s_1 \preceq_{Cs} s_2$ *and* $T_1 \prec_{Cs} T_2 \equiv (T_1 \preceq_{Cs} T_2 \wedge \exists_{s_1 \in T_1, s_2 \in T_2} s_1 \prec_{Cs} s_2)$*.*

The TMSeqs are constructed backwards from the leaf nodes to the initial state of a subgame tree. An empty TMSeq, where Es are evaluations of all local concepts, has to be added for every terminal state of the subgame. Because in general the `terminal` rules cannot be evaluated in a subgame state, an empty TMSeq is added for a state that has no legal move for any player or in which at least one local concept of the `terminal` rules is satisfied. In both cases the subgame state could belong to a terminal state of the game. In each subgame state s a set of TMSeqs is computed for every player in the following way: $TMSeqs(p, s) = \{(p \circ Ts, m \circ Ms, Es)|(Ts, Ms, Es) \in TMSeqs(p', do(p, m, s))\}$. This means that the TMSeqs of a player p in a state s are exactly the TMSeqs of all successor states of s (denoted by $do(p, m, s)$) with the turn sequence Ts augmented by p and the move sequence Ms augmented by the move m that leads to the successor state.

In each state, a set of turn move sequences T_1 obtained from a move of player p is removed if there is a set T_2 from other moves of p such that $\exists_v(T_1 \prec_{Cs} T_2 \wedge \forall_{v'>v} T_1 \preceq_{Cs'} T_2)$ or $\forall_v T_1 \preceq_{Cs} T_2$ where: v and v' are goal values defined for player p in the game rules, Cs are local concepts of $\text{goal}(p, v)$ and `terminal`, and Cs' are local concepts of $\text{goal}(p, v')$ and `terminal`. For example, the two paths with dashed lines in figure 3 have the same evaluation under local concepts of $\text{goal}(x, 100)$ (`line1(x)`) and `terminal` (`terminal1`) as the other two paths, but under local concepts of $\text{goal}(x, 75)$ (`line1(x)`, `not line1(x)`, `not line1(o)`, and `not open1`) and `terminal`, the dashed path with turn sequence (xo) is strongly dominated by the corresponding solid path because it does not entail `not line1(o)`. Thus the two paths with dashed lines can be removed.

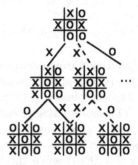

Fig. 3.   Subtree   of   one subgame of Double-Tictactoe

Global Game Search (GGS). GGS is based on standard search techniques (e.g., Minimax, MaxN) but uses TMSeqs returned from SGS instead of the game's `legal` rules to determine the moves in each state. Because of the removal of dominated TMSeqs in SGS the number of moves from the TMSeqs is typically smaller than the number of legal moves. This results in a much smaller game tree compared to full search. Algorithm 4.1 shows the basic idea of DS. We applied iterative-deepening depth-first search (IDDFS) in DS, which finds the shortest solution first and prevents SGS from spending too much time on big subgames.

Algorithm 4.1. Decomposition Search

Input: State: global game state, Player: for which the best move is searched
Output: BestMove
TMSeqs $\leftarrow \varnothing$;
foreach $Subgame \in Subgames$ **do**
    SubState $\leftarrow$ subgame state of Subgame in State;
    SubgameTMSeqs $\leftarrow$ SGS(SubState);
    TMSeqs $\leftarrow$ TMSeqs $\cup$ SubgameTMSeqs;
end
BestMove $\leftarrow$ GGS(TMSeqs,Player);

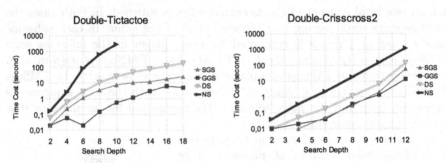

Fig. 4. DS and normal search (NS) testing results

Complexity Comparison and Experiments. The complexity of SGS is the complexity of depth-first search (DFS) plus the complexity of TMSeq simplification. The complexity of GGS depends on the number of TMSeqs returned from SGS. The more TMSeqs can be removed in SGS the less time GGS will use. Assuming a game has n subgames and the average number of states for each subgame is SV, the time complexity of normal search with DFS or IDDFS is $O(|SV|^n + |E_N|)$ while the time complexity of DS is $O(n * |SV| + |E_D| + C)$ where E_N and E_D are edges of the game tree in normal and decomposition search, respectively; $E_N \geq E_D$; and C is the time complexity of GGS. Moreover, the strategies found by DS are just as good as the ones found by normal search.

We have implemented and integrated DS for alternating move games in *Flux-player* [3]. Figure 4 shows the time costs of DS and normal search with different search depths (the depth is related to the global game tree, which is the sum of all SGS depths) for two alternating move games. TMSeq simplification works very well for those games; e.g., only 0.0016% (912 of 58242432) and 0.0661% (10448 of 15864465) of the TMSeqs are returned by subgame search for Double-Tictactoe to depth 9 and for Double-Crisscross2 to depth 6, respectively.

5   Impartial Games

Definition 4. *A game G is **impartial** if G is an alternating move game, in each state of G the player whose turn it is has the same legal moves, and the effects of each move are independent on who is making the move.*

Impartial games allow for a special DS that is more efficient than the general method. Before discussing DS for impartial games, let us describe how a general game player can check whether a game is impartial.

Checking Impartiality. According to the definition of impartial games, we would need to check every state of the game to know if the game is impartial. Since this is not feasible in general, we only do a syntactic analysis of the game rules. This yields a sound but incomplete method. The main idea is to verify that the `legal` and `next` rules for all players are equivalent by checking that for each legal and next rule that is defined for one player there is a correspondent rule for every other player.

Definition 5. *Given two rules R_1 and R_2 of a multi-player game, R_1 and R_2 are **correspondent** for players P_1 and P_2 iff simultaneously substituting P_1 for P_2 and P_2 for P_1 in* control *fluents and* does *predicates in R_1 yields a variant $(R_1[P_1/P_2, P_2/P_1])$ of R_2, that is, $R_1[P_1/P_2, P_2/P_1]$ and R_2 are equal up to renaming of variables and reordering of literals in the bodies of the rules.*

As an example, the legal and next rules of Nim (example 2) are correspondent to themselves for all players.

Decomposition Search (DS). It is easy to prove that a game G is impartial iff its independent subgames are impartial. Another important theorem used in this section is the Sprague-Grundy theorem, which says that each impartial game with normal play convention is equivalent to some Nim heap. Nim is a typical impartial game, which has been mathematically solved. Thus each subgame is actually a Nim heap. If we have the size (called *nimber*) of each subgame, we can use *Nim-sum* to obtain the size of the global game and solve the impartial game by using the strategies used for Nim. More information about Nim and nimber can be found in [12]; explaining the full theory behind impartial games goes beyond the scope of our paper.

Subgame Search (SGS). SGS uses depth-first search to search each subgame with only one player to compute the nimber of the subgame. As all players have the same legal moves in each state, it is sufficient to consider one player's legal moves instead of all players. The nimber of terminal states is 0. For intermediate state, the nimber is the minimal excludent of the nimbers of its successors. For example, if the successors of a state have nimbers 0, 1 and 3, the nimber for the state will be 2.

Global Game Search (GGS). The nimber of the global game can be easily computed as Nim-sum of the nimbers of all subgames, and then the winning strategies for Nim are used to do the rest.

Complexity Comparison and Experiments. For a subgame of size n, the worst-case time complexity of SGS is $O(2^n)$. For an impartial game with m heaps (subgames) of sizes $n_1, n_2, ..., n_m$, the time complexity of DS is $O(\sum 2^{n_i})$, whereas the worst-case time complexity of standard search is $O(2^{\sum n_i})$. In practical play, however, the time complexities are much lower if transposition tables are used in the search; e.g., for Nim this reduces the complexity of SGS to $O(n)$.

The following table shows the time cost of DS and normal search for game Nim with 4 heaps (4 subgames) in *Fluxplayer*:

Time Cost(s)	Normal Play			Misère
	Heaps Size			
	1,5,4,2	2,2,10,10	11,12,15,25	12,12,20,20
Normal Search	0.4	3.5	6607	10797
Decomposition Search	0.01	0.01	0.07	0.06

From the results it is easy to see that the time cost for DS is linear in terms of the biggest heap size, while the one for normal search is exponentially growing in terms of the sum of all heap sizes.

6 Conclusion

We have developed a method by which general game playing systems can search for a decomposition of a multi-player game into independent sub-games in order to significantly improve game tree search. Our result generalises an established method from AI Planning [6,7,8] to General Game Playing. A different, preliminary approach to the decomposition of multi-player games has been independently developed by [13], but there the authors did not address the issue of how a general game player can actually exploit such a reduction during play.

References

1. Kuhlmann, G., Dresner, K., Stone, P.: Automatic heuristic construction in a complete general game player. In: AAAI, pp. 1457–1462 (2006)
2. Clune, J.: Heuristic evaluation functions for general game playing. In: AAAI, pp. 1134–1139 (2007)
3. Schiffel, S., Thielscher, M.: Fluxplayer: A successful general game player. In: AAAI, pp. 1191–1196 (2007)
4. Finnsson, H., Björnsson, Y.: Simulation-based approach to general game playing. In: AAAI, pp. 259–264 (2008)
5. Genesereth, M., Love, N., Pell, B.: General game playing: Overview of the AAAI competition. AI Magazine 26(2), 62–72 (2005)
6. Amir, E., Engelhardt, B.: Factored planning. In: IJCAI, pp. 929–935 (2003)
7. Brafman, R., Domshlak, C.: Factored planning: How, when and when not. In: AAAI, pp. 809–814 (2006)
8. Kelareva, E., Buffet, O., Huang, J., Thiébaux, S.: Factored planning using decomposition trees. In: IJCAI, pp. 1942–1947 (2007)
9. Günther, M., Schiffel, S., Thielscher, M.: Factoring general games. In: Proceedings of the IJCAI Workshop on General Intelligence in Game-Playing Agents (GIGA), pp. 27–34 (2009)
10. Love, N., Hinrichs, T., Haley, D., Schkufza, E., Genesereth, M.: General game playing: Game description language specification. Technical report (2008)
11. Schiffel, S., Thielscher, M.: A multiagent semantics for the Game Description Language. In: ICAART. Springer, Heidelberg (2009)
12. Conway, J.H.: On Numbers and Games. London Mathematical Society Monographs, vol. 6. Academic Press, London (1976)
13. Cox, E., Schkufza, E., Madsen, R., Genesereth, M.: Factoring general games using propositional automata. In: Proceedings of the IJCAI Workshop on General Intelligence in Game-Playing Agents (GIGA), pp. 13–20 (2009)

Extended Full Computation-Tree Logic with Sequence Modal Operator: Representing Hierarchical Tree Structures

Norihiro Kamide[1] and Ken Kaneiwa[2]

[1] Waseda Institute for Advanced Study
1-6-1 Nishi Waseda, Shinjuku-ku, Tokyo 169-8050, Japan
logician-kamide@aoni.waseda.jp
[2] National Institute of Information and Communications Technology
3-5 Hikaridai, Seika, Soraku, Kyoto 619-0289, Japan
kaneiwa@nict.go.jp

Abstract. An extended full computation-tree logic, CTLS*, is introduced as a Kripke semantics with a sequence modal operator. This logic can appropriately represent hierarchical tree structures where sequence modal operators in CTLS* are applied to tree structures. An embedding theorem of CTLS* into CTL* is proved. The validity, satisfiability and model-checking problems of CTLS* are shown to be decidable. An illustrative example of biological taxonomy is presented using CTLS* formulas.

1 Introduction

Full computation-tree logic, CTL*[3, 4], is known as one of the most important branching-time temporal logics that use computation-trees to specify and verify concurrent systems. CTL* is sufficiently expressive to represent almost all the important temporal properties such as liveness, fairness, and safety of concurrent systems. CTL* is more expressive than *computation-tree logic* (CTL) and *linear-time temporal logic* (LTL). CTL [2] is a useful subsystem of CTL*, but it cannot express some important properties such as strong fairness. LTL [5] is also a useful subsystem of CTL*, but it cannot express the properties that verify the existence of a path. An important feature of CTL* is that the existence of paths in computation-trees can be specified and verified. A computation-tree is one that represents a non-deterministic computation or unwinding of a Kripke structure. A Kripke structure is a directed graph; hence, it can naturally express tree structures.

However, CTL* is not suitable for representing the highly complex and informative structures of ontologies and hierarchies. This is because "normal" trees are not sufficiently expressive to represent such complex structures. "Hierarchical" trees are better suited for this purpose; hierarchical tree structures are used to represent hierarchies, taxonomies, and ontologies in some computer science applications. *Biomedical ontologies*, which are knowledge representation

A. Nicholson and X. Li (Eds.): AI 2009, LNAI 5866, pp. 485–494, 2009.

models with hierarchies of biomedical vocabularies, are usually represented by is-a (subtype relation), part-of (inclusion relation), located-in (spatial relation), and proceeded-by (temporal relation) using hierarchical trees or directed acyclic graphs. A biological process *pathway* is searched by finding a path in such a hierarchical tree or directed acyclic graph. In order to represent temporal properties in hierarchical tree structures, we require a very expressive branching-time temporal logic with a new modal operator. The aim of this study is to improve CTL* in order to represent "hierarchical" tree structures, i.e. to obtain computation-trees with "additional information". We introduce a *sequence modal operator* $[b]$, which represents a sequence b of symbols, to describe the ordered labels in a hierarchy.

The reason of using the notion of "sequences" in the new modal operator is explained below. The notion of "sequences" is fundamental to practical reasoning in computer science, because it can appropriately represent "data sequences," "program-execution sequences," "action sequences," "time sequences," "word (character or alphabet) sequences," "DNA sequences" etc. The notion of sequences is thus useful to represent the notions of "information," "attributes," "trees," "orders," "preferences," "strings," "vectors," and "ontologies". "Additional information" can be represented by sequences; this is useful because a sequence structure gives a *monoid* $\langle M, ;, \emptyset \rangle$ with *informational interpretation* [6]:

1. M is a set of pieces of (ordered or prioritized) information (i.e., a set of sequences),
2. ; is a binary operator (on M) that combines two pieces of information (i.e., a concatenation operator on sequences),
3. $\emptyset$ is the empty piece of information (i.e., the empty sequence).

The sequence modal operator $[b]$ represents labels as "additional information". A formula of the form $[b_1 ; b_2 ; \cdots ; b_n]\alpha$ intuitively means that "α is true based on a sequence $b_1 ; b_2 ; \cdots ; b_n$ of (ordered or prioritized) information pieces." Further, a formula of the form $[\emptyset]\alpha$, which coincides with α, intuitively means that "α is true without any information (i.e., it is an eternal truth in the sense of classical logic)." Simple and intuitive consequence relations called *sequence-indexed consequence relations* are required to formalize the sequence modal operator. These consequence relations are regarded as natural extensions of the standard two-valued consequence relation of classical logic. The sequence-indexed consequence relations, denoted as $\models^{\hat{d}}$, are indexed by a sequence $\hat{d}$, and the special case $\models^{\emptyset}$ corresponds to the classical two-valued consequence relation. Then, $\models^{\hat{d}} \alpha$ means that "α is true based on a sequence $\hat{d}$ of information pieces" and $\models^{\emptyset} \alpha$ means that "α is eternally true without any information."

The contents of this paper are summarized as follows: An extended CTL* with the sequence modal operator is introduced as a Kripke semantics with the sequence-indexed consequence relations. An embedding theorem that explains the introduction of CTLS* into CTL* is described. The validity, satisfiability and model checking problems of CTLS* are shown to be decidable. An illustrative example of biological taxonomy using CTLS* formulas is presented.

2   Extended Full Computation-Tree Logic with Sequence Modal Operator

Formulas of CTLS*, which are defined by combining two types of formulas *state formulas* and *path formulas*, are constructed from atomic formulas, $\top$ (truth constant), $\lor$ (disjunction), $\neg$ (negation), E (some computation path), X (next), U (until) and $[b]$ (sequence modal operator) where b is constructed from atomic sequence, the empty sequence $\emptyset$ and ; (concatenation). The other connectives $\bot$ (falsity constant), $\rightarrow$ (implication), $\land$ (conjunction), A (all computation paths), G (always) and F (eventually) can be defined using the connectives displayed above.

Definition 1. *Assume that the numbers of atomic formulas and atomic sequences are respectively countable. The symbol $\emptyset$ represents the empty sequence.*

Formulas α, state formulas β, path formulas γ and sequences b are defined by the following grammar, assuming p and e represent atomic formulas and atomic sequences, respectively:

$$\alpha ::= \beta \mid \gamma$$
$$\beta ::= p \mid \top \mid \beta \lor \beta \mid \neg\beta \mid [b]\beta \mid E\gamma$$
$$\gamma ::= \gamma \lor \gamma \mid \neg\gamma \mid [b]\gamma \mid X\gamma \mid \gamma U\gamma \mid \text{path}(\beta)$$
$$b ::= e \mid \emptyset \mid b \,;\, b$$

Remark that the "path" in Definition 1 is regarded as an auxiliary function from the set of state formulas to the set of path formulas. This means that a state formula is a path formula. The set of atomic formulas is denoted as ATOM, and the set of sequences (including the empty sequence $\emptyset$) is denoted as SE. Lower-case letters $b, c, \dots$ are used for sequences, lower-case letters $p, q, \dots$ are used for atomic formulas, and Greek lower-case letters $\alpha, \beta, \dots$ are used for (state/path) formulas. The symbol ω is used to represent the set of natural numbers. Lower-case letters i, j and k are used for any natural numbers. The symbol $\geq$ or $\leq$ is used to represent a linear order on ω, and the symbol $>$ or $<$ is used to represent a strict linear order on ω. An expression $A \equiv B$ indicates the syntactical identity between A and B. An expression $[\emptyset]\alpha$ coincides with α, and expressions $[\emptyset \,;\, b]\alpha$ and $[b \,;\, \emptyset]\alpha$ coincide with $[b]\alpha$. An expression $[\hat{d}]$ is used to represent $[d_0][d_1][d_2]\cdots[d_i]$ with $i \in \omega$ and $d_0 \equiv \emptyset$, i.e., $[\hat{d}]$ can be the empty sequence. Also, an expression $\hat{d}$ is used to represent $d_0 \,;\, d_1 \,;\, d_2 \,;\, \cdots \,;\, d_i$ with $i \in \omega$ and $d_0 \equiv \emptyset$.

The logic CTLS* is then defined as a Kripke structure with an infinite number of consequence relations.

Definition 2. *A Kripke structure for CTLS* is a structure $\langle S, S_0, R, \{L^{\hat{d}}\}_{\hat{d}\in\mathrm{SE}}\rangle$ such that*

1. *S is a (non-empty) set of states,*
2. *S_0 is a (non-empty) set of initial states and $S_0 \subseteq S$,*

3. R *is a binary relation on* S *which satisfies the condition:* $\forall s \in S \ \exists s' \in S \ [(s, s') \in R]$,

4. $L^{\hat{d}}$ $(\hat{d} \in \mathrm{SE})$ *are mappings from* S *to the power set of* AT $(\subseteq \mathrm{ATOM})$.

Definition 3. *A path in a Kripke structure for CTLS* is an infinite sequence of states,* $\pi = s_0, s_1, s_2, \dots$ *such that* $\forall i \geq 0 \ [(s_i, s_{i+1}) \in R]$. *An expression* π^i *means the suffix of* π *starting at* s_i.

Definition 4. *Let* AT *be a nonempty subset of* ATOM. *Let* α_1 *and* α_2 *be state formulas and* β_1 *and* β_2 *be path formulas. Sequence-indexed consequence relations* $\models^{\hat{d}}$ $(\hat{d} \in \mathrm{SE})$ *on a Kripke structure* $M = \langle S, S_0, R, \{L^{\hat{d}}\}_{\hat{d} \in \mathrm{SE}} \rangle$ *for CTLS* are defined as follows (* π *represents a path constructed from* S, *s represents a state in* S, *and e represents an atomic sequence):*

1. *for* $p \in \mathrm{AT}$, $M, s \models^{\hat{d}} p$ *iff* $p \in L^{\hat{d}}(s)$,

2. $M, s \models^{\hat{d}} \top$ *holds*,

3. $M, s \models^{\hat{d}} \alpha_1 \vee \alpha_2$ *iff* $M, s \models^{\hat{d}} \alpha_1$ *or* $M, s \models^{\hat{d}} \alpha_2$,

4. $M, s \models^{\hat{d}} \neg \alpha_1$ *iff* *not-*$[M, s \models^{\hat{d}} \alpha_1]$,

5. *for any atomic sequence e,* $M, s \models^{\hat{d}} [e]\alpha_1$ *iff* $M, s \models^{\hat{d} \ ; \ e} \alpha_1$,

6. $M, s \models^{\hat{d}} [b \ ; \ c]\alpha_1$ *iff* $M, s \models^{\hat{d}} [b][c]\alpha_1$,

7. $M, s \models^{\hat{d}} \mathrm{E}\beta_1$ *iff there exists a path* π *from* s *such that* $M, \pi \models^{\hat{d}} \beta_1$,

8. $M, \pi \models^{\hat{d}} \mathrm{path}(\alpha_1)$ *iff* s *is the first state of* π *and* $M, s \models^{\hat{d}} \alpha_1$,

9. $M, \pi \models^{\hat{d}} \beta_1 \vee \beta_2$ *iff* $M, \pi \models^{\hat{d}} \beta_1$ *or* $M, \pi \models^{\hat{d}} \beta_2$,

10. $M, \pi \models^{\hat{d}} \neg \beta_1$ *iff* *not-*$[M, \pi \models^{\hat{d}} \beta_1]$,

11. *for any atomic sequence e,* $M, \pi \models^{\hat{d}} [e]\beta_1$ *iff* $M, \pi \models^{\hat{d} \ ; \ e} \beta_1$,

12. $M, \pi \models^{\hat{d}} [b \ ; \ c]\beta_1$ *iff* $M, \pi \models^{\hat{d}} [b][c]\beta_1$,

13. $M, \pi \models^{\hat{d}} \mathrm{X}\beta_1$ *iff* $M, \pi^1 \models^{\hat{d}} \beta_1$,

14. $M, \pi \models^{\hat{d}} \beta_1 \mathrm{U} \beta_2$ *iff* $\exists k \geq 0 \ [(M, \pi^k \models^{\hat{d}} \beta_2)$ *and* $\forall j \ (0 \leq j < k$ *implies* $M, \pi^j \models^{\hat{d}} \beta_1)]$.

Proposition 5. *The following clauses hold for any state formula* β, *any path formula* γ *and any sequences* c *and* $\hat{d}$,

1. $M, s \models^{\hat{d}} [c]\beta$ *iff* $M, s \models^{\hat{d} \ ; \ c} \beta$,

2. $M, s \models^{\emptyset} [\hat{d}]\beta$ *iff* $M, s \models^{\hat{d}} \beta$,

3. $M, \pi \models^{\hat{d}} [c]\gamma$ *iff* $M, \pi \models^{\hat{d} \ ; \ c} \gamma$,

4. $M, \pi \models^{\emptyset} [\hat{d}]\gamma$ *iff* $M, \pi \models^{\hat{d}} \gamma$.

Proof. (1) and (3) are proved by induction on c. (2) and (4) are derived using (1) and (3), respectively. We thus show only (1) below.

Case ($c \equiv \emptyset$): Obvious.

Case ($c \equiv e$ for an atomic sequence e): By the definition of $\models^{\hat{d}}$.

Case ($c \equiv b_1 \ ; \ b_2$): $M, s \models^{\hat{d}} [b_1 \ ; \ b_2]\beta$ iff $M, s \models^{\hat{d}} [b_1][b_2]\beta$ iff $M, s \models^{\hat{d} \ ; \ b_1} [b_2]\beta$ (by induction hypothesis) iff $M, s \models^{\hat{d} \ ; \ b_1 \ ; \ b_2} \beta$ (by induction hypothesis). ∎

Definition 6. *A formula α is* valid *(satisfiable) if and only if one of the following clauses holds:*

1. *if α is a path formula, then $M, \pi \models^\emptyset \alpha$ for any (some) Kripke structure M, any (some) sequence-indexed valuations $\models^{\hat{d}}$ and any (some) path π in M,*
2. *if α is a state formula, then $M, \pi \models^\emptyset \mathrm{path}(\alpha)$ for any (some) Kripke structure M, any (some) sequence-indexed valuations $\models^{\hat{d}}$ and any (some) path π in M.*

An expression $\alpha \leftrightarrow \beta$ means $(\alpha \rightarrow \beta) \wedge (\beta \rightarrow \alpha)$.

Proposition 7. *The following formulas are valid: for any formulas α and β, any path formulas γ_1 and γ_2, any state formula δ and any sequences b and c,*

1. *$[b](\alpha \vee \beta) \leftrightarrow ([b]\alpha) \vee ([b]\beta)$,*
2. *$[b](\neg\alpha) \leftrightarrow \neg([b]\alpha)$,*
3. *$[b \; ; \; c]\alpha \leftrightarrow [b][c]\alpha$,*
4. *$[b]\sharp\gamma_1 \leftrightarrow \sharp[b]\gamma_1$ where $\sharp \in \{E, X\}$,*
5. *$[b](\gamma_1 U \gamma_2) \leftrightarrow ([b]\gamma_1)U([b]\gamma_2)$,*
6. *$[b]\mathrm{path}(\delta) \leftrightarrow \mathrm{path}([b]\delta)$.*

Definition 8. *Let M be a Kripke structure $\langle S, S_0, R, \{L^{\hat{d}}\}_{\hat{d} \in SE} \rangle$ for CTLS*, and $\models^{\hat{d}}$ ($\hat{d} \in SE$) be sequence-indexed consequence relations on M. Then, the model checking problem for CTLS* is defined by: for any formula α, find the set $\{s \in S \mid M, s \models^\emptyset \alpha\}$.*

3   Embedding and Decidability

The logic CTL* can be defined as a sublogic of CTLS*.

Definition 9. *Let AT be a nonempty subset of ATOM. A Kripke structure for CTL* is a structure $\langle S, S_0, R, L \rangle$ such that*

1. *L is a mapping from S to the power set of AT,*
2. *$\langle S, S_0, R \rangle$ is the same as that for a Kripke structure for CTLS*.*

The consequence relation $\models$ on a Kripke structure for CTL* is obtained from Definition 4 by deleting the clauses 5–6 and 11–12 and the sequence notation $\hat{d}$.

Expressions $\models^{\hat{d}}_{\mathrm{CTLS}^*}$ and $\models_{\mathrm{CTL}^*}$ are also used for CTLS* and CTL*, respectively. Note that $\models^\emptyset_{\mathrm{CTLS}^*}$ includes $\models_{\mathrm{CTL}^*}$.
   A translation from CTLS* into CTL* is introduced below.

Definition 10. *Let AT be a nonempty subset of ATOM and $AT^{\hat{d}}$ be the set $\{p^{\hat{d}} \mid p \in AT\}$ ($\hat{d} \in SE$) of atomic formulas where $p^\emptyset := p$, i.e., $AT^\emptyset := AT$. The language $\mathcal{L}^S$ (the set of formulas) of CTLS* is defined using AT, $\top, \vee, \neg, [b], E$, path, X and U by the same way as in Definition 1. The language $\mathcal{L}$ of CTL* is obtained from $\mathcal{L}^S$ by adding $\bigcup_{\hat{d} \in SE} AT^{\hat{d}}$ and deleting $[b]$.*

A mapping f from $\mathcal{L}^S$ to $\mathcal{L}$ is defined by:

1. $f([\hat{d}]p) := p^{\hat{d}} \in AT^{\hat{d}}$ *for any* $p \in AT$,
2. $f([\hat{d}]\top) := \top$,
3. $f([\hat{d}](\alpha \circ \beta)) := f([\hat{d}]\alpha) \circ f([\hat{d}]\beta)$ *where* $\circ \in \{\vee, U\}$,
4. $f([\hat{d}]\sharp\alpha) := \sharp f([\hat{d}]\alpha)$ *where* $\sharp \in \{\neg, E, X\}$,
5. $f([\hat{d}]\text{path}(\alpha)) := \text{path}(f([\hat{d}]\alpha))$,
6. $f([\hat{d}][b\,;\,c]\alpha) := f([\hat{d}][b][c]\alpha)$.

Lemma 11. *Let f be the mapping defined in Definition 10. For any Kripke structure* $M = \langle S, S_0, R, \{L^{\hat{d}}\}_{\hat{d} \in SE}\rangle$ *for CTLS*, any sequence-indexed consequence relations* $\models^{\hat{d}}_{CTLS*}$ *on M and any state or path s in M, there exist a Kripke structure* $N = \langle S, S_0, R, L\rangle$ *for CTL* and a consequence relation* $\models_{CTL*}$ *on N such that for any state or path formula* α *in* $\mathcal{L}^S$,

$$M, s \models^{\hat{d}}_{CTLS*} \alpha \text{ iff } N, s \models_{CTL*} f([\hat{d}]\alpha).$$

Proof. Let AT be a nonempty subset of ATOM and $AT^{\hat{d}}$ be the set $\{p^{\hat{d}} \mid p \in AT\}$ of atomic formulas. Suppose that M is a Kripke structure $\langle S, S_0, R, \{L^{\hat{d}}\}_{\hat{d} \in SE}\rangle$ for CTLS* such that

$$L^{\hat{d}} \ (\hat{d} \in SE) \text{ are mappings from } S \text{ to the powerset of AT.}$$

Suppose that N is a Kripke structure $\langle S, S_0, R, L\rangle$ for CTL* such that

$$L \text{ is a mapping from } S \text{ to the powerset of } \bigcup_{\hat{d} \in SE} AT^{\hat{d}}.$$

Suppose moreover that for any $s \in S$,

$$p \in L^{\hat{d}}(s) \text{ iff } p^{\hat{d}} \in L(s).$$

The lemma is then proved by induction on the complexity of α.

- Base step:
  Case ($\alpha \equiv p \in AT$): $M, s \models^{\hat{d}}_{CTLS*} p$ iff $p \in L^{\hat{d}}(s)$ iff $p^{\hat{d}} \in L(s)$ iff $N, s \models_{CTL*} p^{\hat{d}}$ iff $N, s \models_{CTL*} f([\hat{d}]p)$ (by the definition of f).
- Induction step: We show some cases.

  Case ($\alpha \equiv [b]\beta$): $M, s \models^{\hat{d}}_{CTLS*} [b]\beta$ iff $M, s \models^{\hat{d}\,;\,b}_{CTLS*} \beta$ iff $N, s \models_{CTL*} f([\hat{d}\,;\,b]\beta)$ (by induction hypothesis) iff $N, s \models_{CTL*} f([\hat{d}][b]\beta)$ by the definition of f.

  Case ($\alpha \equiv \alpha_1 \vee \alpha_2$): $M, s \models^{\hat{d}}_{CTLS*} \alpha_1 \vee \alpha_2$ iff $M, s \models^{\hat{d}}_{CTLS*} \alpha_1$ or $M, s \models^{\hat{d}}_{CTLS*} \alpha_2$ iff $N, s \models_{CTL*} f([\hat{d}]\alpha_1)$ or $N, s \models_{CTL*} f([\hat{d}]\alpha_2)$ (by induction hypothesis) iff $N, s \models_{CTL*} f([\hat{d}]\alpha_1) \vee f([\hat{d}]\alpha_2)$ iff $N, s \models_{CTL*} f([\hat{d}](\alpha_1 \vee \alpha_2))$ (by the definition of f).

  Case ($\alpha \equiv E\beta$): $M, s \models^{\hat{d}}_{CTLS*} E\beta$ iff $\exists\pi$: path starting from s $(M, \pi \models^{\hat{d}}_{CTLS*} \beta)$ iff $\exists\pi$: path starting from s $(N, \pi \models_{CTL*} f([\hat{d}]\beta))$ (by induction hypothesis) iff $N, s \models_{CTL*} Ef([\hat{d}]\beta)$ iff $N, s \models_{CTL*} f([\hat{d}]E\beta)$ (by the definition of f).

Case ($\alpha \equiv \beta_1 U \beta_2$ and s is a path π): $M, \pi \models^{\hat{d}}_{CTLS*} \beta_1 U \beta_2$ iff $\exists k \geq 0$ [($M, \pi^k \models^{\hat{d}}_{CTLS*} \beta_2$) and $\forall j \ (0 \leq j < k$ implies $M, \pi^j \models^{\hat{d}}_{CTLS*} \beta_1$)] iff $\exists k \geq 0$ [($N, \pi^k \models_{CTL*} f([\hat{d}]\beta_2)$) and $\forall j \ (0 \leq j < k$ implies $N, \pi^j \models_{CTL*} f([\hat{d}]\beta_1)$)] (by induction hypothesis) iff $N, \pi \models_{CTL*} f([\hat{d}]\beta_1) U f([\hat{d}]\beta_2)$ iff $N, \pi \models_{CTL*} f([\hat{d}](\beta_1 U \beta_2))$ (by the definition of f). ∎

Lemma 12. *Let f be the mapping defined in Definition 10. For any Kripke structure $N = \langle S, S_0, R, L \rangle$ for CTL* and any consequence relation $\models_{CTL*}$ on N, and any state or path s in N, there exist a Kripke structure $M = \langle S, S_0, R, \{L^{\hat{d}}\}_{d \in SE} \rangle$ for CTLS*, sequence-indexed consequence relations $\models^{\hat{d}}_{CTLS*}$ on M such that for any state or path formula α in $\mathcal{L}^S$,*

$$N, s \models_{CTL*} f([\hat{d}]\alpha) \ \textit{iff} \ M, s \models^{\hat{d}}_{CTLS*} \alpha.$$

Proof. Similar to the proof of Lemma 11. ∎

Theorem 13 (Embedding). *Let f be the mapping defined in Definition 10. For any formula α, α is valid in CTLS* iff $f(\alpha)$ is valid in CTL*.*

Proof. By Lemmas 11 and 12. ∎

Theorem 14 (Decidability). *The model checking, validity and satisfiability problems for CTLS* are decidable.*

Proof. By the mapping f defined in Definition 10, a formula α of CTLS* can finitely be transformed into the corresponding formula $f(\alpha)$ of CTL*. By Lemmas 11 and 12 and Theorem 13, the model checking, validity and satisfiability problems for CTLS* can be transformed into those of CTL*. Since the model checking, validity and satisfiability problems for CTL* are decidable, the problems for CTLS* are also decidable. ∎

Since the mapping f is a polynomial-time translation, the complexity results for CTLS* are the same as those for CTL*, i.e., the validity, satisfiability and model-checking problems for CTLS* are 2EXPTIME-complete, deterministic 2EXPTIME-complete and PSPACE-complete, respectively.

4   Illustrative Example

Let us consider an example of biological taxonomy (Linnaean taxonomy [1]) as shown in Fig 1. We model such a case of biological taxonomy using CTLS* formulas in a manner such that the biological classes are partially ordered by sequence modal operators and the life cycles of each class are characterized by temporal operators.

First, the most general taxonomic class *livingThing* is formalized by CTLS* formulas as follows:

[*livingThing*]AF(*living* $\wedge$ AF($\neg living$) $\wedge$ A(*living*U(AG$\neg living$)))

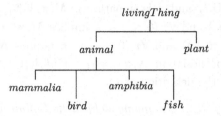

Fig. 1. A biological taxonomy

This formula implies that every living thing is alive until it dies.

> [*animal*; *livingThing*]AF(*motile* ∨ *sentient*)
> [*plant*; *livingThing*]AG¬(*motile* ∨ *sentient*)

These formulas classify living things into animals and plants by characterizing their life cycles. Every animal is motile or sentient, while plants are neither motile nor sentient. The orders of symbols in sequence modal operators are useful in representing a class hierarchy. This is because the orders differentiate their validities; in other words, the two formulas [*b*; *c*]α and [*c*; *b*]α are not equivalent. In addition, the abovementioned example includes both temporal operators and sequence modal operators; CTL* and other logics do not include both these types of operators.

Moreover, several subcategories of animals are defined by the following formulas.

> [*mammalia*; *animal*; *livingThing*]AF(¬*egg* ∧ AX*child* ∧ AXAX*adult*)
> [*bird*; *animal*; *livingThing*](AG(¬*inWater*)∧AF(*egg*∧AX(*child*∧¬*canFly*)∧
> AXAX(*adult* ∧ *canFly*)))
> [*amphibia*; *animal*; *livingThing*]AF((*egg*∧*inWater*)∧AX(*child*∧*inWater*)∧
> AXAX*adult*)
> [*fish*; *animal*; *livingThing*](AG(*inWater*)∧AF(*egg*∧AX*child*∧AXAX*adult*))

Every mammal is viviparous. In the course of its life cycle, it grows from a juvenile into an adult. Every bird is oviparous and can fly when it grows into an adult. In addition, amphibians and fishes are modeled by the changes occurring during growth.

Fig 2 shows a hierarchical tree structure of the biological taxonomy with life cycles (the structure has been simplified to make it easy to understand). We define a Kripke structure $M = \langle S, S_0, R, L^{\hat{d}} \rangle$ that corresponds to biological taxonomy as follows:

1. $S = \{s_0, s_1, s_2, s_3, s_4, s_5, s_6, s_7\}$,
2. $S_0 = \{s_0\}$,
3. $R = \{(s_0, s_1), (s_1, s_2), (s_2, s_3), (s_3, s_4), (s_4, s_4), (s_0, s_5), (s_5, s_6), (s_6, s_7), (s_7, s_4)\}$,
4. $L^{livingThing}(s_0) = L^{animal}(s_0) = \emptyset$,

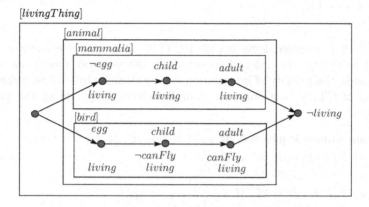

Fig. 2. A biological taxonomy with life cycles

5. $L^{livingThing}(s_1) = L^{animal}(s_1) = L^{mammalia}(s_1) = \{living\}$,
6. $L^{livingThing}(s_2) = L^{animal}(s_2) = L^{mammalia}(s_2) = \{child, living\}$,
7. $L^{livingThing}(s_3) = L^{animal}(s_3) = L^{mammalia}(s_3) = \{adult, living\}$,
8. $L^{livingThing}(s_4) = L^{livingThing}(s_4) = L^{mammalia}(s_4) = L^{bird}(s_4) = \emptyset$,
9. $L^{livingThing}(s_5) = L^{animal}(s_5) = L^{bird}(s_5) = \{egg, living\}$,
10. $L^{livingThing}(s_6) = L^{animal}(s_6) = L^{bird}(s_6) = \{child, living\}$,
11. $L^{livingThing}(s_8) = L^{animal}(s_7) = L^{bird}(s_7) = \{canFly, adult, living\}$,
12. $L^{mammalia}(s_0) = L^{mammalia}(s_5) = L^{mammalia}(s_6) = L^{mammalia}(s_7) = \emptyset$,
13. $L^{bird}(s_0) = L^{bird}(s_1) = L^{bird}(s_2) = L^{bird}(s_3) = \emptyset$.

We can verify the existence of a path that represents required information in the structure M. For example, we can verify: "Is there an animal that hatches from an egg and can fly?" This statement is expressed as:

$[animal](EFegg \wedge (EFcanFly))$

The above statement is true because we have a path $s_0 \rightarrow s_5 \rightarrow s_6 \rightarrow s_7$ with $egg \in L^{animal}(s_5)$ and $canFly \in L^{animal}(s_7)$.

5   Concluding Remarks

An extended full computation-tree logic CTLS* with the sequence modal operator $[b]$ was introduced. This logic could be used to appropriately represent hierarchical tree structures, which are useful in formalizing ontologies. The embedding theorem of CTLS* into CTL* was proved. The validity, satisfiability, and model-checking problems of CTLS* were shown to be decidable. The embedding and decidability results allow us to use the existing CTL*-based algorithms to test the satisfiability. Thus, it was shown that CTLS* can be used as an executable logic to represent hierarchical tree structures.

In the following paragraph, we explain that the applicability of CTLS* can be extended by adding an operator $-$ (converse), which can satisfy the following axiom schemes:

1. $[--b]\alpha \leftrightarrow [b]\alpha,$
2. $[-(b \; ; \; c)]\alpha \leftrightarrow [-c \; ; \; -b]\alpha.$

The resulting extended logic is called CTLS^*_-. The converse operator $-$ can be used to change the order of sequences in the sequence modal operator; in other words, the priority of information can be changed by $-$. The consequence relations of CTLS^*_- are obtained from Definition 4 by adding the following conditions:

1. for any atomic sequence e, $M, s \models^{\hat{d}} [-e]\alpha_1$ iff $M, s \models^{\hat{d} \; ; \; -e} \alpha_1,$
2. $M, s \models^{\hat{d}} [--b]\alpha_1$ iff $M, s \models^{\hat{d}} [b]\alpha_1,$
3. $M, s \models^{\hat{d} \; ; \; --b} \alpha_1$ iff $M, s \models^{\hat{d} \; ; \; b} \alpha_1,$
4. $M, s \models^{\hat{d}} [-(b \; ; \; c)]\alpha_1$ iff $M, s \models^{\hat{d}} [-c \; ; \; -b]\alpha_1,$
5. $M, s \models^{\hat{d} \; ; \; -(b \; ; \; c)} \alpha_1$ iff $M, s \models^{\hat{d} \; ; \; -c \; ; \; -b} \alpha_1,$
6. for any atomic sequence e, $M, \pi \models^{\hat{d}} [-e]\beta_1$ iff $M, \pi \models^{\hat{d} \; ; \; -e} \beta_1,$
7. $M, \pi \models^{\hat{d}} [--b]\beta_1$ iff $M, \pi \models^{\hat{d}} [b]\beta_1,$
8. $M, \pi \models^{\hat{d} \; ; \; --b} \beta_1$ iff $M, \pi \models^{\hat{d} \; ; \; b} \beta_1,$
9. $M, \pi \models^{\hat{d}} [-(b \; ; \; c)]\beta_1$ iff $M, \pi \models^{\hat{d}} [-c \; ; \; -b]\beta_1,$
10. $M, \pi \models^{\hat{d} \; ; \; -(b \; ; \; c)} \beta_1$ iff $M, \pi \models^{\hat{d} \; ; \; -c \; ; \; -b} \beta_1.$

In a similar manner as in the previous sections (with some appropriate modifications), we can obtain the embedding theorem of CTLS^*_- into CTL^* and the decidability theorem for the validity, satisfiability, and model-checking problems of CTLS^*_-.

Acknowledgments. N. Kamide was supported by the Alexander von Humboldt Foundation and the Japanese Ministry of Education, Culture, Sports, Science and Technology, Grant-in-Aid for Young Scientists (B) 20700015. K. Kaneiwa has been partially supported by the Japanese Ministry of Education, Science, Sports and Culture, Grant-in-Aid for Young Scientists (B) 20700147.

References

1. http://en.wikipedia.org/wiki/Linnaean_taxonomy
2. Clarke, E.M., Emerson, E.A.: Design and synthesis of synchronization skeletons using branching time temporal logic. LNCS, vol. 131, pp. 52–71. Springer, Heidelberg (1981)
3. Emerson, E.A., Halpern, J.Y.: "Sometimes" and "not never" revisited: on branching versus linear time temporal logic. Journal of the ACM 33(1), 151–178 (1986)
4. Emerson, E.A., Sistla, P.: Deciding full branching time logic. Information and Control 61, 175–201 (1984)
5. Pnueli, A.: The temporal logic of programs. In: Proceedings of the 18th IEEE Symposium on Foundations of Computer Science, pp. 46–57 (1977)
6. Wansing, H.: The Logic of Information Structures. LNCS (LNAI), vol. 681, 163 p. Springer, Heidelberg (1993)

A Data Model for Fuzzy Linguistic Databases with Flexible Querying

Van Hung Le, Fei Liu, and Hongen Lu

Department of Computer Science and Computer Engineering
La Trobe University, Bundoora, VIC 3086, Australia
vh2le@students.latrobe.edu.au, f.liu@latrobe.edu.au, h.lu@latrobe.edu.au

Abstract. Information to be stored in databases is often fuzzy. Two important issues in research in this field are the representation of fuzzy information in a database and the provision of flexibility in database querying, especially via including linguistic terms in human-oriented queries and returning results with matching degrees. Fuzzy linguistic logic programming (FLLP), where truth values are linguistic, and hedges can be used as unary connectives in formulae, is introduced to facilitate the representation and reasoning with linguistically-expressed human knowledge. This paper presents a data model based on FLLP called fuzzy linguistic Datalog for fuzzy linguistic databases with flexible querying.

Keywords: Fuzzy database, querying, fuzzy logic programming, hedge algebra, linguistic value, linguistic hedge, Datalog.

1 Introduction

Humans mostly use words to characterise and to assess objects and phenomena in the real world. Thus, it is a natural demand for formalisms that can represent and reason with human knowledge expressed in linguistic terms. FLLP [1] is such a formalism. In FLLP, each fact or rule (a many-valued implication) is associated with a linguistic truth value, e.g., *Very True* or *Little False*, taken from a linear hedge algebra (HA) of a linguistic variable *Truth* [2,3], and linguistic hedges (modifiers), e.g., *Very* and *Little*, can be used as unary connectives in formulae. The latter is motivated by the fact that humans often use hedges to state different levels of emphasis, e.g., *very close* and *quite close*.

Deductive databases combine logic programming and relational databases to construct systems that are powerful (e.g., can handle recursive queries), still fast and able to deal with very large volumes of data (e.g., utilizing set-oriented processing instead of one tuple at a time). In this work, we develop an extension of Datalog [4] called *fuzzy linguistic Datalog* (FLDL) by means of FLLP. Features of FLDL are: (i) It enables to find answers to queries over a fuzzy linguistic database (FLDB) using a fuzzy linguistic knowledge base (FLKB) represented by an FLDL program in which all components (except usual connectives) can be expressed in linguistic terms; (ii) Results are returned with a *comparative* linguistic truth value and thus can be ranked accordingly.

A. Nicholson and X. Li (Eds.): AI 2009, LNAI 5866, pp. 495–505, 2009.

The paper is organised as follows: Section 2 gives an overview of FLLP while Section 3 presents FLDL; Section 4 gives discussions and concludes the paper.

2   Preliminaries

2.1   Linguistic Truth Domains and Operations

Values of the linguistic variable *Truth*, e.g., *True, VeryTrue, VeryLittleFalse*, can be regarded to be generated from a set of *primary terms* $G = \{False, True\}$ using *hedges* from a set $H = \{Very, Little, ...\}$ as unary operations. There exists a natural ordering among these terms, with $a \leq b$ meaning that a indicates a degree of truth not greater than b, e.g., $True < VeryTrue$. Hence, the term domain is a partially ordered set and can be characterised by an HA $\underline{X} = (X, G, H, \leq)$, where X is a term set, and $\leq$ is the *semantically ordering relation* on X [2,3]. Hedges either increase or decrease the meaning of terms they modify, i.e., $\forall h \in H, \forall x \in X$, either $hx \geq x$ or $hx \leq x$. The fact that a hedge h modifies terms more than or equal to another hedge k, i.e., $\forall x \in X, hx \leq kx \leq x$ or $x \leq kx \leq hx$, is denoted by $h \geq k$. The primary terms $False \leq True$ are denoted by c^- and c^+, respectively. For an HA $\underline{X} = (X, \{c^-, c^+\}, H, \leq)$, H can be divided into disjoint subsets H^+ and H^- defined by $H^+ = \{h | hc^+ > c^+\}, H^- = \{h | hc^+ < c^+\}$. An HA is said to be *linear* if both H^+ and H^- are linearly ordered. It is shown that the term domain X of a linear HA is also linearly ordered. An *l-limited* HA, where l is a positive integer, is a linear HA in which every term has a length of at most $l + 1$. A *linguistic truth domain* is a finite and linearly ordered set $\overline{X} = X \cup \{0, W, 1\}$, where X is the term domain of an l-limited HA, and 0 (*AbsolutelyFalse*), W (the *middle truth value*), and 1 (*AbsolutelyTrue*) are the least, the neutral and the greatest elements of $\overline{X}$, respectively [1]. Operations are defined on $\overline{X}$ as follows: (*i*) *Conjunction*: $x \wedge y = \min(x, y)$; (*ii*) *Disjunction*: $x \vee y = \max(x, y)$; (*iii*) *An inverse mapping of a hedge*: the idea is that if we modify a predicate by a hedge h, its truth value will be changed by the inverse mapping of h, denoted h^-, e.g., if $young(john) = VeryTrue$, then $Very\ young(john) = Very^-(VeryTrue) = True$; (*iv*) *Many-valued modus ponens* states that from (B, α) and $(A \leftarrow_i B, \beta)$ (i.e., truth values of B and the rule $A \leftarrow_i B$ are at least α and β, respectively), one obtains $(A, \mathcal{C}_i(\alpha, \beta))$, where $\mathcal{C}_i$ is the t-norm, whose *residuum* is the truth function of the implication $\leftarrow_i$, evaluating the modus ponens [5]. Note that our rules can use any of the Łukasiewicz and Gödel implications. Given a linguistic truth domain $\overline{X}$ consisting of $v_0 \leq v_1 \leq ... \leq v_n$, where $v_0 = 0, v_n = 1$, Łukasiewicz and Gödel t-norms are defined as:

$$\mathcal{C}_L(v_i, v_j) = \begin{cases} v_{i+j-n} & \text{if } i + j - n > 0 \\ v_0 & \text{otherwise} \end{cases}$$

$$\mathcal{C}_G(v_i, v_j) = min(v_i, v_j)$$

2.2   Fuzzy Linguistic Logic Programming

Language. The language is a many-sorted predicate language without function symbols. Let $\mathcal{A}$ denote the set of all attributes. For each sort of variables $A \in \mathcal{A}$, there is a set $\mathcal{C}^A$ of constant symbols, which are names of elements in the domain of A. Connectives can be: conjunctions $\wedge$ (also called Gödel) and $\wedge_L$ (Łukasiewicz); the disjunction $\vee$; implications $\leftarrow_L$ (Łukasiewicz) and $\leftarrow_G$ (Gödel); and hedges as unary connectives. For any connective c different from hedges, its truth function is denoted by $c^\bullet$, and for a hedge connective h, its truth function is its inverse mapping h^-. The only quantifier allowed is $\forall$.

A *term* is either a constant or a variable. An *atom* (or *atomic formula*) is of the form $p(t_1, ..., t_n)$, where p is an n-ary predicate symbol, and $t_1, ..., t_n$ are terms of corresponding attributes. A *body formula* is defined inductively as follows: (i) An atom is a body formula; (ii) If B_1 and B_2 are body formulae, then so are $\wedge(B_1, B_2)$, $\vee(B_1, B_2)$ and hB_1, where h is a hedge. A *rule* is a graded implication $(A \leftarrow B.r)$, where A is an atom called *rule head*, B is a body formula called *rule body*, and r is a truth value different from 0; $(A \leftarrow B)$ is called the *logical part* of the rule. A *fact* is a graded atom $(A.b)$, where A is an atom called the logical part of the fact, and b is a truth value different from 0. A *fuzzy linguistic logic program* (program, for short) is a finite set of rules and facts such that there are no two rules (facts) having the same logical part, but different truth values. A program P can be represented as a partial mapping $P : Formulae \rightarrow \overline{X} \setminus \{0\}$, where the domain of P, denoted $dom(P)$, is finite and consists only of logical parts of rules and facts, and $\overline{X}$ is a linguistic truth domain; for a rule $(A \leftarrow B.r)$ (resp. fact $(A.b)$), $P(A \leftarrow B) = r$ (resp. $P(A) = b$). A *query* is an atom $?A$.

Example 1. We take the linguistic truth domain of a 2-limited HA $\underline{X} = (X, \{F, T\}, \{V, M, Q, L\}, \leq)$, where V, M, Q, L, F and T stand for *Very, More, Quite, Little, False* and *True*, i.e., $\overline{X} = \{0, VVF,MVF,VF,QVF,LVF,VMF,$ $MMF, MF,$ $QMF,LMF,F,VQF,MQF,QF,QQF,LQF,LLF,QLF,LF,MLF,VLF,W,VLT,MLT,$ $LT, QLT, LLT,LQT,QQT,QT,MQT,VQT,T,LMT,QMT,MT,MMT,VMT, LVT,$ $QVT,VT,MVT,VVT, 1\}$ (truth values are in an ascending order). Assume that we have the following piece of knowledge: (i) "A car is considered *good* if it is *quite comfortable* and consumes *very less* fuel" is *VeryTrue*; (ii) "A Toyota is *comfortable*" is *QuiteTrue*; (iii) "A Toyota consumes *less* fuel" is *MoreTrue*; (iv) "A BMW is *comfortable*" is *VeryTrue*; (v) "A BMW consumes *less* fuel" is *QuiteTrue*. The knowledge can be represented by the following program:

$$(good(X) \leftarrow_G \wedge(Q\ comfort(X), V\ less\_fuel(X)).VT)$$
$$(comfort(toyota).QT)$$
$$(less\_fuel(toyota).MT)$$
$$(comfort(bmw).VT)$$
$$(less\_fuel(bmw).QT)$$

It can be seen that since it is difficult to give precise numerical assessments of the criteria, the use of qualitative assessments is more realistic and appropriate.

We assume the underlying language of a program P is defined by constants and predicate symbols appearing in P. Thus, we can refer to the *Herbrand base* of P, which consists of all ground atoms, by B_P [6].

Declarative Semantics. Given a program P, let $\overline{X}$ be the linguistic truth domain; a *fuzzy linguistic Herbrand interpretation* (interpretation, for short) f is a mapping $f : B_P \to \overline{X}$. Interpretation f can be extended to all formulae, denoted $\overline{f}$, as follows: (i) $\overline{f}(A) = f(A)$, if A is a ground atom; (ii) $\overline{f}(c(B_1, B_2)) = c^{\bullet}(\overline{f}(B_1), \overline{f}(B_2))$, where B_1, B_2 are ground formulae, and c is a binary connective; (iii) $\overline{f}(hB) = h^{-}(\overline{f}(B))$, where B is a ground body formula, and h is a hedge; (iv) $\overline{f}(\varphi) = \overline{f}(\forall\varphi) = inf_{\vartheta}\{\overline{f}(\varphi\vartheta)|\varphi\vartheta$ is a ground instance of $\varphi\}$. An interpretation f is a *model* of P if for all $\varphi \in dom(P)$, $\overline{f}(\varphi) \geq P(\varphi)$.

Given a program P, let $\overline{X}$ be the linguistic truth domain. A pair $(x; \theta)$, where $x \in \overline{X}$, and θ is a substitution, is called a *correct answer* for P and a query $?A$ if for every model f of P, we have $\overline{f}(A\theta) \geq x$.

Fixpoint Semantics. Let P be a program. An immediate consequence operator T_P is defined as: for an interpretation f and every ground atom A, $T_P(f)(A) = max\{sup\{\mathcal{C}_i(\overline{f}(B), r) : (A \leftarrow_i B.r)$ is a ground instance of a rule in $P\}, sup\{b : (A.b)$ is a ground instance of a fact in $P\}\}$. It is shown in [1] that the Least Herbrand model of the program P is exactly the least fixpoint of T_P and can be obtained by finitely iterating T_P from the bottom interpretation, mapping every ground atom into 0.

3   Fuzzy Linguistic Datalog

According to [4], a *data model* is a mathematical formalism with two parts: (i) A notation for describing data, and (ii) A set of operations used to manipulate that data. Furthermore, model-theoretic, proof-theoretic, fixpoint semantics and their relationship are considered as important parts of a formal data model.

3.1   Language

Our FLDL is an extension of Datalog [4] without negation and possibly with recursion in the same spirit as the one in [7]. The underlying mathematical model of data for FLDL is the notion of *fuzzy linguistic relation* (fuzzy relation, for short): a fuzzy predicate $r(A_1, ..., A_n)$ is interpreted as a fuzzy relation $R : \mathcal{C}^{A_1} \times ... \times \mathcal{C}^{A_n} \to \overline{X}$ and is represented in the form of a (crisp) relation with the relation scheme $R(A_1, ..., A_n, TV)$, where $\overline{X}$ is a linguistic truth domain and is the domain of the *truth-value attribute* TV. Thus, our FLDB is a relational database in which a truth-value attribute is added to every relation to store a linguistic truth value for each tuple. The relations are in the set-of-lists sense, i.e., components appear in a fixed order, and reference to a column is only by its position among the arguments of a given predicate symbol [4]. All notions are the same as those in FLLP. Moreover, we also have some restrictions on logic programs as in the classical case. A rule is said to be *safe* if every variable occurring in the head also occurs in the body. An FLDL program consists of finite safe rules and facts. A predicate appearing in logical parts of facts is called an *extensional database* (EDB) predicate, whose relation is stored in the database and called *EDB relation*, while one defined by rules is called an *intensional database* (IDB) predicate, whose relation is called *IDB relation*, but not both.

3.2 Model-Theoretic Semantics

Let P be an FLDL program; we denote the schema of P by $sch(P)$. For an interpretation f of P, the fact that $f(r(a_1, ..., a_n)) = \alpha$, where $r(a_1, ..., a_n)$ is a ground atom, is denoted by a tuple $(a_1, ..., a_n, \alpha)$ in the relation R for the predicate r. Hence, f can be considered as a database instance over $sch(P)$. As in the classical case [4,8], the *semantics* of P is the least model of P.

3.3 Fuzzy Linguistic Relational Algebra

We extend a *monotone subset*, consisting of Cartesian product, equijoin, projection and union, of relational algebra [4] for the case of our relations and create a new operation called *hedge-modification*. We call the collection of these operations and the classical selection *fuzzy linguistic relational algebra* (FLRA).

Cartesian Product. Predicates can be combined by conjunctions or disjunctions in rule bodies, thus we have two kinds of Cartesian product called *conjunction* and *disjunction Cartesian product*. Let R and S be fuzzy relations of arity k_1 and k_2, respectively. The conjunction (resp. disjunction) Cartesian product of R and S, denoted $R \times^\wedge S$ or $\times^\wedge(R, S)$ (resp. $R \times^\vee S$ or $\times^\vee(R, S)$), is the set of all possible $(k_1 + k_2 - 1)$-tuples of which the first $k_1 - 1$ and the next $k_2 - 1$ components are from a tuple in R and a tuple in S excluding the truth values, respectively, and the new truth value is $\wedge^\bullet(TV_r, TV_s)$ (resp. $\vee^\bullet(TV_r, TV_s)$), where TV_r, TV_s are the truth values of the tuples in R and S, respectively.

Equijoin. We also have two kinds of equijoin called *conjunction* and *disjunction equijoin*. The conjunction (resp. disjunction) equijoin of R and S on column i and j, written $R \bowtie^\wedge_{\$i=\$j} S$ or $\bowtie^\wedge_{\$i=\$j} (R, S)$ (resp. $R \bowtie^\vee_{\$i=\$j} S$ or $\bowtie^\vee_{\$i=\$j} (R, S)$), is those tuples in the conjunction (resp. disjunction) product of R and S such that the ith component of R equals the jth component of S.

Hedge-Modification. Let R be the relation for a predicate r. The relation for formula kr, where k is a hedge, is computed by a *hedge-modification* of R, denoted $\mathcal{H}_k(R)$, as follows: for every tuple in R with a truth value α, there is the same tuple in $\mathcal{H}_k(R)$ except that the truth value is $k^-(\alpha)$.

Projection. Given a relation for the body of a rule, a projection is used to obtain the relation for the IDB predicate in the head. Due to semantics of our rules, the truth value of each tuple in the projected relation is computed using the expression $\mathcal{C}(\_, \rho)$, where $\mathcal{C}$ is the t-norm corresponding to the implication used in the rule, ρ is the truth value of the rule, and the first argument of $\mathcal{C}$ is the truth value of the corresponding tuple in the body relation. More precisely, if R is a relation of arity k, we let $\Pi_{i_1,i_2,\ldots,i_m}^{\mathcal{C}(-,\rho)}(R)$, where the i_j's are distinct integers in the range 1 to $k-1$, denote the projection of R w.r.t. $\mathcal{C}(-,\rho)$ onto components $i_1, i_2, \ldots, i_m$, i.e., the set of $(m+1)$-tuples $(a_1, \ldots, a_m, \alpha)$ such that there is some k-tuple $(b_1, \ldots, b_{k-1}, \beta)$ in R for which $a_j = b_{i_j}$ for $j = 1, \ldots, m$, and $\alpha = \mathcal{C}(\beta, \rho)$.

Union. For the case there is more than one rule with the same IDB predicate in their heads, the relation for the predicate is the union of all projected relations of such rules. The union of relations $R_1, \ldots, R_n$ of the same arity $k+1$, denoted $\bigcup_{i=1}^{n} R_i$, is the set of tuples such that for all tuples $(a_1, \ldots, a_k, \alpha_i)$ in R_i, there is one and only one tuple $(a_1, \ldots, a_k, max\{\alpha_i\})$ in $\bigcup_{i=1}^{n} R_i$.

Remark 1. Clearly, it would not be efficient if we store all possible tuples in the EDB relations; instead we want to store only tuples with non-zero truth values, called *non-zero tuples*, i.e., those have actual meaning, and tuples not appearing in the database are regarded to have a truth value 0. Nevertheless, there are situations where we have to store all possible tuples in some EDB relations. Our language allows rule bodies to be built using the disjunction. Thus, we can have a non-zero tuple in the relation for a disjunction formula even if one of the two tuples in the relations for its disjuncts is missing. However, by disjunction Cartesian product or equijoin, the absence of a tuple in a relation for one of the disjuncts will lead to the absence of the tuple in the relation for the formula. Thus, to ensure that all non-zero tuples will not be lost, all relations for the disjuncts need to consist of all possible tuples that can be formed out of constants (of corresponding sorts of variables) appearing in the program such that tuples which are not explicitly associated with a truth value will have a truth value 0. Moreover, in order for the relations for the operands of a disjunction to consist of all possible tuples, relations for all predicates occurring in the operands must also consist of all their possible tuples; for the relation of a predicate in the head of a rule to consist of all possible tuples, all relations for predicates in its body also contain all their possible tuples. In summary, we can say that the relation for a predicate *involved* in a disjunction must contain all possible tuples; conversely, the relation just needs to consist of only non-zero tuples.

3.4 Translation of FLDL Rules into FLRA Expressions

From the proof-theoretic point of view, we would like to compute relations for IDB predicates from relations for EDB predicates using rules. To that end, we first translate every rule in an FLDL program into an FLRA expression which yields a relation for the IDB predicate in the head of the rule. A *translation algorithm* to do this is adapted from the algorithm for the classical case in [9]. The algorithm requires the following: (a) Function $corr(i)$ returns, for any variable v_i in the head of the rule, the index j of the first occurrence of the same variable in the body; (b) Function $const(w)$ returns *true* if argument w is a constant, and *false* otherwise; (c) Procedure $newvar(x)$ returns a new variable name in x; (d) Let L be a string, and x, y symbols. Then $L < x, y >$ (resp., $L[x,y]$) denotes a new string obtained by replacing the first occurrence (resp., all occurrences) of x in L with y; (e) An artificial predicate $eq(x,y)$ whose relation EQ contains one tuple $(c,c,1)$ for each constant c appearing in the program. EQ is used to express the equality of two components in the crisp sense, i.e., the truth values of such tuples are 1.

TRANSLATION ALGORITHM

INPUT: A rule $r : (H \leftarrow_i B.\rho)$, where H is a predicate $p(v_1, ..., v_n)$, and B is a formula of predicates $q_1, ..., q_m$ whose arguments u_j are indexed consecutively. During the execution, B denotes whatever follows the symbol $\leftarrow$ except ρ.

OUTPUT: An FLRA expression of relations $Q_1, ..., Q_m$.

BEGIN $T(r)$ **END**

RULE 1: $T(r)$
    BEGIN
        IF $\exists i : const(v_i)$ THEN /*case (a)*/
            BEGIN $newvar(x)$; RETURN $T(H < v_i, x > \leftarrow \wedge (B, eq(x, v_i)).\rho)$ END
        ELSEIF $\exists i, j : v_i = v_j, i < j$ THEN /*case (b)*/
            BEGIN $newvar(x)$; RETURN $T(H < v_i, x > \leftarrow \wedge (B, eq(x, v_j)).\rho)$ END
        ELSE RETURN $\Pi^{C_i(\_,\rho)}_{corr(1),...,corr(n)} T'(B)$
    END

RULE 2: $T'(B)$
    BEGIN
        IF $\exists i : const(u_i)$ THEN /*case (a)*/
            BEGIN $newvar(x)$; RETURN $\sigma_{\$i=u_i} T'(B < u_i, x >)$ END
        ELSEIF $\exists i, j : u_i = u_j, i < j$ THEN /*case (b)*/
            BEGIN $newvar(x)$; RETURN $\sigma_{\$i=\$j} T'(B < u_i, x >)$ END
        ELSE
            BEGIN $B[\wedge, \times^{\wedge}]$; $B[\vee, \times^{\vee}]$; $B[h, \mathcal{H}_h]$; $B[eq(...), EQ]$;
                FOR $i := 1$ TO m DO $B[q_i(...), Q_i]$; RETURN B
            END
    END

After the application of rule T', the selection conditions are replaced by those of relations and join conditions as follows: (i) For each condition $\$i = a$, we find relation Q that contains the component of position i. Let l be the total number of components of all relations preceding Q; we put a selection condition $\$j = a$ for Q, where $j = i - l$. (ii) For each condition $\$i = \$j$, we find the innermost expression, which is either a relation, a product $\times^c$, or a join $\bowtie^c$, where $c \in \{\wedge, \vee\}$, that contains both components of positions i and j. Let l be the total number of components of all relations preceding the expression; we put $i' = i - l$, $j' = j - l$, and: (ii.1) If the expression is a relation Q, we put a selection condition $\$i' = \$j'$ for Q. (ii.2) If it is a product $E = \times^c(E_1, E_2)$ or a join $E = \bowtie^c_F (E_1, E_2)$, the components of positions i and j are contained in E_1 and E_2, respectively (otherwise, E is not the innermost expression containing both components). Let $j'' = j' - l_1$, where l_1 is the total number of components of all relations in E_1; we replace E by a join $\bowtie^c_{i'=j''} (E_1, E_2)$ or $\bowtie^c_{F \wedge i' = j''} (E_1, E_2)$.

Example 2. Given a rule $(p(X, X, Z) \leftarrow \vee(r(X, Y), \wedge(s(Y, a, Z), V \ q(X, Z))).\rho)$, by Rule T, case (b), we have (1) in the following. Then, by T, last recursive call, we have (2). After that, by T', cases (a) and (b), we have (3). Then, by T', last recursive call, we have (4). After pushing the selections into relation selections and join conditions, we have (5).

$$T(p(V_1, X, Z) \leftarrow \wedge(\vee(r(X, Y), \wedge(s(Y, a, Z), V \ q(X, Z))), eq(V_1, X)).\rho) \quad (1)$$

$$\Pi_{8,1,5}^{\mathcal{C}(-,\rho)} T'(\wedge(\vee(r(X, Y), \wedge(s(Y, a, Z), V \ q(X, Z))), eq(V_1, X))) \quad (2)$$

$$\Pi_{8,1,5}^{\mathcal{C}(-,\rho)} \sigma_{\$1=\$6 \wedge \$2=\$3 \wedge \$4=a \wedge \$5=\$7 \wedge \$6=\$9} T'(\wedge(\vee(r(V_2, V_3), \wedge(s(Y, V_4, V_5),$$
$$V \ q(V_6, Z))), eq(V1, X))) \quad (3)$$

$$\Pi_{8,1,5}^{\mathcal{C}(-,\rho)} \sigma_{\$1=\$6 \wedge \$2=\$3 \wedge \$4=a \wedge \$5=\$7 \wedge \$6=\$9}(\times^{\wedge}(\times^{\vee}(R, \times^{\wedge}(S, \mathcal{H}_V(Q))), EQ)) \quad (4)$$

$$\Pi_{8,1,5}^{\mathcal{C}(-,\rho)}(\bowtie_{6=2}^{\wedge} (\bowtie_{1=4 \wedge 2=1}^{\vee} (R, \bowtie_{3=2}^{\wedge} (\sigma_{\$2=a} S, \mathcal{H}_V(Q))), EQ)) \quad (5)$$

The translated expression of a rule r with an IDB predicate p in its head will be denoted by $E(p, r)$. Since p can appear in the heads of more than one rule, the relation for p is the union of all such translated expressions, denoted $\mathcal{E}(p) = \bigcup_r E(p, r)$. The collection of all $\mathcal{E}(p)$ is denoted by $\mathcal{E}$.

3.5   Equivalence between $\mathcal{E}$ and T_P

We show the equivalence between $\mathcal{E}$ and T_P in the sense that for every IDB predicate p, if there exists a tuple $(a_1, ..., a_n, \alpha)$ in the relation produced by $\mathcal{E}(p)$, the truth value of the ground atom $p(a_1, ..., a_n)$ computed by the immediate consequence operator $T_P(f)$ is also α, otherwise, the truth value of $p(a_1, ..., a_n)$ is 0, where f is the interpretation corresponding to current values of relations.

Consider a rule $r : (A \leftarrow B.\rho)$ of a program P with predicate p in the head. For each subformula B_j of B, including itself, we denote the subexpression of $E(p, r)$ corresponding to B_j by $E(B_j)$, and the concatenation of arguments of all predicates appearing in B_j by $\mathcal{A}(B_j)$. Moreover, we denote the part of $E(p, r)$ excluding the projection Π, which is a combination of $E(B)$ and occurrences of the relation EQ in the form of conjunction joins or products, by $E_b(r)$.

Example 3. Consider the rule r in Example 2; if $B_1 = \wedge(s(Y, a, Z), V \ q(X, Z))$, $E(B_1) = \bowtie_{3=2}^{\wedge} (\sigma_{\$2=a} S, \mathcal{H}_V(Q))$ and $\mathcal{A}(B_1) = (Y, a, Z, X, Z)$; $E_b(r) = \bowtie_{6=2}^{\wedge} (\bowtie_{1=4 \wedge 2=1}^{\vee} (R, \bowtie_{3=2}^{\wedge} (\sigma_{\$2=a} S, \mathcal{H}_V(Q))), EQ)$.

Let $Q_1, ..., Q_m$ be the relations that have been already computed for all predicates $q_1, ..., q_m$ in B, and f an interpretation such that $f(q_i(b_1^i, ..., b_{k_i}^i)) = \beta_i$ if there is a tuple $(b_1^i, ..., b_{k_i}^i, \beta_i) \in Q_i$, and $f(q_i(b_1^i, ..., b_{k_i}^i)) = 0$, otherwise. For each ground instance $(A' \leftarrow B'.\rho)$ of r, let θ be the ground substitution such that $A' = A\theta$ and $B' = B\theta$. Also, let $B_j' = B_j\theta$ for all j. It can be seen that the substitution θ identifies at most one tuple $(b_1^j, ..., b_{n_j}^j, \beta^j)$, which corresponds to B_j', in each $E(B_j)$. More precisely, $(b_1^j, ..., b_{n_j}^j)$ is obtained from $\mathcal{A}(B_j)$ by replacing each occurrence of a variable by its substitute value in θ. We denote the subset of $E(B_j)$ corresponding to B_j' by $E(B_j')$; thus, $E(B_j')$ is either empty or a tuple $(b_1^j, ..., b_{n_j}^j, \beta^j)$. It can be proved by induction on the structure of B_j' that for all j, if there exists such a tuple, then $\beta^j = \overline{f}(B_j')$, otherwise $\overline{f}(B_j') = 0$. In particular, if $E(B') = \{(b_1, ..., b_k, \beta)\}$, then $\beta = \overline{f}(B')$; otherwise, $E(B')$ is empty and $\overline{f}(B') = 0$. Moreover, if $E(B') = \{(b_1, ..., b_k, \beta)\}$, since there is only one tuple $(c, c, 1)$ in EQ for each constant c, there is exactly one tuple in $E_b(r)$ which is formed by the tuple in $E(B')$ and occurrences of EQ via conjunction joins or products (with one component of EQ being restricted to a constant due to Cases (a) of Rule T and T'). Because $\wedge^\bullet(\beta, 1) = \beta$, the truth value of the tuple in $E_b(r)$ is also β. By projection, the truth value of the tuple corresponding to A' in $E(p, r)$ is $\mathcal{C}(\beta, \rho) = \mathcal{C}(\overline{f}(B'), \rho)$. Otherwise, $E(B')$ is empty, and there is no tuple for A' in $E(p, r)$. Finally, if there exist tuples for A' in $E(p, r)$'s, the truth value of the tuple for A' in $\mathcal{E}(p)$ is $max\{\mathcal{C}(\overline{f}(B'), \rho) | (A' \leftarrow B'.\rho)$ is a ground instance of a rule in P$\}$. On the other hand, since an IDB predicate cannot be an EDB predicate simultaneously, $T_P(f)(A') = max\{\mathcal{C}(\overline{f}(B'), \rho) | (A' \leftarrow B'.\rho)$ is a ground instance of a rule in $P\}$. In the case there is no tuple for A' in $\mathcal{E}(p)$, for all B', $E(B')$ is empty, and $\overline{f}(B') = 0$; hence, $T_P(f)(A') = 0$.

3.6 Fixpoint Semantics

Due to the equivalence between $\mathcal{E}$ and $T_P(f)$, the semantics of the program P can be obtained by repeatedly iterating the expressions in $\mathcal{E}$, obtained from the rules in P, from a set of the relations for the EDB predicates. More concretely, we can write $\mathcal{E}(p)$ as $\mathcal{E}(p, R_1, ..., R_k, P_1, ..., P_m)$, where R_i's are all EDB relations including EQ, and P_i's are all IDB relations. The semantics of program P is the least fixpoint of the equations $P_i = \mathcal{E}(p_i, R_1, ..., R_k, P_1, ..., P_m)$, for $i = 1 ... m$. Similar to [4], we have a *naive evaluation* algorithm:

INPUT: A collection of FLRA expressions $\mathcal{E}$ obtained from rules by the translation algorithm, and lists of IDB and EDB relations $P_1, ..., P_m$ and $R_1, ..., R_k$.

OUTPUT: The least fixpoint of equations $P_i = \mathcal{E}(p_i, R_1, ..., R_k, P_1, ..., P_m)$.

BEGIN
    FOR $i := 1$ TO k DO
        IF r_i involves a disjunction THEN
            $R_i := R_i +$ a set of other possible tuples with a truth value 0;
    FOR $i := 1$ TO m DO
        IF p_i involves a disjunction THEN
            $P_i :=$ a set of all possible tuples with a truth value 0
        ELSE $P_i := \emptyset$;
    REPEAT FOR $i := 1$ TO m DO $Q_i := P_i$; /*save old values of P_i*/
            FOR $i := 1$ TO m DO $P_i := \mathcal{E}(p_i, R_1, ..., R_k, Q_1, ..., Q_m)$
    UNTIL $P_i = Q_i$ for all $i = 1 ... m$;
END

For example, given the program in Example 1, applying the above algorithms, we obtain tuples $\{(toyota, QT), (bmw, VLT)\}$ in the relation for predicate *good* (with $Q^-(QT) = T$, $V^-(MT) = QT$, $Q^-(VT) = VVT$, $V^-(QT) = VLT$).

Since the least model of a program P can be obtained by finitely iterating T_P from the bottom interpretation, the following theorem follows immediately.

Theorem 1. *Every query over an FLKB represented by an FLDL program can be exactly computed by finitely iterating the expressions in $\mathcal{E}$, obtained from the rules in the program, from a set of relations for the EDB predicates.*

4   Discussions and Conclusion

We can utilize some optimization techniques such as *incremental evaluation* and *magic-sets* for evaluation of FLDL as for the classical case [4]. Nevertheless, incremental tuples can be used for only relations of the predicates that do not involve any disjunction. The procedural semantics of FLLP can be used to find answers to queries w.r.t. a threshold as discussed in [1]. However, it may not terminate for recursive programs. In this paper, we have presented a data model for fuzzy data in which every tuple has a linguistic truth value; the data can be stored in a crisp relational database with an extra truth-value attribute added to every relation. Queries to the database are made over an FLKB expressed by an FLDL program. We define FLRA as the set of operations to manipulate the data. Logical rules whose heads have the same IDB predicate can be converted into an expression of FLRA which yields a relation for the predicate. The semantics of the FLDL program can be obtained by finitely iterating the expressions from the database. Concerning related works, the many-valued logic extension of Datalog in [7] can also handle fuzzy similarity and enable threshold computation. The probabilistic Datalog in [10] uses a magic sets method as the basic evaluation strategy for modularly stratified programs. The method transforms a probabilistic Datalog program into a set of relational algebra equations, then the fixpoint of these equations is computed.

References

1. Le, V.H., Liu, F., Tran, D.K.: Fuzzy linguistic logic programming and its applications. Theory and Practice of Logic Programming 9, 309–341 (2009)
2. Nguyen, C.H., Wechler, W.: Hedge algebras: An algebraic approach to structure of sets of linguistic truth values. Fuzzy Sets and Systems 35, 281–293 (1990)
3. Nguyen, C.H., Wechler, W.: Extended hedge algebras and their application to fuzzy logic. Fuzzy Sets and Systems 52, 259–281 (1992)
4. Ullman, J.D.: Principles of database and knowledge-base systems, vol. I, II. Computer Science Press, Inc., New York (1988,1990)
5. Hájek, P.: Metamathematics of Fuzzy Logic. Kluwer, Dordrecht (1998)
6. Lloyd, J.W.: Foundations of logic programming. Springer, Berlin (1987)
7. Pokorný, J., Vojtáš, P.: A data model for flexible querying. In: Caplinskas, A., Eder, J. (eds.) ADBIS 2001. LNCS, vol. 2151, pp. 280–293. Springer, Heidelberg (2001)

8. Abiteboul, S., Hull, R., Vianu, V.: Foundations of databases. Addison-Wesley, USA (1995)
9. Ceri, S., Gottlob, G., Tanca, L.: Logic programming and databases. Springer, Berlin (1990)
10. Fuhr, N.: Probabilistic datalog: implementing logical information retrieval for advanced applications. J. Am. Soc. Inf. Sci. 51(2), 95–110 (2000)

Modelling Object Typicality in Description Logics

Katarina Britz[1,3], Johannes Heidema[2], and Thomas Meyer[1,3]

[1] KSG, Meraka Institute, CSIR, South Africa
[2] Dept of Mathematical Sciences, University of South Africa
[3] School of Computing, University of South Africa
arina.britz@meraka.org.za,
johannes.heidema@gmail.com,
tommie.meyer@meraka.org.za

Abstract. We present a semantic model of typicality of concept members in description logics (DLs) that accords well with a binary, globalist cognitive model of class membership and typicality. We define a general preferential semantic framework for reasoning with object typicality in DLs. We propose the use of feature vectors to rank concept members according to their defining and characteristic features, which provides a modelling mechanism to specify typicality in composite concepts.

1 Introduction

The study of natural language concepts in cognitive psychology has led to a range of hypotheses and theories regarding cognitive constructions such as concept inclusion, composition, and typicality. Description logics (DLs) have been very successful in modelling some of these cognitive constructions, for example IS-A and PART-OF. In this paper, we focus on the semantic modelling of typicality of concept members in such a way that it accords well with empirically well-founded cognitive theories of how people construct and reason about concepts involving typicality. We do not attempt to survey all models of concept typicality, but briefly outline some aspects of the debate:

According to the *unitary model* of concept typicality and class membership, variations in both graded class membership and typicality of class members reflect differences in similarity to a concept prototype. Class membership and typicality are determined by placing some criterion on the similarity of objects to the concept prototype [10,11]. According to the *binary model* of concept typicality and class inclusion, typicality and concept membership reflect essentially different cognitive processes. Concepts have *defining features* providing necessary and sufficient conditions for class membership, as well as *characteristic features* indicating typicality within that class [4,17,18]. According to the *localist* view of concepts, the meaning of a compound concept is a function of the meanings of its semantic constituents. According to the *globalist* view, the meanings of concepts are entrenched in our world knowledge, which is context-dependent and cannot be decomposed into, or composed from, our understanding of basic building blocks [12,16]. Concept typicality can therefore not be determined from concept definition alone, but requires a world view to provide context relative to which typicality may be determined.

A. Nicholson and X. Li (Eds.): AI 2009, LNAI 5866, pp. 506–516, 2009.

Description logics cannot resolve any of these debates, but we can use DLs to model some aspects of them. In particular, we can model typicality of concept members based on their characteristic features. We can also model compositional aspects of typicality. Other aspects, such as the graded class membership that underpins the unitary model, and non-compositionality of compound class membership in the globalist view, cannot be modelled using DLs, or at least not in an intuitively natural way. In [21] a model of graded concept membership was proposed, but this presented a marked departure from classical DL reasoning. We therefore restrict our attention to the binary model, with a compositional model of class membership, where being a member of a class is an all-or-nothing affair, and membership of compound concepts are determined by membership of their atomic constituents or defining features, while characteristic features contribute to induce degrees of typicality within a class.

DLs have gained wide acceptance as underlying formalism in intelligent knowledge systems over complex structured domains, providing an unambiguous semantics to ontologies, and balancing expressive power with efficient reasoning mechanisms [1]. The nature of DL reasoning has traditionally been *deductive*, but there have been a fair number of proposals to extend DLs to incorporate some form of *defeasible* reasoning, mostly centered around the incorporation of some form of default rules, e.g. [5].

In a previous paper [3], we presented a general preferential semantic framework for defeasible subsumption in DLs, analogous to the KLM preferential semantics for propositional entailment [2,13]. We gave a formal semantics of defeasible subsumption, as well as a translation of defeasible subsumption to classical subsumption within a suitably rich DL language. This was done by defining a preference order on objects in a knowledge base, which allowed for defeasible terminological statements of the form "All the most preferred objects in C are also in D".

In practice, an ontology may call for different preference orders on objects, and correspondingly, multiple defeasible subsumption relations within a single knowledge base. An object may be typical (or preferable) with respect to one property, but not another. For example, a guppy may be considered a typical pet fish, even though it is neither a typical fish, nor a typical pet [17]. So we may want a pet typicality order on pets, a fish typicality order on fish, and some way of combining these orders, or other relevant characteristics, into a pet fish typicality order. That is, we want to order objects in a given class according to their typicality with respect to the chosen features of that class. The subjective world view adopted in the fish shop may be different from that adopted in an aquarium, or a pet shop, hence the features deemed relevant may differ in each case, and this has to be reflected in the respective typicality orders.

Relative to a particular interpretation of a DL, any concept C partitions all objects in the domain according to their class membership into those belonging to C, and those not belonging to C. This yields a two-level preference order, with all objects in C preferred to all those not in C. This order may be refined further to distinguish amongst objects in C, but even the basic two-level order suffices to define an important class of preferential subsumption relations, namely those characterising the stereotypical reasoning of [14].

A preference order on objects may be employed to obtain a notion of defeasible subsumption that relaxes the deductive nature of classical subsumption. To this end, we introduce a parameterised defeasible subsumption relation $\sqsubseteq_j$ to express terminological statements of the form $C \mathrel{\sqsubseteq_j} D$, where C and D are arbitrary concepts, and $\mathrel{\sqsubseteq_j}$ is induced by a preference order $\leq_j$. If $\leq_j$ prefers objects in A to objects outside of A, we say that C is preferentially subsumed by D relative to A iff all objects in C that are typical in A (i.e. preferred by the typicality order corresponding to A), are also in D. When translated into DL terminology, the proposal of [14] reads as follows: Given concepts C, D and S such that S represents a best stereotype of C, C is preferentially subsumed by D relative to S if all stereotypical objects in C also belong to D.

The rest of the paper is structured as follows: We first fix some standard semantic terminology on DLs that will be useful later on. After giving some background on rational preference orders, we introduce the notion of an ordered interpretation, and present a formal semantics of parameterised defeasible subsumption. This is a natural extension of the work presented in [3], and provides a way of reasoning defeasibly with the IS-A relationship between concepts relative to a given concept. We then put forward two approaches to the definition of a derived typicality order on concepts, namely *atomic composition* and *feature composition*. We argue that feature composition is the more general approach, and is not as vulnerable to arguments against compositionality as is the case with atomic composition. We show how feature vectors may be used to determine typicality compositionally, taking into account semantic context.

2 Preliminaries

2.1 DL Terminology

In the standard set-theoretic semantics of concept descriptions, concepts are interpreted as subsets of a domain of interest, and roles as binary relations over this domain. An interpretation I consists of a non-empty set Δ^I (the *domain* of I) and a function $\cdot^I$ (the *interpretation function* of I) which maps each atomic concept A to a subset A^I of Δ^I, and each atomic role R to a subset R^I of $\Delta^I \times \Delta^I$. The interpretation function is extended to arbitrary concept descriptions (and role descriptions, if complex role descriptions are allowed in the language) in the usual way.

A DL knowledge base consists of a *Tbox* which contains *terminological axioms*, and an *Abox* which contains *assertions*, i.e. facts about specific named objects and relationships between objects in the domain. Depending on the expressive power of the DL, a knowledge base may also have an *Rbox* which contains *role axioms*. Tbox statements are *concept inclusions* of the form $C \sqsubseteq D$, where C and D are (possibly complex) concept descriptions. $C \sqsubseteq D$ is also called a *subsumption statement*, read "C is subsumed by D". An interpretation I *satisfies* $C \sqsubseteq D$, written $I \Vdash C \sqsubseteq D$, iff $C^I \subseteq D^I$. $C \sqsubseteq D$ is *valid*, written $\models C \sqsubseteq D$, iff it is satisfied by all interpretations. Rbox statements include *role inclusions* of the form $R \sqsubseteq S$, and assertions used to define *role properties* such as asymmetry. Objects named in the Abox are referred to by a finite number of *individual names*. These names may be used in two types of assertional statements – *concept assertions* of the form $C(a)$ and *role assertions* of the form $R(a, b)$, where C is a concept description, R is a role description, and a and b are individual names. To provide a semantics for Abox statements it is necessary to add to every interpretation a *denotation function* which satisfies the unique names assumption, mapping each individual name a to a different element a^I of the domain of interpretation Δ^I. An interpretation I satisfies the assertion $C(a)$ iff $a^I \in C^I$; it satisfies $R(a,b)$ iff $(a^I, b^I) \in R^I$. An interpretation I satisfies a DL knowledge base $\mathcal{K}$ iff it satisfies every statement in $\mathcal{K}$. A DL knowledge base $\mathcal{K}$ *entails* a DL statement ϕ, written as $\mathcal{K} \models \phi$, iff every interpretation that satisfies $\mathcal{K}$ also satisfies ϕ.

2.2 Preferential Semantics

In a preferential semantics for a propositional language, one assumes some order relation on propositional truth valuations (or on interpretations or worlds or, more generally, on states) to be given. The intuitive idea captured by the order relation is that interpretations higher up (greater) in the order are more typical in the context under consideration, than those lower down. For any given class C, we assume that all objects in the application domain that are in (the interpretation of) C are more typical of C than those not in C. This is a technical construction which allows us to order the entire domain, instead of only the members of C. This leads us to take as starting point a finite set of preference orders $\{\leq_j : j \in \mathcal{J}\}$ on objects in the application domain, with index set $\mathcal{J}$. If $\leq_j$ prefers any object in C to any object outside of C, we call $\leq_j$ a C-order.

To ensure that the subsumption relations generated are *rational*, i.e. satisfy a weak form of strengthening on the left, the *rational monotonicity* postulate (see [6,15], we assume the preference orders to be *modular partial orders*, i.e. reflexive, transitive, anti-symmetric relations such that, for all a, b, c in Δ^I, if a and b are incomparable and a is strictly below c, then b is also strictly below c.

Modular partial orders have the effect of stratifying the domain into layers, with any two elements in the same layer being unrelated to each other, and any two elements in different layers being related to each other. (We could also have taken the preference order to be a total preorder, i.e. a reflexive, transitive relation such that, for all a, b in Δ^I, a and b are comparable. Since there is a bijection between modular partial orders and total preorders on Δ^I, it makes no difference here which formalism we choose.)

We further assume that the order relations have no infinite chains (and hence, in Shoham's terminology [20, p.75], are bounded, which is the dual of well-founded, which in turn implies, in the terminology of [13], that the order relations are smooth). In the presence of transitivity, this implies that, for any $j \in \mathcal{J}$, nonempty $X \subseteq \Delta^I$ and $a \in X$, there is an element $b \in X$, $\leq_j$-maximal in X, with $a \leq_j b$.

3 Preferential Subsumption

We now develop a formal semantics for preferential subsumption in DLs. We assume a DL language with a finite set of preference orders $\{\preceq_j : j \in \mathcal{J}\}$ in its signature. We make the preference orders on the domain of interpretation explicit through the notion of an *ordered interpretation*: $(I, \{\leq_j : j \in \mathcal{J}\})$ is the interpretation I with preference orders $\{\leq_j : j \in \mathcal{J}\}$ on the domain Δ^I. The preference orders on domain elements may be constrained by means of role assertions of the form $a \preceq_j b$ for $j \in \mathcal{J}$, where the interpretation of $\preceq_j$ is $\leq_j$, that is, $\preceq_j^I = \leq_j$:

Definition 1. *An ordered interpretation* $(I, \{\leq_j : j \in \mathcal{J}\})$ *consists of an interpretation* I *and finite, indexed set of modular partial orders* $\{\leq_j : j \in \mathcal{J}\}$ *without infinite chains over their domain* Δ^I.

Definition 2. *An ordered interpretation* $(I, \{\leq_j : j \in \mathcal{J}\})$ *satisfies an assertion* $a \preceq_j b$ *iff* $a^I \leq_j b^I$.

We do not make any further assumptions about the DL language, but assume that concept and role assertions and constructors, and classical subsumption are interpreted in the standard way, ignoring the preference orders of ordered interpretations.

We first introduce the notion of satisfaction by an ordered interpretation, thereafter we relax the semantics of concept inclusion to arrive at a definition of satisfaction of a parameterised preferential subsumption relation $\succcurlyeq_j$ by an ordered interpretation. Finally, we define what it means for a preferential subsumption statement to be entailed by a knowledge base.

3.1 Satisfaction of Preferential Subsumption Statements

Definition 3. *An ordered interpretation* $(I, \{\leq_j : j \in \mathcal{J}\})$ *satisfies* $C \sqsubseteq D$, *written* $(I, \{\leq_j : j \in \mathcal{J}\}) \Vdash C \sqsubseteq D$, *iff* I *satisfies* $C \sqsubseteq D$.

The preferential semantics of $\succcurlyeq_j$ is then defined as follows:

Definition 4. *An ordered interpretation* $(I, \{\leq_j : j \in \mathcal{J}\})$ *satisfies the preferential subsumption* $C \succcurlyeq_j D$, *written* $(I, \{\leq_j : j \in \mathcal{J}\}) \Vdash C \succcurlyeq_j D$, *iff* $C_j^I \subseteq D^I$, *where*

$$C_j^I = \{x \in C^I : \text{there is no } y \in C^I \text{ such that } x \leq_j y \text{ and } x \neq y\}.$$

For brevity, we shall at times write $\leq_{\mathcal{J}}$ instead of $\{\leq_j : j \in \mathcal{J}\}$. Preferential subsumption satisfies the following three properties:

Supraclassicality: If $(I, \leq_{\mathcal{J}}) \Vdash C \sqsubseteq D$ then $(I, \leq_{\mathcal{J}}) \Vdash C \succcurlyeq_j D$ for all $j \in \mathcal{J}$.
Nonmonotonicity: $(I, \leq_{\mathcal{J}}) \Vdash C \succcurlyeq_j D$ does not necessarily imply
$(I, \leq_{\mathcal{J}}) \Vdash C \sqcap C' \succcurlyeq_j D$ for any $j \in \mathcal{J}$.
Defeasibility: $(I, \leq_{\mathcal{J}}) \Vdash C \succcurlyeq_j D$ does not necessarily imply $(I, \leq_{\mathcal{J}}) \Vdash C \sqsubseteq D$ for any $j \in \mathcal{J}$.

It also satisfies the familiar properties of rational preferential entailment [13,15] (when expressible in the DL under consideration): Reflexivity, And, Or, Left Logical Equivalence, Left Defeasible Equivalence, Right Weakening, Cautious Monotonicity, Rational Monotonicity, and Cut.

3.2 Entailment of Preferential Subsumption Statements

Satisfaction for defeasible subsumption is defined relative to a fixed, ordered interpretation. We now take this a step further, and develop a general semantic theory of entailment relative to a knowledge base using ordered interpretations. Note that, although the knowledge base may contain preferential subsumption statements, entailment from the knowledge base is classical and monotonic.

Definition 5. *The preferential subsumption statement* $C \mathrel{\sqsubseteq_j} D$ *is valid, written* $\models$ $C \mathrel{\sqsubseteq_j} D$, *iff it is satisfied by all ordered interpretations* $(I, \{\leq_j : j \in \mathcal{J}\})$.

Definition 6. *A DL knowledge base* $\mathcal{K}$ *entails the preferential subsumption statement* $C \mathrel{\sqsubseteq_j} D$, *written* $\mathcal{K} \models C \mathrel{\sqsubseteq_j} D$, *iff every ordered interpretation that satisfies* $\mathcal{K}$ *also satisfies* $C \mathrel{\sqsubseteq_j} D$.

The following properties of $\mathrel{\sqsubseteq_j}$ are direct consequences of its corresponding properties relative to a fixed, ordered interpretation:

$\mathrel{\sqsubseteq_j}$ is supraclassical: If $\mathcal{K} \models C \sqsubseteq D$ then also $\mathcal{K} \models C \mathrel{\sqsubseteq_j} D$.
$\mathrel{\sqsubseteq_j}$ is nonmonotonic: $\mathcal{K} \models C \mathrel{\sqsubseteq_j} D$ does not necessarily imply that $\mathcal{K} \models C \sqcap C' \mathrel{\sqsubseteq_j} D$.
$\mathrel{\sqsubseteq_j}$ is defeasible: $\mathcal{K} \models C \mathrel{\sqsubseteq_j} D$ does not necessarily imply that $\mathcal{K} \models C \sqsubseteq D$.

The other properties of $\mathrel{\sqsubseteq_j}$ mentioned earlier relative to a fixed, ordered interpretation extend analogously in the context of entailment relative to a knowledge base. For example, reflexivity of $\mathrel{\sqsubseteq_j}$ relative to $\mathcal{K}$ reads $\mathcal{K} \models C \mathrel{\sqsubseteq_j} C$.

4 Derived Typicality of Concept Membership

In the previous section we presented a semantic framework to model typicality of concept membership: $\leq_j$ is a C-order if it ranks any object in C higher than any object outside of C. In a DL with value restrictions, we can write this as: $C \sqsubseteq \forall \preceq_j . C$. We now address the question of derived typicality C-orders. We distinguish between two possible approaches to resolve this problem:

1. *Atomic composition*: Here we use the atomic constituents or defining features of the compound concept C as building blocks. We combine their respective typicality orders recursively, depending on the operators used in the syntactic construction of C. Say $C \equiv A \sqcap B$, and typicality orders $\leq_j$ and $\leq_k$ are defined such that $\leq_j$ is an A-order and $\leq_k$ is a B-order respectively. We may then form a new typicality order for C by composing $\leq_j$ and $\leq_k$ according to some composition rule for $\sqcap$.
2. *Feature composition*: Here we identify the relevant features of the concept C. For each object a belonging to C, we form a feature vector characterising a. These feature vectors are then used to determine the typicality of a in C.

Irrespective of the composition rules applied, atomic composition is vulnerable to the same criticisms that have been levied against localist, compositional cognitive models of typicality of concept membership [16].

Feature composition is also compositional, but, in contrast with atomic composition, it is not localist. That is, the typicality of a member of a concept may be influenced by characteristic features that do not constitute part of the definition of the concept. For example, the diet of penguins may be a relevant characteristic feature in determining their typicality, but atomic composition cannot take this into account when determining typicality unless this feature forms part of the definition of a penguin.

Atomic composition may be viewed as a restricted version of feature composition, since any defining feature may be considered a relevant feature. Hence, we will only consider feature composition further. We consider the definition of feature vectors, their normalisation, and their composition.

4.1 Feature Vectors

The features of a concept come in two guises: They are either *characteristic features*, co-determining typicality of objects in the concept, or they are *defining features* of the concept. In a DL extended with suitable preferential subsumption relations, characteristic features may be introduced on the right-hand side of preferential subsumption statements. For example, in the axioms given below, if $\precsim_1$ is derived from the *Penguin*-order $\leq_1$, then $\forall eats.Fish$ is a characteristic feature of *Penguin*. Defining features are introduced on the right hand-side of classical subsumption statements. For example, in the following axioms, *Seabird* is a defining feature of *Penguin*, so are *Bird* and $\exists eats.Fish$. Similarly, *Bird* and $\exists eats.Fish$ are both defining features of *Seabird*: $Seabird \equiv Bird \sqcap \exists eats.Fish;\ Penguin \sqsubseteq Seabird;\ Penguin \precsim_1 \forall eats.Fish.$

The question arises whether relevant features should be determined algorithmically through some closure operator, or whether their identification is a modelling decision. While defining features can easily be derived from the knowledge base, this is not obvious in the case of characteristic features. We therefore view the choice of relevant features as a modelling decision, in accordance with a globalist view of concepts as context sensitive. The choice of features relevant for a particular concept, and their respective preference orders, are therefore determined by a subjective world view and have to be re-evaluated in each new context. The following development assumes a fixed ordered interpretation, even when some order is defined in terms of others.

Definition 7. *A feature vector is an n-tuple of concepts* $\langle C_1^I, \ldots, C_n^I \rangle$ *with corresponding preference vector* $\langle \leq_1, \ldots, \leq_n \rangle$ *such that* $\leq_j$ *is a* C_j*-order, for* $1 \leq j \leq n$, *and weight vector* $\langle w_1, \ldots, w_n \rangle$ *such that* $w_j \in \mathbb{Z}$, *for* $1 \leq j \leq n$.

We do not place any formal relevance restriction on the choice of elements of a feature vector, as this is a modelling decision. We may even, for example, have two feature vectors for *Fish*, one for use in the fish shop, and one for the pet shop. We may also define different preference orders for the same concept, for use in different contexts. For example, miniature, colourful fish may be typical in a pet shop, but not even relevant in a fish shop.

Next, we consider the normalisation of preference orders, which paves the way for their composition.

Definition 8. *Let* $\langle C_1^I, \ldots, C_n^I \rangle$ *be a feature vector with corresponding preference vector* $\langle \leq_1, \ldots, \leq_n \rangle$. *The level of an object* $x \in \Delta^I$ *relative to preference order* $\leq_j$, *written* $level_j(x)$, *is defined recursively as follows:*

$$level_j(x) := \begin{cases} 1 \text{ if } x \text{ is } \leq_j \text{ -minimal in } C_j^I; \\ 0 \text{ if } x \text{ is } \leq_j \text{ -maximal in } \Delta^I \backslash C_j^I; \\ max\{level_j(y) : y <_j x\} + 1 \text{ for non-minimal objects in } C_j^I; \\ min\{level_j(y) : x <_j y\}) - 1 \text{ for non-maximal objects in } \Delta^I \backslash C_j^I. \end{cases}$$

Definition 8 maps objects in the domain to integers. We note that the absence of infinite $\leq_j$-chains ensures that $level_j$ is defined on the whole of Δ^I. Given any feature C_j^I in the feature vector, Definition 8 assigns a positive level to all objects in C_j, and a non-positive level to all objects not in C_j^I. In the case where $\leq_j$ is a two-level order, $level_j(x) = 1$ for $x \in C_j^I$, and $level_j(x) = 0$ for $x \notin C_j^I$.

It is not difficult to see (given the modularity of the preference orders) that this mapping preserves the relative order of elements in the corresponding preference order:

Proposition 1. *For any* $x, y \in \Delta^I$, $x \leq_j y$ *iff* $level_j(x) \leq level_j(y)$.

We now have the required apparatus to compose the chosen preference orders of a feature vector. We define the typicality of objects relative to a given concept, based on its relevant features. The weight vector may be used in two ways – to normalise the preference orders so that they have the same range, or to adjust the relative importance of each feature. Normalisation can be done without intervention from the modeller, and resonates better with the qualitative approach to typicality followed so far in the paper.

The intuition of Definition 9 is that it ranks those objects that conform better to the features of C in terms of typicality on a higher level. The function f first maps each object in the domain to a non-negative integer. This induces a modular C-order, say $\leq_k$, on objects in the domain.

Definition 9. *Given concept* C *with feature vector* $\langle C_1^I, \ldots, C_n^I \rangle$, *preference vector* $\langle \leq_1, \ldots, \leq_n \rangle$ *and weight vector* $\langle w_1, \ldots, w_n \rangle$, *let* $f : \Delta^I \to \mathbb{Z}_0^+$, *such that* $f(a) := Max\{1, \sum_{j=1}^n (level_j(a) \times w_j)\}$ *if* $a \in C^I$ *and 0 otherwise, for any object* $a \in \Delta^I$. *The associated preference relation* $\leq_k$ *on* Δ^I *given by:* $a \leq_k b$ *iff* $f(a) \leq f(b)$, *for some* $k \in \mathcal{J}$, *is the typicality* C-*order induced by the features, preferences and weights.*

Our choice for f is not arbitrary, but there are alternatives, such as taking the maximum of the input preferences instead of their sum. By choosing different functions for different connectives, atomic composition can be simulated using feature vectors.

4.2 Example

We conclude this section with an illustrative example. Suppose we have the following terminological statements:

$$Penguin \sqsubseteq Bird \sqcap Flightless \sqcap Aquatic \qquad (1)$$

$$Penguin \sqsubseteq_1 \forall habitat.Southern \qquad (2)$$

$$Southern \sqsubseteq \neg Equatorial \qquad (3)$$

$$GalapagosPenguin \sqsubseteq Penguin \qquad (4)$$

$$Penguin \sqsubseteq \forall \preceq_1 .Penguin \qquad (5)$$

$$\exists habitat.Equatorial \sqsubseteq \forall \preceq_2 .\exists habitat.Equatorial \qquad (6)$$

Line (2) of the TBox states that the habitat of typical penguins is restricted to the southern regions. Note that we cannot derive from (2) and (4) that the habitat of typical Galapagos penguins is restricted to the southern regions. Lines (5-6) ensure that $\preceq_1$ and $\preceq_2$ are indeed, respectively, a $Penguin$-order and an $\exists habitat.Equatorial$-order. In the ordered interpretation I satisfying this Tbox, and where $\preceq_1^I$ partitions objects into typical penguins, atypical penguins, and non-penguins, we have that:

$$level_1(a) := \begin{cases} 2 \text{ if } a \text{ is typical in } Penguin^I; \\ 1 \text{ if } a \text{ is atypical in } Penguin^I; \\ 0 \text{ otherwise.} \end{cases}$$

Suppose further that $\preceq_2^I$ is the modular default $\exists habitat.Equatorial$-order that partitions this concept into two classes. Then $level_2(a) := 1$ if $a \in \exists habitat.Equatorial^I$, 0 otherwise.

We now construct a feature vector for $GalapagosPenguin$. We choose $Penguin$ as relevant defining feature, and $\exists habitat.Equatorial$ as relevant characteristic feature. That is, a Galapagos penguin is a penguin whose distinctive characteristic is that it occurs in the equatorial region. The feature vector for $GalapagosPenguin$ is therefore $\langle Penguin^I, \exists habitat.Equatorial^I \rangle$. Its preference vector is $\langle \leq_1, \leq_2 \rangle$, and as weight vector we choose $\langle 1, 2 \rangle$ in order to normalise the ranges of $\leq_1$ and $\leq_2$. The resulting derived $GalapagosPenguin$-order is $\leq_3$, obtained from:

$$f_3(a) := \begin{cases} 4 \text{ if } a \text{ is typical in } Penguin^I \text{ and } a \in \exists habitat.Equatorial^I; \\ 3 \text{ if } a \text{ is atypical in } Penguin^I \text{ and } a \in \exists habitat.Equatorial^I; \\ 2 \text{ if } a \text{ is typical in } Penguin^I \text{ and } a \in \forall habitat.(\neg Equatorial)^I; \\ 1 \text{ if } a \text{ is atypical in } Penguin^I \text{ and } a \in \forall habitat.(\neg Equatorial)^I; \\ 0 \text{ otherwise.} \end{cases}$$

Note that the first case, i.e. where $f_3(a) = 4$, does not hold for any object a, as it contradicts terminological axiom (2) in the knowledge base. The following preferential subsumption statement holds in I: $GalapagosPenguin \sqsubseteq_3 \exists habitat.Equatorial$.

So, typically, Galapagos penguins are found in the equatorial region, not exclusively in the southern regions. Of course, in this example we could simply have stated this, but the point is that defining and characteristic features may be used to derive compositionally the typicality of objects in a class based on chosen relevant features. Our example gives a simple illustration of this claim.

5 Related Work

Notions of typicality have been studied in a wide variety of contexts, most of them beyond the scope of this paper. In the context of ontologies, Yeung and Leung [21] proposed a model of graded membership, but their representation is not directly in terms of DLs. Giordano et al. [7,8] define a nonmonotonic extension of the description logic $\mathcal{ALC}$ to reason about typicality, while Grossi et al. [9] use contexts, modelled as sets of DL models, to describe a version of typicality. In order to be able to determine similarity between objects, Sheremet et al. [19] extend a DL with the constructors of the similarity logic $\mathcal{SL}$.

6 Conclusion

We presented a semantic framework for modelling object typicality in description logics. In [3] we showed how reasoning with a single typicality order on the domain of interpretation (and the induced defeasible subsumption relation) can be reduced to reasoning in a sufficiently expressive DL. This translation is also applicable when reasoning with typicality of individual concept members, as presented in this paper.

We also presented a proposal for deriving new typicality orders from existing ones using feature vectors. Our proposal is compositional, and rooted in a globalist cognitive stance on the semantics of typicality. The determination of compositional rules is therefore a modelling decision, unlike compound class membership, the meaning of which can be completely determined from the meanings of its atomic constituents. Implementation of feature vectors in a DL setting is a topic for further research.

References

1. Baader, F., Horrocks, I., Sattler, U.: Description logics. In: van Harmelen, F., Lifschitz, V., Porter, B. (eds.) Handbook of Knowledge Representation, pp. 135–180. Elsevier, Amsterdam (2008)
2. Britz, K., Heidema, J., Labuschagne, W.: Semantics for dual preferential entailment. Journal of Philosophical Logic 38, 433–446 (2009)
3. Britz, K., Heidema, J., Meyer, T.: Semantic preferential subsumption. In: Proceedings of KR 2008, pp. 476–484. AAAI Press, Menlo Park (2008)
4. Chater, N., Lyon, K., Meyers, T.: Why are conjunctive categories overextended? Journal of Experimental Psychology: Learning, Memory, and Cognition 16(3), 497–508 (1990)
5. Donini, F.M., Nardi, D., Rosati, R.: Description logics of minimal knowledge and negation as failure. ACM Transactions on Computational Logic 3(2), 177–225 (2002)
6. Freund, M., Lehmann, D., Morris, P.: Rationality, transitivity and contraposition. Artificial Intelligence 52(2), 191–203 (1991)
7. Giordano, L., Gliozzi, V., Olivetti, N., Pozzato, G.L.: Preferential description logics. In: Dershowitz, N., Voronkov, A. (eds.) LPAR 2007. LNCS (LNAI), vol. 4790, pp. 257–272. Springer, Heidelberg (2007)
8. Giordano, L., Gliozzi, V., Olivetti, N., Pozzato, G.L.: Reasoning about typicality in preferential description logics. In: Hölldobler, S., Lutz, C., Wansing, H. (eds.) JELIA 2008. LNCS (LNAI), vol. 5293, pp. 192–205. Springer, Heidelberg (2008)
9. Grossi, D., Dignum, F., Meyer, J.-J.C.: Context in categorization. In: Serafini, L., Bouquet, P. (eds.) Proceedings of CRR 2005. CEUR Workshop Proceedings, vol. 136. CEUR-WS (2005)
10. Hampton, J.A.: Overextension of conjunctive concepts: Evidence for a unitary model of concept typicality and class inclusion. Journal of Experimental Psychology: Learning, Memory, and Cognition 14(1), 12–32 (1988)
11. Hampton, J.A.: Concepts as prototypes. The Psychology of Learning and Motivation 46, 79–113 (2006)
12. Kamp, H., Partee, B.: Prototype theory and compositionality. Cognition 57, 129–191 (1995)
13. Kraus, S., Lehmann, D., Magidor, M.: Nonmonotonic reasoning, preferential models and cumulative logics. Artificial Intelligence 44, 167–207 (1990)
14. Lehmann, D.: Stereotypical reasoning: Logical properties. Logic Journal of the IGPL 6(1), 49–58 (1998)
15. Lehmann, D., Magidor, M.: What does a conditional knowledge base entail? Artificial Intelligence 55, 1–60 (1992)
16. Lyon, K., Chater, N.: Localist and globalist approaches to concepts. In: Gilhooly, K.J., Keane, M.T.G., Logie, R.H., Erdos, G. (eds.) Lines of Thinking. John Wiley & Sons Ltd., Chichester (1990)

17. Osherson, D.N., Smith, E.E.: Gradedness and conceptual combination. Cognition 12, 299–318 (1982)
18. Osherson, D.N., Smith, E.E.: On typicality and vagueness. Cognition 64, 189–206 (1997)
19. Sheremet, M., Tishkovsky, D., Wolter, F., Zakhharyaschev, M.: A logic for concepts and similarity. Journal of Logic and Computation 17(3), 415–452 (2007)
20. Shoham, Y.: Reasoning about Change: Time and Causation from the Standpoint of Artificial Intelligence. MIT Press, Cambridge (1988)
21. Yeung, C., Leung, H.: Ontology with likeliness and typicality of objects in concepts. In: Embley, D.W., Olivé, A., Ram, S. (eds.) ER 2006. LNCS, vol. 4215, pp. 98–111. Springer, Heidelberg (2006)

Efficient SAT Techniques for Relative Encoding of Permutations with Constraints

Miroslav N. Velev* and Ping Gao

Aries Design Automation
Chicago, IL 60660, U.S.A.
miroslav.velev@aries-da.com
http://www.aries-da.com

Abstract. We present new techniques for relative SAT encoding of permutations with constraints, resulting in improved scalability compared to the previous approach by Prestwich, when applied to searching for Hamiltonian cycles. We observe that half of the ordering variables and two-thirds of the transitivity constraints can be eliminated. We exploit minimal enumeration of transitivity, based on 12 triangulation heuristics, and 11 heuristics for selecting the first node in the Hamiltonian cycle. We propose the use of inverse transitivity constraints. We achieve 3 orders of magnitude average speedup on satisfiable random graphs from the phase transition region, 2 orders of magnitude average speedup on unsatisfiable random graphs, and up to 4 orders of magnitude speedup on satisfiable structured graphs from the DIMACS graph coloring instances.

1 Introduction

We investigate efficient SAT techniques for solving of problems that can be reformulated as permutations with constraints, based on the relative SAT encoding [21], thus exploiting the tremendous advances in the speed and capacity of SAT solvers without reimplementing the same optimizations in specialized tools for specific problems. Particularly, we do an in-depth study of Hamiltonian Cycle Problems (HCPs)—where the goal is to find a route in a graph by visiting each node exactly once and returning to the starting node—a known class of hard combinatorial problems, classified as NP-complete (p. 199 of [9]). Another hard combinatorial problem, quasigroup completion [1, 11, 16, 28], can be reformulated as multiple permutations with constraints; it has applications to design of experiments, and wavelength routing in switches on optical networks. Other combinatorial problems that can be viewed as permutations with constraints are discussed in [2, 12]. At NASA, problems that can be reformulated as permutations of tasks, subject to additional constraints, arise in preparing sites for human habitation [8]. The efficient encoding of real-world problems as equivalent SAT formulas is a challenge identified by Kautz and Selman [17].

The first method proposed for HCP solving by translation to SAT was based on the absolute SAT encoding of permutations [15], and was used in [2, 13, 21]. However,

---

* Corresponding author.

A. Nicholson and X. Li (Eds.): AI 2009, LNAI 5866, pp. 517–527, 2009.

those papers present results for graphs with at most 24 nodes. Recently, we did an in-depth study of techniques to improve the scalability of the absolute SAT encoding for permutations with constraints, when applied to solving HCPs [29]. We found that the above method for absolute SAT encoding of permutations does not scale for a suite with 100 graphs of 30 nodes each, generated in the phase-transition region [3, 7, 27]. We proposed new techniques that improved the scalability of the absolute encoding, and resulted in at least 4 orders of magnitude average speedup when solving HCPs for both satisfiable and unsatisfiable benchmarks. However, the new techniques scaled only for graphs with up to 95 nodes.

In this paper, we extend the relative SAT encoding of permutations with constraints [21], and present techniques to improve its scalability when solving HCPs. The paper makes four contributions: 1) the observation that half of the ordering variables and two-thirds of the transitivity constraints can be eliminated in the relational encoding of permutations with constraints [21]; 2) the use of minimal enumeration of transitivity, based on 12 triangulation heuristics, and 11 heuristics for selecting the first node in the Hamiltonian cycle; 3) the concept of inverse transitivity constraints; and 4) experimental results showing 3 orders of magnitude average speedup on satisfiable random graphs from the phase transition region, 2 orders of magnitude average speedup on unsatisfiable random graphs, and up to 4 orders of magnitude speedup on satisfiable structured graphs from the DIMACS graph coloring instances [5].

2 Background

In the *relative encoding* [21], represented are the relative positions of objects with respect to each other in a permutation. The motivation for this encoding is the observation that in the absolute encoding of permutations, if node v is to be moved forward (backward) in a permutation, then all nodes between the old and the new positions of v will also have to be moved, which will result in value changes for the Boolean variables encoding the placement of all those nodes. In contrast, in the relative encoding, we only need to change the relative position of node v with respect to the rest of the nodes by changing the values of some or all of the Boolean variables used for node v, while keeping the rest of the nodes in their original relative positions, i.e., keeping the values assigned to the Boolean variables encoding the relative placement of those nodes, thus significantly reducing the number of Boolean variables whose values have to be changed, and so significantly reducing the corresponding work of the SAT solver. Since Prestwich [21] presented constraints for Hamiltonian paths, while our focus is on Hamiltonian cycles that require additional constraints to ensure that the first and last nodes in the permutation are neighbors in the graph, we present next an extension of his formulation for Hamiltonian cycles.

Two types of Boolean variables are introduced—successor and ordering variables. The *successor variables* encode the constraints that exactly one out of a node's neighbors is selected to be its successor in the permutation, and exactly one is selected to be its predecessor, based on the direct encoding [4]. A successor Boolean variable s_{ij} is introduced for every ordered pair of nodes (v_i, v_j) that are neighbors in the graph, i.e., 2 such variables are required for each edge—one variable for each of the two orderings of the two nodes of the edge—such that s_{ij} is true iff node v_j appears immediately after v_i in the permutation that represents the ordering of the nodes in the Hamiltonian cycle.

The constraints for the successor variables are:

- each node v_i has at least one successor from its neighbors: $s_{ij} \vee s_{ik} \vee \dots$, where the neighbors of v_i are nodes v_j, v_k, ... ;
- each node v_i has at most one successor from its neighbors: $\neg s_{ij} \vee \neg s_{ik}$, for every pair of neighbors v_j and v_k of node v_i;
- each node v_i is the successor of at least one of its neighbors: $s_{ji} \vee s_{ki} \vee \dots$, where the neighbors of v_i are nodes v_j, v_k, ... ; and
- each node v_i is the successor of at most one of its neighbors: $\neg s_{ji} \vee \neg s_{ki}$, for every pair of neighbors v_j and v_k of node v_i.

The *ordering variables* encode the relative ordering of the nodes in the permutation that represents the Hamiltonian cycle, starting from the node selected to be first, v_f, and ending with the first node's neighbor selected to be the last node in the cycle, v_l. An ordering Boolean variable o_{ij} is introduced only for ordered pairs of two different nodes (v_i, v_j), where $i \neq j$. The ordering variable o_{ij} is defined to be true iff node v_i appears before node v_j in the permutation. The ordering variables satisfy several properties:

- *transitivity*: $o_{ij} \wedge o_{jk} \Rightarrow o_{ik}$, i.e., $\neg o_{ij} \vee \neg o_{jk} \vee o_{ik}$, for all permutations of 3 nodes v_i, v_j, and v_k, such that $i \neq j$, $j \neq k$, and $i \neq k$;
- *antisymmetry*: $\neg(o_{ij} \wedge o_{ji})$, or equivalently $\neg o_{ij} \vee \neg o_{ji}$, i.e., it is impossible for node v_i to be before node v_j and for node v_j to be before node v_i simultaneously;
- the first node v_f precedes all others: o_{fi}, for all $i \neq f$; and
- the first node's neighbor v_l selected to be the last node in the cycle succeeds all others: $s_{lf} \Rightarrow o_{il}$, or equivalently $\neg s_{lf} \vee o_{il}$, for all $i \neq f$, $i \neq l$, and all neighbors v_l of the first node v_f.

Finally, the relationship between successor and ordering variables is: $s_{ij} \Rightarrow o_{ij}$, or equivalently $\neg s_{ij} \vee o_{ij}$, i.e., if node v_i has its neighbor v_j selected to be v_i's successor along the Hamiltonian cycle, then the ordering variable o_{ij} is true.

Prestwich [21] enforced transitivity for all possible permutations of three nodes. However, this results in $n \times (n - 1)$ ordering variables and $n \times (n - 1) \times (n - 2)$ transitivity constraints for a graphs with n nodes, i.e., in a prohibitive number of transitivity constraints for large graphs. We refer to this method as *exhaustive enumeration of transitivity*, and will present improvements to it in Sect. 3 and 4.

3 Eliminating Half of the Ordering Variables and Two-Thirds of the Transitivity Constraints

We exploit the observation that the ordering variables o_{ij} and o_{ji} are complements of each other, since when node v_i precedes node v_j then o_{ij} is true and o_{ji} is false by definition, and vice versa—when node v_j precedes node v_i then o_{ij} is false and o_{ji} is true by definition. Thus, we can introduce an ordering variable o_{ij} only if the node index i is less than the node index j, and replace the ordering variable o_{ji} with $\neg o_{ij}$. This leads to the following lemma:

LEMMA 1. *For a triple of nodes* $\{v_i, v_j, v_k\}$, *whose indices are ordered* $1 \leq i < j < k \leq n$, *where n is the number of nodes in the graph, after half of the ordering variables are eliminated by introducing an ordering variable* o_{pq} *only for an ordered pair of nodes* (v_p, v_q) *where the node indices satisfy* $1 \leq p < q \leq n$, *and replacing the ordering variable* o_{qp} *with* $\neg o_{pq}$, *then the six transitivity constraints that are introduced for the triple of nodes* $\{v_i, v_j, v_k\}$ *in exhaustive enumeration of transitivity will reduce to two transitivity constraints:* (1) $\neg o_{ij} \lor \neg o_{jk} \lor o_{ik}$; *and* (2) $o_{ij} \lor o_{jk} \lor \neg o_{ik}$.

Thus, half of the ordering variables, two-thirds of the transitivity constraints, and all the antisymmetry constraints (that evaluate to true now) are eliminated. This results in fewer decisions to be made by the SAT solver when solving the resulting CNF formula, as well as reduces significantly the *Boolean Constraint Propagation* (BCP)—the iterative process of assigning values to literals as implied by clauses that have become unit, i.e., where all literals but one have assigned values, and those values are false, so that the only literal without a value has to be set to true in order for the clause to be satisfied. BCP takes up to 90% of the execution time of SAT solvers [19].

4 Minimal Enumeration of Transitivity

Exhaustive transitivity requires $n \times (n - 1)$ ordering variables and $n \times (n - 1) \times (n - 2)$ transitivity constraints, where n is the number of nodes in the graph—see Sect. 2. Eliminating half of the ordering variables as described in Sect. 3 and applying Lemma 1 results in $n \times (n - 1) / 2$ ordering variables and $n \times (n - 1) \times (n - 2) / 3$ transitivity constraints. In this section we aim to further reduce the number of ordering variables and transitivity constraints by introducing only a minimal set of such variables and constraints in a way that the property of transitivity will be enforced for the *required ordering variables*—those that are used in the constraints in Sect. 2, except for the transitivity constraints.

We introduce the concept of a *relational graph* that has the same set of nodes as the original graph that we want to find a Hamiltonian cycle in, but an extended set of edges, such that a pair of nodes is connected with an edge in the relational graph iff an ordering variable is introduced for that pair of nodes. We will use the relational graph to efficiently enforce the property of transitivity for the required ordering variables. These variables determine three types of edges that must be present in the relational graph:

1. each edge from the original graph, since the relationship between successor and ordering variables requires that each edge in the original graph should have an associated ordering variable;
2. given a choice for a first node, v_f, an edge between v_f and each of the other $n - 1$ nodes in the relational graph, because of the constraints that v_f must precede all other nodes in the permutation; and

3. given a choice for a first node, v_f, an edge between each of v_f's neighbors and the other $n-2$ nodes in the relational graph besides v_f, because of the conditions that when that neighbor is selected to be the last node in the permutation, then the other nodes must precede that neighbor in the permutation (that v_f precedes the last node is already ensured with constraints that resulted in edges of type 2 above).

We refer to the resulting graph as *base relational graph*, and designate it R_B. Because of the second and third types of edges that depend on the choice of the first node, v_f, the base relational graph must be constructed after the first node in the Hamiltonian cycle is selected. Note that the base relational graph R_B is different from the original graph for which we are searching for a Hamiltonian cycle, and is used for an efficient relative SAT encoding of the node permutation.

Besides the three types of required edges in R_B that were described, we want to introduce additional edges such that the resulting graph becomes *chordal* [22]. That is, the graph has the property that every cycle of length greater than three has a chord—an edge joining two nodes that are not adjacent in the cycle. Chordal graphs are also called *triangulated graphs*. We can then generate a sufficient set of transitivity constraints by enforcing transitivity for each triangle in the resulting graph. Computing a minimal triangulation consists of embedding a given graph into a triangulated graph by adding a set of edges called *a fill*. Finding a fill that is minimum is NP-complete [30]. However, there are good heuristic solutions. The ones that we explored are described next.

Our triangulation method proceeds as a series of elimination steps, starting with graph $G_0 = R_B$. On elimination step i, we create a new graph G_i that is identical to G_{i-1}, except that some vertex v_m and its incident edges are removed, and new edges are possibly added in the following way: for every pair of distinct vertices v_j and v_k such that G_{i-1} contains the edges (v_m, v_j) and (v_m, v_k), we add the edge (v_j, v_k) to G_i if this edge does not already exist. This process continues until we reach an empty graph G_n. Let R_T be the relational graph obtained from R_B after extending it with all edges added during the elimination process. It can be shown that R_T is a chordal graph [22]. To choose which node v_m to eliminate on step i, we implemented the following 12 triangulation heuristics that select the node with the current minimum degree, and if there are several nodes with such degree, the tie is broken as follows: *t1*—break the tie based on a lesser sum of neighbors' degrees; *t2*—break the tie based on a greater sum of neighbors' degrees; *t3*—break the tie based on a lesser fill (i.e., additional edges) that will be added at this step of the elimination; *t4*—break the tie based on a greater fill that will be added at this step of the elimination; *t5*—break the tie based on a lesser original degree of the node in R_B; *t6*—break the tie based on a greater original degree of the node in R_B; *t7*—break the tie based on a lesser number of additional triangles created in R_T that include node v_m, and edges from G_{i-1} and those added at step i; *t8*—break the tie based on a greater number of additional triangles created in R_T that include node v_m, and edges from G_{i-1} and those added at step i; *t9*—select the node that will result in a minimum fill at this step of the elimination, and break ties by selecting the first such node in the graph description; *t10*—select the node that will result in a minimum fill at this step of the elimination, and break ties randomly; *t11*—select the node that will result in a minimal number of additional triangles created in R_T that include node v_m, and edges from G_{i-1} and those added at step i, and break ties by selecting the first such node in the graph description; and $t12$—select the node that will result in a minimal number of additional triangles created in R_T that include node v_m, and edges from G_{i-1} and those added at step i, and break ties randomly.

We use the triangles in R_T to enforce transitivity of the ordering variables. When alhe optimizations from Sect. 3, we only introduce one ordering variable per edge in R_T and the two transitivity constraints per triangle—see Lemma 1.

Triangulation heuristic $t10$ was also explored by Kjaerulff [18], who found it to performance when triangulating belief networks in knowledge representation. He called that heuristic minimum fill. Rose [23] refers to it as minimum deficiency heuristic, and references other papers [25, 26] in which it is assumed that this heuristic produces a near optimal node-elimination ordering.

5   Selecting the First Node for the Hamiltonian Cycle

We exploit the observation that when the search is for a Hamiltonian cycle, as opposed to a Hamiltonian path, the first node in the cycle (i.e., permutation) can be selected in any way, since all nodes have to be included in the cycle, and if a cycle exists then all nodes are symmetrical in it. Thus, if a cycle does not exist when a particular node is selected as a first node, then a cycle would not exist when any other node is selected as a first node. That is, the first node can be chosen in any way because of the symmetry of a solution, and the general formulation of the relative encoding in Sect. 2.

We implemented the following 11 heuristics for selecting the first node: $f1$—choose the first node in the graph description; $f2$—choose the first node of max. degree in the graph description; $f3$—choose the first node of min. degree in the graph description; $f4$—choose the first node of average degree in the graph description; $f5$—random choice; $f6$—choose a node of max. degree, and break ties based on a lesser sum of neighbors' degrees; $f7$—choose a node of max. degree, and break ties based on a greater sum of neighbors' degrees; $f8$—choose a node of average degree, and break ties based on a lesser sum of neighbors' degrees; $f9$—choose a node of average degree, and break ties based on a greater sum of neighbors' degrees; $f10$—choose a node of min. degree, and break ties based on a lesser sum of neighbors' degrees; and $f11$—choose a node of min. degree, and break ties based on a greater sum of neighbors' degrees. In contrast, Prestwich [21] selected the first node for a Hamiltonian path (he was searching for Hamiltonian paths, as opposed to cycles) to be the first node in the graph description.

6   Inverse Transitivity Constraints

We consider triples of nodes v_i, v_j, and v_k, such that v_i and v_j are neighbors in the original graph. Then, if node v_j is selected to be the successor of node v_i in the permutation, as indicated by the successor Boolean variable s_{ij} being true, and node v_i precedes node v_k in the permutation, as indicated by the ordering Boolean variable o_{ik} being true, it follows that node v_j must also precede node v_k in the permutation, i.e., o_{jk} must be true, since v_j is the closest node that is after v_i and thus v_k must be after v_j, i.e., $s_{ij} \wedge o_{ik} \Rightarrow o_{jk}$, or equivalently $\neg s_{ij} \vee \neg o_{ik} \vee o_{jk}$ (see Fig. 1.a). We call these constraints *forward inverse transitivity constraints*. Note that they are different from the regular transitivity constraints that state that if v_i is before v_j, and v_j is before v_k, then v_i is before v_k.

We can also define a variant of these constraints for the case when node v_k is before node v_j, as indicated by the ordering Boolean variable o_{kj} being true, such that node v_j is selected to be the successor of node v_i in the permutation, as indicated by the successor Boolean variable s_{ij} being true. Then, it follows that node v_k must be before v_i, i.e., o_{ki} must be true, since v_i is the closest node that is before v_j by being the predecessor of v_j and thus v_k must be before v_i, i.e., $s_{ij} \wedge o_{kj} \Rightarrow o_{ki}$, or equivalently $\neg s_{ij} \vee \neg o_{kj} \vee o_{ki}$ (see Fig. 1.b). We call these constraints *backward inverse transitivity constraints*.

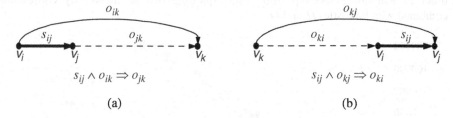

(a)   (b)

Fig. 1. Illustration of: (a) forward inverse transivity; (b) backward inverse transivity

When the optimizations from Sect. 3 are also applied, only two of the above four ordering variables will actually be introduced, while the other two will be each replaced with the negation of the introduced ordering variable for the same pair of nodes. In our tool, we imposed the inverse transitivity constraints for every ordered pair of neighboring nodes (v_i, v_j) from the original graph, and for every node v_k that is different from v_i and v_j, such that the ordering variables for v_k with respect to both v_i and v_j have been introduced by corresponding edges in the triangulated relational graph R_T (see Sect. 4). Note that these constraints are optional, and that we can use only a subset of them.

7 Results

The experiments were conducted on a Dell Precision T7400 workstation with two 3.2-GHz quad-core Intel Xeon processors, 32 GB of 800-MHz memory, and running Red Hat Enterprise Linux v5.3. Only one CPU core was used for each experiment. We started with satisfiable benchmarks (guaranteed to have Hamiltonian cycles) by generating suites of 100 graphs from the phase-transition region [3] that satisfy the ratio $e / (n \log n) = 1$, where e is the number of edges and n the number of nodes in the graph, since that ratio was shown to result in the hardest instances for backtrack search algorithms [7, 27]. We used the Hamiltonian-graph generator by Vandegriend and Culberson[1] [27]—that was also used in all previous papers on SAT-based solving

---

[1] http://web.cs.ualberta.ca/~joe/Theses/HCarchive/main.html

of HCPs [2, 13, 21, 29]. We produced 9 suites of 100 random graphs each from the phase-transition region, with the following number of nodes in the graphs of a suite, respectively—20, 30, 50, 100, 150, 200, 250, 300, and 350—for a total of 900 graphs.

We compared the SAT solvers siege [24], BerkMin [10], minisat v1.14 [6], tinisat [14], and rsat v3.1 [20], of which rsat v3.1 outperformed the rest by 2× or more, and so was used for the experiments discussed next. Fig. 2 presents the total CPU times for the 9 suites, plotted on a logarithmic scale, for the previous translation, and the best strategy for each of the proposed optimizations. As can be seen, the speedup is increasing with the size of the graphs. For the suite with the largest graphs of 350 nodes, eliminating half of the ordering variables and two-thirds of the transitivity constraints resulted in approximately an order of magnitude speedup, followed by another order of magnitude speedup from minimal transitivity with heuristic *t4* and selecting the first node in the Hamiltonian cycle with heuristic *f4*, and then another order of magnitude speedup from also exploiting inverse transitivity constraints combined with heuristics *f1* and *t9*.

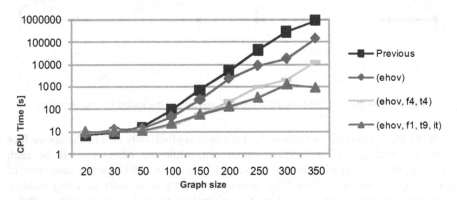

Fig. 2. Total CPU time [s] on a logarithmic scale for all 100 graphs in each of the 9 suites of satisfiable benchmarks from the phase-transition region

Table 1 shows detailed statistics for the four strategies from Fig. 2 for the suite of 100 graphs with 350 nodes. Strategy (*ehov, f1, t9, it*) resulted in efficient pruning of the solution space, reduced the number of conflicts and decisions by 3 orders of magnitude, and produced an average speedup of 1,071×, i.e., *3 orders of magnitude average speedup* computed based on the total time for the suite of 100 graphs with 350 nodes, relative to the previous translation. The individual speedup was up to 4 orders of magnitude. *Note that the reduction in the number of clauses and file size by approximately 30× will allow scalability for significantly larger instances.*

Although inverse transitivity constraints increased the number of clauses by 19% on average, they produced speedup due to the resulting pruning of the solution space, and eliminated wasteful work of the SAT solver by preventing it from exploring assignments that are guaranteed to not be part of solutions. This coincides with conclusions from previous work that adding redundant constraints to a CNF formula may make it easier to solve, e.g., see [1, 11, 16, 28].

Table 1. Details of the four strategies in Fig. 2 on the suite of 100 graphs with 350 nodes

Strategy	CNF Formula Average Statistics			SAT Solving Average Statistics		Average CPU Time per Graph [s]				Total CPU Time for Entire Suite [s]	Speedup
	Variables	Clauses	File Size [MB]	Conflicts	Decisions	Translate	SAT Solve	Check	Total for 3 Steps		
Previous	124,818	42,657×10³	905	119,758	1,402,382	14.2	9,425	0.03	9,439	943,924	——
(ehov)	63,743	14,196×10³	284	53,748	629,956	5.4	1,424	0.02	1,429	142,919	6.6×
(ehov, f4, t4)	18,596	1,339×10³	25	46,638	508,685	6.5	124	0.4	131	13,053	72×
(ehov, f1, t9, it)	18,544	1,593×10³	30	34	875	7.1	1.3	0.4	8.8	882	1,071×

To conduct experiments with unsatisfiable benchmarks that do not have Hamiltonian cycles, we generated such instances by implementing a program to rewrite each of the graphs with Hamiltonian cycles used for the presented experiments by injecting structures that make it impossible to find a Hamiltonian cycle in the resulting graph. We used two such transformations. First, selecting a random node of degree greater than 2, and modifying each of its neighbors to make them of degree 2, thus ensuring that only two of the neighbors can be on a path that includes the selected node, and so making it impossible to find a Hamiltonian cycle in the resulting graph. Second, selecting a random number of nodes up to 60% of the nodes in the graph, but at least 7 nodes, and converting them to form a chain of k cycles of 4 nodes each, where each of these cycles is a sequence of four nodes (n_1, n_2, n_3, n_4), such that node n_1 for the first cycle in the chain is connected to a random node in the portion of the graph that was not used for the chain, and node n_3 in each cycle is used as node n_1 for the next cycle in the chain, except in the last cycle of the chain, where node n_3 is connected to a random node in the portion of the graph that was not used for the chain. Since nodes n_2 and n_4 in each cycle are of degree 2, it is impossible to include all four nodes in such a cycle in a Hamiltonian cycle for the entire graph. Furthermore, this chain has a total of 2^k paths between node n_1 in the first cycle and node n_3 in the last cycle, so that a backtrack search algorithm may be affected negatively by the exponential number of such paths. Thus, each of the resulting 1,800 graphs was guaranteed to not have a Hamiltonian cycle. We found that strategy $(ehov, f1, t9, it)$ resulted in up to 2 orders of magnitude of average speedup, compared to the previous translation. The speedup is again increasing with the size of the graphs.

To evaluate the benefit from these optimizations on structured benchmarks, we ran experiments with the DIMACS graph coloring instances [5]—see Table 2. The best was strategy $(ehov, f4, t4, it)$, resulting in up to *4 orders of magnitude speedup*. After we filtered the graphs that were trivially unsatisfiable because of nodes of degree 0 or 1, all of the remaining DIMACS graphs had Hamiltonian cycles.

Table 2. Results for structured benchmarks from the DIMACS graph-coloring suite

DIMACS Graph	Nodes	Edges	Has Hamiltonian Cycle?	Total CPU Time [s]		Speedup	Orders of Magnitude Speedup
				Previous	$(ehov, f4, t4, it)$		
myciel7	191	2,360	Yes	>100,000	2.8	>35,714×	4
queen15_15	225	10,360	Yes	23,102	35	660×	2
queen16_16	256	12,640	Yes	>100,000	86	>1,163×	3
miles500	128	1,170	Yes	6,590	2.0	3,295×	3
miles750	128	2,113	Yes	>50,000	1.2	>41,667×	4
miles1000	128	3,216	Yes	>50,000	2.5	>20,000×	4

8 Conclusion

We presented new techniques for relative SAT encoding of permutations with constraints, resulting in improved scalability compared to the previous translation [21] when applied to HCPs. We achieved 3 orders of magnitude average speedup on satisfiable random graphs from the phase transition region, 2 orders of magnitude average speedup on unsatisfiable random graphs, and up to 4 orders of magnitude speedup on satisfiable structured graphs from the DIMACS graph coloring instances. We believe that these techniques can be used to solve other types of permutations with constraints.

References

[1] Ansótegui, C., del Val, A., Dotú, I., Fernández, C., Manyà, F.: Modeling Choices in Quasigroup Completion: SAT vs. CSP. In: 19th National Conference on Artificial Intelligence, July 2004, pp. 137–142 (2004)
[2] Cadoli, M., Schaerf, A.: Compiling Problem Specifications into SAT. Artificial Intelligence 162(1-2), 89–120 (2005)
[3] Cheeseman, P., Kanefsky, B., Taylor, W.M.: Where the Really Hard Problems Are. In: 12th International Joint Conference on AI (IJCAI 1991), pp. 331–337 (1991)
[4] de Kleer, J.: A Comparison of ATMS and CSP Techniques. In: 11th Int'l. Joint Conference on Artificial Intelligence (IJCAI 1989), August 1989, pp. 290–296 (1989)
[5] DIMACS Graph Coloring Instances,
    http://mat.gsia.cmu.edu/COLOR/instances.html
[6] Eén, N., Sörensson, N.: MiniSat—a SAT solver with conflict-clause minimization. In: Bacchus, F., Walsh, T. (eds.) SAT 2005. LNCS, vol. 3569. Springer, Heidelberg (2005)
[7] Frank, J., Martel, C.U.: Phase Transitions in the Properties of Random Graphs. In: Principles and Practice of Constraint Programming, CP 1995 (1995)
[8] Frank, J.: Personal Communication (April 2009)
[9] Garey, M.R., Johnson, D.S.: Computers and Intractability: A Guide to the Theory of NP-Completeness. W. H. Freeman, New York (1979)
[10] Goldberg, E., Novikov, Y.: BerkMin: A Fast and Robust Sat-Solver. In: Design, Automation, and Test in Europe (DATE 2002), March 2002, pp. 142–149 (2002)

[11] Gomes, C.P., Shmoys, D.B.: The Promise of LP to Boost CSP Techniques for Combinatioral Problems. In: International Conference on Integration of AI and OR Techniques in Constraint Programming for Combinatorial Optimization Problems (CP-AI-OR 2002), March 2002, pp. 291–305 (2002)

[12] Hnich, B., Walsh, T., Smith, B.M.: Dual Modelling of Permutation and Injection Problems. Journal of Artificial Intelligence Research (JAIR) 21, 357–391 (2004)

[13] Hoos, H.H.: SAT-Encodings, Search Space Structure, and Local Search Performance. In: Int'l. Joint Conference on Artificial Intelligence (IJCAI 1999), August 1999, pp. 296–303 (1999)

[14] Huang, J.: A Case for Simple SAT Solvers. In: Bessière, C. (ed.) CP 2007. LNCS, vol. 4741, pp. 839–846. Springer, Heidelberg (2007)

[15] Iwama, K., Miyazaki, S.: SAT-Varible Complexity of Hard Combinatorial Problems. In: IFIP 13th World Computer Congress, August–September 1994, vol. (1), pp. 253–258 (1994)

[16] Kautz, H.A., Ruan, Y., Achlioptas, D., Gomes, C.P., Selman, B., Stickel, M.E.: Balance and Filtering in Structured Satisfiable Problems. In: 17th Int'l. Joint Conference on Artificial Intelligence (August 2001)

[17] Kautz, H.A., Selman, B.: The State of SAT. Discrete Applied Mathematics 155(12) (June 2007)

[18] Kjaerulff, U.: Triangulation of Graphs—Algorithms Giving Small Total State Space. Technical Report R 90-09, Institute for Electronic Systems, Denmark (March 1990)

[19] Moskewicz, M.W., Madigan, C.F., Zhao, Y., Zhang, L., Malik, S.: Chaff: Engineering an Efficient SAT Solver. In: 38th Design Automation Conference (DAC 2001) (June 2001)

[20] Pipatsrisawat, K., Darwiche, A.: A New Clause Learning Scheme for Efficient Unsatisfiability Proofs. In: Twenty-Third AAAI Conference on Artificial Intelligence (AAAI 2008), July 2008, pp. 1481–1484 (2008)

[21] Prestwich, S.D.: SAT Problems with Chains of Dependent Variables. Discrete Applied Mathematics 130(2), 329–350 (2003)

[22] Rose, D.: Triangulated Graphs and the Elimination Process. Journal of Mathematical Analysis and Applications 32, 597–609 (1970)

[23] Rose, D.: A Graph-Theoretic Study of the Numerical Solution of Sparse Positive Definite Systems of Linear Equations. In: Read, R.C. (ed.) Graph Theory and Computing, pp. 183–217. Academic Press, NY (1973)

[24] Ryan, L.: Siege SAT Solver v.4, http://www.cs.sfu.ca/~loryan/personal/

[25] Sato, N., Tinney, W.F.: Techniques for Exploiting the Sparsity or the Network Admittance Matrix. IEEE Transactions on Power Apparatus and Systems 82(69), 944–950 (1963)

[26] Tinney, W.F., Walker, J.W.: Direct Solutions of Sparse Network Equations by Optimally Ordered Triangular Factorization. Proceedings of the IEEE 55(11), 1801–1809 (1967)

[27] Vandegriend, B., Culberson, J.: The $G_{n,m}$ Phase Transition Is Not Hard for the Hamiltonian Cycle Problem. Journal of Artificial Intelligence Research (JAIR) 9, 219–245 (1998)

[28] Velev, M.N., Gao, P.: Efficient SAT-Based Techniques for Design of Experiments by Using Static Variable Ordering. In: 10th International Symposium on Quality Electronic Design (ISQED 2009) (March 2009)

[29] Velev, M.N., Gao, P.: Efficient SAT Techniques for Absolute Encoding of Permutation Problems: Application to Hamiltonian Cycles. In: 8th Symposium on Abstraction, Reformulation and Approximation (2009)

[30] Yanakakis, M.: Computing the Minimum Fill-in Is NP-Complete. SIAM Journal of Algebraic and Discrete Mathematics 2, 77–79 (1981)

Uniform Interpolation for $\mathcal{ALC}$ Revisited*

Zhe Wang[1], Kewen Wang[1], Rodney Topor[1], Jeff Z. Pan[2], and Grigoris Antoniou[3]

[1] Griffith University, Australia
[2] University of Aberdeen, UK
[3] University of Crete, Greece

Abstract. The notion of uniform interpolation for description logic $\mathcal{ALC}$ has been introduced in [9]. In this paper, we reformulate the uniform interpolation for $\mathcal{ALC}$ from the angle of forgetting and show that it satisfies all desired properties of forgetting. Then we introduce an algorithm for computing the result of forgetting in concept descriptions. We present a detailed proof for the correctness of our algorithm using the Tableau for $\mathcal{ALC}$. Our results have been used to compute forgetting for $\mathcal{ALC}$ knowledge bases.

1 Introduction

The Web Ontology Language (OWL) [15] provides a construct *owl:imports* for importing and merging Web ontologies by referencing axioms contained in another ontology that may be located somewhere else on the Web. This construct is very limited in that it can only merge some linked ontologies together but is unable to resolve conflicts among merged ontologies or to filter redundant parts from those ontologies. However, an ontology is often represented as a logical theory, and the removal of one term may influence other terms in the ontology. Thus, more advanced methods for dealing with large ontologies and reusing existing ontologies are desired.

It is well-known that OWL is built on description logics (DLs) [1]. Recent efforts show that the notions of *uniform interpolation* and *forgetting* are promising tools for extracting modular ontologies from a large ontology. In a recent experiment reported in [5], uniform interpolation and forgetting have been used for extracting modular ontologies from two large medical ontologies SNOMED CT [16] and NCI [17]. SNOMED CT contains around 375,000 concept definitions while NCI Thesaurus has 60,000 axioms. The experiment result is promising. For instance, if 2,000 concepts definitions are forgotten from SNOMED CT, the success rate is 93% and if 5,000 concepts definitions are forgotten from NCI, the success rate is 97%.

Originally, *interpolation* is proposed and investigated in pure mathematical logic, specifically, in proof theory. Given a theory T, *ordinary interpolation* for T says that if $T \vdash \phi \rightarrow \psi$ for two formulas ϕ and ψ, then there is a formula $I(\phi, \psi)$ in the language containing only the shared symbols, say S, such that $T \vdash \phi \rightarrow I(\phi, \psi)$ and $T \vdash I(\phi, \psi) \rightarrow \psi$. *Uniform interpolation* is a strengthening of ordinary interpolation in that the interpolant can be obtained from either ϕ and S or from ψ and S. Uniform

---

* This work was partially supported by the Australia Research Council (ARC) Discovery Project 0666107.

A. Nicholson and X. Li (Eds.): AI 2009, LNAI 5866, pp. 528–537, 2009.

interpolation for various propositional modal logics have been investigated by Visser [10] and Ghilardi [3]. A definition of uniform interpolation for the description logic $\mathcal{ALC}$ is given in [9] and it is used in investigating the definability of TBoxes for $\mathcal{ALC}$.

On the other hand, (semantic) forgetting is studied by researchers in AI [8,7,2]. Informally, given a knowledge base K in classical logic or nonmonotonic logic, we may wish to forget about (or discard) some redundant predicates but still preserve certain forms of reasoning. Forgetting has been investigated for DL-Lite and extended $\mathcal{EL}$ in [11,6,5] but not for expressive DLs. Forgetting for modal logic is studied in [13].

Forgetting and uniform interpolation have different intuitions behind them and are introduced by different communities. Uniform interpolation is originally investigated as a syntactic concept and forgetting is a semantic one. However, if the axiom system is sound and complete, they can be characterized by each other.

In this paper, we first reformulate the notion of uniform interpolation for $\mathcal{ALC}$ studied in [9] from the angle of forgetting and show that all desired properties of forgetting are satisfied. We introduce an algorithm for computing the result of forgetting in concept descriptions and, a novel and detailed proof for the correctness of the algorithm is developed using the Tableau for $\mathcal{ALC}$. We note that a similar algorithm for uniform interpolation is provided in [9] in which it is mentioned that the correctness of their algorithm can be shown using a technique called *bisimulation* that is widely used in modal logic[1]. In a separate paper [12], we use the results obtained in this paper to compute forgetting for $\mathcal{ALC}$ knowledge bases.

Due to space limitation, proofs are omitted in this paper but can be found at http://www.cit.gu.edu.au/~kewen/Papers/alc_forget_long.pdf.

2 Preliminaries

We briefly recall some basics of $\mathcal{ALC}$. Further details of $\mathcal{ALC}$ and other DLs can be found in [1].

First, we introduce the syntax of *concept descriptions* for $\mathcal{ALC}$.

Elementary concept descriptions consist of both *concept names* and *role names*. So a concept name is also called *atomic concept* while a role name is also called *atomic role*. Complex concept descriptions are built inductively as follows: A (atomic concept); $\top$ (universal concept); $\bot$ (empty concept); $\neg C$ (negation); $C \sqcap D$ (conjunction); $C \sqcup D$ (disjunction); $\forall R.C$ (universal quantification) and $\exists R.C$ (existential quantification). Here, A is an (atomic) concept, C and D are concept descriptions, and R is a role.

An interpretation $\mathcal{I}$ of $\mathcal{ALC}$ is a pair $(\Delta^{\mathcal{I}}, \cdot^{\mathcal{I}})$ where $\Delta^{\mathcal{I}}$ is a non-empty set called the *domain* and $\cdot^{\mathcal{I}}$ is an interpretation function which associates each (atomic) concept A with a subset $A^{\mathcal{I}}$ of $\Delta^{\mathcal{I}}$ and each atomic role R with a binary relation $R^{\mathcal{I}} \subseteq \Delta^{\mathcal{I}} \times \Delta^{\mathcal{I}}$. The function $\cdot^{\mathcal{I}}$ can be naturally extended to complex descriptions:

$$\top^{\mathcal{I}} = \Delta^{\mathcal{I}} \qquad\qquad \bot^{\mathcal{I}} = \emptyset$$
$$(\neg C)^{\mathcal{I}} = \Delta^{\mathcal{I}} - C^{\mathcal{I}} \qquad (C \sqcap D)^{\mathcal{I}} = C^{\mathcal{I}} \cap D^{\mathcal{I}}$$
$$(C \sqcup D)^{\mathcal{I}} = C^{\mathcal{I}} \cup D^{\mathcal{I}}$$

---

[1] An email communication with one of the authors of [9] shows that they have not got a complete proof yet.

$$(\forall R.C)^{\mathcal{I}} = \{a \in \Delta^{\mathcal{I}} \; : \; \forall b.(a, b) \in R^{\mathcal{I}} \text{ implies } b \in C^{\mathcal{I}}\}$$
$$(\exists R.C)^{\mathcal{I}} = \{a \in \Delta^{\mathcal{I}} \; : \; \exists b.(a, b) \in R^{\mathcal{I}} \text{ and } b \in C^{\mathcal{I}}\}$$

An interpretation $\mathcal{I}$ satisfies an *inclusion* (or *subsumption*) of the form $C \sqsubseteq D$ where C, D are concept descriptions, denoted $\mathcal{I} \models C \sqsubseteq D$, if $C^{\mathcal{I}} \subseteq D^{\mathcal{I}}$. We write $\models C \sqsubseteq D$ if $\mathcal{I} \models C \sqsubseteq D$ for all $\mathcal{I}$. Similarly, $\models C \equiv D$ is an abbreviation of $\models C \sqsubseteq D$ and $\models D \sqsubseteq C$. $\mathcal{I}$ satisfies an assertion of the form form $C(a)$ or $R(a, b)$, where a and b are individuals, C is a concept description and R is a role name, if, respectively, $a^{\mathcal{I}} \in C^{\mathcal{I}}$ and $(a^{\mathcal{I}}, b^{\mathcal{I}}) \in R^{\mathcal{I}}$.

The signature of a concept description C, written $\text{sig}(C)$, is the set of all concept and role names in C.

3   Forgetting in Concept Descriptions

In this section, we discuss the problem of forgetting about a concept/role in description logic $\mathcal{ALC}$. In the rest of this paper, by a *variable* we mean either a concept name or a role name. Intuitively, the result C' of forgetting about a variable from a concept description C may be weaker than C (w.r.t. subsumption) but it should be as close to C as possible. For example, after the concept *Male* is forgotten from a concept description for "Male Australian student" $C = Australians \sqcap Students \sqcap Male$, then we should obtain a concept description $C' = Australians \sqcap Students$ for "Australian student".

3.1   Semantic Forgetting

Let C be a concept description that contains a variable V. If we want to forget (or discard) V from C, intuitively, the result of forgetting about V in C will be a concept description C' that satisfies the condition: C' defines a minimal concept among all concepts that subsumes C and is irrelevant to V (i.e. V does not appear in the concept description).

Definition 3.1. *(Forgetting for concept descriptions) Let C be a concept description in $\mathcal{ALC}$ and V be a variable. A concept description C' on the signature $\text{sig}(C) \setminus \{V\}$ is a result of forgetting about V in C if the following conditions are satisfied:*

(RF1) $\models C \sqsubseteq C'$.
(RF2) *For all concept description C'' on $\text{sig}(C) \setminus \{V\}$, $\models C \sqsubseteq C''$ and $\models C'' \sqsubseteq C'$ implies $\models C'' \equiv C'$, i.e., C' is the strongest concept description weaker than C that does not contain V.*

Notice that we can forget about a set of concept names and role names in a concept description by a straightforward generalization of Definition 3.1. The above (RF1) and (RF2) correspond to the conditions (2) and (3) of Theorem 8 in [9][2], and generalize them by allowing V to be a role name.

A fundamental property of forgetting in $\mathcal{ALC}$ concept descriptions is that the result of forgetting is unique under concept description equivalence.

---

[2] The correspondence between (RF2) and (3) of Theorem 8 can be seen from this: $\models C \sqsubseteq C'$ and $\models C \sqsubseteq C''$ implies $\models C \sqsubseteq C' \sqcap C''$. It implies by (RF2), $\models C' \sqcap C'' \equiv C'$, which equals $\models C' \sqsubseteq C''$ in (3).

Theorem 3.1. *For any concept description C and variable V, if C_1 and C_2 are results of forgetting about V in C, then $\models C_1 \equiv C_2$.*

As all results of forgetting are equivalent, we write $\mathsf{forget}(C, V)$ to denote an arbitrary result of forgetting about V in C.

We use the following examples of concept descriptions to illustrate our semantic definitions of forgetting for $\mathcal{ALC}$. We will introduce an algorithm later and explain how we can compute a result of forgetting through a series of syntactic transformations of concept descriptions.

Example 3.1. Suppose the concept "Research Student" is defined by $C = Student \sqcap (Master \sqcup PhD) \sqcap \exists supervised.Professor$ where "Master", "PhD" and "Professor" are all concepts; "supervised" is a role and $supervised(x, y)$ means that x is supervised by y. If the concept description C is used only for students, we may wish to forget about $Student$: $\mathsf{forget}(C, Student) = (Master \sqcup PhD) \sqcap \exists supervised.Professor$. If we do not require that a supervisor for a research student must be a professor, then the filter "Professor" can be forgotten: $\mathsf{forget}(C, Professor) = Student \sqcap (Master \sqcup PhD) \sqcap \exists supervised.\top$.

3.2 Properties of Semantic Forgetting

The semantic forgetting for description logic possesses several important properties. The following result, which is not obvious, shows that forgetting distributes over union $\sqcup$.

Proposition 3.1. *Let $C_1, \ldots, C_n$ be concept descriptions in $\mathcal{ALC}$. For any variable V, we have*

$$\mathsf{forget}(C_1 \sqcup \cdots \sqcup C_n, V) = \mathsf{forget}(C_1, V) \sqcup \cdots \sqcup \mathsf{forget}(C_n, V)$$

However, forgetting for $\mathcal{ALC}$ does not distribute over intersection $\sqcap$. For example, if the concept description $C = A \sqcap \neg A$, then $\mathsf{forget}(C, A) = \bot$, since $\models C \equiv \bot$. But $\mathsf{forget}(A, A) \sqcap \mathsf{forget}(\neg A, A) = \top$. So $\mathsf{forget}(A \sqcap \neg A, A) \neq \mathsf{forget}(A, A) \sqcap \mathsf{forget}(\neg A, A)$.

On the other hand, forgetting for $\mathcal{ALC}$ does preserve the subsumption relation between concept descriptions.

Proposition 3.2. *Let C_1 and C_2 be concept descriptions in $\mathcal{ALC}$, and $\models C_1 \sqsubseteq C_2$. For any variable V, we have $\models \mathsf{forget}(C_1, V) \sqsubseteq \mathsf{forget}(C_2, V)$.*

As a corollary of Proposition 3.2, it is straightforward to show that semantic forgetting also preserves the equivalence of concept descriptions.

Proposition 3.3. *Let C_1 and C_2 be concept descriptions in $\mathcal{ALC}$, and $\models C_1 \equiv C_2$. For any variable V, we have $\mathsf{forget}(C_1, V) \equiv \mathsf{forget}(C_2, V)$.*

Satisfiability is key reasoning task in description logics. We say a concept C is satisfiable if $C^{\mathcal{I}} \neq \emptyset$ for some interpretation $\mathcal{I}$. C is unsatisfiable if $\models C \equiv \bot$. Forgetting also preserves satisfiability of concept descriptions.

Proposition 3.4. *Let C be a concept description in $\mathcal{ALC}$, and V be a variable. Then C is satisfiable iff* forget(C,V) *is satisfiable.*

When we want to forget about a set of variables, they can be forgotten one by one since the ordering of forgetting is irrelevant to the result.

Proposition 3.5. *Let C be a concept description in $\mathcal{ALC}$ and let $\mathcal{V} = \{V_1,\ldots,V_n\}$ be a set of variables. Then* forget$(C,\mathcal{V}) =$ forget(forget(forget$(C,V_1),V_2),\ldots),V_n)$.

The next result shows that forgetting distributes over quantifiers.

Proposition 3.6. *Let C be a concept description in $\mathcal{ALC}$, R be a role name and V be a variable. Then*
(1) forget$(\forall V.C, V) = \top$ *where V is a role name, and* forget$(\forall R.C, V) = \forall R.$forget$(C, V)$ *if $V \neq R$;*
(2) forget$(\exists V.C, V) = \top$ *where V is a role name, and* forget$(\exists R.C, V) = \exists R.$forget$(C, V)$. *if $V \neq R$.*

4   Computing the Result of Forgetting

In this section we introduce an intuitive algorithm for computing the result of forgetting through rewriting of concept descriptions (syntactic concept transformations). Given a concept description C and a variable V, this algorithm consists of two stages: (1) C is first transformed into an equivalent disjunctive normal form (DNF), which is a disjunction of conjunctions of simple concept descriptions; (2) the result of forgetting about V in each such simple concept description is obtained by removing some parts of the conjunct.

We call an atomic concept A or its negation $\neg A$ a *literal concept* or simply *literal*. A *pseudo-literal* with role R is a concept description of the form $\exists R.F$ or $\forall R.F$, where R is a role name and F is an arbitrary concept description. A *generalized literal* is either a literal or a pseudo-literal.

First, each arbitrary concept description can be equivalently transformed into a disjunction of conjunctions of generalized literals. This basic DNF for $\mathcal{ALC}$ is well known in the literature [1] and thus the details of the transformation are omitted here.

Definition 4.1. *(Basic DNF) A concept description D is in* basic disjunctive normal form *or* basic DNF *if $D = \bot$ or $D = \top$ or D is a disjunction of conjunctions of generalized literals $D = D_1 \sqcup \cdots \sqcup D_n$, where each $D_i \not\equiv \bot$ and D_i is of the form*
$$\bigsqcap L \sqcap \bigsqcap_{R \in \mathcal{R}} [\forall R.U_R \sqcap \bigsqcap_k \exists R.E_R^{(k)}]$$
where each L is a literal, $\mathcal{R}$ is the set of role names that occur in D_i, $k \geq 0$, and each U_R or $E_R^{(k)}$ is a concept description in basic DNF.

The reason for transforming a concept description into its basic DNF is that forgetting distributes over $\sqcup$ (Proposition 3.1). When a concept description is in its basic DNF, we only need to compute the result of forgetting in a conjunction of generalized literals. It can be shown that the result of forgetting about an atomic concept A in a conjunction B of literals can be obtained just by extracting A (or $\neg A$) from the conjuncts (extracting a conjunct equals replacing it by $\top$). Unlike classical logics and DL-Lite, the basic DNF in $\mathcal{ALC}$ is not clean in that a generalized literal (i.e. pseudo-literal) can still be very complex. When C is a conjunction containing pseudo-literals, it is not straightforward to compute the result of forgetting about A in C. For example, let $C = \forall R.A \sqcap \forall R.\neg A$. Through simple transformation we can see $C \equiv \forall R.\bot$ is the result of forgetting about A in C, while simply extracting A and $\neg A$ results in $\top$. A similar example is when $C = \forall R.(A \sqcup B) \sqcap \exists R.(\neg A \sqcup B)$, the result of forgetting is $\exists R.B$ rather than $\exists R.\top$ (note that $\forall R.C_1 \sqcap \exists R.C_2 \sqsubseteq \exists R.(C_1 \sqcap C_2)$). For this reason, the following key step in obtaining a DNF is required for computing forgetting:

$$\forall R.C_1 \sqcap \exists R.C_2 \rightsquigarrow \forall R.C_1 \sqcap \exists R.(C_1 \sqcap C_2)$$

By applying this transformations to a concept description in its basic DNF, each $\mathcal{ALC}$ concept description can be transformed into the DNF as defined below.

Definition 4.2. *(Disjunctive Normal Form or DNF)*

A concept description D is in disjunctive normal form *if $D = \bot$ or $D = \top$ or D is a disjunction of conjunctions of generalized literals $D = D_1 \sqcup \cdots \sqcup D_n$, where each $D_i \not\equiv \bot$ $(1 \le i \le n)$ is a conjunction $\bigsqcap L$ of literals or in the form of*

$$\bigsqcap L \sqcap \bigsqcap_{R \in \mathcal{R}} \left[\forall R.U_R \sqcap \bigsqcap_k \exists R.(E_R^{(k)} \sqcap U_R) \right]$$

where each L is a literal, $\mathcal{R}$ is the set of role names that occur in D_i, $k \ge 0$, and each U_R or $E_R^{(k)} \sqcap U_R$ is a concept description in DNF.

For convenience, each D_i is called a *normal conjunction* in this paper. The disjunctive normal form is a bit different from the normal form in [1] but they are essentially the same.

Once a concept D is in the normal form, the result of forgetting about a variable V in D can be obtained from D by simple symbol manipulations.

Obviously, the major cost of Algorithm 1 is from transforming the given concept description into its normal form. For this reason, the algorithm is exponential time in the worst case. However, if the input concept description C is in DNF, Algorithm 1 takes only linear time (w.r.t. the size of C).

Example 4.1. Given a concept $D = (A \sqcup \exists R.\neg B) \sqcap \forall R.(B \sqcup C)$, we want to forget about concept name B in D. In Step 1 of Algorithm 1, D is firstly transformed into its DNF $D' = [A \sqcap \forall R.(B \sqcup C)] \sqcup [\forall R.(B \sqcup C) \sqcap \exists R.(\neg B \sqcap C)]$. Note that $\exists R.(\neg B \sqcap C)$ is transformed from $\exists R.[\neg B \sqcap (B \sqcup C)]$. Then in Step 2, each occurrence of B in D' is replaced by $\top$, and $\forall R.(\top \sqcup F)$ is replaced with $\top$. We obtain $\mathsf{CForget}(D, B) = A \sqcup \exists R.C$. To forget about role R in D, Algorithm 1 replaces each pseudo-literals in D' of the form $\forall R.F$ or $\exists R.F$ with $\top$, and returns $\mathsf{CForget}(D, R) = \top$.

Indeed we can prove that Algorithm 1 is sound and complete w.r.t. the semantic definition of forgetting for $\mathcal{ALC}$ in Definition 3.1.

Theorem 4.1. *Let V be a variable and C a concept description in $\mathcal{ALC}$. Then*

$$\models \mathsf{CForget}(C, V) \equiv \mathsf{forget}(C, V).$$

Algorithm 1
Input: A concept description C in $\mathcal{ALC}$ and a variable V in C.
Output: The result of forgetting about V in C.
Method CForget(C, V):
Step 1. Transform C into its DNF D. If D is $\top$ or $\bot$, return D; otherwise, let $D = D_1 \sqcup \cdots \sqcup D_n$ as in Definition 4.2.
Step 2. For each conjunct E in each D_i, perform the following transformations:

- if (V is a concept name and) E is a literal equals V or $\neg V$, replace E with $\top$;
- if (V is a role name and) E is a pseudo-literal of the form $\forall V.F$ or $\exists V.F$, replace E with $\top$;
- if E is a pseudo-literal in the form of $\forall R.F$ or $\exists R.F$ where $R \neq V$, replace F with CForget(F, V), and replace each resulting $\forall S.(\top \sqcup G)$ with $\top$.

Step 3. Return the resulting concept description as CForget(C, V).

Fig. 1. Forgetting in concept descriptions

Theorem 4.1 and Proposition 3.6 can be immediately derived from some lemmas in the next section and the validity of these lemmas is established by using the Tableau for $\mathcal{ALC}$.

5 Proof of Theorem 4.1

Proofs of Proposition 3.6 and Theorem 4.1 (i. e. the correctness of Algorithm 1) heavily rely on the Tableau theory for $\mathcal{ALC}$ and thus we first briefly introduce it.

Given two concept descriptions C_0 and D_0, the Tableau theory states that $\models C_0 \sqsubseteq D_0$ iff no (finite) interpretation $\mathcal{I}$ can be constructed such that $\mathcal{I}$ satisfies concept $C_0 \sqcap \neg D_0$, i.e., there is an individual x_0 with $x_0^{\mathcal{I}} \in (C_0 \sqcap \neg D_0)^{\mathcal{I}}$. Equivalently, ABox $\mathcal{A}_0 = \{(C_0 \sqcap \neg D_0)(x_0)\}$ must be inconsistent for an arbitrary individual x_0. And by transforming $C_0 \sqcap \neg D_0$ into its Negation Normal Form (NNF) and applying Tableau transformation rules to the ABox, *clashes* of the form $A(x), \neg A(x)$ must occur.

The $\mathcal{ALC}$ Tableau transformation rules are as follows:

- $\{(C_1 \sqcap C_2)(x), \ldots\} \rightarrow_{\sqcap} \{(C_1 \sqcap C_2)(x), C_1(x), C_2(x), \ldots\}$.
- $\{(C_1 \sqcup C_2)(x), \ldots\} \rightarrow_{\sqcup} \{(C_1 \sqcup C_2)(x), C_1(x), \ldots\}$ and $\{(C_1 \sqcup C_2)(x), C_2(x), \ldots\}$.
- $\{(\exists R.C)(x), \ldots\} \rightarrow_{\exists} \{(\exists R.C)(x), R(x, y), C(y), \ldots\}$.
- $\{(\forall R.C)(x), R(x, y)\} \rightarrow_{\forall} \{(\forall R.C)(x), R(x, y), C(y)\}$.

Conversely, if clashes occur in each resulting ABox after applying Tableau rules, then $\mathcal{A}_0$ must be inconsistent and $\models C_0 \sqsubseteq D_0$ holds.

Before presenting the proofs, we first show a useful lemma, whose correctness can be immediately obtained using the Tableau.

Lemma 5.1. *Let C_i's be concepts and R a role name. Then, for every $j(1 \leq j \leq m)$ or $C_j = \bot$, $\models \forall R.(\bigsqcup_{i=1}^{m} C_i) \sqsubseteq \forall R.C_j \sqcup \bigsqcup_{i \neq j}^{m} \exists R.C_i$.*

Lemma 5.2. *Let U, E_i's be concepts with $\models E_i \sqsubseteq U$, and R a role name. Denote $D = \forall R.U \sqcap \bigsqcap \exists R.E_i$ with $\not\models D \equiv \perp$. Suppose $D' = \bigsqcup \forall R.C_j \sqcup \exists R.C \sqcup \bigsqcup L_k$, where C and C_j's are concepts, and L_k's are generalized literals not containing R. Then $\models D \sqsubseteq D'$ implies that at least one of the following three holds:*

(1) $\models D' \equiv \top$; or
(2) $\models E_i \sqsubseteq C$ for some i, and $\models D \sqsubseteq \exists R.C \sqsubseteq D'$; or
(3) $\models U \sqsubseteq C \sqcup C_j$, and $\models D \sqsubseteq \forall R.(C \sqcup C_j) \sqsubseteq D'$ for some j.

Proof. For simplicity, we discuss the case of $m = 2$ and $n = 2$, that is, $D = \forall R.U \sqcap \exists R.E_1 \sqcap \exists R.E_2$ and $D' = \forall R.C_1 \sqcup \forall R.C_2 \sqcup \exists R.C \sqcup \bigsqcup L_k$. Other cases can be proved in the same way.

According to the Tableau, the ABox $\{D(x), \neg D'(x)\}$ is inconsistent, which can be transformed through Tableau rules into $\{ (\forall R.U)(x), (\exists R.E_1)(x), (\exists R.E_2)(x), (\exists R.\neg C_1)(x), (\exists R.\neg C_2)(x), (\forall R.\neg C)(x), \neg(\bigsqcup L_k)(x) \}$

It can be further transformed into

$$\mathcal{A} = \{ R(x, y_1), E_1(y_1), R(x, y_2), E_2(y_2), R(x, z_1), \neg C_1(z_1), R(x, z_2),$$
$$\neg C_2(z_2), (\forall R.U)(x), U(y_1), U(y_2), U(z_1), U(z_2), (\forall R.\neg C)(x),$$
$$\neg C(y_1), \neg C(y_2), \neg C(z_1), \neg C(z_2), \neg(\bigsqcup L_k)(x), \dots \}.$$

Note that L_k's do not contain any pseudo-literal of the form $\forall R.F$ or $\exists R.F$. Thus there is no way to generate any new assertions about y_i or z_j from $\neg(\bigsqcup L_k)(x)$ $(i, j = 1, 2)$. Neither can $R(x, v)$ with $v \neq y_i$ and $v \neq z_j$ be generated from $\mathcal{A}$. This means no Tableau rule is applicable to $R(x, y_i), R(x, z_j), (\forall R.U)(x)$ or $(\forall R.\neg C)(x)$ any more. Thus we can ignore those assertions.

According to the Tableau, $\mathcal{A}$ must be inconsistent for arbitrary instances x, y_1, y_2, z_1, z_2. Thus it is safe to assume that x, y_1, y_2, z_1, z_2 represent different individuals. Thus $\mathcal{A}$ can be written as

$$\{ \neg(\bigsqcup L_k)(x) \} \cup \{ E_1(y_1), U(y_1), \neg C(y_1) \} \cup \{ E_2(y_2), U(y_2), \neg C(y_2) \}$$
$$\cup \{ U(z_1), \neg C_1(z_1), \neg C(z_1) \} \cup \{ U(z_2), \neg C_2(z_2), \neg C(z_2) \}.$$

Consider three cases:
Case 1. $\{\neg(\bigsqcup L_k)(x)\}$ is inconsistent, then we have $\models \bigsqcup L_k \equiv \top$. That is, $\models D' \equiv \top$.
Case 2. $\{E_i(y_i), U(y_i), \neg C(y_i)\}$ $(i = 1$ or $2)$ is inconsistent, then $\models E_i \sqcap U \sqsubseteq C$. From $\models E_i \sqsubseteq U$, it follows that $\models E_i \sqsubseteq C$. Thus $\models D \sqsubseteq \exists R.E_i \sqsubseteq \exists R.C \sqsubseteq D'$.
Case 3. $\{U(z_j), \neg C_j(z_j), \neg C(z_j)\}$ $(j = 1$ or $2)$ is inconsistent, then $\models U \sqsubseteq C \sqcup C_j$. Thus, by Lemma 5.1, we have $\models D \sqsubseteq \forall R.(C \sqcup C_j) \sqsubseteq \forall R.C_j \sqcup \exists R.C \sqsubseteq D'$.   ∎

Now we can show a general property of forgetting w.r.t. quantifiers. Proposition 3.6 is just a special case of the following lemma.

Lemma 5.3. *Let U, E_i's be concepts with $\models E_i \sqsubseteq U$, R be a role name, and V be a variable. Denote $C = \forall R.U \sqcap \bigsqcap_{i \in M} \exists R.E_i$ where $M = \{1, \dots, m\}$ is a set of natural numbers.*

Suppose $\not\models C \equiv \perp$. If $V = R$, then $\mathsf{forget}(C, V) = \top$. Otherwise, we have $\mathsf{forget}(C, V) = \forall R.\mathsf{forget}(U, V) \sqcap \bigsqcap_{i \in M} \exists R.\mathsf{forget}(E_i, V)$.

Proof. Set $C' = \forall R.\text{forget}(U, V) \sqcap \prod_{i \in M} \exists R.\text{forget}(E_i, V)$.

(CF1): Obviously, $\models C \sqsubseteq \top$. And we have $\models C \sqsubseteq C'$, since $\models \forall R.U \sqsubseteq \forall R.\text{forget}(U, V)$ and $\models \exists R.E_i \sqsubseteq \exists R.\text{forget}(E_i, V)$ for all i.

(CF2): Suppose that D is a concept not containing V and $\models C \sqsubseteq D$. We want to prove $\models \top \sqsubseteq D$ for $V = R$, and otherwise, $\models C' \sqsubseteq D$.

Let $D = \prod_{k \in N} D_k$, and every D_k is of the form $\bigsqcup \forall R.U'_j \sqcup \exists R.E' \sqcup B'$ where E' and U'_j's are concepts, B' is a disjunction of generalized literals not containing R. From $\models C \sqsubseteq D$, we have $\models C \sqsubseteq D_k$ for all k.

If $V = R$, then each D_k contains no disjunct of $\forall R.U'_j$, and $\models E' \equiv \bot$. By Lemma 5.2, we have $\models D_k \equiv \top$ for each k. In this case, $\models \top \sqsubseteq D$. We have shown in this case, $\text{forget}(C, V) = \top$.

Otherwise, suppose for some D_k, it does not contain any occurrence of R. In this case, $\models D_k \equiv \top$, and we can remove D_k from the conjunction. In what follows, we assume D_k contains R and $\not\models D_k \equiv \top$ for each $k \in N$.

By Lemma 5.2, for some D_k's in D (denoted as $k \in K \subseteq N$), we always have some E_i ($i \in M$) in C such that $\models E_i \sqsubseteq F_{\mathcal{E}_k}$ where $\exists R.F_{\mathcal{E}_k}$ is the existential quantified disjunct of D_k, and thus $\models C \sqsubseteq \exists R.F_{\mathcal{E}_k} \sqsubseteq D_k$. For the other D_k's with $k \in N - K$, we always have $\models U \sqsubseteq F_{\mathcal{U}_k}$ and $\models C \sqsubseteq \forall R.F_{\mathcal{U}_k} \sqsubseteq D_k$ for some concept $F_{\mathcal{U}_k}$ not containing V. This is to say, we can always find
$$D' = \prod_{k \in K} \exists R.F_{\mathcal{E}_k} \sqcap \prod_{l \in N - K} \forall R.F_{\mathcal{U}_l}$$
such that D' does not contain V, and $\models C \sqsubseteq D' \sqsubseteq D$.

By the definition of K, for each $F_{\mathcal{E}_k}$ ($k \in K$), we always have some E_i ($i \in M$) in C such that $\models E_i \sqsubseteq F_{\mathcal{E}_k}$. By the definition of c-forgetting, $\models \text{forget}(E_i, V) \sqsubseteq F_{\mathcal{E}_k}$. That is, for each $k \in K$, there always exists some $i \in M$ such that $\models \exists R.\text{forget}(E_i, V) \sqsubseteq \exists R.F_{\mathcal{E}_k}$. This implies $\models \prod_{i \in M} \exists R.\text{forget}(E_i, V) \sqsubseteq \prod_{k \in K} \exists R.F_{\mathcal{E}_k}$. Similarly, we have $\models \forall R.\text{forget}(U, V) \sqsubseteq \forall R.\prod_{l \in N - K} F_{\mathcal{U}_l}$. Thus, we can conclude that
$$\models \forall R.\text{forget}(U, V) \sqcap \prod_{i \in M} \exists R.\text{forget}(E_i, V) \sqsubseteq \forall R.\prod_{l \in N - K} F_{\mathcal{U}_l} \sqcap \prod_{k \in K} \exists R.F_{\mathcal{E}_k}.$$
which is, $\models C' \sqsubseteq D'$. Hence, we have $\models C' \sqsubseteq D$. $\blacksquare$

Similar to the above lemma, we can show the following result. For a literal L, we use L^+ to denote the concept name in L.

Lemma 5.4. *Let C be a disjunct in DNF such that $C = \prod L_i \sqcap \prod_{R \in \mathcal{R}} C_R$, where each L is a literal, concept C_R is of the form $\forall R.U \sqcap \prod \exists R.E_k$ with $\models E_k \sqsubseteq U$ for each k. Then we have, $\text{forget}(C, V) = \prod_{L_i^+ \neq V} L_i \sqcap \prod_{R \in \mathcal{R}} \text{forget}(C_R, V)$.*

Since forgetting is distributive over disjunction, Theorem 4.1 is proven.

6   Conclusion

We have looked into the concept of uniform interpolation for $\mathcal{ALC}$ from the angle of variable forgetting. As a result, a theory of forgetting in $\mathcal{ALC}$ concept descriptions is developed, in which forgetting can be done for both concepts and roles. As well as several important properties, we have developed algorithms for computing results of forgetting and provide a novel proof for the correctness of the algorithm w.r.t. the semantic definition of forgetting. Forgetting for $\mathcal{ALC}$ concept descriptions has been implemented in C^{++} as a new component of the DL reasoner FaCT++ [14] and it is available at http://www.cit.gu.edu.au/~kewen/DLForget/. Such a forgetting component can be used by an ontology editor to enhance its ability to partially reuse existing ontologies and thus provides a flexible tool for tailoring large ontologies. Although semantic forgetting can be easily adapted to most DLs, it is not straightforward to generalize the algorithms for computing forgetting to other expressive DLs.

References

1. Baader, F., Calvanese, D., McGuinness, D., Nardi, D., Patel-Schneider, P.: The Description Logic Handbook. Cambridge University Press, Cambridge (2002)
2. Eiter, T., Wang, K.: Semantic forgetting in answer set programming. Artificial Intelligence 172(14), 1644–1672 (2008)
3. Ghilardi, S.: An algebraic theory of normal forms. Annals Pure Appl. Logic 71(3), 189–245 (1995)
4. Konev, B., Walther, D., Wolter, F.: The logical difference problem for description logic terminologies. In: Armando, A., Baumgartner, P., Dowek, G. (eds.) IJCAR 2008. LNCS (LNAI), vol. 5195, pp. 259–274. Springer, Heidelberg (2008)
5. Konev, B., Walther, D., Wolter, F.: Forgetting and uniform interpolation in large-scale description logic terminologies. In: Proc. IJCAI 2009 (2009)
6. Kontchakov, R., Wolter, F., Zakharyaschev, M.: Can you tell the difference between DL-Lite ontologies? In: Proc. KR 2008, pp. 285–295 (2008)
7. Lang, J., Liberatore, P., Marquis, P.: Propositional independence: Formula-variable independence and forgetting. J. Artif. Intell. Res. 18, 391–443 (2003)
8. Lin, F., Reiter, R.: Forget it. In: Proc. AAAI Fall Symposium on Relevance, pp. 154–159 (1994)
9. ten Cate, B., Conradie, W., Marx, M., Venema, Y.: Definitorially complete description logics. In: Proc. KR 2006, pp. 79–89 (2006)
10. Visser, A.: Uniform interpolation and layered bisimulation. In: Proc. of Gödel 1996, pp. 139–164 (1996)
11. Wang, Z., Wang, K., Topor, R., Pan, J.Z.: Forgetting concepts in DL-Lite. In: Bechhofer, S., Hauswirth, M., Hoffmann, J., Koubarakis, M. (eds.) ESWC 2008. LNCS, vol. 5021, pp. 245–257. Springer, Heidelberg (2008)
12. Wang, Z., Wang, K., Topor, R., Pan, J.Z., Antoniou, G.: Concept and Role Forgetting in $\mathcal{ALC}$ Ontologies. In: Proc. ISWC 2009, pp. 666–681 (2009)
13. Zhang, Y., Zhou, Y.: Knowledge Forgetting: Properties and Applications. Artificial Intelligence 173, 1525–1537 (2009)
14. Tsarkov, D., Horrocks, I.: Description logic reasoner: System description. In: Furbach, U., Shankar, N. (eds.) IJCAR 2006. LNCS (LNAI), vol. 4130, pp. 292–297. Springer, Heidelberg (2006)
15. Bechhofer, S., et al.: OWL (Web Ontology Language) Reference (2004), http://www.w3.org/TR/2004/REC-owl-ref-20040210/
16. Systematized Nomenclature of Medicine-Clinical Terms, http://www.ihtsdo.org/snomed-ct/
17. NCI Wiki, http://ncicb.nci.nih.gov/

Modeling Abstract Behavior: A Dynamic Logic Approach

Yi Zhou and Yan Zhang

Intelligent Systems Lab
School of Computing and Mathematics
University of Western Sydney
Locked Bag 1797, Penrith South DC, NSW 1797, Australia
{yzhou,yan}@scm.uws.edu.au

Abstract. Modeling abstract behavior is essential for intelligent agents under incomplete and uncertain environments. In this paper, we extend Propositional Dynamic Logic (PDL) to Propositional Abstract Dynamic Logic (PADL) for modeling abstract behavior in two aspects. On the one hand, we treat the task of finding a plan to achieve a certain formula as an abstract action. On the other hand, we explicitly represent the subsumption relation between two actions as a formula in the language. We propose the semantics for the two operators and discuss some important related properties.

1   Introduction

Modeling rational behavior of agents is one of the most fundamental problems for intelligent agents. In the AI literature, behavior, or actions, of the agents are usually represented simply as atomic actions or sequences of atomic actions called plans. However, little attention has been paid to the problem of modeling more flexible and complex forms of actions.

Propositional Dynamic Logic (PDL for short) [1,2,3,4,5] is an elegant and powerful logic for modeling various kinds of compound actions. It can represent not only sequences of actions, but also nondeterministic choices of actions, query actions and iteration actions. More importantly, PDL relates actions with formulas so that it can be used for reasoning about properties of actions. Indeed, as successfully shown in the area of theoretical computer science [1], PDL is able to represent a large variety of programs and to verify the partial and total correctness of them.

However, PDL is not powerful enough for modeling abstract behavior for intelligent agents. By abstract behavior, we mean those actions that are not fully specified. In other words, abstract actions are relatively general in contrast to more specific actions. An abstract action is maybe a high level description of the agent about what to do; it gives some but not all information. Thus, it can be further refined or elaborated to more specific actions by fixing some of the details, which might be given in many different ways.

Let us consider an example first. Suppose that Alice wants to present her research at a conference. A reasonable solution for this problem is to divide it into two steps, a) to get her paper accepted by the conference and b) to attend the conference. Here, both the

A. Nicholson and X. Li (Eds.): AI 2009, LNAI 5866, pp. 538–546, 2009.

two steps are abstract actions. The former is a subgoal of Alice, which can be achieved by various specific plans. The latter is a general description of behavior, which needs to be further refined by other specific actions such as to register the conference, to book a flight and so on.

Abstract behavior is essential for intelligent agents for several reasons. Firstly, the agent may only have incomplete information about the environment. Thus, sometimes it is impossible for the agent to make a very specific plan due to the lack of information. However, it is still possible to make abstract plans (behavior) according to its incomplete information. Secondly, abstract actions are more reliable than specific actions. In realistic domains, environments are essentially unpredictable and uncertain. Therefore, actions often lead to unexpected results. Abstract actions are usually more reliable since they have less details. Thirdly, abstract actions are more flexible and robust than specific actions. Abstract actions can be further refined into various specific actions. Once one of them is failed, we can trace back and choose another refinement according to the abstract action. However, if no abstract action is allowed, then the agent may need to replan the whole picture. Finally, abstract behavior is beneficial to the resource bounded barriers, including computational resources. To consider problems in an abstract level can reduce the cost of resources (e.g. computational cost) because representation in the abstract level is usually more succinct.

In this paper, we extend Propositional Dynamic Logic into Propositional Abstract Dynamic Logic (PADL for short) for modeling abstract behavior in two aspects. Firstly, we treat the task of finding a plan to achieve a certain formula as an abstract action. Technically, given a proposition ϕ, we introduce a new operator $\#$, the achievement operator, in front of ϕ as an abstract action $\#\phi$, meaning that the action to find an (arbitrary) action (or plan) to achieve ϕ. Secondly, we explicitly represent the subsumption relation between two actions as a formula in the language. Technically, we introduce a new operator $\rhd$, the subsumption operator, between two actions α and β. $\alpha \rhd \beta$ is a well defined formula in PADL, meaning that α is more specific than β.

This paper is organized as follows. In the next section, we recall some basic definitions of PDL. In section 3, provide the syntax as well as the semantics for PADL. We extensively study the properties of PADL in Section 4. Finally, we draw our conclusions.

2   Propositional Dynamic Logic

Propositional Dynamic Logic (PDL for short) [1] is a logic for representing and reasoning about the interactions between propositions (or properties, formulas) and programs (or actions, plans, events, behavior). It inherits three classical components: propositional logic, modal logic and regular expressions. The basic idea of PDL is to represent each (regular) program as a modal operator in propositional modal logic. Therefore, PDL is essentially a multi-modal logic.

The syntax of PDL has two kinds of expressions, namely, *formulas* and *actions*. They are defined recursively from a set Π_0 of *events* (or *atomic actions*), a set Φ_0 of *atoms* and the following operators:

Propositional operators: $\rightarrow$ (*implication*) and $\perp$ (*falsity*),
Behavioral operators: ; (*composition*), $\cup$ (*choice*) and $*$ (*iteration*),
Mixed operators: $[]$ (*necessity*) and ? (*test*).

Given a set Π_0 of primitive actions and a set Φ_0 of atoms, we define the set Π of all
actions and the set Φ of all formulas to be the smallest set such that

S1 $\Phi_0 \subseteq \Phi$ and $\perp \in \Phi$,
S2 $\Pi_0 \subseteq \Pi$,
S3 if $\phi, \psi \in \Phi$, then $\phi \rightarrow \psi \in \Phi$,
S4 if $\alpha, \beta \in \Pi$, then $\alpha; \beta$, $\alpha \cup \beta$ and $\alpha^* \in \Pi$,
S5 if $\alpha \in \Pi$ and $\phi \in \Phi$, then $[\alpha]\phi \in \Phi$,
S6 if $\phi \in \Phi$ then $\phi? \in \Pi$.

The other propositional operators $\top$ (*truth*), $\neg$ (*negation*), $\wedge$ (*conjunction*), $\vee$ (*disjunction*) and $\leftrightarrow$ (*equivalence*) are defined as usual. The *possibility operator* $<>$ is the dual
of the necessity operator $[]$. Given a formula ϕ and an action α, $< \alpha > \phi$ is defined as
$\neg[\alpha]\neg\phi$.

The semantics of PDL inherits from standard Kripke semantics for modal logic. A
(Kripke) frame is a pair $\mathcal{F} = \langle W, M_{\mathcal{F}} \rangle$, where W is a set of elements called *worlds* or
states and $M_{\mathcal{F}}$ is a *meaning function* assigning a subset of W to each atom and a binary
relation on W to each atomic action. That is,

$$M_{\mathcal{F}}(p) \subseteq W, p \in \Phi_0$$
$$M_{\mathcal{F}}(a) \subseteq W \times W, a \in \Pi_0.$$

We extend the meaning function $M_{\mathcal{F}}$ to all actions and formulas inductively as follows

M1 $M_{\mathcal{F}}(\perp) = \emptyset$,
M2 $M_{\mathcal{F}}(\phi \rightarrow \psi) = (W - M_{\mathcal{F}}(\phi)) \cup M_{\mathcal{F}}(\psi)$,
M3 $M_{\mathcal{F}}([\alpha]\phi) = \{u \mid \forall v \in W, \text{ if } (u, v) \in M_{\mathcal{F}}(\alpha), \text{ then } v \in M_{\mathcal{F}}(\phi)\}$,
M4 $M_{\mathcal{F}}(\alpha; \beta) = \{(u, v) \mid \exists w \in W, \text{ such that } (u, w) \in M_{\mathcal{F}}(\alpha) \text{ and } (w, v) \in M_{\mathcal{F}}(\beta)\}$,
M5 $M_{\mathcal{F}}(\alpha \cup \beta) = M_{\mathcal{F}}(\alpha) \cup M_{\mathcal{F}}(\beta)$,
M6 $M_{\mathcal{F}}(\alpha^*) = \bigcup_{n \geq 0} M_{\mathcal{F}}(\alpha^n)$,
M7 $M_{\mathcal{F}}(\phi?) = \{(u, u) \mid u \in M_{\mathcal{F}}(\phi)\}$.

Therefore,

- $M_{\mathcal{F}}(\top) = W$,
- $M_{\mathcal{F}}(\phi \vee \psi) = M_{\mathcal{F}}(\phi) \cup M_{\mathcal{F}}(\psi)$,
- $M_{\mathcal{F}}(\phi \wedge \psi) = M_{\mathcal{F}}(\phi) \cap M_{\mathcal{F}}(\psi)$,
- $M_{\mathcal{F}}(\neg\phi) = W - M_{\mathcal{F}}(\phi)$,
- $M_{\mathcal{F}}(< \alpha > \phi) = \{u \mid \exists v \in W, \text{ and } v \in M_{\mathcal{F}}(\phi)\}$.

We say that a world u *satisfies* a formula ϕ in a frame $\mathcal{F}$, or that ϕ is *true* at u in $\mathcal{F}$, if
$u \in M_{\mathcal{F}}(\phi)$, also written $\mathcal{F}, u \models \phi$. We say that a formula ϕ is satisfiable in a frame
$\mathcal{F} = \langle W, M_{\mathcal{F}} \rangle$ if there exists some $u \in W$ such that $\mathcal{F}, u \models \phi$.

We say that ϕ is *valid* in $\mathcal{F}$, written $\mathcal{F} \models \phi$, if for all $u \in W$, $\mathcal{F}, u \models \phi$. We say
that ϕ is *valid*, denoted by $\models \phi$, if for all frames $\mathcal{F}$, $\mathcal{F} \models \phi$. Given a set Σ of formulas
and a frame $\mathcal{F}$, we write $\mathcal{F} \models \Sigma$ if for all $\phi \in \Sigma$, $\mathcal{F} \models \phi$. We say that a formula ψ
is a *logical consequence* of a formula set Σ if for all frames $\mathcal{F}$, $\mathcal{F} \models \Sigma$ implies that
$\mathcal{F} \models \psi$. Note that this notion of validity defined for PDL is defined on frames. That is,
this is not the same as saying that $\mathcal{F}, u \models \Sigma$ implies that $\mathcal{F}, u \models \psi$ for all pairs $\mathcal{F}, u$.

3   Propositional Abstract Dynamic Logic

In this section, we extend Propositional Dynamic Logic (PDL) to Propositional Abstract Dynamic Logic (PADL) for formalizing abstract behavior in two aspects. Firstly, we treat the task of finding a plan to achieve a certain formula as an abstract action. Technically, we use a new mixed operator #, the *achievement* operator, to denote the task of achieving a certain formula. Given a formula ϕ, $\#\phi$, reading as "to achieve ϕ", is an action in the language of PADL. Performing this action intuitively means to find a(n) (arbitrary) action (plan) to achieve the formula ϕ. Particularly, we simply write $\#\top$ as U, which denotes the *universal relation*.

Example 1. Recall the example proposed in the introduction section. Alice's first step of *getting her paper accepted by the conference* can be represented as an abstract action $\#paper - accepted$, where $paper - accepted$ is the formula saying that her paper is accepted by the conference. Performing this action $\#paper - accepted$ means to find an arbitrary action (or plan) to achieve the formula $paper - accepted$. There might be many specific actions that can accomplish this task. However, the abstract action itself does not care what the specific action really is. It represents the set of all specific actions that can achieve the formula $paper - accepted$. In addition, Alice's original idea of *presenting her research at the conference* can also be considered as another abstract action $\#research - presented$.

Secondly, we consider the *subsumption* relation between two (abstract) actions. We use a new mixed operator $\triangleright$, the *subsumption* operator, to denote the subsumption relation between two actions. Given two actions α and β, $\alpha \triangleright \beta$, reading as "$\alpha$ subsumes β", is a formula in the language of PADL. Intuitively, this formula is true if α always yields more consequences than β. In other words, α is more specific than β.

The subsumption operator can be further explained by considering the *equivalence* operator $\bowtie$ between two actions. Given two actions α and β, we write $\alpha \bowtie \beta$, reading as "α is equivalent to β", as an abbreviation of $(\alpha \triangleright \beta) \wedge (\beta \triangleright \alpha)$. Intuitively, $\alpha \bowtie \beta$ means that α and β have the same consequences in every situation. In other words, α and β have the same ability.

Example 2. Again, recall the example in the introduction section. Alice's second step *to attend the conference* can be further refined into small pieces, for instance, 1) to register the conference, 2) to book a flight, 3) to book a hotel, 4) to take the trip to the conference and 5) to present the paper. Thus, this procedure of refinement can be represented in PADL as an assertion (PADL formula) $register - conference; book - flight; book - hotel; trip - conference; present - paper \triangleright attend - conference$, meaning that the abstract action to attend the conference is subsumed by a sequence of more specific actions as mentioned above. In fact, Alice's solution of dividing her original idea into two steps can also be represented in PADL as a subsumption relation $\#paper - accepted; attend - conference \triangleright \#research - presented$.

Hence, the syntax of Propositional Abstract Dynamic Logic (PADL) is an extension of the syntax of PDL with two more mixed operators $\triangleright$ (for subsumption) and $\#$ (for achievement). As PDL, the language of PADL has two components, namely formulas as well as actions. Based on the syntactical composition rules of PDL, given a set Π_0 of atomic actions and a set Φ_0 of atoms, the set Π of all actions and the set Φ of all formulas in PADL can be obtained by two more composition rules

S7 if $\alpha, \beta \in \Pi$, then $\alpha \triangleright \beta \in \Phi$,
S8 if $\phi \in \Phi$ then $\#\phi \in \Pi$.

In particular, we write $\alpha \bowtie \beta$ (the equivalence relation between two actions) and U (the universal operator) as the shorthand of $(\alpha \triangleright \beta) \wedge (\beta \triangleright \alpha)$ and $\#\top$ respectively.

We still adopt the Kripke semantics for PADL. Again, a (Kripke) frame is a pair consisting of a set of worlds W and a meaning function $M_{\mathcal{F}}$ assigning a subset of W to each atom and a binary relation on W to each atomic action. We extend the meaning function $M_{\mathcal{F}}$ to all actions and formulas in a similar way. In addition, we need two additional explanations for the two new mixed operators $\triangleright$ and $\#$ respectively.

M8 $M_{\mathcal{F}}(\alpha \triangleright \beta) = W - \{u \mid \exists v, (u, v) \in M_{\mathcal{F}}(\alpha), (u, v) \notin M_{\mathcal{F}}(\beta)\}$,
M9 $M_{\mathcal{F}}(\#\phi) = \{(u, v) \mid v \in M_{\mathcal{F}}(\phi)\}$.

Therefore, the semantics of the operator $\bowtie$ can be induced from the above definitions:

$$M_{\mathcal{F}}(\alpha \bowtie \beta) = W - \{u \mid \exists v, (u, v) \in (M_{\mathcal{F}}(\alpha) \backslash M_{\mathcal{F}}(\beta)) \cup (M_{\mathcal{F}}(\beta) \backslash M_{\mathcal{F}}(\alpha))\}.$$

Similarly, the semantics of the operator U can be induced as well.

$$M_{\mathcal{F}}(\mathsf{U}) = \{(u, v) \mid v \in M_{\mathcal{F}}(\top)\} = W \times W.$$

Similarly, we say that a world u *satisfies* a formula ϕ in a frame $\mathcal{F}$, or that ϕ is *true* at u in $\mathcal{F}$, if $u \in M_{\mathcal{F}}(\phi)$, also written $\mathcal{F}, u \models \phi$.

In fact, we can rewrite the semantics of PADL equivalently as follows:

M1' $\mathcal{F}, u \not\models \bot$,
M2' $\mathcal{F}, u \models \phi \to \psi$ iff $\mathcal{F}, u \models \phi$ implies that $\mathcal{F}, u \models \psi$,
M3' $\mathcal{F}, u \models [\alpha]\phi$ iff $\forall v$, if $(u, v) \in M_{\mathcal{F}}(\alpha)$, then $\mathcal{F}, v \models \phi$,
M4' $(u, v) \in M_{\mathcal{F}}(\alpha; \beta)$ iff $\exists w$ such that $(u, w) \in M_{\mathcal{F}}(\alpha)$ and $(w, v) \in M_{\mathcal{F}}(\beta)$,
M5' $(u, v) \in M_{\mathcal{F}}(\alpha \cup \beta)$ iff $(u, v) \in M_{\mathcal{F}}(\alpha)$ or $(u, v) \in M_{\mathcal{F}}(\beta)$,
M6' $(u, v) \in M_{\mathcal{F}}(\alpha^*)$ iff $\exists n \geq 0$, and $\exists u_0, u_1 \ldots, u_n$ such that a) $u_0 = u$, b) $u_n = v$ and c) $(u_i, u_{i+1}) \in M_{\mathcal{F}}(\alpha)$ for all $i, (0 \leq i \leq n - 1)$,
M7' $(u, v) \in M_{\mathcal{F}}(\phi?)$ iff $u = v$ and $\mathcal{F}, u \models \phi$,
M8' $\mathcal{F}, u \models \alpha \triangleright \beta$ iff $\forall v, (u, v) \in M_{\mathcal{F}}(\alpha)$ implies that $(u, v) \in M_{\mathcal{F}}(\beta)$,
M9' $(u, v) \in M_{\mathcal{F}}(\#\phi)$ iff $v \in M_{\mathcal{F}}(\phi)$.

The notions of satisfiability, validity and logical consequence in PADL are defined in the same way as those for PDL. We say that ϕ is *valid* in $\mathcal{F}$, written $\mathcal{F} \models \phi$, if for all $u \in W, \mathcal{F}, u \models \phi$. We say that ϕ is *valid*, denoted by $\models \phi$, if for all frames $\mathcal{F}, \mathcal{F} \models \phi$. Given a set Σ of formulas and a frame $\mathcal{F}$, we write $\mathcal{F} \models \Sigma$ if for all $\phi \in \Sigma, \mathcal{F} \models \phi$. We say that a formula ψ is a *logical consequence* of a formula set Σ if for all frames $\mathcal{F}, \mathcal{F} \models \Sigma$ implies that $\mathcal{F} \models \psi$.

In fact, the two operators are not exactly new in the literature. The notion of subsumption relation and equivalence relation between actions (programs) are considerably studied in the area of theoretical computer science [1,2]. However, in most approaches, subsumption and equivalence relation are treated in the meta level but not in the logic language. Passay and Tinchev [2] discussed the idea of explicitly introducing subsumption relation into PDL. Also, they considered to introduce the universal relation into PDL. Note that the achievement operator and the universal relation can be defined interchangeably, i.e., $\#\phi \bowtie U; \phi?$ is valid in PADL. Therefore, mathematically, PADL actually has the same expressive power as Passay and Tinchev's CPDL^C without the constant set. Some researchers used the universal relation for representing the any action or arbitrary action of agents [4], whilst others believe that it is too strong. Giacomo and Lenzerini [3] used a pre-defined subset of the universal relation to represent the any action. Broersen [6] considered to use some accessible relations instead.

4 Properties

In this section, we study the properties of PADL. We first show that both the achievement operator and subsumption operator cannot be represented in standard PDL. Due to the space limitation, we only outline the proofs in this section.

Theorem 1. *The operator $\#$ cannot be represented in PDL.*

Proof. Consider the logical consequence relation $\Gamma \models \phi$. It can be represented in PADL as a formula $[U] \bigwedge \Gamma \to [U]\phi$, where $\bigwedge \Gamma$ is the conjunction of all formulas in Γ. In other words, ϕ is a logical consequence of Γ iff $[U] \bigwedge \Gamma \to [U]\phi$ is valid. On the other hand, logical consequence cannot be represented as a formula in PDL itself [1]. This shows that PDL with the achievement operator is strictly more expressive than PDL.

Theorem 2. *The operator $\triangleright$ cannot be represented in PDL.*

Proof. As shown in [1], by induction on the structure of formulas, we have that given a frame $\mathcal{F}$ and two worlds u and v in $\mathcal{F}$ that agree on all atoms, for all PDL formulas ϕ, $\mathcal{F}, u \models \phi$ iff $\mathcal{F}, v \models \phi$.

Now we construct a frame which contains three worlds u, v, w, where u and v agree the same on all atoms. Let $M_{\mathcal{F}}(\alpha) = \{(u, w), (v, w)\}$ and $M_{\mathcal{F}}(\beta) = \{(u, w)\}$. We have that $\mathcal{F}, u \models \alpha \triangleright \beta$ but $\mathcal{F}, v \not\models \alpha \triangleright \beta$. This shows that there does not exist a PDL formula equivalent to $\alpha \triangleright \beta$.

Theorem 1 and Theorem 2 show that PDL<PADL. That is, PADL is a strict extension of PDL. Interestingly, the following proposition shows that the PDL modal operator $[\alpha]\phi$ can be represented by the new achievement operator together with the subsumption operator.

Theorem 3. *Let α be an action and ϕ be a formula. We have that $\models [\alpha]\phi \leftrightarrow \alpha \triangleright \#\phi$.*

Proof. For any $M_{\mathcal{F}}$,
$M_{\mathcal{F}}(\alpha \triangleright \#\phi)$
$= W - \{u \mid \exists v, (u, v) \in M_{\mathcal{F}}(\alpha), (u, v) \notin M_{\mathcal{F}}(\#\phi)\}$
$= W - \{u \mid \exists v, (u, v) \in M_{\mathcal{F}}(\alpha), v \notin M_{\mathcal{F}}(\phi)\}$
$= \{u \mid \forall v \in W, \text{ if } (u, v) \in M_{\mathcal{F}}(\alpha), \text{ then } v \in M_{\mathcal{F}}(\phi)\}$
$= M_{\mathcal{F}}([\alpha]\phi).$

Thus, this assertion holds.

Theorem 3 coincides with the intuitions. The left side means that after executing α, ϕ should be true. On the other hand, the right side means that the action α always yields more consequences than the action $\#\phi$. That is, all the consequences of the action $\#\phi$ (e.g., ϕ itself) are also consequences of α. Clearly, this is same as saying that after executing α, ϕ is true.

Theorem 4. *Let ϕ be a formulas and α and β two actions. We have that $\models \alpha \triangleright \beta \rightarrow ([\alpha]\phi \rightarrow [\beta]\phi)$. Also, $\models \alpha \triangleright \beta \rightarrow (<\alpha>\phi \rightarrow <\beta>\phi)$.*

Proof. We prove the latter. The former can be proved similarly. For any frame $\mathcal{F} = \langle W, M_{\mathcal{F}} \rangle$ and world $u \in W$, if $\mathcal{F}, u \models \alpha \triangleright \beta$ and $\mathcal{F}, u \models <\alpha>\phi$, then there exists $v \in W$, such that $(u, v) \in M_{\mathcal{F}}(\alpha)$, and $v \in M_{\mathcal{F}}(\phi)$. Thus, $(u, v) \in M_{\mathcal{F}}(\beta)$. Therefore, $\mathcal{F}, u \models <\beta>\phi$. This shows that the latter assertion holds.

Theorem 4 shows that if an action α subsumes another action β, then for any formula ϕ, if ϕ must hold after executing α, then ϕ also must hold after executing β; if ϕ possibly holds after executing α, then ϕ also possibly holds after executing β. This shows that all consequences of α are also consequences of β. This coincides with our intuition on subsumption, as we discussed in Section 2.

Theorem 5. *Let ϕ be a formula. We have that $\models \#\phi \bowtie [\mathsf{U}]; \phi?$.*

Proof. For any $M_{\mathcal{F}}$,
$M_{\mathcal{F}}([\mathsf{U}]; \phi?)$
$= \{(u, v) \mid \exists w \in W, \text{ such that } (u, w) \in M_{\mathcal{F}}(\mathsf{U}) \text{ and } (w, v) \in M_{\mathcal{F}}(\phi?)\}$
$= \{(u, v) \mid \exists w \in W, \text{ such that } w = v \text{ and } v \in M_{\mathcal{F}}(\phi)\}$
$= \{(u, v) \mid v \in M_{\mathcal{F}}(\phi)\}$
$= M_{\mathcal{F}}(\#\phi)$.
Thus, this assertion holds holds.

Theorem 5 shows that the achievement operator can be defined by the universal operator. Together with the fact that the universal operator U can be defined as $\#\top$, we have that the two operators can indeed be defined from each other.

Theorem 6. *Let α, β and γ three actions. We have that $\models (\alpha \triangleright \beta) \wedge (\beta \triangleright \gamma) \rightarrow \alpha \triangleright \gamma$.*

Proof. For any frame $\mathcal{F}$ and a world u in it. Suppose that $\mathcal{F}, u \models \alpha \triangleright \beta$ and $\mathcal{F}, u \models \beta \triangleright \gamma$. Then, for any v, if $(u, v) \in M_{\mathcal{F}}(\alpha)$, then $(u, v) \in M_{\mathcal{F}}(\beta)$. Therefore, $(u, v) \in M_{\mathcal{F}}(\gamma)$. This shows that $\mathcal{F}, u \models \alpha \triangleright \gamma$ as well.

Theorem 6 shows that the subsume operator satisfies transitivity.

Theorem 7. *Let ϕ be a formula. We have that $\models \#\phi; \#\phi \bowtie \#\phi$. Also, $\models (\#\phi)^* \bowtie \#\phi$.*

Proof. We only prove the former. The latter holds similarly. For any $M_{\mathcal{F}}$, $M_{\mathcal{F}}(\#\phi; \#\phi)$
$= \{(u, v) \mid \exists w, (u, w) \in M_{\mathcal{F}}(\#\phi), (w, v) \in M_{\mathcal{F}}(\#\phi)\}$
$= \{(u, v) \mid \exists w, w \in M_{\mathcal{F}}(\phi), v \in M_{\mathcal{F}}(\phi)\}$
$= \{(u, v) \mid v \in M_{\mathcal{F}}(\phi)\}$
$= M_{\mathcal{F}}(\#\phi)$.
This shows that $\models \#\phi; \#\phi \bowtie \#\phi$.

Theorem 7 shows that the achievement operator is closed under iteration. That is, repeating to achieve a formula twice or many times is the same as to achieve it just once.

Theorem 8. *Let ϕ and ψ be two formulas. We have that $\models [\#\phi]\psi \rightarrow (\phi \rightarrow \psi)$.*

Proof. For any frame $\mathcal{F}$ and a world u in it. Suppose that $\mathcal{F}, u \models [\#\phi]\psi$. Then, for all v, if $\mathcal{F}, v \models \phi$, then $\mathcal{F}, v \models \psi$. Hence, if $\mathcal{F}, u \models \phi$, then $\mathcal{F}, u \models \models \psi$. This shows that $\mathcal{F}, u \models [\#\phi]\psi \rightarrow (\phi \rightarrow \psi)$.

Theorem 8 shows that if a formula ψ holds in company with achieving another formula ϕ, then ψ must be a logical consequence of ϕ.

Theorem 9. *Let ϕ and ψ be two formulas and α an action. We have that $\models [\#\phi]\psi \rightarrow [\alpha](\phi \rightarrow \psi)$.*

Proof. For any frame $\mathcal{F}$ and a world u in it. Suppose that $\mathcal{F}, u \models [\#\phi]\psi$. Then, for all v, if $\mathcal{F}, v \models \phi$, then $\mathcal{F}, v \models \psi$. Therefore, for any world v in $\mathcal{F}$, $\mathcal{F}, v \models \phi \rightarrow \psi$. Therefore, for any α, $\mathcal{F}, v \models \alpha(\phi \rightarrow \psi)$.

Theorem 9 shows that if a formula ψ holds in company with achieving another formula ϕ, then after executing any actions α, ψ always holds in company with ϕ i.e., if ϕ holds, then ψ holds as well.

There are of course many other valid formulas in PADL, other than the valid formulas in PDL. Due to the space limitation, we are not able to list more. Here, we propose an additional one.

Theorem 10. *Let ϕ and ψ be two formulas. We have that $\models (\phi \rightarrow \psi) \rightarrow [\#\phi] < \#\phi > \psi$.*

Proof. The proof of this assertion is similar to the above techniques of proving valid formulas in PADL, we leave it to the readers.

An important task is to axiomatize the logic PADL. However, this might not be an easy task. One reason comes from the studies of another extended version of PDL with intersection and negation on actions, which can also represent the subsumption operator. Researcher have found that it is very difficult to axiomatize this logic.

Is PADL expressive enough for capturing the essence of abstract behavior. Here, we argue that PADL is indeed powerful for this purpose. Firstly, PADL makes it possible to treat both formulas and actions as the same objects by the achievement operator. Secondly, using the subsumption operator, one can reason about the abilities of actions explicitly. Hence, the two new operators themselves and the combination of them, together with the original operators in PDL, offer adequate expressive power for modeling abstract behavior for intelligent agents. In fact, it can be shown that some complex action formalisms, such as conditional planning and HTN planning, can be represented in PADL. Nevertheless, there are other features that cannot be modeled by PADL, for instance, temporal operators and joint actions in multi-agent environment.

The term abstract behavior (or actions, plans) has been used elsewhere in the AI literature. However, the basic ideas of these work are different from PADL in nature. In HTN planning [7], abstract plan is introduced to denote those plans on a higher abstract level, whilst concrete plan is those plans that can be really executed. A partial order relation among plans is used to represent the refinement relation (i.e., subsumption relation) among plans in the meta level. In 3APL agent language [8], abstract plan has to be transformed into basic actions according to some rules. A set AP is used to denote the set of all abstract plans. Schmidt and Tishkovsky [9] used abstract actions to represent those actions that can be performed by any agent, whilst concrete actions can only performed by some particular agents.

5   Conclusion

The main contributions of this paper are summarized as follows. Firstly, we argued that abstract behavior plays an essential role for intelligent agent decision making under incomplete and uncertain environments. Secondly, we extended PDL into PADL for modeling abstract behavior by adding two mixed operators # (for achievement) and ▷ (for subsumption). The action $\#\phi$ is an abstract action meaning that to find a specific plan to achieve the goal ϕ; the formula $\alpha \rhd \beta$ states that the action α always yields more consequences that the action β in any cases. We defined the semantics for them and showed that both operators are strict extensions of PDL. We also discussed other important related properties of PADL.

For future work, as we mentioned, one important task is to find a sound and complete axiom system for PADL. Also, another work worth pursuing is to further extend PADL for modeling abstract behavior in multi-agent systems. Finally, it is crucial to apply PADL to real agent programming languages, e.g. 3APL [8]. We leave these work to our future investigations.

References

1. Harel, D., Kozen, D., Tiuryn, J.: Dynamic Logic. MIT Press, Cambridge (2000)
2. Passay, S., Tinchev, T.: An essay in combinatory dynamic logic. Information and Computation 93(2), 263–332 (1991)
3. Giacomo, G., Lenzerini, M.: PDL-based framework for reasoning about actions. In: AI*IA 1995, pp. 103–114 (1995)
4. Prendinger, H., Schurz, G.: Reasoning about action and change. a dynamic logic approach. Journal of Logic, Language and Information 5(2), 209–245 (1996)
5. Meyer, J.J.C.: Dynamic logic for reasoning about actions and agents, pp. 281–311 (2000)
6. Broersen, J.: Relativized action complement for dynamic logics. In: Advances in Modal Logic, pp. 51–70 (2002)
7. Erol, K., Hendler, J., Nau, D.S.: Semantics for hierarchical task-network planning. Technical report (1994)
8. Dastani, M., van, R.B., Dignum, F., Meyer, J.J.C.: A programming language for cognitive agents goal directed 3APL. In: Dastani, M.M., Dix, J., El Fallah-Seghrouchni, A. (eds.) PRO-MAS 2003. LNCS (LNAI), vol. 3067, pp. 111–130. Springer, Heidelberg (2004)
9. Schmidt, R.A., Tishkovsky, D.: Multi-agent logics of dynamic belief and knowledge. In: Flesca, S., Greco, S., Leone, N., Ianni, G. (eds.) JELIA 2002. LNCS (LNAI), vol. 2424, pp. 38–49. Springer, Heidelberg (2002)

Restoring Punctuation and Casing in English Text

Timothy Baldwin[1,2] and Manuel Paul Anil Kumar Joseph[1]

[1] Department of Computer Science and Software Engineering
University of Melbourne, VIC 3010, Australia
[2] NICTA Victoria Laboratories
University of Melbourne, VIC 3010, Australia
tb@ldwin.net, mjoseph@students.csse.unimelb.edu.au

Abstract. This paper explores the use of machine learning techniques to restore punctuation and case in English text, as part of which it investigates the co-dependence of case information and punctuation. We achieve an overall F-score of .619 for the task using a variety of lexical and contextual features, and iterative retagging.

1 Introduction

While digitised text data is growing exponentially in volume, the majority is of low quality. Such text is often produced automatically (e.g. via speech recognition or optical character recognition [OCR]) or in a hurry (e.g. instant messaging or web user forum data), and hence contains noise. Normalisation of case and punctuation in such text can greatly improve its consistency and accessibility to natural language processing methods [10].

This research is focused on the restoration of case and punctuation. To illustrate the task, given the following input text:

(1) ... club course near mount fuji marnie mcguire of new zealand winner of the mitsukoshi cup ladies in april had a ...

we would hope to restore it to:

(1') ... club course near Mount Fuji. Marnie McGuire of New Zealand, winner of the Mitsukoshi Cup Ladies in April, had a ...

There are some notable effects taking place, such as *mcguire*, where the first and third letters are capitalised, and *april*, which is both capitalised and has a comma attached to it. While *cup* and *ladies* would not standardly be capitalised, they require capitalisation in this context because of their occurrence in the proper name *Mitsukoshi Cup Ladies*. Similarly, words such as *prime* and *minister*, and *new* and *york*, need to be capitalised primarily when they co-occur. The above example illustrates the complexities involved in the task of case and punctuation restoration.

A. Nicholson and X. Li (Eds.): AI 2009, LNAI 5866, pp. 547–556, 2009.

This research has applications over the output of speech dictation systems and automatic speech recognition (ASR) systems, which tend to have difficulties predicting where to insert punctuation and sentence boundaries, and also over noisy web data (e.g. as found in web user forums), where case and punctuation are often haphazard [2].

In the process of exploring the complexity of this task and proposing a restoration method, we have developed a benchmark dataset for research on case restoration and punctuation restoration.

2   Related Work

Case and punctuation restoration is a relatively unexplored field. CYBERPUNC [2] is a lightweight method for automatic insertion of intra-sentence punctuation into text. It uses a simple hidden Markov model with trigram probabilities to model the comma restoration problem, restoring the punctuation of 54% of the sentences correctly. [15] tackle the problem of comma restoration using syntactic information, and improve on this to achieve an accuracy of 58%. In both of these cases, sentence boundaries are assumed to be given. In our case, we assume no punctuation whatsoever, including sentence boundaries, and hence direct comparison with our work is not possible.

The above-mentioned methods deal with punctuation restoration at the sentence level, i.e., the input to both systems is a single sentence. For instance, the following input instances:

(2) the golf tournament was at the country club course near mount fuji
(3) marnie mcguire of new zealand winner of the mitsukoshi cup ladies in april
     had a 72 for 212

would be converted to:

(2') The golf tournament was at the country club course near Mount Fuji.
(3') Marnie Mcguire of New Zealand, winner of the Mitsukoshi Cup Ladies in
      April, had a 72 for 212.

This simplifies the task significantly, as the sentence boundaries are explicitly specified. This is not the case in our system, where the input is a stream of words, thus requiring the system to detect sentence boundaries (explicitly or implicitly). Hence it is not possible to apply these systems over our data or compare the results directly.

In ASR systems, researchers have made use of prosodic information, disfluencies and overlapping speech to predict punctuation, which they have then supplemented with language models [15].

[10] look into the task of truecasing, or case restoration of text. They propose a language model-based truecaser, which achieves a *word* accuracy of around 98% on news articles. The high accuracy reported here can be used as an indication that the case restoration task is simpler in comparison to punctuation restoration. Note that a direct comparison of the accuracy of punctuation methods

mentioned above and the truecasing task is misleading: the accuracy reported for the punctuation tasks is at the *sentence* level, whereas, in case of the truecasing task it is at the *word* level.

3 Task Description

To generate our dataset, we randomly selected 100 articles each (roughly 65K words) from the AP Newswire (APW) and New York Times (NYT) sections of the English Gigaword Corpus, as training, development and test data. We then tokenised the data, before stripping off all punctuation and converting all the text to lower case. The only punctuation that was left in the text was hyphenation, apostrophes and in-word full stops (e.g. *U.S* and *trade-off*).[1] Each of these Table 1 shows the number of words in each of the datasets.

Each token is treated as a single instance and annotated with a class indicating the punctuation and case restoration that needs to take place, in the form of a capitalisation class, indicating the character indices requiring capitalisation, and a list of zero or more punctuation labels, each representing a punctuation mark to be appended to the end of the word. For example, CAP1+FULLSTOP+COMMA applied to *corp* would restore it to *Corp.,*. The class ALLCAPS is used to represent that all letters in the word need to be converted to uppercase, irrespective of the character length of the word.

Unsurprisingly, the distribution of classes in the data is heavily skewed as detailed in Table 2, with the vast majority of instances belonging to the NOCHANGE class.

In addition to the token instances, we fashioned a set of base features for each word as part of the data release. The base features consist of:

1. the lemma of the word, based on MORPH [13];
2. Penn part-of-speech (POS) tags [12] based on FNTBL 1.0 [14];
3. CLAWS7 POS tags [16] based on the RASP tagger [4]; and
4. CoNLL-style chunk tags based on FNTBL 1.0.

Table 1. Size of the training, development and test datasets

Dataset	Number of tokens
Training	66371
Development	65904
Test	64072

---

[1] The decision to leave in-word full stops and hyphens in the data is potentially controversial. In future work, we intend to explore their impact on classification performance by experimenting with data which contains literally no punctuation information.

Table 2. The top-8 classes in the training data, with a description of the corresponding change to the token

Class	Description	%	Example
NOCHANGE	no change	75.8%	*really → really*
CAP1	capitalise first letter	12.4%	*thursday → Thursday*
NOCHANGE+COMMA	append comma	4.1%	*years → years,*
NOCHANGE+FULLSTOP	append full stop	3.8%	*settlement → settlement.*
CAP1+COMMA	capitalise first letter and append comma	1.7%	*thursday → Thursday,*
CAP1+FULLSTOP	capitalise first letter and append full stop	0.9%	*thursday → Thursday.*
ALLCAPS	capitalise all letters	0.7%	*tv → TV*
ALLCAPS+FULLSTOP	capitalise all letters and append a full stop	0.2%	*u.s → U.S.*

We generate all of these features over the case-less, punctuation-less text, meaning that we don't have access to sentence boundaries in our data. For both POS taggers and the full text chunker [1], therefore, we process 5-token sequences, generated by running a sliding window over the text. For a given token, therefore, 5 separate tags are generated for each preprocessor, at 5 discrete positions in the sliding window; all 5 tags are included as features.[2] The total number of base features is thus 16 per instance.

This dataset is available for download at `http://www.csse.unimelb.edu.au/research/lt/resources/casepunct/`.

4   Feature Engineering

While the data release includes a rich array of features, we chose to optimise classifier performance via feature engineering, modifying the feature description in various ways. In all cases, feature engineering was performed over the development data, holding out the test data for final evaluation.

4.1   Lemma and POS/Chunk Tag Normalisation

First, we converted all 4-digit numbers (most commonly years, e.g. *2008*) into a single lemma, and all other sequences of all digits into a second number lemma. Similarly, we converted all month and day of the week references into separate lemmas. The primary reason for this was the high frequency of date strings such as *14 Jan, 2007* which require comma restoration; in this case, the string would be lemmatised into the three tokens *non-4digit-num-ersatz month-ersatz 4digit-num-ersatz*, respectively. In the case of these filters successfully matching with

---

[2] For tokens at the very start or end of a dataset which do not feature in all 5 positions, we populate any missing features with the value _.

the wordform, the resultant lemma substituted for that in the original dataset. For example, the lemma for *2007* is *2007* in the original dataset, but this would be replaced with *4digit-num-ersatz*. As such, this processing doesn't generate any new features, but simply modifies the existing lemma feature column.

Rather than include all 5 POS and chunk tags for a given token, we select the POS and chunk tags which are generated when the token is at the left extremity in the 5-word sliding window. That is, we remove all but the leftmost POS and chunk tags from the features provided in the dataset. Surprisingly, this simple strategy of taking the first out of the 5 POS and chunk tags provided in the dataset was superior to a range of more motivated disambiguation strategies trialled, and also superior to preserving all 5 tags.

4.2 Lexical Features

We capture hyphens, apostrophes and in-word full stops by way of a vector of three Boolean lexical features per instance.

In an attempt to capture the large number of acronyms (e.g. *dvd*) and proper nouns in the data, we fashioned a list of acronyms from GCIDE and WordNet 2.1 [7]. We used the British National Corpus [5] to determine which capitalisation form had the highest frequency for a given lemma. For lemmas where the word form with the highest prior involves capitalisation, we encode the capitalisation schema (e.g. ALLCAPS+FULLSTOP for *u.s.a*) via a fixed set of Boolean features, one each for the different schemas. We additionally encode the conditional probabilities for a given lemma being lower case, all caps, having its first letter capitalised, or having the first and third letters capitalised (e.g. *mcarthur*); these were discretised into three values using the unsupervised equal frequency algorithm algorithm as implemented in NLTK [3].

4.3 Context Information

Punctuation and capitalisation are very dependent on context. For example, *prime* is most readily capitalised when to the immediate left of *minister*. To capture context, we include the lemma, Penn POS tag, CLAWS7 POS tag and chunk tag (disambiguated as described in Section 4.1) for the immediately preceding and proceeding words, for each target word. That is, we copy across a sub-vector from the preceding and proceeding words. We also include: (1) bigrams of the target word and proceeding word, in the form of each of word, Penn POS tag and CLAWS7 POS tag bigrams; and (2) trigrams of the Penn and CLAWS7 POS tags of the preceding, target and proceeding words; and (3) trigrams of the CLAWS7 tags of the target word and two preceding words.

5 Classifier Architecture

We experimented with a range of classifier architectures, decomposing the task into different sub-tasks and combining the results differently.

5.1 Class Decomposition

We first experimented with a 3-way class decomposition, performing dedicated classification for each of: (1) acronym detection, (2) case restoration, and (3) punctuation restoration. We then take the predictions of the three classifiers for a given instance, and combine them into a fully-specified class. Note that acronym detection cross-cuts both punctuation and case restoration, but is a well-defined standalone sub-task.

For the acronym detection task, we focus exclusively on the three classes of CAP1+FULLSTOP, ALLCAPS+FULLSTOP and NOCHANGE (the three most frequent classes). This was achieved through simple class translation over the training/development instances, by stripping off any extra classes from the original data to form a modified class set.

To perform case restoration, we again strip all but the case information from the class labels. There will inevitably be some overlap with the acronym classifier, so we exclude ALLCAPS+FULLSTOP instances from classification with this classifier (i.e. transform all ALLCAPS+FULLSTOP instances into NOCHANGE instances).

Finally, for the punctuation restoration sub-task, we strip off any case information from class labels, leaving only the punctuation-related labels.

To combine the predictions for the classifiers associated with each of the three sub-tasks, we tested two approaches: (1) using heuristics to directly combine the class labels of the three classifiers, and (2) performing meta-classification across the classifier outputs. In the heuristic approach, the class label produced by the abbreviation sub-task overrides the predictions of the other two classifiers if both predict that case restoration is necessary. For example, if ALLCAPS+FULLSTOP was predicted by the abbreviation classifier and CAP1 was predicted by the case restoration classifier, we would accept ALLCAPS+FULLSTOP as the final case prediction. If the punctuation classifier then predicted COMMA, the final class would be ALLCAPS+FULLSTOP+COMMA. If, on the other hand, the abbreviation classifier predicted NOCHANGE, the prediction from the case restoration classifier would be accepted.

In the meta-classification approach, the three classifiers are run over both the test and development datasets, and the outputs over the development data are used to train a meta-classifier. The outputs from the three classifiers for each test instance are fed into the meta-classifier to generate the final class.

5.2 Retagging

As stated in Section 3, the base features were generated using a sliding window approach (without case or punctuation information). We expect the performance of the preprocessors to improve with correct case and punctuation information, and sentence-tokenised inputs. We thus experiment with a feedback mechanism, where we iteratively: (a) classify the instances, and restore the (training, development and test) text on the basis of the predicted classes; and (b) sentence tokenise, re-tag, lemmatise and chunk the data, and then feed the updated tags back into the data as new feature values. As our sentence tokeniser, we used the

NLTK implementation of PUNKT [9]; the lemma, POS and chunk tags are generated in the same way as mentioned in Section 3. We stop the iterative process when the relative change in tags from one iteration to the next becomes sufficiently small (gauged as a proportion of test instances which undergo change in their predicted class).

6 Results

All evaluation was carried out using token-level accuracy, precision, recall and F-score ($\beta = 1$).

As our baseline classifier, we ran the TiMBL implementation of the IB1 learner [6] over the base set of features (which actually outperformed an SVM learner over the same features). All other classifiers are based a multi-class support vector machine, implemented as part of the BSVM toolkit [8].[3] We used a linear kernel in all experiments described in this paper, because we found that it performed better than the Radial Basis Function (RBF) and polynomial kernels.

The results for all the experiments are presented in Table 3.

First, we can see that the strategy of using only the first POS and chunk tag improves accuracy and precision, but actually leads to a drop in recall and F-score over the baseline. The addition of lexical features (Section 4.2) appreciably improved results in all cases, and had the greatest impact of any one of the feature sets described in Section 4. The incorporation of all the extra features brought precision down slightly, but improved recall and led to an overall improvement in F-score.

Using the same set of expanded features with the 3-way task decomposition and either direct heuristic combination or meta-classification, actually led to a slight drop in F-score in both cases relative to the monolithic classification strategy. The meta-classifier generated the highest precision and equal-best accuracy of all the classifiers using only automatically-generated features, but precision dropped considerably.

The retagging method, in combination with the monolithic classifier architecture, resulted in the best accuracy, recall and F-score of all automatic methods tried. The indicated F-score is based on three iterations, as the number of changes dropped exponentially across iterations to only 335 at this point. Error analysis of this final classifier revealed that the performance over case restoration actually deteriorated (to below-baseline levels for the class ALLCAPS+FULLSTOP, e.g.), but the performance over punctuation restoration picks up considerably. Results for the top-10 classes (based on F-score[4]) are presented in Table 4.

To investigate the potential for the retagging method to improve results, we separately ran the lemmatiser, taggers and chunker over the original text (with correct case and punctuation information, and sentence tokenisation), and re-ran the classifier. This caused the F-score to jump up to .740, suggesting that this

---

[3] http://mloss.org/software/view/62/
[4] Excluding the NOCHANGE class.

Table 3. Classification results (*italicised* numbers indicate gold-standard data used; **bold** numbers are the best achieved without gold-standard data)

Classifier Description	Accuracy	Precision	Recall	F-score
Baseline (IB1)	.784	.516	.301	.381
Base features, with only first tag	.790	.571	.282	.378
+ extra lexical features	.837	.637	.534	.581
+ all extra features	.839	.620	.604	.611
Heuristic combination	.834	.596	.612	.604
Meta-classifier	**.840**	**.639**	.554	.594
Iterative retagging	**.840**	.615	**.622**	**.619**
Retagging (based on original text)	*.885*	*.715*	*.766*	*.740*
With gold-standard punct labels	*.926*	*.887*	*.793*	*.837*
With gold-standard case/abbrev labels	*.912*	*.793*	*.813*	*.803*

Table 4. Best-10 performing classes for the iterative retagger (ranked based on F-score)

Class	Accuracy	Precision	Recall	F-score
ALLCAPS+FULLSTOP	.657	.787	.799	.793
ALLCAPS	.561	.853	.621	.719
CAP1	.523	.719	.658	.687
CAP1+FULLSTOP	.312	.607	.391	.476
CAP1+COMMA	.276	.523	.369	.433
NOCHANGE+FULLSTOP	.251	.450	.361	.401
NOCHANGE+COMMA	.191	.351	.294	.320
CAP1-3	.143	.750	.150	.250
CAP1+FULLSTOP+COMMA	.100	.667	.105	.182
CAP1+COLON	.082	1.000	.082	.151

approach could lead to much greater improvement given higher performance of the base classifier.

Finally, we investigated the co-dependency of the case and punctuation restoration tasks in the context of the meta-classification approach, by combining gold-standard case labels with automatically-generated punctuation labels, and vice versa. This resulted in the final two lines of Table 3, which clearly show that if we can get one of the two tasks correct, the other becomes considerably easier.

7   Future Work

Our research focussed on a small sub-set of punctuation. Punctuation such as question marks and colons was not explored here, and features targeting these could be considered to further improve the performance of the classifier.

Another area for future investigation is instance selection [11]. The distribution of instances over the set of classes is skewed heavily in favour of the

NOCHANGE class. Instance filtering could have helped in alleviating this bias, forcing the classifier to look at the other classes. This could help especially when looking at the sub-tasks, where the number of NOCHANGE instances increased because of the stripping off of the case or punctuation class information.

The original motivation for this research was in applications such as ASR and OCR, but all of our results are based on the artificially-generated dataset, which lacks case and punctuation but is otherwise clean. We are keen to investigate the applicability of the proposed method to noisy outputs from ASR and OCR in more realistic settings.

8   Conclusion

We have explored the task of case and punctuation restoration over English text. First, we established a benchmark dataset for the task, complete with a base feature set, and then we proposed an expanded feature set, and a range of classifier architectures based on decomposition of the overall task. The best results were achieved with the expanded feature set, a monolithic classifier architecture and iterative retagging of the text.

Acknowledgements

NICTA is funded by the Australian Government as represented by the Department of Broadband, Communications and the Digital Economy and the Australian Research Council through the ICT Centre of Excellence program.

References

1. Abney, S.P.: Parsing by chunks. In: Berwick, R.C., Abney, S.P., Tenny, C. (eds.) Principle-Based Parsing: Computation and Psycholinguistics, pp. 257–278. Kluwer, Dordrecht (1991)
2. Beeferman, D., Berger, A., Lafferty, J.: Cyberpunc: A lightweight punctuation annotation system for speech. In: Proceedings of 1998 IEEE International Conference on Acoustics, Speech, and Signal Processing (ICASSP 1998), Seattle, USA (1998)
3. Bird, S., Klein, E., Loper, E.: Natural Language Processing with Python — Analyzing Text with the Natural Language Toolkit. O'Reilly Media, Sebastopol (2009)
4. Briscoe, E., Carroll, J., Watson, R.: The second release of the RASP system. In: Proceedings of the COLING/ACL 2006 Interactive Poster System, Sydney, Australia, pp. 77–80 (2006)
5. Burnard, L.: User Reference Guide for the British National Corpus. Technical report, Oxford University Computing Services (2000)
6. Daelemans, W., Zavrel, J., van der Sloot, K., van den Bosch, A.: TiMBL: Tilburg Memory Based Learner, version 5.1, Reference Guide. ILK Technical Report 04-02 (2004)
7. Fellbaum, C.: Wordnet: An Electronic Lexical Database. MIT Press, Cambridge (1998)

8. Hsu, C.-W., Chang, C.-C., Lin, C.-J.: A practical guide to support vector classification. Technical report, Department of Computer Science National Taiwan University (2008)
9. Kiss, T., Strunk, J.: Unsupervised multilingual sentence boundary detection. Computational Linguistics 32(4), 485–525 (2006)
10. Lita, L.V., Ittycheriah, A., Roukos, S., Kambhatla, N.: tRuEcasIng. In: Proceedings of the 41st Annual Meeting of the Association for Computational Linguistics, Sapporo, Japan, pp. 152–159 (2003)
11. Liu, H., Motoda, H.: Feature Extraction, Construction and Selection: A Data Mining Perspective. Kluwer Academic Publishers, Dordrecht (1988)
12. Marcus, M.P., Santorini, B., Marcinkiewicz, M.A.: Building a large annotated corpus of English: the Penn treebank. Computational Linguistics 19(2), 313–330 (1993)
13. Minnen, G., Carroll, J., Pearce, D.: Applied morphological processing of English. Natural Language Engineering 7(3), 207–223 (2001)
14. Ngai, G., Florian, R.: Transformation-based learning in the fast lane. In: Proceedings of the 2nd Annual Meeting of the North American Chapter of Association for Computational Linguistics (NAACL 2001), Pittsburgh, USA, pp. 40–47 (2001)
15. Shieber, S.M., Tao, X.: Comma restoration using constituency information. In: Proceedings of the 3rd International Conference on Human Language Technology Research and 4th Annual Meeting of the NAACL (HLT-NAACL 2003), Edmonton, Canada, pp. 142–148 (2003)
16. Wynne, M.: A post-editor's guide to CLAWS7 tagging. UCREL University of Lancaster, Lancaster, England (1996)

A Novel Connectionist Network for Solving Long Time-Lag Prediction Tasks

Keith Johnson and Cara MacNish

School of Computer Science and Software Engineering
The University of Western Australia, Australia
{keith,cara}@csse.uwa.edu.au

Abstract. Traditional Recurrent Neural Networks (RNNs) perform poorly on learning tasks involving long time-lag dependencies. More recent approaches such as LSTM and its variants significantly improve on RNNs ability to learn this type of problem. We present an alternative approach to encoding temporal dependencies that associates temporal features with nodes rather than state values, where the nodes explicitly encode dependencies over variable time delays. We show promising results comparing the network's performance to LSTM variants on an extended Reber grammar task.

1 Introduction

An intelligent agent must be able to make sense of temporally distributed observations, learn to recognise temporally correlated occurrences and patterns, and make use of these for future predictions. Connectionist approaches to this problem can be roughly divided into two camps.

In Time-delay Neural Networks (TDNN) [1], each feature (observation) is supplied to the network not only in the time step in which it occurs, but for a fixed number of additional time steps. The network therefore has access to a finite history of observations on which to base classifications or predictions. While this approach has proved successful in a range of practical applications, it is limited in its generality. It cannot solve problems with arbitrarily long time delays, and there is a trade off between the length of history supplied, and the number of input nodes that must be incorporated to access each feature at each time point. This in turn has a bearing on training requirements.

A more general and intuitively appealing approach makes use of recurrent connections to create memory elements within the network. In these approaches the activation state of the recurrent nodes encodes a history of the observations presented to the network. Conceptually this history may be arbitrarily long. In practice, however, traditional recurrent neural networks (RNNs) have performed poorly on learning tasks involving long time-lag dependencies (more than about 5 to 10 time steps), due to the tendency of back-propagated error flows to vanish or explode [2,3]. This has led to the development of a range of extended recurrent networks designed to deal with long time-lags. We discuss some of the key approaches and their limitations in Section 2.

A. Nicholson and X. Li (Eds.): AI 2009, LNAI 5866, pp. 557–566, 2009.

In this paper we propose an alternative approach to encoding temporal dependencies, where the responsibility of storing temporal patterns is shifted from an implicit representation within the state space of the recurrent network, to the explicit responsibility of nodes within the network. That is, nodes themselves encode dependencies over variable time delays. At first sight, this would appear to suggest an explosion in the number of nodes, not unlike the TDNN approach. The key to our approach, however, is the recognition that, while there are infinitely many *potential* temporal relationships, relatively few of those will turn out to play a significant role in any problem of interest.

This thinking is not new to artificial intelligence and machine learning, which rely heavily on abstracting away the infinite detail of a problem domain to identify key features on which problem solving or classification might depend. Humans do this naturally as a way of dealing with complexity. It is also central to techniques such as principal component analysis, discriminant analysis and information gain, as well as constructive connectionist approaches. Our work seeks to identify and exploit the important *temporal features* in a problem.

Since our approach associates temporal features with nodes rather than state values, it is necessary that some of the dynamicism of the network is also shifted from activation values to the network structure and connections. To achieve this we employ a novel, highly dynamic, Hebbian-style network, where new nodes are created to reflect temporally correlated occurrences, evidence for these relationships is accumulated, and less useful nodes are pruned. The resulting *Constructive Hypothesise and Test Network (CHTN)* is described in Section 3.

In order to compare our approach with existing networks we evaluate it in Section 4 on an extended Reber grammar task that has been extensively used in the literature. We show that, unlike recurrent approaches targeted at long time-lag problems, our approach is able to consistently find exact solutions, with significantly less training. At the same time, we provide evidence to show that the networks' size remains manageable. The paper is concluded in Section 5.

2 Background

2.1 The Embedded Reber Grammar Problem

A benchmark long time-lag problem used for comparison in the literature is the embedded Reber grammar (ERG) [3,4], illustrated in Figure 1. This is a popular choice because the grammar generates strings with time dependencies that are just beyond the capabilities of traditional networks. The ERG is an extension of the far simpler standard Reber grammar (SRG) which is not a long time-lag problem and can be learned successfully by most RNNs.

The standard Reber grammar (SRG) and its derivatives are a set of rules describing sequences of symbols. Sequences are generated by starting at the left-most node and traversing the directed edges until the right-most node is reached. At each traversal, the symbol associated with the edge is appended to the sequence. Nondeterministic choices are made with equal probability.

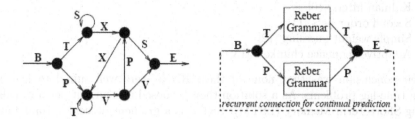

Fig. 1. Left: Transition diagram for SRG. Right: Transition diagram for ERG and CERG. The dashed line indicates the recurrent connection that distinguishes the CERG.

The embedded Reber grammar (ERG) extends the SRG and is a significantly harder problem to learn. Two types of string can be generated by the ERG; BT<Reber string>TE and BP<Reber string>PE. To learn this grammar, a system must remember the first T or P to correctly predict the penultimate symbol. Furthermore, the system must remember this symbol over the duration of the SRG, which is variable in length.

RNNs that are applied to learning the SRG and ERG use an external "teacher", which knows *a priori* about the ends of each string, to reset the network's internal state at the end of each sequence. The continual embedded Reber grammar (CERG) [5] is a more stringent form of the ERG that does not use an external teacher. This is of potential importance in cases where the start and end of repetitive patterns (in this case sequences) within a body of data is not available and must be learned by the network. The CERG generates a contiguous stream of ERG sequences, where the boundaries between each ERG are not known to the learning system, making it a more difficult problem to solve.

2.2 Previous Approaches

Tradional neural networks, such as TDNNs [1], and RNNs trained using Back-Propagation Through Time [6] are well suited to learning problems with short time-lag dependencies. The SRG is one such example, as it only requires the network to remember information for up to one time step. The same networks however, struggle to learn problems with time-lags over about 5-10 time steps. When applied to the ERG, a long time-lag problem, these networks fail to perform well because they are required to remember information over arbitrary time periods, often greater than 10 time steps.

Hochreiter and Schmidhuber in [3] provide a review of a large variety of existing RNNs, including the following types of networks:

- Gradient-descent variants [7,4,6]
- Time delays [1]
- Time constants [8]
- Ring's approach [9]
- Bengio *et al*'s approaches [2]

- Kalman filters [10]
- Second order nets [11]
- Simple weight guessing [12]
- Adaptive sequence chunkers [13]

Their conclusion was that none of these RNNs were well suited to learning long time-lag problems. As a solution they proposed a recurrent network called Long Short-Term Memory (LSTM). LSTM is a gradient-descent variant but is designed to overcome the difficulties faced by existing variants with respect to learning long time-lag patterns.

Gradient-descent approaches use back-propagation of error to adjust network parameters such that the network output more closely matches the desired output. The error flow, however, tends to either vanish or explode quickly over a short number of steps [2,3]. LSTM overcomes this problem by using input and output gates to control the error flow. Each input and output gate learns to open and close access to the error flow when appropriate, allowing it to bridge arbitrarily long time-lags. Experimental results on long time-lag problems showed that the network was the first to perform well on problems of this type.

LSTM did not perform well however on the more difficult variant of the Reber grammar task. Gers *et al* [5] applied LSTM to the CERG and their results showed that LSTM performed poorly on this problem, unable to find a perfect solution in any trial. They reported that this was because without external resets, a continual input stream eventually may cause the internal values of the cells to grow without bound, even if the repetitive nature of the problem suggests they should be reset occasionally. Their solution was *extended LSTM* with forget gates (ELSTM). The forget gates allowed the network to learn when to reset its own internal values. As a result, the network offered a significant improvement over the LSTM for learning the CERG.

While the LSTM and ELSTM provide promising approaches to long time-lag problems, previously unsolvable by tradition RNNs, these networks still cannot achieve 100% success on the CERG problem and training times are long. The following section describes our alternative approach to encoding temporal dependencies. Experimental results show that this approach is successful in finding solutions to the CERG in every trial with relatively few training examples.

3   The CHTN Architecture

The goal of the CHTN network is to recognize temporal patterns in a sequence of observations (or state descriptions), and use these to predict future observations. More specifically, each observation in a discrete-time sequence can be represented as a feature vector, and the goal of the network is to predict the feature vector at the subsequent time step.

The network is constructive: it begins with input nodes only, one for each feature in the input sequence. It then dynamically builds *connections* and hypothesis *nodes* which attempt to encode the temporal relationships in the sequence. The survival of the nodes depends on how useful they turn out to be.

In the following we say that an input node is *activated* when the corresponding feature is seen in the current observation.

3.1   Connections

Connections are unidirectional links from a source node to a destination node. They play two complementary roles. The first is to *learn* about correlated activity between nodes from past observations; the second is to *predict* future activity.

Correlation Learning

Hebbian learning is nature's way of recording correlated activity by strengthening neuronal connections. It is often paraphrased as "neurons that fire together, wire together". This notion is modelled in our network by associating a simple frequency-based probability with connections.

Each connection in our network has two properties associated with it: a signal *delay*, and a signal *spread*. When the source node of a connection is activated, it emits a "learning" signal. The signal reaches the destination node after the specified delay, and then remains active for the duration of the specified spread (the "active period"). If the destination node is activated at any time during that duration, they're considered to have fired together and the connection is strengthened (positive correlation), otherwise it is weakened (negative correlation). The observed probability of the connection's pattern is then simply:

$$observed\ probability = \frac{\#positive\ correlations}{\#positive\ correlations\ +\ \#negative\ correlations}$$

This observed probability is subsequently used in making predictions.

Note that the observed probability is not time-weighted towards more recent events. This encourages stability. In order to learn patterns that change over time the system must discover appropriate features that correlate with the changes over time.

In the implementation of the architecture used for the experiments in this paper, only two types of connection instances are used for learning and prediction. Both have a one step time delay. The first has a spread of one time instant, while the second has an infinite spread. We will refer to these as *instantaneous* and *persistent* connections respectively.

Making Predictions

Connections predict future activity on the basis of past observations. When the source node of a connection is activated, it predicts that the destination node will become active during the connection's active period at the probability determined from previous observations. If the active period includes more than one point in discrete time, the probability at each point is the measured probability divided by the number of points within the active period. (Persistent connections therefore do not contribute directly to predictions. We will see their role later in node construction.)

Before a connection is allowed to make predictions about node activity, the number of observations on which its prediction is based must reach a minimum threshold or *maturity*. This is a parameter of the network that acts much like a traditional learning rate, trading off rate of adaptation against stability.

A destination node may receive multiple prediction signals from different sources at any one time, and must resolve these to give a single prediction for the node's activity. This is achieved by selecting the prediction signal with the *largest positive or negative activity correlation.*

Forming New Connections

Connections are freely formed within the network. Whenever a node becomes active, connections are formed with any input node that is active or incorrectly predicted within the active durations specified above. More specifically, if A_t is the set of all nodes active at time t, and E_t is the set of all nodes incorrectly predicted at time t, then:

- Instantaneous connections are formed between all nodes A_t and $A_{t+1} \cup E_{t+1}$; that is, between any active node and active or error node that form a contigous sequence.
- Persistent connections are formed between all nodes A_t and $A_n \cup E_n$, where $n >= t+1$. Because most nodes become active at some point, this will result in a persistent connection between every node combination.

3.2   Nodes

While connections allow the network to learn correlations and make predictions based on simple patterns of temporal activity between input nodes, more complex patterns require a way of compounding these relationships. This is achieved by constructing additional "intermediary" nodes that represent compound features. These nodes can be regarded as hypotheses for *useful* temporal relationships or features. To maintain a workable network size, the hypothesised nodes must be pruned when they do not turn out to be useful.

Construction of New Nodes

The construction of new nodes is targeted at improving predictions. If a node's activity is imperfectly predicted, the network will seek to combine nodes to improve the prediction. Candidate nodes for combining are sought from those which already contribute to the node's prediction. The idea is that if a node is already providing some (though imperfect) predictive information, then combining it with other nodes (thus incorporating additional features or temporal combinations) may be able to improve the prediction.

A node with an imperfect prediction will only motivate the construction of a new node if it is considered useful. Similarly, only nodes that are considered useful are regarded as candidates for combining to form the new nodes. A node is deemed useful if, via one of its outgoing connections, it is successfully contributing to improving the prediction of any node. In the implemention used for this article, "successfully contributing to improving the prediction" is restricted to a perfect prediction, that is, one with a probability of 0 or 1.

New nodes are constructed by combining signals from two existing (useful) nodes, which we will refer to as the *primary* and *secondary* nodes, and specifying a time range that defines their temporal combination. The nodes that contribute to an imperfect prediction form a pool of candidates for secondary nodes. The signals that arrived at these secondary nodes at the time they made their prediction, and the connections that carried those signals, are in turn considered. The connection's source and destination nodes and active period form the primary and secondary nodes and active period for the new node. The new node activates (or "fires") if the secondary node becomes active during the active period after the primary node activates.

Once a node has been created from a given connection, no further nodes will be constructed from it. This ensures that duplicate nodes are not created.

Pruning

The CHTN network works according to the "hypothesise and test" principle. The constructed nodes can be regarded as hypotheses, and their usefulness as the test. Nodes that prove to be useful are kept, while the others are pruned. The pruning process is peformed at each time step. Any node whose outgoing connections have all reached maturity, yet is not useful, is pruned. Pruning serves to reduce the computing complexity of the network, but does not affect the prediction.

4   CERG Experiment and Results

In the CERG experiment, the network is presented with a sequence of symbols generated by the continual embedded Reber grammar. This is encoded as a sequence of feature vectors with a component for each of the seven symbols, one of which will be active (1) at each time point. The network initially consists of seven corresponding input nodes.

The task of the network is to predict the next symbol, or the next two possible symbols where there is a nondeterministic choice. In traditional RNNs, LSTM, and its variants, the network expresses predictions with a real value ($value \in [0, 1]$) for each output node. Where there is only one possible next symbol in the grammar, correct prediction requires that the output node corresponding to that symbol be the one with the highest value. Where there are two possible symbols, the two appropriate output nodes must have the highest values.

The CHTN network expresses prediction by assigning an activation probability for each node that corresponds to a symbol (the input nodes). Correct prediction in this network requires that when there is one possible next symbol, the corresponding node is predicted to be active with 100% probability, while all others are 0%. When there are two possible symbols, the two corresponding nodes must be predicted with some positive probability, while all others are 0%.

After Gers *et al* [5] we say the network has "solved" the problem if it is able to correctly predict 1 million consecutive symbols.

Table 1. Results for ERG (left) and CERG (right) showing percentage of trials resulting in a perfect solution and average number of symbols to reach the perfect solutions. ERG results taken from [3]. CERG results for LSTM and ELSTM taken from [5].

ERG	% perfect	# symbols est.
RTRL	"some fraction"	287,500
ELM	0	>2,300,000
RCC	50	2,093,000
LSTM	100	97,060

CERG	% perfect	# symbols
Standard LSTM	0	-
ELSTM	18	>18,889
ELSTM α decay	62	>14,087
CHTN	100	2127

4.1   Comparative Results

The CHTN network, when applied to the CERG problem, was able to find a perfect solution in every trial. Averaged over a run of 20 tests, using a maturity threshold of 5, the network was able find the solution after only 2127 symbols. In each case, the following 1,000,000 symbols were predicted correctly.

Table 1 shows the comparative results on the CERG problem. For completeness, we also include results of earlier recurrent networks on the simpler ERG problem. CERG results for LSTM and ELSTM are taken from Gers *et al* [5], where the experimental conditions differ slightly to ours. Gers *et al* use input streams of 100,000 symbols that are stopped as soon as the network makes an incorrect prediction or the stream ends. They alternate learning and testing — after each stream, weights are frozen and the network is fed 10 streams of 100,000 symbols. Performance is measured by the average number of test symbols fed on each stream before an incorrect prediction is made.

Gers *et al* only report on how many training streams were given to each network, not the number of sequences or symbols seen by the network, so it is not possible to give a direct comparison of training time to our own results. We do know, however, that the number of symbols used for training is at least as many as the number of streams, giving us a lower bound for comparison. We can, on the other hand, directly compare percentage of successful trials as this requirement is 1,000,000 consecutively correct predictions, the same used in our experiments. LSTM and the two variants of ELSTM find perfect solutions in up to 62% of trials. CHTN is the only network that finds perfect solutions every time.

4.2   Solution Size and Maturity Threshold

It is important in a constructive network that its size (number of nodes) does not grow uncontrollably. Figure 2(a) shows the number of nodes being hypothesized, kept and pruned over a single CERG trial. It can be seen that the node count increases from the original 7 to a maximum of 519 before starting to decline. By the end of training, the network has found 66 useful nodes and pruned 2176 nodes. This suggests the hypothesise and test approach is working to restrict attention to key temporal features.

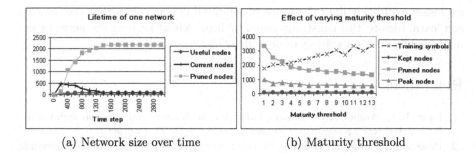

(a) Network size over time   (b) Maturity threshold

Fig. 2. (a) Network size over the lifetime of one network: Useful nodes and total current nodes at each time step and the cumulative count of pruned nodes over time. (b) Effects of changing the maturity threshold: Training symbols required to reach a perfect solution, the number of useful nodes kept in the final solution, the number of nodes pruned whilst training and peak node count of a network at any given time.

Figure 2(b) shows the effect of the maturity threshold on the learning time and node counts. It can be seen that the maturity threshold provides a trade-off between the rate of network growth and the time to find a solution. Lower maturity leads to quicker solutions, but at the expense of more nodes being hypothesized and pruned. The lower maturity threshold networks find solutions faster by searching more hypotheses simultaneously.

5 Conclusion

Traditional RNNs have difficulty learning problems with long time-lag dependencies. LSTM and its variants provide the first set of solutions that perform well in this domain. On a benchmark long time-lag problem, CERG, the LSTM variants outperform previous RNNs but are not able to find perfect solutions in every trial and require many training samples.

In this paper we presented a novel network in which the responsibility for encoding temporal relationships is shifted from an implicit representation in the network's activation space, to an explicit encoding in the network's nodes. In order for the network to identify key temporal features and manage the number of features, a hypothesise and test approach is used. The network hypothesises new nodes that represent potentially important temporal relationships, measures their usefulness, and prunes those that do not contribute to successful outcomes. The CHTN network was shown to find perfect solutions in every trial on the CERG problem with far fewer training samples than the existing networks.

While the experimentation with the CERG problem did not reveal problems with network growth, it is nevertheless possible that problems may arise with larger scale problems. We are currently investigating larger applications with promising preliminary results. A further limitation of the algorithm in its current form is the restriction of connections to two active duration types. This constraint, while improving the training speed and limiting the potential of growth explosion, will likely limit the scope of temporal dependencies that the network can learn. Ideally, this constraint could be lifted whilst keeping the network size manageable, and is a subject of ongoing work.

References

1. Lang, K.J., Waibel, A.H., Hinton, G.E.: A time-delay neural network architecture for isolated word recognition. Neural Networks 3(1), 23–43 (1990)
2. Bengio, Y., Simard, P., Frasconi, P.: Learning long-term dependencies with gradient descent is difficult. IEEE Transactions on Neural Networks 5(2), 157–166 (1994)
3. Hochreiter, S., Schmidhuber, J.: Long short-term memory. Neural Computation 9(8), 1735–1780 (1997)
4. Fahlman, S.E.: The recurrent cascade-correlation architecture. In: Advances in Neural Information Processing Systems, vol. 3, pp. 190–196 (1991)
5. Gers, F.A., Schmidhuber, J., Cummins, F.: Learning to Forget: Continual Prediction with LSTM (2000)
6. Williams, R.J., Zipser, D.: Gradient-based learning algorithms for recurrent networks and their computational complexity. Backpropagation: Theory, architectures, and applications, 433–486 (1995)
7. Elman, J.L.: Finding structure in time. Cognitive science 14(2), 179–211 (1990)
8. Mozer, M.C.: Induction of multiscale temporal structure. In: Advances in Neural Information Processing Systems, vol. 4, pp. 275–282 (1992)
9. Ring, M.B.: Learning Sequential Tasks by Incrementally Adding Higher Orders. In: Advances in Neural Information Processing Systems, pp. 115–122 (1992)
10. Puskorius, G.V., Feldkamp, L.A.: Neurocontrol of nonlinear dynamical systems with Kalman filtertrained recurrent networks. IEEE Transactions on Neural Networks 5(2), 279–297 (1994)
11. Watrous, R.L., Kuhn, G.M.: Induction of finite-state languages using second-order recurrent networks. Neural Computation 4(3), 406–414 (1992)
12. Schmidhuber, J., Hochreiter, S.: Guessing can outperform many long time lag algorithms (1996)
13. Schmidhuber, J.: Netzwerkarchitekturen, Zielfunktionen und Kettenregel. Habilitation, Technische Universitat Munchen 1(1), 1 (1993)

An Abstraction-Based Data Model for Information Retrieval

Richard A. McAllister and Rafal A. Angryk*

Montana State University Department of Computer Science
Bozeman, MT 59717-3880
{mcallis, angryk}@cs.montana.edu

Abstract. Language ontologies provide an avenue for automated lexical analysis that may be used to supplement existing information retrieval methods. This paper presents a method of information retrieval that takes advantage of WordNet, a lexical database, to generate paths of abstraction, and uses them as the basis for an inverted index structure to be used in the retrieval of documents from an indexed corpus. We present this method as a entree to a line of research on using ontologies to perform word-sense disambiguation and improve the precision of existing information retrieval techniques.

1 Introduction

In the creation of an information retrieval system, an ambiguity must be addressed in finding the relationships among documents and queries. Word to word or word-to-synonym comparisons may address this ambiguity. But methods relying on such comparisons fail to address the issue that relevance may elude such comparisons. Documents seen in comparison with one another may refer to relative supersets or subsets of the words used to describe the subject matter [1]. In this paper we present a method for fusing the two worlds of word-sense disambiguation and concept abstraction to create a new, abstraction path-based data model for documents for use in an information retrieval system.

An illustration of this ambiguity involves the problem of polysemy. A polysemous word has more than one meaning. For example, the word "sign" is polysemous in that it may mean [2]:

1. n: any object, action, event, pattern, etc., that conveys a meaning.
2. v: to engage by written agreement.
3. v: to obligate oneself by signature.

It is clear from the distinctness of each definition that a system that compares documents in the dimension of the word "sign" will need to be aware of the difference among the referred definitions and provide a method for disambiguating such words. In addition to the polysemy problem, the panoply of lexica that may

* This work was supported in part by the RightNow Technologies' Grant under MSU's Award No. RIGNOW-W1774.

A. Nicholson and X. Li (Eds.): AI 2009, LNAI 5866, pp. 567–576, 2009.

be used to refer to any singular topic may lead to a truncation of relevant results. This may occur, for example, as time advances and different jargon is used to refer to a topic. Such a problem may be mitigated through an abstraction-based search, since an assemblage of broader concepts may imply specificity with respect to the desired information need.

Such dimensional considerations would arguably be even more useful for a corpus of specialized documents, such as a database of medical papers in which a specialized lexicon is used. Documents related in such a manner will tend to address the same issues repeatedly, using similar language in each repetition. This would allow a greater degree of confidence in the way that a set of words is being used [3].

2   Related Works

Pedersen, et al. [4] presented a method of word-sense disambiguation based on assigning a target word the sense that is most related to the senses of its neighboring words. Their methodology was based on finding paths in concept networks composed of information derived from the corpus and word definitions. Wan and Angryk [5] proposed using WordNet to create context vectors to judge relationships between semantic concepts. Their measure involves creating a geometric vector representation of words and their constituent concepts and using the cosine of the angle between concepts to measure the similarity between them. Perhaps one of the most relevant ideas comes from Resnik [6] who created a measure of semantic similarity in an IS-A taxonomy based on shared information content. The relevance comes from the IS-A taxonomy idea, since this betrays the use of subclass/superclass relationships within the measure. Measuring semantic distance between network nodes for word-sense disambiguation was addressed by Sussna [1], who also clarified the perils of inaccuracy in keyword-based search. Jiang and Conrath [3] combined lexical taxonomy structures with statistical information hewn from the corpus. In doing so they were not reliant on either of the methods for a measure of semantic similarity, but rather both. An essential idea to them was that words used in a subject-specific corpus would be more likely to mean some things based on how they are normally used in relation to the subject address in the corpus. Lin [7] proposed a measure of similarity of words based on the probability that they contain common independent features. Widdows and Dorow [8] presented a graph method for recognizing polysemy. Word-sense disambiguation provided motivation for the technique, which is based on creating graphs of words, using the words' equivalence relations as the connections.

3   Proposed System

3.1   Overview of Frequent Abstraction Path Discovery

Our approach consists of the following steps:

1. Obtain a word-vector representation of the entire corpus under consideration.
2. Use WordNet to create document graph representations of each document's word vector and a master document graph representation of the entire corpus.

3. Use the FP-Growth algorithm for association rule mining to obtain corpus-wide frequent paths.
4. Use these frequent abstraction paths to create an inverted index for document retrieval.

The word-vector representation we have chosen to use for the primary step is the *tfidf* word vector. *tfidf* is a measure based upon a term count that is normalized with respect to both the document in which a term appears and also with respect to the entire corpus in which the document appears [9].

The concept of an *abstraction path* is central to our approach. An abstraction path is a path of term derivation, from most general to most specific, that corresponds to an objective ontological hierarchy of terms. For our purposes, WordNet is this ontology though other domain-specific ontologies exist.

WordNet [10] is a lexical database that depicts relationships among words [4]. In WordNet, words are organized by several standard relationships. Among these are *synonomy*, which describes words having the same meaning; *polysemy*, which describes words that have many meanings; and *hypernymy/hyponymy* which is the property that describes abstracts of a term (hypernyms) or terms that are more concrete than a term (hyponyms).

Figure 1 shows two abstraction paths for the term 'basketball'. As an example of polysemy, the two paths correspond to 'basketball' the game and 'basketball' the ball. Starting from the bottom, each term above the current term is a more general term whose meaning embodies some subset of the current term. The term 'entity', the most general noun in WordNet is where all abstraction paths terminate.

Our process makes use of the hypernym/hyponym relationship in WordNet. There are many other relationships in WordNet that may also be useful in this type of analysis. However, for the demonstration of the validity of our method, we chose to restrict our analysis to the hypernym/hyponym relationship in the interest of saving processing time and memory.

For exploiting this hypernym/hyponym relationship among words we use document graphs, which depict original *keywords* (words that are found in a document) as well as *implicit words* (words whose presence is implied by being the hypernyms of those keywords) in a hierarchy that is created according to the ontology in use [11]. For each keyword in a document's vector representation, a query is made to WordNet to obtain all hypernyms of this keyword recursively, all the way to the the word *entity*, the most abstract hypernym in WordNet 3.0. The result of this is a subgraph of WordNet's hypernym ontology that depicts the document's ontological footprint. During this process, the keyword weight in the word vector is distributed to its direct hypernyms (and then further up the hypernyms' graph) by dividing the weight of the keyword evenly among them.

Figure 2 shows how the time and memory usage changes as the number of documents for which we created document graphs increases from 1,000 through

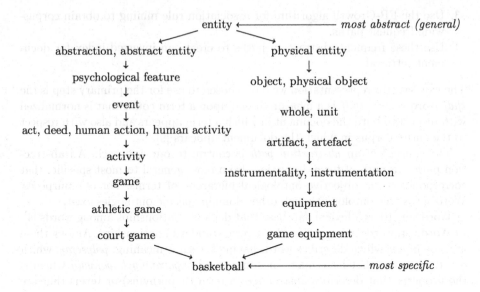

Fig. 1. Examples of Abstraction Paths for the word 'basketball'

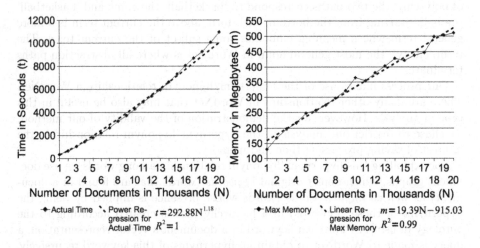

Fig. 2. Time to create document graphs and maximal memory usage for creating document graphs

20,000. As is apparent, the time is polynomial, but close to linear and memory usage is approximately linear. This supports our claim that the construction of this data model can be achieved in a time and space-efficient manner.

A *document tree* is a tree representation of a document graph. It is created by continually splitting each vertex that has an in-degree greater than 1 into as many nodes as its in-degree until all vertices have an in-degree of 1 (or less in the case of the root vertex).

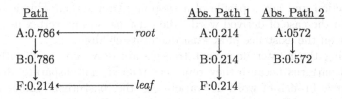

Fig. 3. An Example of Abstraction Path Beginning Criteria

A *document abstraction path* is a path of term derivation, from most general to most specific. A singular abstraction path will contain all of the words in a unique path from a keyword to the word "entity" (in WordNet). For our purposes, a vertex in a document tree must fulfill one of two criteria to be the beginning of an abstraction path:

1. The vertex is a leaf.
2. The vertex has a greater weight than any of its children.

In the path shown in figure 3 the abstraction path from the leaf vertex would produce path F, B, A because F is a leaf vertex and will therefore have a corresponding abstraction path. It would also produce path B, A because the support of B is greater than that of F, indicating that B has more of a contribution to the document than it does just as a matter of being a hypernym of F. These criteria guarantee that the first vertex in any abstraction path corresponds to either a keyword a *junction word* (a word that is at the junction of two or more hyponym branches).

An aggregate document graph representation for the entire corpus is created during the creation of each individual document graph. This is called the *master document graph*, and it includes every type of relationship that is depicted in each document graph. Its corresponding tree representation is called the *master document tree*. We will use this structure to discover abstraction paths that, as we believe, provide much better representation of the content of the corpus than the original keywords.

FP-Growth [12] is a method of mining frequent patterns in a dataset without generating candidates, as is done with the popular Apriori method [11] [12] [13]. Instead of generating frequent pattern candidates and checking them against the entire dataset, as is done with Apriori, the FP-Growth traverses the dataset exactly twice to create its representation in the form of a tree structure, the FP-Tree, that is monotonic in its weight values for each vertex. It then mines this tree by recursively decomposing the original FP-Tree into itemset-dependent FP-Trees and pruning items that are no longer frequent (for a detailed description of this process, see Han et al. [12]).

An important characteristic of this master document tree is that it is monotonic, since vertex weights accumulate as vertices get closer to the root. This makes the use of the FP-Growth algorithm possible. However, the master document tree structure does differ in that each path from root to leaf represents an abstraction and not a profile of a shopping transaction. This is important because, in the case of derived words, the existence of any vertex in the tree is dependent on the existence of at least one of its children.

In mining the master document tree we are only concerned with directly connected patterns because they represent related, non-disparate abstractions from a single branch of word derivation, e.g. the 'basketball' game equipment path from figure 1 as opposed to the game path. We extended the FG-Growth algorithm with this heuristic, since the original FP-Growth algorithm would generate patterns of words without checking if the words are directly linked via a hypernym relationship.

The frequent patterns that are hewn from the master document tree are called frequent paths, as they are all connected. To make them compatible for comparison with the document abstraction paths we convert these frequent paths into *frequent abstraction paths*, which are frequent paths augmented with their entire hypernym traces all the way to the most abstract hypernym, (*entity*). The *frequent abstraction paths* are used to create the inverted index. Documents are related to an intersection of the document abstraction paths and the frequent abstraction paths.

3.2   Query Handling

Upon acceptance of a plain-text query, several actions proceed in sequence:

1. The query undergoes the same preprocessing as the documents (creation of the query graph and query tree and harvesting the query abstraction paths).
2. Selection of the related documents from the inverted index is performed using the query abstraction paths.
3. Ranking of documents and presentation take place.

A *query tree* is the same as a document tree, except that it is constructed for a query. It allows us to perform standard comparisons between queries and documents as if we were comparing only documents. To facilitate the search through the collection of abstraction paths that came from the document analysis, only the same structures are obtained from the query tree.

The query abstraction paths' hash codes are used to query the inverted index for relevant postings. If any of the paths is found to have a document in its postings list, that document is included in the result set.

The ranking procedure uses the cosine similarity measure [14], but now calculated in *frequent abstraction path* space, not in regular keyword space. Each document in the result set, generated by the inverted index, is compared against the query using this measure. The results are then sorted by descending order of cosine similarity and the results are presented to the user. The high dimensionality of the frequent abstraction path space is reduced using the inverted index, as only the frequent abstraction paths that occur in the query are used for document ranking (i.e. cosine-based ordering).

3.3   Preliminary Experiments

For our experimentation we used a 10 newsgroup subset of the 20 Newsgroups dataset [15], a dataset that contains almost 20,000 newsgroup documents from 20 different newsgroups, about 1,000 per newsgroup. A template for our preliminary experimentation was taken from the work of Cohn and Gruber [16]. The main purpose of this truncation to 10 newsgroups was to pick the newsgroups that were most distinct from each other [17], and to take advantage of benchmarks with queries freely available on the Internet [16].

The results from our procedure are influenced by several parameters that can be manipulated to adjust the efficacy of our approach. These are:

1. *Path length upper limit*: How short a path must be in order to be considered. This parameter limits the specificity of the paths to a low abstraction level.
2. *Path length lower limit*: How long a path must be.
3. *Path length range*: Limits the abstraction level to a length that is within a range.
4. *Path popularity upper limit*: How many documents in which a particular path may be found.
5. *Path popularity lower limit*: How few documents in which a particular path may be found.
6. *Path popularity range*: Limits the number of document in which an abstraction path may be found to a range.

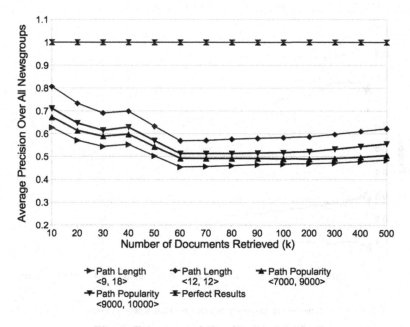

Fig. 4. Experimental Results: Precision at k

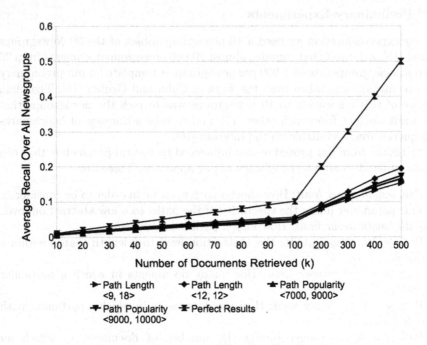

Fig. 5. Experimental Results: Recall at k

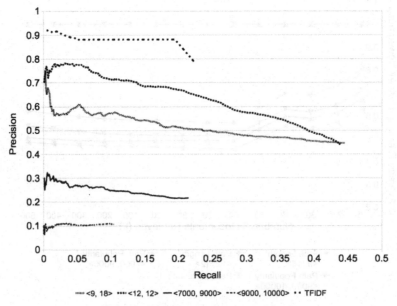

Fig. 6. Results: Precision vs. Recall

We ran several experiments using several different scenarios. Figures 4 and 5 depict the results of the best of these configurations. Each point represents the precision (figure 4) and recall (figure 5) for a specific number of documents retrieved. Specifically, using a path length range of 1 through 9, using a path length of 12, and using the path popularities of 0 through 5,000, 9,000 through 10,000, and 7,000 through 9,000. Our best results were achieved when we used a path length of 12. At this point the 10-document precision was above 0.8.

Figure 6 shows an approximation of Cohn and Gruber's TFIDF results [16] as well as the results from our frequent abstraction path experiments (FAP). As can be seen, Cohn and Gruber's TFIDF results achieve a precision just below 0.9 for a recall level of 0.1. The chart shows a gradual waning of this efficacy for higher recall levels. In comparison, our process achieved a precision high that was just below 0.8 at a recall level of 0.09. A possible explanation for the difference between our best results and the results achieved by Cohn and Gruber's TFIDF process is that they do not mention using an inverted index.

4   Conclusions and Future Work

In this paper we have shown how the consideration of word abstractions, found using objective (non corpus-based) lexical ontologies, can be leveraged to provide an effective and efficient way to construct an inverted index. We have also demonstrated a framework for further work involving ontological abstractions in information retrieval.

In the future, we intend to investigate the ramifications of each of the parametric considerations in all of their permutations to see how each parameter affects the efficacy of the system. The parameters we have identified are those of abstraction path popularity thresholds and ranges, abstraction path length thresholds and ranges, minimum support threshold, and dimension collapse via a common hypernym substitution strategy.

Another avenue of future work that we wish to pursue is the creation of a framework by which other ontologies may be adapted. Since there exist many domains that have produced these ontologies, e.g. the aerospace, medical, and pharmaceutical industries, we hope to use this system to resolve disparate lexica that differ with respect to time or region. An example of this may be if someone is performing a search for "potentiometer," which also may be referred to as a "voltage divider," or a "potential divider." We hope to discover whether or not such vernaculars may be resolved through the use of these ontologies.

References

1. Sussna, M.: Word sense disambiguation for free-text indexing using a massive semantic network. In: Proceedings of the 2nd International Conference on Information and Knowledge Management, pp. 67–74 (1993)
2. Dictionary.com, http://www.dictionary.com

3. Jiang, J.J., Conrath, D.W.: Semantic similarity based on corpus statistics and lexical taxonomy. In: Proceedings of International Conference Research on Computational Linguistics (1997)
4. Pedersen, T., Banerjee, S., Padwardhan, S.: Maximizing semantic relatedness to perform word sense disambiguation (February 2009), citeseer.ist.psu.edu/pedersen03maximizing.html
5. Wan, S., Angryk, R.: Measuring semantic similarity using wordnet-based context vectors. In: Proceedings of the IEEE International Conference on Systems, Man & Cybernetics (2007)
6. Resnik, P.: Semantic similarity in a taxonomy: An information-based measure and its application to problems of ambiguity in natural language. Journal of Artificial Intelligence Research 11, 95–130 (1999)
7. Lin, D.: An information-theoretic definition of similarity. In: Proceedings of the 15th International Conference on Machine Learning, pp. 296–304 (1998)
8. Widdows, D., Dorow, B.: A graph model for unsupervised lexical acquisition. In: 19th International conference on Computational Linguistics, pp. 1093–1099 (2002)
9. Feldman, R., Sanger, J.: The Text Mining Handbook: Advanced Approaches in Analyzing Unstructured Data. Cambridge University Press, Cambridge (2006)
10. Wordnet: a lexical database for the english language, http://wordnet.princeton.edu/
11. Hossain, M.S., Angryk, R.A.: Gdclust: A graph-based document clustering technique. In: ICDM Workshops, pp. 417–422. IEEE Computer Society, Los Alamitos (2007), http://dblp.uni-trier.de/db/conf/icdm/icdmw2007.html#HossainA07
12. Han, J., Pei, J., Yin, Y.: Mining frequent patterns without candidate generation: a frequent-pattern tree approach. SIGMOD Rec. 29(2), 1–12 (2000), http://dx.doi.org/10.1145/335191.335372
13. Tan, P.-N., Steinbach, M., Kumar, V.: Introduction to Data Mining. Addison-Wesley Publi., Reading (2006)
14. Manning, C.D., Raghavan, P., Schütze, H.: Introduction to Information Retrieval. Cambridge University Press, Cambridge (2008)
15. Home page for 20 newsgroups data set (May 2009), http://people.csail.mit.edu/jrennie/20Newsgroups/
16. Cohn, I., Gruber, A.: Information retrieval experiments (May 2009), http://www.cs.huji.ac.il/~ido_cohn
17. Slonim, N., Tishby, N.: Document clustering using word clusters via the information bottleneck method. In: ACM SIGIR 2000, pp. 208–215. ACM Press, New York (2000)

Vietnamese Document Representation and Classification

Giang-Son Nguyen, Xiaoying Gao, and Peter Andreae

School of Engineering and Computer Science
Victoria University of Wellington, New Zealand
{nguyenson, Xiaoying.Gao, Peter.Andreae}@ecs.vuw.ac.nz

Abstract. Vietnamese is very different from English and little research has been done on Vietnamese document classification, or indeed, on any kind of Vietnamese language processing, and only a few small corpora are available for research. We created a large Vietnamese text corpus with about 18000 documents, and manually classified them based on different criteria such as topics and styles, giving several classification tasks of different difficulty levels. This paper introduces a new syllable-based document representation at the morphological level of the language for efficient classification. We tested the representation on our corpus with different classification tasks using six classification algorithms and two feature selection techniques. Our experiments show that the new representation is effective for Vietnamese categorization, and suggest that best performance can be achieved using syllable-pair document representation, an SVM with a polynomial kernel as the learning algorithm, and using Information gain and an external dictionary for feature selection.

Keywords: Vietnamese language processing, text categorization, classification, machine learning.

1 Introduction

Since the beginning of the World Wide Web, the need to organize and structure the text-based information available electronically has constantly increased. Automated text categorization is one of the important techniques for arranging and finding relevant information. Many applications have been built for such tasks as classifying news by subjects or newsgroups, sorting and filtering e-mail messages, etc. Many Vietnamese documents are available online but little research has been done on Vietnamese document categorization.

Text categorization is the task of assigning a natural language document to one or more predefined categories based on its content. Formally, text categorization can be represented by a function $\Phi: D \times C \rightarrow \{True, False\}$, where $D = \{d_1...d_N\}$ is a set of documents d_i to be classified, and $C = \{c_1...c_M\}$ is a set of predefined labels c_j of each category [1]. A number of approaches to automatic text categorization have been proposed such as Naïve Bayes, K-Nearest Neighbour (K-NN), and Support Vector Machines (SVM). They differ in their representation of documents and their algorithms to decide the category of documents. Some approaches are good only for a particular data set, while others can be applied to many different tasks.

A. Nicholson and X. Li (Eds.): AI 2009, LNAI 5866, pp. 577–586, 2009.

In most research, the techniques are applied to English documents. How to apply these techniques on Vietnamese documents has not been explored and the performance of these techniques on this very different language is unknown. There are many differences between English and Vietnamese, for example, in Vietnamese, a sentence is written as a sequence of syllables rather than words; each word in Vietnamese may have two or three syllables, and word segmentation is not trivial. Therefore, the standard bag-of-words model for English document categorization and the techniques based on this model is hard to apply to Vietnamese documents.

In fact, there has been little research on Vietnamese Language Processing of any kind. Among the few active researchers in this area, Dien *et al* [2], [3] mainly do research on English-Vietnamese translation; Nguyen *et al* [4] are trying to build and standardize Part of Speech (POS) tagging of Vietnamese corpora. Some others focus on building spoken and written language resources for Vietnamese [5]. However, to our knowledge, no one has used machine learning techniques for Vietnamese text categorization yet.

This paper describes our efforts to build an effective classifier for Vietnamese documents. We created a large corpus with about 18000 documents and investigated document representation, feature extraction, and learning methods that are suitable for Vietnamese document categorization. Our research questions include the following:

- What are the special languages features of this language that might be useful for classification? How can we represent the documents effectively?
- Among all the different learning algorithm and feature selection techniques, which ones are suitable for Vietnamese document categorization?

The paper is organized as follows. Section 2 describes the important features of Vietnamese language. Section 3 introduces a syllable-level representation of Vietnamese documents. Section 4 describes our Vietnamese text corpus and the testing environment. Section 5 presents empirical results and Section 6 concludes the paper.

2   Vietnamese Language Features

This section summarizes the main characteristics of Vietnamese language and then details its alphabet, tones, syllables and words. Related research is found [5,6,7] that discusses Vietnamese features at syntactic and semantic linguistic levels. This paper focuses only on the features that are related to the document representation and feature extraction techniques.

Vietnamese belongs to the Austro-Asiatic language family. Its vocabulary is based on Sino-Vietnamese and enriched from the French language. Like other isolating languages, Vietnamese has the following common characteristics:

- It is a tonal, monosyllabic language;
- In contrast to Latin languages, Vietnamese word forms never change when making a sentence. For example, a noun does not have different forms for singular or plural; a verb does not change with tense, person, number, voice and mood.

- Word order and auxiliary words are necessary for building grammatical relations in a sentence;
- Sentence and paragraph structure most commonly has a "topic-elaboration" pattern.

The Vietnamese alphabet is based on the Latin alphabet with some digraphs and the addition of nine accent marks or diacritics — four of them to create additional sounds, and the other five to indicate the tone of each word. The Vietnamese alphabet has 29 letters (12 vowels and 17 consonants). Its vowels form 24 diphthongs and 4 triphthongs, and its consonants form 9 digraphs and 1 trigraph.

Vietnamese is a tonal language so that syllables with different tones have different meanings. There are six distinct tones; the first one ("level tone") is not marked, but the other five are indicated by diacritics applied to the vowel part of the syllable. In syllables where the vowel part consists of more than one vowel (such as diphthongs and triphthongs), the placement of the tone is still a matter of debate. For example, "hóa" and "hoá" are the same and both are acceptable.

In contrast to words in English, Vietnamese is written as a sequence of syllables, separated by spaces and/or punctuation. Each syllable consists of at most three parts, in the following order from left to right:

1. An optional beginning consonant part
2. A required vowel part and the tone mark applied above or below it if needed.
3. An optional ending consonant part, which must be one of the following: *c, ch, m, n, ng, nh, p, t.*

For example, *"trường"* consists of 3 parts: the beginning consonant part: *"tr"*; the vowel part: *"ươ"* with the falling tone `; and the ending part: *"ng"*.

There are three types of syllables:
1. Syllables with a complete meaning. Such syllables are monosyllable words;
2. Syllables with a meaning, but also used to form compound words (especially sino-vietnamese words); and
3. Syllables with no meaning of their own.

The semantically meaningful vocabulary words in Vietnamese do not necessarily correspond to single grams (syllables) in Vietnamese documents – each word consists of 1, 2 or 3 grams, separated by space characters and/or punctuations. For example, the word *"doanh nghiệp"* /company is written as two separated grams *"doanh"* and *"nghiệp"*.

According to Cao X. Hao [8], the Vietnamese vocabulary can be partitioned into the following categories:

1. Monosyllabic words, (e.g. *đi*/go, *ở*/live, *ăn*/eat), constituting about 15% of a general dictionary.
2. Reduplicated words composed of phonetic reduplication (e.g. *trăng trắng*/whitish)
3. Compound words composed by semantic coordination (e.g. *quần áo*/clothes from *quần*/ pant + *áo*/shirt)
4. Compound words composed by semantic subordination (e.g. *xe đạp*/bicycle from *xe*/vehicle + *đạp*/pedal, or *máy bay*/airplane from *máy*/machine + *bay*/fly)

5. Complex words whose phonetics are transcribed from foreign languages, especially French (e.g. *cà phê*/coffee).

3 Syllable-Based Document Representation

Text can be analyzed at five different levels:

1. Morphological level: the components of words
2. Lexical level: complete words
3. Syntactic level: the structure of sentences
4. Semantic level: the meaning of text
5. Pragmatic level: the meaning of text with respect to context and situation

We investigated a number of different representations at different levels. The paper introduces a new representation at morphological level, where documents are represented either as a sequences of syllables (unigram) or syllable pairs (bigrams).

For English document categorization, text is typically represented at the lexical level, which is easy because words are separated by spaces or punctuations in English documents. The most common methods use a bag-of-words representation which throws away the structure of the document but captures the semantic information present at the lexical level. However, this approach can not be directly applied to Vietnamese documents since complete words in Vietnamese are not identified and may consist of multiple separated syllables.

It is not trivial to segment words in Vietnamese [7]. For example, the sentence "*học sinh ra từ hành*" may segment to "*học*/learning | *sinh ra*/become | *từ*/from | *hành*/practice" or to "*học sinh*/pupil | *ra*/go out | *từ*/of | *hành*/practice". All the unigrams (single syllable) and bigrams (two syllables) from this example are meaningful and exist in a Vietnamese dictionary. However, only the first segmentation is semantically correct. In one of our projects, we are working on an algorithm that learns word segmentation from unlabelled documents with the help of a dictionary, but this paper describes an alternative approach.

This paper analyzes text primarily at the morphological level, treating a document as a sequence of syllables. We represent a Vietnamese document at the syllable level as a bag of syllables. The bag-of-syllables model is simple to construct, and to our surprise, it is reasonably effective for easy classification tasks as shown in our experiments. One problem is that it is poor at capturing the semantic properties of a document because the majority of meaningful Vietnamese words consist of more than one syllable. So we introduce a bag-of-syllable-pairs model and represent documents as a sequence of syllable pairs. Since more than 80% of complete Vietnamese words consist of two syllables, we expect that the bag-of-syllable-pairs model is likely to capture many of the semantic properties of the document. Our experiments show it is effective and sufficient for our classification tasks.

We also investigated an even lower level representation that breaks down each syllable into three parts: onset, rhyme and tone in order to capture the phonological structure of Vietnamese language. Our preliminary results showed that this representation is very effective for classifying documents into different styles, for example, to classify documents into ancient vs. modern, and to identify poems, songs and dialogues. However, this low level representation is not very effective for general categorization tasks and these results are not included in this paper.

4 Constructing the Corpus and the Testing Environment

We created a corpus with about 18000 Vietnamese documents and made it publically available on the Internet. We constructed a test environment and plan to make it publically available too, so users can plug in their own algorithms and compare it with our approach.

4.1 Vietnamese Document Corpus

To our knowledge, there is no standard Vietnamese document corpus for text categorization. Therefore, we had to build our corpus from scratch. In order to create a good corpus, we collected documents from different sources (e.g. websites, emails, newspapers, etc), and made sure that the documents had a good range of document sizes, from hundreds to thousands of words. We manually labelled them by category and ensured that the categories were balanced, so the numbers of documents in each category were approximately equal. Our corpus allows ambiguous words and lexical errors. These documents are manually labeled using different criteria such as topics and styles, which construct multiple data sets with different classification tasks at different difficulty levels. The first two data sets used in this paper are detailed in this section.

Our first data set, D1, contains modern Vietnamese documents from four general categories: Vietnamese Society, Sciences, Economy and Sport. Each category has approximately 4000 documents. The average document size is about 800 syllables or 500 words. Each document has about 20 sentences divided into 5-6 paragraphs. The spelling error ratio is nearly 0.5%.

Our second data set, D2, is a subset of 1743 documents from the first data set, but classified into two categories based on the type of document rather than on the topic of the document. The two categories were *"Tu-lieu/Documentary"* and *"Phan-tich/* Analysis". Both categories contain documents on a range of topics, but "Documentary" contains documents which give information about a fact or event in the past, whereas "Analysis" contains documents that discuss or analyse an issue. Of the 1743 documents in D2, 825 documents were manually classified into the "Analysis" category, and the other 918 documents into the "Documentary" category. About 15% of them are hard to classify even for a human expert.

Data set D2 constitutes a more challenging classification task than D1. Documents in one category of D1 are likely to share vocabulary associated with the topic of the category that is distinct from the vocabulary associated with the other topics. However, documents in both categories of D2 may be dealing with the same topic, facts and events and therefore the vocabularies of the documents of the two categories are likely to by much less distinct than in D1; many of the distinguishing features of the two categories in D2 are in the logical and semantic structure of the document that is lost in a morphological (or even lexical) model.

4.2 The Testing Environment

Our current test environment compares six typical classification methods: Naïve Bayes(NB), K-Nearest Neighbour(KNN) and Support Vector Machine(SVM) with four different kernel functions including Linear (LSVM), Polynomial (PSVM), Gaussian (GSVM), and Sigmoid (SSVM). NB is a simple method that gives a useful baseline. For English text categorization, SVM-based and k-NN classifiers deliver top performance on many different experiments and data sets, e.g. Reuter-21578, Reuter-22173(ModApte), or OHSUMED [1]. Our experiments tested whether they work well for Vietnamese text categorization.

In our experiments, we use a small validation set Va (5% of the training set Tr) to optimize the parameters of these learning methods. For our text corpus, the number of neighbors in the KNN method is 32, and the parameters for SVM kernels are $d=3$ for the polynomial kernel, $c=16$ (ball $radius = 4$) for the Gaussian, and $\theta = 1.0$ for the sigmoid kernel.

Our current testing environment also supports two feature selection techniques: using information gain and an external dictionary for filtering the vocabulary.

In our experiments, we use classification accuracy to evaluate the classification performance of the different algorithms on the data sets: Accuracy = TP / N where TP is the number of correctly classified documents and N is the number of test documents. In all the experiments, each classification algorithm was run ten times using a randomly selected 10% of the documents as training data using the remainder of the documents for testing. The average classification accuracy over the ten runs was used for the final results.

5 Experimental Results

Our document representation is tested on two data sets using six algorithms and two feature selection techniques.

5.1 Bag-of-Syllables *vs.* Bag-of-Syllable-Pairs

The first experiment was designed to compare the two document representations: bag-of-syllables ("unigram") and bag-of-syllable-pairs ("bigram"). Both models are based on a bag-of-tokens model, but we expected the bigram model to lead to better classification accuracy than the unigram model because it captures more of the semantically meaningful elements of the documents. On the other hand, because the bigram representation involves a much larger vocabulary, we also expected it to be less efficient. We tested each representation of the documents in the D1 data set using six different algorithms. We measured both the quality (accuracy) and the efficiency (time consumed); the results are presented in Tables 1 and 2 below.

As expected, the bigram representation achieves a higher accuracy for all algorithms but also takes considerable more time. However, for this classification task, the accuracy is extremely good for all the algorithms (regardless of the model), and there appears to be little difference between the accuracy of the algorithms. This demonstrates that representing Vietnamese documents at the morphological level is adequate for straightforward classification into topics, even when using the unigram

Table 1. Accuracy (%): unigram *vs* bigram

	NB	KNN	LSVM	PSVM	GSVM	SSVM
syllable	97.8	97.5	98.1	98.1	98.0	98.1
syllable pair	98.1	98.4	98.5	98.6	98.6	98.5

Table 2. Time consumed (s): unigram *vs* bigram

	NB	KNN	LSVM	PSVM	GSVM	SSVM
syllable	51	45	68	81	89	96
syllable pair	235	228	305	655	832	925

model. For this task, there appears to be no need to pursue the more expensive route of segmenting Vietnamese text into lexical words.

5.2 Task Difficulty

The second experiment was designed to compare the different algorithms on a harder classification task. We used the same six algorithms as in the previous experiment, but ran them on the harder data set D2, as well as on D1. We only used the bigram model since it showed slightly better performance in experiment 1. We expected the accuracy to be lower for D2, and perhaps a wider spread of accuracies across the algorithms. Table 3 shows the average accuracy of the six algorithms on the two data sets.

Table 3. Accuracy of six methods on two clasification tasks

	NB	KNN	LSVM	PSVM	GSVM	SSVM
D1	98.1	98.4	98.5	98.6	98.6	98.5
D2	84.0	73.2	86.7	89.8	88.6	89.3

The results confirmed our expectations, with a drop in accuracy of at least 9%, though we were surprised how well the best algorithms performed on the harder classification task, given that the human expert (who had access to the full semantic content of the document) considered that about 15% of the documents were very hard to classify.

KNN was clearly the worst algorithm at the harder task – 16% worse than the best algorithm, and the SVM algorithms were the best. The polynomial kernel (PSVM) was the best of the SVM algorithms by a small margin.

With an accuracy of almost 90% on this difficult classification task, it is clear that the bigram model, combined with SVM, can be successfully applied to Vietnamese document classification.

However, the size of the bigram models means that the time cost of the classification algorithms is high. The remaining experiments explore various approaches to reducing the cost of the classification.

5.3 Dictionary Based Filtering

The third experiment explored a method for reducing the model size which exploited semantic information. In this case, we used the harder classification task of data set D2 in order to get a wider range of accuracies and used both the unigram and bigram models. We used a Vietnamese dictionary to identify and prune out grams that did not correspond to complete words. This reduced the total size of the unigram models to 17% (from 30,000 to 5,000 grams), and the size of the bigram models to 10% (from 300,000 to 30,000 bigrams).

Again, we ran the six algorithms on the two different models with and without the dictionary based filtering. Table 4 shows the results. The four different SVM methods had very similar results so only the results for Linear SVM are presented.

Table 4. Accuracy (%) with and without a dictionary

Model	Vocabulary	NB	KNN	LSVM
Unigram	30K	83.6	65.5	84.9
	5K (dictionary)	82.1	74.3	82.1
Bigram	300K	84.0	73.2	86.7
	30K (dictionary)	85.1	78.4	85.9

Interestingly, the accuracy increased for KNN in both unigrams and bigrams. We believe that this is because KNN is very sensitive to irrelevant features, such as the presence of semantically meaningless elements in the morphological-level models. We suggest that the dictionary is able to remove many of these features in a simple way. It also increased for NB in the bigram case, though not as dramatically. For SVM (and NB unigrams), the accuracy decreased very slightly, although the much smaller vocabulary represents a significant increase in efficiency. For SVM, which is more able to ignore irrelevant features than KNN, pruning non-word grams and bigrams actually loses a small amount of useful information.

This experiment demonstrates that dictionary based filtering is a useful technique for feature selection in morphological-level models of Vietnamese text documents.

5.4 Selection with Information Gain

The fourth experiment uses a commonly used information theoretic approach to reduce the size of the document models. As in the previous experiment, we used the harder classification task of data set D2 and used both unigram and bigram models. We calculated the *infogain* measure for each term (unigram or bigram) in the data set, and then pruned from the document models all but the top M terms from the list of terms ordered by *infogain*.

We ran the classification algorithms on the pruned documents using seven different values of $M=\{3000, 2000, 1500, 1000, 500, 100, 50\}$. The accuracy of the unigram models was similar to that of the bigram models, and all the SVM algorithms also had similar results. Table 5 shows the results on the bigram models of NB, KNN, and LSVM.

Table 5. Using infogain for feature selection

Number of terms	NB	KNN	LSVM
Full(300K)	84.0	73.2	86.7
M=3000	85.5	77.3	83.4
M=2000	85.1	79.8	82.1
M=1500	84.8	80.6	82.4
M=1000	84.9	81.5	80.8
M=500	84.3	80.0	78.7
M=100	72.7	72.1	71.5
M=50	69.5	71.8	70.7

As with the dictionary based pruning, KNN accuracy improves when the less informative features are pruned, with a maximum accuracy when the vocabulary is pruned to 0.33%. NB accuracy also improves, though less dramatically, but enough to make NB the best classifier for all but levels of pruning at the most extreme. The *infogain* pruning can be far more extreme than the dictionary based pruning for the same loss of accuracy.

The experiment shows that *infogain* is a very effective feature selection method for morphological-level models of Vietnamese documents. The size of the vocabulary can be significantly reduced from 300K to 100 with minimal impact on accuracy, and significant reduction in cost.

6 Conclusions

The paper has introduced a morphological level representation of Vietnamese text documents using bags of syllables and bags of syllable pairs. Our experiments show that it is possible to build highly effective classifiers using syllable based documents representations, particularly using the syllable pairs model. The SVM with a polynomial kernel appeared to be the best classification algorithm with this model. The experiments also demonstrated that using a dictionary and the *InfoGain* measure for feature selection is able to reduce the vocabulary size considerably with little loss in accuracy. We created a large text corpus with about 18000 documents and a testing environment with different learning algorithms and feature selection techniques which can be used in future research.

This paper is one of the earliest attempts at text categorization for Vietnamese documents. We investigate a number of language features and a number of different classification tasks. But there are a lot of areas yet to be explored. In future work, we will explore further ways to enrich the bag-of-syllables models, for example, by taking into account the similarities between syllables, and the relatedness of bigrams. We are also working on word segmentation techniques, both to enable document representations at the lexical level and to provide better feature selection for the bigram model. Finally, we are investigating ways of improving our classifier by adding further linguistic knowledge, for example, developing a new representation to capture the phonological structure of Vietnamese language and enriching document representations using part-of-speech annotations.

References

1. Sebastiani, F.: Machine learning in automated text categorization. ACM Computing Surveys 34(1), 1–47 (2002)
2. Dinh, D., Hoang, K.: Vietnamese Word Segmentation. In: Proceedings of Sixth Natural Language Processing Pacific Rim Symposium (NLPRS 2001), Tokyo (2001)
3. Dinh, D.: POS-Tagger for English – Vietnamese Bilingual Corpus. In: Workshop: Building and Using Parallel Texts: Data Driven Machine Translation and Beyond, Edmonton, CA (2003)
4. Nguyen, T.M.H., Vu, X.L.: Une etude de cas pour l'etiquetage morpho-syntaxique de texts vietnamiens. In: The TALN Conference, Batz-sur-mer, France (2003)
5. Nguyen, T.B.: Lexical descriptions for Vietnamese language processing. In: The 1st International Joint Conference on Natural Language Processing, Workshop on Asian Language Resource, Sanya, Hainan Island, China (2004)
6. Vu, D.: The Vietnamese and contemporary languages – an overview of syntax (in Vietnamese), Viet Stuttgart, Germany (2004)
7. Nguyen, T.C.: Vietnamese Grammar. NXB Đại học Quốc Gia, Hanoi (1998)
8. Cao, X. H.: The Vietnamese – some problems about phonetics, grammar, and semantics, (in Vietnamese). NXB Giáo dục, Hanoi (2000)

Can Shallow Semantic Class Information Help Answer Passage Retrieval?

Bahadorreza Ofoghi and John Yearwood

Centre for Informatics and Applied Optimization
University of Ballarat, Ballarat, Victoria 3350, Australia

Abstract. In this paper, the effect of using *semantic class overlap* evidence in enhancing the passage retrieval effectiveness of question answering (QA) systems is tested. The semantic class overlap between questions and passages is measured by evoking FrameNet semantic frames using a shallow term-lookup procedure. We use the semantic class overlap evidence in two ways: i) fusing passage scores obtained from a baseline retrieval system with those obtained from the analysis of semantic class overlap (*fusion-based* approach), and ii) revising the passage scoring function of the baseline system by incorporating semantic class overlap evidence (*revision-based* approach). Our experiments with the TREC 2004 and 2006 datasets show that the revision-based approach significantly improves the passage retrieval effectiveness of the baseline system.

1 Introduction

Having received natural language questions, question answering (QA) systems perform various processes (question analysis, information retrieval, and answer processing) to return actual direct answers to the information requests eliminating the burden of query formulation and reading lots of irrelevant documents to reach the exact desired information. This is because users usually seek not the entire document texts but brief text snippets to specific questions like: "How old is the President? Who was the second person on the moon? When was the storming of the Bastille?" [1].

In the pipelined architecture of information retrieval-based QA systems, three main processes are carried out: i) question processing to find the answer type of the given question and formulate the best representative query, ii) information (document and/or passage) retrieval using the query formed in the first step, and iii) answer processing (answer extraction and scoring) in the most related texts retrieved.

There are empirical studies in the domain of QA which show that the answer processing task can be handled more effectively on the passage-level information in documents rather than document-level texts [2][3][13][14][15]. In effectively answering questions, the successful processing of candidate and actual answers can be achieved by analyzing the passages which are most similar to the queries. However, in many cases in the context of QA, *relatedness* of the passages is not enough and *specificity* is required to have short texts containing real and

A. Nicholson and X. Li (Eds.): AI 2009, LNAI 5866, pp. 587–596, 2009.
© Springer-Verlag Berlin Heidelberg 2009

potentially correct answers. This introduces new limitations in passage retrieval and requires more precise retrieval processes to improve QA effectiveness.

The effectiveness of passage retrieval can be affected by two aspects: *a*) the performance of the passage scoring and ranking algorithm, and *b*) the formulation of representative queries. In this paper, we focus on the first aspect and analyze the impact of using semantic class information in scoring and ranking retrieved passages. We use FrameNet [4] frames to capture the semantic class of question and passage predicates and utilize it in two ways: i) fusing the scoring evidence from the baseline passage retrieval method (MultiText [5][6][16][17]) with the score that is obtained after semantic class overlap analysis between the question and passage predicates, and ii) revising the passage scoring function of the baseline retrieval system to incorporate the semantic class overlap evidence in scoring retrieved passages. We call the first method a *fusion-based* method and the second one a *revision-based* method.

Our fusion functions in the fusion-based method are based on the mathematical aggregation of the scores that are obtained from each type of evidence, while in the revision-based method we modify the internal passage scoring function of the baseline retrieval system by replacing a part of the function with the semantic class overlap evidence. The main question in this paper is whether semantic class overlap evidence, when fused or incorporated with other types of evidence, can enhance the answer passage retrieval performance.

2   FrameNet Semantic Classes

Frame semantics, that has been developed from Charles Fillmore's Case Structure Grammar [8][9], emphasizes the continuities between language and human experience [10][11][12]. The main idea behind frame semantics is that the meaning of a single word is dependent on the essential knowledge related to that word. The required knowledge about each single word is stored in a semantic class. In order to encapsulate frame semantics in such classes, the FrameNet project [4] has been developing a network of interrelated semantic classes that is now a lexical resource for English being used in many natural language applications.

The main entity in FrameNet is the *frame* which encapsulates the semantic relation between concepts based on the scenario of an event. Each frame (semantic class) contains a number of *frame elements* (FEs) which represent different semantic and syntactic roles regarding a target word. For instance, in the frame *Personal_relationship*, *partner_1* and *partner_2* are two core FEs which participate in the scenario of having personal relationships. "Adultery.n", "affair.n", "date.v", and "wife.n" are examples of the terms (Lexical Units (LUs)) which are covered by (and can evoke) this frame.

3   Baseline Passage Retrieval System

MultiText interprets all textual documents as a continuous series of words and also interprets passages as any number of words starting and ending at any position in the documents. A document d is treated as a sequence of terms $\{t_1, t_2, \ldots, t_{|d|}\}$ and the query is translated to an unordered set of terms $Q = \{q_1, q_2, \ldots, q_{|Q|}\}$. There are two definitions necessary:

- An *extent* over a document d is a sequence of words in d which contains a subset of Q. It is denoted by the pair (p, q) where $1 \leq p \leq q \leq |d|$. This is translated to the interval of texts in document d from t_p to t_q. An extent (p, q) satisfies a term set $T \subseteq Q$ if it includes all of the terms in T.
- An extent (p, q) is a *cover* for the term set T if it satisfies T and there is no shorter extent $(\acute{p}, \acute{q})$ over the document d which satisfies T. A shorter extent $(\acute{p}, \acute{q})$ is a nested extent in (p, q) where $p < \acute{p} \leq \acute{q} \leq q$ or $p \leq \acute{p} \leq \acute{q} < q$. In any document d there may be different covers for T which are represented in the cover set C for the term set T.

The passages retrieved by MultiText are identified by covers and scored based on the length of the passages and the weight of the query terms covered in the passages. Each term t gets the IDF[1]-like weight w_t as shown in Equation 1, where f_t is the frequency of the term t in the corpus and N is the total length of the unique string constructed over the document set.

$$w_t = log(\frac{N}{f_t}) \tag{1}$$

A passage containing a set T of the terms is assigned a score according to the formula in Equation 2 where p and q are the start and end points of the passage in the unique string of words in the document set.

$$Score(T, p, q) = \sum_{t \in T} w_t - |T| log(q - p + 1) \tag{2}$$

The high performance of MultiText [7], as well as its frequent participation in TREC [5], is the main reason for choosing MultiText as the baseline passage retrieval method in our experiments.

4   Semantic Class Overlap Evidence for Passage Specificity

4.1   Theory of Semantic Class Overlap

The existing passage retrieval algorithms which rely on surface structures of passages and questions are dependent on the occurrences of exact matches of syntactical features. As a result, their highest performance of retrieval cannot reach very high levels due to the limitations imposed by syntactic structures. Example 1 shows a case where surface structures fail to resolve the connection between the answer-bearing passage and the question. The predicate "discover" appears in the question whereas the answer-containing passage is formulated using an alternative predicate "spot".

---

[1] Inverse Document Frequency.

Example 1

> *Who discovered Hale-Bopp?*
> *The comet, one of the brightest comets this century, was first spotted by Hale*
> *and Bopp, both astronomers in the United States, on July 23, 1995.*

These types of mismatches are tackled by other passage retrieval methods which incorporate linguistic information. However, there are other types of mismatches which are more complicated.

Example 2

> *Who is his [Horus's] mother?*
> *Osiris, the god of the underworld, his wife, Isis, the goddess of fertility, and*
> *their son, Horus, were worshiped by ancient Egyptians.*

In this example, there is no direct relationship between the terms "mother" and "son". To overcome such mismatches, we use the frame semantics encapsulated in FrameNet. In the case of Example 2, for instance, the semantic class evoked by "mother" in the question is *Kinship* which is the same as the semantic class invoked by the predicate "son" in the passage. Such shared semantic classes imply scenario-based similarities between questions and passages that may help retrieval systems improve their performance. This can be carried out by considering semantic class similarity evidence in scoring and ranking retrieved passages.

4.2 Procedure

We use two types of evidence for scoring and ranking retrieved passages by the baseline passage retrieval method explained in section 3:

- Passage relatedness scores obtained from the baseline passage retrieval system (MultiText) on the basis of query terms coverage and passage length
- Semantic class similarity score obtained by comparing the semantic classes in questions and retrieved passages

In order to measure the semantic class similarity of a given question q and a passage p, we take the following steps:

- For each question term t_q in question q, we evoke all frames from FrameNet which include t_q as a LU. The invocation of frames is based on a shallow term-lookup in all FrameNet frames. A frameset $frames_q$ is formed by the union of the frames evoked for the whole question term set.
- For each passage term t_p in passage p, we evoke all frames from FrameNet which include t_p as a LU. A frameset $frames_p$ is formed by the union of the frames evoked for the whole passage term set.
- We form the intersection of the two framesets $frames_q$ and $frames_p$ as the semantic class overlap evidence ($|frames_q \cap frames_p|$).

The two types of evidence (MultiText and semantic class overlap) are then used to score retrieved passages by the MultiText passage retrieval system in two ways.

```
foreach passage in retrieved passages{
        mscore ← get passage score from MultiText}
shift mscores from [-x,+y] to [0,+z]
foreach passage in retrieved passages{
        if mscore/max(mscores) < threshold{
                fscore ← semantic_class_overlap(question, passage)
                if fscore>0{
                        passage_score ← (fscore+1) Δ mscore}
                else{
                        passage_score ← mscore}}
        else{
                passage_score ← mscore}}
foreach passage in retrieved passages{
        passage_score ← passage_score/max(passage_scores)
```

Fig. 1. Fusion of the MultiText evidence and semantic class similarity evidence for passage scoring

Fusion-Based: We fuse the results obtained from the two above-mentioned types of evidence using different *operators* and *threshold* values.

A fusion process in our experiments is defined by the pair $(\triangle, threshold)$. The $\triangle$ operator is a mathematical operator, either *addition* or *multiplication* of the scores obtained using the two types of evidence ($\triangle \in \{+, \times\}$). The *threshold*[2] factor is the minimum MultiText score that is required for each passage in order to not use any semantic class overlap evidence. The threshold factor in our experiments includes two values $threshold \in \{0.5, 0.75\}$.

Figure 1 shows the pseudo code of the fusion process with a generic fusion operator $\triangle$ and a generic threshold factor *threshold*.

Revision-Based: As an alternative approach, we change the passage scoring function of MultiText (see Equation 2) to incorporate semantic class overlap evidence in its cover scoring process. Equation 3 shows the new scoring function that we use in MultiText where *fscore(Psg)* is the score obtained according to the semantic class similarity of the question and passage *Psg* (this is measured using the procedure explained in section 4.2). This approach generalizes the MultiText coverage concept. In MultiText only exact matches are considered to calculate the coverage rate. This is encapsulated in the $|T|$ factor of Equation 2. By using $fscore(Psg) = |frames_q \cap frames_p|$ instead of $|T|$, it is possible to consider pairs of question and passage terms t_q and t_p as exact matches if they appear in the same semantic class (FrameNet frame). However, it is important to notice that in this approach, we are not expanding T itself and so the $\sum_{t \in T} w_t$ part of Equation 2 remains intact. The only concept that we change is the coverage rate of passages formulated in $|T|$ by developing it to $|frames_q \cap frames_p|$.

---

[2] We use this threshold to avoid fusion where there is already much evidence for passages to be containing answers.

$$Score(Psg, T, p, q) = \sum_{t \in T} w_t - fscore(Psg) \times log(q - p + 1) \qquad (3)$$

5 Empirical Analysis

5.1 Datasets

The datasets under experiment are the TREC 2004 and TREC 2006 factoid question lists and their corresponding text collection - AQUAINT. The TREC 2004 question list contains 230 factoid questions. We have run the retrieval methods on the subset of 208 questions for which there exists an answer in the document collection. In the TREC 2006 track, there are 403 factoid questions. We have run the experiments on 386 factoid questions in this set for which there is an answer in the document collection.

5.2 Results

We have performed two types of retrieval effectiveness analysis for each passage retrieval method:

Table 1. Analysis of the top 100 passages: the TREC 2004 dataset

	1st answer passage rank					
Method	1-25	25-50	50-75	75-100	Total	Avg.
MultiText	127	15	7	2	151	12.27
Best fusion-based	120	19	8	2	149	14.29
Revision-based	134	9	5	4	152	11.08

Long-List Analysis: An analysis based on the top 100 passages retrieved to observe the effect of each method in terms of the ranking of the first answer passages. Table 1 and Table 2 show the results of this analysis for the TREC 2004 and TREC 2006 datasets. The *Avg.* column in these tables shows the average rank of the first answer passage calculated over the total number of questions for which an answer passage is found in the top 100 passages. From a set of experiments with different fusion parameters (the addition and multiplication operators) and two threshold values {0.5,0.75}, we only report the best fusion method that uses the multiplication operator and selects 0.75 for the threshold value formulated as the fusion task ($\times$,0.75).

Short-List Analysis: An analysis based on the top 10, 15, and 20 passages retrieved by each method to test the effect of the methods regarding a small number of retrieved passages. This analysis is more dedicated to information retrieval-based QA systems where the final answer processing task can be more efficiently and effectively carried out on a small number of retrieved text passages. Here,

Table 2. Analysis of the top 100 passages: the TREC 2006 dataset

Method	1st answer passage rank				Total	Avg.
	1-25	25-50	50-75	75-100		
MultiText	176	32	18	10	236	19.25
Best fusion-based	160	42	24	8	234	21.21
Revision-based	186	31	11	10	238	15.72

Table 3. Accuracy analysis of the top 10, 15, and 20 passages: the TREC 2004 and TREC 2006 datasets

Method	trec 04			trec 06		
	%@10	%@15	%@20	%@10	%@15	%@20
MultiText	48.07	53.36	57.21	28.75	36.01	40.15
Best fusion-based	41.82	47.11	52.88	28.23	33.67	37.56
Revision-based	50.48	58.17*	62.50*	37.04‡	43.26†	45.33*

we report the *accuracy*[3], mean reciprocal rank (mrr)[4], and *F-measure*[5] of the retrieval methods. For calculating F-measure, we do not calculate *precision* values at standard *recall* levels; instead, the precision values are evaluated at the three levels of top 10, 15, and 20 passages retrieved. The main reason for this is the importance of measuring the appearance of answer-containing passages at high ranks. Therefore, our focus is on a limited number of top-ranked passages instead of the distribution of precision at a range of standard recall levels. Table 3, Table 4, and Table 5 show the results regarding accuracy, mrr, and F-measure respectively for both the TREC 2004 and TREC 2006 datasets. The symbols $*$, $†$, and $‡$ show the statistical significance with $p < 0.25$, $p < 0.05$, and $p < 0.01$ respectively[6]. The statistical tests have been carried out between the results obtained by the MultiText and revision-based methods, leaving aside those obtained by the best fusion-based method that do not show any improvement.

5.3 Discussion

From the results in Table 1 and Table 2, it can be seen that the fusion-based method does not assist with retrieval of a greater number of answer passages in either of the datasets. With regard to the high ranks (1-25 and 25-50), the fusion-based method shifts a number of passages from the range of 1-25 to 25-50. As a result, the average ranking of the first answer passage increases from 12.27 (by

---

[3] Accuracy is calculated as the rate of questions with at least a single answer passage to the total number of questions in each dataset.

[4] $mrr = \frac{1}{n_q} \sum_{i=1}^{n_q} \frac{1}{ar_i}$, where n_q is the total number of questions and ar_i stands for the rank of the first answer-bearing passage for the question q_i.

[5] $F_1 = 2 \times \frac{recall \times precision}{recall + precision}$.

[6] The t-test has been carried out to measure the statistical significance level of the results.

Table 4. mrr analysis of the top 10, 15, and 20 passages: the TREC 2004 and TREC 2006 datasets

Method	trec 04			trec 06		
	mrr@10	mrr@15	mrr@20	mrr@10	mrr@15	mrr@20
MultiText	0.29	0.29	0.29	0.16	0.17	0.17
Best fusion-based	0.24	0.24	0.24	0.12	0.13	0.13
Revision-based	0.30	0.30	0.30	$0.25^{\ddagger}$	$0.25^{\ddagger}$	$0.25^{\ddagger}$

Table 5. F-measure analysis of the top 10, 15, and 20 passages: the TREC 2004 and TREC 2006 datasets

Method	trec 04			trec 06		
	F_1@10	F_1@15	F_1@20	F_1@10	F_1@15	F_1@20
MultiText	0.098	0.076	0.061	0.058	0.049	0.042
Best fusion-based	0.085	0.067	0.056	0.054	0.044	0.038
Revision-based	0.103	0.081^{*}	0.068^{*}	$0.070^{\dagger}$	$0.058^{\dagger}$	0.047^{*}

MultiText) to 14.29 by the best fusion-based method. The results summarized in Table 3, Table 4, and Table 5 also show that not only is the best fusion-based method not able to improve the results of the baseline MultiText method, but also it damages the retrieval effectiveness of MultiText in terms of accuracy, mrr, and F-measure in the both TREC 2004 and TREC 2006 datasets.

We have carried out a set of experiments using the fusion-based methodology (only the best of which reported here). Our experiments consisted of different settings for the fusion operation including the mathematical addition and multiplication of the scores obtained from each type of evidence ($\triangle \in \{+, \times\}$, see section 4.2). We also experimented with two threshold values $threshold \in \{0.5, 0.75\}$ for using the semantic class overlap evidence. Surprisingly, none of the fusion-based runs could improve the baseline effectiveness of MultiText. We are planning to carry out an accurate failure analysis to possibly learn ways of improving the fusion-based method.

The revision-based method, however, outperforms MultiText in the experiments on both datasets shown in Table 1 and Table 2. The revision-based method retrieves 7 and 10 more answer passages in the range of 1-25 passages in the TREC 2004 and TREC 2006 datasets respectively. On average, the first answer passage by the revision-based method has the rank of 11.08 and 15.72 in the top 100 passages (for the TREC 2004 and TREC 2006 datasets). This is better than the average ranks 12.27 and 19.25 achieved by MultiText.

Considering the results shown in Table 3, Table 4, and Table 5, the revision-based method quite significantly outperforms MultiText at the levels of top 15 and top 20 passages in terms of accuracy and F-measure in the TREC 2004 dataset. In terms of mrr, however, there is no significant improvement achieved using the revision-based method in the same dataset.

With respect to the TREC 2006 dataset, the accuracy, mrr, and F-measure of the results obtained by the revision-based method are significantly higher than those of the results obtained by MultiText at all retrieval levels.

All our results (which are promising regarding the revision-based method) are based on a term-lookup approach to identifying FrameNet frames. The frame evocation task can be further elaborated by using a shallow semantic parser that can perform word sense disambiguation and focus on the exact semantic classes of terms. This would, however, introduce more complexity to the process of measuring semantic class overlap in terms of efficiency. On the other hand, existing shallow semantic parsers are still far from the best parsing performance that could be achieved.

6   Conclusion

We have tested the effect of using semantic class overlap evidence in enhancing the passage retrieval effectiveness for QA systems. To measure the semantic class overlap between questions and passages, we have used a shallow term-lookup procedure in FrameNet frames to evoke a set of semantic classes related to each question/passage term. We have exploited the semantic class overlap evidence in a fusion-based method and a revision-based method in an effort to enhance the retrieval effectiveness of a baseline passage retrieval method (MultiText).

In our experiments with the TREC 2004 and TREC 2006 datasets, the fusion-based method fails in improving the baseline MultiText method and we are planning to conduct a failure analysis task to understand the underpinning reasons for this.

The revision-based approach that incorporates semantic class overlap evidence in the passage scoring function of MultiText has, however, significantly improved the answer passage retrieval effectiveness of the baseline system. This has been achieved at high retrieval levels in terms of the accuracy, mrr, and F-measure evaluation metrics in the both TREC 2004 and TREC 2006 datasets. This is more significant when considering our shallow frame invocation process that does not involve any shallow semantic parsing stage which is recognized as a real challenge in natural language processing.

References

1. Hovy, E., Gerber, L., Hermjakob, U., Junk, M., Lin, C.-Y.: Question answering in Webclopedia. In: The Ninth Text Retrieval Conference (TREC-9), pp. 655–665 (2000)
2. Clarke, C.L.A., Cormack, G.V., Lynam, T.R., Li, C.M., McLearn, G.L.: Web reinforced question answering (MultiText experiments for TREC 2001). In: Tenth Text Retrieval Conference (TREC 2001), pp. 673–679 (2001)
3. Clarke, C.L.A., Terra, E.L.: Passage retrieval vs. document retrieval for factoid question answering. In: The 26th Annual International ACM-SIGIR Conference on Research and Development in Information Retrieval. Toronto, Canada, pp. 427–428 (2003)

4. Baker, C.F., Fillmore, C.J., Lowe, J.B.: The Berkeley FrameNet project. In: The 17th International Conference on Computational Linguistics (COLING 1998), pp. 86–90. Universite de Montreal, Montreal (1998)

5. Clarke, C.L.A., Cormack, G.V., Kisman, D., Lynam, T.: Question answering by passage selection (MultiText experiments for TREC-9). In: The Ninth Text Retrieval Conference (TREC-9), pp. 673–684 (2000)

6. Cormack, G., Palmer, C., Biesbrouck, M., Clarke, C.: Deriving very short queries for high precision and recall (MultiText experiments for TREC-7). In: The Seventh Text Retrieval Conference (TREC-7), pp. 121–133 (1998)

7. Tellex, S., Katz, B., Lin, J., Fernandes, A., Marton, G.: Quantitative evaluation of passage retrieval algorithms for question answering. In: The 26th Annual International ACM-SIGIR Conference on Research and Development in Information Retrieval. Toronto, Canada, pp. 41–47 (2003)

8. Fillmore, C.J.: The case for case. Universals in Linguistic Theory, 1–88 (1968)

9. Cook, W.A.: Case grammar theory. Georgetown University Press (1989)

10. Fillmore, C.J.: Frame semantics and the nature of language. In: The Annals of the New York Academy of Sciences: Conference on the Origin and Development of Language and Speech, vol. 280, pp. 20–32 (1976)

11. Lowe, J.B., Baker, C.F., Fillmore, C.J.: A frame-semantic approach to semantic annotation. In: SIGLEX Workshop on Tagging Text with Lexical Semantics: Why, What, and How? (1997)

12. Petruck, M.R.L.: Frame semantics. Handbook of Pragmatics. John Benjamins, Philadelphia (1996)

13. Harabagiu, S., Maiorano, S.J.: Finding answers in large collections of texts: Paragraph indexing + abductive inference. In: AAAI 1999, pp. 63–71 (1999)

14. Lee, Y.-S., Hwang, Y.-S., Rim, H.-C.: Variable length passage retrieval for Q&A system. In: The 14th Hangul and Korean Information Processing, pp. 259–266 (2002)

15. Moldovan, D., Pasca, M., Harabagiu, S., Surdeanu, M.: Performance issue and error analysis in an open-domain question answering system. The ACM Transactions on Information Systems (TOIS) 21, 113–154 (2003)

16. Clarke, C.L.A., Cormack, G.V., Burkowski, F.: Shortest substring ranking (MultiText experiments for TREC-4). In: The Fourth Text Retrieval Conference (TREC-4), pp. 295–304 (1995)

17. Clarke, C.L.A., Cormack, G.V., Tudhope, E.A.: Relevance ranking for one to three term queries. Information Processing and Management 36, 291–311 (2000)

English Article Correction System Using Semantic Category Based Inductive Learning Rules

Hokuto Ototake and Kenji Araki

Graduate School of Information Science and Technology, Hokkaido University,
Kita-14, Nishi-9, Kita-ku, Sapporo, 060-0814 Japan
{hokuto, araki}@media.eng.hokudai.ac.jp

Abstract. In this paper, we describe a system for automatic correction of English. Our system uses rules based on article context features, and generates new abstract rules by Semantic Category Based Inductive Learning that we proposed before. In the experiments, we achieve 93% precision with the best set of parameters. This method scored higher than our previous system, and is competitive with a related method for the same task.

1 Introduction

Using articles correctly is difficult for Japanese learners of English. Lee[1] reported that native speakers of languages which do not have any articles often have difficulty in choosing appropriate English articles, and tend to underuse them. Also Kawai et al.[2] reported there are many article errors in English written by Japanese.

In view of these circumstances, various methods for automatic detecting or correcting article errors have been proposed in the past. Kawai et al.[2] proposed a rule-based method which uses rules made by hand based on linguistic knowledge. To make such rules much effort and a lot of expenses are needed, and it is very difficult to achieve good coverage of articles usage. Izumi et al.[3] trained a maximum entropy classifier to recognize various errors including articles using contextual features. They reported the results for different error types (omission - precision 76%, recall 46%; replacement - precision 31%, recall 8%). Yi et al.[4] proposed a web count-based system for article error correction achieving precision 62% and recall 41%. De Felice et al[5] proposed a classifier-based approach to correct article and preposition errors (accuracy 92% for articles).

We also proposed an article error correction system[6] which corrects errors using rules extracted and generated automatically by Inductive Learning[7]. Apart from the performance, it has an advantage of being more understandable while investigating why a given article was erroneous because of using rules. However, its performance needs to be improved.

To improve the performance of our system, we propose an introduction of WordNet category information into rules for article error correction and Semantic

A. Nicholson and X. Li (Eds.): AI 2009, LNAI 5866, pp. 597–606, 2009.

Category Based Inductive Learning. Considering semantic category information, the learning process of our system can generate more reliable rules even though the system does not have any rules with the same head noun as user input. We aim to improve precision of article error correction.

The rest of the paper is organized as follows. In section 2, we describe details of the proposed system. In section 3, we evaluate the system and finally conclusions are given in section 4.

2   System

Fig. 1 shows the process flow of our system. When the system gets English sentences including article errors, it extracts feature slots. Using the resultant feature slots, the system extracts moderate amounts of rules from the rule database in order to generate abstract rules by Inductive Learning that is described in **2.3**. The system corrects article errors by the correction algorithm using the resultant rules. In the correction algorithm, the system calculates scores of rules to rank them by reliability.

We describe the details for each part of the process below.

2.1   Feature Slots and Rules

We use similar feature slots and rules as our previous system[6]. Consulting literature concerning usage of articles[8,9], we define feature slots that can be easily extracted automatically for the context within a sentence. A difference between the proposed system and our previously proposed system[6] is whether slots for nouns and verbs have WordNet category information or not. The proposed system extracts category information from WordNet 3.0 when it extracts feature

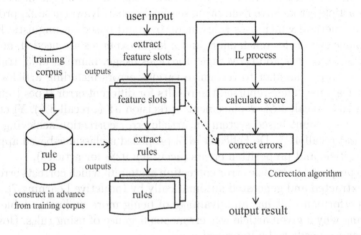

Fig. 1. Process flow

(i) This is the only *soccer ball* which I bought yesterday.

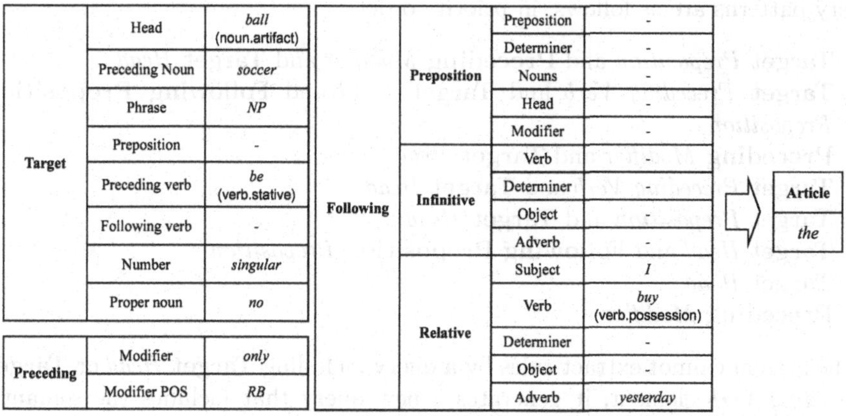

Target		
	Head	*ball* (noun.artifact)
	Preceding Noun	*soccer*
	Phrase	*NP*
	Preposition	-
	Preceding verb	*be* (verb.stative)
	Following verb	-
	Number	*singular*
	Proper noun	*no*
Preceding	Modifier	*only*
	Modifier POS	*RB*

Following			
Preposition	Preposition	-	
	Determiner	-	
	Nouns	-	
	Head	-	
	Modifier	-	
Infinitive	Verb	-	
	Determiner	-	
	Object	-	
	Adverb	-	
Relative	Subject	*I*	
	Verb	*buy* (verb.possession)	
	Determiner	-	
	Object	-	
	Adverb	*yesterday*	

⇨ **Article** *the*

*) Slot element "-" means that no corresponding element is present.

Fig. 2. An example of a rule

slots. There are 26 semantic categories for nouns (e.g. 'noun.act', 'noun.food'), and 15 categories for verbs (e.g. 'verb.change', 'verb.creation'). A rule consists of a combination of an article and such feature slots. They are extracted from POS-tagged and chunked sentences automatically.

The feature slots and rule extracted from "soccer ball" in sentence (i) is as shown in Fig. 2. Feature slots consists of three categories, **Target**, **Preceding** and **Following**. **Target** category has context information about a target noun. Since "soccer ball" is a compound noun, "ball" is put into the *Head* slot and "soccer" is put into *Preceding Noun*. *Head* "ball" has category "noun.artifact" and *Preceding verb* has category "verb.stative". These categories are also put into *Head* and *Preceding verb* slots respectively. The **Preceding** category means modifiers preceding the target noun phrase, such as an adjective or adverb. The **Following** category contains information about modifiers following the target noun, such as prepositions, infinitive and relative clauses.

A requirement when applying rules to new text is that the rule feature slots and the context of the target noun must agree. If the slot element is "-", it counts as agreeing with any other element.

2.2 Rule Extraction

After extracting feature slots of user input, the system tries to extract rules that have feature slots similar to the user input from a rule database. These extracted rules are used in next learning process. The rule database is constructed from a training corpus in advance. The reason why we try to extract moderate amounts of rules for every user input is that it is impossible for the system to generate rules from the whole large corpus (say over 100 million words) because of the exponential growth of the recursive Inductive Learning.

When the system receives feature slots of user input, it searches rules from the rule database using queries based on feature slot elements of the input. The query patterns are as follows, in priority order:

1. **Target** *Preposition* and **Preceding** *Modifier* and **Target** *Head*
2. **Target** *Preceding Verb* and **Target** *Head* and **Following Preposition** *Preposition*
3. **Preceding** *Modifier* and **Target** *Head*
4. **Target** *Preceding Verb* and **Target** *Head*
5. **Target** *Preposition* and **Target** *Head*
6. **Target** *Head* and **Following Preposition** *Preposition*
7. **Target** *Head*
8. **Preceding** *Modifier*

If the system cannot extract rules by a query including **Target** *Head* or **Target** *Preceding Verb* element, it generates a new query that includes the semantic category information of the element instead of the element value. Using such queries with semantic category information, the system can get appropriate rules to be used in the learning process even though the rule database has no identical words as head nouns or verbs of user input.

2.3 Rule Abstraction

The system generates new abstract rules from extracted rules based on Inductive Learning (IL) that is proposed by us originally. IL is defined as discovering inherent regularities from actual examples by extracting their common and different parts recursively[7]. In our case, the actual examples are represented as the feature slots extracted from training data.

In the IL process, by extracting common and different elements recursively from comparison of feature slots of two rules, abstract rules are generated one after another. The abstract rules have the common parts of elements and another parts are abstracted.

However, by IL, abstract rules have only two kinds of elements, concrete element and full abstract element. Full abstract element is allowed to agree with any element. There is a possibility that many abstract rules which are too generalized and possess lower reliability are generated because of ease of full abstraction.

In this paper, we propose Semantic Category Based IL (SCB-IL). SCB-IL supplies additional granularity for rule abstraction, which abstract rules can have abstract elements with category information. We aim both to enhance accuracy of error correction and to carry on versatility of abstract rules using SCB-IL.

Fig. 3 shows an example of rule abstraction by SCB-IL. In Fig. 3, rule 1 is the same as Fig. 2 from sentence (i). Rule 2 and rule 4 are extracted from sentence (ii) and (iii) respectively.

Rule 3 is an abstract rule generated from rule 1 and rule 2. These two upper rules have many common slot elements: all of **Preceding**, a portion of **Target** and the article. The other elements differ. **Target** *Head* elements also differ, but

(i) This is the only *soccer ball* which I bought yesterday.

(ii) I have the only *guitar*.

(iii) Bobby is the only *child* in his family.

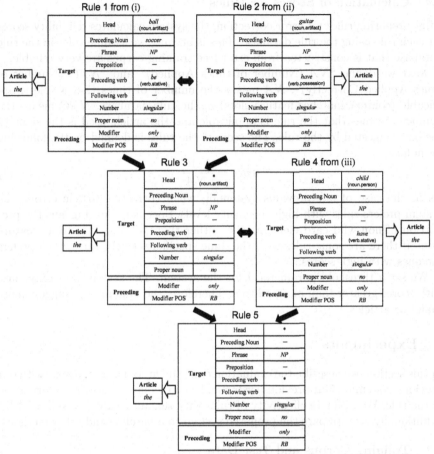

*) For clarity, the **Following** category is not included.

Fig. 3. An example of when the SCB-IL process generates new abstract rules

these categories are the same. SCB-IL thus generates rule 3 as a new abstract rule which has these common slot elements, an abstract element with category information and full abstract elements at the differing slots. If slots of source rules have different kinds of categories as **Target** *Head* slots of rule 3 and rule 4, the slot of the resultant rule is fully abstracted as rule 5. In Fig. 3, elements with "*" similarly to "-" mean the slot is allowed to agree with any element.

If the only requirement for SCB-IL was that two rules must have common slot elements, many rules abstracted too much would be generated. Thus, we add the requirements that the article elements of the rules must agree and that

Target *Head* or **Preceding** *Modifier* must also agree. If **Target** *Head* elements disagree but their categories agree, the system can continue SCB-IL. SCB-IL then generates rules recursively until no rules meeting these requirements remain.

2.4   Calculation of Scores for Rules

After preparing rules for error correction, the system calculates reliability scores for each rule using the rule database. This is because we assume rules in the rule database that is constructed from error free training corpus are very reliable.

Next we define "NA (Number of times Applied)" and "NAC (Number of times Applied Correctly)". NA means the number of times that a rule is applicable (context matches feature slots) in the rule database. NAC means the number of times that the rule is applicable and the article used is the same as the one suggested by the rule. The score of the rules is defined by the following formula:

$$score = \frac{NAC}{NA} \tag{1}$$

As in the case of our previous system[6], when correcting article errors, the system prefers rules with high specificity levels. Specificity level means the proportion of non-wild card elements (i.e., other than "-" and "*") to all feature slots. If there are some rules with the same specificity levels, the system prefers the ones with high scores.

We set a threshold parameter θ for the system. The system uses only rules with scores greater than or equal to θ. Therefore, the system may suggest more than one articles.

3   Experiments

In this section, we describe the evaluation experiment and a comparison with our previous system[6]. Our previous system[6] has no semantic category information in the rule. We evaluate effectivity of rules with semantic category and SCB-IL. Additionally, we compare the proposed system to a baseline and other methods.

3.1   Training Corpus and Test Data

We use the Reuters Corpus[10] (about 188 million words) as a source of the rule database. Constructing the rule database from the corpus in advance, it has 46,140,689 rules. We use Brill's Tagger[11] to extract feature slots from tagged sentences. The test data is sentences that include 2,269 article instances from the Reuters Corpus separate from the rule database. Since the test data does not include article errors, we evaluate whether the system suggests the same article as in the test data.

3.2   Experimental Procedures

We compare the proposed system and our previous system[6] results to the test data for $\theta = 0$ to 1 with 0.2 increments in between, and evaluate its precision (P) and recall (R) that are defined as follows:

$$P = \frac{\text{the number of suggesting articles correctly}}{\text{articles for which at least one rule matches the context}} \qquad (2)$$

$$R = \frac{\text{the number of suggesting articles correctly}}{\text{all existing articles in the test data}} \qquad (3)$$

And we also define F (F-measure) as follows:

$$F = \frac{2 \cdot P \cdot R}{P + R} \qquad (4)$$

A requirement of suggesting articles correctly is that the rule with the highest score suggests the correct article.

Our proposed and previous systems need many rule extraction and abstraction processes if input has many noun phrases. Furthermore, these are not very lightweight processes. Therefore, in this experiment, we set 20 as the upper limit of the number of extracting rules from the rule database per noun phrase.

3.3 Results and Discussion

Fig. 4, 5 and 6 show the results of precision, recall and F-measure respectively. The threshold θ has a little effect on the change of the precision and recall of both systems, except for $\theta = 1.0$ in Fig. 5. The reason why only $\theta = 1.0$ degrades the recall is that there are very fewer rules with score 1.0 than others. In the course of the scores calculation, many rules have scores less than 1.0 even if the rule is reliable. Many reliable rules have scores close to 1.0 but not equal to 1.0.

Comparing both systems, the proposed system achieves better precision than the previous system in every θ as Fig. 4. The best precision of the proposed system is 0.93 with $\theta = 0.8$. That of the previous system is 0.87 with $\theta = 1.0$. On the other hand, the previous system has better recall performance than

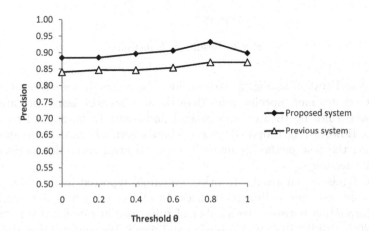

Fig. 4. Results of precision

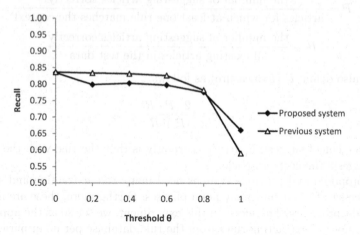

Fig. 5. Results of recall

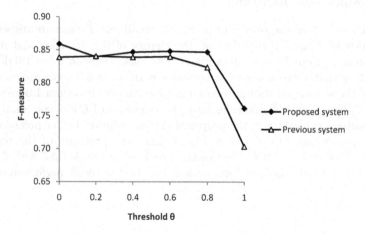

Fig. 6. Results of F-measure

the proposed system as Fig. 5. We consider the reason is the effect of SCB-IL which generates more specific rules than IL, as rules with semantic categories. Conversely, rules by IL are more general and easier to be applied to various contexts. However, the proposed system achieves better F-measure for about 2%. Therefore, the new method is more effective for error correction performance, especially accuracy.

Table 1 relates our results to other system[5] reported in the literature on comparable tasks for individual articles. In our system, it is not possible to find a correlation between the number of rules seen in rule database ("a":8%, "the":20%, "null":72%) and precision and recall. We consider that the lowest performance for "the" is attributed to a lack of information of a rule. Choice of a definite article depends not only on context within a processed sentence but

Table 1. Individual articles results

System	Article	Precision	Recall
Baseline	a	-	0.00
	the	-	0.00
	null	0.72	1.00
De Felice et al.[5]	a	0.71	0.54
	the	0.85	0.92
	null	0.99	0.99
Our system ($\theta = 0.8$)	a	0.94	0.82
	the	0.88	0.69
	null	0.94	0.79

also on context of the preceding sentence or semantic information. A currently used rule has no information about the preceding sentence and sense except for WordNet category. We believe that the rule should have such information for performance improvement about definite articles.

In Table 1, The baseline refers to the most frequent article, "null". Our system outperforms baseline in the comparison of precision. And in comparison of our system and De Felice et al.[5], our system gets over 20% better performance of both precision and recall for "a". For other articles, our system can achieve similar precision to theirs. However, recall of our system is about 20% lower than theirs except for "a". Our system suggests no article when it cannot generate rules that matches input context and we consider it as one of the factors of lower recall.

4   Conclusions and Future Work

In this paper, we proposed a system for correcting article errors based on automatic rules generation by Semantic Category Based Inductive Learning. Our proposed system achieved better precision performance than our previous system. We confirmed that the introduction of WordNet category information into rules for article error correction and SCB-IL have positive effect on our system.

However, our experiments are not sufficient in the following respects. First, the size of test data is too small. For example, De Felice et al.[5] used over 300,000 article contexts for test data. Second, our experiments have no error correction for sentences that are written by second learners of English. We try to evaluate the system with larger test data and with sentences including article errors by the second learners of English in the near future.

References

1. Lee, J.: Automatic Article Restoration. In: Proc. HLT/NAACL Student Research Workshop, Boston USA, pp. 195–200 (2004)
2. Kawai, A., Sugihara, K., Sugie, N.: ASPEC-I: An error detection system for English composition (in Japanese). Transactions of Information Processing Society of Japan 25(6), 1072–1079 (1984)

3. Izumi, E., Uchimoto, K., Isahara, H.: SST speech corpus of Japanese learners' English and automatic detection of learners' errors. ICAME Journal 28, 31–48 (2004)

4. Yi, X., Gao, J., Dolan, W.B.: A Web-based English Proofing System for English as a Second Language Users. In: Proc. IJCNLP 2008, Hyderabad, India, pp. 619–624 (2008)

5. Felice, R.D., Pulman, S.G.: A classifier-based approach to preposition and determiner error correction in L2 English. In: Proc. 22nd International Conference on Computational Linguistics (Coling 2008), Manchester, UK, pp. 169–176 (2008)

6. Ototake, H., Araki, K.: Extraction of Useful Training Data for Article Correction Depending on User Inputs. In: Proc. the 2008 Empirical Methods for Asian Languages Processing Workshop (EMALP 2008) at The Tenth Pacific Rim International Conference on Artificial Intelligence (PRICAI 2008), Hanoi, Vietnam, pp. 92–101 (2008)

7. Araki, K., Tochinai, K.: Effectiveness of natural language processing method using inductive learning. In: Proc. IASTED International Conference Artificial Intelligence and Soft Computing, Mexico, pp. 295–300 (2001)

8. Cole, T.: The Article Book. The University of Michigan Press, Michigan (2000)

9. Swan, M.: Practical English Usage, 2nd edn. Oxford University Press, Oxford (1995)

10. Reuters Corpus, http://trec.nist.gov/data/reuters/reuters.html

11. Brill, E.: Some Advances in Transformation-Based Part of Speech Tagging. In: Proc. the twelfth National Conference on Artificial Intelligence, Seattle, Washington, USA, vol. 1, pp. 722–727 (1994)

Towards Interpreting Task-Oriented Utterance Sequences

Patrick Ye and Ingrid Zukerman

Faculty of Information Technology, Monash University,
Clayton, VICTORIA 3800, Australia
ye.patrick@gmail.com, ingrid@infotech.monash.edu.au

Abstract. This paper describes a probabilistic mechanism for the interpretation of utterance sequences in a task-oriented domain. The mechanism receives as input a sequence of sentences, and produces an interpretation which integrates the interpretations of individual sentences. For our evaluation, we collected a corpus of hypothetical requests to a robot, which comprise different numbers of sentences of different length and complexity. Our results are promising, but further improvements are required in our algorithm.

1   Introduction

DORIS (Dialogue Oriented Roaming Interactive System) is a spoken dialogue system under development, which will eventually be mounted on a household robot. In this paper, we describe our most recent work on *Scusi?*, *DORIS*'s language interpretation module, focusing on its mechanism for the interpretation of a sequence of utterances.

People often utter several separate sentences to convey their requests, rather than producing a single sentence that contains all the relevant information. For instance, people are likely to say "Go to my office. Get my mug. It is on the table", instead of "Get my mug on the table in my office". This observation, which was validated in our corpus study (Section 4), motivates the mechanism for the interpretation of a sequence of utterances presented in this paper.

Our previous work focused on interpreting single-sentence utterances, where each sentence is a command to *DORIS* [1]; and two-sentence utterances, where one sentence is a command, and the other further specifies the command [2]. In this paper, we extend our previous work, offering a probabilistic mechanism for interpreting multi-sentence utterances.[1] This mechanism combines sentence mode classification (declarative or imperative), sequential coreference resolution, and sequential sentence interpretation to produce integrated interpretations of multi-sentence requests, and estimate the probability of these interpretations. Specifically, the sentence mode classification determines whether the interpretation of a sentence in a request should be combined with the interpretations of other sentences; and the coreference resolution determines which portions of the individual interpretations should be combined.

Our evaluation demonstrates that our mechanism exhibits creditable performance for requests comprising different numbers of sentences of different length and level

---

[1] Since all of our test utterances are requests (Section 4), we henceforth refer to them as requests.

A. Nicholson and X. Li (Eds.): AI 2009, LNAI 5866, pp. 607–616, 2009.

of complexity, and highlights particular aspects of our algorithms that require further inspection (Section 4).

This paper is organized as follows. In the next section, we describe our mechanism for interpreting an utterance sequence. In Section 3, we present our formalism for estimating the probability of an interpretation. The performance of our system is evaluated in Section 4, followed by related research and concluding remarks.

2   Interpreting a Sequence of Utterances

We reported in previous work [1,2] that *Scusi?* follows a pipeline architecture, where a speech wave is first transformed to text, which is then parsed with the Charniak parser[2] [3]. The resultant parse tree is then converted to an *Uninstantiated Concept Graph (UCG)* — a meaning representation based on Concept Graphs [4], which is finally grounded to a virtual world in the form of an *Instantiated Concept Graph (ICG)*. Figure 1 shows the pictorial and textual form of a UCG sequence generated for the request "Go to the desk near the computer. The mug is on the desk near the phone. Fetch it for me." (this request is typical of our corpus).[3]

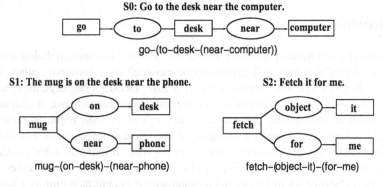

Fig. 1. Sample UCGs for a 3-sentence sequence

In this paper, we extend our previous work to interpret multi-sentence requests. To avoid issues arising from speech recognition error, and focus on the integration of single-sentence interpretations, we assume that we have the correct text for the speech wave. Further, we focus on the interpretation process up to the UCG stage. This process is broken down into four tasks: (1) generating candidate UCG sequences from individual UCGs, (2) determining the sentence mode of the individual sentences, (3) resolving the coreferences in a sentence sequence, and (4) merging the UCGs in a sequence.

2.1   Generating Candidate UCG Sequences from Individual UCGs

The interpretation of a multi-sentence request starts by composing candidate UCG sequences from the candidate UCGs generated for each sentence. The process of

---

[2] Version of August 2009 using a language model trained on the Brown Corpus.

[3] The textual form is a short-hand representation which will be used in the rest of the paper.

Sentence	UCG	Prob.	Prob.	Candidate UCG sequence
S0	go-(to-desk-(near-computer))	0.6		go-(to-desk-(near-computer))
	*go-(to-desk)-(near-computer)	0.4	0.264	mug-(on-desk)-(near-phone)
S1	mug-(on-desk)-(near-phone)	0.55		fetch-(object-it)-(for-me)
	mug-(on-desk-(near-phone))	0.45		go-(to-desk-(near-computer))
S2	fetch-(object-it)-(for-me)	0.8	0.216	mug-(on-desk-(near-phone))
	*fetch-(object-it-(for-me))	0.2		fetch-(object-it)-(for-me)

(a) Top two UCGs for the individual sentences (b) Top two candidate UCG sequences

Fig. 2. Example of candidate UCG sequence generation

transforming parse trees to UCGs has been described in [1]. Since the intended parse tree for a sentence is often not top ranked by the Charniak parser, we decided to consider up to 50 parse trees for each sentence during the interpretation process, which yields multiple UCGs for each sentence. The candidate UCG sequences are generated by combining the most probable UCGs for the individual sentences (the calculation of the probability of a UCG is described in [1] and outlined in Section 3). For example, Figure 2(a) shows the top two UCGs generated for each sentence in Figure 1, together with their probabilities. The top two UCG sequences produced from these UCGs appear in Figure 2(b). Given the combinatorial nature of the process for generating candidate UCG sequences, we applied a thresholding technique to limit the total number of candidate UCG sequences. Using this technique, a candidate UCG sequence is retained for further processing only if its probability matches or exceeds a threshold relative to the probability of the top-ranked candidate UCG sequence.[4]

2.2 Determining the Sentence Mode for Individual Sentences

Since a user's request contains commands (imperative sentences) and specifications to the commands (declarative sentences), it is necessary to distinguish between these two types of sentences so that we can determine how the information in these UCGs should be combined. This is the task of sentence-mode classification. At present, we consider only two modes: imperative and declarative.

We trained a Maximum Entropy based classifier[5] [5] to assign mode to the sentences in our corpus. The features used (obtained from the highest probability parse tree for a sentence) are: (1) top parse-tree node; (2) types of the top level phrases under the top parse-tree node combined with their positions, e.g., (0, NP), (1, VP), (2, PP); (3) top phrases under the top parse-tree node reduced to a regular expression, e.g., VP-NP$^+$ to represent, say, VP NP NP; (4) top VP head – the head word of the first top level VP; (5) top NP head – the head word of the first top level NP; (6) first three tokens in the sentence; and (7) last token in the sentence. Using leave-one-out cross validation, this classifier has an accuracy of 99.2% on the test data — a 33% improvement over the majority class (imperative) baseline. As for the generation of candidate UCG sequences, we applied a threshold to retain the most probable sentence-mode sequences.

---

[4] We use an empirically determined threshold of 80% for all stages of the interpretation process. The interpretations whose probability is below this threshold are marked with an asterisk (*).

[5] http://homepages.inf.ed.ac.uk/s0450736/maxent_toolkit.html

2.3   Resolving the Coreferences in a Sentence Sequence

UCGs obtained from declarative sentences usually contain descriptions of objects. In order to produce a fully specified UCG sequence, these descriptions must be appropriately incorporated in the (declarative or imperative) UCGs which mention these objects. A first step in this process is the resolution of the coreferences in the noun phrases of the UCGs.

Scusi? handles pronouns, one-anaphora (e.g., "the blue one") and NP identifiers (e.g., "the book"). At present, we consider only exact matches between NP identifiers and referents, e.g., "the cup" does not match "the dish". In the future, we will incorporate similarity scores, e.g., Leacock and Chodorow's WordNet-based scores for approximate lexical matches [6]; such matches occurred in 4% of our corpus (Section 4).

To reduce the complexity of reference resolution across a sequence of sentences, and the amount of data required to reliably estimate probabilities (Section 3), we separate the coreference resolution problem into two parts: (1) identifying the sentence being referred to, and (2) determining the referent within that sentence.

Identifying a sentence. Most referents in our corpus appear in the *current, previous* or *first* sentence in a sequence, with a few referents appearing in *other* sentences. Hence, we have chosen the sentence classes {*current, previous, first, other*}. The probability of referring to a sentence of a particular class from a sentence in position i is estimated from our corpus, where $i = 1, \ldots, 5, > 5$ (there are only 13 sequences with more than 5 sentences). This distribution is estimated for each fold of the leave-one-out cross validation (Section 4).

Determining a referent. We use heuristics based on those described in [7] to classify pronouns, and heuristics based on the results obtained in [8] to classify one-anaphora. If a term is classified as a pronoun or one-anaphor, then a list of potential referents is constructed using the head nouns in the target sentence. We use the values in [7] to assign a score to each anaphor-referent pair according to the grammatical role of the referent in the target UCG (obtained from the highest probability parse tree that is a parent of this UCG). These scores are then converted to probabilities using a linear mapping function.

After we have calculated the coreference scores for every pair of noun phrases in a request, we generate candidate coreference sequences using the same thresholding method as for the UCG and sentence-mode sequences.

2.4   Merging the UCGs in a Sequence

After the candidate sequences of UCGs (*US*), sentence modes (*MS*) and coreferences (*CS*) have been determined, we generate the most probable combinations of these candidate sequences. For example, Figures 3(a) and 3(b) respectively show the top two candidate sentence-mode sequences and coreference sequences for the 3-sentence request in Figure 1, together with their probabilities (the (desk(s1),desk(s0)) coreference pair is omitted for clarity of exposition). The resultant top two combinations of UCG, mode and coreference sequences and their probabilities appear in Figure 3(c). The UCG sequences differ in the attachment of the prepositional phrase (PP) "near the phone" (boldfaced).

Prob.	Sentence mode sequence	Prob.	Coref. resolution sequence for pronouns
0.8	Imperative, Declarative, Imperative	0.48	[(it, mug)]
0.1	*Imperative, Imperative, Declarative	0.384	[(it, phone)]
(a) Candidate sentence-mode sequences		(b) Candidate coreference resolution sequences	

Prob.	Candidate combinations
0.1014	{Sentence mode: [Imperative, Declarative, Imperative] }, {Coref.: [(it, mug)]} {UCG: [go-(to-desk-(near-computer)), **mug-(on-desk)-(near-phone)**, fetch-(object-it)-(for-me)]}
0.0829	{Sentence mode: [Imperative, Declarative, Imperative]}, {Coref.: [(it, mug)]} {UCG: [go-(to-desk-(near-computer)), **mug-(on-desk-(near-phone))**, fetch-(object-it)-(for-me)]}

(c) Top two candidate combinations

Fig. 3. Example of combined candidate sequences

Given a combination of US_i, MS_j and CS_k, we first replace the pronouns in the UCGs of US_i with their antecedents. We then aggregate the information about the objects mentioned in the declarative UCGs by merging these UCGs (the algorithm for merging two UCGs is described in [2]). The resultant (possibly merged) declarative UCGs are combined with imperative UCGs that mention the objects described in the declarative UCGs. For example, the third UCG in the top-ranked candidate in Figure 3(c) yields fetch-(object-**mug-(on-desk)-(near-phone)**)-(for-me). The generation of a fully specified UCG sequence takes under 1 second on a typical desktop.

A limitation of this merging process is that the information about the objects specified in an imperative UCG is not aggregated with the information about these objects in other imperative UCGs. This may cause some imperative UCGs to be under-specified.

3   Estimating the Probability of a Merged Interpretation

We now present a formulation for estimating the probability of a sequence of UCGs, which supports the selection of the sequence with the highest probability.

One sentence. The probability of a UCG generated from a sentence T is estimated as described in [1].

$$\Pr(U|T) \propto \sum_P \Pr(P|T) \cdot \Pr(U|P) \qquad (1)$$

where T, P and U denote text, parse tree and UCG respectively. The Charniak parser returns an estimate of $\Pr(P|T)$; and $\Pr(U|P)=1$, since the process of generating a UCG from a parse tree is deterministic.

A sentence sequence. An interpretation of a sequence of sentences $T_1, \ldots, T_n$ comprises a sequence of (possibly merged) UCGs $\mathbf{U_1}, \ldots, \mathbf{U_m}$ obtained by combining the UCGs for individual sentences $U_1, \ldots, U_n$ as prescribed by their mode sequence $M_1, \ldots, M_n$ and coreference sequence $R_1, \ldots, R_n$. Thus, the probability of an interpretation of a sequence of sentences is

$$\Pr(\mathbf{U_1}, \ldots, \mathbf{U_m}) = \Pr(U_1, \ldots, U_n, M_1, \ldots, M_n, R_1, \ldots, R_n | T_1, \ldots, T_n)$$

By making judicious conditional independence assumptions, and incorporating parse trees into the formulation, we obtain

$$\Pr(\mathbf{U_1}, \ldots, \mathbf{U_m}) = \prod_i^n \Pr(U_i|T_i) \cdot \Pr(M_i|P_i, T_i) \cdot \Pr(R_i|P_1, \ldots, P_i) \qquad (2)$$

This formulation is independent of the number of UCGs (m) in the merged sequence, thereby supporting the comparison of sequences of different lengths (which are produced when different numbers of mergers are performed).

$\Pr(U_i|T_i)$ is obtained from Equation 1, and $\Pr(M_i|P_i, T_i)$ is obtained as described in Section 2.2, where P_i is the highest-probability parse tree for sentence i (recall that the input features to the classifier depend on the parse tree and the sentence).

To estimate $\Pr(R_i|P_1, \ldots, P_i)$ we assume conditional independence between the identifiers in a sentence, yielding

$$\Pr(R_i|P_1, \ldots, P_i) = \prod_{j=1}^{k_i} \Pr(R_{ij}|P_1, \ldots, P_i)$$

where k_i is the number of identifiers in sentence i, R_{ij} is the referent for anaphor j in sentence i, denoted A_{ij}, and $P_1, \ldots, P_i$ are the highest probability parse trees that are parents of $U_1, \ldots, U_i$ respectively. As mentioned in Section 2.3, this factor is separated into determining a sentence, and determining a referent in that sentence. We also include in our formulation the Type of anaphor A_{ij} (pronoun, one-anaphor or NP) and sentence position i, yielding

$$\Pr(R_{ij}|P_1, \ldots, P_i) = \Pr(A_{ij} \text{ refer to } NP_a \text{ in sent } b, \text{Type}(A_{ij})|i, P_1, \ldots, P_i)$$

After making conditional independence assumptions we obtain

$$\Pr(R_{ij}|P_1, \ldots, P_i) = \Pr(A_{ij} \text{ refer to } NP_a|A_{ij} \text{ refer to sent } b, \text{Type}(A_{ij}), P_i, P_b) \times$$
$$\Pr(A_{ij} \text{ refer to sent } b|\text{Type}(A_{ij}), i) \times \Pr(\text{Type}(A_{ij})|P_i)$$

As stated in Section 2.3, $\Pr(A_{ij} \text{ refer to } NP_a|A_{ij} \text{ refer to sent } b, \text{Type}(A_{ij}), P_i, P_b)$ and $\Pr(\text{Type}(A_{ij})|P_i)$ are estimated in a rule-based manner, while statistics obtained from the corpus are used to estimate $\Pr(A_{ij} \text{ refer to sent } b|\text{Type}(A_{ij}), i)$ (recall that we distinguish between sentence classes, rather than specific sentence positions).

4   Evaluation

We conducted a web-based survey to collect a corpus comprising multi-sentence utterances. To this effect, we presented participants with a scenario where they are in a meeting room, and they ask a robot to fetch something from their office. The idea is that if people cannot see a scene, their instructions would be more segmented than if they can view the scene. The participants were free to decide which object to fetch, what was in the office, and how to phrase their requests.

We collected 115 requests mostly from different participants (a few people did the survey more than once). These requests contain between 1 and 9 sentences. Figure 4(a) shows the frequency of the number of sentences in a request, e.g., 51 requests consist of 2 sentences; around 75% of the requests consist of 1 to 3 sentences. Our evaluation

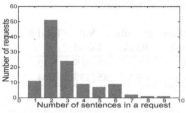
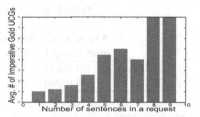

(a) Frequency of number of sentences in a request

(b) Avg. # of imperative Gold UCGs for a request of N sentences ($N = 1, \ldots, 9$)

Fig. 4. Distribution of the requests and Gold Standard UCGs in the corpus

Table 1. Original and modified texts

Original	Get my book *"The Wizard of Oz"* from my office. It's green *and yellow. It has a picture of a dog and a girl on it.* It's in my *desk* drawer on the right *side* of my desk, *the second drawer down. If it's not there, it's somewhere on my shelves that are on the left side of my office as you face the window.*
Modified	Get my book from my office. It's green. It's in my drawer on the right of my desk.
Original	*DORIS, I left* my mug in my office *and I want a coffee. Can you* go into my office *and* get my mug? It is on top of the cabinet *that is* on the left *side* of my desk.
Modified	My mug is in my office. Go into my office. Get my mug. It is on top of the cabinet on the left of my desk.

focuses on imperative UCG sequences, as they contain the actions the robot is expected to perform. Figure 4(b) shows the average number of imperative Gold Standard UCGs for requests of N sentences, e.g., the average number of command Gold UCGs for 3-sentence requests is 1.7. These Gold Standard UCG sequences were manually constructed by a tagger and the authors through consensus-based annotation [9]. As seen in Figure 4(b), many requests of 1 to 3 sentences can be transformed to a single command UCG, but requests with more sentences tend to include several commands.

Many of the sentences in our corpus had grammatical requirements which exceeded the capabilities of our system. In order to focus our attention on the interpretation and merging of an arbitrary number of UCGs, we made systematic manual changes to produce sentences that meet our system's grammatical restrictions (in the future, we will relax these restrictions, as required by a deployable system). The main types of changes we made are: (1) indirect speech acts in the form of questions were changed to commands; (2) sentences with relative clauses were changed to two sentences; (3) conjoined verb phrases or sentences were separated into individual sentences; (4) composite verbs were simplified, e.g., "I think I left it on" was changed to "it is on"; (5) composite nouns were replaced by simple nouns or adjective-noun NPs; and (6) conditional sentences were removed. Table 1 shows two original texts compared with the corresponding modified texts (the changed portions in the originals have been italicized).

We tested *Scusi?*'s performance using leave-one-out cross validation. An interpretation generated by *Scusi?* was deemed successful if it correctly represented the speaker's intention, which was encoded by a sequence of imperative Gold Standard UCGs.

Table 2. *Scusi?*'s interpretation performance

	# Gold with prob. in top 1	Average Rank	Median Rank	75%-ile Rank	Not found	Total #
UCG seqs.	59 (51%)	3.14	0	1	36 (31%)	115
UCGs	146 (62%)	NA	NA	NA	55 (23%)	234

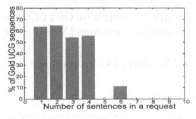   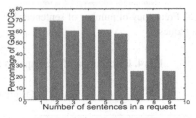

(a) % of Top-1 Gold UCG sequences   (b) % of Top-1 Gold UCGs

Fig. 5. *Scusi?*'s performance broken down by request length

4.1   Results

Our results are summarized in Table 2. The first row shows the performance of *Scusi?*
on generating UCG sequences for the requests in our corpus. The **top 1** column indicates that for about half of the test instances, the most probable UCG sequence generated by *Scusi?* is the correct interpretation of the user's request. *Scusi?* ranks the UCG
sequences that it generates by their probabilities, with the most probable sequence having a rank of 0. The median and 75%-ile ranks show that whenever *Scusi?* generates
the correct interpretation, this interpretation tends to be highly ranked ("not found"
Gold UCGs are excluded from these three statistics). The average rank of the correct
UCG sequences generated by *Scusi?* is above 3 because there are a few requests for
which *Scusi?* ranked the correct UCG sequence above 30, thereby worsening the overall average. The second row of Table 2 shows how many correct individual UCGs were
generated by *Scusi?* in the top-1 ranked UCG sequence. This result indicates that when
Scusi? cannot fully interpret a user's request, it can sometimes still generate a partially
correct interpretation.

Figure 5 shows a break-down of the results in Table 2 by request length (in number
of sentences). Figure 5(a) depicts the percentage of requests of length L for which the
most probable UCG sequence generated by *Scusi?* is the Gold UCG sequence, e.g., 65%
of 2-sentence requests were interpreted correctly. Figure 5(b) depicts the percentage of
top-ranked individual Gold UCGs generated by *Scusi?* for requests of length L, e.g.,
69% of the top-ranked UCGs generated for 2-sentence requests were Gold standard.
These results indicate that although very few requests comprising 5 or more sentences
were correctly interpreted in their entirety, *Scusi?* still found some Gold Standard UCGs
for each of these requests.

Table 2 shows that 31% of Gold Standard UCG sequences and 23% of Gold Standard UCGs were not found. Most of these cases (as well as the poorly ranked UCGs
and UCG sequences) were due to (1) several imperatives with object specifications (19
sequences), (2) wrong anaphora resolution (6 sequences), and (3) wrong PP-attachment

(6 sequences). As mentioned in Section 2.4, we will refine the merging process to address the first problem. The second problem occurs mainly when there are multiple anaphoric references in a sequence. We propose to include this factor in our estimation of the probability of referring to a particular sentence. The PP-attachment problem is expected from a purely semantic interpreter. In order to alleviate this problem, we propose to interleave semantic and pragmatic interpretation of prepositional phrases to improve the rank of candidates which are pragmatically more plausible.

5   Related Research

This research extends our mechanism for interpreting stand-alone utterances [1] and utterance pairs [2] to the interpretation of utterance sequences. Our approach may be viewed as an *information state* approach [10,11], in the sense that utterances may update different informational aspects of other utterances, without requiring a particular "legal" set of dialogue acts. However, unlike these information state approaches, ours is probabilistic.

Several researchers have investigated probabilistic approaches to the interpretation of spoken utterances in dialogue systems, e.g., [12,13,14]. However, these systems do not handle utterance sequences. Additionally, the first two systems employ semantic grammars, while the third system performs word spotting. In contrast, *Scusi?* uses generic, syntactic tools, and incorporates semantic- and domain-related information only in the final stage of the interpretation process. This approach is supported by the findings reported in [15] for relatively unconstrained utterances by users unfamiliar with the system, such as those expected by *DORIS*.

Our mechanism is well suited for processing replies to clarification questions [16,17], as this is a special case of the problem addressed in this paper — the interpretation of spontaneously volunteered, rather than prompted, information. Further, our probabilistic output can be used by a utility-based dialogue manager [16].

6   Conclusion

We have extended *Scusi?*, our spoken language interpretation system, to interpret sentence sequences. Specifically, we have proposed a procedure that combines the interpretations of the sentences in a sequence, and presented a formalism for estimating the probability of a merged interpretation. This formalism supports the comparison of interpretations comprising different numbers of UCGs obtained from different mergers. Additionally, our mechanism can be readily used to process replies to clarification questions — a special case of the situation considered in this paper.

Our empirical evaluation shows that *Scusi?* performs well for textual input at the sentence level, but several issues pertaining to the integration of UCGs still need to be considered. Thereafter, we propose to extend the interpretation process to ICG sequences, and investigate the influence of speech recognition performance on *Scusi?*'s performance. In the future, we intend to expand *Scusi?*'s grammatical capabilities.

Acknowledgments. This research was supported in part by grant DP0878195 from the Australian Research Council. The authors thank Kapil K. Gupta for his assistance with tagging.

References

1. Zukerman, I., Makalic, E., Niemann, M., George, S.: A probabilistic approach to the interpretation of spoken utterances. In: Ho, T.-B., Zhou, Z.-H. (eds.) PRICAI 2008. LNCS (LNAI), vol. 5351, pp. 581–592. Springer, Heidelberg (2008)
2. Zukerman, I., Makalic, E., Niemann, M.: Interpreting two-utterance requests in a spoken dialogue system. In: Proceedings of the 6th IJCAI Workshop on Knowledge and Reasoning in Practical Dialogue Systems, Pasadena, California, pp. 19–27 (2009)
3. Charniak, E.: Maximum-entropy-inspired parser. In: The 2nd Meeting of the North American Chapter of the Association for Computational Linguistics, Seattle, USA, pp. 132–139 (2000)
4. Sowa, J.: Conceptual Structures: Information Processing in Mind and Machine. Addison-Wesley, Reading (1984)
5. Berger, A.L., Pietra, V.J.D., Pietra, S.A.D.: A maximum entropy approach to natural language processing. Computational Linguistics 22(1), 39–71 (1996)
6. Leacock, C., Chodorow, M.: Combining local context and WordNet similarity for word sense identification. In: Fellbaum, C. (ed.) WordNet: An Electronic Lexical Database, pp. 265–285. MIT Press, Cambridge (1998)
7. Lappin, S., Leass, H.: An algorithm for pronominal anaphora resolution. Computational Linguistics 20, 535–561 (1994)
8. Ng, H., Zhou, Y., Dale, R., Gardiner, M.: A machine learning approach to identification and resolution of one-anaphora. In: IJCAI 2005 – Proceedings of the 19th International Joint Conference on Artificial Intelligence, Edinburgh, Scotland, pp. 1105–1110 (2005)
9. Ang, J., Dhillon, R., Krupski, A., Shriberg, E., Stolcke, A.: Prosody-based automatic detection of annoyance and frustration in human-computer dialog. In: ICSLP 2002 – Proceedings of the 7th International Conference on Spoken Language Processing, Denver, Colorado, pp. 2037–2040 (2002)
10. Larsson, S., Traum, D.: Information state and dialogue management in the TRINDI dialogue move engine toolkit. Natural Language Engineering 6, 323–340 (2000)
11. Becker, T., Poller, P., Schehl, J., Blaylock, N., Gerstenberger, C., Kruijff-Korbayová, I.: The SAMMIE system: Multimodal in-car dialogue. In: Proceedings of the COLING/ACL 2006 Interactive Presentation Sessions, Sydney, Australia, pp. 57–60 (2006)
12. He, Y., Young, S.: A data-driven spoken language understanding system. In: ASRU 2003 – Proceedings of the IEEE Workshop on Automatic Speech Recognition and Understanding, St. Thomas, US Virgin Islands, pp. 583–588 (2003)
13. Gorniak, P., Roy, D.: Probabilistic grounding of situated speech using plan recognition and reference resolution. In: ICMI 2005 – Proceedings of the 7th International Conference on Multimodal Interfaces, Trento, Italy, pp. 138–143 (2005)
14. Hong, J.H., Song, Y.S., Cho, S.B.: Mixed-initiative human-robot interaction using hierarchical Bayesian networks. IEEE Transactions on Systems, Man and Cybernetics, Part A 37(6), 1158–1164 (2007)
15. Knight, S., Gorrell, G., Rayner, M., Milward, D., Koeling, R., Lewin, I.: Comparing grammar-based and robust approaches to speech understanding: A case study. In: Proceedings of Eurospeech 2001, Aalborg, Denmark, pp. 1779–1782 (2001)
16. Horvitz, E., Paek, T.: DeepListener: Harnessing expected utility to guide clarification dialog in spoken language systems. In: ICSLP 2000 – Proceedings of the 6th International Conference on Spoken Language Processing, Beijing, China, pp. 226–229 (2000)
17. Bohus, D., Rudnicky, A.: Constructing accurate beliefs in spoken dialog systems. In: ASRU 2005 – Proceedings of the IEEE Workshop on Automatic Speech Recognition and Understanding, San Juan, Puerto Rico, pp. 272–277 (2005)

Fuzzy Rank Linear Regression Model

Jin Hee Yoon[1] and Seung Hoe Choi[2]

[1] School of Economics, Yonsei University, Seoul 120-749, South Korea
[2] School of Liberal Arts and Sciences, Korea Aerospace University,
Koyang 412-791, South Korea

Abstract. In this paper, we construct a fuzzy rank linear regression model using the rank transform (RT) method and least absolute deviation (LAD) method based on the α-level sets of fuzzy numbers. The rank transform method is known to be efficient when the error distribution does not satisfy the conditions for normality and the method is not sensitive to outliers in the regression analysis. Some examples are given to compare the effectiveness of the proposed method with other existing methods.

1 Introduction

Tanaka et al. [14] introduced fuzzy regression models to explain the functional relationship among the variables that are vaguely expressed. Numerical and statistical methods have been introduced to develop the parametric fuzzy regression models. The numerical methods estimate the fuzzy regression model by minimizing the sum of the spreads of the estimated dependent variable. Many authors have studied fuzzy regression models using linear or nonlinear programming methods [3,6,8,9,11,12,13,14]. Statistical methods minimize the squares of the differences between observed fuzzy data and predicted fuzzy data to construct the parametric fuzzy regression models. Many authors have used the statistical methods that employ the method of least squares to estimate the fuzzy regression model [1,2,7,10]. The least squares method is suited to cases where the errors of the estimated regression model have a normal distribution; however, in cases where the errors do not have a normal distribution, the method of least squares is not the best method. This shows that the accuracy of the fuzzy regression model using this method, which is sensitive to the outliers, may decline in some cases. Moreover, the errors in fuzzy regression models arise from the inaccuracy of the regression models. Therefore, it is necessary to consider another method that is both independent of the distributions of errors and insensitive to outliers. The regression model with fuzzy outliers was developed to increase the efficiency of such models [1].

In this paper, we introduce the rank transform (RT) method that uses the rank of the modes and endpoints of the α-level sets of the fuzzy number to develop the fuzzy regression model. We compare the efficiency of the proposed regression model with those developed by the method of least squares and the method of least absolute deviation (LAD) using some data that have outliers.

A. Nicholson and X. Li (Eds.): AI 2009, LNAI 5866, pp. 617–626, 2009.

2   Fuzzy Regression Model

In order to explain the functional relationship among the variables that are vaguely expressed, Tanaka et al. [14] first introduced the fuzzy regression model that can be expressed as follows:

$$Y(\mathbf{X_i}) = A_0 + A_1 X_{i1} + \cdots + A_p X_{ip}, \tag{1}$$

where X_{ij}, A_j, and $Y(\mathbf{X_i})$ are LR-fuzzy numbers. One of the purposes of the fuzzy regression analysis is to determine the regression coefficients that minimize the difference between the observed fuzzy numbers and predicted fuzzy numbers based on the observed data $\{(X_{i1}, \cdots, X_{ip}, Y_i) : i = 1, \cdots, n\}$.

The membership function of the LR-fuzzy number $A = (a, l_a, r_a)_{LR}$ is

$$\mu_A(x) = \begin{cases} L_A\left((a-x)/l_a\right) & \text{if } 0 \le a - x \le l_a, \\ R_A\left((x-a)/r_a\right) & \text{if } 0 \le x - a \le r_a, \\ 0 & \text{otherwise,} \end{cases}$$

where L_A and R_A are monotonic decreasing functions. Further, they satisfy $L_A(0) = R_A(0) = 1$ and $L_A(1) = R_A(1) = 0$. Here, a denotes the mode of A, l_a and r_a denote the left and right spreads of the fuzzy number A, respectively. If $L_A(x) = R_A(x) = 1 - x$, then the LR-fuzzy number A is called the triangular fuzzy number and represented by $(a, l_a, r_a)_T$. In particular, we express the LR-fuzzy number as $(a, s_a)_{LR}$, when the fuzzy number is symmetric, that is, the left and right spreads are identical.

The α-level set of the LR-fuzzy number $A = (a, l_a, r_a)_{LR}$ is

$$A(\alpha) = \begin{cases} \overline{\{x : \mu_A(x) > \alpha\}} & \text{if } \alpha = 0 \\ \{x : \mu_A(x) \ge \alpha\} & \text{if } 0 < \alpha \le 1, \end{cases}$$

where $\bar{A}$ denotes the closure of A. The α-level set of the fuzzy number A is the closed interval with mode a, left spread $(l_a L_A^{-1}(\alpha))$, and right spread $(r_a R_A^{-1}(\alpha))$, respectively. Hence, we can represent the α-level set of the fuzzy number as follows:

$$A(\alpha) \doteq [a - l_a L_A^{-1}(\alpha), a + r_a R_A^{-1}(\alpha)].$$

Thus, the α-level set of the observed fuzzy number $Y_i = (y_i, l_{y_i}, r_{y_i})_{LR}$ is

$$Y_i(\alpha) \doteq [y_i - l_{y_i} L_{Y_i}^{-1}(\alpha), y_i + r_{y_i} R_{Y_i}^{-1}(\alpha)]$$

and the α-level set of the predicted fuzzy numbers $Y(\mathbf{X}_i)$ is

$$\sum_{k=0}^{p} [l_{A_k}(\alpha), r_{A_k}(\alpha)] \cdot [l_{X_{ik}}(\alpha), r_{X_{ik}}(\alpha)], \tag{2}$$

where $l_{A_k}(\alpha) = a_k - l_k L_{A_i}^{-1}(\alpha), r_{A_k}(\alpha) = a_k + r_k R_{A_i}^{-1}(\alpha), l_{X_{ik}}(\alpha) = x_{ik} - l_{X_{ik}} L_{X_{ik}}^{-1}(\alpha),$ and $r_{X_{ik}}(\alpha) = x_{ik} + r_{X_{ik}} R_{X_{ik}}^{-1}(\alpha).$

The extension principle and resolution identity theorem developed by Zadeh [15] play an important role in fuzzy set theory. The extension principle implies that an arithmetic of fuzzy numbers depends on a calculating operation of intervals. The resolution identity states that the fuzzy number may be decomposed into its level sets. As such, we can apply the α-level regression model (2) to derive the predicted fuzzy output. In this paper, in order to increase the effectiveness of the estimated fuzzy regression model, we first apply the RT method, which is not affected by the distribution of the errors, to the α-level regression model. We then construct the fuzzy regression model by using a robust method of fitting the reference function to the estimated α-level sets.

3   Fuzzy Rank Linear Regression Model

In this section, the three types of fuzzy regression models classified by the conditions of regression coefficients and independent variables are considered and the RT method is applied to construct the proposed regression models in this section. In the regression analysis, the RT method was recommended and used when the assumptions of normal distribution are violated [4]. Iman and Conover [5] applied the RT method to regression analysis, whose high efficiency was verified through experimental analysis. They proved that the estimators obtained by the RT method are robust and not sensitive to the outliers in the regression model, including monotonic increasing data. To apply the RT method to the fuzzy regression model, we use following two types of simple linear fuzzy regression models:

$$Y(x_{i1}) = A_0 + A_1 x_{i1}, \tag{3}$$

where $x_{i1}(i = 1, \cdots, n)$ are crisp numbers, and A_0, A_1, and $Y(x_{i1})$ are LR-fuzzy numbers.

$$Y(X_{i1}) = A_0 + A_1 X_{i1}, \tag{4}$$

where X_{i1}, A_0, A_1, and $Y(X_{i1})$ are LR-fuzzy numbers.

3.1   Fuzzy Rank Linear Regression with Crisp Independent Variables

We first consider the fuzzy regression model where input data are ordinal numbers with zero spreads and regression coefficients are fuzzy numbers. The left-hand(right-hand) side $l_{y_i}(\alpha)$ $(r_{y_i}(\alpha))$ of the α-level sets of the proposed model are given by

$$l_{y_i}(\alpha) = l_0(\alpha) + l_1^*(\alpha)x_{i1} \quad \text{and} \quad r_{y_i}(\alpha) = r_0(\alpha) + r_1^*(\alpha)x_{i1}, \tag{5}$$

where the points $l_1^*(\alpha)$ and $r_1^*(\alpha)$ satisfy the following.

$$l_1^*(\alpha) = \begin{cases} l_1(\alpha) & \text{if } x_{i1} \geq 0 \\ r_1(\alpha) & \text{if } x_{i1} < 0 \end{cases} \quad and \quad r_1^*(\alpha) = \begin{cases} r_1(\alpha) & \text{if } x_{i1} \geq 0 \\ l_1(\alpha) & \text{if } x_{i1} < 0. \end{cases}$$

The result $L_{Y_i}^{-1}(\alpha) = R_{Y_i}^{-1}(\alpha) = L_{A_k}^{-1}(1) = R_{A_k}^{-1}(1) = 0 (k = 0, 1)$ yields that the equation (3) is equivalent to an ordinary regression model

$$y_i = a_0 + a_1 x_{i1}$$

and that $y_i = l_{y_i}(1) = r_{y_i}(1)$. After constructing the regression model based on the set of modes of the fuzzy data, we can obtain the predicted value $Y(X_i)(\alpha)$ by estimating the endpoints $l_{y_i}(\alpha)$ $(r_{y_i}(\alpha))$ for $\alpha \in [0, 1)$.

Now we consider the application of the RT method to the regression model (3) in the following four steps:

(i) Estimate the mode y_i based on the RT method and the set $\{(x_{i1}, y_i) : i = 1, \cdots, n\}$.

(ii) Estimate the left(the right) endpoints $l_{y_i}(\alpha_o)$ $(r_{y_i}(\alpha_o))$ using the RT procedures and the set $\{(x_{i1}, l_{y_i}(\alpha)) : i = 1, \cdots, n\}$ for some $\alpha_o \in (0, 1)$.

(iii) Estimate the endpoints $l_{y_i}(\alpha)$ $(r_{y_i}(\alpha))$ using the RT method and the results given in (ii) for any $\alpha \in (0, 1)$ and $\alpha \neq \alpha_o$.

(iv) Find the membership function $\mu_{\hat{Y}_i}$ using a robust method of fitting an estimated membership function $L_{\hat{Y}_i}(x)$ $(R_{\hat{Y}_i}(x))$ to the left-hand(right-hand) side $(l_{y_i}(\alpha_i), \alpha_i)$ $((r_{y_i}(\alpha_i), \alpha_i))$ given in (i) and (iii).

We now introduce a method that applies the RT method, which uses the ranks of the components of the α-level sets of the observed fuzzy numbers, to estimate the endpoints of the predicted value $Y(x_i)(\alpha^*)$ for some $\alpha^* \in [0, 1]$. The procedures of the RT method to construct the regression model

$$l_{y_i}(\alpha^*) = l_0(\alpha^*) + l_1(\alpha^*) x_{i1}$$

are as follows:

1. Determine the rank $R(x_{i1})$ of the observed value x_{i1} among $\{x_{11}, \cdots, x_{n1}\}$ and $R(l_{y_i}(\alpha^*))$ of $l_{y_i}(\alpha^*)$. If the ranks of the two values are the same, we take the average.

2. We use the following regression model:

$$R(l_{y_i}(\alpha^*)) = \frac{(n+1)}{2} + \beta_1 \left(R(x_{i1}) - \frac{(n+1)}{2} \right).$$

3. Estimate the predicted ranks $\hat{R}(l_{y_i}(\alpha^*))$ using the best regression model (3) and the set of the rank of the observed value.

4. Calculate the predicted value corresponding to the input x_i by the estimated rank $\hat{R}(l_{y_i}(\alpha^*))$ given in the third step as follows:

$$\tilde{l}_{y_i}(\alpha^*) = \begin{cases} l_{y_{(1)}}(\alpha^*) & \text{if } \hat{R}(l_{y_i}(\alpha^*)) < R(l_{y_{(1)}}(\alpha^*)), \\ l_{y_{(n)}}(\alpha^*) & \text{if } \hat{R}(l_{y_i}(\alpha^*)) > R(l_{y_{(n)}}(\alpha^*)), \\ l_{y_{(j)}}(\alpha^*) & \text{if } \hat{R}(l_{y_i}(\alpha^*)) = R(l_{y_{(j)}}(\alpha^*)). \end{cases}$$

If $R(l_{y_{(j)}}(\alpha^*)) < \hat{R}(l_{y_i}(\alpha^*)) < R(l_{y_{(j+1)}}(\alpha^*))$, then

$$\tilde{l}_{y_i}(\alpha^*) = l_{y_{(j)}}(\alpha^*) + (l_{y_{(j+1)}}(\alpha^*) - l_{y_{(j)}}(\alpha^*)) \cdot \frac{\hat{R}(l_{y_i}(\alpha^*)) - R(l_{y_{(j)}}(\alpha^*))}{R(l_{y_{(j+1)}}(\alpha^*)) - R(l_{y_{(j)}}(\alpha^*))}, \quad (6)$$

where $l_{y_{(j)}}(\alpha^*)$ denotes the j-th largest value among $\{l_{y_1}(\alpha^*), \cdots, l_{y_n}(\alpha^*)\}$. Since the left-hand side of the α-level set of the fuzzy number has to be less than the mode, the estimated left endpoint of $Y(X_i)(\alpha)$ is given by

$$\bar{l}_{y_i}(\alpha^*) = \min\{\tilde{l}_{y_i}(\alpha^*), \hat{y}_i\}.$$

Here, if the explanatory variable x^* with respect to the response variable $l_{y^*}(\alpha)$ satisfies $x^* \neq x_{i1}(i = 1, \cdots, n)$, it is necessary to know the rank of x^* in order to predict the value using the regression model estimated in the second step. For this, we use the following method.

$$R(x^*) = \begin{cases} 1 & \text{if } x^* < x^{(1)}, \\ n & \text{if } x^* > x^{(n)}, \\ R(x^{(i)}) & \text{if } x^* = x^{(i)}. \end{cases}$$

Further if $x^{(i)} < x^* < x^{(i+1)}$, then

$$R(x^*) = R(x^{(i)}) + (R(x^{(i+1)}) - R(x^{(i)})) \cdot \frac{x^* - x^{(i)}}{x^{(i+1)} - x^{(i)}}, \tag{7}$$

where $x^{(i)}$ denotes the i-th largest value among $\{x_{11}, \cdots, x_{n1}\}$. Similarly, we can obtain the predicted value of the right-hand side $(\hat{r}_{y_i}(\alpha^*))$ and the mode $(\hat{y}_i)$ by applying the same procedure to $\{(x_i, r_{y_i}(\alpha^*)) : i = 1, \cdots, n\}$, and then construct the parametric fuzzy regression model.

As the third step, we suggest the procedure to estimate the endpoints of the α-level set of $Y(x_{i1})(\alpha)$ when $\alpha \neq \alpha^*$. For each α, the estimated right (left) endpoints based on the values $\tilde{r}_{y_i}(\alpha^*)$ $(\tilde{l}_{y_i}(\alpha^*))$, derived in the previous step, are given by

$$\hat{r}_{y_i}(\alpha) = \begin{cases} \max\{\max_{\{\alpha \leq s < \alpha^*\}}\{\bar{r}_{y_i}(s)\}, \hat{y}_i\} & \text{if } \alpha < \alpha^* \\ \max\{\min_{\{\alpha^* < s \leq \alpha\}}\{\bar{r}_{y_i}(s)\}, \hat{y}_i\} & \text{if } \alpha* < \alpha \end{cases}$$

and

$$\hat{l}_{y_i}(\alpha) = \begin{cases} \min\{\max_{\{\alpha^* \leq s < \alpha\}}\{\bar{l}_{y_i}(s)\}, \hat{y}_i\} & \text{if } \alpha* < \alpha \\ \min\{\min_{\{\alpha < s \leq \alpha^*\}}\{\bar{l}_{y_i}(s)\}, \hat{y}_i\} & \text{if } \alpha < \alpha^*. \end{cases}$$

We know that the value $\hat{l}_{y_i}(\alpha)$ $(\hat{r}_{y_i}(\alpha))$ is decreasing (increasing) as α is decreasing. The maximum and minimum value of the function $\hat{l}_{y_i}(\alpha)$ are $\hat{l}_{y_i}(0)$ and $\hat{y}_i = \hat{l}_{y_i}(1)$, respectively.

Finally, the membership function of the estimated fuzzy output $\hat{Y}_i$ is obtained using a parametric estimation method of fitting the reference function to the data $\{(\hat{l}_{y_i}(\alpha_1), \alpha_1 k))\}$ $(\{(\hat{r}_{y_i}(\alpha_1), \alpha_1))\})$ subject to $L_{\hat{Y}_i}(\hat{l}_{y_i}(1)) = R_{\hat{Y}_i}(\hat{r}_{y_i}(1)) = 1$. The predicted fuzzy output given by the proposed method is represented as follows:

$$\hat{Y}_i = \left(\hat{y}_i, \hat{y}_i - L_{\hat{Y}_i}^{-1}(0), R_{\hat{Y}_i}^{-1}(0) - \hat{y}_i\right)_{LR}.$$

3.2 Fuzzy Rank Linear Regression with Fuzzy Input and Output

In this subsection, we study a method estimating the fuzzy regression using the RT method and the resolution identity theorem when independent and dependent variables are fuzzy numbers. If the regression coefficients are crisp numbers, the α-level set of the proposed fuzzy model is

$$Y(X_i)(\alpha) = [l_{y_i}(\alpha), r_{y_i}(\alpha)] = \left[a_0 + a_1 l^*_{X_{i1}}(\alpha), \quad a_0 + a_1 r^*_{X_{i1}}(\alpha)\right], \quad (8)$$

where

$$l^*_{X_{i1}}(\alpha) = \begin{cases} l_{X_{i1}}(\alpha) & \text{if } a_1 \geq 0 \\ r_{X_{i1}}(\alpha) & \text{if } a_1 < 0 \end{cases} \quad and \quad r^*_{X_{i1}}(\alpha) = \begin{cases} r_{X_{i1}}(\alpha) & \text{if } a_1 \geq 0 \\ l_{X_{i1}}(\alpha) & \text{if } a_1 < 0. \end{cases}$$

The above fuzzy regression model that employs crisp parameters has been developed by many authors [2,7,10]. Further, in the case where the regression parameters are both fuzzy numbers, the α-level set of the fuzzy model is given by

$$Y(X_i)(\alpha) = [l_0(\alpha), r_0(\alpha)] + [l_1(\alpha), r_1(\alpha)] \cdot [l_{X_{i1}}(\alpha), r_{X_{i1}}(\alpha)].$$

Although a multiplication of interval numbers have complicated formula to analyze, we can convert the multiplication in the above regression into simple forms under some conditions. That is, the multiplication of the intervals $[l_1(\alpha), r_1(\alpha)]$ and $[l_{X_{i1}}(\alpha), r_{X_{i1}}(\alpha)]$ equals

$$\begin{cases} [l_1(\alpha) l_{X_{i1}}(\alpha), r_1(\alpha) r_{X_{i1}}(\alpha)] & \text{if } l_1(\alpha), l_{X_{i1}}(\alpha) \geq 0 \\ [r_1(\alpha) r_{X_{i1}}(\alpha), l_1(\alpha) l_{X_{i1}}(\alpha)] & \text{if } r_1(\alpha), r_{X_{i1}}(\alpha) < 0. \end{cases}$$

Thus, if all the values of the independent variable $X_{i1}(i = 1, \cdots, n)$ are all positive (or negative) fuzzy numbers, the α-level set of the fuzzy regression model can be written as follows:

$$Y(X_i)(\alpha) = [l_0(\alpha), r_0(\alpha)] + [l_1(\alpha) l_{X_{i1}}(\alpha), r_1(\alpha) r_{X_{i1}}(\alpha)]. \quad (9)$$

Without loss of generality, in this paper, we assume that control variables $X_{i1}(i = 1, \cdots, n)$ are all positive (negative). Then, since model (7) is a special case of equation (8), we can construct the fuzzy linear regression model with fuzzy input and output by applying the RT method and parametric estimation to the data

$$\{(l_{X_{i1}}(\alpha), l_{y_i}(\alpha)) : i = 1, \cdots, n\} \quad (\{r_{X_{i1}}(\alpha), r_{y_i}(\alpha) : i = 1, \cdots, n\}).$$

The same third procedure given in the previous subsection gives the values $\hat{l}_{y_i}(\alpha)$ $(\hat{r}_{y_i}(\alpha))$ and $\tilde{l}_{y_i}(\alpha)$ $(\tilde{r}_{y_i}(\alpha))$, which are the estimates of $l_{y_i}(\alpha)$ $(r_{y_i}(\alpha))$. Next, the conditions of reference function and parametric estimation provide the membership function of the estimated fuzzy output $\hat{Y}_i$. In the following section, we show the development of a fuzzy regression model by using the RT method and a suitable estimation, and evaluate the effectiveness of the proposed fuzzy regression model.

4   Numerical Examples

In this section, we provide a measure of performance and fuzzy outliers to compare the efficiency of the fuzzy regression model estimated by the RT and LAD methods. Further, we compare the accuracy of the fuzzy regression model estimated by the method proposed in this paper with those of other fuzzy regression models that were constructed using different methods.

Kim and Bishu [9] used an integration of the membership functions to compare the accuracy of the developed fuzzy regression models. The difference between the membership values of the observed fuzzy number Y_i and estimated fuzzy number $\widehat{Y}_i$ is defined by

$$d\left(Y_i, \widehat{Y}_i\right) = \frac{\int_{-\infty}^{\infty} |\mu_{Y_i}(x) - \mu_{\widehat{Y}_i}(x)| dx}{\int_{-\infty}^{\infty} \mu_{Y_i}(x) dx}. \tag{10}$$

The smaller the difference between the two fuzzy numbers, the closer is the value in equation (10) to zero. Hence, we can assume that when the value in equation (10) is the smallest, the accuracy of the method is the highest. Thus, we can also conclude that when the value in equation (10) is the smallest, the accuracy of the developed fuzzy regression model is the highest.

On the other hand, Choi and Buckley [1] introduced a fuzzy outlier that is numerically distant from the other fuzzy data. They defined outliers in the fuzzy regression model with respect to modes and spreads as M-type and S-type fuzzy outliers.

In the following example, we consider a fuzzy rank linear regression model to illustrate the use of the RT and LAD methods, and compare them with other methods.

Example 4.1. The data in Table 1(a) were first used by Tanaka et al. [13] and many authors have referred to this data. Tanaka et al. [13] and Kao and Chyu [6] applied a method that minimizes the spreads of the estimated fuzzy numbers to the data. Further, Kim and Bishu [9] and Nasrabadi [11] used the least squares methods with the same data. Simple calculations show that the ranks for the endpoints of the observed variable do not change with the values of α. That is, $R(l_{y_i}(\alpha)) = R(l_{y_i}(\alpha^*))$ for $\alpha \neq \alpha^*$. The RT method implies that the estimated regression model for the ranks of the value $l_{y_i}(\alpha)$ derived from the data given in Table 1(a) are

$$\widehat{R}(l_{y_i}(\alpha)) = 2.5 + 0.9(R(x_i) - 2.5).$$

The predicted α-level sets calculated by the RT method are given in the right side of Table 1(a).

Further, the least squares method yields that the third data item in Table 1(a) is almost an S-type outlier ($k = 1.01$) [1]. The fuzzy regression model using the least squares method may be affected by this data. From the estimated level set given in Table 1(a), we get the fuzzy out in Table 1(b) using the LAD method. The errors between the observed value and the predicted value obtained by the several methods are also given in Table 1(b). The result in Table 1(b)

Table 1. (a) Numerical Data and Estimates for Example 4.1. (b) Estimation Errors for Example 4.1.

(a)

Input	Output		Estimated level sets			
x_i	$(y_i, s_i)_T$	$\widehat{Y}_i(1)$	$\widehat{Y}_i(0.75)$	$\widehat{Y}_i(0.5)$	$\widehat{Y}_i(0.25)$	$\widehat{Y}_i(0)$
1	(8, 1.8)$_T$	6.81	[6.31, 7.32]	[5.81, 7.83]	[5.31, 8.34]	[4.83, 8.86]
2	(6.4, 2.2)$_T$	8.19	[7.72, 8.66]	[7.25, 9.12]	[6.78, 9.59]	[6.31, 10.05]
3	(9.5, 2.6)$_T$	9.5	[8.85, 10.15]	[8.2, 10.8]	[7.55, 11.45]	[6.9, 12.1]
4	(13.5, 2.6)$_T$	12.55	[11.92, 13.18]	[11.29, 13.8]	[10.66, 14.42]	[10.02, 15.04]
5	(13, 2.4)$_T$	13.37	[12.73, 14.02]	[12.09, 14.66]	[11.45, 15.3]	[10.81, 15.95]

(b)

Input	Predicted output	$m(Y_i, \widehat{Y}_i^T)$	$m(Y_i, \widehat{Y}_i^{KC})$	Errors in estimation $m(Y_i, \widehat{Y}_i^{KB})$	$m(Y_i, \widehat{Y}_i^N)$	$m(Y_i, \widehat{Y}_i)$
1	$(6.81, 2, 2.04)_T$	3.356	2.789	2.207	2.564	1.991
2	$(8.19, 1.87, 1.87)_T$	2.85	2.589	3.025	2.813	2.792
3	$(9.5, 2.6, 2.61)_T$	1.522	0.553	1.042	0.718	0.005
4	$(12.55, 2.52, 2.49)_T$	2.257	3.363	2.902	3.062	1.737
5	$(13.37, 2.57, 2.57)_T$	2.414	0.385	0.85	0.614	0.72
	Total errors	12.399	9.679	10.026	9.771	7.245

shows that the total error of the regression model estimated by the RT method is smaller than that using the least squares method.

The estimates $\widehat{Y}_i^T$, $\widehat{Y}_i^{KC}$, $\widehat{Y}_i^{KB}$, $\widehat{Y}_i^N$ and $\widehat{Y}_i$ are the results of [13], [6], [9], [11] and proposed method, respectively.

Example 4.2. Sakawa and Yano [12] used the data in Table 2(a) to develop a fuzzy regression model with fuzzy coefficients. Furthermore, Kao and Chyu [6] estimated the fuzzy regression model for the data in Table 2(a) and compared their two-stage method with other methods. Nasrabadi [11] used the method of least squares for the same data.

In Table 2(a), we know that the ranks of explanatory variables accord with those of response variables. That is, $s_{y_i} = s_{x_i}(i = 1, \cdots, 8)$. Thus, the regression coefficients for the spreads in equation (2) equal 1. The regression model for the ranks of the modes is

$$\widehat{R}(y_i) = 4.5 + 0.974(R(x_i) - 4.5).$$

The LAD method gives the predicted fuzzy output in Table 2(b) based on the estimated α-level sets in Table 2(a). Further, a simple calculation reveals that the sixth data item in Table 2(a) is an M-type outlier ($k = 1.3$) [1].

Hence, we can confirm that the estimators using the RT method are more efficient than those using the least squares method through the total error given in Table 2(b). The estimates $\widehat{Y}_i^{SY}$, $\widehat{Y}_i^{KC}$, $\widehat{Y}_i^N$ and $\widehat{Y}_i$ are the results of [12], [6], [11] and proposed method, respectively.

Table 2. (a) Numerical Data and Estimates for Example 4.2. (b) Estimation Errors for Example 4.2.

(a)

Input	Output	Estimated level sets		
X_i	Y_i	$\widehat{Y}_i(1)$	$\widehat{Y}_i(0.5)$	$\widehat{Y}_i(0)$
$(2.0, 0.5)_T$	$(4.0, 0.5)_T$	4.45	[4.20, 4.70]	[3.96, 4.94]
$(3.5, 0.5)_T$	$(5.5, 0.5)_T$	5.71	[5.46, 5.96]	[5.22, 6.21]
$(5.5, 1.0)_T$	$(7.5, 1.0)_T$	6.63	[6.35, 6.91]	[6.07, 7.19]
$(7.0, 0.5)_T$	$(6.5, 0.5)_T$	7.52	[7.02, 8.02]	[6.52, 8.52]
$(8.5, 0.5)_T$	$(8.0, 0.5)_T$	7.98	[7.48, 8.48]	[6.98, 8.98]
$(10.5, 1.0)_T$	$(8.0, 1.0)_T$	8.44	[8.15, 8.72]	[7.87, 9.79]
$(11.0, 0.5)_T$	$(10.5, 0.5)_T$	9.29	[9.04 9.54]	[8.78 9.79]
$(12.5, 0.5)_T$	$(9.5, 0.5)_T$	10.20	[9.95, 10.45]	[9.70, 10.70]

(b)

Input	Predicted output	Errors in estimation			
		$m\left(Y_i, \widehat{Y}_i^{SY}\right)$	$m\left(Y_i, \widehat{Y}_i^{KC}\right)$	$m\left(Y_i, \widehat{Y}_i^{N}\right)$	$m\left(Y_i, \widehat{Y}_i\right)$
$(2.0, 0.5)_T$	$(4.45, 0.5, 0.5)_T$	0.633	0.848	0.891	0.698
$(3.5, 0.5)_T$	$(5.71, 0.5, 0.51)_T$	0.453	0.208	0.019	0.38
$(5.5, 1.0)_T$	$(6.63, 0.57, 0.56)_T$	1.613	1.489	1.413	1.26
$(7.0, 0.5)_T$	$(7.52, 1, 1)_T$	1.165	0.910	0.991	1.346
$(8.5, 0.5)_T$	$(7.98, 1, 1)_T$	0.770	0.760	0.476	0.86
$(10.5, 1.0)_T$	$(8.44, 0.57, 0.78)_T$	1.977	1.449	1.767	0.862
$(11.0, 0.5)_T$	$(9.29, 0.5, 0.5)_T$	1.368	1.000	0.992	1.00
$(12.5, 0.5)_T$	$(10.2, 0.5, 0.5)_T$	1.452	0.806	0.992	0.91
Total errors		9.431	7.470	7.541	7.316

5   Conclusions

In this paper, we construct a fuzzy rank linear regression model using the RT method and LAD method based on the α-level sets of fuzzy numbers, and then obtained the predicted value based on the equation for the ranks of the dependent variables. We investigated the efficiency of the fuzzy regression model with the data that have outliers through some examples. We confirm that the proposed method may be a robust estimator in the fuzzy regression analysis through some examples.

References

1. Choi, S.H., Buckley, J.J.: Fuzzy regression using least absolute deviation estimators. Soft Computing 12, 257–263 (2008)
2. Diamond, P.: Fuzzy Least Squares. Inform. Sci. 46, 141–157 (1988)
3. Diamond, P., Tanaka, H.: Fuzzy regression analysis. Fuzzy sets in decision analysis, operations research and statistics. Kluwer Academic Publishers, Norwell (1999)

4. Headrick, T.C., Rotou, O.: An investigation of the rank transformation in multiple regression. Computational Statistics and Data Analysis 38, 203–215 (2001)
5. Iman, R.L., Conover, W.J.: The use of the rank transform in regression. Technomerics 21, 499–509 (1979)
6. Kao, C., Chyu, C.: A fuzzy linear regression model with better explanatory power. Fuzzy Sets and Systems 126, 401–409 (2002)
7. Kao, C., Chyu, C.: Least-squares estimates in fuzzy regression analysis. European J. of Operational Research 148, 426–435 (2003)
8. Kao, C., Lin, P.: Entropy for fuzzy regression analysis. Int. Journal of Sys. Sci. 36, 869–876 (2005)
9. Kim, B., Bishu, R.R.: Evaluation of fuzzy linear regression models by comparing membership functions. Fuzzy Sets and Systems 100, 343–352 (1998)
10. Kim, H.K., Yoon, J.H., Li, Y.: Asymptotic properties of least squares estimation with fuzzy observations. Inform. Sci. 178, 439–451 (2008)
11. Nasrabadi, M.M., Nasrabadi, E.: A mathematical programming approach to fuzzy linear regression analysis. Applied Mathematical and Computation 155, 873–881 (2004)
12. Sakawa, M., Yano, H.: Multiobjective fuzzy linear regression analysis for fuzzy input-output data. Fuzzy Sets and Systems 47, 173–181 (1992)
13. Tanaka, H., Hayashi, I., Watada, J.: Possibilistic linear regression analysis for fuzzy data. European Journal of Operational Research 40, 389–396 (1989)
14. Tanaka, H., Uejima, S., Asai, K.: Linear regression analysis with fuzzy model. IEEE Trans. Syst., Man Cybernet. 12, 903–907 (1982)
15. Zadeh, L.A.: Fuzzy sets. Information and Control 8, 338–353 (1965)

Numerical *versus* Analytic Synchronization in Small-World Networks of Hindmarsh-Rose Neurons

Mahdi Jalili

Department of Computer Engineering, Sharif University of Technology, Tehran, Iran
Mjalili@sharif.edu

Abstract. Neuronal temporal synchronization is one of the key issues in studying binding phenomenon in neural systems. In this paper we consider identical Hindmarsh-Rose neurons coupled over Newman-Watts small-world networks and investigate to what extent the numerical and analytic synchronizing coupling strengths are different. We use the master-stability-function approach to determine the unified coupling strength necessary for analytic synchronization. We also solve the network's differential equations numerically and track the synchronization error and consequently determine the numerical synchronizing coupling parameters. Then, we compare these two values and investigate the influence of various network parameters on the gap between them. We find that this gap is almost not influenced by network size. The only parameter that affects the gap between the analytic and numerical synchronizing parameters is the average degree, i.e. average connection per node in the connection graph. In networks with higher average degree this gap is larger than those with lower average degree.

1 Introduction

Synchronization activity of two or many interconnected dynamical units is believed to play an important role in information processing in the brain both in macroscopic and cellular levels [1, 2]. It is hypothesized that synchronous brain activity is the most likely mechanism for many cognitive functions such as attention and feature binding, as well as learning, development and memory formation [3]. Neurons in a population synchronize their activity using electrical and chemical synapses with other neurons in the same population as well as with neurons from other populations. However, synchronization is not useful all the time; high levels of synchrony may proceed to epileptic behaviors. In other words, brain disorders such as Epilepsy, Schizophrenia, Alzheimer's disease and Parkinson influence functional synchronization maps of different brain areas [4-9]. Thus, understanding the mechanisms behind the neural synchronization is of high importance in order to understand various brain functions.

Studies of neuronal synchronization based on different neuronal models can be separated into two categories; those using threshold models of integrate-and-fire type and those with conductance-based realization such as various Hodgkin-Huxley type models. There are a number of simplified versions of the Hodgkin-Huxley model; Hindmarsh-Rose (HR) model is one, which consists of three first-order differential equations [10]. It has been shown to be capable of producing many of observed

A. Nicholson and X. Li (Eds.): AI 2009, LNAI 5866, pp. 627–635, 2009.

neuronal behaviors such as regular bursting [11]. Without being biophysically meaningful, the HR model exhibits all of the behaviors that the Hodgkin-Huxley model is able to show, but with about a 10-fold increase in computational efficiency [11]. Therefore, one could get a great advance in understanding the collective behavior of real-world neuronal populations by studying the synchronization in meaningful networks of the HR systems.

Here, we study the synchronization phenomenon in ensembles of HR neurons whose connection topology is Newman-Watts (NW) small-world network [12, 13]. We investigate the gap between the analytic and numerical synchronizing coupling parameters. In order to determine the coupling strength necessary for the analytic stability of the synchronization manifold, we use the master-stability-function (MSF) formalism [14], while the numerical synchronizing parameter is determined by numerically solving the network's differential equations and tracking the synchronization error.

2 Newman-Watts Small-World Networks

It has been shown that lots of real-world networks have small-world property, including those for cortical neurons, i.e. their connection graph has a structure that is neither purely random nor a regular one [3, 5, 15]. In such networks, average characteristic path length scales almost logarithmically with network size, like random networks, while the clustering coefficient is as high as regular networks [16, 17]. In this work as connection network model, we consider the one proposed by Newman and Watts [18], which guarantees connectedness of the network. The NW small-world networks are constructed as follows. Starting with a ring graph with N nodes each connected to its k–nearest neighbors by undirected links, the unconnected nodes get connected with probability P. The graphical representation of the Newman-Watts networks with 20 nodes each connected to their first and second nearest neighbors and with different values for the probability of shortcuts ($P = 0$, $P = 0.05$, $P = 0.15$, and $P = 1$) is shown in Fig. 1.

3 Hindmarsh-Rose Neuron Model and Neuronal Bursting Behavior

In order to understand the underlying mechanisms of neural systems along with *in vivo* and *in vitro* measurements, computer simulations using model neurons should also be performed. In many works, different models of the family of Hodgkin-Huxley neuron model have been used, which are expensive to solve. A number of reduced models have been proposed in the literature and HR model is one [10], which is well-known for its chaotic behavior and different types of bursts [19]. The model consists of three first-order ordinary differential equations and takes a form as follows

$$F(X) = \begin{cases} \dot{x} = y + ax^2 - x^3 - z + I \\ \dot{y} = 1 - dx^2 - y \\ \dot{z} = \mu\left(b\left(x - x_0\right) - z\right) \end{cases}, \qquad (1)$$

Where $X = (x,y,z)$ is the state variables of the system; x represents the membrane potential (dimensionless), y and z are virtual states representing the fast and slow current dynamics, respectively. I is the external input current injected to the neuron and a governs the qualitative behavior of the model. μ is a small parameter that governs the bursting and adaptation behavior of the model. b governs adaptation in which small b (values around b = 1) results in fast spiking behavior without accommodation and subthershold adaptations, whereas values around $b = 4$ gives strong accommodation[20]. x_0 sets the resting potential of the system and d is a positive value. We adopted the parameters of the model as $\mu = 0.01$, $b = 4$, $d = 5$, $x_0 = -1.6$, $a = 2.6$, and $I = 4$, which produces bursting behavior [20]. Fig. 2 shows the time history of the membrane potential (x-component) of such bursting neurons.

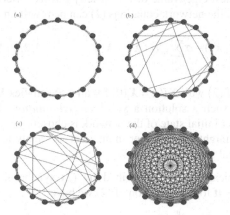

Fig. 1. Graphical representation of NW networks with 20 nodes in a ring each connected to their first and second nearest neighbors and randomly with the other nodes with probability a) $P = 0$, b) $P = 0.05$, c) $P = 0.15$, and d) $P = 1$

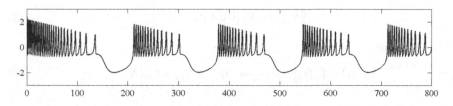

Fig. 2. The time history of the x-component of the Hindmarsh-Rose neuron exhibiting regular bursting behavior. We have considered the parameters of the model described by equations (1) as as $\mu = 0.01$, $b = 4$, $d = 5$, $x_0 = -1.6$, $a = 2.6$, and $I = 4$.

4   Numerical and Analytic Synchronization

Let us consider an undirected and unweighted network with N nodes. On each node a dynamical system sits and the equations of the motion of the dynamical network read

$$\dot{X}_i = F(X_i) - \sigma \sum_{j=1}^{N} l_{ij} H X_j \quad ; \quad i = 1, 2, ..., N , \tag{2}$$

where $X_i \in \mathbb{R}^d$ are the state vectors, $F : \mathbb{R}^d \to \mathbb{R}^d$ defines the individual system's dynamical equation (equations (1)). These dynamical systems are coupled via a unified coupling strength σ and coupling matrix $L = (l_{ij})$. L that is called *Laplacian* is a symmetric matrix with vanishing row-sums and negative off-diagonal entries, i.e. $l_{ij} = l_{ji}$ for all pairs of (i,j), $l_{ij} < 0$ for $i \neq j$, and $\sum_{j=1}^{N} l_{ij} = 0$ for all i. The nonzero elements of 3×3 matrix H determines the coupled elements of the oscillators.

In the following we suppose that the connection graph is connected, which implies that the second smallest eigenvalue of L is strictly positive. Because of the zero row-sums of the matrix L, the network equations (2) can be rewritten as

$$\frac{dX_i}{dt} = F(X_i) - \sigma \sum_{\substack{j=1 \\ j \neq i}}^{N} l_{ij} H \left(X_j - X_i \right) \quad ; \quad i = 1, 2, ..., N . \tag{3}$$

Thus, any solution of (2) with $X_i(0) = X_j(0)$ for all (i,j) satisfies $X_i(t) = X_j(t)$ for all (i,j) and $t \geq 0$. We call such a solution a *synchronized solution*. The question is what happens when another initial state of the network is chosen.

Synchronization might be whether numerical or analytic and we distinguish analytic and numerical synchronization.

a) The dynamical network described by (2) synchronizes *numerically* (and completely), if for any solution of (2) we have

$$\left\| X_i(t) - X_j(t) \right\| \xrightarrow[t \to \infty]{} 0 \quad \forall \ i, j = 1, ... N . \tag{4}$$

b) The dynamical network (2) synchronizes *analyticly*, if there exists an $\varepsilon > 0$ such that for any solution with

$$\left\| X_i(0) - X_j(0) \right\| < \varepsilon , \tag{5}$$

we have

$$\left\| X_i(t) - X_j(t) \right\| \xrightarrow[t \to \infty]{} 0 \quad \forall \ i, j = 1, ... N . \tag{6}$$

4.1 Determining the Analytic Synchronizing Coupling Parameter

We have used the MSF formalism proposed by Pecora and Carroll to determine the analytic synchronization of the network [14]. The MSF gives necessary conditions for the analytic stability of the synchronization manifold $X_1(t) = X_2(t) = ... = X_N(t) = s(t)$ [14]. Considering the dynamical network (2), the stability of the synchronization manifold can be determined by the variational equations, i.e. each dynamical system is considered to have extremely small perturbation from the synchronous state. The variational equations are written as

$$\dot{\eta}_i = DF(s)\eta_i - \sigma \lambda_i H \eta_i \quad ; \quad i = 1, 2, ..., N , \tag{7}$$

where D stands for Jacobian. λ_i's are the eigenvalues of L, ordered as $0 = \lambda_1 \leq \lambda_2 \leq \ldots$ $\leq \lambda_N$, in which $\lambda_1 = 0$ is associated with the synchronized manifold $s(t)$.

The largest Lyapunov exponent of the variational equation expressed by (7), $\Lambda(a = \sigma\lambda_i)$ called MSF [14] and accounts for the linear stability of the synchronization manifold, i.e. if $\Lambda(a) < 0$, the synchronized state is linearly stable. The MSF depends only on the coupling configuration expressed by H and the dynamics of the individual dynamical systems expressed by $F(\cdot)$. In this way, a necessary condition for the analytic stability of the synchronization manifold is obtained. It is worth mentioning that the MSF is computed for a dynamical system with specific H once and one only needs to compute λ_2 and (for some systems) λ_N of the Laplacian L for determining the synchronization conditions of the dynamical network (2). Fig. 3 shows the MSF x-coupled HR neurons. It shows the border of the analytic synchrony, i.e. above the red line where the MSF is positive the synchronization manifold is unstable, whereas below the line it is stable. Therefore, for providing synchronization in the network, the coupling parameter must be larger than the one needed for making the MSF negative.

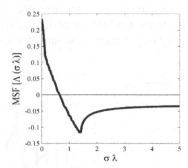

Fig. 3. The MSF for x-coupled bursting HR neurons. The graph shows the border of the analytic synchrony, i.e. above the red line where the MSF is positive the synchronization manifold is unstable, whereas below the line it is stable.

4.2   Determining the Numerical Synchronizing Coupling Parameter

Although we can study the (linear) synchronization in an analytic way, the most precise way for determining the (global) synchronizing coupling strength is numerically solving the differential equations and monitoring the synchronization error. In order to have the complete synchronization in the network, the error between the dynamical systems must distinctly goes to zero as time goes to infinity. The average synchronization error E of the network (2) in a time-interval $(0,T)$ is defined as [21]

$$E(t) = \frac{2}{N(N-1)} \sum_{i<j} \left\| X_i(t) - X_j(t) \right\|^2 \; ; \; E = \frac{1}{T} \sum_{t=0}^{T} E(t). \tag{8}$$

In order to determine the synchronization error and hence the synchronizing coupling strength, the initial state of the trajectory, $X_i(0)$, is randomly chosen, with the constraint $E(0) = 1$. By starting the coupling strength from a value (usually the one obtained through the MSF method), it gradually increases until E gets a value less than ε, i.e. $E < \varepsilon$ ($\varepsilon = 10^{-5}$ in this work).

5 Simulation Results

The analytic synchronizing coupling strength was determined through the MSF formalism (Fig. 3). In order to determine the numerical synchronizing coupling strength through tracking the average synchronization error, 20 different dynamical networks were randomly generated and a trajectory $X_i(t)$; $i = 1,...,N$ was computed, and then the error expressed by (8) was computed and the corresponding synchronizing coupling strength was determined. Fig. 4 shows the numerical and analytic synchronizing coupling strengths ($\sigma_{numerical}$ and $\sigma_{analytic}$, respectively) and the normalized gap between them (($\sigma_{numerical} - \sigma_{analytic}$) / $\sigma_{analytic}$) as a function of the network size N. In this experiment the nodes of the networks were connected at least to their $k = 2$ nearest neighbors and the probability of shortcuts was $P = 0.01$. It is seen that as the number of nodes increases the synchronizing coupling strength decreases. However, the normalized gap between the numerical and analytic synchronizing parameters is almost not influenced by the network size. Note that in this experiment the average degree per node is the same for all of the cases.

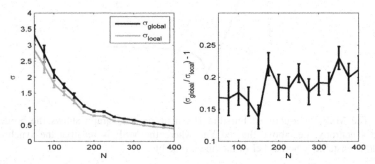

Fig. 4. The numerical and analytic synchronizing coupling strengths ($\sigma_{numerical}$ and $\sigma_{analytic}$, respectively) and the normalized gap between them (($\sigma_{numerical} - \sigma_{analytic}$) / $\sigma_{analytic}$) as a function of the network size N. The nodes of the networks are connected at least to their $k = 2$ nearest neighbors with $P = 0.01$. The graphs show averages and standard errorbars over 20 realizations.

We further investigated the influence of the average degree on the gap between σ_{Slobal} and $\sigma_{analytic}$. First, in networks with $N = 200$ the shortcut probability was fixed at $P = 0.01$ and the average degree changed by increasing k. The results are shown in Fig. 5 in which the influence of the average degree on the gap between these two synchronizing parameters is clear; the more the average degree the larger the gap. In another experiment we fixed $k = 2$ and the average degree changed by controlling P. The results of this experiment are shown in Fig. 6 where again by increasing P (increasing the average degree per node) the gap also increases. In other words, as the number of links per node increases, the strength providing the numerical synchronization in the network gets farther from the one necessary for the analytic synchronization, i.e. the one obtained through the MSF method.

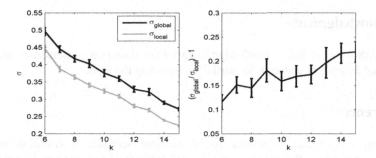

Fig. 5. The numerical and analytic synchronizing coupling strengths and the normalized gap between them as a function of k. The network size is $N = 200$ and $P = 0.01$ and data refer to averages over 20 realizations with corresponding errorbars for standard error.

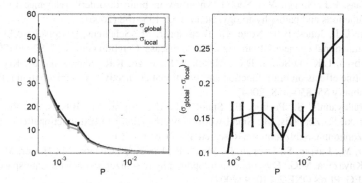

Fig. 6. The numerical and analytic synchronizing coupling strengths and the normalized gap between them as a function of P. The network size is $N = 200$ and $k = 2$ and data refer to averages over 20 realizations with corresponding errorbars for standard error.

6 Conclusions

In this paper we investigated the numerical and analytic synchronization behavior of HR neurons connected over NW networks. The analytic synchronizing parameters were obtained through the MSF formalism, while the parameters providing global stability of the synchronization manifold were obtained through numerically solving the network's differential equation and tracking the average synchronization error. We found that the gap between the analytic and numerical synchronizing parameters was almost independent of the networks size. However, the average degree was important in this context; the more the average degree of the network the larger the gap. Indeed, in networks with high average degree per node, the parameter obtained through the MSF approach, which is much simpler and cheaper than the numerical approaches, is far from the real synchronizing one.

Acknowledgments

The Author would like to thank Martin Hasler for stimulating discussions and Swiss National Science Foundation for partially supporting the work.

References

1. Gray, C.M., Singer, W.: Stimulus-specific neuronal oscillations in orientation columns of cat visual cortex. Proceedings of the National Academy of Science of the United States of America 86, 1698–1702 (1989)
2. Glass, L.: Synchronization and rhythmic processes in physiology. Nature 410, 277–284 (2001)
3. Buzsaki, G.: Rhythms of the brain. Oxford University Press, New York (2006)
4. Uhhaas, P.J., Singer, W.: Neural synchrony in brain disorders: relevance for cognitive dysfunctions and pathophysiology. Neuron 52, 155–168 (2006)
5. Stam, C.J., Jones, B.F., Nolte, G., Breakspear, M., Scheltens, P.: Small-world networks and functional connectivity in Alzheimer's disease. Cerebral Cortex 17, 92–99 (2006)
6. Jacobsen, L.K., D'Souza, D.C., Mencl, W.E., Pugh, K.R., Skudlarski, P., Krystal, J.H.: Nicotine effects on brain function and functional connectivity in schizophrenia. Biological Psychiatry 55, 850–858 (2004)
7. Micheloyannis, S., Pachou, E., Stam, C.J., Breakspear, M., Bitsios, P., Vourkas, M., Erimaki, S., Zervakis, M.: Small-world networks and disturbed functional connectivity in schizophrenia. Schizophrenia Research 87, 60–66 (2006)
8. Jalili, M., Lavoie, S., Deppen, P., Meuli, R., Do, K.Q., Cuenod, M., Hasler, M., De Feo, O., Knyazeva, M.G.: Dysconnection topography in schizophrenia with state-space analysis of EEG. PLoS ONE 2, e1059 (2007)
9. Knyazeva, M.G., Jalili, M., Brioschi, A., Bourquin, I., Fornari, E., Hasler, M., Meuli, R., Maeder, P., Ghika, J.: Topography of EEG multivariate phase synchronization in early Alzheimer's disease. Neurobiology of Aging (2008)
10. Hindmarsh, J.L., Rose, R.M.: A model of neuronal bursting using three coupled first order differential equations. Proceedings of the Royal Society of London Series B-Biological Sciences 221, 87–102 (1984)
11. Izhikevich, E.M.: Which model to use for cortical spiking neurons? IEEE Transactions on Neural Networks 15, 1063–1070 (2004)
12. Jalili, M.: Synchronizing Hindmarsh-Rose neurons over Newman-Watts networks. Chaos 19, 33103 (2009)
13. Jalili, M.: Neuronal Synchronization over Networks with Small-world Property. In: International Conference on Information Management and Engineering, pp. 17–21. IEEE Computer Society, Kuala Lumpur (2009)
14. Pecora, L.M., Carroll, T.L.: Master stability functions for synchronized coupled systems. Physical Review Letters 80, 2109–2112 (1998)
15. Sporns, O., Zwi, J.D.: The small world of the cerebral cortex. Neuroinformatics 2, 145–162 (2004)
16. Newman, M.E.J.: The structure and function of complex networks. SIAM Review 45, 167–256 (2003)
17. Newman, M.E.J., Moore, C., Watts, D.J.: Mean-field solution of the small-world network model. Physical Review Letters 84, 3201–3204 (2000)

18. Newman, M.E.J., Watts, D.J.: Renormalization group analysis of the small-world network model. Physics Letters A 263, 341–346 (1999)
19. Storace, M., Linaro, D., de Lange, E.: The Hindmarsh-Rose neuron model: Bifurcation analysis and piecewise-linear approximations. Chaos 18, 33128 (2008)
20. de Lange, E.: Neuron models of the generic bifurcation type: network analysis and data modeling. PhD Thesis. Ecole Polytechnique Federal de Lausanne, Lausanne (2006)
21. Jalili, M., Ajdari Rad, A., Hasler, M.: Enhancing synchronizability of dynamical networks using the connection graph stability method. International Journal of Circuit Theory and Applications 35, 611–622 (2007)

Outline Capture of Images by Multilevel Coordinate Search on Cubic Splines

Muhammad Sarfraz

Department of Information Science, Adailiya Campus, Kuwait University,
P.O. Box 5969, Safat 13060, Kuwait
prof.m.sarfraz@gmail.com

Abstract. An optimization technique is proposed for the outline capture of planar images. The overall technique has various phases including extracting outlines of images, detecting corner points from the detected outline, and curve fitting. The idea of multilevel coordinate search has been used to optimize the shape parameters in the description of the generalized cubic spline introduced. The spline method ultimately produces optimal results for the approximate vectorization of the digital contour obtained from the generic shapes.

Keywords: Optimization, multilevel coordinate search, Generic shapes, curve fitting, spline.

1 Introduction

Capturing and vectorizing outlines of images is one of the important problems of computer graphics, vision, and imaging. Various mathematical and computational phases are involved in the whole process. This is usually done by computing a curve close to the data point set. Computationally economical and optimally good solution is an ultimate objective to achieve the vectorized outlines of images for planar objects.

Curve modeling [21-23] plays significant role in various applications The representation of planar objects in terms of curves has many advantages. For example, scaling, shearing, translation, rotation and clipping operations can be performed without any difficulty. Although a good amount of work has been done in the area [8-20], it is still desired to proceed further to explore more advanced and interactive strategies. Most of the up-to-date research has tackled this kind of problem by curve subdivision or curve segmentation.

This work is inspired by a global optimization algorithm based on multilevel coordinate search (MCS) by Huyer and Neumaier [24-25]. It motivates the authors to a global optimization technique proposed for the outline capture of planar images. In this paper, the data point set represents any generic shape whose outline is required to be captured. We present an iterative process to achieve our objective. The algorithm comprises of various phases to achieve the target. First of all, it finds the contour of the gray scaled bitmap image [26-27]. Secondly, it uses the idea of corner points [1-7]

A. Nicholson and X. Li (Eds.): AI 2009, LNAI 5866, pp. 636–645, 2009.

to detects corners. These phases are considered as preprocessing steps. The next phase detects the corner points on the digital contour of the generic shape under consideration. The idea of multilevel coordinate search (MCS) is then used to fit a generalized cubic spline which passes through the corner points. It globally optimizes the shape parameters in the description of the generalized cubic spline to provide a good approximation to the digital curve.

The organization of the paper is as follows, Section 2 discusses about preprocessing step which includes finding the boundary of planar object and corner detection algorithm for finding the significant points. Section 3 is about the interpolant form of cubic spline curves and computation of its associated tangents. The process of multilevel coordinate search is explained in Section 4. Overall methodology of curve fitting is explained in Section 5, it includes the idea of knot insertion as well as the algorithm design for the proposed vectorization scheme. Demonstration of the proposed scheme is presented in Section 6. Finally, the paper is concluded in Section 7.

2 Preprocessing

The proposed scheme starts with first finding the boundary of the generic shape and then using the output to find the corner points. The image of the generic shape can be acquired either by scanning or by some other mean. The aim of boundary detection is to produce an object's shape in graphical or non-scalar representation. Chain codes [27], in this paper, have been used for this purpose. Demonstration of the method can be seen in Figure 1(b) which is the contour of the bitmap image shown in Figure 1(a).

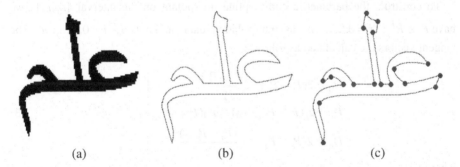

(a)   (b)   (c)

Fig. 1. Pre-processing Steps: (a) Original Image, (b) Outline of the image, (c) Corner points achieved

Corners in digital images give important clues for the shape representation and analysis. These are the points that partition the boundary into various segments. The strategy of getting these points is based on the method proposed in [1]. The demonstration of the algorithm is made on Figure 1(b). The corner points of the image are shown in Figure 1(c).

3 Curve Fitting and Spline

The motive of finding the corner points, in Section 2, was to divide the contours into pieces. Each piece contains the data points in between two subsequent corners inclusive. This means that if there are m corner points cp_1, cp_2, ..., cp_m then there will be m pieces pi_1, pi_2, ..., pi_m. We treat each piece separately and fit the spline to it. In general, the ith piece contains all the data points between cp_i and cp_{i+1} inclusive. After breaking the contour of the image into different pieces, we fit the spline curve to each piece.

The curve fitted by an ordinary Hermite cubic spline is a candidate of best fit, but it may not be a desired fit. This leads to the need of introducing some shape parameters in the description of the cubic spline. This section deals with the generalized form of cubic spline. It introduces two parameters v and w in the description of cubic spline defined as follows:

$$P(t) = (1-\theta)^3 F_i + 3\theta(1-\theta)^2 V_i + 3\theta^2(1-\theta)W_i + \theta^3 F_{i+1} \qquad (1)$$

where

$$V_i = F_i + h_i v_i D_i, \ W_i = F_{i+1} - h_i w_i D_{i+1}, \qquad (2)$$

F_i and F_{i+1} are corner points of the ith piece. D_i and D_{i+1} are the corresponding tangents at corner points.

Obviously, the parameters v_i's and w_i's, when equal to 1/3, provide the special case of cubic spline. Otherwise, these parameters can be used to loose or tight the curve. This paper proposes an evolutionary technique, namely multilevel coordinate search (MCS), to optimize these parameters so that the curve fitted is optimal.

To construct the parametric cubic spline interpolant on the interval $[t_0, t_n]$, we have $F_i \in R^m$, $i = 0,1,......,n$, as interpolation data, at knots t_i, $i = 0,1,......,n$. The tangent vectors are calculated as follows:

$$\left. \begin{array}{l} D_0 = 2(P_1 - P_0) - \dfrac{(P_2 - P_0)}{2} \\[2mm] D_i = a_i(P_i - P_{i-1}) + (1 - a_i)(P_{i+1} - P_i) \\[2mm] D_n = 2(P_n - P_{n-1}) - \dfrac{(P_n - P_{n-2})}{2} \end{array} \right\}, \qquad (3)$$

where

$$a_i = \frac{\|P_{i+1} - P_i\|}{\|P_{i+1} - P_i\| + \|P_i - P_{i-1}\|}.$$

Since, the objective of the paper is to come up with an optimal technique which can provide a decent curve fit to the digital data. Therefore, the interest would be to compute the curve in such a way that the sum square error of the computed curve with the actual curve (digitized contour) is minimized. Mathematically, the sum squared distance is given by:

$$S_i = \sum_{j=1}^{m_i} \left[P_i(u_{i,j}) - P_{i,j} \right]^2, \ t_{i,j} \in \left[t_i, t_{i+1} \right], \ i = 0,1,...,n-1, \tag{4}$$

where

$$P_{i,j} = (x_{i,j}, y_{i,j}), \quad j = 1,2,...,m_i, \tag{5}$$

are the data points of the ith segment on the digitized contour. The parameterization over t's is in accordance with the chord length parameterization. Thus the curve fitted in this way will be a candidate of best fit.

4   Multilevel Coordinate Search (MCS)

Multi-level coordinate search (MCS) is an optimization technique [24]. It guarantees to converge to the optimal solution if the function is continuous in the neighborhood of a global minimizer. It works by combining two types of searching: global searching and local searching. The advantage is that if the optimal value is somewhere near the current position, local search makes sure that the algorithm does not divert to distant locations in the solution space. It also reduces the time to reach the exact optimal value after reaching near it. A detailed description of the mapping of the MCS technique on our problem is given in the next section.

5   Proposed Approach for Vectorization

The proposed approach to the curve problem is described here in detail. It includes the phases of problem matching with MCS using cubic spline, description of parameters used for MCS, curve fitting, and the overall algorithm design.

5.1   Problem Mapping

Our interest is to optimize the values of curve parameters v and w such that the defined curve fits as close to the original contour segment as possible. We use MCS for this optimization of these two variables for the fitted curve. Hence the dimensionality of the solution space is 2, and each point in MCS represents a pair of values for v and w. We start with an initial set of points that are taken to be the corner points of the 2-dimensional solution space and the midpoints along the two directions. Since the solution space is bounded, with boundary values as -1 and 1 for both the dimensions, the initial points are chosen at these corners. Then we make boxes of the solution spaces using these points. For each point, we also compute and store the objective function value and associate each with one of the boxes. Now each box corresponds to a range of values of v and w. From all these boxes (ranges of v and w values), we first select the one having an associated point with the lowest function value. In this box, we apply local search and try to find the optimum in the determined direction of minimization within the box. If the v and w pair found in this box is not the optimal solution, then this box is split. That is, the range of v and w

values within this box is further split into smaller mutually exclusive ranges. Each new range is associated with a new representative point in the solution space and its fitness value. The shopping basket is hence kept updated with these ranges and fitness values.

Note that we apply MCS independently for each segment of a contour between two consecutive corner points that we have identified using corner point algorithm. MCS is applied sequentially on each of the segments, generating an optimized fitted curve for each segment. The algorithm is run until the maximum level of allowed splitting is reached, or an optimal value is reached. Once, all the contour segments are exhausted and still the desired global optimum solution is not achieved, MCS is applied again.

5.2 Initialization and Curve Fitting

Once we have the bitmap image of a generic shape, the boundary of the image can be extracted using the method described in Section 2. After the boundary points of the image are found, the next step is to detect corner points as explained in Section 2. This corner detection technique assigns a measure of 'corner strength' to each of the points on the boundary of the image. This step helps to divide the boundary of the image into n segments. Each of these segments is then approximated by interpolating spline described in Section 3. The initial solution of spline parameters (v and w) are randomly selected within the range [-1, 1].

After an initial approximation for the segment is obtained, better approximations are obtained through MCS to reach the optimal solution. We experiment with our system by approximating each segment of the boundary using the generalized cubic spline of Section 3. Each boundary segment is approximated by the spline. The shape parameters. v and w, in the cubic spline provide greater flexibility over the shape of the curve. These parameters are adjusted using MCS to get the optimal fit. Here, we try to minimize the sum squared error.

5.3 MCS Parameters Used

Although MCS sets default values of the algorithm variables, but it also gives the option of manipulating some parameters that define various factors affecting its performance. One of the factors is that how much weight MCS should give to global searching as opposed to local searching. The higher this value, the more global level search will be done. Similarly, another parameter that defines how much local search to do is also specified.

An initial set of starting solution points have to be specified for the system to start with. MCS requires an initial guess for the solution. It is the starting state parameters that affect the performance of the algorithm. If the starting solution is very near to the optimal solution, it is more likely to find the optimal solution readily than if the starting solution is distant from the optimal solution. An acceptable error value has to be defined, so that if the system comes within this error range from the optimal value, it terminates with the found solution.

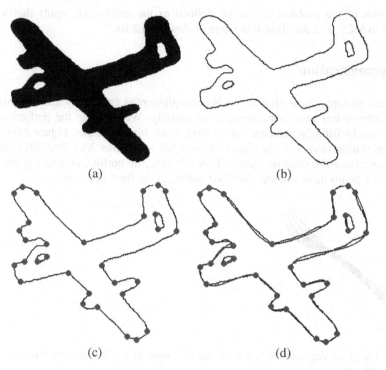

Fig. 2. Pre-processing Steps: (a) Original Image, (b) Outline of the image, (c) Corner points achieved, (d) Fitted Outline of the image

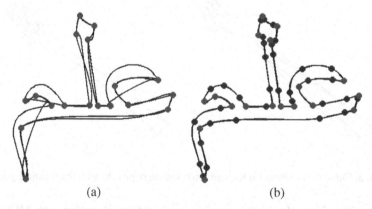

Fig. 3. (a) Fitted Outline of the image, (b) Fitted Outline of the image with intermediate points

An overall constraining factor is the maximum number of epochs that the algorithm may run. In this way, it does not run indefinitely if it is not reaching a stable solution. The direction of optimization of the fitness function has to be specified i.e. specific value that has to be attained. The default value is negative infinity and it can be used for our problem since the lowest value for our objective function is zero. The dimension of the problem has to be defined as the number of inputs that will be passed to MCS, and the allowable range of these variables.

6 Demonstration

The methodology, in Section 5, has been implemented practically and the proposed curve scheme has been implemented successfully. We evaluate the performance of our system by fitting parametric curves to different binary images. Figure 2 shows the implementation results of the algorithm with MCS: Figures 2(a), 2(b), 2(c) and 2(d) are respectively the original image of an Airplane, its outline, outline together with the corner points detected, and the fitted outline at the final iteration.

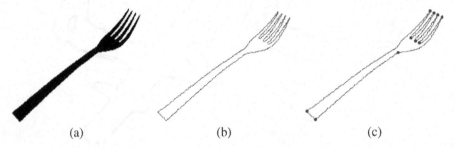

(a)   (b)   (c)

Fig. 4. Pre-processing steps for curve fitting (a) Image of a plane, (b) Extracted outline (c) Initial corner points

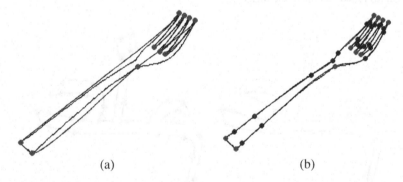

(a)   (b)

Fig. 5. Cubic curve fitting (a) without intermediate points (b) with intermediate points

Figure 3(a) shows the implementation results of the algorithm with MCS for the original image of an Arabic language word "Ilm" in Figure 1. One can see that the approximation is not satisfactory, this is specifically due to those segments which are bigger in size and highly curvy in nature. Thus, some more treatment is required for such outlines. One of the idea is to insert some intermediate points, this is demonstrated in Figure 3(b) where excellent result has been achieved. The idea of how to insert intermediate points is not explained here due to limitation of space. It will be explained in a subsequent paper.

Table 1. Names and contour details of images

Image	Name	# of Contours	# of Contour Points
	Ilm	1	[1641]
	Plane	3	[1106+61+83]
	Fork	1	[693]

Table 2. Comparison of number of initial corner points, intermediate points and total time taken (in seconds) for cubic interpolation approaches

Image	# of Initial Corner Points	# of Intermediate Points in Cubic Interpolation	Total Time Taken For Cubic Interpolation	
			Without Intermediate Points	With Intermediate Points
Ilm.bmp	18	34	46.312	164.17
Plane.bmp	31	13	56.766	100.58
Fork.bmp	10	22	18.438	70.297

Table 3. Comparison of number of epochs taken by MCS cubic interpolation approach with and without intermediate points

Image	# of Epochs taken by MCS	
	Cubic Interpolation Without Intermediate Points	Cubic Interpolation With Intermediate Points
Ilm.bmp	2459	8915
Plane.bmp	4726	7613
Fork.bmp	1035	4690

Another experiment is made on an image of Fork in Figure 4(a). Its outline is detected in Figure 4(b), and the corner points are shown in 4(c). Figures 5(a) and 5(b) demonstrate the fitted curves to the outline of Figure 4(b) corresponding to the scheme without and with insertion points respectively. It can be noticed that the fitted curve in Figure 5(a) has a good approximation, without inserting extra points, except at two segments. However, inserting extra points, has highly refined the approximation every where in Figure 5(b).

Tables 1 to 3 summarize the experimental results for different bitmap images. These results highlight various information including contour details of images (Table 1), intermediate points (Table 2), and number of iterations (Table 3).

7 Conclusion and Future Work

A global optimization technique, based on multilevel coordinate search, is proposed for the outline capture of planar images. The proposed technique uses the multilevel coordinate search to optimize a cubic spline to the digital outline of planar images. By starting a search from certain good points (initially detected corner points), an improved convergence result is obtained. The overall technique has various phases including extracting outlines of images, detecting corner points from the detected outline, curve fitting, and addition of extra knot points if needed. The idea of multilevel coordinate search has been used to optimize the shape parameters in the description of the generalized cubic spline introduced. The spline method ultimately produces optimal results for the approximate vectorization of the digital contours obtained from the generic shapes. It provides an optimal fit with an efficient computation cost as far as curve fitting is concerned. The proposed algorithm is fully automatic and requires no human intervention. The author is also thinking to apply the proposed methodology for another model curve namely conic. It might improve the approximation process. This work is in progress to be published as a subsequent work.

Acknowledgments

The author is grateful for constructive comments of the anonymous reviewers towards the improvement of the paper. This work was supported by Kuwait University, Research Grant No. [WI 01/09].

References

1. Chetrikov, D., Zsabo, S.: A Simple and Efficient Algorithm for Detection of High Curvature Points in Planar Curves. In: Proceedings of the 23rd Workshop of the Australian Pattern Recognition Group, pp. 1751–2184 (1999)
2. Goshtasby, A.A.: Grouping and Parameterizing Irregularly Spaced Points for Curve Fitting. ACM Transactions on Graphics 19(3), 185–203 (2000)
3. Reche, P., Urdiales, C., Bandera, A., Trazegnies, C., Sandoval, F.: Corner Detection by Means of Contour Local Vectors. Electronic Letters 38(14) (2002)
4. Marji, M., Siv, P.: A New Algorithm for Dominant Points Detection and Polygonization of Digital Curves. Pattern Recognition, 2239–2251 (2003)
5. Hu, W.-C.: Multiprimitive Segmentation Based on Meaningful Breakpoints for Fitting Digital Planar Curves with Line Segments and Conic Arcs. Image and Vision Computing, 783–789 (2005)
6. Freeman, H., Davis, L.S.: A corner finding algorithm for chain-coded curves. IEEE Trans. Computers 26, 297–303 (1977)

7. Richard, N., Gilbert, T.: Extraction of Dominant Points by estimation of the contour fluctuations. Pattern Recognition (35), 1447–1462 (2002)
8. Sarfraz, M.: Representing Shapes by Fitting Data using an Evolutionary Approach. International Journal of Computer-Aided Design & Applications 1(1-4), 179–186 (2004)
9. Sarfraz, M., Khan, M.A.: An Automatic Algorithm for Approximating Boundary of Bitmap Characters. Future Generation Computer Systems, 1327–1336 (2004)
10. Sarfraz, M.: Some Algorithms for Curve Design and Automatic Outline Capturing of Images. International Journal of Image and Graphics, 301–324 (2004)
11. Hou, Z.J., Wei, G.W.: A New Approach to Edge Detection. Pattern Recognition, 1559–1570 (2002)
12. Sarfraz, M.: Computer-Aided Reverse Engineering using Simulated Evolution on NURBS. International Journal of Virtual & Physical Prototyping 1(4), 243–257 (2006)
13. Sarfraz, M., Riyazuddin, M., Baig, M.H.: Capturing Planar Shapes by Approximating their Outlines. International Journal of Computational and Applied Mathematics 189(1-2), 494–512 (2006)
14. Sarfraz, M., Rasheed, A.: A Randomized Knot Insertion Algorithm for Outline Capture of Planar Images using Cubic Spline. In: The Proceedings of The 22th ACM Symposium on Applied Computing (ACM SAC 2007), Seoul, Korea, pp. 71–75. ACM Press, New York (2007)
15. Kano, H., Nakata, H., Martin, C.F.: Optimal Curve Fitting and Smoothing using Normalized Uniform B-Splines: A Tool for Studying Complex Systems. Applied Mathematics and Computation, 96–128 (2005)
16. Yang, Z., Deng, J., Chen, F.: Fitting Unorganized Point Clouds with Active Implicit B-Spline Curves. Visual Computer, 831–839 (2005)
17. Lavoue, G., Dupont, F., Baskurt, A.: A New Subdivision Based Approach for Piecewise Smooth Approximation of 3D Polygonal Curves. Pattern Recognition 38, 1139–1151 (2005)
18. Yang, H., Wang, W., Sun, J.: Control Point Adjustment for B-Spline Curve Approximation. Computer Aided Design 36, 639–652 (2004)
19. Horng, J.H.: An Adaptive Smoothing Approach for Fitting Digital Planar Curves with Line Segments and Circular Arcs. Pattern Recognition Letters 24(1-3), 565–577 (2003)
20. Sarkar, B., Singh, L.K., Sarkar, D.: Approximation of Digital Curves with Line Segments and Circular Arcs using Genetic Algorithms. Pattern Recognition Letters 24, 2585–2595 (2003)
21. Yang, X.: Curve Fitting and Fairing using Conic Splines. Computer Aided Design 6(5), 461–472 (2004)
22. Yang, X.N., Wang, G.Z.: Planar Point Set Fairing and Fitting by Arc Splines. Computer Aided Design, 35–43 (2001)
23. Sarfraz, M.: Designing Objects with a Spline. International Journal of Computer Mathematics 85(7) (2008)
24. Huyer, W., Neumaier, A.: Global Optimization by Multilevel Coordinate Search. Journal of Global Optimization 14, 331–355 (1999)
25. Multi-level Coordinate Search,
   http://www.mat.univie.ac.at/~neum/software/mcs/
26. Gonzalez, R.C., Woods, R.E., Eddins, S.L.: Digital Image Processing Using MATLAB, 2nd edn. Gatesmark Publishing (2009)
27. Nixon, M.S., Aguado, A.S.: Feature extraction and image processing. Elsevier, Amsterdam (2008)

Aggregation Trade Offs in Family Based Recommendations

Shlomo Berkovsky, Jill Freyne, and Mac Coombe

CSIRO Tasmanian ICT Center,
GPO Box 1538, Hobart, 7001, Australia
firstname.lastname@csiro.au

Abstract. Personalized information access tools are frequently based on collaborative filtering recommendation algorithms. Collaborative filtering recommender systems typically suffer from a data sparsity problem, where systems do not have sufficient user data to generate accurate and reliable predictions. Prior research suggested using group-based user data in the collaborative filtering recommendation process to generate group-based predictions and partially resolve the sparsity problem. Although group recommendations are less accurate than personalized recommendations, they are more accurate than general non-personalized recommendations, which are the natural fall back when personalized recommendations cannot be generated. In this work we present initial results of a study that exploits the browsing logs of real families of users gathered in an eHealth portal. The browsing logs allowed us to experimentally compare the accuracy of two group-based recommendation strategies: aggregated group models and aggregated predictions. Our results showed that aggregating individual models into group models resulted in more accurate predictions than aggregating individual predictions into group predictions.

1 Introduction

The quantity of potentially interesting information services available online has been growing rapidly and exceeds human processing capabilities. The vast amount of online information necessitates Web sites and portals to provide users with intelligent and personalized navigation support tools. These tools help users to identify the information items most relevant to them and filter out the rest by predicting the level of interest of users in particular information items. Collaborative filtering [9] is a statistical recommendation technique that can be applied to predict the interest level of a user in unvisited Web pages.

Collaborative filtering is commonly used in many online recommender systems to support users in selecting news items [2], courses [3], and many more [13]. The input for a collaborative filtering algorithm is a two dimensional matrix consisting of user models describing their preferences, interests, and information needs in the form of a feature vector. Collaborative filtering is based on the assumption that users with similar interests prefer similar information items [15]. In order

A. Nicholson and X. Li (Eds.): AI 2009, LNAI 5866, pp. 646–655, 2009.

to generate a recommendation, collaborative filtering initially compares the user models to identify users with the highest similarity to the current user and then generates predictions on items by calculating the normalized and weighted average of the opinions of the similar users[1].

One of the emerging practical problems of collaborative filtering recommender systems is the sparsity of user data [7], *i.e.*, the lack of sufficient information about users, which prevents the system from generating accurate and reliable predictions of interest in yet unseen information items. To partially resolve this problem and increase the accuracy of the generated recommendations, [8] proposed to aggregate the sparse individual user data into group-based data and then use the aggregated data in the collaborative recommendation process. Although in most conditions group-based recommendations cannot be as accurate as the personalized recommendations, they have the potential to be more accurate than general non-personalized recommendations, which are the natural fall back if the sparsity problem prevents the system from generating the personalized recommendations.

In this work we analyze family-based collaborative filtering recommendations – a particular case of group recommendations – using real life browsing data gathered in a study involving the users of an experimental eHealth family portal. We implemented several strategies that aggregated individual browsing logs into group-based data, generated collaborative filtering recommendations using the aggregated data, and then evaluated them against the observed browsing logs of the users.

The obtained experimental results demonstrate that group recommendations are superior to global and inferior to personalized recommendations. Also, we compared two aggregation strategies. The first aggregated the individual user models into group models and then applied collaborative filtering to the aggregated models. The second applied collaborative filtering algorithm to the individual user models and then aggregated the individual predictions into group predictions. The results show that aggregating the user models allows generating more accurate recommendations than aggregating the predictions.

Hence, the main contributions of this work are two-fold. Firstly, we evaluate the accuracy of collaborative filtering group recommendations and compare it to the accuracy of personalized and general recommendations. Secondly, we compare two strategies for the data aggregation: aggregation of browsing models and aggregation of predictions.

The rest of this paper is structured as follows. In section 2 we discuss related work on collaborative filtering and group recommendations. In section 3 we present and formulate the two aggregated group-based recommendation strategies. In section 4 we present our experimental settings, results and findings. Finally, in section 5 we conclude this work and present our future research directions.

---

[1] This presentation of collaborative filtering is narrowed down to user-to-user memory based approach. For a recent through survey of collaborative filtering algorithms the reader is referred to [14].

2 Collaborative Filtering and Group Recommendations

Collaborative filtering is one of the most popular and widely-used recommendation algorithms. It is based on the notion of *word of mouth* [15], which assumes that users who agreed in the past will agree in the future. In other words, it uses opinions of similar users to generate future predictions for a target user. The opinions of users on the items are expressed either as explicit ratings given by users according to a predefined scale or as implicit ratings accumulated and inferred through logging users' interactions with the system.

The main stages of the collaborative filtering recommendation generation process are: (1) recognizing commonalities between users by computing their inter-user similarities; (2) selecting the most similar users; and (3) generating recommendations by aggregating the opinions of the most similar users [9]. As it is being based on the similarities of users, the collaborative filtering process is sometimes referred to *people-to-people correlation*. In comparison with other recommendation algorithms, the main advantage of collaborative filtering over other algorithms is that it is not domain specific and independent of the representation of users and items. That is, a single collaborative filtering recommender systems can generate recommendations for any type of items (movies, images, or text) regardless of their content. As such, it is considered a universal technique applicable to a wide variety of domains and applications [13].

Collaborative filtering recommender systems suffer from the well-known sparsity problem [7]. It prevents the system from generating accurate predictions due to the insufficient data available about the users. Two particular cases of the sparsity problem can be differentiated: new user problem – the number of user ratings is insufficient for the identification of similar users and reliable generation of recommendations for that user [10], and new item problem – the number of item ratings is insufficient for a reliable generation of recommendations for that item [5].

In recent years the focus of collaborative filtering recommendation algorithms shifted from predictions for individuals to the more complex task of predictions for groups. To date, group recommendations were generated using one of the following three strategies: merging recommendations generated for individuals (very rare occurrence; will not be considered in this work), aggregating individual user models into group-based models, or aggregating them predictions for individual users into group-based recommendations [8].

The group modeling and aggregated predictions strategies differ in the timing of the aggregation of information in the recommendation process as illustrated in Figure 1. Specifically, group modeling strategy [4,16] aggregates individual user models of the group members *before* the prediction computation and then generates recommendations basing on the aggregated group model. Alternatively, aggregated predictions [11,12] treats group members as individuals for the prediction computation and *afterwards* aggregates the individual predictions to generate group recommendations.

As discussed in [8], the selection between the group modeling and aggregated prediction strategies depends on external factors, such as the ability to examine

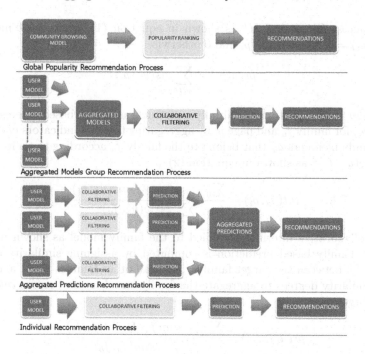

Fig. 1. Recommendation generation process

or negotiate group preferences, coverage of the recommendations, privacy considerations, and ability to explain the recommendations. However, to the best of our knowledge no prior work compared the *accuracy* of collaborative filtering recommendations generated using the above two strategies.

3   Prediction Strategies

The aim of this work is to determine which of the above two strategies for aggregating individual data and generating group-based recommendations is more appropriate when dealing with coherent groups consisting of individuals within a nuclear family structure. We concentrate on the following four recommendation strategies (see Figure 1). Our baseline strategy, *global popularity*, exploits the wisdom of the crowd at large and recommends the same most frequently visited items to all users. Our second and third strategies, *group modeling* and *aggregated predictions*, examine group-based recommendation algorithms and focus, respectively, on the group modeling and aggregated predictions strategies. Our fourth strategy is a standard *personalized collaborative filtering* recommendation algorithm. We will elaborately present these four strategies.

The *global popularity* strategy implements a simple social navigation mechanism [1], which guides users to areas of global interest. Each page p_i is assigned a predicted popularity score $pred(p_i)$ based on the number of times that it was visited across all users $u_x \epsilon U$ as shown in equation (1), where visit indicator $vis(u_x, p_i) = 1$ if u_x visited p_i and 0 otherwise.

$$pred(p_i) = \sum_{x \epsilon U} vis(u_x, p_i) \tag{1}$$

The *group modeling* strategy initially constructs a family based interest score $int(f_a, p_i)$ for family f_a and page p_i by aggregating the visit indicators $vis(u_x, p_i)$ of all family members u_x that belong to the family f_a according to their relative weight $\omega(u_x, f_a)^2$ as shown in equation (2).

$$int(f_a, p_i) = \frac{\sum_{u_x \epsilon f_a} \omega(u_x, f_a) \, vis(u_x, p_i)}{\sum_{u_x \epsilon f_a} \omega(u_x, f_a)} \tag{2}$$

Then, collaborative filtering is applied to the family model as shown in equation (3). Family based prediction is computed by assigning similarity degrees $sim(f_a, f_b)$ between the target family f_a and all other families $f_b \epsilon F$, and using these similarity degrees to aggregate the family based interest scores $int(f_b, p_i)$ in the target page p_i.

$$pred(f_a, p_i) = \frac{\sum_{f_b \epsilon F} sim(f_a, f_b) \, int(f_b, p_i)}{\sum_{f_b \epsilon F} sim(f_a, f_b)} \tag{3}$$

Finally, the computed family based prediction $pred(f_a, p_i)$ is assigned to all the users u_x that belong to the family f_a, i.e., $pred(u_x, p_i \mid u_x \epsilon f_a) = pred(f_a, p_i)$.

The *aggregated prediction* strategy maintains an individual model for each user and initially generates individual predictions using the standard collaborative filtering recommendation algorithm as shown in equation (4). Prediction $pred(u_x, p_i)$ for user u_x and page p_i is computed by assigning similarity degrees $sim(u_x, u_y)$ between the target user u_x and all other users $u_y \epsilon U$, and using these similarity degrees to aggregate the individual visit indicators $vis(u_y, p_i)$ for the target page p_i.

$$pred(u_x, p_i) = \frac{\sum_{u_y \epsilon U} sim(u_x, u_y) \, vis(u_y, p_i)}{\sum_{u_y \epsilon U} sim(u_x, u_y)} \tag{4}$$

Then, the process becomes group focused. To generate a family based prediction $pred(f_a, p_i)$, the individual predictions $pred(u_x, p_i)$ for the family members are aggregated according to their relative weight $\omega(u_x, f_a)$ as shown in equation (5).

$$pred(f_a, p_i) = \frac{\sum_{u_x \epsilon f_a} \omega(u_x, f_a) \, pred(u_x, p_i)}{\sum_{u_x \epsilon f_a} \omega(u_x, f_a)} \tag{5}$$

Similarly to the previous strategy, the computed family based prediction $pred(f_a, p_i)$ is assigned to all the users u_x that belong to the family f_a.

---

[2] Uniform weighting is currently used to assign equal weight $\omega(u_x, f_a) = 1$ to all the users. Other weighting strategies will be investigated in the future.

The *personalized collaborative filtering* recommendation strategy examines the browsing patterns of individual users regardless of their membership in a family. For each user u_x, each page p_i is assigned a prediction score $pred(u_x, p_i)$ using the standard collaborative filtering algorithm as shown in equation (4) presented in the previous strategy.

In all four strategies we simplify the recommendation generation and recommend k unvisited pages having the highest prediction scores, *i.e.*, k pages that maximize the product $\prod_{i=1}^{k} pred(u_x, p_i)$. Note that the global popularity strategy generates one list of recommendations for all users, the two group-based strategies generate one list of recommendations for each family, and the personalized collaborative filtering produces one list for each family member.

4   Evaluation

The presented analysis was carried out through the browsing logs gathered as part of an eHealth family portal study. The aim of the analysis was to determine which strategy would be best to implement in a family based recommender in future versions of the portal. Specifically, we aimed to ascertain the differences (1) between the simple global popularity model, the aggregated family based models, and the personalized recommendation model, and (2) between the combined group model and the aggregated predictions strategies.

4.1   Experimental Setting

The data used was gathered over a two week period in March 2009. Members of the general public (families to be specific) were invited to take part in a study of family engagement with an eHealth application. The task for each family member was to visit the experimental eHealth portal, possibly browse the healthy living content, and submit suggestions for improving their lifestyles. A by product of the study was the capture of browsing activity for all the members of the involved families over the 23 portal pages.

In total, 64 users from 40 families took part in the trial. In 24 families only one person interacted with the portal, in 8 families two members interacted with the portal, in 2 families three members interacted with the portal, and in 6 families all four members interacted with the portal. In total 188 individual page visits and 151 aggregated family based interest levels were logged, yielding an individual matrix having 87.23% sparsity and a denser family based matrix having 83.59% sparsity[3]. Each user visited on average 2.94 pages (stdev=2.77) and each page was visited on average by 8.17 users (stdev=4.33).

The distribution of page visits across the users demonstrates a typical long tail distribution. Only 2 users visited more than 10 pages, 6 users visited between 5 and 10 pages, and 56 users visited less than 5 pages. Conversely, the distribution

---

[3] We disregard the families in which only one member interacted with the portal and exclude them from the evaluation. However, we do use these users' browsing logs as sources of recommendation content in the training set.

of page visits across the pages is more balanced. 7 pages were visited by more than 10 users[4], 7 pages were visited by between 5 and 10 users, and 9 pages were visited by less than 5 users.

For each user or family, a one off similarity matrix with other users or families was created using Pearson's Correlation similarity metric [9][5]. Using this similarity matrix, four recommendation lists were produced for each user using the four prediction strategies detailed in Section 3 (global, group modeling, aggregated prediction, and personalized collaborative filtering). A leave one out experimental evaluation was carried out to evaluate the performance of the algorithms. In particular, the accuracy of the recommendations was evaluated using the classification accuracy metrics of precision, recall, and F1 by comparing the recommendation lists with the actual logs of the users [6].

Let us denote by $\mathcal{V}$ the set of pages that were visited by a user (will be considered as the relevant pages) and by $\mathcal{R}$ the set of pages that were recommended by the system to the user. In this context, *precision* of the recommendations is computed by $\frac{|\mathcal{V} \cap \mathcal{R}|}{|\mathcal{R}|}$ and *recall* by $\frac{|\mathcal{V} \cap \mathcal{R}|}{|\mathcal{V}|}$. When the size of the recommended set $\mathcal{R}$ is limited to k, the computed precision metric is referred to as *precision@k*. Combining the two metrics of precision and recall yields a single metric, *F1 score*, which represents their harmonic mean assigning them equal weights. The F1 score is computed as

$$F1 = \frac{2 \times precision \times recall}{precision + recall}$$

4.2 Experimental Results

The first question we posed related to the accuracy of recommendations based on the *global* strategy of all users versus smaller groups of users in *aggregated models* and *aggregated predictions* strategies versus individual activity in *personalized collaborative filtering* strategy. Table 1 shows the average precision, recall, and F1 scores obtained for each of the above recommendation strategies.

Table 1. Precision, Recall, and F-measure

	global	aggregated models	aggregated predictions	personalized
precision	0.219	0.300	0.235	0.534
recall	0.552	0.689	0.609	0.779
F1	0.314	0.418	0.339	0.633

It can be seen that, as expected, the personalized recommendation strategy outperformed all other strategies in terms of accuracy, returning the highest

---

[4] One of the pages was an outlier – it was visited by 20 users.
[5] Similar experimental results were obtained for the Cosine Similarity metric.

precision, recall, and F1 scores. Statistical significance tests[6] showed that the personalized strategy significantly outperformed both the group-based strategies. For precision we obtained $p = 4.36 \times 10^{-4}$ vs. aggregated models and $p = 1.44 \times 10^{-5}$ vs. aggregated predictions, and for recall we obtained $p = 2.32 \times 10^{-2}$ vs. aggregated models and $p = 2.72 \times 10^{-2}$ vs. aggregated predictions. Both group-based strategies outperformed the global strategy. For precision we obtained $p = 5.20 \times 10^{-2}$ vs. aggregated models and not statistically significant difference vs. aggregated predictions, and for recall we obtained $p = 5.45 \times 10^{-2}$ vs. aggregated models and not statistically significant difference vs. aggregated predictions.

Examining the whole recommendation lists and their accuracy is only one dimension of the recommendations' success. Precision@k measure analyzes the position of the visited pages within the recommendation lists. Figure 2 depicts the precision@k of the four recommendation strategies for gradually increasing from 1 to 7 values of k. Precision@k curves showed that the personalized strategy outperformed both the group-based strategies and the global strategy. For example, for $k = 1$ (the most strict metric focusing on the first recommended page) the personalized strategy achieved a precision of 74% in comparison to 38% and 44% for the two group-based strategies, and only 29% for the global non-personalized strategy. This observation was valid also for other values of k.

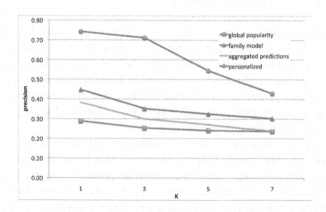

Fig. 2. Precision@k for various values of k

The second question we posed related to the comparative accuracy of the two group-based strategies. Both Table 1 and Figure 2 showed that the aggregated model strategy consistently outperformed the aggregated predictions strategy. For the overall precision and recall scores, the differences were statistically significant: $p = 1.79 \times 10^{-2}$ for precision and $p = 1.84 \times 10^{-2}$ for recall. We explain these findings by observing that aggregation of individual models yields a reasonably dense and accurate group model, which allows the system to generate

---

[6] All statistical significance results hereafter refer to a two-tailed t-test assuming equal variances.

reasonably accurate recommendations. Conversely, recommendations generated using the individual sparse models are inaccurate, such that their aggregation does not allow to improve their accuracy.

5   Conclusions and Future Work

The sparsity of user data is a well known problem of collaborative filtering recommender systems. To resolve it, the sparse individual user data can be aggregated into group-based data and these can then be used in the recommendation process. In this work we analyzed collaborative filtering family based recommendations using the browsing logs of the users of an experimental eHealth family portal. We implemented several strategies that aggregated individual data into group-based data, generated family based collaborative filtering recommendations using the aggregated data, and evaluated their accuracy against the observed browsing logs of the users.

Our empirical results showed that group recommendations were, as expected, more accurate than non-personalized one-size-fits-all recommendations at determining relevant Web pages. However, personalized collaborative filtering recommendations still outperformed group recommendations when comparing precision, recall, F1, and precision@k scores.

While previous works analyzed conditions when one group recommendation strategy would be preferred over another, this work experimentally compared the accuracy of two group-based aggregation strategies with real families of users of an eHealth portal. Our results consistently showed that aggregating individual browsing models into group models resulted in more accurate recommendations than aggregating the predicted interest levels. That is, generating recommendations using a dense group model was more accurate than aggregating the predictions generated for individual users.

In this work the users were assigned uniform weights when aggregating individual models into the family models. However, this is not reflective of the real setting, where different users may have different browsing patterns and frequencies. In the future we will evaluate the impact of various weighting heuristics on the accuracy of the recommendations. The results presented in this work are preliminary as they are supported by a reasonably small data set. In the future we will conduct a larger scale user study of online group recommendations, which will allow us to determine which strategies perform best under varying conditions such as richer user models, larger and more heterogeneous groups of users, and different content domains.

Acknowledgments

This research is jointly funded by the Australian Government through the Intelligent Island Program and CSIRO. The Intelligent Island Program is administered by the Tasmanian Department of Economic Development, Tourism and the Arts. The authors thank Nathalie Colineau, Cecile Paris, Dipak Bhandari, and Greg

Smith for their help with the development of the experimental eHealth family portal.

References

1. Brusilovsky, P., Chavan, G., Farzan, R.: Social adaptive navigation support for open corpus electronic textbooks. In: De Bra, P.M.E., Nejdl, W. (eds.) AH 2004. LNCS, vol. 3137, pp. 24–33. Springer, Heidelberg (2004)
2. Claypool, M., Gokhale, A., Miranda, T., Murnikov, P., Netes, D., Sartin, M.: Combining content-based and collaborative filters in an online newspaper. In: Workshop on Recommender Systems: Algorithms and Evaluation (1999)
3. Farzan, R., Brusilovsky, P.: Social navigation support in a course recommendation system. In: Proceedings of the International Conference on Adaptive Hypermedia and Adaptive Web-Based Systems, pp. 91–100 (2006)
4. Freyne, J., Smyth, B.: Cooperating search communities. In: Proceedings of the International Conference on Adaptive Hypermedia and Adaptive Web-Based Systems, pp. 101–110 (2006)
5. Gokhale, A., Claypool, M.: Correlation thresholds for more accurate collaborative filtering. In: Proceedings of the IASTED International Conference Artificial Intelligence and Soft Computing (1999)
6. Herlocker, J.L., Konstan, J.A., Terveen, L.G., Riedl, J.T.: Evaluating collaborative filtering recommender systems. ACM Transactions on Information Systems 22(1), 5–53 (2004)
7. Huang, Z., Chen, H., Zeng, D.: Applying associative retrieval techniques to alleviate the sparsity problem in collaborative filtering. ACM Transactions on Information Systems 22(1), 116–142 (2004)
8. Jameson, A., Smyth, B.: Recommendation to groups. In: Brusilovsky, P., Kobsa, A., Nejdl, W. (eds.) Adaptive Web 2007. LNCS, vol. 4321, pp. 596–627. Springer, Heidelberg (2007)
9. Konstan, J.A., Miller, B.N., Maltz, D., Herlocker, J.L., Gordon, L.R., Riedl, J.: Grouplens: applying collaborative filtering to usenet news. Commun. ACM 40(3), 77–87 (1997)
10. Linden, G., Smith, B., York, J.: Amazon.com recommendations: item-to-item collaborative filtering. IEEE Internet Computing 7(1), 76–80 (2003)
11. Masthoff, J.: Group modeling: Selecting a sequence of television items to suit a group of viewers. User Model. User-Adapt. Interact. 14(1), 37–85 (2004)
12. O'Connor, M., Cosley, D., Konstan, J.A., Riedl, J.: Polylens: a recommender system for groups of users. In: Proceedings of the European Conference on Computer Supported Cooperative Work, pp. 199–218 (2001)
13. Schafer, B.J., Konstan, J.A., Riedl, J.: E-commerce recommendation applications. Data Mining and Knowledge Discovery 5(1), 115–153 (2001)
14. Schafer, J., Frankowski, D., Herlocker, J., Sen, S.: Collaborative filtering recommender systems. In: Brusilovsky, P., Kobsa, A., Nejdl, W. (eds.) Adaptive Web 2007. LNCS, vol. 4321, pp. 291–324. Springer, Heidelberg (2007)
15. Shardanand, U., Maes, P.: Social information filtering: Algorithms for automating "word of mouth". In: Proceedings of the International Conference on Human Factors in Computing Systems, pp. 210–217 (1995)
16. Yu, Z., Zhou, X., Hao, Y., Gu, J.: Tv program recommendation for multiple viewers based on user profile merging. User Modeling and User-Adapted Interaction 16(1), 63–82 (2006)

Using Keyword-Based Approaches to Adaptively Predict Interest in Museum Exhibits

Fabian Bohnert and Ingrid Zukerman

Faculty of Information Technology, Monash University,
Clayton, VIC 3800, Australia
{fabian.bohnert,ingrid.zukerman}@infotech.monash.edu.au

Abstract. Advances in mobile computing and user modelling have enabled technologies that help museum visitors select personally interesting exhibits to view. This is done by generating personalised exhibit recommendations on the basis of non-intrusive observations of visitors' behaviour in the physical museum space. We describe a simple methodology for manually annotating museum exhibits with bags of keywords (viewed as item features), and present two personalised keyword-based models for predicting a visitor's viewing times of unseen exhibits from his/her viewing times at visited exhibits (viewing time is indicative of interest). Our models were evaluated with a real-world dataset of visitor pathways collected by tracking visitors in a museum. Both models achieve a higher predictive accuracy than a non-personalised baseline, and perform at least as well as a nearest-neighbour collaborative filter.

1 Introduction

Cultural heritage spaces such as museums provide a wealth of information. However, a visitor's receptivity and time are typically limited, posing the challenge of selecting personally interesting exhibits to view within the available time. Advances in mobile computing and user modelling provide the opportunity to assist a visitor in this selection process by means of suitable technologies. Such technologies can (1) utilise non-intrusive observations of a visitor's behaviour in the physical space to learn a model of his/her interests, and (2) generate personalised exhibit recommendations based on interest predictions.

In this paper, we describe a simple methodology for manually annotating museum exhibits with bags of keywords, which we view as item features. We then present two personalised keyword-based models for predicting a visitor's viewing times of unseen exhibits from his/her viewing times at visited exhibits (we use viewing time to measure interest in exhibits): (1) a memory-based nearest-neighbour *Content-Based Filter (CBF)*, and (2) a model-based *Normalised Least Mean Squares Filter (NLMSF)*. Our models were evaluated with a dataset we collected by manually tracking visitors at Melbourne Museum (Melbourne, Australia). We compared the performance of our models with that of a non-personalised baseline, and a memory-based nearest-neighbour collaborative filter — the traditional approach for domains where features of items are not readily apparent [1,2]. Both models attain a higher predictive accuracy than the baseline, and perform at least as well as the collaborative filter (with *NLMSF* outperforming *CBF* in the realistic *Progressive Visit* setting, Section 5).

A. Nicholson and X. Li (Eds.): AI 2009, LNAI 5866, pp. 656–665, 2009.

This paper is organised as follows. In Section 2, we summarise related research. Section 3 describes our domain, our dataset of visitor pathways, and our approach for annotating museum exhibits, and Section 4 discusses our models for predicting a visitor's viewing times. We then summarise the results of our evaluation (Section 5), before we conclude in Section 6.

2   Related Research

Our research lies at the intersection of statistical user modelling and personalised guide systems for physical domains. Personalised guide systems for physical domains have often employed adaptable user models, which require visitors to explicitly state their interests in some form, e. g., [3,4]. Less attention has been paid to predicting preferences from non-intrusive observations, and to utilising adaptive user models that do not require explicit user input. In the museum domain, adaptive user models have usually been updated from the user's interactions with the system, with a focus on adapting content presentation [5,6,7] rather than predicting or recommending exhibits to be viewed.

The above systems, like most systems in the museum domain, rely on knowledge-based user models in some way, and hence, require an explicit, a-priori engineered representation of the domain knowledge. In contrast, our research investigates non-intrusive statistical user modelling and recommendation techniques that do not require such an explicit domain knowledge representation.

3   Domain and Datasets

This section introduces our domain. We briefly discuss the dataset of visitor pathways we collected at Melbourne Museum (Section 3.1), and our methodology for annotating and representing museum exhibits (Section 3.2).

3.1   Visitor Pathways

Our dataset of visitor pathways was obtained by manually tracking visitors at Melbourne Museum from April to June 2008, using a custom-made tracking tool running on laptop computers [2]. We only shadowed first-time adult visitors travelling on their own, to ensure that neither prior knowledge about the museum nor other accompanying visitors' interests influenced a visitor's decisions about which exhibits to view. In total, we recorded 158 visitor pathways in the form of time-annotated sequences of visited exhibit areas,[1] obtaining data which provides information of the type that may be inferred from sensing data. The dataset (described in detail in [2]) contains 8327 viewing durations at the 126 exhibit areas of Melbourne Museum, yielding an average of 52.7 exhibit areas per visitor (41.8% of the exhibit areas). Hence, on average 58.2% of the exhibit areas were not viewed by a visitor, which indicates that there is potential for pointing a visitor to relevant but unvisited exhibits.[2]

---

[1] Prior to collecting the data, we grouped the individual exhibits of Melbourne Museum into 126 semantically coherent and spatially confined *exhibit areas*.

[2] We use the terms 'exhibit' and 'exhibit area' synonymously in the remainder of this paper.

Table 1. Annotator statistics

	A1	A2	A3	A4
Annotated exhibit areas	126	122	85	79
Phrases	1242	772	403	324
Tokens	1578	806	889	906
Distinct tokens	764	328	495	447
Phrases / exhibit area	9.86	6.33	4.74	4.10
Tokens / exhibit area	12.52	6.61	10.46	11.47
Tokens / phrase	1.27	1.04	2.21	2.80

3.2 Exhibit Representation

Museum exhibits are complex, as they are usually composed of a collection of items (including text labels and multimedia content). Hence, it is difficult to determine features and feature weights that accurately describe the content being presented. In this section, we outline the methodology we used to annotate the exhibit areas of Melbourne Museum. The idea is that subjective, manual annotations comprising keywords associated with exhibits can be used to represent the content of the exhibits.

The 126 exhibit areas were annotated by four annotators — two were familiar with the museum ($A1, A2$) and two were not ($A3, A4$) (the annotators operated in pairs $A1$-$A4$ and $A2$-$A3$, so that exhibit area boundaries were clear to all annotators). The annotators were instructed to engage with an exhibit for a length of time equal to the expected viewing time of the exhibit (estimated from our pathway dataset), before writing down any keywords or phrases. This enabled us to take into account exhibit complexity. We did not restrict the vocabulary that the annotators could use, and asked them to use as many keywords or multi-word phrases they considered appropriate. In total, 77 exhibit areas were annotated by all four annotators, and overall the annotators used 1381 distinct tokens (we split phrases into one-word tokens). Table 1 summarises annotator-specific statistics, which highlight the differences between the annotators' styles.

In this paper, we explore two content-based models based on a *bag of words* representation of the exhibits (we remove stopwords in both models):

- **Simple.** We simply collect the tokens and phrases given by our annotators in an exhibit-specific bag of words. We then remove tokens that occur in less than two bags of words or have a corpus frequency of less than 3.[3]
- **Populated.** We proceed as in the *Simple* model, and use Wordnet [8] to further populate the bags of words. This is done by including Wordnet topics and Wordnet synonyms from the most likely synset of a term (i. e., the first synset). Both are included in proportion to how often the corresponding term was used by the annotators. The inclusion of topics and synonyms is expected to alleviate the differences between the words used by the annotators, and hence improve predictive performance. We then remove tokens that occur in less than three bags of words or have a corpus frequency of less than 5.

---

[3] These thresholds, and those for the *Populated* model, were empirically determined.

Equivalently to its bag of words, an exhibit i is represented by an f-dimensional term vector x_i (where f is the number of distinct tokens used across all exhibit areas). For each model (*Simple* and *Populated*), we consider the following variants:

- **Binary.** The exhibit-specific bags of words are transformed into binary term vectors x_i by setting to 1 those elements of x_i for which the term occurs in exhibit i's bag of words.

- **Frequency.** The bags of words are transformed into term vectors x_i whose elements are set to word frequencies, i.e., the number of times a term occurs in exhibit i's bag of words.

- **Tf-idf.** The elements of x_i are *term frequency inverse document frequency (tf-idf)* term weights, where the *term frequency* $\text{tf}_{t,i}$ of a term t for an exhibit i, and the *inverse document frequency* idf_t for a term t are defined as follows:

$$\text{tf}_{t,i} = \frac{n_{t,i}}{\sum_s n_{s,i}} \quad \text{and} \quad \text{idf}_t = 1 + \log \frac{|I|}{|\{i \in I : n_{t,i} > 0\}|}$$

where $n_{t,i}$ is the number of times term t occurs in exhibit i's bag of words, s is a term in this bag of words, $|I|$ is the total number of exhibits, and $|\{i \in I : n_{t,i} > 0\}|$ is the number of exhibits annotated with term t. The *tf-idf* weight for term t and exhibit i is $\text{tf-idf}_{t,i} = \text{tf}_{t,i} \times \text{idf}_t$.

4 Adaptive, Content-Based Viewing Time Prediction from Non-intrusive Observations

In this section, we first describe how we use viewing time to quantify interest in exhibits (Section 4.1). We then propose two keyword-based approaches for predicting a visitor's (log) viewing times (viewed as interest in exhibits) from non-intrusive observations of his/her (log) viewing times at visited exhibits (Sections 4.2 and 4.3).

4.1 From Viewing Time to Exhibit Interest

In an information-seeking context, people usually spend more time on relevant information than on irrelevant information, as viewing time correlates positively with preference and interest [9]. Hence, viewing time can be used as an indirect measure of interest. We propose to use log viewing time (instead of raw viewing time), due to the following reasons. When examining our dataset of visitor pathways (Section 3.1), we found the distributions of viewing times at exhibits to be positively skewed. Thus, the usual assumption of a Gaussian model did not seem appropriate. To select a more appropriate family of probability distributions, we used the *Bayesian Information Criterion (BIC)* [10]. We tested exponential, gamma, normal, log-normal and Weibull distributions. The log-normal family fitted best, with respect to both number of best fits and average BIC score (averaged over all exhibits). Hence, we transformed all viewing times to their log-equivalent to obtain approximately normally distributed data. This transformation fits well with the idea that for high viewing times, an increase in viewing time indicates a smaller increase in the modelled interest than a similar increase in the context of low viewing times.

The basis for our predictive models is the current visitor u's implicit rating vector $\boldsymbol{r}_u$, where an element r_{ui} of $\boldsymbol{r}_u$ is the implicit rating given by visitor u to exhibit i. To build $\boldsymbol{r}_u$, for each element r_{ui}, we first calculate the log-equivalent t_{ui} of an observed viewing time. We then normalise this log viewing time by calculating an exhibit-specific z-score, thereby ensuring that varying exhibit complexity does not affect the predictions of our models (viewing time increases with exhibit complexity). This is done by subtracting from the log viewing time t_{ui} the exhibit's log viewing time mean $\bar{t}_{\cdot i}$, and dividing by the sample standard deviation σ_i.[4] Finally, we add the resultant normalised log viewing time r_{ui} to $\boldsymbol{r}_u$.

4.2 Nearest-Neighbour Content-Based Filter

Our *Content-Based Filter (CBF)* for adaptively predicting a visitor's viewing times of unseen exhibits is a nearest-neighbour content-based filter, e.g., [11]. We start by calculating $\tilde{r}_{ui}$, a similarity-weighted personalised prediction of a current visitor u's unobserved normalised log viewing time r_{ui} for a current exhibit i, from the normalised log viewing times r_{uj} in visitor u's rating vector $\boldsymbol{r}_u$ as follows:

$$\tilde{r}_{ui} = \frac{\sum_{j \in N(u,i)} s_{ij} r_{uj}}{\sum_{j \in N(u,i)} s_{ij}}$$

where $N(u,i)$ is the set of nearest neighbours, and s_{ij} is the similarity between exhibits i and j (calculated using the cosine similarity measure on the feature vectors $\boldsymbol{x}_i$ and $\boldsymbol{x}_j$ of exhibits i and j, Section 3.2).[5] The set of nearest neighbours $N(u,i)$ for the current visitor u and exhibit i is constructed by (1) calculating s_{ij} for all exhibits j which were viewed by current visitor u, and (2) selecting up to k exhibits that are most similar to the current exhibit i. Finally, we employ *shrinkage to the mean* [12] to calculate $\hat{r}_{ui}$, a shrunken personalised prediction of r_{ui} (we unnormalise $\hat{r}_{ui}$ afterwards to obtain a log viewing time prediction $\hat{t}_{ui}$):

$$\hat{r}_{ui} = \tilde{r}_{\cdot i} + \omega \left(\tilde{r}_{ui} - \tilde{r}_{\cdot i} \right)$$

where $\tilde{r}_{\cdot i}$ is a (normalised, non-personalised) mean prediction of r_{ui}, i.e., $\tilde{r}_{\cdot i} = 0$, and $\omega \in [0, 1]$ is the shrinkage weight. Setting $\tilde{r}_{\cdot i} = 0$ corresponds to using the log viewing time mean $\bar{t}_{\cdot i}$ as a non-personalised mean prediction of exhibit i's log viewing time t_{ui}.

Whenever a similarity-weighted personalised prediction is not possible (e.g., when the set of nearest neighbours is empty), we estimate r_{ui} using simply $\tilde{r}_{\cdot i} = 0$. We proceed in the same fashion for less than m observations in visitor u's rating vector $\boldsymbol{r}_u$.

The maximum number of nearest neighbours k, minimum number of viewed exhibits m and shrinkage weight ω are chosen so that an error measure of choice is minimised — in our case, the *mean absolute error (MAE)* (Section 5.1).

---

[4] We use the other visitors' log viewing times to compute estimates for $\bar{t}_{\cdot i}$ and σ_i. Hence, our predictive models are not content-based in the strict sense, as this normalisation procedure uses information acquired from visitors other than the current visitor.

[5] The similarities s_{ij} can be pre-computed, as they do not depend on visitor ratings.

4.3 Normalised Least Mean Squares Filter

In addition to the memory-based approach discussed in Section 4.2, we present a model-based approach for predicting a visitor's (log) viewing times of unseen exhibits, called *Normalised Least Mean Squares Filter (NLMSF)*. For *NLMSF*, the current visitor u's user model is an f-dimensional weight vector $\boldsymbol{w}_u$ (recall that f is the number of tokens, Section 3.2). A personalised prediction $\tilde{r}_{ui}$ of a current visitor u's unobserved normalised log viewing time r_{ui} for a current exhibit i is computed as the inner product of the weight vector $\boldsymbol{w}_u$ and the feature vector $\boldsymbol{x}_i$ of the exhibit:

$$\tilde{r}_{ui} = \boldsymbol{w}_u \cdot \boldsymbol{x}_i$$

As for *CBF*, we use shrinkage to the mean to improve the predictive accuracy of *NLMSF*, i.e., $\hat{r}_{ui} = \tilde{r}_{\cdot i} + \omega\,(\tilde{r}_{ui} - \tilde{r}_{\cdot i})$, where as above, $\tilde{r}_{\cdot i} = 0$ is a (normalised, non-personalised) mean prediction of r_{ui} (we unnormalise $\hat{r}_{ui}$ afterwards to obtain a log viewing time prediction $\hat{t}_{ui}$). If we have less than m observations in the current visitor u's rating vector $\boldsymbol{r}_u$, we estimate r_{ui} using $\tilde{r}_{\cdot i} = 0$.

Let $\boldsymbol{w}_{u,t}$ be the weight vector at time t. We initialise $\boldsymbol{w}_{u,0} = 0$ ($t = 0$ corresponds to having an empty rating vector $\boldsymbol{r}_u$), and use the normalised version of the *Widrow-Hoff rule* [13] (also called *gradient descent rule*) to adaptively update the weight vector $\boldsymbol{w}_u$, whenever a normalised log viewing time r_{ui} is added to the current visitor's rating vector $\boldsymbol{r}_u$:

$$\boldsymbol{w}_{u,t+1} = \boldsymbol{w}_{u,t} - \frac{\eta}{\|\boldsymbol{x}_i\|^2} \underbrace{(\boldsymbol{w}_{u,t} \cdot \boldsymbol{x}_i - r_{ui})}_{\tilde{r}_{ui}}\boldsymbol{x}_i$$

where $\eta > 0$ is the *learning rate*, which controls the degree to which an observation affects the update (we normalise the learning rate by dividing it by $\|\boldsymbol{x}_i\|^2$). This update rule moves the current weight vector in the direction of the negative gradient of the squared error of the observation (being the direction in which the squared error decreases most rapidly) — hence the name *Normalised Least Mean Squares Filter*.

The learning rate η, minimum number of viewed exhibits m and shrinkage weight ω are chosen so that the MAE is minimised (Section 5.1).

5 Evaluation

This section reports on the results of an evaluation performed with our datasets (Section 3), including a comparison with a nearest-neighbour collaborative filter [1,2].

5.1 Experimental Setup

To evaluate the performance of our predictive models *CBF* and *NLMSF* (Section 4), we implemented two additional models: *Mean Model (MM)* and *Collaborative Filter (CF)*. *MM*, which we use as a baseline, predicts the log viewing time of an exhibit area i to be its (non-personalised) mean log viewing time $\bar{t}_{\cdot i}$. For *CF* (described in detail in [2]), we implemented a nearest-neighbour collaborative filtering algorithm, and added modifications from the literature that improve its performance, such as significance weighting [1] and shrinkage to the mean [12]. Additionally, as for *CBF* and *NLMSF*, we transformed the log viewing times into exhibit-specific z-scores, to ensure that varying exhibit area complexity does not affect the similarity computation for selecting the nearest neighbours. Visitor-to-visitor differences with respect to their mean viewing durations were removed by transforming predictions to the current visitor's viewing-time scale [1].

Due to the relatively small dataset, we used leave-one-out cross validation to evaluate the performance of the different models. That is, for each visitor, we trained the models with a reduced dataset containing the data of 157 of the 158 visit trajectories, and used the withheld visitor pathway for testing.[6] As discussed in Section 3.2, we explored three exhibit-representation variants for the *Simple* model and three for the *Populated* model. For each of the six variants, we tested several thousand different parametrisations of *CBF* and *NLMSF* (we proceeded similarly for *CF*). For *CBF*, we varied the parameters k, m and ω, and for *NLMSF*, the parameters η, m and ω. In this paper, we report only on the performance obtained with the best configurations.

We performed two types of experiments: *Individual Exhibit* and *Progressive Visit*.

- **Individual Exhibit (IE).** *IE* evaluates predictive performance for a single exhibit. For each observed visitor-exhibit pair (u, i), we first removed the implicit rating r_{ui} from the vector r_u of visitor u's normalised log viewing durations. We then computed a prediction $\hat{r}_{ui}$ from the other observations, and unnormalised $\hat{r}_{ui}$ to obtain a log viewing time prediction $\hat{t}_{ui}$. This experiment is lenient in the sense that all available observations except the observation for exhibit area i are kept in a visitor's rating vector r_u.

- **Progressive Visit (PV).** *PV* evaluates performance as a museum visit progresses, i. e., as the number of viewed exhibit areas increases. For each visitor, we started with an empty visit, and iteratively added each viewed exhibit area to the visit history, together with its normalised log viewing time. We then predicted the normalised log viewing times of all yet unvisited exhibit areas, and unnormalised these predictions to obtain log viewing times.

For both experiments, we used the *mean absolute error (MAE)* with respect to log viewing times to measure predictive accuracy as follows:

$$\text{MAE} = \frac{1}{\sum_{u \in U} |I_u|} \sum_{u \in U} \sum_{i \in I_u} |t_{ui} - \hat{t}_{ui}|$$

where U is the set of all visitors, and I_u denotes a visitor u's set of exhibit areas for which predictions were computed. For *IE*, we calculated the total MAE for all valid visitor-exhibit pairs; and for *PV*, we computed the MAE for the yet unvisited exhibit areas for all visitors at each time fraction of a visit (to account for different visit lengths, we normalised all visits to a length of 1).

5.2 Results

Table 2 shows the results for the *IE* experiment, where the best *CBF* variant achieves an MAE of 0.7673 (stderr 0.0067), and the best *NLMSF* variant achieves an MAE of

---

[6] For our experiments, we ignored travel time between exhibit areas, and collapsed multiple viewing events of one area into one event.

Table 2. Model performance for the *IE* experiment (MAE)

											MAE
MM											0.8618
CF											0.7868

CBF	Binary				Frequency				Tf-idf		
	MAE	k	m	ω	MAE	k	m	ω	MAE	k	m	ω
Simple	0.7714	22	20	0.8	**0.7673**	20	21	0.8	0.7705	20	15	0.7
Populated	0.7758	35	15	0.8	0.7704	14	15	0.7	0.7699	14	15	0.7

NLMSF	Binary				Frequency				Tf-idf		
	MAE	η	m	ω	MAE	η	m	ω	MAE	η	m	ω
Simple	0.7829	0.3	1	1.0	0.7850	0.3	1	0.9	0.7780	0.4	1	1.0
Populated	0.7860	0.3	1	0.9	0.7875	0.2	1	1.0	**0.7772**	0.3	1	1.0

0.7772 (stderr 0.0067) (typeset in boldface in Table 2). Both models outperform *MM* and *CF* (statistically significantly with $p \ll 0.01$).[7] For the *Binary* and *Frequency* variants, *Simple* performs marginally better than *Populated*. For *Tf-idf* on the other hand, the *Populated* variants perform better than *Simple*. However, these differences are not very pronounced. Further experiments are necessary to make more conclusive statements regarding the effects of the different ways of representing exhibits by bags of words. Interestingly, the minimum number of viewed exhibits m is 1 for all *NLMSF* variants, meaning that *NLMSF* produces personalised predictions that are more accurate than the non-personalised mean with only little information about a visitor. Additionally, the learning rates η are relatively small (the optimal learning rate for a normalised least mean squares filter can be shown to be 1.0 in the case of perfect observations), and the shrinkage weights ω are close to 1.0. In contrast, the shrinkage weights for *CBF* are smaller. This can be explained by the fact that for *NLMSF*, small learning rates have the same effect on performance as small shrinkage weights (i. e., weights close to 0.0).

The performance of our models for the *PV* experiment is depicted in Figure 1 (we show only the results for the *Populated Tf-idf* variants of our models, as they yielded the best performance). *CF* outperforms *MM* slightly (statistically significantly for visit fractions 0.191 to 0.374 and for several shorter intervals later on, $p < 0.05$). *CBF* performs at least as well as *CF* (statistically significantly better for 0.0701 to 0.2513 and for several shorter intervals later on, $p < 0.05$), and *NLMSF* performs significantly better than both *MM* and *CF* for visit fractions 0.0160 to 0.9039 with gaps around 0.3 and 0.8 ($p < 0.05$). Additionally, *NLMSF* outperforms *CBF* (statistically significantly for 0.0160 to 0.9199 with gaps from 0.1982 to 0.3674 and 0.8078 to 0.8308, $p < 0.05$).

Drawing attention to the initial portion of the visits, *NLMSF*'s MAE decreases instantly, whereas the MAE for *MM*, *CF* and *CBF* remains at a higher level. Generally, the faster a model adapts to a visitor's interests, the more likely it is to quickly deliver (personally) useful recommendations (alleviating the *new-user problem*, which is typical for content-based user modelling approaches). Such behaviour in the early stages of

---

[7] Throughout this paper, the statistical tests performed are paired two-tailed t-tests.

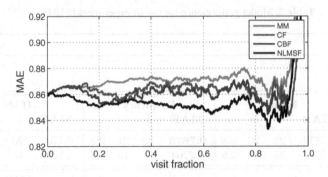

Fig. 1. Model performance for the *PV* experiment (MAE)

a museum visit is essential in order to build trust in the system, and to guide a visitor in a phase of his/her visit where such guidance is most likely needed.

As expected, *MM* performs at a relatively constant MAE level. For the other models, we expected to see an improvement in performance (relative to *MM*) as the number of visited exhibit areas increases. However, this trend is rather subtle. Additionally, for all four models, there is a performance drop towards the end of a visit. We postulate that these phenomena may be explained, at least partially, by the increased influence of outliers on the MAE, as the number of exhibit areas remaining to be viewed is reduced with the progression of a visit. This influence in turn offsets potential gains in performance obtained from additional observations. Our hypothesis is supported by a widening in the standard error bands for all models as a visit progresses, in particular towards the end (not shown in Figure 1 for clarity of presentation).

6   Conclusions and Future Work

We have offered a simple methodology for manually annotating museum exhibits with bags of keywords. We then presented two content-based models for predicting a visitor's viewing times of unseen exhibits (used to measure interest in exhibits) from his/her viewing times at visited exhibits: memory-based nearest-neighbour *Content-Based Filter (CBF)*, and model-based *Normalised Least Mean Squares Filter (NLMSF)*. For both models, we considered six variants for the keyword-based exhibit representation. The results of our evaluation favour slightly the *Populated Tf-idf* representation, but this outcome requires further investigation. More importantly, both content-based models attain a higher predictive accuracy than a non-personalised baseline, and perform at least as well as a nearest-neighbour collaborative filter (with *NLMSF* outperforming *CBF* in the realistic *Progressive Visit* setting). Additionally, in the *Progressive Visit* setting, *NLMSF* rapidly adapts to observed visitor behaviour, alleviating the *new-user problem* of content-based approaches.

In the future, we plan to hybridise collaborative and content-based models, e. g., in an ensemble fashion [14]. We also plan to investigate whether more sophisticated approaches for representing museum exhibits improve the predictive performance of our models. Further, the exhibits predicted by our interest-based models to be personally interesting (Section 4) will be combined with the most likely pathway through the museum (predicted by a location-based model [15]), in order to recommend exhibits that a visitor may be interested in but is likely to overlook.

Acknowledgments. This research was supported in part by grant DP0770931 from the Australian Research Council. The authors thank Carolyn Meehan and her team from Museum Victoria for fruitful discussions and their support; and Karl Grieser, Kapil K. Gupta and Patrick Ye for their assistance with annotating exhibits.

References

1. Herlocker, J.L., Konstan, J.A., Borchers, A., Riedl, J.T.: An algorithmic framework for performing collaborative filtering. In: Proc. of the 22nd Annual Intl. ACM SIGIR Conf. on Research and Development in Information Retrieval (SIGIR 1999), pp. 230–237 (1999)
2. Bohnert, F., Zukerman, I.: Non-intrusive personalisation of the museum experience. In: Houben, G.-J., et al. (eds.) UMAP 2009. LNCS, vol. 5535, pp. 197–209. Springer, Heidelberg (2009)
3. Cheverst, K., Mitchell, K., Davies, N.: The role of adaptive hypermedia in a context-aware tourist GUIDE. Communications of the ACM 45(5), 47–51 (2002)
4. Aroyo, L., Stash, N., Wang, Y., Gorgels, P., Rutledge, L.: CHIP demonstrator: Semantics-driven recommendations and museum tour generation. In: Aberer, K., Choi, K.-S., Noy, N., Allemang, D., Lee, K.-I., Nixon, L.J.B., Golbeck, J., Mika, P., Maynard, D., Mizoguchi, R., Schreiber, G., Cudré-Mauroux, P. (eds.) ASWC 2007 and ISWC 2007. LNCS, vol. 4825, pp. 879–886. Springer, Heidelberg (2007)
5. Petrelli, D., Not, E.: User-centred design of flexible hypermedia for a mobile guide: Reflections on the HyperAudio experience. User Modeling and User-Adapted Interaction 15(3-4), 303–338 (2005)
6. Hatala, M., Wakkary, R.: Ontology-based user modeling in an augmented audio reality system for museums. User Modeling and User-Adapted Interaction 15(3-4), 339–380 (2005)
7. Stock, O., Zancanaro, M., Busetta, P., Callaway, C., Krüger, A., Kruppa, M., Kuflik, T., Not, E., Rocchi, C.: Adaptive, intelligent presentation of information for the museum visitor in PEACH. User Modeling and User-Adapted Interaction 18(3), 257–304 (2007)
8. Fellbaum, C.: Wordnet: An Electronic Lexical Database. Bradford Books (1998)
9. Parsons, J., Ralph, P., Gallager, K.: Using viewing time to infer user preference in recommender systems. In: Proc. of the AAAI Workshop on Semantic Web Personalization (SWP 2004), pp. 52–64 (2004)
10. Schwarz, G.: Estimating the dimension of a model. The Annals of Statistics 6(2), 461–464 (1978)
11. Pazzani, M., Billsus, D.: Content-Based Recommendation Systems. In: Brusilovsky, P., Kobsa, A., Nejdl, W. (eds.) Adaptive Web 2007. LNCS, vol. 4321, pp. 325–341. Springer, Heidelberg (2007)
12. James, W., Stein, C.: Estimation with quadratic loss. In: Proc. of the Fourth Berkeley Symp. on Mathematical Statistics and Probability, vol. 1, pp. 361–379 (1961)
13. Kivinen, J., Warmuth, M.: Exponentiated gradient versus gradient descent for linear predictors. Information and Computation 132(1), 1–63 (1997)
14. Burke, R.: Hybrid recommender systems: Survey and experiments. User Modeling and User-Adapted Interaction 12(4), 331–370 (2002)
15. Bohnert, F., Zukerman, I., Berkovsky, S., Baldwin, T., Sonenberg, L.: Using interest and transition models to predict visitor locations in museums. AI Communications 21(2-3), 195–202 (2008)

Behaviour Recognition from Sensory Streams in Smart Environments

Sook-Ling Chua, Stephen Marsland, and Hans W. Guesgen

School of Engineering and Advanced Technology, Massey University,
Palmerston North 4442, New Zealand
{s.l.chua,s.r.marsland,h.w.guesgen}@massey.ac.nz
http://muse.massey.ac.nz/

Abstract. One application of smart homes is to take sensor activations from a variety of sensors around the house and use them to recognise the particular behaviours of the inhabitants. This can be useful for monitoring of the elderly or cognitively impaired, amongst other applications. Since the behaviours themselves are not directly observed, only the observations by sensors, it is common to build a probabilistic model of how behaviours arise from these observations, for example in the form of a Hidden Markov Model (HMM). In this paper we present a method of selecting which of a set of trained HMMs best matches the current observations, together with experiments showing that it can reliably detect and segment the sensor stream into behaviours. We demonstrate our algorithm on real sensor data obtained from the MIT PlaceLab. The results show a significant improvement in the recognition accuracy over other approaches.

Keywords: Behaviour Recognition, Hidden Markov Models (HMMs), Activity Segmentation, Smart Home.

1 Introduction

It is a well-reported fact that the populations of the Western world are aging. In Europe, for example, the number of people aged 65 and over is projected to increase from 10% of the entire population in 1950 to more than 25% in 2050. Older adults are more frequently subject to physical disabilities and cognitive impairments than younger people. It is clearly impossible to rely solely on increasing the number of caregivers, since even now it is difficult and expensive to find care. Additionally, many people are choosing to stay in their own homes as long as possible, and hope to remain independent. This has lead to a large number of monitoring systems (also known as 'smart homes', or 'ubiquitous computing systems') that aim to assist in the Activities of Daily Living (ADLs) such as bathing, grooming, dressing, eating and so on [11], either directly through involvement with the person, or by alerting carers when a problem arises.

As the majority of the ADLs involve using physical objects, such as washing machines, cooking utensils, refrigerators, televisions and so forth, it is possible to infer the inhabitant's behaviour [9], [13]. As a result, behaviour recognition has been drawing significant attention from the research community. The idea behind behaviour recognition is to infer the inhabitant's behaviours from a series of observations acquired through sensors.

A. Nicholson and X. Li (Eds.): AI 2009, LNAI 5866, pp. 666–675, 2009.

One of the main challenges in behaviour recognition is that the exact activities are not directly observed. The only information provided are the sensor observations, which could be that the kitchen light is on, the oven is turned on and the burner is on; the inference that therefore somebody is cooking is left to the intelligent part of the system. Two challenges of behaviour recognition are that many of the same sensor activations will be involved in multiple behaviours, and that the number of observations in a behaviour can vary between activities, and within different instances of the same activity. For example, making breakfast could involve sensors on the fridge, toaster, and cabinet one day, and also the kettle the next day when the person decides to have coffee as well. Making lunch will also involve the fridge and cabinet, and other unrelated sensors.

One common approach to recognising behaviours is to use Hidden Markov Models (HMMs), which are probabilistic graphical models where sensor observations give rise to latent variables which represent the behaviours. To use HMMs there are a few problems that have to be solved. One is to break the token sequence into appropriate pieces that represent individual behaviours (i.e., segmentation), and another is to classify the behaviours using the HMM. Most current approaches assume that the activities have been segmented, and use a fixed window length to partition the input stream. With each behaviour produces different numbers of sensor actions, it is inappropriate to rely on fixed window length, as activity segmentation can be biased in this way. Thus, an intelligent method is required to self-determine the window size based on the data. This paper presents a prototype system that performs the behaviour recognition and segmentation by using a set of HMMs that each recognise different behaviours and that compete to explain the current observations. In this paper we propose a variable window length that moves over the sequence of observations and use hand-labelled data to demonstrate the efficacy of the system.

2   Related Work

There has been a lot of work on activity segmentation in smart homes. Within smart home research it is common to use more complicated variants of the HMM, such as the Hierarchical Hidden Markov Model [7], or Switching Hidden Semi-Markov Model [1]. In both of these models, a top-level representation of behaviours (e.g., cooking or making coffee) is built up from a set of recognised activities that arise from the individual sensor values. A variant of these methods uses a three level Dynamic Bayesian Network [5] (the HMM is one of the simplest dynamic Bayesian network). These models can be seen as adding complexity to the HMM in order to represent one complete model of behaviours arising from sensor activations. The difficulty with these methods is that more complex models require more data for training, and higher computational cost.

There are many other places where time series of activities are recognised and classified into 'behaviours', and our method owes more to other areas of temporal signal analysis, such as recognising activities from posture information from video [3] and motion patterns [8,4,14]. In common with our algorithm, Kellokumpu, Pietikäinen and Heikkilä [3] use a set of HMMs, one for each activity, and apply the forward algorithm in the same way that we do to monitor likelihood values. However, they do not use a sliding time window, preferring multiple window sizes and thresholding in order to separate out the activities. Niu and Abdel-Mottaleb [8] merge the outputs of the different

HMMs using majority voting. A vote is assigned to each window and activity is classified based on the most common classification from the set of HMMs. A similar method is used in [12] for the identification of housekeeping activities using RFID data.

Kim, Song and Kim [4] turn the problem around and perform segmentation before classification, in this case for gesture recognition. The starting point of gestures is detected, and then a window is slid across the observation sequence until an end point is reached. The extracted gestures are then fed to HMMs for gesture recognition, with the final gesture type being determined by majority vote. An attempt to simultaneously detect the sequences and train the HMMs was described by Yin, Shen, Yang and Li [14]. A window is moved over the observation sequence to construct a linear dynamic system, and the likelihood of each model is computed based on these linear systems. A modified EM algorithm is used to simultaneously update these estimates. High-level goals can then be inferred from these sequences of consecutive motion patterns. Another method that is closely related to our approach is the work of Govindaraju and Veloso [2], which attempts to recognise activities from a stream of video. They use a set of HMMs, but maintain a single fixed window size, which is determined by averaging the lengths of the training segments used.

3   Behaviour Recognition

In our work we use the Hidden Markov Model as the basic representation of a behaviour. We posit that a typical behaviour is a sequence of activities that occur close to one another in time, in one location. While this is not always true, for now we are focussing on these types of behaviour, which includes activities such as cooking and preparing beverages. It would not include common activities such as laundry, which may well be separated in time (while waiting for the washer to finish) and in space (for example, if clothes are hung outside rather than using a dryer).

The Hidden Markov Model (HMM) [10] is very commonly used for these types of problem. It is a probabilistic model that uses a set of hidden (unknown) states to classify a sequence of observations over time. The HMM uses three sets of probabilities, which form the model parameters: (1) state transition probability distribution $A = a_{ij}$, the probability of transition from state i to state j conditional on current state i, (2) observation probability distribution $B = b_j(O)$, which illustrates the probability of observing observation O given that current state is j and (3) initial state distribution $\pi = \pi_i$. The HMM is a special case of the Dynamical Bayesian Network or Graphical Model [6], and unlike most graphical models, HMMs admit tractable algorithms for learning and prediction without the need for sampling or approximation. We use a separate HMM to recognise each behaviour. This is allows for variation in the activity, such as different orders of sensor activation, the fact that certain sensor activations can be shared by multiple behaviours, and the fact that the algorithm is probabilistic and can hence deal with 'noise' in the data.

Given a set of HMMs trained on different behaviours, we present data from the sensor stream to all of the HMMs, and use the forward algorithm [10] to compute the likelihood of this sequence of activities according to the model of each behaviour, i.e. $P(O_1, O_2, \ldots, O_T|\lambda)$, for HMM λ and observation sequence $O_1, O_2, \ldots, O_T$ using:

$$P(O_1, O_2, \ldots, O_T | \lambda) = \sum_{i=1}^{N} \alpha_T(i)$$

which can be recursively computed by:

$$\alpha_1(i) = \pi_i b_i(O_1)$$

$$\alpha_{t+1}(j) = \sum_{i=1}^{N} \alpha_t(i) a_{ij} b_j(O_{t+1})$$

The α_t values are quantised into the set $\{0, 1\}$ using a winner-takes-all approach to simplify the calculations at each subsequent step. The forward algorithm fits well into the context of our study because it determines how well the 'winning' HMM explains the observed sequences. This can be determined by monitoring the forward variable (α) for each observation. A change in the quantised α value signifies a 'change' of activity from the observation stream. The data that is presented to the HMMs is chosen from the sensor stream using a window that moves over the sequence. The choice of the size of this window is important, because it is unlikely that all of the activities in the sequence belong to one behaviour, and so the HMM chosen to represent it will, at best, represent only some of the activities in the sequence. Many of the methods described in related work used multiple sizes of window to try and deal with this fact, which arises because sequences of different behaviours (or indeed, the same behaviour in different instances) can be of different lengths. We present an alternative solution to this problem. To see the importance of the problem, consider the three different cases shown in Fig. 1. In each, a behaviour w takes up much of the window and is the winning behaviour. However, the location of it in the window differs, and we want to ensure that other behaviours in the window are also recognised.

The solid line shown in Fig. 1 illustrates how the quantised α values computed by the forward algorithm applied to one particular HMM, the one selected as the 'winner' for this window. If the quantised α values are high (that is, $\alpha = 1$) at the beginning of the observation sequence then it is likely that case (Fig. 1(a)) is occurring. Following Fig. 1(a) we see that there is a drop in α value between observations O_5 and O_6, which suggests that the behaviour has changed. We can therefore classify $O_1, O_2, \ldots, O_5$ as belonging to the winning behaviour, w, and then initialise a new window of default size (D_2) at O_6. When D_2 is initialized, all the observations within D_2 will then be fed to HMMs for competition and the process iterates. The second case occurs when the winning behaviour best describes observations that fall in the middle of the window, e.g., $O_4, O_5, \ldots, O_{10}$ in Fig. 1(b). Since the winning behaviour (w) does not describe observations O_1, O_2 and O_3, the probability for these three observations is low and we observe a jump in the α value at O_4. When this is observed, a new window (D_2) is initialized that contains only the three observations that are not explained by behaviour w. The whole process is then recursively computed on this window. With regard to the remaining sequence (O_4 and onwards) it would be possible to use HMM w and continue to monitor the α values. However, it was observed that sometimes there may be an overlap in individual sensor activations between the first and second behaviours, which can confuse things. For this reason, a new window of default size (D_3) is started at O_4 and the HMM competition is rerun on this sequence.

Since these two cases have ensured that the winning behaviour is at the beginning of the window, the only possibilities are that the behaviour stops during the window (Fig. 1(a)) or does not (Fig. 1(c)). The first case is already dealt with, and in the second case, we could simply classify the activities in the window as w and start a new one at the end of the current window. However, instead we extend the size of the window (shown as a dashed arrow in Fig. 1(c)) and continue to calculate the α value for each observation until the α drops. Fig. 2 summarises the overall procedures of the proposed method.

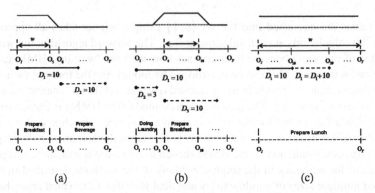

(a)   (b)   (c)

Fig. 1. An activity w does not need to take up the entire window. Even assuming that the actions in a behaviour are contiguous, it could be (a) at the start of the window, (b) in the middle, or (c) at the end. If the entire window is classified as one behaviour, then a potentially large number of behaviours are missed. $O_1, O_2, \ldots, O_T$ is the observation sequence, D is the window size and the initial default window size is 10. The solid line above the observation sequence shows the possible representations of a winning sequence using the α values. The long dash below the observation sequence shows the original observation sequence. For details, see the text.

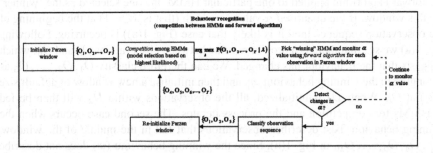

Fig. 2. Summary of our algorithm. When no changes is observed in α value, the algorithm will continue to monitor the α value based on the winning HMM (shown in dashed line). The recognition process is recursively computed until it reaches the end of the observation stream.

4   Experiment and Results

In order to demonstrate our algorithm, we took a dataset from the MIT PlaceLab [13]. They designed a system based on a set of simple and easily installed state-change sensors that were placed in two different apartments with real people living in them. The subjects kept a record of their activities that form a set of annotations for the data, meaning that there is a 'ground-truth' segmentation of the dataset. We trained the HMMs using this hand-segmented and labelled data. While this is a simplification of the overall aims of the project, it enables us to evaluate the method properly; future work will consider the problems of training with noisy and unlabelled data.

The actual dataset consists of state changes in objects within the home (such as the washing machine, TV, coffee machine, and toaster). For the first of the two subjects there were 77 sensors and data was collected for 16 consecutive days. It is this dataset that will form the basis for most of the experiment reported here. Further details on these datasets and PlaceLab architecture can be found in [13]. We assume for now that activities take place in one room, and that the location of the sensors is known *a priori*. For this reason, we concentrated on just one room, namely the kitchen, which contained more behaviours than any other. The behaviours that were originally labelled in the kitchen were (i) prepare breakfast, (ii) prepare beverage, (iii) prepare lunch, and (iv) do the laundry. We split behaviour (i) into two different ones, prepare toast and prepare cereal. This made two relatively similar behaviours, which is important to test recognition accuracy to distinguish activities and to avoid bias classification.

In order to train the HMMs, a subset of the data was required. We partitioned the data into a training set consisting of the first few days, followed by a test set consisting of the remainder. From the total of 16 days of data, we tried different splits of the data, from 15 days for training (and 1 for testing) through 11 days, 8 days, and 5 days for training. There were approximately 5-6 activities each day, made up of around 90-100 sensor observations. The HMMs were each trained on the relevant labelled data in the training set using the standard Expectation-Maximization (EM) algorithm [10]. We conducted three separate experiments using these five trained HMMs. In the first, we compared the algorithm with fixed window length, while in the second we looked at the effects of window size on the efficiency and accuracy of the algorithm. In the third experiment, we looked at how much training data was required for accurate results.

We defined two separate measurements of accuracy for our results:

Behaviour-level recognition accuracy: This simply compares the behaviour output by the algorithm with that of the label whenever the behaviour changed (e.g. behaviour such as 'doing laundry', 'preparing lunch', etc.).

Observation-level recognition accuracy: This compares the behaviour output by the algorithm with that of the label for every observation. This is particularly sensitive to times when two behaviours that occur one after the other share the same observations (e.g. observation such as 'oven is turned on' should be classified as 'preparing lunch' rather than 'doing laundry').

4.1   Experiment 1: Comparison between the Algorithm with Fixed Window Length

The first experiment is designed to compare the algorithm with the fixed window length. In this experiment, we used a fixed window length of size 10, with 5 days of training and 11 days of testing, and ran the algorithm over the sensory stream. Table 1 shows

Table 1. Comparison results between the variable window length and fixed window length

Recognition Accuracy	Variable Window Length	Fixed Window Length
Behaviour-level	87.80%	78%
Observation-level	98.35%	86.93%

the comparison results between the algorithm with the fixed window length. Results are based on contiguous manner, so there is no variation when the experiment is rerun, hence averages and standard deviations are not reported.

4.2   Experiment 2: Competition among HMMs

Before beginning this experiment, we used a window size of 10, with 5 days of training and 11 days of testing, and ran the entire algorithm over the sensor observations. The results of sliding this window over the data is shown in Fig. 3, which displays the outputs of the algorithm, with the winning behaviour at each time being clearly visible. As the figure shows, we can determine that the subject is doing laundry at observation

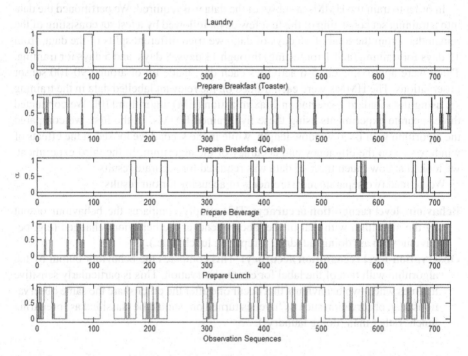

Fig. 3. Illustration of competition between HMMs based on a testing set of 727 sensor observations for five different behaviours: laundry, preparing toaster, preparing cereal, preparing beverage and preparing lunch. Since behaviours may share the same sensor observation, this explains why the $\alpha = 1$ is seen in multiple behaviours (e.g. between observation 120 and 140 in the last two behaviours). The 'preparing lunch' is selected as winner because it appears in a continuous manner.

150 and preparing breakfast (toaster) at observation 550. The classification accuracy of this experiment was high enough to encourage us to look further.

Table 2 shows the results of using different lengths of window. The different in the results is not significantly different across the different sizes, and therefore a shorter window length is preferred in order to keep the computational costs low.

Table 2. The results of using different initial window length on different training–test sets

Training Set (days)	Test Set (days)	Initial Window Length	% Accuracy (Behaviour)	% Accuracy (Observation)
11	5	10	86.96	97.29
		20	86.96	97.29
		50	86.96	97.29
		100	86.96	97.29
8	8	10	88.23	98.14
		20	82.35	97.80
		50	82.35	97.80
		100	88.23	98.14
5	11	10	87.80	98.35
		20	82.93	98.07
		50	82.93	98.07
		100	82.93	98.35

4.3 Experiment 3: Size of Training Data

The objective of this experiment is to analyze the amount of training data needed to train the HMMs. The most important thing is that every behaviour is seen several times in the training set to ensure that the HMM acquires a good representation of that behaviour. The results on recognition accuracy on both behaviour-level and observation-level are presented in Table 3. As the table shows, the size of training data does not have much significant impact on recognition accuracy. Even when only 5 days of training and 11 days of testing with window size 10 are used, we are still able to achieve 87.80% recognition accuracy on behaviour-level and 98.35% on observation-level. It seems that the proposed method does not need a significant large amount of training data for this dataset, although this may not be true for more complicated behaviours.

Table 3. Behaviour-level and observation-level recognition accuracy using window length of size 10

Training Datasets	Test Datasets	Behaviour-level		Observation-level	
		Total Activities	Accuracy	Total Observations	Accuracy
15 Days	1 Day	5	100%	99	100%
11 Days	5 Days	23	86.96%	369	97.29%
8 Days	8 Days	34	88.23%	591	98.14%
5 Days	11 Days	41	87.80%	727	98.35%

5   Discussion

On this relatively simple dataset our algorithms have worked very well, producing over 98% accuracy at the observation-level. However, it is still instructive to see if there are consistent reasons for the misclassifications that did occur.

We identified one main reason for misclassification, which is that individual sensor observations can be in several behaviours. There are two places where this can be a problem. The first is when the end of one behaviour contains observations that could be in the start of the next. This will not pose a problem if the second behaviour happens immediately after the first. However, if the second behaviour happened two hours after the first, that would be a totally different unrelated behaviours. The second place that this can be seen is where the winning behaviour is not at the start of the window, but those activities at the start could be interpreted as being part of that behaviour. It was experimentally observed that this was more likely to happen where the size of the window was large, because more behaviours were observed.

One way to reduce the misclassification is by adding extra information in order to improve the classification accuracy. This can be achieved by augmenting the current algorithm with spatio-temporal information. If spatio-temporal information is included, then places where two behaviours abut one another can be reduced, since there could be other non-kitchen behaviours inbetween.

6   Conclusions

We have presented a simple system that performs behaviour recognition based on competition between trained Hidden Markov Models, and demonstrated that the method works on labelled data. Our experimental results show that the method works effectively, with an average of around 90.75% behaviour-level recognition accuracy and 98.45% observation-level recognition accuracy (by averaging the accuracy percentage from table 3) based on relatively small amount of training data. We have investigated the size of window required, and found that relatively small ones work best, which reduces the amount of training data required even further. As the model is relatively simple and based on recursive computation, the computational costs are significantly lower than many other methods. We have also shown that a comparison between variable window length and fixed window length and that the variable window length works best.

It is important to note that this study is purely performed on labelled data and have proven the ability to distinguish activities given a series of observations. The encouraging results highlight the need to test on unlabelled data, resulting in a system that can be built up from nothing when sensors are placed into a new environment, and allowing on-line recognition. The MIT PlaceLab dataset is very clean, in that there is little sensor noise or inaccuracy. This may well not be the case with other datasets, since sensors can be 'twitchy' or fail, there may be other people or animals in the house, etc. It is possible that smoothing the sensor stream will deal with this, e.g., by using a median filter. It may also be that behaviours are interleaved: a person may well make a beverage at the same time preparing lunch, which could be done while the laundry was running. Our current system will not deal with these behaviours in any sensible way, highlighting all of the separate parts of the behaviour as different instances of that behaviour. This is left for future work.

Acknowledgments. This study is conducted as part of PhD research in the School of Engineering and Advanced Technology at Massey University. The financial support from Massey University is gratefully acknowledged. We also acknowledge the support of the other members of the MUSE group (http://muse.massey.ac.nz). We would like to thank MIT PlaceLab for providing access to their dataset.

References

1. Duong, T.V., Bui, H.H., Phung, D.Q., Venkatesh, S.: Activity recognition and abnormality detection with the switching hidden semi-markov model. In: CVPR 2005: Proc. of the 2005 IEEE Computer Society Conference on Computer Vision and Pattern Recognition (CVPR 2005), vol. 1, pp. 838–845. IEEE Computer Society, Los Alamitos (2005)
2. Govindaraju, D., Veloso, M.: Learning and recognizing activities in streams of video. In: Proc. of the AAAI Workshop on Learning in Computer Vision (2005)
3. Kellokumpu, V., Pietikäinen, M., Heikkilä, J.: Human activity recognition using sequences of postures. In: MVA, pp. 570–573 (2005)
4. Kim, D., Song, J., Kim, D.: Simultaneous gesture segmentation and recognition based on forward spotting accumulative hmms. Pattern Recognition 40(11), 3012–3026 (2007)
5. Liao, L., Patterson, D.J., Fox, D., Kautz, H.: Learning and inferring transportation routines. Artificial Intelligence 171(5-6), 311–331 (2007)
6. Marsland, S.: Machine Learning: An Algorithmic Introduction. CRC Press, New Jersey (2009)
7. Nguyen, N., Phung, D., Venkatesh, S., Bui, H.: Learning and detecting activities from movement trajectories using the hierarchical hidden markov model. In: IEEE Computer Society Conference on Computer Vision and Pattern Recognition (CVPR 2005), vol. 2, pp. 955–960 (2005)
8. Niu, F., Abdel-Mottaleb, M.: HMM-based segmentation and recognition of human activities from video sequences. In: IEEE International Conference on Multimedia and Expo. (ICME 2005), pp. 804–807 (2005)
9. Philipose, M., Fishkin, K., Perkowitz, M., Patterson, D., Fox, D., Kautz, H., Hahnel, D.: Inferring activities from interactions with objects. IEEE Pervasive Computing 3(4), 50–57 (2004)
10. Rabiner, L.: A tutorial on hidden Markov models and selected applications in speech recognition. Proc. of the IEEE 77(2), 257–286 (1989)
11. Robert, N., Taewoon, K., Mitchell, l., Stephen, K.: Living quarters and unmet need for personal care assistance among adults with disabilities. Journal of Gerontology: Social Sciences 60B(4), S205–S213 (2005)
12. Stikic, M., Huỳnh, T., Van Laerhoven, K., Schiele, B.: ADL recognition based on the combination of RFID and accelerometer sensing. In: Second International Conference on Pervasive Computing Technologies for Healthcare, PervasiveHealth 2008, pp. 258–263 (2008)
13. Tapia, E.M., Intille, S.S., Larson, K.: Activity recognition in the home using simple and ubiquitous sensors. In: Ferscha, A., Mattern, F. (eds.) PERVASIVE 2004. LNCS, vol. 3001, pp. 158–175. Springer, Heidelberg (2004)
14. Yin, J., Shen, D., Yang, Q., Li, Z.-N.: Activity recognition through goal-based segmentation. In: Proc. of the 19th AAAI Conference on Artificial Intelligence (AAAI 2005), pp. 28–33 (2005)

Probabilistic Seeking Prediction in P2P VoD Systems

Weiwei Wang, Tianyin Xu, Yang Gao, and Sanglu Lu

State Key Laboratory for Novel Software Technology,
Nanjing University, Nanjing 210093, PRC
ww.wang.cs@gmail.com

Abstract. In P2P VoD streaming systems, user behavior modeling is critical to help optimise user experience as well as system throughput. However, it still remains a challenging task due to the dynamic characteristics of user viewing behavior. In this paper, we consider the problem of user seeking prediction which is to predict the user's next seeking position so that the system can proactively make response. We present a novel method for solving this problem. In our method, frequent sequential patterns mining is first performed to extract abstract states which are not overlapped and cover the whole video file altogether. After mapping the raw training dataset to state transitions according to the abstract states, we use a simpel probabilistic contingency table to build the prediction model. We design an experiment on the synthetic P2P VoD dataset. The results demonstrate the effectiveness of our method.

Keywords: User seeking prediction, State abstraction, Contingency table, P2P VoD systems, User behavior modeling, PrefixSpan.

1 Introduction

With the proliferation of emerging applications, including Internet TV, online video, and distance education, media streaming service over the Internet has become immensely popular and generated a large percentage of today's Internet traffic. Peer-to-peer (P2P) technology has been proved as a successful solution which can effectively alleviate server workload, save server bandwidth consumed and thus bring high system resilience and scalability[1,2,3,4,5]. In P2P media streaming systems, the users do not need to download the complete video files before playback which introduces long startup delay. Instead, "play-as-download" streaming service is provided to let the users watch videos while downloading. P2P live streaming, a typical media streaming service designed for all peers receiving streamed video at the same playback position, has been widely deployed to provide "play-as-download" service for a large number of users[4,5,6]. However, P2P video-on-demand (VoD) streaming is more difficult to design and deploy than P2P live streaming. Unlike live streaming, VoD systems allow users' interactive behaviors, i.e., users can seek forward or backward freely when watching video streams. If not handled properly, such seeking requests may lead to long response latency, which severely deteriorates users' viewing quality, e.g., playback freezing or even blackout.

To improve user viewing experience, user behavior understanding is critical. If a VoD system could detect or predict user seeking patterns, it could proactively make response to them in order to optimise media content delivery[7]. On the server side, the media

A. Nicholson and X. Li (Eds.): AI 2009, LNAI 5866, pp. 676–685, 2009.

server could use spare bandwidth to push out the appropriate media contents to a user before being requested. On the client side, a peer could prefetch media contents that are likely to be requested by upcoming seeking events. This can effectively reduce the response latency and maximize system throughput[8].

User behavior modeling has been already studied for a few years. Some researchers have studied single genres like sports videos[7] and education videos[9] while others have studied a range of video types[10,11]. Brampton et al.[7] analyzed the user inter-activity characteristics for sports VoD systems and derived some statistical distributions for user behavior which we employ to generate synthetic P2P VoD dataset for experiments as currently the original viewer logs are not available to us. Both He et al.[12] and Huang et al.[13] performed association rule mining to learn user seeking patterns used to do prediction. Zheng et al.[14] analyzed the statistical pattern hidden in the VoD dataset and applied the optimal quantization theory to learn user seeking behavior. Vilas et al.[11] proposed a user behavior model on the observed dataset but no evaluation or utilization of that model was presented. A survey paper on probabilistic human behavior prediction models by Albrecht and Zukerman can be found in [15], which is a great material to get a thorough understanding of probabilistic approaches on human behavior modeling. However, user behavior modeling still remains a challenging task due to the dynamic characteristics of the viewing behavior which is always changing over time.

In this paper, we employ a simple probabilistic contingency table to solve the problem of user seeking prediction which is to predict the user's next seeking position so that the P2P VoD system can proactively make response. In the design of our method, frequent sequential patterns mining[16] is first performed to extract abstract states which are not overlapped and cover the whole video file altogether. After mapping the raw training dataset to state transitions according to the abstract states, we simple count the number of each seeking operation to build a transition model. We evaluate the prediction model on a synthetic P2P VoD dataset containing 4000 user viewing logs. The results demonstrate the effectiveness of our proposed method.

The rest of this paper is organized as follows. In Section 2, we state the user seeking prediction problem in P2P VoD systems. Then our method is presented in Section 3. Next in Section 4, our method is validated by the experiments on synthetic P2P VoD dataset. Section 5 concludes and discusses some future work.

2   Problem Statement

In this section, we present some related terminology and define the problem of user seeking prediction.

Terminology. Since the most basic user activity is the continuous viewing of a section of a video, a peer maintains such basic activity in a *user viewing record* (UVR) in playback time. The important parts of an UVR format is shown as follows.

(UID, MID, Start Position, Inter-Seek Duration, Jump Position)

where UID refers to the user's identifier while MID refers to the movie's identifier. The *inter-seek duration* is described as the number of segments contained in the section of the video the user watched before seeking to a new position. The *start position* points to the first watched segment in the current inter-seek duration while the *jump position* points to the first segment in the next inter-seek duration.

In most cases, as soon as a user finishes recording one UVR, a new UVR is initialized to record the next inter-seek duration. A sequence of UVRs forms a *user viewing log* which represents a complete user viewing history called a *session*. For example, $\{(U1, V1, 1, 6, 73), (U1, V1, 73, 3, 4), (U1, V1, 4, 3, \text{End})\}$ depicts a session as follows: a user $U1$ first views the video $V1$ from the 1-st segment to the 6-th segment, then seeks forward to the 73-rd segment and views until the 75-th segment, and finally seeks backward to the 4-th segment, re-views for 3 segments and finishes playback.

User viewing logs can also be represented in *sequence format* as $\{s_0, s_1, \cdots, s_i, \cdots, s_{n-1}\}$, where s_i denotes that the user has viewed the s_i-th segment of the video. In this example, the corresponding sequence format is $\{1, 2, 3, 4, 5, 6, 73, 74, 75, 4, 5, 6\}$. Notice that the UVR format can be easily transferred into the sequential format which is used for mining frequent patterns in the state abstraction stage.

Problem Statement. Given a database of user viewing logs, the problem of *user seeking prediction* is to predict the next seeking position according to the user's viewing history in the current session.

Still use the example mentioned above. Given large volumes of user viewing logs of movie $V1$ and suppose the viewing history of user $U1$'s current session to movie $V1$ is $\{1, 2, 3, 4, 5\}$, we hope to predict $U1$'s next seeking position 73 and pre-fetch it in advance. As a result, when $U1$ finishes viewing segment 6 and requests to seek to segment 73, the client-side software can directly satisfy $U1$ with little response latency.

3   Learning User Seeking Behavior

To learn the seeking behavior, we first do frequent sequential pattern mining on the collected P2P VoD dataset and split the patterns into abstract states. Then we map the raw data to state transitions according to the abstract states. Finally, a prediction model is built using a simple contingency table.

3.1   State Abstraction

In typical P2P streaming systems, a video stream is divided into segments of uniform length and each segment contains 1 second video content [6,17]. Typical video stream on the Internet such as movies and sports videos take more than 1.5 hours (5400 seconds/segments) long. If we simply use "segment" as the unit to do learning, it would generate too many fine grained intermediates that bring difficulties for learning the prediction model. As a result, state abstraction is essential. In this problem, we extract the strongly associated segments into *abstract states* by distilling large volumes of user logs and then maps the raw user viewing logs into *state transitions*.

Mine Frequent Sequential Patterns. According to the measurements of real deployed media streaming systems [7,18,14], there are always some popular segments called *highlights* which attracts far more viewing times than other segments. This indicates that users are much willing to watch some interesting scenes while skip boring scenes. Fig.1 is the segment popularity statistics in [7], which is collected from a real deployed 8200-second sports video, a football match between Argentina vs. Serbia and Montenegro in World Cup 2006. We can see that the match has about 10 highlights, either of which is a kick-off or a goal. Fig.2 is the segment popularity of the synthetic P2P VoD dataset generated according to the statistic distributions in [7]. We will explain in detail the generation process in section 4.1.

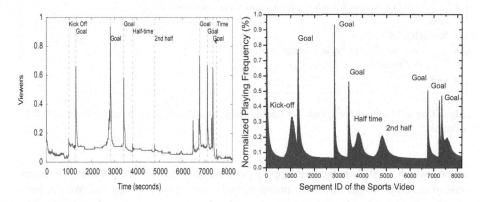

Fig. 1. Segment popularity of the dataset in [7] **Fig. 2.** Segment popularity of synthetic dataset

Directly cutting the 8200 seconds into equally length time series would not work well as it does not take the viewers' watching patterns into consideration. For example, if most viewers watched from 1000 to 1300 as a goal happened in that period, we should try to aggregate these seconds as a state or some contiguous states. From this point of view, we employ the frequent sequential pattern mining method **Prefix-projected Sequential pattern mining (PrefixSpan)**[16] to mining frequent patterns from the dataset. The general ideal of *PrefixSpan* is to examine the prefix subsequences and project their corresponding postfix subsequences into projected databases. In each projected database, sequential patterns are grown by exploring only local frequent patterns. This method is considerably fast than the Apriori-based algorithms and **Frequent pattern projected Sequential pattern mining (FreeSpan)**[19]. We do some modification to the *PrefixSpan* method as we aim at finding the contiguous sequential patterns. For example, the original *PrefixSpan* will find patterns like $\langle 1, 2, 4, 5, 6, 7, 10 \rangle$ which are not allowed in our result. We need patterns like $\langle 4, 5, 6, 7 \rangle$ which are not only sequential but also contiguous. For this reason, we modified the *PrefixSpan* to generate only frequent sequential and contiguous patterns. A procedural form of *PrefixSpan* is given in Algorithm 1. We follow the code by Yasuo Tabei[20] in our implementation.

Split Sequential Patterns into States. As the patterns found are largely overlapped, e.g., $\langle 1, 2, 3, 4, 5, 6, 7 \rangle$ and $\langle 5, 6, 7, 8, 9, 10, 11, 12 \rangle$ may both exist in the mining result,

---

Input   : A sequence database S, and the minimum support threshold min_sup
Output   : The complete set of sequential patterns
Method   : Call PrefixSpan($\emptyset, 0, S$)
Subroutine: PrefixSpan($\alpha, l, S|_\alpha$)
Parameters: α: asequential pattern; l: the length of α; $S|_\alpha$: the $\alpha -$ projected database, if
          $\alpha \neq \emptyset$; otherwise, the sequence database S
1 call $S|_\alpha$ once, find the set of frequent items b such that: b can be assembled to the last
   element of α to form a sequential pattern; or $\langle b \rangle$ can be appended to α to form a
   sequential pattern;
2 **foreach** *frequent item b and if b is **contiguous after** α* **do**
3   Append it to α to form a sequential pattern α', and output α';
4 **end**
5 **foreach** α' **do**
6   construct $\alpha' -$ projected database $S|_{\alpha'}$;
7   call **PrefixSpan**($\alpha', l + 1, S|_{\alpha'}$);
8 **end**

Algorithm 1. Modified *PrefixSpan* for mining frequent and contiguous time series in P2P VoD systems

among which $\langle 5, 6, 7 \rangle$ is the overlapping part. We need to split the patterns into intervals of different length which are not overlapped and remain contiguous. We design a simple splitting algorithm which scans over the sequential patterns and cuts them into intervals without overlapping, e.g., $\langle 1, 2, 3, 4, 5, 6, 7 \rangle$ and $\langle 5, 6, 7, 8, 9, 10, 11, 12 \rangle$ will be cut into intervals $[1, 7]$ and $[8, 12]$. For the intervals which do not exist in the mined sequential patterns, we take each of them as a separate interval. After that, we split the contiguous intervals into appropriate granularity intervals, called *states*, in order to fit the pre-fetching buffer. A too large interval makes no sense for pre-fetching because the buffer is size-limited. Here, we set two tunable parameters *min-state-len* and *max-state-len* which are the minimum and maximum state length allowed according to the condition of the client-side peer. The *min-state-len* avoids splitting an isolated segment as a state while the *max-state-len* is set as the pre-fetching buffer size in our experiment. Thus, the whole video stream can be represented by these abstract states.

Map Raw Dataset into State Transitions. With the abstract states generated from the above step, we can easily map user viewing logs into state transitions. The mapped results are in the following form:

$$\langle s,\ s' \rangle$$

where s is the state the current playback position is in while s' means the next state the viewer will seek into. For a single inter-seek duration, several contiguous state transitions may be generated as the duration may be very large for a single state.

3.2   Probabilistic Seeking Model Building

We assume the user seek operation satisfies the Markov property, that is, the next seek position is dependent on the current position and independent on the previous positions before the current position. With this assumption, we employ a simple contingency table to build a prediction model for predicting user behavior, that is the seeking operation.

Model Building using State Transitions. For the prediction task in this paper, we use a simple contingency table to represent the transition probability. The table is shown in Fig.3, in which s represents the current state in the mapped trainning data and s' represents the next state. By simply counting the number of seeking operation of each transition pair $\langle s, s' \rangle$, we can build this simple model in an efficient and incremental way.

s'	s	P(s'|s)
S_0	S_0	P_{00}
S_1	S_0	P_{01}
...	...	...
S_0	S_1	P_{10}
S_1	S_1	P_{11}
...	...	...

Fig. 3. A simple probabilistic contingency table for predicting user seeking behavior

After the training process, a model is built and can be used to do predictions. Given current state s, we can infer $P(s'|s)$ from the learnt transition table. According to this distribution, we can predict the next seeking, e.g. we can employ roulette wheel section or softmax selection. In our approach, we simply apply roulette wheel section strategy.

4 Performance Evaluation

In this section, we evaluate our method on the user seeking prediction problem. The data used here is the state transitions generated in the above steps.

4.1 Data Generation

In the experiment, we generated a synthetic P2P VoD dataset of user viewing logs according to the statistical distribution in [7]. The chosen video is the 8200-second football match described in Section 3.1. In [7], the segment popularity, the session length as well as the inter-seek duration follows some probability distribution, see Table 1.

For the generation, we modified the GISMO streaming generator [10] to produce 4000 user viewing logs in UVR format. We set most parameters of GISMO generator to the values in Table 1. Moreover, we modified the jump sub-routine in GISMO using a log-normal distribution to let users trend to jump around highlights. The segment popularity of the synthetic dataset is shown in Fig.2 which is similar with the real popularity statistics in [7]. So we believe our dataset can reflects the user behavior well.

Table 1. Metrics with their corresponding distribution used in data generation

Metric	Distribution	R-Square
Segment popularity	Log-normal, $\mu=0.016$, $\sigma=1.35$	0.0941
Session length	Log-normal, $\mu=4.835$, $\sigma=1.704$	0.127
Inter-seek duration	Log-normal, $\mu=1.4796$, $\sigma=2.2893$	0.0358

4.2 Problem Analysis

The average number of seeking behavior is about 7 times in a whole session[14][7]. Most of the state transitions, $\langle s, s' \rangle$, are contiguous, that is, $s' = s + 1$ as shown in Table 2. These state transitions are useless for our prediction as the playback buffer has already done this job. For this reason, we skip all the contiguous state transitions in the training data, i.e., state transitions like $\langle s, s + 1 \rangle$ will be just skipped.

Table 2. Statistical data of the user behavior prediction problem

Description	Statistics
Forward seek ratio	99.28%
One step forward seek ratio	80.54%
Backward seek ratio	0.72%
One step backward seek ratio	31.9%

In our experiment, the streaming rate S of the video is set as 256 Kbps (most video stream over the Internet today is encoded in the 200-400 Kpbs range [2]). The total available downloading bandwidth of each peer is randomly distributed in $[1.5S, 5S]$. The length of the client-side buffer is 30Mbytes which can be easily accommodated in state-of-art personal computers, i.e., each peer can cache 120 segments. The client-side buffer is split into two parts: the playback buffer with 25Mbytes and the pre-fetching buffer with 5Mbytes. In each time slot, Peers download urgent segments in playback buffer in high priority of using bandwidth to guarantee continuous normal playback. If there is still residual bandwidth, peers pre-fetch the segments in the predictive states into the pre-fetching buffer for supporting user seeking behavior. Both of the two parts use LRU (Least Recently Used) as a default buffer replacement policy.

4.3 Experimental Results

We use our data generation method to produce 4000 user viewing logs. The threshold value used in *PrefixSpan* is $1/10$ of the population, that is 400, and *min-state-len* is set to 5 while *max-state-len* is 20. After preprocessing, we split the whole data into a training dataset and a validation dataset with a split ratio of 0.7. We run all the experiments 10 times and average the results. Fig.4 shows the learnt user behavior model, in which each color represent a transition probability $P(s'|s)$ for a specific state s. We can see that our learnt transition model seems very similar to the segment popularity in Fig.2,

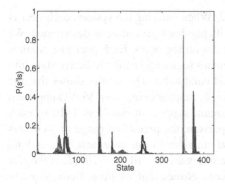

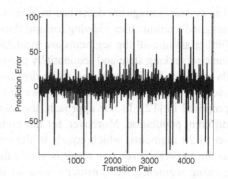

Fig. 4. Learnt user behavior model   **Fig. 5.** Prediction curve

which should not be surprising as we can expect users are more willing to seeking to the highlights and this also validates our method. The prediction error curve of the validation set is shown in Fig.5, in which, the x-coordinate is the transition pair and the y-coordinate represents the prediction error between the predictive next state and the actual next state in the validation data.

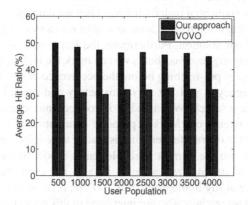

Fig. 6. Comparison of VOVO and our approach

We apply our learnt model to do prediction-based per-fetching and evaluate the pre-fetching performance in terms of hit ratio which is calculated using (1). When a user imposes a seeking request on a peer, the peer checks its local buffer (both the playback buffer and the pre-fetch buffer). If its local buffer contains segments in the requested state, the seeking is considered as a hit event and the peer can continue playback without jitter. Otherwise, it is a miss event and the peer must try to search and download the requested segment from other peers, which leads to long latency.

$$Average\ hit\ ratio = \frac{Total\ number\ of\ hit\ events}{Total\ number\ of\ seeking\ requests}. \tag{1}$$

The system settings are described in Sect. 4.2. When entering the system, each peer is assigned a unique user viewing log and thus its playback procedure is determined. We further assume all the segments are available for all the peers. Each peer just predicts the next seeking state and sequentially pre-fetches segments in the predictive state into its pre-fetch buffer according to its download bandwidth. The results shows that the average hit ratio is about 45% as shown in Fig. 6, more accurate than VOVO approach proposed in [12]. In our approach, the minimum support threshold is $1/10$ of each different population. Moreover, for fair comparison, the prediction range of VOVO is set to 20 segments, which equals to the *max-state-len* in our approach, i.e., as long as the VOVO's predictive result is within the interval $[R - 10, R + 10]$ (R is the real seeking segment), we consider it as a hit event. Notice that we do not consider any collaboration between peers in this paper. Thus, we can prospect a much higher hit ratio with the help of peer collaboration from which peers can exchange their buffer contents with neighbors and are more likely to find appropriate contents. Consequently, the results demonstrate the effectiveness of our approach.

5   Conclusion and Future Work

In this paper, we propose a new method on the user seeking prediction problem and get good results. In data preprocessing, we extract abstract states from the raw user viewing logs through frequent sequential pattern mining. Then we employ a simple contingency table to build a state transition model. State abstraction as a step of preprocessing plays an important role in our solution. Furthermore, the learnt user seeking model can be used to do pre-fetching suggestions, that is we can mark the highlights beside the video and offer suggestions for pre-fetching before the occurrence of seeking operations.

However, the accuracy is still not very satisfactory and much improvement could be done in the future research. We intend to introduce time series analysis approach into this problem to release our Markov property assumption of the state transitions. Besides, model transfer or transfer learning is also a very important research for our future work so as to use the current available learnt model to build new model for new videos instead of starting from scratch. Finally, our code and synthetic dataset are publicly available at my homepage http://cs.nju.edu.cn/rl/people/weiweiwang to all researchers who are interested in this novel problem.

Acknowledgments

This work is partially supported by the National Natural Science Foundation of China under Grant No. 60775046, 90718031, 60721002 and 60903025; the National Basic Research Program of China (973) under Grant No. 2009CB320705; the Natural Science Foundation of Jiangsu Province under Grant No. SBK200921645.

Moreover, we would like to thank Yinghuan Shi, Yongyan Cui and Liangdong Shi for their helpful comments.

References

1. Sripanidkulchai, K., Maggs, B., Zhang, H.: The feasible of supporting large-scale live streaming applications with dynamic application end-points. In: Proc. of the ACM SIG-COMM 2004 (August 2004)
2. Huang, C., Li, J., Ross, K.W.: Can internet video-on-demand profitable? In: Proc. of ACM SIGCOMM 2007 (August 2007)
3. Huang, Y., Fu, T.Z.J., Chiu, D.M., Liu, J.C.S., Huang, C.: Challenges, design and analysis of a large-scale p2p-vod system. In: Proc. of ACM SIGCOMM 2008 (August 2008)
4. PPLive: A free p2p internet tv software, http://www.pplive.com/(September 2007)
5. Joost: A website to watch videos, music, tv, movies and more over the internet, http://www.joost.com/ (June 2006)
6. Zhang, X., Liu, J., Li, B., Yum, T.S.P.: CoolStreaming/DoNet: A data-driven overlay network for efficient live media streaming. In: Proc. of IEEE INFOCOM 2005 (March 2005)
7. Brampton, A., MacQuire, A., Rai, I.A., Race, N.J.P., Mathy, L., Fry, M.: Characterising user interactivity for sports video-on-demand. In: Proc. of ACM NOSSDAV 2007 (April 2007)
8. Annapureddy, S., Guha, S., Gkantisdis, C., Gunawardena, D., Rodriguez, P.: Is high-quality vod feasible using p2p swarming? In: Proc. of ACM WWW 2007 (May 2007)
9. Costa, C., Cunha, I., Borges, A., Ramos, C., Rocha, M., Almeida, J., Riberio-Neto, B.: Analyzing client interactivity in streaming media. In: Proc. of ACM WWW 2004 (May 2004)
10. Jin, S., Bestavros, A.: GISMO: A generator of internet streaming media objects and workloads. In: Proc. of ACM SIGMETRICS 2001 (June 2001)
11. Vilas, M., Paneda, X.G., Garcia, R., Melendi, D., Garcia, V.G.: User behaviour analysis of a video-on-demand service with a wide variety of subjects and lengths. In: Proc. of IEEE EUROMICRO-SEAA 2005 (August 2005)
12. He, Y., Liu, Y.: VOVO: VCR-oriented video-on-demand in large-scale peer-to-peer networks. IEEE Transactions on Parallel and Distributed Systems 20(4), 528–539 (2009)
13. Huang, C.M., Hsu, T.H.: A user-aware prefetching mechanism for video streaming. World Wide Web: Internet and Web Information Systems 6(4), 353–374 (2003)
14. Zheng, C., Shen, G., Li, S.: Distributed prefetching scheme for random seek support in peer-to-peer streaming applications. In: Proc. of the ACM P2PMMS 2005 (November 2005)
15. Zukerman, I., Albrecht, D.W.: Predictive statistical models for user modeling. In: User Modeling and User-adapted Interaction, pp. 5–18. Kluwer Academic Publishers, Dordrecht (2001)
16. Pei, J., Han, J., Mortazavi-Asl, B., Wang, J., Pinto, H., Chen, Q., Dayal, U., Hsu, M.C.: Mining sequential patterns by pattern-growth: The PrefixSpan approach. IEEE Transactions on Knowledge and Data Engineering 16(10), 1–17 (2004)
17. Xu, T., Chen, J., Li, W., Lu, S., Guo, Y., Hamdi, M.: Supporting VCR-like operations in derivative tree-based P2P streaming systems. In: Proc. of IEEE ICC 2009 (June 2009)
18. Yu, H., Zheng, D., Zhao, B.Y., Zheng, W.: Understanding user behavior in large-scale video-on-demand systems. SIGOPS Oper. Syst. Rev. 40(4), 333–344 (2006)
19. Han, J., Pei, J., Mortazavi-Asl, B., Chen, Q., Dayal, U., Hsu, M.C.: Freespan: frequent pattern-projected sequential pattern mining. In: Proc. of the ACM SIGKDD 2000 (August 1999)
20. Tabei, Y.: PrefixSpan: An implementation of prefixspan (prefix-projected sequential pattern mining) (December 2008), http://www.cb.k.u-tokyo.ac.jp/asailab/tabei/prefixspan/prefixspan.html

Author Index